The Concise Encyclopedia
of Statistics

T0178007

Yadolah Dodge

The Concise Encyclopedia
of Statistics

 Springer

Prof. Yadolah Dodge
Université Neuchâtel
Fac. Droit et Sc. Économiques
Dépt. Statistique
espace de l'Europe 4
2002 Neuchâtel
Switzerland
yadolah.dodge@unine.ch

ISBN 978-1-4419-1390-6   ISBN 978-0-387-32833-1 (eBook)

Springer New York Dordrecht Heidelberg London

Library of Congress Control Number: 2009941851

Printed on acid-free paper

Springer is part of Springer Science+Business Media (www.springer.com)

Dedicated to the memory of
my lovely wife K, my mother and my father

Preface

With this concise volume we hope to satisfy the needs of a large scientific community previously served mainly by huge encyclopedic references. Rather than aiming at a comprehensive coverage of our subject, we have concentrated on the most important topics, but explained those as deeply as space has allowed. The result is a compact work which we trust leaves no central topics out.

Entries have a rigid structure to facilitate the finding of information. Each term introduced here includes a definition, history, mathematical details, limitations in using the terms followed by examples, references and relevant literature for further reading. The reference is arranged alphabetically to provide quick access to the fundamental tools of statistical methodology and biographies of famous statisticians, including some currents ones who continue to contribute to the science of statistics, such as Sir David Cox, Bradley Efron and T.W. Anderson just to mention a few. The critera for selecting these statisticians, whether living or absent, is of course rather personal and it is very possible that some of those famous persons deserving of an entry are absent. I apologize sincerely for any such unintentional omissions.

In addition, an attempt has been made to present the essential information about statistical tests, concepts, and analytical methods in language that is accessible to practitioners and students and the vast community using statistics in medicine, engineering, physical science, life science, social science, and business/economics.

The primary steps of writing this book were taken in 1983. In 1993 the first French language version was published by Dunod publishing company in Paris. Later, in 2004, the updated and longer version in French was published by Springer France and in 2007 a student edition of the French edition was published at Springer.

In this encyclopedia, just as with the *Oxford Dictionary of Statistical Terms*, published for the International Statistical Institute in 2003, for each term one or more references are given, in some cases to an early source, and in others to a more recent publication. While some care has been taken in the choice of references, the establishment of historical priorities is notoriously difficult and the historical assignments are not to be regarded as authoritative. For more information on less central terms not found in this encyclopedia short articles can be found in the following encyclopedias and dictionaries:

International Encyclopedia of Statistics, eds. William Kruskal and Judith M. Tanur (The Free Press, 1978).

Encyclopedia of Statistical Sciences, eds. Samuel Kotz, Norman L. Johnson and Cambell Reed (John Wiley and Sons, 1982).

The Encyclopedia of Biostatistics, eds. Peter Armitage and Ted Colton (Chichester: John Wiley and Sons, 1998).

The Encyclopedia of Environmetrics, eds. A.H. El-Sharaawi and W.W. Paregoric (John Wiley and Sons, 2001).

The Encyclopedia of Statistics in Quality and Reliability, eds. F. Ruggeri, R.S. Kenett and F.W. Faltin (John Wiley and Sons, 2008).

Dictionnaire- Encylopédique en Statistique, Yadolah Dodge, Springer 2004

In between the publication of the first version of the current book in French in 1993 and the later edition in 2004 to the current one, the manuscript has undergone many corrections. Special care has been made in choosing suitable translations for terms in order to achieve sound meaning in both the English and French languages. If in some cases this has not happen, I apologize. I would be very grateful to readers for any comments regarding inaccuracies, corrections, and suggestions for the inclusion of new terms, or any matter that could improve the next edition. Please send your comments to Springer-Verlag.

I wish to thank many people who helped me throughout these many years to bring this manuscript to its current form. Starting with my former assistants from 1983 to 2004, Nicole Rebetez, Sylvie Gonano-Weber, Maria Zegami, Jurg Schmid, Severine Pfaff, Jimmy Brignony Elisabeth Pasteur, Valentine Rousson, Alexandra Fragnieire, and Theiry Murrier. To my colleagues Joe Whittaker of University of Lancaster, Ludevic Lebart of France Telecom, and Bernard Fisher, University of Marseille, for reading parts of the manuscript. Special thanks go to Gonna Serbinenko and Thanos Kondylis for their remarkable cooperation in translating some of terms from the French version to English. Working with Thanos, my former Ph.D. student, was a wonderful experience. To my colleague Shahriar Huda whose helpful comments, criticisms, and corrections contributed greatly to this book. Finally, I thank the Springer-Verlag, especially John Kimmel, Andrew Spencer, and Oona Schmid for their meticulous care in the production of this encyclopedia.

January 2008

Yadolah Dodge
Honorary Professor
University of Neuchâtel
Switzerland

About the Author

Founder of the Master in Statistics program in 1989 for the University of Neuchâtel in Switzerland, Professor Yadolah Dodge earned his Master in Applied Statistics from the Utah State University in 1970 and his Ph.D in Statistics with a minor in Biometry from the Oregon State University in 1973. He has published numerous articles and authored, co-authored, and edited several books in the English and French languages, including *Mathematical Programming in Statistics* (John Wiley 1981, Classic Edition 1993), *Analysis of Experiments with Missing Data* (John Wiley 1985), *Alternative Methods of Regression* (John Wiley 1993), *Premier Pas en Statistique* (Springer 1999), *Adaptive Regression* (Springer 2000), *The Oxford Dictionary of Statistical Terms* (2003), *Statistique: Dictionnaire encyclopédique* (Springer 2004), and *Optimisation appliquée* (Springer 2005). Professor Dodge is an elected member of the International Statistical Institute (1976) and a Fellow of the Royal Statistical Society.

Acceptance Region

The acceptance region is the **interval** within the **sampling distribution** of the test **statistic** that is consistent with the **null hypothesis** H_0 from **hypothesis testing**.
It is the complementary region to the **rejection region**.
The acceptance region is associated with a **probability** $1 - \alpha$, where α is the **significance level** of the test.

MATHEMATICAL ASPECTS
See **rejection region**.

EXAMPLES
See **rejection region**.

FURTHER READING
► **Critical value**
► **Hypothesis testing**
► **Rejection region**
► **Significance level**

Accuracy

The general meaning of accuracy is the proximity of a **value** or a **statistic** to a reference value. More specifically, it measures the proximity of the **estimator** T of the unknown **parameter** θ to the true value of θ.

The accuracy of an **estimator** can be measured by the **expected value** of the squared deviation between T and θ, in other words:

$$E\left[(T - \theta)^2\right].$$

Accuracy should not be confused with the term **precision**, which indicates the degree of exactness of a measure and is usually indicated by the number of decimals after the comma.

FURTHER READING
► **Bias**
► **Estimator**
► **Parameter**
► **Statistics**

Algorithm

An algorithm is a process that consists of a sequence of well-defined steps that lead to the solution of a particular type of problem. This process can be iterative, meaning that it is repeated several times. It is generally a numerical process.

HISTORY
The term algorithm comes from the Latin pronunciation of the name of the ninth century mathematician al-Khwarizmi, who lived in Baghdad and was the father of algebra.

DOMAINS AND LIMITATIONS

The word algorithm has taken on a different meaning in recent years due to the advent of computers. In the field of computing, it refers to a process that is described in a way that can be used in a computer program.

The principal goal of statistical software is to develop a programming language capable of incorporating statistical algorithms, so that these algorithms can then be presented in a form that is comprehensible to the user. The advantage of this approach is that the user understands the results produced by the algorithm and trusts the precision of the solutions. Among various statistical reviews that discuss algorithms, the *Journal of Algorithms* from the *Academic Press* (New York), the part of the *Journal of the Royal Statistical Society Series C (Applied Statistics)* that focuses on algorithms, *Computational Statistics* from *Physica-Verlag* (Heidelberg) and *Random Structures and Algorithms* edited by *Wiley* (New York) are all worthy of special mention.

EXAMPLES

We present here an algorithm that calculates the absolute value of a nonzero number; in other words $|x|$.
Process:

Step 1. Identify the algebraic sign of the given number.

Step 2. If the sign is negative, go to step 3. If the sign is positive, specify the absolute value of the number as the number itself:

$$|x| = x$$

and stop the process.

Step 3. Specify the absolute value of the given number as its opposite number:

$$|x| = -x$$

and stop the process.

FURTHER READING

▶ **Statistical software**
▶ **Yates' algorithm**

REFERENCES

Chambers, J.M.: Computational Methods for Data Analysis. Wiley, New York (1977)

Khwarizmi, Musa ibn Meusba (9th cent.). Jabr wa-al-muqeabalah. The algebra of Mohammed ben Musa, Rosen, F. (ed. and transl.). Georg Olms Verlag, Hildesheim (1986)

Rashed, R.: La naissance de l'algèbre. In: Noël, E. (ed.) Le Matin des Mathématiciens. Belin-Radio France, Paris (1985)

Alternative Hypothesis

An alternative hypothesis is the hypothesis which differs from the hypothesis being tested.
The alternative hypothesis is usually denoted by H_1.

HISTORY

See **hypothesis** and **hypothesis testing**.

MATHEMATICAL ASPECTS

During the **hypothesis testing** of a **parameter** of a **population**, the **null hypothesis** is presented in the following way:

$$H_0: \quad \theta = \theta_0,$$

where θ is the parameter of the population that is to be estimated, and θ_0 is the presumed **value** of this parameter. The alternative hypothesis can then take three different forms:

1. $H_1 : \theta > \theta_0$
2. $H_1 : \theta < \theta_0$
3. $H_1 : \theta \neq \theta_0$

In the first two cases, the **hypothesis test** is called the **one-sided**, whereas in the third case it is called the **two-sided**.

The alternative hypothesis can also take three different forms during the **hypothesis testing** of **parameters** of two **populations**. If the **null hypothesis** treats the two parameters θ_1 and θ_2 equally, then:

$$H_0 : \theta_1 = \theta_2 \text{ or}$$
$$H_0 : \theta_1 - \theta_2 = 0.$$

The alternative hypothesis could then be

- $H_1 : \theta_1 > \theta_2$ or $H_1 : \theta_1 - \theta_2 > 0$
- $H_1 : \theta_1 < \theta_2$ or $H_1 : \theta_1 - \theta_2 < 0$
- $H_1 : \theta_1 \neq \theta_2$ or $H_1 : \theta_1 - \theta_2 \neq 0$

During the comparison of more than two **populations**, the **null hypothesis** supposes that the **values** of all of the **parameters** are identical. If we want to compare k populations, the null hypothesis is the following:

$$H_0 : \theta_1 = \theta_2 = \ldots = \theta_k.$$

The alternative hypothesis will then be formulated as follows:

H_1: the values of $\theta_i (i = 1, \ldots, k)$ are not all identical.

This means that only one **parameter** needs to have a different value to those of the other parameters in order to reject the **null hypothesis** and accept the alternative hypothesis.

EXAMPLES

We are going to examine the alternative hypotheses for three examples of **hypothesis testing**:

1. *Hypothesis testing on the percentage of a population*

   An election candidate wants to know if he will receive more than 50% of the votes. The **null hypothesis** for this problem can be written as follows:

   $$H_0 : \pi = 0.5,$$

   where π is the **percentage** of the **population** to be estimated.

   We carry out a **one-sided test** on the right-hand side that allows us to answer the candidate's question. The alternative hypothesis will therefore be:

   $$H_1 : \pi > 0.5.$$

2. *Hypothesis testing on the mean of a population*

   A bolt maker wants to test the precision of a new machine that should make bolts 8 mm in diameter.

   We can use the following **null hypothesis**:

   $$H_0 : \mu = 8,$$

   where μ is the **mean** of the **population** that is to be estimated.

   We carry out a **two-sided test** to check whether the bolt diameter is too small or too big.

   The alternative hypothesis can be formulated in the following way:

   $$H_1 : \mu \neq 8.$$

3. *Hypothesis testing on a comparison of the means of two populations*

   An insurance company decided to equip its offices with microcomputers. It wants to buy these computers from two different companies so long as there is no significant difference in durability between the two brands. It therefore tests the time that passes before the first breakdown on a **sample** of microcomputers from each brand.

According to the **null hypothesis**, the **mean** of the elapsed time before the first breakdown is the same for each brand:

$$H_0: \mu_1 - \mu_2 = 0.$$

Here μ_1 and μ_2 are the respective means of the two **populations**.

Since we do not know which mean will be the highest, we carry out a **two-sided test**. Therefore the alternative hypothesis will be:

$$H_1: \mu_1 - \mu_2 \neq 0.$$

FURTHER READING
▶ **Analysis of variance**
▶ **Hypothesis**
▶ **Hypothesis testing**
▶ **Null hypothesis**

REFERENCE
Lehmann, E.I., Romann, S.P.: Testing Statistical Hypothesis, 3rd edn. Springer, New York (2005)

Analysis of Binary Data

The study of how the probability of success depends on expanatory variables and grouping of materials.

The analysis of binary data also involves **goodness-of-fit** tests of a sample of binary variables to a theoretical distribution, as well as the study of 2×2 **contingency tables** and their subsequent analysis. In the latter case we note especially **independence tests** between attributes, and **homogeneity tests**.

HISTORY
See **data analysis**.

MATHEMATICAL ASPECTS
Let Y be a binary **random variable** and $X_1, X_2, \ldots, X_k$ be supplementary binary variables. So the **dependence** of Y on the variables $X_1, X_2, \ldots, X_k$ is represented by the following models (the coefficients of which are estimated via the **maximum likelihood**):

1. *Linear model:* $P(Y = 1)$ is expressed as a linear function (in the parameters) of X_i.
2. *Log-linear model:* $\log P(Y = 1)$ is expressed as a linear function (in the parameters) of X_i.
3. *Logistic model:* $\log\left(\frac{P(Y=1)}{P(Y=0)}\right)$ is expressed as a linear function (in the parameters) of X_i.

Models 1 and 2 are easier to interpret. Yet the last one has the advantage that the quantity to be explained takes all possible values of the linear models. It is also important to pay attention to the extrapolation of the model outside of the domain in which it is applied. It is possible that among the independent variables $(X_1, X_2, \ldots, X_k)$, there are categorical variables (eg. binary ones). In this case, it is necessary to treat the nonbinary categorical variables in the following way: let Z be a random variable with m categories. We enumerate the categories from 1 to m and we define $m - 1$ random variables $Z_1, Z_2, \ldots, Z_{m-1}$. So Z_i takes the value 1 if Z belongs to the category represented by this index. The variable Z is therefore replaced by these $m - 1$ variables, the coefficients of which express the influence of the considered category. The reference (used in order to avoid the situation of **collinearity**) will have (for the purposes of comparison with other categories) a parameter of zero.

FURTHER READING
► Binary data
► Data analysis

REFERENCES
Cox, D.R., Snell, E.J.: The Analysis of Binary Data. Chapman & Hall (1989)

Analysis of Categorical Data

The analysis of **categorical data** involves the following methods:

(a) A study of the **goodness-of-fit test**;

(b) The study of a **contingency table** and its subsequent analysis, which consists of discovering and studying relationships between the attributes (if they exist);

(c) An **homogeneity test** of some populations, related to the distribution of a binary qualitative categorical variable;

(d) An examination of the **independence** hypothesis.

HISTORY
The term "contingency", used in the relation to cross tables of **categorical data** was probably first used by **Pearson, Karl** (1904). The **chi-square test**, was proposed by Barlett, M.S. in 1937.

MATHEMATICAL ASPECTS
See **goodness-of-fit** and **contingency table**.

FURTHER READING
► Data
► Data analysis
► Categorical data
► Chi-square goodness of fit test
► Contingency table
► Correspondence analysis
► Goodness of fit test
► Homogeneity test
► Test of independence

REFERENCES
Agresti, A.: Categorical Data Analysis. Wiley, New York (1990)

Bartlett, M.S.: Properties of sufficiency and statistical tests. Proc. Roy. Soc. Lond. Ser. A **160**, 268–282 (1937)

Cox, D.R., Snell, E.J.: Analysis of Binary Data, 2nd edn. Chapman & Hall, London (1990)

Haberman, S.J.: Analysis of Qualitative Data. Vol. I: Introductory Topics. Academic, New York (1978)

Pearson, K.: On the theory of contingency and its relation to association and normal correlation. Drapers' Company Research Memoirs, Biometric Ser. I., pp. 1–35 (1904)

Analysis of Residuals

An analysis of **residuals** is used to test the validity of the statistical model and to control the assumptions made on the error term. It may be used also for outlier detection.

HISTORY
The analysis of residuals dates back to Euler (1749) and Mayer (1750) in the middle of

the eighteenth century, who were confronted with the problem of the **estimation** of **parameters** from **observations** in the field of astronomy. Most of the methods used to analyze residuals are based on the works of Anscombe (1961) and Anscombe and Tukey (1963). In 1973, Anscombe also presented an interesting discussion on the reasons for using graphical methods of analysis. Cook and Weisberg (1982) dedicated a complete book to the analysis of residuals. Draper and Smith (1981) also addressed this problem in a chapter of their work *Applied Regression Analysis*.

MATHEMATICAL ASPECTS

Consider a general model of **multiple linear regression**:

$$Y_i = \beta_0 + \sum_{j=1}^{p-1} \beta_j X_{ij} + \varepsilon_i, \quad i = 1, \ldots, n,$$

where ε_i is the nonobservable random **error** term.

The **hypotheses** for the **errors** ε_i are generally as follows:

- The errors are independent;
- They are normally distributed (they follow a **normal distribution**);
- Their **mean** is equal to zero;
- Their **variance** is constant and equal to σ^2.

Regression analysis gives an **estimation** for Y_i, denoted $\hat{Y}_i$. If the chosen **model** is adequate, the distribution of the **residuals** or "observed **errors**" $e_i = Y_i - \hat{Y}_i$ should confirm these hypotheses.

Methods used to analyze residuals are mainly graphical. Such methods include:
1. Representing the residuals by a frequency chart (for example a **scatter plot**).

2. Plotting the residuals as a function of time (if the chronological order is known).
3. Plotting the residuals as a function of the estimated **values** $\hat{Y}_i$.
4. Plotting the residuals as a function of the **independent variables** X_{ij}.
5. Creating a **Q–Q plot** of the residuals.

DOMAINS AND LIMITATIONS

To validate the analysis, some of the hypotheses need to hold (like for example the normality of the residuals in estimations based on the **mean square**).

Consider a plot of the **residuals** as a function of the estimated **values** $\hat{Y}_i$. This is one of the most commonly used graphical approaches to verifying the validity of a **model**. It consists of placing:

- The **residuals** $e_i = Y_i - \hat{Y}_i$ in increasing **order**;
- The estimated values $\hat{Y}_i$ on the abscissa.

If the chosen **model** is adequate, the **residuals** are uniformly distributed on a horizontal band of points.

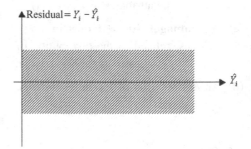

However, if the **hypotheses** for the **residuals** are not verified, the shape of the plot can be different to this. The three figures below show the shapes obtained when:
1. The **variance** σ^2 is not constant. In this case, it is necessary to perform a **transformation** on the **data** Y_i before tackling the **regression analysis**.

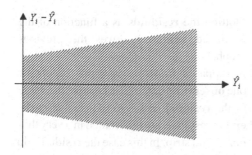

2. The chosen **model** is inadequate (for example, the model is linear but the constant term was omitted when it was necessary).

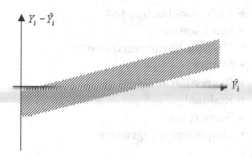

3. The chosen **model** is inadequate (a parabolic tendency is observed).

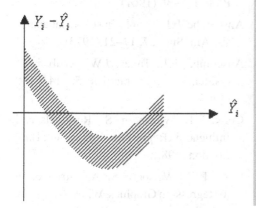

Different **statistics** have been proposed in order to permit numerical measurements that are complementary to the visual techniques presented above, which include those given by Anscombe (1961) and Anscombe and Tukey (1963).

EXAMPLES

In the nineteenth century, a Scottish physicist named Forbe, James D. wanted to estimate the altitude above sea level by measuring the boiling point of water. He knew that the altitude could be determined from the atmospheric pressure; he then studied the relation between pressure and the boiling point of water. Forbe suggested that for an **interval** of observed **values**, a plot of the logarithm of the pressure as a function of the boiling point of water should give a straight line. Since the logarithm of these pressures is small and varies little, we have multiplied these values by 100 below.

X boiling point	Y 100 · log (pressure)
194.5	131.79
194.3	131.79
197.9	135.02
198.4	135.55
199.4	136.46
199.9	136.83
200.9	137.82
201.1	138.00
201.4	138.06
201.3	138.05
203.6	140.04
204.6	142.44
209.5	145.47
208.6	144.34
210.7	146.30
211.9	147.54
212.2	147.80

The **simple linear regression model** for this problem is:

$$Y_i = \beta_0 + \beta_1 X_i + \epsilon_i, \quad i = 1, \ldots, 17.$$

Using the **least squares** method, we can find the following estimation function:

$$\hat{Y}_i = -42.131 + 0.895X_i$$

where $\hat{Y}_i$ is the estimated **value** of **variable** Y for a given X.

For each of these 17 values of X_i, we have an estimated value $\hat{Y}_i$. We can calculate the **residuals**:

$$e_i = Y_i - \hat{Y}_i .$$

These results are presented in the following table:

i	X_i	Y_i	$\hat{Y}_i$	$e_i =$ $Y_i - \hat{Y}_i$
1	194.5	131.79	132.037	−0.247
2	194.3	131.79	131.857	−0.067
3	197.9	135.02	135.081	−0.061
4	198.4	135.55	135.529	0.021
5	199.4	136.46	136.424	0.036
6	199.9	136.83	136.872	−0.042
7	200.9	137.82	137.768	0.052
8	201.1	138.00	137.947	0.053
9	201.4	138.06	138.215	−0.155
10	201.3	138.05	138.126	−0.076
11	203.6	140.04	140.185	−0.145
12	204.6	142.44	141.081	1.359
13	209.5	145.47	145.469	0.001
14	208.6	144.34	144.663	−0.323
15	210.7	146.30	146.543	−0.243
16	211.9	147.54	147.618	−0.078
17	212.2	147.80	147.886	−0.086

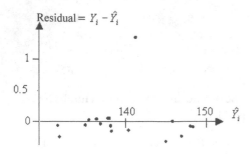

Residual $= Y_i - \hat{Y}_i$

Plotting the **residuals** as a function of the estimated **values** $\hat{Y}_i$ gives the previous graph.

It is apparent from this graph that, except for one **observation** (the 12th), where the value of the **residual** seems to indicate an **outlier**, the residuals are distributed in a very thin horizontal strip. In this case the **residuals** do not provide any reason to doubt the validity of the chosen **model**. By analyzing the standardized residuals we can determine whether the 12th observation is an **outlier** or not.

FURTHER READING
▶ **Anderson–Darling test**
▶ **Least squares**
▶ **Multiple linear regression**
▶ **Outlier**
▶ **Regression analysis**
▶ **Residual**
▶ **Scatterplot**
▶ **Simple linear regression**

REFERENCES
Anscombe, F.J.: Examination of residuals. Proc. 4th Berkeley Symp. Math. Statist. Prob. **1**, 1–36 (1961)

Anscombe, F.J.: Graphs in statistical analysis. Am. Stat. **27**, 17–21 (1973)

Anscombe, F.J., Tukey, J.W.: Analysis of residuals. Technometrics **5**, 141–160 (1963)

Cook, R.D., Weisberg, S.: Residuals and Influence in Regression. Chapman & Hall, London (1982)

Cook, R.D., Weisberg, S.: An Introduction to Regression Graphics. Wiley, New York (1994)

Cook, R.D., Weisberg, S.: Applied Regression Including Computing and Graphics. Wiley, New York (1999)

A

Draper, N.R., Smith, H.: Applied Regression Analysis, 3rd edn. Wiley, New York (1998)

Euler, L.: Recherches sur la question des inégalités du mouvement de Saturne et de Jupiter, pièce ayant remporté le prix de l'année 1748, par l'Académie royale des sciences de Paris. Republié en 1960, dans Leonhardi Euleri, Opera Omnia, 2ème série. Turici, Bâle, 25, pp. 47–157 (1749)

Mayer, T.: Abhandlung über die Umwälzung des Monds um seine Achse und die scheinbare Bewegung der Mondflecken. Kosmographische Nachrichten und Sammlungen auf das Jahr 1748 1, 52–183 (1750)

Analysis of Variance

The analysis of variance is a technique that consists of separating the total variation of **data** set into logical components associated with specific sources of variation in order to compare the **mean** of several **populations**. This analysis also helps us to test certain **hypotheses** concerning the parameters of the model, or to estimate the components of the **variance**. The sources of variation are globally summarized in a component called **error variance**, sometime called within-treatment mean square and another component that is termed "effect" or treatment, sometime called between-treatment mean square.

HISTORY

Analysis of variance dates back to **Fisher, R.A.** (1925). He established the first fundamental principles in this field. Analysis of variance was first applied in the fields of biology and agriculture.

MATHEMATICAL ASPECTS

The analysis of **variance** compares the **means** of three or more random **samples** and determines whether there is a significant difference between the **populations** from which the samples are taken. This technique can only be applied if the random samples are independent, if the population distributions are approximately normal and all have the same variance σ^2.

Having established that the **null hypothesis**, assumes that the **means** are equal, while the **alternative hypothesis** affirms that at least one of them is different, we fix a **significant level**. We then make two **estimates** of the unknown **variance** σ^2:

- The first, denoted s_E^2, corresponds to the mean of the variances of each sample;
- The second, s_{Tr}^2, is based on the variation between the means of the samples.

Ideally, if the **null hypothesis** is verified, these two estimations will be equal, and the F ratio ($F = s_{Tr}^2 / s_E^2$, as used in the **Fisher test** and defined as the quotient of the second estimation of σ^2 to the first) will be equal to 1. The **value** of the F ratio, which is generally more than 1 because of the variation from the **sampling**, must be compared to the value in the **Fisher table** corresponding to the fixed **significant level**. The decision rule consists of either rejecting the null hypothesis if the calculated value is greater than or equal to the tabulated value, or else the **means** are equal, which shows that the **samples** come from the same **population**.

Consider the following **model**:

$$Y_{ij} = \mu + \tau_i + \varepsilon_{ij},$$
$$i = 1, 2, \ldots, t, \quad j = 1, 2, \ldots, n_i.$$

Here

Y_{ij} represents the **observation** j receiving the **treatment** i,

μ is the general **mean** common to all treatments,

τ_i is the actual effect of treatment i on the observation,

ε_{ij} is the experimental error for observation Y_{ij}.

In this case, the **null hypothesis** is expressed in the following way:

$$H_0: \tau_1 = \tau_2 = \ldots = \tau_t,$$

which means that the t treatments are identical.

The **alternative hypothesis** is formulated in the following way:

$$H_1: \text{the } \textbf{values} \text{ of } \tau_i(i = 1, 2, \ldots, t)$$
$$\text{are not all identical}.$$

The following formulae are used:

$$SS_{Tr} = \sum_{i=1}^{t} n_i(\bar{Y}_{i.} - \bar{Y}_{..})^2, \quad s_{Tr}^2 = \frac{SS_{Tr}}{t-1},$$

$$SS_E = \sum_{i=1}^{t} \sum_{j=1}^{n_i} (Y_{ij} - \bar{Y}_{i.})^2, \quad s_E^2 = \frac{SS_E}{N-t},$$

and

$$SS_T = \sum_{i=1}^{t} \sum_{j=1}^{n_i} (Y_{ij} - \bar{Y}_{..})^2$$

or

$$SS_T = SS_{Tr} + SS_E.$$

where

$$\bar{Y}_{i.} = \sum_{j=1}^{n_i} \frac{Y_{ij}}{n_i} \quad \text{is the \textbf{mean} of the } i\text{th set}$$

$$\bar{Y}_{..} = \frac{1}{N} \sum_{i=1}^{t} \sum_{j=1}^{n_i} Y_{ij} \quad \text{is the global mean taken on all the \textbf{observations}, and}$$

$$N = \sum_{i=1}^{t} n_i \quad \text{is the total number of observations.}$$

and finally the value of the F ratio

$$F = \frac{s_{Tr}^2}{s_E^2}.$$

It is customary to summarize the information from the analysis of variance in an analysis of variance table:

Source of variation	Degrees of freedom	Sum of squares	Mean of squares	F
Among treatments	$t - 1$	SS_{Tr}	s_{Tr}^2	$\dfrac{s_{Tr}^2}{s_E^2}$
Within treatments	$N - t$	SS_E	s_E^2	
Total	$N - 1$	SS_T		

DOMAINS AND LIMITATIONS

An analysis of variance is always associated with a **model**. Therefore, there is a different analysis of variance in each distinct case. For example, consider the case where the analysis of variance is applied to **factorial experiments** with one or several **factors**, and these factorial experiments are linked to several **designs of experiment**.

We can distinguish not only the number of **factors** in the **experiment** but also the type of **hypotheses** linked to the effects of the **treatments**. We then have a **model** with fixed effects, a model with variable effects and a model with mixed effects. Each of these requires a specific analysis, but whichever model is used, the basic assumptions of additivity, normality, homoscedasticity and independence must be respected. This means that:

1. The experimental **errors** of the **model** are **random variables** that are independent of each other;

2. All of the errors follow a **normal distribution** with a **mean** of zero and an unknown **variance** σ^2.

All **designs of experiment** can be analyzed using analysis of variance. The most common designs are **completely randomized designs**, **randomized block designs** and **Latin square designs**.

An analysis of variance can also be performed with simple or multiple **linear regression**.

If during an analysis of variance the null hypothesis (the case for equality of means) is rejected, a **least significant difference test** is used to identify the **populations** that have significantly different means, which is something that an analysis of variance cannot do.

EXAMPLES

See **two-way analysis of variance**, **one-way analysis of variance**, **linear multiple regression** and **simple linear regression**.

FURTHER READING

▶ **Design of experiments**
▶ **Factor**
▶ **Fisher distribution**
▶ **Fisher table**
▶ **Fisher test**
▶ **Least significant difference test**
▶ **Multiple linear regression**
▶ **One-way analysis of variance**
▶ **Regression analysis**
▶ **Simple linear regression**
▶ **Two-way analysis of variance**

REFERENCES

Fisher, R.A.: Statistical Methods for Research Workers. Oliver & Boyd, Edinburgh (1925)

Rao, C.R.: Advanced Statistical Methods in Biometric Research. Wiley, New York (1952)

Scheffé, H.: The Analysis of Variance. Wiley, New York (1959)

Anderson, Oskar

Anderson, Oskar (1887–1960) was an important member of the Continental School of Statistics; his contributions touched upon a wide range of subjects, including correlation, time series analysis, nonparametric methods and sample survey, as well as econometrics and statistical applications in social sciences.

Anderson, Oskar received a bachelor degree with distinction from the Kazan Gymnasium and then studied mathematics and physics for a year at the University of Kazan. He then entered the Faculty of Economics at the Polytechnic Institute of St. Petersburg, where he studied mathematics, statistics and economics.

The publications of Anderson, Oskar combine the traditions of the Continental School of Statistics with the concepts of the English Biometric School, particularly in two of his works: "Einführung in die mathematische Statistik" and "Probleme der statistischen Methodenlehre in den Sozialwissenschaften".

In 1949, he founded the journal *Mitteilungsblatt für Mathematische Statistik* with Kellerer, Hans and Münzner, Hans.

Some principal works of Anderson, Oskar:

1935 Einführung in die Mathematische Statistik. Julius Springer, Wien

1954 Probleme der statistischen Methodenlehre in den Sozialwissenschaften. Physica-Verlag, Würzberg

Anderson, Theodore W.

Anderson, Theodore Wilbur was born on the 5th of June 1918 in Minneapolis, in the state of Minnesota in the USA. He became a Doctor of Mathematics in 1945 at the University of Princeton, and in 1946 he became a member of the Department of Mathematical Statistics at the University of Columbia, where he was named Professor in 1956. In 1967, he was named Professor of Statistics and Economics at Stanford University. He was, successively: Fellow of the Guggenheim Foundation between 1947 and 1948; Editor of the *Annals of Mathematical Statistics* from 1950 to 1952; President of the Institute of Mathematical Statistics in 1963; and Vice-President of the American Statistical Association from 1971 to 1973. He is a member of the American Academy of Arts and Sciences, of the National Academy of Sciences, of the Institute of Mathematical Statistics and of the Royal Statistical Society. Anderson's most important contribution to statistics is surely in the domain of multivariate analysis. In 1958, he published the book entitled *An Introduction to Multivariate Statistical Analysis*. This book was the reference work in this domain for over forty years. It has been even translated into Russian.

Some of the principal works and articles of Theodore Wilbur Anderson:

1952   (with Darling, D.A.) Asymptotic theory of certain goodness of fit criteria based on stochastic processes. Ann. Math. Stat. 23, 193–212.

1958   An Introduction to Multivariate Statistical Analysis. Wiley, New York.

1971   The Statistical Analysis of Time Series. Wiley, New York.

1989   Linear latent variable models and covariance structures. J. Econometrics, 41, 91–119.

1992   (with Kunitoma, N.) Asymptotic distributions of regression and autoregression coefficients with Martingale difference disturbances. J. Multivariate Anal., 40, 221–243.

1993   Goodness of fit tests for spectral distributions. Ann. Stat. 21, 830–847.

FURTHER READING
▶ Anderson–Darling test

Anderson–Darling Test

The Anderson–Darling test is a goodness-of-fit test which allows to control the hypothesis that the distribution of a random variable observed in a sample follows a certain theoretical distribution. In particular, it allows us to test whether the empirical distribution obtained corresponds to a **normal distribution**.

HISTORY
Anderson, Theodore W. and Darling D.A. initially used Anderson–Darling statistics, denoted A^2, to test the conformity of a distribution with perfectly specified parameters (1952 and 1954). Later on, in the 1960s and especially the 1970s, some other authors (mostly Stephens) adapted the test to a wider range of distributions where some of the parameters may not be known.

MATHEMATICAL ASPECTS
Let us consider the **random variable** X, which follows the normal distribution with an expectation μ and a variance σ^2, and has a distribution function $F_X(x; \theta)$, where θ is a parameter (or a set of parameters) that

determine, F_X. We furthermore assume θ to be known.

An observation of a sample of size n issued from the variable X gives a distribution function $F_n(x)$. The Anderson–Darling statistic, denoted by A^2, is then given by the weighted sum of the squared deviations $F_X(x; \theta) - F_n(x)$:

$$A^2 = \frac{1}{n} \left(\sum_{i=1}^{n} (F_X(x; \theta) - F_n(x))^2 \right) .$$

Starting from the fact that A^2 is a random variable that follows a certain distribution over the interval $[0; +\infty[$, it is possible to test, for a significance level that is fixed a priori, whether $F_n(x)$ is the realization of the random variable $F_X(X; \theta)$; that is, whether X follows the probability distribution with the distribution function $F_X(x; \theta)$.

Computation of A^2 Statistic

Arrange the observations $x_1, x_2, \ldots, x_n$ in the sample issued from X in ascending order i.e., $x_1 < x_2 < \ldots < x_n$. Note that $z_i = F_X(x_i; \theta)$, $(i = 1, 2, \ldots, n)$. Then compute, A^2 by:

$$A^2 = -\frac{1}{n} \left(\sum_{i=1}^{n} (2i - 1) \left(\ln(z_i) + \ln(1 - z_{n+1-i}) \right) \right) - n .$$

For the situation preferred here (X follows the normal distribution with expectation μ and variance σ^2), we can enumerate four cases, depending on the known parameters μ and σ^2 (F is the distribution function of the standard normal distribution):

1. μ and σ^2 are known, so $F_X(x; (\mu, \sigma^2))$ is perfectly specified. Naturally we then have $z_i = F(w_i)$ where $w_i = \frac{x_i - \mu}{\sigma}$.
2. σ^2 is known but μ is unknown and is estimated using $\bar{x} = \frac{1}{n} \left(\sum_i x_i \right)$, the mean of

the sample. Then, let $z_i = F(w_i)$, where $w_i = \frac{x_i - \bar{x}}{\sigma}$.

3. μ is known but σ^2 is unknown and is estimated using $s'^2 = \frac{1}{n} \left(\sum_i (x_i - u)^2 \right)$. In this case, let $z_i = F(w_i)$, where $w_i = \frac{x_{(i)} - \mu}{s'}$.

4. μ and σ^2 are both unknown and are estimated respectively using $\bar{x}$ and $s^2 = \frac{1}{n-1} \left(\sum_i (x_i - \bar{x})^2 \right)$. Then, let $z_i = F(w_i)$, where $w_i = \frac{x_i - \bar{x}}{s}$.

Asymptotic distributions were found for A^2 by Anderson and Darling for the first case, and by Stephens for the next two cases. For last case, Stephens determined an asymptotic distribution for the transformation: $A^* = A^2(1.0 + \frac{0.75}{n} + \frac{2.25}{n^2})$.

Therefore, as shown below, we can construct a table that gives, depending on the case and the significance level (10%, 5%, 2.5% or 1% below), the limiting values of A^2 (and A^* for the case 4) beyond which the normality hypothesis is rejected:

	Significance level			
Case:	0.1	0.050	0.025	0.01
1: $A^2 =$	1.933	2.492	3.070	3.857
2: $A^2 =$	0.894	1.087	1.285	1.551
3: $A^2 =$	1.743	2.308	2.898	3.702
4: $A^* =$	0.631	0.752	0.873	1.035

DOMAINS AND LIMITATIONS

As the distribution of A^2 is expressed asymptotically, the test needs the sample size n to be large. If this is not the case then, for the first two cases, the distribution of A^2 is not known and it is necessary to perform a transformation of the type $A^2 \longmapsto A^*$, from which A^* can be determined. When $n > 20$, we can avoid such a transformation and so the data in the above table are valid.

The Anderson–Darling test has the advantage that it can be applied to a wide range of distributions (not just a **normal distribution** but also exponential, logistic and gamma distributions, among others). That allows us to try out a wide range of alternative distributions if the initial test rejects the null hypothesis for the distribution of a random variable.

EXAMPLES

The following data illustrate the application of the Anderson–Darling test for the normality hypothesis:

Consider a sample of the heights (in cm) of 25 male students. The following table shows the observations in the sample, and also w_i and z_i. We can also calculate $\bar{x}$ and s from these data: $\bar{x} = 177.36$ and $s = 4.98$. Assuming that F is a standard normal distribution function, we have:

Obs:	x_i	$w_i = \frac{x_i - \bar{x}}{s}$	$z_i = F(w_i)$
1	169	−1.678	0.047
2	169	−1.678	0.047
3	170	−1.477	0.070
4	171	−1.277	0.100
5	173	−0.875	0.191
6	173	−0.875	0.191
7	174	−0.674	0.250
8	175	−0.474	0.318
9	175	−0.474	0.318
10	175	−0.474	0.318
11	176	−0.273	0.392
12	176	−0.273	0.392
13	176	−0.273	0.392
14	179	0.329	0.629
15	180	0.530	0.702
16	180	0.530	0.702
17	180	0.530	0.702
18	181	0.731	0.767
19	181	0.731	0.767
20	182	0.931	0.824
21	182	0.931	0.824

Obs:	x_i	$w_i = \frac{x_i - \bar{x}}{s}$	$z_i = F(w_i)$
22	182	0.931	0.824
23	185	1.533	0.937
24	185	1.533	0.937
25	185	1.533	0.937

We then get $A^2 \cong 0.436$, which gives

$$A^* = A^2 \cdot \left(1.0 + \frac{0.75}{25} + \frac{0.25}{625}\right)$$
$$= A^2 \cdot (1.0336) \cong 0.451.$$

Since we have case 4, and a significance level fixed at 1%, the calculated value of A^* is much less then the value shown in the table (1.035). Therefore, the normality hypothesis cannot be rejected at a significance level of 1%.

FURTHER READING
▸ **Goodness of fit test**
▸ **Histogram**
▸ **Nonparametric statistics**
▸ **Normal distribution**
▸ **Statistics**

REFERENCES
Anderson, T.W., Darling, D.A.: Asymptotic theory of certain goodness of fit criteria based on stochastic processes. Ann. Math. Stat. **23**, 193–212 (1952)

Anderson, T.W., Darling, D.A.: A test of goodness of fit. J. Am. Stat. Assoc. **49**, 765–769 (1954)

Durbin, J., Knott, M., Taylor, C.C.: Components of Cramer-Von Mises statistics, II. J. Roy. Stat. Soc. Ser. B **37**, 216–237 (1975)

Stephens, M.A.: EDF statistics for goodness of fit and some comparisons. J. Am. Stat. Assoc. **69**, 730–737 (1974)

Arithmetic Mean

The arithmetic **mean** is a **measure of central tendency**. It allows us to characterize the center of the **frequency distribution** of a **quantitative variable** by considering all of the observations with the same weight afforded to each (in contrast to the **weighted arithmetic mean**).

It is calculated by summing the observations and then dividing by the number of observations.

HISTORY

The arithmetic mean is one of the oldest methods used to combine observations in order to give a unique approximate value. It appears to have been first used by Babylonian astronomers in the third century BC. The arithmetic mean was used by the astronomers to determine the positions of the sun, the moon and the planets. According to Plackett (1958), the concept of the arithmetic mean originated from the Greek astronomer Hipparchus.

In 1755 Thomas Simpson officially proposed the use of the arithmetic mean in a letter to the President of the Royal Society.

MATHEMATICAL ASPECTS

Let $x_1, x_2, \ldots, x_n$ be a set of n quantities or n observations relating to a **quantitative variable** X.

The arithmetic mean $\bar{x}$ of $x_1, x_2, \ldots, x_n$ is the sum of these observations divided by the number n of observations:

$$\bar{x} = \frac{\sum_{i=1}^{n} x_i}{n} .$$

When the observations are ordered in the form of a **frequency distribution**, the arithmetic mean is calculated in the following way:

$$\bar{x} = \frac{\sum_{i=1}^{k} x_i \cdot f_i}{\sum_{i=1}^{k} f_i} ,$$

where x_i are the different values of the **variable**, f_i are the frequencies associated with these values, k is the number of different values, and the sum of the frequencies equals the number of observations:

$$\sum_{i=1}^{k} f_i = n .$$

To calculate the **mean** of a frequency distribution where values of the quantitative variable X are grouped in classes, we consider that all of the observations belonging to a certain class take the central value of the class, assuming that the observations are uniformly distributed inside the classes (if this **hypothesis** is not correct, the arithmetic mean obtained will only be an approximation.)

Therefore, in this case we have:

$$\bar{x} = \frac{\sum_{i=1}^{k} x_i \cdot f_i}{\sum_{i=1}^{k} f_i} ,$$

where the x_i are the class centers, the f_i are the frequencies associated with each class, and k is the number of classes.

Properties of the Arithmetic Mean

- The algebraic sum of deviations between every value of the set and the arithmetic mean of this set equals 0:

$$\sum_{i=1}^{n} (x_i - \bar{x}) = 0 .$$

- The sum of square deviations from every value to a given number "a" is smallest when "a" is the arithmetic mean:

$$\sum_{i=1}^{n} (x_i - a)^2 \geq \sum_{i=1}^{n} (x_i - \bar{x})^2 .$$

Proof:

We can write:

$$x_i - a = (x_i - \bar{x}) + (\bar{x} - a) .$$

Finding the squares of both members of the equality, summarizing them and then simplifying gives:

$$\sum_{i=1}^{n} (x_i - a)^2$$

$$= \sum_{i=1}^{n} (x_i - \bar{x})^2 + n \cdot (\bar{x} - a)^2 .$$

As $n \cdot (\bar{x} - a)^2$ is not negative, we have proved that:

$$\sum_{i=1}^{n} (x_i - a)^2 \geq \sum_{i=1}^{n} (x_i - \bar{x})^2 .$$

- The arithmetic mean $\bar{x}$ of a **sample** $(x_1, \ldots, x_n)$ is normally considered to be an **estimator** of the **mean** μ of the **population** from which the sample was taken.
- Assuming that x_i are independent random variables with the same **distribution function** for the mean μ and the **variance** σ^2, we can show that
  1. $E[\bar{x}] = \mu$,
  2. $\mathrm{Var}(\bar{x}) = \frac{\sigma^2}{n}$,

  if these moments exist.

Since the **mathematical expectation** of $\bar{x}$ equals μ, the arithmetic mean is an estimator without **bias** of the mean of the population.

- If the x_i result from the **random sampling** without replacement of a finite population with a mean μ, the identity

$$E[\bar{x}] = \mu$$

is still valid, but the variance of $\bar{x}$ must be adjusted by a factor that depends on the size N of the population and the size n of the sample:

$$\mathrm{Var}(\bar{x}) = \frac{\sigma^2}{n} \cdot \left[\frac{N - n}{N - 1} \right] ,$$

where σ^2 is the variance of the population.

Relationship Between the Arithmetic Mean and Other Measures of Central Tendency

- The arithmetic mean is related to two principal measures of central tendency: the **mode** M_0 and the **median** M_d.

  If the distribution is symmetric and unimodal:

$$\bar{x} = M_d = M_0 .$$

  If the distribution is unimodal, it is normally true that:

  $\bar{x} \geq M_d \geq M_0$ if the distribution is stretched to the right,

  $\bar{x} \leq M_d \leq M_0$ if the distribution is stretched to the left.

  For a unimodal, slightly asymmetric distribution, these three measures of the central tendency often approximately satisfy the following relation:

$$(\bar{x} - M_0) = 3 \cdot (\bar{x} - M_d) .$$

- In the same way, for a unimodal distribution, if we consider a set of positive numbers, the **geometric mean** G is always smaller than or equal to the **arithmetic mean** $\bar{x}$, and is always greater than or equal to the **harmonic mean** H. So we have:

$$H \leq G \leq \bar{x}.$$

These three means are identical only if all of the numbers are equal.

DOMAINS AND LIMITATIONS

The arithmetic mean is a simple measure of the central **value** of a set of quantitative observations. Finding the mean can sometimes lead to poor data interpretation:

If the monthly salaries (in Euros) of 5 people are 3000, 3200, 2900, 3500 and 6500, the arithmetic mean of the salary is $\frac{19100}{5} = 3820$. This **mean** gives us some idea of the sizes of the salaries sampled, since it is situated between the biggest and the smallest one. However, 80% of the salaries are smaller then the mean, so in this case it is not a particularly good representation of a typical salary.

This case shows that we need to pay attention to the form of the distribution and the reliability of the observations before we use the arithmetic mean as the **measure of central tendency** for a particular set of values. If an absurd observation occurs in the distribution, the arithmetic mean could provide an unrepresentative value for the central tendency. If some observations are considered to be less reliable then others, it could be useful to make them less important. This can be done by calculating a **weighted arithmetic mean**, or by using the **median**, which is not strongly influenced by any absurd observations.

EXAMPLES

In company A, nine employees have the following monthly salaries (in Euros):

3000 3200 2900 3440 5050

4150 3150 3300 5200

The arithmetic mean of these monthly salaries is:

$$\bar{x} = \frac{(3000 + 3200 + \cdots + 3300 + 5200)}{9}$$

$$= \frac{33390}{9} = 3710 \, \text{Euros}.$$

We now examine a case where the data are presented in the form of a **frequency distribution**.

The following **frequency table** gives the number of days that 50 employees were absent on sick leave during a period of one year:

x_i: Days of illness	f_i: Number of employees
0	7
1	12
2	19
3	8
4	4
Total	50

Let us try to calculate the mean number of days that the employees were absent due to illness.

The total number of sick days for the 50 employees equals the sum of the product of each x_i by its respective **frequency** f_i:

$$\sum_{i=1}^{5} x_i \cdot f_i = 0 \cdot 7 + 1 \cdot 12 + 2 \cdot 19 + 3 \cdot 8$$

$$+ 4 \cdot 4 = 90.$$

The total number of employees equals:

$$\sum_{i=1}^{5} f_i = 7 + 12 + 19 + 8 + 4 = 50.$$

L The arithmetic mean of the number of sick days per employee is then:

$$\bar{x} = \frac{\sum_{i=1}^{5} x_i \cdot f_i}{\sum_{i=1}^{5} f_i} = \frac{90}{50} = 1.8$$

which means that, on average, the 50 employees took 1.8 days off for sickness per year.

In the following example, the data are grouped in classes.

We want to calculate the arithmetic mean of the daily profits from the sale of 50 types of grocery. The frequency distribution for the groceries is given in the following table:

Classes (profits in Euros)	Mid-points x_i	Frequencies f_i (number of groceries)	$x_i \cdot f_i$
500–550	525	3	1575
550–600	575	12	6900
600–650	625	17	10625
650–700	675	8	5400
700–750	725	6	4350
750–800	775	4	3100
Total		50	31950

The arithmetic mean of the profits is:

$$\bar{x} = \frac{\sum_{i=1}^{6} x_i \cdot f_i}{\sum_{i=1}^{6} f_i} = \frac{31950}{50} = 639,$$

which means that, on average, each of the 50 groceries provide a daily profit of 639 Euros.

FURTHER READING

▶ **Geometric mean**
▶ **Harmonic mean**
▶ **Mean**
▶ **Measure of central tendency**
▶ **Weighted arithmetic mean**

REFERENCES

Plackett, R.L.: Studies in the history of probability and statistics. VII. The principle of the arithmetic mean. Biometrika **45**, 130–135 (1958)

Simpson, T.: A letter to the Right Honorable George Earl of Macclesfield, President of the Royal Society, on the advantage of taking the mean of a number of observations in practical astronomy. Philos. Trans. Roy. Soc. Lond. **49**, 82–93 (1755)

Simpson, T.: An attempt to show the advantage arising by taking the mean of a number of observations in practical astronomy. In: Miscellaneous Tracts on Some Curious and Very Interesting Subjects in Mechanics, Physical-Astronomy, and Speculative Mathematics. Nourse, London (1757). pp. 64–75

Arithmetic Triangle

The arithmetic triangle is used to determine **binomial** coefficients $(a + b)^n$ when calculating the number of possible combinations of k objects out of a total of n objects (C_n^k).

HISTORY

The notion of finding the number of combinations of k objects from n objects in total has been explored in India since the ninth century. Indeed, there are traces of it in the

Meru Prastara written by Pingala in around 200 BC.

Between the fourteenth and the fifteenth centuries, al-Kashi, a mathematician from the Iranian city of Kashan, wrote *The Key to Arithmetic*. In this work he calls **binomial** coefficients "exponent elements".

In his work *Traité du Triangle Arithmétique*, published in 1665, Pascal, Blaise (1654) defined the numbers in the "arithmetic triangle", and so this triangle is also known as Pascal's triangle.

We should also note that the triangle was made popular by Tartaglia, Niccolo Fontana in 1556, and so Italians often refer to it as Tartaglia's triangle, even though Tartaglia did not actually study the arithmetic triangle.

MATHEMATICAL ASPECTS

The arithmetic triangle has the following form:

$$
\begin{array}{ccccccccccccc}
 & & & & & & 1 & & & & & & \\
 & & & & & 1 & & 1 & & & & & \\
 & & & & 1 & & 2 & & 1 & & & & \\
 & & & 1 & & 3 & & 3 & & 1 & & & \\
 & & 1 & & 4 & & 6 & & 4 & & 1 & & \\
 & 1 & & 5 & & 10 & & 10 & & 5 & & 1 & \\
1 & & 6 & & 15 & & 20 & & 15 & & 6 & & 1 \\
\end{array}
$$

$$\cdots$$

Each element is a **binomial** coefficient

$$
\begin{aligned}
C_n^k &= \frac{n!}{k! \, (n-k)!} \\
&= \frac{n \cdot (n-1) \cdot \ldots \cdot (n-k+1)}{1 \cdot 2 \cdot \ldots \cdot k} .
\end{aligned}
$$

This coefficient corresponds to the element k of the line $n+1$, $k = 0, \ldots, n$.

Any particular number is obtained by adding together its neighboring numbers in the previous line.

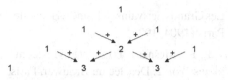

For example:

$$C_6^4 = C_5^3 + C_5^4 = 10 + 5 = 15 .$$

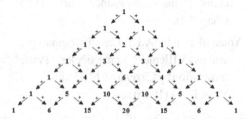

More generally, we have the **relation**:

$$C_n^k + C_n^{k+1} = C_{n+1}^{k+1} ,$$

because:

$$
\begin{aligned}
C_n^k + C_n^{k+1} &= \frac{n!}{(n-k)! \cdot k!} \\
&\quad + \frac{n!}{(n-k-1)! \cdot (k+1)!} \\
&= \frac{n! \cdot [(k+1) + (n-k)]}{(n-k)! \cdot (k+1)!} \\
&= \frac{(n+1)!}{(n-k)! \cdot (k+1)!} \\
&= C_{n+1}^{k+1} .
\end{aligned}
$$

FURTHER READING

▶ **Binomial**

▶ **Binomial distribution**

▶ **Combination**

▶ **Combinatory analysis**

REFERENCES

Pascal, B.: Traité du triangle arithmétique (publ. posthum. in 1665), Paris (1654)

Pascal, B.: Œuvres, vols. 1–14. Brunschvicg, L., Boutroux, P., Gazier, F. (eds.)

Les Grands Ecrivains de France. Hachette, Paris (1904–1925)

Pascal, B.: Mesnard, J. (ed.) Œuvres complètes. Vol. 2. Desclée de Brouwer, Paris (1970)

Rashed, R.: La naissance de l'algèbre. In: Noël, E. (ed.) Le Matin des Mathématiciens. Belin-Radio France, Paris (1985) Chap. 12).

Youschkevitch, A.P.: Les mathématiques arabes (VIIIème-XVème siècles). Partial translation by Cazenave, M., Jaouiche, K. Vrin, Paris (1976)

ARMA Models

ARMA models (sometimes called Box-Jenkins models) are autoregressive moving average models used in time series analysis. The autoregressive part, denoted AR, consists of a finite linear combination of previous observations. The moving average part, MA, consists of a finite linear combination in t of the previous values for a white noise (a sequence of mutually independent and identically distributed random variables).

MATHEMATICAL ASPECTS

1. AR model (autoregressive)

In an autoregressive process of order p, the present observation y_t is generated by a weighted mean of the past observations up to the pth period. This takes the following form:

$$AR(1): y_t = \theta_1 y_{t-1} + \varepsilon_t,$$
$$AR(2): y_t = \theta_1 y_{t-1} + \theta_2 y_{t-2} + \varepsilon_t,$$
$$\vdots$$

$$AR(p): y_t = \theta_1 y_{t-1} + \theta_2 y_{t-2} + \dots + \theta_p y_{t-p} + \varepsilon_t,$$

where $\theta_1, \theta_2, \dots, \theta_p$ are the positive or negative parameters to be estimated and ε_t is the error factor, which follows a normal distribution.

2. MA model (moving average)

In a moving average process of order q, each observation y_t is randomly generated by a **weighted arithmetic mean** until the qth period:

$$MA(1): y_t = \varepsilon_t - \alpha_1 \varepsilon_{t-1}$$
$$MA(2): y_t = \varepsilon_t - \alpha_1 \varepsilon_{t-1} - \alpha_2 \varepsilon_{t-2}$$
$$\dots$$

$$MA(p): y_t = \varepsilon_t - \alpha_1 \varepsilon_{t-1} - \alpha_2 \varepsilon_{t-2} - \dots - \alpha_q \varepsilon_{t-q},$$

where $\alpha_1, \alpha_2, \dots, \alpha_q$ are positive or negative parameters and ε_t is the Gaussian random error.

The MA model represents a time series fluctuating about its mean in a random manner, which gives rise to the term "moving average", because it smoothes the series, subtracting the white noise generated by the randomness of the element.

3. $ARMA$ model (autoregressive moving average model)

$ARMA$ models represent processes generated from a combination of past values and past errors. They are defined by the following equation:

$$ARMA(p, q):$$
$$y_t = \theta_1 y_{t-1} + \theta_2 y_{t-2} + \dots + \theta_p y_{t-p} + \varepsilon_t - \alpha_1 \varepsilon_{t-1} - \alpha_2 \varepsilon_{t-2} - \dots - \alpha_q \varepsilon_{t-q},$$

with $\theta_p \neq 0, \alpha_q \neq 0$, and $(\varepsilon_t, t \in Z)$ is a weak white noise.

FURTHER READING
▶ Time series
▶ Weighted arithmetic mean

REFERENCES
Box, G.E.P., Jenkins, G.M.: Time Series Analysis: Forecasting and Control (Series in Time Series Analysis). Holden Day, San Francisco (1970)

Arrangement

Arrangements are a concept found in **combinatory analysis**.

The number of arrangements is the number of ways drawing k objects from n objects where the order in which the objects are drawn is taken into account (in contrast to **combinations**).

HISTORY
See **combinatory analysis**.

MATHEMATICAL ASPECTS
1. *Arrangements without repetitions*
   An arrangement without repetition refers to the situation where the objects drawn are not placed back in for the next drawing. Each object can then only be drawn once during the k drawings.
   The number of arrangements of k objects amongst n without repetition is equal to:

$$A_n^k = \frac{n!}{(n-k)!} \,.$$

2. *Arrangements with repetitions*
   Arrangements with repetition occur when each object pulled out is placed back in for the next drawing. Each object can then be drawn r times from k drawings, $r = 0, 1, \ldots, k$.

The number of arrangements of k objects amongst n with repetitions is equal to n to the power k:

$$A_n^k = n^k \,.$$

EXAMPLES
1. *Arrangements without repetitions*
   Consider an urn containing six balls numbered from 1 to 6. We pull out four balls from the urn in succession, and we want to know how many numbers it is possible to form from the numbers of the balls drawn. We are then interested in the number of arrangements (since we take into account the order of the balls) without repetition (since each ball can be pulled out only once) of four objects amongst six. We obtain:

$$A_n^k = \frac{n!}{(n-k)!} = \frac{6!}{(6-4)!} = 360$$

possible arrangements. Therefore, it is possible to form 360 different numbers by drawing four numbers from the numbers 1,2,3,4,5,6 when each number can appear only once in the four-digit number formed.

As a second example, let us investigate the arrangements without repetitions of two letters from the letters A, B and C. With $n = 3$ and $k = 2$ we have:

$$A_n^k = \frac{n!}{(n-k)!} = \frac{3!}{(3-2)!} = 6 \,.$$

We then obtain:
AB, AC, BA, BC, CA, CB.

2. *Arrangements with repetitions*
   Consider the same urn as described previously. We perform four successive drawings, but this time we put each ball drawn back in the urn.

We want to know how many four-digit numbers (or arrangements) are possible if four numbers are drawn.

In this case, we are investigating f the number of arrangements with repetition (since each ball is placed back in the urn before the next drawing). We obtain

$$A_n^k = n^k = 6^4 = 1296$$

different arrangements. It is possible to form 1296 four-digit numbers from the numbers 1,2,3,4,5,6 if each number can appear more than once in the four-digit number.

As a second example we again take the three letters A, B and C and form an arrangement of two letters with repetitions. With $n = 3$ and $k = 2$, we have:

$$A_n^k = n^k = 3^2 = 9.$$

We then obtain:

AA, AB, AC, BA, BB, BC, CA, CB, CC.

FURTHER READING
▶ **Combination**
▶ **Combinatory analysis**
▶ **Permutation**

REFERENCES
See **combinatory analysis**.

Attributable Risk

The attributable risk is the difference between the **risk** encountered by individuals exposed to a particular factor and the risk encountered by individuals who are not exposed to it. This is the opposite to **avoidable risk**. It measures the absolute effect of a cause (that is, the excess **risk** or cases of illness).

HISTORY
See **risk**.

MATHEMATICAL ASPECTS
By definition we have:

attributable risk = risk for those exposed
− risk for those not exposed .

DOMAINS AND LIMITATIONS
The confidence interval of an attributable risk is equivalent to the confidence interval of the difference between the proportions p_E and p_{NE}, where p_E and p_{NE} represent the risks encountered by individuals exposed and not exposed to the studied factor, respectively. Take n_E and n_{NE} to be, respectively, the size of the exposed and nonexposed populations. Then, for a confidence level of $(1 - \alpha)$, is given by:

$$(p_E - p_{NE}) \pm z_\alpha \sqrt{\tfrac{p_E \cdot (1 - p_E)}{n_E} + \tfrac{p_{NE} \cdot (1 - p_{NE})}{n_{NE}}},$$

where z_α the **value** obtained from the **normal table** (for example, for a confidence interval of 95%, $\alpha = 0.05$ and $z_\alpha = 1.96$). The confidence interval for $(1 - \alpha)$ for an avoidable risk has bounds given by:

$$(p_{NE} - p_E) \pm z_\alpha \sqrt{\tfrac{p_E \cdot (1 - p_E)}{n_E} \cdot \tfrac{p_{NE} \cdot (1 - p_{NE})}{n_{NE}}}.$$

Here, n_E and n_{NE} need to be large. If the confidence interval includes zero, we cannot rule out an absence of attributable risk.

EXAMPLES
As an example, we consider a study of the risk of breast cancer in women due to smoking:

Group	Incidence rate (/100000 /year)	Attributable to risk from smoking (A) (/100000 /year)
Nonex-posed	57.0	57.0 − 57.0 = 0
Passive smokers	126.2	126.2 − 57.0 = 69.2
Active smokers	138.1	138.1 − 57.0 = 81.1
Total	114.7	114.7 − 57.0 = 57.7

The risks attributable to passive and active smoking are respectively 69 and 81 (/100000 year). In other words, if the exposure to tobacco was removed, the incidence rate for active smokers (138/100000 per year) could be reduced by 81/100000 per year and that for passive smokers (126/100000 per year) by 69/100000 per year. The incidence rates in both categories of smokers would become equal to the rate for nonexposed women (57/100000 per year). Note that the incidence rate for nonexposed women is not zero, due to the influence of other factors aside from smoking.

Group	No. indiv. observed over two years	Cases attrib. to smoking (for two-year period)	Cases attrib. to smoking (per year)
Nonex-posed	70160	0.0	0.0
Passive smok-ers	110860	76.7	38.4
Active smok-ers	118636	96.2	48.1
Total	299656	172.9	86.5

We can calculate the number of cases of breast cancer attributable to tobacco exposure by multiplying the number of individuals observed per year by the attributable risk. By dividing the number of incidents attributable to smoking in the two-year period by two, we obtain the number of cases attributable to smoking per year, and we can then determine the risk attributable to smoking in the population, denoted PAR, as shown in the following example. The previous table shows the details of the calculus.

We describe the calculus for the passive smokers here. In the two-year study, 110860 passive smokers were observed. The risk attributable to the passive smoking was 69.2/100000 per year. This means that the number of cases attributable to smoking over the two year period is (110860 · 69.2)/100000 = 76.7. If we want to calculate the number of cases attributable to passive smoking per year, we must then divide the last value by 2, obtaining 38.4. Moreover, we can calculate the risk attributable to smoking per year simply by dividing the number of cases attributable to smoking for the two-year period (172.9) by the number of individuals studied during these two years (299656 persons). We then obtain the risk attributable to smoking as 57.7/100000 per year. We note that we can get the same result by taking the difference between the total incidence rate (114.7/100000 per year, see the examples under the entries for **incidence rate**, **prevalence rate**) and the incidence rate of the nonexposed group (57.0/100000 per year).

The risk of breast cancer attributable to smoking in the population (PAR) is the ratio of the number of the cases of breast cancer attributable to exposure to tobacco and the number of cases of breast cancer diagnosed in the population (see the above table). The attributable risk in the population is 22.3% (38.4/172) for passive smoking and 28% (48.1/172) for active smoking. For both forms of exposure, it is 50.3% (22.3% + 28%). So, half of the cases of breast cancer diagnosed each year in this population are attributable to smoking (active or passive).

Group	Case attrib. to smoking (for a two-year period)	Case attrib. to smoking (per year)	PAR
Nonex-posed	0.0	20	0.0
Passive smok-ers	38.4	70	22.3
Active smok-ers	48.1	82	28.0
Total	86.5	172	50.3

FURTHER READING
- ▶ Avoidable risk
- ▶ Cause and effect in epidemiology
- ▶ Incidence rate
- ▶ Odds and odds ratio
- ▶ Prevalence rate
- ▶ Relative risk
- ▶ Risk

REFERENCES
Cornfield, J.: A method of estimating comparative rates from clinical data. Applications to cancer of the lung, breast, and cervix. J. Natl. Cancer Inst. **11**, 1269–75 (1951)

Lilienfeld, A.M., Lilienfeld, D.E.: Foundations of Epidemiology, 2nd edn. Clarendon, Oxford (1980)

MacMahon, B., Pugh, T.F.: Epidemiology: Principles and Methods. Little Brown, Boston, MA (1970)

Morabia, A.: Epidemiologie Causale. Editions Médecine et Hygiène, Geneva (1996)

Morabia, A.: L'Épidémiologie Clinique. Editions "Que sais-je?". Presses Universitaires de France, Paris (1996)

Autocorrelation

Autocorrelation, denoted ρ_k, is a measure of the correlation of a particular **time series** with the same time series delayed by k lags (the distance between the observations that are so correlated). It is obtained by dividing the **covariance** between two observations, separated by k lags, of a time series (autocovariance) by the **standard deviation** of y_t and y_{t-k}. If the autocorrelation is calculated for all values of k we obtain the **autocorrelation function**. For a time series that does not change over time, the autocorrelation function decreases exponentially to 0.

HISTORY
The first research into autocorrelation, the partial autocorrelation and the correlogram was performed in the 1920s and 1930s by Yule, George, who developed the theory of autoregressive processes.

MATHEMATICAL ASPECTS
We define the autocorrelation of time series Y_t by:

$$\rho_k = \frac{\text{cov}(y_t, y_{t-k})}{\sigma_{y_t}\sigma_{y_{t-k}}}$$

$$= \frac{\displaystyle\sum_{t=k+1}^{T} (y_t - \bar{y})(y_{t-k} - \bar{y})}{\sqrt{\displaystyle\sum_{t=k+1}^{T} (y_t - \bar{y})^2} \sqrt{\displaystyle\sum_{t=k+1}^{T} (y_{t-k} - \bar{y})^2}} .$$

Here $\bar{y}$ is the mean of the series calculated on $T - k$ lags, where T is the number of observations.

We find out that:

$$\rho_0 = 1 \quad \text{and}$$

$$\rho_k = \rho_{-k} .$$

It is possible to estimate the autocorrelation (denoted $\widehat{\rho}_k$) provided the number of observations is large enough $(T > 30)$ using the following formula:

$$\rho_k = \frac{\displaystyle\sum_{t=k+1}^{T} (y_t - \bar{y})(y_{t-k} - \bar{y})}{\displaystyle\sum_{t=1}^{T} (y_t - \bar{y})^2}$$

The partial autocorrelation function for a delay of k lags is defined as the autocorrelation between y_t and y_{t-k}, the influence of other variables is moved by k lags $(y_{t-1}, y_{t-2}, \ldots, y_{t-k+1})$.

Hypothesis Testing

When analyzing the autocorrelation function of a time series, it can be useful to know the terms ρ_k that are significantly different from 0. Hypothesis testing then proceeds as follows:

$$H_0 : \rho_k = 0$$

$$H_1 : \rho_k \neq 0 .$$

For a large sample $(T > 30)$, the coefficient ρ_k tends asymptotically to a **normal distribution** with a mean of 0 and a standard deviation of $\frac{1}{\sqrt{T}}$. The **Student test** is based on the comparison of an empirical t and a theoretical t.

The confidence interval for the coefficient ρ_k is given by:

$$\rho_k = 0 \pm t_{\alpha/2} \frac{1}{\sqrt{T}} .$$

If the calculated coefficient $\widehat{\rho}_k$ does not fall within this confidence interval, it is significantly different from 0 at the level α (generally $\alpha = 0.05$ and $t_{\alpha/2} = 1.96$).

DOMAINS AND LIMITATIONS

The partial autocorrelation function is principally used in studies of time series and, more specifically, when we want to adjust an ARMA model. These functions are also used in **spatial statistics**, although in the context of spatial autocorrelation, where we investigate the correlation of a variable with itself in space. If the presence of a phenomenon in a particular spatial region affects the probability of the phenomenon being present in neighboring regions, the phenomenon displays spatial autocorrelation. In this case, positive autocorrelation occurs when the neighboring regions tend to have identical properties or similar values (examples include homogeneous regions and regular gradients). Negative autocorrelation occurs when the neighboring regions have different qualities, or alternate between strong and weak values for the phenomenon. Autocorrelation measures depend on the scaling of the variables which are used in the analysis as well as on the grid that registers the observations.

EXAMPLES

We take as an example the national average wage in Switzerland from 1950 to 1994, measured every two years.

We calculate the autocorrelation function between the data; we would like to find a positive autocorrelation. The following figures show the presence of this autocorrelation.

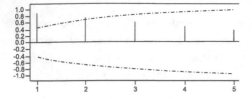

We note that the correlation significance peaks between the observation at time t and the observation at time $t-1$, and also between the observation at time t and the observation at time $t-2$. This data configuration is typical of an autoregressive process. For two first values, we can see that this autocorrelation is significant, because the Student statistic t for the $T = 23$ observations gives:

$$\rho_k = 0 \pm 1.96 \frac{1}{\sqrt{23}} \, .$$

Year	National average wage
50	11999
52	12491
54	13696
56	15519
58	17128
60	19948
62	23362
64	26454
66	28231
68	30332
70	33955
72	40320
74	40839
76	37846
78	39507
80	41180

Year	National average wage
82	42108
84	44095
86	48323
88	51584
90	55480
92	54348
94	54316

Source: Swiss Federal Office of Statistics

FURTHER READING
▶ **Student test**
▶ **Time series**

REFERENCES

Bourbonnais, R.: Econométrie, manuel et exercices corrigés, 2nd edn. Dunod, Paris (1998)

Box, G.E.P., Jenkins, G.M.: Time Series Analysis: Forecasting and Control (Series in Time Series Analysis). Holden Day, San Francisco (1970)

Chatfield, C.: The Analysis of Time Series: An Introduction, 4th edn. Chapman & Hall (1989)

Avoidable Risk

The avoidable risk (which, of course, is avoidable if we neutralize the effect of exposure to a particular phenomenon) is the opposite to the **attributable risk**. In other words, it is the difference between the **risk** encountered by nonexposed individuals and that encountered by individuals exposed to the phenomenon.

HISTORY
See **risk**.

MATHEMATICAL ASPECTS

By definition we have:

$$\text{avoidable risk} = \text{risk if not exposed} - \text{risk if exposed} .$$

DOMAINS AND LIMITATIONS

The avoidable risk was introduced in order to avoid the need for defining a negative **attributable risk**. It allows us to calculate the number of patients that will need to be treated, because:

$$\frac{\text{Number of patients}}{\text{to be treated}} = \frac{1}{\text{Avoidable risk}} .$$

See also **attributable risk**.

EXAMPLES

As an example, consider a study of the efficiency of a drug used to treat an illness.

The 223 patients included in the study are all at risk of contracting the illness, but they have not yet done so. We separate them into two groups: patients in the first group (114 patients) received the drug; those in the second group (109 patients) were given a placebo. The study period was two years. In total, 11 cases of the illness are diagnosed in the first group and 27 in the placebo group.

Group	Cases of illness	Number of patients in the group	Risk for the two-year period
	(A)	(B)	(A/B in %)
1st group	11	114	9.6%
2nd group	27	109	24.8%

So, the avoidable risk due to the drug is $24.8 - 9.6 = 15.2\%$ per two years.

FURTHER READING

► **Attributable risk**
► **Cause and effect in epidemiology**
► **Incidence rate**
► **Odds and odds ratio**
► **Prevalence rate**
► **Relative risk**
► **Risk**

REFERENCES

Cornfield, J.: A method of estimating comparative rates from clinical data. Applications to cancer of the lung, breast, and cervix. J. Natl. Cancer Inst. **11**, 1269–75 (1951)

Lilienfeld, A.M., Lilienfeld, D.E.: Foundations of Epidemiology, 2nd edn. Clarendon, Oxford (1980)

MacMahon, B., Pugh, T.F.: Epidemiology: Principles and Methods. Little Brown, Boston, MA (1970)

Morabia, A.: Epidemiologie Causale. Editions Médecine et Hygiène, Geneva (1996)

Morabia, A.: L'Épidémiologie Clinique. Editions "Que sais-je?". Presses Universitaires de France, Paris (1996)

Bar Chart

Bar chart is a type of **quantitative graph**. It consists of a series of vertical or horizontal bars of identical width but with lengths relative to the represented quantities.

Bar charts are used to compare the **categories** of a **categorical qualitative variable** or to compare sets of **data** from different years or different places for a particular **variable**.

HISTORY

See **graphic representation**.

MATHEMATICAL ASPECTS

A vertical axis and a horizontal axis must be defined in order to construct a vertical bar chart.

The horizontal axis is divided up into different **categories**; the vertical axis shows the value of each category.

To construct a horizontal bar chart, the axes are simply inverted.

The bars must all be of the same width since only their lengths are compared.

Shading, hatching or color can be used to make it easier to understand the the graphic.

DOMAINS AND LIMITATIONS

A bar chart can also be used to represent negative category values. To be able to do this, the **scale** of the axis showing the category values must extend below zero.

There are several types of bar chart. The one described above is called a simple bar chart. A multiple bar chart is used to compare several **variables**.

A composite bar chart is a multiple bar chart where the different sets of **data** are stacked on top of each other. This type of diagram is used when the different data sets can be combined into a total population, and we would like to compare the changes in the data sets and the total population over time.

There is another way of representing the subsets of a total population. In this case, the total population represents 100% and value given for each subset is a **percentage** of the total (also see **pie chart**).

EXAMPLES

Let us construct a *bar chart divided into percentages* for the **data** in the following **frequency table**:

Marital status in a sample of the Australian female population on the 30th June 1981 (in thousands)

Marital status	Fre-quency	Relative fre-quency	Percent-age
Bachelor	6587.3	0.452	45.2
Married	6836.8	0.469	46.9
Divorced	403.5	0.028	2.8
Widow	748.7	0.051	5.1
Total	14576.3	1.000	100.0

Source: ABS (1984) Australian Pocket Year Book. Australian Bureau of Statistics, Canberra, p. 11

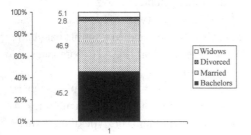

FURTHER READING
► **Graphical representation**
► **Quantitative graph**

Barnard, George A.

Barnard, George Alfred was born in 1915, in Walthamstow, Essex, England. He gained a degree in mathematics from Cambridge University in 1936. Between 1942 and 1945 he worked in the Ministry of Supply as a scientific consultant. Barnard joined the Mathematics Department at Imperial College London from 1945 to 1966. From 1966 to 1975 he was Professor of Mathematics in the University of Essex, and from 1975 until his retirement in 1981 he was Professor of Statistics at the University of Waterloo, Canada.

Barnard, George Alfred received numerous distinctions, including a gold medal from the Royal Statistical Society and from the Institute of Mathematics and its Applications. In 1987 he was named an Honorary Member of the International Statistical Institute. He died in 2002 in August.

Some articles of Barnard, George Alfred:

1954 Sampling inspection and statistical decisions. J. Roy. Stat. Soc. Ser. B 16, 151–174.

1958 Thomas Bayes – A biographical note. Biometrika 45, 293–315.

1989 On alleged gains in power from lower p-values. Stat. Mcd., 8, 1469–1477.

1990 Must clinical trials be large? The interpretation of p-values and the combination of test results. Stat. Med., 9, 601–614.

Bayes' Theorem

If we consider the set of the "reasons" that an **event** occurs, Bayes' theorem gives a formula for the **probability** that the event is the direct result of a particular reason.
Therefore, Bayes' theorem can be interpreted as a formula for the **conditional probability** of an event.

HISTORY
Bayes' theorem is named after **Bayes, Thomas,** and was developed in the middle of eighteenth century. However, Bayes did not publish the theorem during his lifetime; instead, it was presented by Price, R. on the 23rd December 1763, two years after his death, to the Royal Society of London,

which Bayes was a member of during the last twenty last years of his life.

MATHEMATICAL ASPECTS

Let $\{A_1, A_2, \ldots, A_k\}$ be a partition of the **sample space** Ω. We suppose that each event $A_1, \ldots, A_k$ has a nonzero **probability**. Let E be an event such that $P(E) > 0$.

So, for every $i (1 \le i \le k)$, Bayes' theorem (for the discrete case) gives:

$$P(A_i|E) = \frac{P(A_i) \cdot P(E|A_i)}{\sum_{j=1}^{k} P(A_j) \cdot P(E|A_j)} .$$

In the continuous case, where X is a **random variable** with **density function** $f(x)$, said also to be an a priori density function, Bayes' theorem gives the density a posteriori according to

$$f(x|E) = \frac{f(x) \cdot P(E|X = x)}{\int_{-\infty}^{\infty} f(t) \cdot P(E|X = t) \, d_t} .$$

DOMAINS AND LIMITATIONS

Bayes' theorem has been the object of much controversy, relating to the ability to use it when the values of the probabilities used to determine the **probability function** a posteriori are not generally established in a precise way.

EXAMPLES

Three urns contain red, white and black balls:
- Urn A contains 5 red balls, 2 white balls and 3 black balls;
- Urn B contains 2 red balls, 3 white balls and 1 black balls;
- Urn C contains 5 red balls, 2 white balls and 5 black balls.

Randomly choosing an urn, we draw a ball at random: it is white. We wish to determine the **probability** that it was taken from urn A. Let A_1 correspond to the **event** where we "choose urn A", A_2 be the event where we "choose urn B," and A_3 be the event where we "choose urn C." $\{A_1, A_2, A_3\}$ forms a partition of the **sample space**.

Let E be the event where "the ball taken is white," which has a strictly positive probability.

We have:

$$P(A_1) = P(A_2) = P(A_3) = \tfrac{1}{3},$$
$$P(E|A_1) = \tfrac{2}{10}, \quad P(E|A_2) = \tfrac{3}{6},$$
$$\text{and} \quad P(E|A_3) = \tfrac{2}{12}.$$

Bayes' formula allows us to determine the probability that the drawn white ball comes from the urn A:

$$P(A_1|E) = \frac{P(A_1) \cdot P(E|A_1)}{\sum_{i=1}^{3} P(A_i) \cdot P(E|A_i)}$$
$$= \frac{\tfrac{1}{3} \cdot \tfrac{2}{10}}{\tfrac{1}{3} \cdot \tfrac{2}{10} + \tfrac{1}{3} \cdot \tfrac{3}{6} + \tfrac{1}{3} \cdot \tfrac{2}{12}} = \tfrac{3}{13} .$$

FURTHER READING
▶ **Conditional probability**
▶ **Probability**

REFERENCE

Bayes, T.: An essay towards solving a problem in the doctrine of chances. Philos. Trans. Roy. Soc. Lond. **53**, 370–418 (1763). Published, by the instigation of Price, R., 2 years after his death. Republished with a biography by Barnard, George A. in 1958 and in Pearson, E.S., Kendall, M.G.: Studies in the History of Statistics and Probability. Griffin, London, pp. 131–153 (1970)

Bayes, Thomas

Bayes, Thomas (1702–1761) was the eldest son of Bayes, Joshua, who was one of the first six Nonconformist ministers to be ordained in England, and was a member of the Royal Society. He was privately schooled by professors, as was customary in Nonconformist families. In 1731 he became reverend of the Presbyterian chapel in Tunbridge Wells, a town located about 150 km south-west of London. Due to some religious publications he was elected a Fellow of the Royal Society in 1742.

His interest in mathematics was well-known to his contemporaries, despite the fact that he had not written any technical publications, because he had been tutored by **De Moivre, A.**, one of the founders of the theory of **probability**. In 1763, Price, R. sorted through the papers left by Bayes and had his principal work published:

1763 An essay towards solving a problem in the doctrine of chances. Philos. Trans. Royal Soc. London, 53, pp. 370–418. Republished with a biography by Barnard, G.A. (1958). In: Pearson, E.S. and Kendall, M (1970). Studies in the History of Statistics and Probability. Griffin, London, pp. 131–153.

FURTHER READING
▶ **Bayes' theorem**

Bayesian Statistics

Bayesien statistics is a large domain in the field of statistics that differs due to an axiomatization of the statistics that gives it a certain internal coherence.

The basic idea is to interpret the **probability** of an **event** as it is commonly used; in other words as the uncertainty that is related to it. In contrast, the classical approach considers the probability of an event to be the limit of the relative frequency (see **probability** for a more formal approach).

The most well-known aspect of Bayesian **inference** is the probability of calculating the **joint probability distribution** (or **density function**) $f(\theta, X = x_1, \ldots, X = x_n)$ of one or many parameters θ (one parameter or a vector of parameters) having observed the data $x_1, \ldots, x_n$ sampled independently from a random variable X on which θ depends. (It is worth noting that it also allows us to calculate the probability distribution for a new observation x_{n+1}).

Bayesian statistics treat the unknown parameters as random variables not because of possible variability (in reality, the unknown parameters are considered to be fixed), but because of our ignorance or uncertainty about them.

The posterior distribution $f(\theta | X = x_1, \ldots, X = x_n)$ is direct to compute since it is the prior $(f(\theta))$ times the likelihood $f(X = x_1, \ldots, X = x_n | \theta)$.

posterior $\propto$ prior x likelihood

The second expression does not cause problems, because it is a function that we often use in classical statistics, known as the likelihood (see **maximum likelihood**).

In contrast, the first part supposes a prior distribution for θ. We often use the initial distribution of θ to incorporate possible supplementary information about the parameters of interest. In the absence of this information, we use a reference function that maximizes the lack of information (which is then the most "objective" or "noninformative" function, following the common but not precise usage).

Once the distribution $f(\theta \,|x_1,\ldots,x_n)$ is calculated, all of the information on the parameters of interest is available. Therefore, we can calculate plausible values for the unknown parameter (the **mean**, the **median** or some other **measure of central tendency**), its **standard deviation**, confidence intervals, or perform hypothesis testing on its value.

HISTORY

See **Bayes, Thomas** and **Bayes' theorem**.

MATHEMATICAL ASPECTS

Let D be the set of data $X = x_1,\ldots,X = x_n$ independently sampled from a random variable X of unknown distribution. We will consider the simple case where there is only one interesting parameter, θ, which depends on X.

Then a standard Bayesian procedure can be expressed by:

1. Identify the known quantities $x_1,\ldots,x_n$.
2. Specify a model for the data; in other words a parametric family $f(x\,|\theta)$ of distributions that describe the generation of data.
3. Specify the uncertainty concerning θ by an initial distribution function $f(\theta)$.
4. We can then calculate the distribution $f(\theta\,|D)$ (called the *final distribution*) using **Bayes' theorem**.

The first two points are common to every statistical **inference**.

The third point is more problematic. In the absence of supplementary information about θ, the idea is to calculate a *reference distribution* $f(\theta)$ by maximizing a function that specifies the missing information on the

parameter θ. Once this problem is resolved, the fourth point is easily tackled with the help of Bayes' theorem.

Bayes' theorem can be expressed, in its continuous form, by:

$$f(\theta\,|D) = \frac{f(D\,|\theta)\cdot f(\theta)}{f(D)}$$
$$= \frac{f(D\,|\theta)\cdot f(\theta)}{\int f(D\,|\theta)f(\theta)\,d\theta}.$$

Since the x_i are independent, we can write:

$$f(\theta\,|D) = \frac{\displaystyle\prod_{i=1}^{n} f(x_i\,|\theta)\cdot f(\theta)}{\displaystyle\int \prod_{i=1}^{n} f(x_i\,|\theta)f(\theta)\,d\theta}.$$

Now we have the means to calculate the **density function** of a new (independent) observation x_{n+1}, given $x_1,\ldots,x_n$:

$$f(X - x_{n+1}\,|D)$$
$$= \frac{f(X = x_1,\ldots,X = x_{n+1})}{f(X = x_1,\ldots,X = x_n)}$$
$$= \frac{\int f(X = x_1,\ldots,X = x_{n+1}\,|\theta)\cdot f(\theta)\,d\theta}{\int f(D\,|\theta)\cdot f(\theta)\,d\theta}$$
$$= \frac{\int \prod_{i=1}^{n+1} f(X = x_i\,|\theta)\cdot f(\theta)\,d\theta}{\int \prod_{i=1}^{n} f(X = x_i\,|\theta)\cdot f(\theta)\,d\theta}$$
$$= \int f(X = x_{n+1}\,|\theta)\cdot f(\theta\,|D)\,d\theta.$$

We will now briefly explain the methods that allow us to:

• Find a value for the estimated parameter that is more probable than the others;
• Find a confidence interval for θ, and;
• Perform hypothesis testing.

These methods are strictly related to decision theory, which plays a considerable role in Bayesian statistics.

Point Estimation of the Parameter

To get a **point estimation** for the parameter θ, we specify a loss function $l(\hat{\theta}, \theta)$ that comes from using $\hat{\theta}$ (the estimated value) instead of the true value (the unknown) θ. Then we minimize this function of $\hat{\theta}$. For example, if θ is real and "the loss is quadratic" (that is, $l(\hat{\theta}, \theta) = (\hat{\theta}-\theta)^2$), then the point estimation of θ will be the mean of the calculated distribution.

Confidence Intervals

The concept of a **confidence interval** is replaced in bayesian statistics by the concept of an α-credible region (where α is the "confidence level"), which is simply defined as an interval I such that:

$$\int_I f(\theta \mid D)\, d\theta = \alpha .$$

Often, we also require that the width of the interval is minimized.

Hypothesis Testing

The general approach to **hypothesis testing**:

$$H_0: \ \theta \in I \qquad \text{versus} \qquad H_1: \ \theta \notin I .$$

is related to decision theory. Based on this, we define a loss function $l(a_0, \theta)$ that accepts H_0 (where the true value of the parameter is θ) and a loss function $l(a_1, \theta)$ that rejects H_0. If the value for the true value obtained by accepting H_0, that is,

$$\int l(a_0, \theta)\, d\theta ,$$

is smaller than to the one obtained by rejecting H_0, then we can accept H_0.

Using this constraint, we reject the restrictions imposed on θ by the null hypothesis ($\theta \in I$).

EXAMPLES

The following example involves estimating the parameter θ from the **Bernoulli distribution** X with the help of n independent observations $x_1, \ldots, x_n$, taking the value 1 in the case of success and 0 in the case of failure. Let r be the number of successes and $n - r$ be the number of failures among the observations.

We have then:

$$L(\theta) = P(X = x_1, \ldots, X = x_n \mid \theta)$$

$$= \prod_{i=1}^{n} P(X = x_i \mid \theta)$$

$$= \theta^r \cdot (1 - \theta)^{n-r} .$$

An estimation of the **maximum likelihood** of θ, denoted by $\hat{\theta}_{\text{mle}}$, maximizes this function. To do this, we consider the logarithm of this function, in other words the log-likelihood:

$$\log L(\theta) = r \log(\theta) + (n - r) \log(1 - \theta).$$

We maximize this by setting its derivative by θ equal to zero:

$$\frac{\partial (\log L)}{\partial \theta} = \frac{r}{\theta} - \frac{n - r}{1 - \theta} = 0 .$$

This is equivalent to $r(1 - \theta) = (n - r)\theta$, and it simplifies to $\hat{\theta}_{\text{mle}} = \frac{r}{n}$. The estimator for the maximum likelihood of θ is then simply the proportion of observed successes.

Now we return to the bayesian method. In this case, the Bernoulli distribution, the a priori reference distribution of the parameter θ, is expressed by:

$$f(\theta) = c \cdot (h(\theta))^{\frac{1}{2}} ,$$

where c is an appropriate constant (such as $\int f(\theta)\, d\theta = 1$) and where $h(\theta)$ is called the

Fisher information of X:

$$h(\theta) = -E\left[\frac{\partial^2 L}{\partial \theta^2}\right]$$

$$= -E\left[-\frac{X}{\theta^2} - \frac{1-X}{(1-\theta)^2}\right]$$

$$= \frac{E[X]}{\theta^2} + \frac{1-E[X]}{(1-\theta)^2}$$

$$= \frac{\theta}{\theta^2} + \frac{1-\theta}{(1-\theta)^2}$$

$$= \frac{1}{\theta} + \frac{1}{1-\theta} = \frac{1}{\theta(1-\theta)} .$$

The distribution will then be:

$$f(\theta) = c \cdot \theta^{-\frac{1}{2}} \cdot (1-\theta)^{-\frac{1}{2}} .$$

The distribution function of θ, given the observations $x_1, \ldots, x_n$, can then be expressed by:

$$f(\theta \mid X = x_1, \ldots, X = x_n)$$

$$= \frac{1}{d}\prod_{i=1}^{n} P(X = x_i \mid \theta) \cdot f(\theta)$$

$$= \frac{1}{d} \cdot \theta^r \cdot (1-\theta)^{n-r} \cdot c \cdot \theta^{-\frac{1}{2}} \cdot (1-\theta)^{-\frac{1}{2}}$$

$$= \frac{c}{d} \cdot \theta^{r-\frac{1}{2}} \cdot (1-\theta)^{n-k-\frac{1}{2}}$$

$$= \frac{c}{d} \cdot \theta^{(r+\frac{1}{2})-1} \cdot (1-\theta)^{(n-r+\frac{1}{2})-1} .$$

which is a **beta distribution** with parameters

$$\alpha = r + \frac{1}{2} \quad \text{and} \quad \beta = n - r + \frac{1}{2},$$

and with a constant of

$$\frac{c}{d} = \frac{\Gamma(\alpha + \beta)}{\Gamma(\alpha)\,\Gamma(\beta)},$$

where Γ is the gamma function (see **gamma distribution**).

We now consider a concrete case, where we want to estimate the proportion of HIV-positive students. We test 200 students and none of them is HIV-positive. The proportion of HIV-positive students is therefore estimated to be 0 by the maximum likelihood. Confidence intervals are not very useful in this case, because (if we follow the usual approach) they are calculated by:

$$0 \pm 1.96 \cdot \sqrt{p(1-p)} .$$

In this case, as p is the proportion of observed successes, the confidence interval reduces to 0.

Following bayesian methodology, we obtain the distribution, based only on the data, that describes the uncertainty about the parameters to be estimated. The larger n is, the more sure we are about θ; the final reference distribution for θ is then more concentrated around the true value of θ.

In this case, we find as a final reference distribution a beta distribution with parameters $\alpha = 0.5$ and $\beta = 200.5$, which summarizes the information about θ (the values that correspond to the spikes in the distribution are the most probable). It is often useful to graphically represent such results:

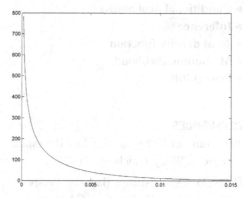

We can see from this that:

- The probability that the proportion of interest is smaller then 0.015 is almost 1;

• The probability that the proportion of interest is smaller then 0.005 is approximately 0.84, and;
• The median, which is chosen as the best measure of the central tendency due to the strong asymmetry of the distribution, is 0.011.

We remark that there is a qualitative difference between the classical result (the proportion can be estimated as zéro) and the bayesian solution to the problem, which allows us to calculate the probability distribution of the parameter, mathematically translating the uncertainty about it. This method tell us in particular that, given certain information, the correct estimation for the proportion of interest is 0.011. Note that the bayesian estimation depends on (like the uncertainty) the number of the observed cases, and it is equivalent to observing 0 cases of HIV among 2, 200 or 20.0 students in the classical case.

FURTHER READING
▶ **Bayes' theorem**
▶ **Bayes, Thomas**
▶ **Conditional probability**
▶ **Inference**
▶ **Joint density function**
▶ **Maximum likelihood**
▶ **Probability**

REFERENCES
Bernardo, J.M., Smith, A.F.M.: Bayesian Theory. Wiley, Chichester (1994)

Berry, D.A.: Statistics, a Bayesian Perspective. Wadsworth, Belmont, CA (1996)

Box, G.E.P., Tiao, G.P.: Bayesian Inference in Statistical Analysis. Addison-Wesley, Reading, MA (1973)

Carlin, B., Louis, T.A.: Bayes & Empirical Bayes Methods. Chapman & Hall /CRC, London (2000)

Finetti, B. de: Theory of Probability; A Critical Introductory Treatment. Trans. Machi, A., Smith, A. Wiley, New York (1975)

Gelman, A., Carlin, J.B., Stern H.S., Rubin, D.B.: Bayesian Data Analysis. Chapman & Hall /CRC, London (2003)

Lindley, D.V.: The philosophy of statistics. Statistician **49**, 293–337 (2000)

Robert, C.P.: Le choix Bayesien. Springer, Paris (2006)

Bernoulli Distribution

A **random variable** X follows a Bernoulli distribution with **parameter** p if its **probability function** takes the form:

$$P(X = x) = \begin{cases} p & \text{for } x = 1 \\ q = 1 - p & \text{for } x = 0 \end{cases}.$$

where p and q represent, respectively, the **probabilities** of "success" and "failure," symbolized by the **values** 1 and 0.

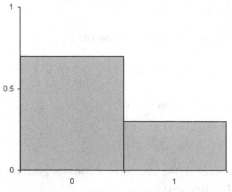

Bernoulli's law, $p = 0.3$, $q = 0.7$

The Bernoulli distribution is a **discrete probability distribution**.

HISTORY
See **binomial distribution**.

MATHEMATICAL ASPECTS
The **expected value** of the Bernoulli distribution is by definition:

$$E[X] = \sum_{x=0}^{1} x \cdot P(X = x)$$
$$= 1 \cdot p + 0 \cdot q = p.$$

The **variance** of the Bernoulli distribution is by definition:

$$\text{Var}(X) = E[X^2] - (E[X])^2$$
$$= 1^2 \cdot p + 0^2 \cdot q - p^2$$
$$= p - p^2 = pq.$$

DOMAINS AND LIMITATIONS
The Bernoulli distribution is used when a **random experiment** has only two possible results: "success" or "failure." These results are usually symbolized by 0 and 1.

FURTHER READING
▶ **Binomial distribution**
▶ **Discrete probability distribution**

Bernoulli Family

Originally from Basel, the Bernoulli family contributed several mathematicians to science. Bernoulli, Jacques (1654–1705) studied theology at the University of Basel, according to the will of his father, and then traveled for several years, teaching and continuing his own studies. Having resolutely steered himself towards mathematics, he became a professor at the University of Basel in 1687. According to Stigler (1986),

Bernoulli, Jacques (called Bernoulli, James in most the English works) is the father of the quantification of uncertainty.

It was only in 1713, seven years after his death, and on the instigation of his nephew Nicolas, that his main work *Ars Conjectandi* was published. This work is divided into four parts: in the first the author comments upon the famous treatise of Huygens; the second is dedicated to the theory of **permutations** and **combinations**; the third to solving diverse problems about games of chance; and finally the fourth discusses the application of the theory of **probability** to questions of moral interest and economic science.

A great number of the works of Bernoulli, Jacques were never published.

Jean Bernoulli (1667–1748), who was more interested in mathematics than the medical career his father intended for him, also became a university professor, first in Groningen in the Netherlands, and then in Basel, where he took over the chair left vacant by the death of his brother.

The two brothers had worked on differential and integral calculus and minimization problems and had also studied functions.

REFERENCES
Stigler, S.: The History of Statistics, the Measurement of Uncertainty Before 1900. Belknap, London (1986) pp. 62–71

Bernoulli, Jakob

Bernoulli, Jakob (or Jacques or Jacob or James) (1655–1705) and his brother Jean were pioneers of Leibniz calculus. Jakob reformulated the problem of calculating an expectation into probability calculus. He also formulated the weak law of large numbers, upon which modern probability and statistics are based. Bernoulli, Jakob was nominated *maître ès lettres* in 1671. During his studies in theology, which he terminated in 1676, he studied mathematics and astronomy, contrary to the will of his father (Nikolaus Bernoulli).

In 1682, Leibniz published (in *Acta Eruditorium*) a method that could be used to determine the integrals of algebraic functions, a brief discourse on differential calculus in an algorithmic scheme, and some remarks on the fundamental idea of integral calculus. This paper attracted the attention of Jakob and his brother, and they considerably improved upon the work already done by Leibniz. Leibniz himself recognized that infinitesimal calculus was mostly founded by the Bernoulli brothers rather than himself. Indeed, in 1690 Jakob introduced the term "integral."

In 1687, he was nominated Professor of Mathematics at the University of Basel, where he stayed until his death in 1705.

Ars conjectandi is the title of what is generally accepted as Bernoulli's most original work. It consists of four parts: the first contains a reprint of Huygen's *De Ratiociniis in Ludo Aleae* (published in 1657), which is completed via important modifications. In the second part of his work, Bernoulli addresses combinatory theory. The third part comprises 24 examples that help to illustrate the modified concept of the expected value. Finally, the fourth part is the most interesting and original, even though Bernoulli did not have the time to finish it. It is in this part that Bernoulli distinguishes two ways of defining (exactly or approximately) the classical measure of probability.

Around 1680, Bernoulli, Jakob also became interested in stochastics. The evolution of his ideas can be followed in his scientific journal *Meditations.*

Some of the main works and articles of Bernoulli, Jakob include:

1677   Meditationes, Annotationes, Animadversiones Theologicae & Philosophicae, a me JB. concinnatae & collectae ab anno 1677. Universitätsbibliothek, Basel, L I a 3.

1713   Ars Conjectandi, Opus Posthumum. Accedit Tractatus de Seriebus infinitis, et Epistola Gallice scripta de ludo Pilae recticularis. Impensis Thurnisiorum, Fratrum, Basel.

Bernoulli Trial

The Bernoulli trials are repeated tests of an **experiment** that obey the following rules:

1. Each trial results in either success or failure;
2. The **probability** of success is the same for each trial; the probability of success is denoted by p, and the probability of failure by $q = 1 - p$;
3. The trials are independent.

HISTORY

The Bernoulli trials take their name from the Swiss mathematician **Bernoulli, Jakob** (1713).

Bernoulli, Jakob (1654–1705) was the eldest of four brothers and it was his father's will that he should study theology. When he had finished his studies in theology in Basel in 1676, he briefly left the town only to return in 1680 in order to devote himself to mathematics. He obtained the Chair of Mathematics at the University of Basel in 1687.

EXAMPLES

The most simple example of a Bernoulli trial is the flipping of a coin. If obtaining "heads" is considered to be a success (S) while obtaining "tails" is considered to be a failure (F), we have:

$$p = P(S) = \tfrac{1}{2}$$

and

$$q = P(F) = 1 - p = \tfrac{1}{2} \, .$$

FURTHER READING

▶ Bernoulli distribution
▶ Binomial distribution

REFERENCES

Bernoulli, J.: Ars Conjectandi, Opus Posthumum. Accedit Tractatus de Seriebus infinitis, et Epistola Gallice scripta de ludo Pilae recticularis. Impensis Thurnisiorum, Fratrum, Basel (1713)

Ross, S.M.: Introduction to Probability Models, 8th edn. John Wiley, New York (2006)

Bernoulli's Theorem

Bernoulli's theorem says that the relative frequency of success in a sequence of Bernoulli trials approaches the probability of success as the number of trials increases towards infinity.

It is a simplified form of the law of large numbers and derives from the Chebyshev inequality.

HISTORY

Bernoulli's theorem, sometimes called the "weak law of large numbers," was first described by Bernoulli, Jakob (1713) in his work *Ars Conjectandi*, which was published (with the help of his nephew Nikolaus) seven years after his death.

MATHEMATICAL ASPECTS

If S represents the number of successes obtained during n Bernoulli trials, and if p is the probability of success, then we have:

$$\lim_{n \to \infty} P\left(\left| \frac{S}{n} - p \right| \geq \varepsilon \right) = 0 \, ,$$

or

$$\lim_{n \to \infty} P\left(\left| \frac{S}{n} - p \right| < \varepsilon \right) = 1 \, ,$$

where $\varepsilon > 0$ and arbitrarily small.

In an equivalent manner, we can write:

$$\frac{S}{n} \xrightarrow[n \to \infty]{} p \, ,$$

which means that the relative frequency of success tends to the probability of success when n tends to infinity.

FURTHER READING

▶ Bernoulli distribution
▶ Bernoulli trial
▶ Convergence
▶ Law of large numbers

REFERENCES

Bernoulli, J.: Ars Conjectandi, Opus Posthumum. Accedit Tractatus de Seriebus infinitis, et Epistola Gallice scripta de ludo Pilae recticularis. Impensis Thurnisiorum, Fratrum, Basel (1713)

Rényi, A.: Probability Theory. North Holland, Amsterdam (1970)

Beta Distribution

A **random variable** X follows a beta distribution with **parameters** α and β if its **density function** is of the form:

$$f(x) = \frac{1}{B(\alpha, \beta)}(x-a)^{\alpha-1}(b-x)^{\beta-1}$$
$$\cdot (b-a)^{-(\alpha+\beta-1)},$$
$$a \le x \le b, \quad \alpha > 0 \text{ and } \beta > 0,$$

where:

$$B(\alpha, \beta) = \int_0^1 t^{\alpha-1}(1-t)^{\beta-1}dt$$
$$= \frac{\Gamma(\alpha)\Gamma(\beta)}{\Gamma(\alpha+\beta)},$$

Γ is the gamma function (see **gamma distribution**).

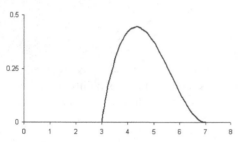

0.5

0.25

0

0   1   2   3   4   5   6   7   8

Beta distribution $\alpha = 2$, $\beta = 3$, $a = 3$, $b = 7$

The beta distribution is a **continuous probability distribution**.

MATHEMATICAL ASPECTS

The **density function** of the standard beta distribution is obtained by performing the **variable** change $Y = \frac{X-a}{b-a}$:

$$f(y) = \begin{cases} \frac{1}{B(\alpha,\beta)}y^{\alpha-1} & \text{if } 0 < y < 1 \\ \cdot(1-y)^{\beta-1} & \\ 0 & \text{if not} \end{cases}.$$

Consider X, a **random variable** that follows the standard beta distribution. The **expected** value and the **variance** of X are, respectively, given by:

$$E[X] = \frac{\alpha}{\alpha + \beta},$$
$$\text{Var}(X) = \frac{\alpha \cdot \beta}{(\alpha + \beta)^2 \cdot (\alpha + \beta + 1)}.$$

DOMAINS AND LIMITATIONS

The beta distribution is one of the most frequently used to adjust empirical distributions where the **range** (or variation **interval**) $[a, b]$ is known.

Here are some particular cases where the beta distribution is used, related to other **continuous probability distributions**:

- If X_1 and X_2 are two independent **random variables** each distributed according to a **gamma distribution** with **parameters** $(\alpha_1, 1)$ and $(\alpha_2, 1)$, respectively, the random variable

$$\frac{X_1}{X_1 + X_2},$$

is distributed according to a beta distribution with parameters (α_1, α_2).

- The beta distribution becomes a **uniform distribution** when

$$\alpha = \beta = 1.$$

- When the **parameters** α and β tends towards infinity, the beta distribution tends towards the standard **normal distribution**.

FURTHER READING

▶ **Continuous probability distribution**
▶ **Gamma distribution**
▶ **Normal distribution**
▶ **Uniform distribution**

Bias

From a statistical point of view, the bias is defined as the difference between the **expected value** of a **statistic** and the true **value** of the corresponding **parameter**. Therefore, the bias is a measure of the systematic error of an **estimator**. If we calculate the mean of a large number of unbiased estimations, we will find the correct value. The bias indicates the distance of the estimator from the true value of the parameter.

HISTORY

The concept of an unbiased **estimator** comes from **Gauss, C.F.** (1821), during the time when he worked on the **least squares** method.

DOMAINS AND LIMITATIONS

We should not confuse the bias of an estimator of a parameter with its degree of precision, which is a measurement of the sampling error.

There are several types of bias, selection bias (due to systematic differences between the groups compared), exclusion bias (due to the systematic exclusion of certain individuals from the study) or analytical bias (due to the way that the results are evaluated).

MATHEMATICAL ASPECTS

Consider a **statistic** T used to estimate a **parameter** θ. If $E[T] = \theta + b(\theta)$ (where $E[T]$ represents the **expected value** of T), then the quantity $b(\theta)$ is called the bias of the statistic T.

If $b(\theta) = 0$, we have $E[T] = \theta$, and T is an unbiased **estimator** of θ.

EXAMPLES

Consider $X_1, X_2, \ldots, X_n$, a sequence of independent **random variables** distributed according to the same **law of probability** with a **mean** μ and a finite **variance** σ^2. We can calculate the bias of the **estimator**

$$S^2 = \frac{1}{n} \cdot \sum_{i=1}^{n} (x_i - \bar{x})^2 ,$$

used to estimate the variance σ^2 of the **population** in the following way:

$$E[S^2] = E\left[\frac{1}{n} \sum_{i=1}^{n} (x_i - \bar{x})^2 \right]$$

$$= E\left[\frac{1}{n} \sum_{i=1}^{n} (x_i - \mu)^2 - (\bar{x} - \mu)^2 \right]$$

$$= \frac{1}{n} E\left[\sum_{i=1}^{n} (x_i - \mu)^2 \right] - E(\bar{x} - \mu)^2$$

$$= \frac{1}{n} \sum_{i=1}^{n} E(x_i - \mu)^2 - E(\bar{x} - \mu)^2$$

$$= \text{Var}(x_i) - \text{Var}(\bar{x})$$

$$= \sigma^2 - \frac{\sigma^2}{n} .$$

The bias of S^2 is then equal to $\frac{-\sigma^2}{n}$.

FURTHER READING

▶ **Estimation**
▶ **Estimator**
▶ **Expected value**

REFERENCES

Gauss, C.F.: Theoria Combinationis Observationum Erroribus Minimis Obnoxiae, Parts 1, 2 and suppl. Werke **4**, 1–108 (1821, 1823, 1826)

Gauss, C.F.: Méthode des Moindres Car-
rés. Mémoires sur la Combinaison des
Observations. Traduction Française par J.
Bertrand. Mallet-Bachelier, Paris (1855)

Bienaymé, Irénée-Jules

Bienaymé,   Irénée-Jules   (1796–1878)
entered the French Ministry of Finance and
became the Inspector General for Finance
in 1836, although he lost his employment
in 1848 for political reasons. Shortly after
this, he began to give lessons in the Faculty
of Science in Paris.

A follower of **Laplace, Pierre Simon
de**, Bienaymé proved the Bienaymé–
Chebyshev inequality some years before
Chebyshev, Pafnutii Lvovich. The reedited version of Bienaymé's paper from 1867
precedes the French version of Chebychev's
proof. This inequality was then used in
Chebyshev's incomplete proof of **central
limit theorem**, which was later finished by
Markov, Andrei Andreevich.

Moreover, Bienaymé correctly formulated
the theorem for branching processes in 1845.
His most famous public work is probably the
corrections he made to the use of Duvillard's
mortality table.

*Some principal works and articles of Bien-
aymé, Irénée-Jules:*

1853 Considérations à l'appui de la découverte de Laplace sur la loi de probabilité dans la méthode des moindres carrés. Comptes Rendus de l'Académie
des Sciences, Paris **37**, 5–13; reedited
in 1867 in the Journal de Liouville
preceding the proof of the Bienaymé–
Chebyshev inequality in J. Math.
Pure. Appl., 12, 158–176.

Binary Data

Binary data occur when the **variable** of interest can only take two **values**. These two values are generally represented by 0 and 1, even
if the variable is not quantitative.

Gender, the presence or absence of a characteristic and the success or failure of an experiment are just a few examples of **variables**
that result in binary data. These variables are
called **dichotomous variables**.

EXAMPLES

A meteorologist wants to know how reliable
his forecasts are. To do this, he studies a **random variable** representing the prediction.
This **variable** can only take two **values**:

$$X = \begin{cases} 0 & \text{if the prediction was incorrect} \\ 1 & \text{if the prediction was correct} \end{cases}.$$

The meteorologist makes predictions for
a period of 50 consecutive days. The prediction is found to be correct 32 times, and incor-
rect 18 times.

To find out whether his predictions are better
than the ones that could have been obtained
by flipping a coin and predicting the weather
based on whether heads or tails are obtained,
he decides to test the **null hypothesis** H_0,
that the proportion p of correct predictions is
equal to 0.5, against the **alternative hypothesis** H_1, that this proportion is different to 0.5:

$$H_0 : \quad p = 0.5$$
$$H_1 : \quad p \neq 0.5 .$$

Let us calculate the **value** of

$$\chi^2 = \sum \frac{\left(\begin{array}{c} \text{observed frequency} \\ -\text{theoretical frequency} \end{array} \right)^2}{\text{theoretical frequency}} ,$$

where the observed **frequencies** are 18 and 32, and the theoretical frequencies are, respectively, 25 and 25:

$$\chi^2 = \frac{(32-25)^2 + (18-25)^2}{25}$$

$$= \frac{49+49}{25} = \frac{98}{25} = 3.92.$$

By assuming that the **central limit theorem** applies, we compare this **value** with the value of the **chi-square distribution** with one **degree of freedom**. For a 'significance level of 5%, we find in the **chi-square table**:

$$\chi^2_{1,0.95} = 3.84.$$

Therefore, since $3.92 > 3.84$, the **null hypothesis** is rejected, which means that the meteorologist predicts the weather better than a coin.

FURTHER READING
▶ **Bernoulli distribution**
▶ **Categorical data**
▶ **Contingency table**
▶ **Data**
▶ **Dichotomous variable**
▶ **Likelihood ratio test**
▶ **Logistic regression**
▶ **Variable**

REFERENCES
Cox, D.R., Snell, E.J.: The Analysis of Binary Data. Chapman & Hall (1989)

Bishop, Y.M.M., Fienberg, S.E., Holland, P.W.: Discrete Multivariate Analysis: Theory and Practice. MIT Press, Cambridge, MA (1975)

Binomial

Algebraic sums containing **variables** are called polynomials (from the Greek "poly," meaning "several"). An expression that contains two terms is called a binomial (from the Latin "bi", meaning "double"). A monomial (from the Greek "mono", meaning "unique") is an expression with one term and a trinomial (from the Latin "tri", meaning "triple") contains three elements.

HISTORY
See **arithmetic triangle**.

MATHEMATICAL ASPECTS
The square of a binomial is easily calculated using

$$(a+b)^2 = a^2 + 2ab + b^2.$$

Binomial formulae for exponents higher than two also exist, such as:

$$(a+b)^3 = a^3 + 3a^2b + 3ab^2 + b^3,$$
$$(a+b)^4 = a^4 + 4a^3b + 6a^2b^2 + 4ab^3 + b^4.$$

We can write a generalized binomial formula in the following manner:

$$(a+b)^n = a^n + na^{n-1}b$$
$$+ \frac{n(n-1)}{2}a^{n-2}b^2$$
$$+ \cdots + \frac{n(n-1)}{2}a^2b^{n-2}$$
$$+ nab^{n-1} + b^n$$
$$= C_n^0 a^n + C_n^1 a^{n-1}b + C_n^2 a^{n-2}b^2$$
$$+ \cdots + C_n^{n-1}ab^{n-1} + C_n^n b^n$$
$$= \sum_{k=0}^{n} C_n^k a^{n-k}b^k.$$

DOMAINS AND LIMITATIONS

The binomial $(p + q)$ raised to the power n lends its name to the **binomial distribution** because it corresponds to the total **probability** obtained after n **Bernoulli trials**.

FURTHER READING

► **Arithmetic triangle**
► **Binomial distribution**
► **Combination**

Binomial Distribution

A **random variable** X follows a binomial distribution with **parameters** n and p if its **probability function** takes the form:

$$P(X = x) = C_n^x \cdot p^x \cdot q^{n-x},$$
$$x = 0, 1, 2, \ldots, n.$$

Therefore if an event comprises x "successes" and $(n - x)$ "failures," where p is the **probability** of "success" and $q = 1 - p$ the probability of "failure," the binomial distribution allows us to calculate the probability of obtaining x successes from n independent trials.

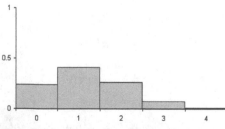
Binomial distribution, $p = 0.3$, $q = 0.7$, $n = 3$

The binomial distribution with **parameters** n and p, denoted $B(n, p)$, is a **discrete probability distribution**.

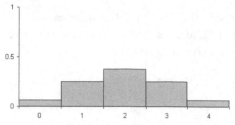
Binomial distribution, $p = 0.5$, $q = 0.5$, $n = 4$

HISTORY

The binomial distribution is one of the oldest known **probability distributions**. It was discovered by **Bernoulli, J.** in his work entitled *Ars Conjectandi* (1713). This work is divided into four parts: in the first, the author comments on the treatise from Huygens; the second part is dedicated to the theory of **permutations** and **combinations**; the third is devoted to solving various problems related to games of chance; finally, in the fourth part, he proposes applying probability theory to moral questions and to the science of economics.

MATHEMATICAL ASPECTS

If $X_1, X_2, \ldots, X_n$ are n independent **random variables** following a **Bernoulli distribution** with a **parameter** p, then the random variable

$$X = X_1 + X_2 + \ldots + X_n$$

follows a binomial distribution $B(n, p)$. To calculate the **expected value** of X, the following property will be used, where Y and Z are two **random variables**:

$$E[Y + Z] = E[Y] + E[Z].$$

We therefore have:

$$E[X] = E[X_1 + X_2 + \ldots + X_n]$$
$$= E[X_1] + E[X_2] + \ldots + E[X_n]$$
$$= p + p + \ldots + p = np.$$

To calculate the **variance** of X, the following property will be used, where Y and Z are two independent **variables**:

$$\text{Var}(Y + Z) = \text{Var}(Y) + \text{Var}(Z).$$

We therefore have:

$$\begin{aligned}\text{Var}(X) &= \text{Var}(X_1 + X_2 + \ldots + X_n) \\ &= \text{Var}(X_1) + \text{Var}(X_2) + \ldots \\ &\quad + \text{Var}(X_n) \\ &= pq + pq + \ldots + pq = npq.\end{aligned}$$

Binomial distribution tables (for the **probability distribution** and the **distribution function**) have been published by Rao et al. (1985), as well as by the National Bureau of Standards (1950). Extended distribution function tables can be found in the *Annals of the Computation Laboratory* (1955).

EXAMPLES

A coin is flipped ten times. Consider the **random variable** X, which represents the number of times that the coin lands on "tails." We therefore have:

number of trials: $n = 10$

probability of one success: $p = \frac{1}{2}$

(tails obtained)

probability of one failure: $q = \frac{1}{2}$

(heads obtained)

The **probability** of obtaining tails x times amongst the ten trials is given by

$$P(X = x) = C_{10}^x \cdot \left(\frac{1}{2}\right)^x \cdot \left(\frac{1}{2}\right)^{10-x}.$$

The **probability** of obtaining tails exactly eight times is therefore equal to:

$$\begin{aligned}P(X = 8) &= C_{10}^8 \cdot p^8 \cdot q^{10-8} \\ &= \frac{10!}{8!(10-8)!} \cdot \left(\frac{1}{2}\right)^8 \cdot \left(\frac{1}{2}\right)^2 \\ &= 0.0439.\end{aligned}$$

The **random variable** X follows the binomial distribution $B(10, \frac{1}{2})$.

FURTHER READING

▶ **Bernoulli distribution**
▶ **Binomial table**
▶ **Discrete probability distribution**
▶ **Negative binomial distribution**

REFERENCES

Bernoulli, J.: Ars Conjectandi, Opus Posthu mum. Accedit Tractatus de Seriebus infinitis, et Epistola Gallice scripta de ludo Pilae recticularis. Impensis Thurnisiorum, Fratrum, Basel (1713)

Harvard University: Tables of the Cumulative Binomial Probability Distribution, vol. 35. Annals of the Computation Laboratory, Harvard University Press, Cambridge, MA (1955)

National Bureau of Standards.: Tables of the Binomial Probability Distribution. U.S. Department of Commerce. Applied Mathematics Series 6 (1950)

Rao, C.R., Mitra, S.K., Matthai, A., Ramamurthy, K.G.: Formulae and Tables for Statistical Work. Statistical Publishing Company, Calcutta, pp. 34–37 (1966)

Binomial Table

The binomial table gives the values for the **distribution function** of a **random**

variable that follows a **binomial distribution**.

HISTORY
See **binomial distribution**.

MATHEMATICAL ASPECTS
Let the **random variable** X follow the **binomial distribution** with parameters n and p. Its **probability function** is given by:

$$P(X = x) = C_n^x \cdot p^x \cdot q^{n-x},$$

$$x = 0, 1, 2, \ldots, n,$$

where C_n^x is the binomial coefficient, equal to $\frac{n!}{x!(n-x)!}$, parameter p is the **probability** of success, and $q = 1-p$ is the complementary probability that corresponds to the probability of failure (see **normal distribution**).
The **distribution function** of the random variable X is defined by:

$$P(X \le x) = \sum_{i=0}^{x} C_n^i \cdot p^i \cdot q^{n-i},$$

$$0 \le x \le n.$$

The binomial table gives the value of $P(X \le x)$ for various combinations of x, n and p.
For large n, this calculation becomes tedious. Thankfully, we can use some very good approximations instead. If $\min(np, n(1-p)) > 10$, we can approximate it with the normal distribution:

$$P(X \le x) = \phi\left(\frac{x + \frac{1}{2} - np}{\sqrt{npq}}\right),$$

where ϕ is the distribution function for the standard normal distribution, and the continuity correction $\frac{1}{2}$ is included.

DOMAINS AND LIMITATIONS
The binomial table is used to perform nonparametric tests on **statistics** that are distributed according to **binomial distribution**, especially the **sign test** and the **binomial test**.
The National Bureau of Standards (1950) published individual and cumulative binomial distribution probabilities for $n \le 49$, while cumulative binomial distribution probabilities for $n \le 1000$ are given in the *Annals of the Computation Laboratory* (1955).

EXAMPLES
See Appendix B.
We can verify that for $n = 2$ and $p = 0.5$:

$$P(X \le 1) = \sum_{i=0}^{1} C_2^i (0.5)^i (0.5)^{2-i}$$

$$= 0.75.$$

or that for $n = 5$ and $p = 0.05$:

$$P(X \le 3) = \sum_{i=0}^{3} C_5^i (0.05)^i (0.95)^{5-i}$$

$$= 1.0000.$$

For an example of the application of the binomial table, see **binomial test**.

FURTHER READING
▶ **Binomial distribution**
▶ **Binomial test**
▶ **Sign test**
▶ **Statistical table**

REFERENCES
National Bureau of Standards.: Tables of the Binomial Probability Distribution. U.S. Department of Commerce. Applied Mathematics Series 6 (1950)

Harvard University: Tables of the Cumulative Binomial Probability Distribution, vol. 35. Annals of the Computation Laboratory, Harvard University Press, Cambridge, MA (1955)

Binomial Test

The binomial test is a **parametric hypothesis test** that applies when the **population** can be divided into two classes: each **observation** of this population will belong to one or the other of these two categories.

MATHEMATICAL ASPECTS

We consider a **sample** of n independent trials. Each trial belongs to either the class C_1 or the class C_2. We note the number of observations n_1 that fall into C_1 and the number of observations n_2 that fall into C_2.

Each trial has a **probability** p of belonging to class C_1, where p is identical for all n trials, and a probability $q = 1 - p$ of belonging to class C_2.

Hypotheses

The binomial test can be either a **two-sided test** or a **one-sided test**. If p_0 is the presumed value of p, $(0 \leq p_0 \leq 1)$, the hypotheses are expressed as follows:

A: Two-sided case

$$H_0: \quad p = p_0,$$
$$H_1: \quad p \neq p_0.$$

B: One-sided case

$$H_0: \quad p \leq p_0,$$
$$H_1: \quad p > p_0.$$

C: One-sided case

$$H_0: \quad p \geq p_0,$$
$$H_1: \quad p < p_0.$$

Decision Rules

Case A

For $n < 25$, we use the **binomial table** with n and p_o as parameters.

Therefore, we initially look for the closest **value** to $\frac{\alpha}{2}$ in this table (which we denote α_1), where α is the **significance level**. We denote the value corresponding to α_1 by t_1.

Next we find the value of $1 - \alpha_1 = \alpha_2$ in the table. We denote the value corresponding to α_2 by t_2.

We reject H_0 at the level α if

$$n_1 \leq t_1 \quad \text{or} \quad n_1 > t_2,$$

where n_1 is the number of observations that fall into the class C_1.

When n is bigger then 25, we can use the **normal distribution** as an approximation for the **binomial distribution**.

The parameters for the binomial distribution are:

$$\mu = n \cdot p_0,$$
$$\sigma = \sqrt{n \cdot p_0 \cdot q_0}.$$

The **random variable** Z that follows the standard normal distribution then equals:

$$Z = \frac{X - \mu}{\sigma} = \frac{X - n \cdot p_0}{\sqrt{n \cdot p_0 \cdot q_0}},$$

where X is a random binomial variable with parameters n and p_0.

The approximations for the values of t_1 and t_2 are then:

$$t_1 = n \cdot p_0 + z_{\alpha_1} \cdot \sqrt{n \cdot p_0 \cdot q_0},$$
$$t_2 = n \cdot p_0 + z_{\alpha_2} \cdot \sqrt{n \cdot p_0 \cdot q_0},$$

where z_{α_1} and z_{α_2} are the values found in the **normal table** corresponding to the levels α_1 and α_2.

Case B

For $n < 25$, we take t to be the value in the binomial table corresponding to $1 - \alpha$, where α is the **significance level** (or the closest value), and n and p_0 are the parameters described previously. We reject H_0 at the level α if

$$n_1 > t,$$

where n_1 is the number of the observations that fall into the class C_1.

For $n \geq 25$ we can make the approximation:

$$t = n \cdot p_0 + z_\alpha \sqrt{n \cdot p_0 \cdot q_0},$$

where z_α can be found in the normal table for $1 - \alpha$.

The decision rule in this case is the same as that for the $n < 25$.

Case C

For $n < 25$, t is the value in the binomial table corresponding to α, where α is the significance level (or the closest value), and with n and p_0 the same parameters described previously.

We reject H_0 at the level α if

$$n_1 \leq t,$$

where n_1 is the number of observations that fall into the class C_1.

For $n \geq 25$, we make the approximation:

$$t = n \cdot p_0 + z_\alpha \sqrt{n \cdot p_0 \cdot q_0},$$

where z_α is found in the normal table for the significance level α. Then, the decision rule is the same as described previously.

DOMAINS AND LIMITATIONS

Two basic conditions that must be respected when performing the binomial test are:

1. The n observations must be mutually independent;
2. Every observation has a **probability** p of falling into the first class. This probability is also the same for all observations.

EXAMPLES

A machine is considered to be operational if a maximum of 5% of the pieces that it produces are defective. The **null hypothesis**, denoted H_0, expresses this situation, while the **alternative hypothesis**, denoted H_1, signifies that the machine is failing:

$$H_0: \quad p \leq 0.05$$
$$H_1: \quad p > 0.05.$$

Performing the test, we take a **sample** of 10 pieces and we note that there are three defective pieces ($n_1 = 3$). As the hypotheses correspond to the **one-tailed test** (*case B*), decision rule B is used. If we choose a **significance level** of $\alpha = 0.05$, the **value** of t in the **binomial table** equals 1 (for $n = 10$ and $p_0 = 0.05$).

We reject H_0 because $n_1 = 3 > t = 1$, and we conclude that the machine is failing.

Then we perform the test again, but on 100 pieces this time. We notice that there are 12 defective pieces. In this case, the value of t can be approximated by:

$$t = n \cdot p_0 + z_\alpha \sqrt{n \cdot p_0 \cdot q_0},$$

where z_α can be found from the **normal table** for $1 - \alpha = 0.95$. We then have:

$$t = 100 \cdot 0.05 + 1.64\sqrt{100 \cdot 0.05 \cdot 0.95}$$
$$= 8.57.$$

We again reject H_0, because $n_1 = 12 > t = 8.57$.

FURTHER READING
- ▶ Binomial table
- ▶ Goodness of fit test
- ▶ Hypothesis testing
- ▶ One-sided test
- ▶ Parametric test
- ▶ Two-sided test

REFERENCES

Abdi, H.: Binomial distribution: Binomical and Sign Tests. In: Salkind, N.J. (ed.) Encyclopedia of Measurement and Statistics. Sage, Thousand Oaks (2007)

Biostatistics

Biostatistics is the scientific field where statistical methods are applied in order to answer questions related to human biology and medicine (the prefix "bio" comes from Greek "bios," which means "life").

The domains of biostatistics are mainly epidemiology, clinical and biological trials, and it is also used when studying the ethics of these trials.

HISTORY

Biostatistics began in the middle of the seventeenth century, when Petty, Sir William (1623–1687) and Graunt, John (1620–1674) created new methods of analyzing the *London Bills of Mortality*. They applied these new methods of analysis to death rate, birthrate and census studies, creating the field of biometrics. Then, in the middle of the nineteenth century, the works of Mendel, Gregor studied inheretance in plants. His observations and results were based on systematically gathered data, and also on the application of numerical methods of describing the regularity of hereditary transmission.

Galton, Francis and Pearson, Karl were two of the most important individuals associated with the development of this science. They used the new concepts and statistical methods when investigating the resemblance in physical, psychological and behavioral data between parents and their children. Pearson, Galton and Weldon, Walter Frank Raphael (1860–1906) cofounded the journal *Biometrika*. Fisher, Ronald Aylmer, during his agricultural studies performed at Rothamsted Experimental Station, proposed a method of random sampling where animals were partitioned into different groups and allocated different treatments, marking the first studies into clinical trials.

DOMAINS AND LIMITATIONS

The domains of biostatistics are principally epidemiology, clinical trials and the ethical questions related to them, as well as biological trials. One of the areas where biostatistical concepts have been used to analyze a specific question was in indirect measurements of the persistence of a substance (for example vitamins and hormones) when administered to a living creature. Biostatistics therefore mainly refer to statistics used to solve problems that appear in the biomedical sciences.

The statistical methods most commonly used in biostatistics include resampling methods (bootstrap, jackknife), multivariate analysis, various regression methods, evaluations of uncertainty related to estimation, and the treatment of missing data.

The most well-known scientific journals related to biostatistics are: *Statistics in Medicine, The Biometrical Journal, Controlled Clinical Trials, The Journal of Biopharmaceutical Statistics, Statistical Methods in Medical Research and Biometrics.*

FURTHER READING
▶ **Demography**
▶ **Epidemiology**

REFERENCES
Armitage, P., Colton, T.: Encyclopedia of Biostatistics. Wiley, New York (1998)

Graunt, J.: Natural and political observations mentioned in a following index, and made upon the bills of mortality: with reference to the government, religion, trade, growth, ayre, diseases, and the several changes of the said city. Tho. Roycroft, for John Martin, James Allestry, and Tho. Dicas (1662)

Greenwood, M.: Medical Statistics from Graunt to Farr. Cambridge University Press, Cambridge (1948)

Petty, W.: Another essay in political arithmetic concerning the growth of the city of London. Kelley, New York (1682) (2nd edn. 1963)

Petty, W.: Observations Upon the Dublin-Bills of Mortality, 1681, and the State of That City. Kelley, New York (1683) (2nd edn. 1963)

Block

Blocks are sets where **experimental units** are grouped in such a way that the units are as similar as possible within each block.

We can expect that the experimental **error** associated with a block will be smaller than that obtained if the same number of units were randomly located within the whole experimental space.

The blocks are generally determined by taking into account both controllable causes related by the **factors** studied and causes that may be difficult or impossible to keep constant over all of the **experimental units**.

The variations between the blocks are then eliminated when we compare the effects of the **factors**.

Several types of regroupings can be used to reduce the effects of one or several sources of **error**. A **randomized block design** results if there is only one source of error.

HISTORY
The block concept used in the field of **design of experiment** originated in studies made by **Fisher, R.A.** (1925) when he was Head of the Rothamsted Experimental Station. When working with agricultural researchers, Fisher realized that the ground(field) chosen for the **experiment** was manifestly heterogenous in the sense that the fertility varies in a systematic way from one point on the ground to another.

FURTHER READING
▶ **Design of experiments**
▶ **Experimental Unit**
▶ **Factor**
▶ **Graeco-Latin square design**
▶ **Latin square designs**
▶ **Randomized block design**

REFERENCES
Fisher, R.A.: Statistical Methods for Research Workers. Oliver & Boyd, Edinburgh (1925)

Bonferroni, Carlo E.

Bonferroni, Carlo Emilio was born in 1892 in Bergamo, Italy. He obtained a degree in mathematics in Turin and completed his education by spending a year at university in

Vienna and at the Eidgenössiche Technische Hochschule in Zurich. He was a Military officer during the First World War, and after the war had finished became an assistant professor at Turin Polytechnic. In 1923, he received the Financial Mathematics Chair at the Economics Institute in Bari, where he was the Rector for seven years. He finally transferred to Florence in 1933, where he held his chair until his death.

Bonferroni tackled various subjects, including actuarial mathemetics, probability and statistical mathematics, analysis, geometry and mechanics. He gave his name to the two Bonferroni inequalities that facilitate the treatment of statistical dependences. These appeared for the first time in 1936, in the article *Teoria statistica delle classi e calcolo delle probabilità*. The development of these inequalities prompted a wide range of new literature in this area.

Bonferroni died in 1960 in Florence, Italy. *Principal article of Carlo Emilio Bonferroni:*

1936 Teoria statistica delle classi e calcolo delle probabilità. Publ. R. Istit. Super. Sci. Econ. Commerc. Firenze, 8, 1–62.

Bootstrap

The term bootstrap describes a family of techniques that are principally used to estimate the standard error, the **bias** and the **confidence interval** of a parameter (or more than one parameter). It is based on n independent observations of a **random variable** with an unknown **distribution function** F, and is particularly useful when the parameters to be estimated relate to a complicated function of F.

The basic idea of bootstrap (which is somewhat similar to the idea behind the jackknife method) is to estimate F using a possible distribution $\hat{F}$ and then to resample from $\hat{F}$. Bootstrap procedures usually require the use of computers, since they can perform a large number of simulations in a relatively short time. In bootstrap methods, automatic simulations take the place of the analytical calculations used in the "traditional" methods of estimation, and in certain cases they can provide more freedom, for example when we do not want to (or we cannot) accept a hypothesis for the structure of the distribution of the data. Certain bootstrap methods are included in statistical software such as S-PLUS, SAS and MATLAB.

HISTORY

The origin of the use of the word "bootstrap" in relation to the methods described here neatly illustrates the reflexive nature of the secondary samples generated by them (the ones constructed from $\hat{F}$). It originates from the literary character Baron Munchausen (from *The Surprising Adventures of Baron Munchausen* by Raspe, R.E.), who fell into a lake, but pulled himself out by his own bootstraps (his laces).

The history of bootstrap, like many self-statistical techniques, starts in 1979 with the publication of an article by Efron (1979), which had a great impact on many researchers. This article triggered the publication of a huge number of articles on the theory and applications of bootstrap.

MATHEMATICAL ASPECTS

Different types of bootstrap have been proposed; the goal here is not to review all of these exhaustively, but instead to give an idea of the types of the methods used.

Let $x_1, x_2, \ldots, x_n$ be n independent observations of a random variable with an unknown distribution function F.

We are interested in the estimation of an unknown parameter δ that depends on F and the reliability (in other words its bias, its **variance** and its confidence interval). The estimation of δ is afforded by the **statistic** $t = t(F)$, which is dependent on F. Since F is unknown, we find an estimation $\hat{F}$ for F based on the sample $x_1, x_2, \ldots, x_n$, and we estimate δ using $t\left(\hat{F}\right)$. Classical examples of t are the **mathematical expectation**, the variance and the quantiles of F. We denote the random variable corresponding to the statistic t (which depends on $x_1, x_2, \ldots, x_n$) by T.

We distinguish the following types of bootstrap:

- *Parametric bootstrap.* If $F = F_\theta$ is a member of the parametric family of distribution functions, we can estimate (the vector) θ using the estimator $\hat{\theta}$ of the **maximum likelihood**. We naturally estimate F by $F_{\hat{\theta}}$.
- *Nonparametric bootstrap.* If F is not a member of the parametric family, it is estimated via an empirical distribution function calculated based on a sample of size n. The empirical distribution function in x is defined by:

$$\hat{F}(x) = \frac{\text{number of observations } x_i \leq x}{n}.$$

The idea of the bootstrap is the following.

1. Consider R independent samples of type $x_1^*, x_2^*, \ldots, x_n^*$ of a random variable of the distribution $\hat{F}$.
2. Calculate the estimated values $t_1^*, t_2^*, \ldots, t_R^*$ for the parameter δ based on the generated bootstrap samples. First calculate the estimation $\hat{F}_i^*$ based on the sample i

($i = 1, \ldots, R$), and then set:

$$t_i^* = t\left(\hat{F}_i^*\right).$$

Following the bootstrap method, the bias b and the variance v of t are estimated by:

$$b_{\text{boot}} = \bar{t}^* - t = \left(\frac{1}{R}\sum_{i=1}^{R} t_i^*\right) - t,$$

$$v_{\text{boot}} = \frac{1}{R-1}\sum_{i=1}^{R}(t_i^* - \bar{t}^*)^2.$$

We should remark that two types of errors intervene here:

- A statistical error arising from the fact that we proceed by considering the bias and variance for $\hat{F}$ and not the real F, which is unknown.
- A simulation error arising from the fact that the number of simulations is not high enough. In reality, it is enough to resample between $R = 50$ and 200 times.

There there are many possible methods of obtaining confidence intervals using bootstrap, and we present only a few here:

1. *Normal intervals.* Suppose that T follows the Gaussian (normal) distribution. Hoping that T^* also follows the normal distribution, we then find the following confidence interval for the (bias) corrected value $t_b = t - b$ of t:

$$\left[t_b - \sqrt{v_{\text{boot}}} \cdot z_{1-\alpha/2},\right.$$

$$\left. t_b + \sqrt{v_{\text{boot}}} \cdot z_{1-\alpha/2}\right],$$

where z_γ is the γ quantile of the Gaussian distribution $N(0, 1)$. For a tolerance limit of α, we use the value from the **normal table** $z_{1-\frac{\alpha}{2}}$, which gives $\alpha = 5\%$: $z_{1-\frac{0.05}{2}} = 1.96$.

2. *Bootstrap intervals.* Suppose that $T - \delta$ does not depend on an unknown variable. In this case we calculate the α-quantile a_α of $T - \delta$. To do this, we arrange the values $t_1^* - t, t_2^* - t, \ldots, t_R^* - t$ in increasing order:

$$t_{(1)}^* - t, t_{(2)}^* - t, \ldots, t_{(R)}^* - t,$$

and then we read the quantiles of interest $a_\alpha = t_{((R+1)\alpha)}^* - t$ and $a_{1-\alpha} = t_{((R+1)(1-\alpha))}^* - t$.
And as far as

$$1 - 2\alpha = \Pr\left(a_\alpha \leq T - \delta \leq a_{1-\alpha}\right)$$

$$= \Pr\left(2t - t_{((R+1)(1-\alpha))}^* \leq \delta\right)$$

$$\leq 2t - t_{((R+1)\alpha)}^*\right),$$

the confidence interval for t at the significance level α is:

$$\left[2t - t_{((R+1)(1-\alpha))}^*, 2t - t_{((R+1)\alpha)}^*\right].$$

3. *Studentized bootstrap intervals.* It is possible to improve the previous estimation by considering

$$Z = \frac{T - \delta}{\sqrt{v}},$$

where v is the variance of t, which must be estimated (normally using known methods like Delta methods or the jackknife method). The confidence interval for t is found in an analogous way to that described previously:

$$\left[t - \sqrt{v} \cdot z_{((R+1)(1-\alpha))}^*,\right.$$

$$\left. t + \sqrt{v} \cdot z_{((R+1)\alpha)}^*\right],$$

where $z_{((R+1)(1-\alpha))}^*$ and $z_{((R+1)\alpha)}^*$ are "empirically" the $(1 - \alpha$ and $\alpha)$ quantiles of $\left(z_i^* = \frac{t_i^* - t}{v^*}\right)_{i=1,\ldots,R}$.

EXAMPLES

We consider a typical example of nonparametric bootstrap (taken from the work of Davison, A. and Hinkley, D.V. (1997)). The data concern the populations of ten American cities in 1920 (U) and 1930 (X):

u	138	93	61	179	48
x	143	104	69	260	75

u	37	29	23	30	2
x	63	50	48	111	50

We are interested in the value δ of the statistic $T = \frac{E(X)}{E(U)}$ which will allow us to deduce the populations in 1930 from those in 1920. We estimate δ using $t = \frac{\bar{x}}{\bar{u}} = 1.52$, but what is the uncertainty in this estimation?
The bootstrap allows us to simulate the values t_l^* ($l = 1, \ldots, R$) of δ sampled from a bivariate distribution $Y = (U, X)$ with an empirical distribution function $\hat{F}$ that gives a weight $\frac{1}{10}$ to each observation.
In practice, this is a matter of resampling with return the ten data values and calculating the corresponding

$$t_i^* = \frac{\sum_{j=1}^{10} x_{ij}^*}{\sum_{j=1}^{10} u_{ij}^*}, \quad i = 1, \ldots, R,$$

where x_{ij}^* and u_{ij}^* are the values of the variables X and U. This allows us to approximate the bias and the variance of t using the formulae given before.
The following table summarizes the simulations obtained with $R = 5$:

	j						
	1	2	3	4	5	6	7
u_{1j}^*	93	93	93	61	61	37	29
x_{1j}^*	104	104	104	69	69	63	50
u_{2j}^*	138	61	48	48	37	37	29
x_{2j}^*	143	69	75	75	63	63	50

			j				
	1	2	3	4	5	6	7
u_{3j}^*	138	93	179	37	30	30	30
x_{3j}^*	143	104	260	63	111	111	111
u_{4j}^*	93	61	61	48	37	29	29
x_{4j}^*	104	69	69	75	63	50	50
u_{5j}^*	138	138	138	179	48	48	48
x_{5j^*}	143	143	143	260	75	75	75

		j		Sum
	8	9	10	
u_{1j}^*	29	30	2	$\sum u_{1j}^* = 528$
x_{1j}^*	50	111	50	$\sum x_{1j}^* = 774$
u_{2j}^*	30	30	2	$\sum u_{2j}^* = 460$
x_{2j}^*	111	111	50	$\sum x_{2j}^* = 810$
u_{3j}^*	30	2	2	$\sum u_{3j}^* = 571$
x_{3j}^*	111	50	50	$\sum x_{3j}^* = 1114$
u_{4j}^*	23	23	2	$\sum u_{4j}^* = 406$
x_{4j}^*	48	48	50	$\sum x_{4j}^* = 626$
u_{5j}^*	29	23	30	$\sum u_{5j}^* = 819$
x_{5j^*}	50	48	111	$\sum x_{5j}^* = 1123$

So, the t_i^* are:

$$t_1^* = 1.466, \quad t_2^* = 1.761, \quad t_3^* = 1.951,$$
$$t_4^* = 1.542, \quad t_5^* = 1.371.$$

From this, it is easy to calculate the bias and the variance of t using the formulae given previously:

$$b = 1.62 - 1.52 = 0.10 \text{ and } v = 0.0553.$$

It is possible to use these results to calculate the normal confidence intervals, supposing that $T - \delta$ is normally distributed, $N(0.1, 0.0553)$. We obtain, for a significance level of 5%, the following confidence interval for t: [1.16, 2.08].
The small number of cities considered and simulations mean that we cannot have much confidence in the values obtained.

FURTHER READING
▶ Monte Carlo method
▶ Resampling
▶ Simulation

REFERENCES
Davison, A.C., Hinkley, D.V.: Bootstrap Methods and Their Application. Cambridge University Press, Cambridge (1997)

Efron, B.: Bootstrap methods: another look at the jackknife. Ann. Stat. **7**, 1–26 (1979)

Efron, B., Tibshirani, R.J.: An Introduction to the Bootstrap. Chapman & Hall, New York (1993)

Hall, P.: The Bootstrap and Edgeworth Expansion. Springer, Berlin Heidelberg New York (1992)

Shao, J., Tu, D.: The Jackknife and Bootstrap. Springer, Berlin Heidelberg New York (1995)

Boscovich, Roger J.

Boscovich, Roger Joseph was born at Ragusa (now Dubrovnik, Croatia) in 1711. He attended the *Collegium Ragusinum*, then he went to Rome to study in the *Collegio Romano*; both colleges were Jesuitic. He died in 1787, in Milano.
In 1760, Boscovich developed a geometric method for finding a simple regression line L_1 (in order to correct errors in stellar observations). **Laplace, Pierre Simon de** disapproved of the fact that Boscovich used geometrical terminology and translated this method into an algebraic version in 1793.

The principal work of Boscovich, Roger Joseph:

1757 De Litteraria Expeditione per Pontificiam ditionem, et Synopsis amplioris Operis, ac habentur plura eius ex exemplaria etiam sensorum impressa. Bononiensi Scientiarium et Artium Instituto Atque Academia Commentarii, Tomus IV, pp. 353–396.

REFERENCES

Whyte, Lancelot Law: Roger Joseph Boscovich, Studies of His Life and Work on the 250th Anniversary of his Birth. Allen and Unwin, London (1961)

Box, E.P. George

Box, George E.P. was born in 1919 in England. He served as a chemist in the British Army Engineers during World War II. After the war he received a degree in mathematics and statistics from University College London. In the 1950s, he worked as a visiting professor in the US at the Institute of Statistics at the University of North Carolina. In 1960 he moved to Wisconsin, where he served as the first chairman of the statistics department. He received the British Empire Medal in 1946, and the Shewhart Medal in 1968. His main interest was in experimental statistics and the design of experiments. His book *Statistics for Experimenters*, coauthored with Hunter, J. Stuart and Hunter, William G., is one of the most highly recommended texts in this field. Box also wrote on time series analysis.

Recommended publication of Box, George:

2005 (with Hunter, William G. and Hunter, J. Stuart) Statistics for Experimenters: Design, Innovation, and Discovery, 2nd edn. Wiley, New York

FURTHER READING
▶ **Design of experiments**

Box Plot

The box plot is a way to represent the following five quantities for a set of data: the **median**; the first **quartile** and the third quartile; the maximum and minimum **values**.

The box plot is a diagram (a box) that illustrates:

- The **measure of central tendency** (in principal the **median**);
- The variability, and;
- The symmetry.

It is often used to compare several sets of **observations**.

HISTORY
The "box-and-whisker plot," or box plot, was introduced by Tukey in 1972, along with other methods of representing data semigraphically, one of the most famous of which is the **stem and leaf diagram**.

MATHEMATICAL ASPECTS
Several ways of representing a box plot exist. We will present it in the following way:

The central rectangle represents 50% of the observations. It is known as the **interquartile range**. The lower limit of this rectangle is fixed at the first **quartile**, and the upper limit at the third quartile. The position of the **median** is indicated by a line through the rectangle.

A line segment (a "whisker") connects each **quartile** to the corresponding extreme (minimum or maximum, unless outliers are present; see below) value on each side of the rectangle.

In this representation, **outliers** are treated in a special way. When the observations are very spread out we define two values called the "internal limits" or "whiskers" by:

int.lim.1 = 1st **quartile**

$\qquad$ − (1.5 · **interquartile range**)

int.lim.2 = 3rd **quartile**

$\qquad$ + (1.5 · **interquartile range**)

For each internal limit, we then select the data value that is closest to the limit but still inside the interval between the internal limits. These two data values are known as adjacent points. Now, when we construct the box plot, the line segments connect the **quartiles** to the adjacent points. Observations outside of the interval between the internal limits (outliers) are represented by stars in the plot.

EXAMPLE

The following example presents the revenue indexes per inhabitant for each Swiss canton (Swiss revenue = 100) in 1993:

Canton	Index	Canton	Index
Zurich	125.7	Schaffhouse	99.2
Bern	86.2	Appenzell Rh.-Ext.	84.2
Luzern	87.9	Appenzell Rh.-Int.	72.6
Uri	88.2	Saint-Gall	89.3

Canton	Index	Canton	Index
Schwyz	94.5	Grisons	92.4
Obwald	80.3	Argovie	98.0
Nidwald	108.9	Thurgovie	87.4
Glaris	101.4	Tessin	87.4
Zoug	170.2	Vaud	97.4
Fribourg	90.9	Valais	80.5
Soleure	88.3	Neuchâtel	87.3
Bâle-City	124.2	Genève	116.0
Bâle-Campaign	105.1	Jura	75.1

Source: Federal Office for Statistics (1993)

We now calculate the different quartiles:
The box plot gives us information on the central tendency, on the dispersion of the distribution and the degree of symmetry:

- Central tendency:
  The **median** equals 90.10.
- Dispersion:
  The **interquartile interval** indicates the **interval** that contains 50% of the observations, and these observations are the closest to the center of distribution.
  In our example, we have:
  − 50% of the cantons of Switzerland have an index that falls in the interval [102.33 − 87.03] (with a width of 15.3).

The limits of the box plot are the following:

lim.int.1 = 1st **quartile**

$\qquad$ − (1.5 · **interquartile interval**)

$\qquad$ = 87.03 − 1.5 · 15.3 = 64.08.

lim.int.2 = 3rd **quartile**

$\qquad$ + (1.5 · **interquartile interval**)

$\qquad$ = 102.33 + 1.5 · 15.3 = 125.28.

The values bigger then 125.28 are therefore outliers. There are two of them: 125.7

(Zurich) and 170.2 (Zoug). The box plot is then as follows:

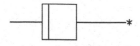

FURTHER READING
▶ **Graphical representation**
▶ **Exploratory data analysis**

▶ **Measure of location**
▶ **Measure of central tendency**

REFERENCES
Tukey, J.W.: Some graphical and semigraphical displays. In: Bancroft, T.A. (ed.) Statistical Papers in Honor of George W. Snedecor. Iowa State University Press, Ames, IA , pp. 293–316 (1972)

B

Categorical Data

Categorical data consists of counts of observations falling into specified classes.

We can distinguish between various types of categorical data:

- Binary, characterizing the presence or absence of a property;
- Unordered multicategorical (also called "nominal");
- Ordered multicategorical (also called "ordinal");
- Whole numbers.

We represent the categorical data in the form of a contingency table.

DOMAINS AND LIMITATIONS

Variables that are essentially continuous can also be presented as categorical variables. One example is "age", which is a continuous variable, but ages can still be grouped into classes so it can still be presented as categorical data.

EXAMPLES

In a public opinion survey for approving or disapproving a new law, the votes cast can be either "yes" or "no". We can represent the results in the form of a contingency table:

	Yes	No
Votes	8546	5455

If we divide up the employees of a business into professions (and at least three professions are presented), the data we obtain is unordered multicategorical data (there is no natural ordering of the professions).

In contrast, if we are interested in the number of people that have achieved various levels of education, there will probably be a natural ordering of the categories: "primary, secondary" and then university. Such data would therefore be an example of ordered multicategorical data.

Finally, if we group employees into categories based on the size of each employee's family (that is, the number of family members), we obtain categorical data where the categories are whole numbers.

FURTHER READING

▶ **Analysis of categorical data**
▶ **Binary data**
▶ **Category**
▶ **Data**
▶ **Dichotomous variable**
▶ **Qualitative categorical variable**
▶ **Random variable**

REFERENCES

See **analysis of categorical data**.

Category

A category represents a set of people or objects that have a common characteristic.

If we want to study the people in a **population**, we can sort them into "natural" categories, by gender (men and women) for example, or into categories defined by other criteria, such as vocation (managers, secretaries, farmers ...).

FURTHER READING
▶ **Binary data**
▶ **Categorical data**
▶ **Dichotomous variable**
▶ **Population**
▶ **Random variable**
▶ **Variable**

Cauchy Distribution

A **random variable** X follows a Cauchy distribution if its **density function** is of the form:

$$f(x) = \frac{1}{\pi \theta} \cdot \left[1 + \left(\frac{x - \alpha}{\theta} \right)^2 \right]^{-1},$$
$$\theta > 0.$$

The **parameters** α and θ are the location and dispersion parameters, respectively.
The Cauchy distribution is symmetric about $x = \alpha$, which represents the **median**. The first **quartile** and the third quartile are given by $\alpha \pm \theta$.
The Cauchy distribution is a **continuous probability distribution**.

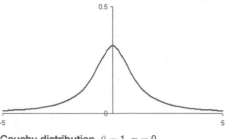

Cauchy distribution, $\theta = 1, \alpha = 0$

MATHEMATICAL ASPECTS
The **expected value** $E[X]$ and the **variance** Var(X) do not exist.
If $\alpha = 0$ and $\theta = 1$, the Cauchy distribution is identical to the **Student distribution** with one **degree of freedom**.

DOMAINS AND LIMITATIONS
Its importance in physics is mainly due to the fact that it is the solution to the differential equation describing force resonance.

FURTHER READING
▶ **Continuous probability distribution**
▶ **Student distribution**

REFERENCES
Cauchy, A.L.: Sur les résultats moyens d'observations de même nature, et sur les résultats les plus probables. C.R. Acad. Sci. **37**, 198–206 (1853)

Causal Epidemiology

The aim of causal epidemiology is to identify how cause is related to effect with regard to human health.
In other words, it is the study of causes of illness, and involves attempting to find statis-

tical evidence for a causal relationship or an association between the illness and the factor proposed to cause the illness.

HISTORY

See **epidemiology**.

MATHEMATICAL ASPECTS

See **cause and effect in epidemiology, odds and odds ratio, relative risk, attributable risk, avoidable risk, incidence rate, prevalence rate**.

DOMAINS AND LIMITATIONS

Studies of the relationship between tobacco smoking and the development of lungial cancers and the relationship between HIV and the AIDS are examples of causal epidemiology. Research into a causal relation is often very complex, requiring many studies and the incorporation and combination of various data sets from biological and animal experiments, to clinical trials.

While causes cannot always be identified precisely, a knowledge of the risk factors associated with an illness and therefore the groups of people at risk allows us to intervene with preventative measures that could preserve health.

EXAMPLES

As an example, we can investigate the relationship between smoking and the development of lung cancer. Consider a study of 2000 subjects: 1000 smokers and 1000 nonsmokers. The age distributions and male/female proportions are identical in both groups. Let us analyze a summary of data obtained over many years. The results are presented in the following table:

	Smokers	Non-smokers
Number of cases of lung cancer (/1000)	50	10
Proportion of people in this group that contracted lung cancer	5%	1%

If we compare the proportion of smokers that contract lung cancer to the proportion of nonsmokers that do, we get $5\%/1\% = 5$, so we can conclude that the risk of developing lung cancer is five times higher in smokers than in nonsmokers. We generally also evaluate the significance of this result computing **p-value**. Suppose that the **chi-square test** gave $p < 0.001$. It normally accepted that if the **p-value** is smaller then 0.05, then the results obtained are statistically significant.

Suppose that we perform the same study, but the dimension of each group (smokers and nonsmokers) is 100 instead of 1000, and we observe the same proportions of people that contract lung cancer:

	Smokers	Non-smokers
Number of cases of lung cancer (/100)	5	1
Proportion of individuals in this group that contracted lungs cancer	5%	1%

If we perform the same statistical test, the **p-value** is found to be 0.212. Since this is greater than 0.05, we can cannot draw any solid conclud that there is about the existence of a significant statistical relation between smoking and lung cancer. This illustrates that, in order to have a statistically signifi-

C

cant level of difference between the results for different populations obtained from epidemiological studies, it is usually necessary to study large samples.

FURTHER READING
See **epidemiology**.

REFERENCES
See **epidemiology**.

Cause and Effect in Epidemiology

In epidemiology, the "cause" is an agent (microbial germs, polluted water, smoking, etc.) that modifies health, and the "effect" describes the the way that the health is changed by the agent. The agent is often potentially pathogenic (in which case it is known as a "risk factor").

The effect is therefore effectively a risk comparison. We can define two different types of risk in this context:

- The absolute effect of a cause expresses the increase in the risk or the additional number of cases of illness that result or could result from exposure to this cause. It is measured by the **attributable risk** and its derivatives.
- The relative effect of a cause expresses the strength of the association between the causal agent and the illness.

A cause that produces an effect by itself is called *sufficient*.

HISTORY
The terms "cause" and "effect" were defined at the birth of **epidemiology**, which occurred in the seventeenth century.

MATHEMATICAL ASPECTS
Formally, we have:

$$\text{Absolute effect} = \text{Risk for exposed} - \text{risk for unexposed}.$$

The absolute effect expresses the excess risk or cases of illness that result (or could result) from exposure to the cause.

$$\text{Relative effect} = \frac{\text{risk for exposed}}{\text{risk for unexposed}}.$$

The relative effect expresses the strength of the association between the illness and the cause. It is measured using the **relative risk** and the **odds ratio**.

DOMAINS AND LIMITATIONS
Strictly speaking, the strength of an association between a particular factor and an illness is not enough to establish a causal relationship between them. We also need to consider:

- The "temporality criterion" (we must be sure that the supposed cause precedes the effect), and;
- Fundamental and experimental research elements that allow us to be sure that the supposed causal factor is not actually a "confusion factor" (which is a factor that is not causal, but is statistically related to the unidentified real causal factor).

Two types of causality correspond to these relative and absolute effects. Relative causality is independent of the clinical or public health impact of the effect; it generally does not allow us to prejudge the clinical or public health impact of the associated effect. It can be strong when the risk of being exposed is very high or when the risk of being unexposed is very low.

Absolute causality expresses the clinical or public health impact of the associated effect, and therefore enables us to answer the question: if we had suppressed the cause, what level of impact on the population (in terms of cases of illness) would have been avoided? If the patient had stopped smoking, what would the reduction in his risk of developing lung cancer or having a myocardial infarction be? A risk factor associated with a high relative effect, but which concerns only a small number of individuals, will cause fewer illnesses and deaths then a risk factor that is associated with a smaller relative effect but where many more individuals are exposed. It is therefore clear that the importance of a causal relation varies depending on whether we are considering relative or absolute causality.

We should also make an important point here about causal interactions. There can be many causes for the same illness. While all of these causes contribute to the same result, they can also interact. The main consequence of this causal interaction is that we cannot prejudge the effect of simultaneous exposure to causes A and B (denoted $A+B$) based on what we know about the effect of exposure to only A or only B. In contrast to the case for independent causes, we must estimate the joint effect, not restrict ourselves with the isolated analyses of the interacting causes.

EXAMPLES

The relative effect and the absolute effect are subject to different interpretations, as the following example shows.

Suppose we have two populations P_1 and P_2, each comprising 100000 individuals. In population P_1, the risk of contracting a given illness is 0.2% for the exposed and 0.1% for the unexposed. In population P_2, the risk for the exposed is 20% and that for the unexposed is 10%, as shown in the following table:

Population	Risk for the exposed (%) A	Risk for the unexposed (%) B
P_1	0.2	0.1
P_2	20	10

Population	Relative effect $\dfrac{A}{B}$	Absolute effect (%) $C = A - B$	Avoidable cases $C \times 100000$
P_1	2.0	0.1	100
P_2	2.0	10	10000

The relative effect is the same for populations P_1 and P_2 (the ratio of the risk for the exposed to the risk for the unexposed is 2), but the impact of the same prevention measures would be very different in the two populations, because the absolute effect is ten times more important in P_2: the number of potentially avoidable cases is therefore 100 in population P_1 and 10000 in population P_2. Now consider the **incidence rate** of lung cancer in a population of individuals who smoke 35 or more cigarettes per day: 3.15/1000/year. While this rate may seem small, it masks the fact that there is a strong relative effect (the risk is 45 times bigger for smokers then for nonsmokers) due to the fact that lung cancer is very rare in nonsmokers (the incidence rate for nonsmokers is 0.07/1000/year).

FURTHER READING

▶ **Attributable risk**

▶ **Avoidable risk**

▶ **Incidence rate**

▶ **Odds and odds ratio**

▶ **Prevalence rate**

▶ **Relative risk**

REFERENCES

Lilienfeld, A.M., Lilienfeld, D.E.: Foundations of Epidemiology, 2nd edn. Clarendon, Oxford (1980)

MacMahon, B., Pugh, T.F.: Epidemiology: Principles and Methods. Little Brown, Boston, MA (1970)

Morabia, A.: Epidemiologie Causale. Editions Médecine et Hygiène, Geneva (1996)

Rothmann, J.K.: Epidemiology. An Introduction. Oxford University Press (2002)

Census

A census is an operation that consists of observing all of the individuals in a **population**. The word census can refer to a population census, in other words a population count, but it can also refer to inquiries (called "exhaustive" inquiries) where we retrieve information about a population by observing all of the individuals in the population. Clearly, such inquiries will be very expensive for very large populations. That is why exhaustive inquiries are rarely performed; **sampling**, which consists of observing of only a portion of the population (called the **sample**), is usually preferred instead.

HISTORY

Censuses originated with the great civilizations of antiquity, when the large areas of empires and complexity associated with governing them required knowledge of the populations involved.

Among the most ancient civilizations, it is known that censuses were performed in Sumeria (between 5000 and 2000 BC), where the people involved reported lists of men and goods on clay tables in cuneiform characters.

Censuses were also completed in Mesopotamia (about 3000 BC), as well as in ancient Egypt from the first dynasty onwards; these censuses were performed due to military and fiscal objectives. Under Amasis II, everybody had to (at the risk of death) declare their profession and source(s) of revenue.

The situation in Israel was more complex: censuses were sometimes compulsary and sometimes forbidden due to the Old Testament. This influence on Christian civilization lasted quite some time; in the Middle Ages, St. Augustin and St. Ambroise were still condemning censuses.

In China, censuses have been performed since at least 200 BC, in different forms and for different purposes. Hecht, J. (1987) reported the main censuses:

1. *Han Dynasty (200 years BC to 200 years AD):* **population** censuses were related to the system of conscription.

2. *Three Kingdoms Period to Five Dynasties (221–959 AD):* related to the system of territorial distribution.

3. *Song and Yuan Dynasties (960–1368 AD):* censuses were performed for fiscal purposes.

4. *Ming Dynasty (1368–1644 AD):* "yellow registers" were established for ten-year censuses. They listed the name, profession, gender and age of every person.

5. *Qing Dynasty (since 1644 AD):* censuses were performed in order to survey population migration.

In Japan during the Middle Age, different types of census have been used. The first census was probably performed under the rule of Emperor Sujin (86 BC).

Finally, in India, a political and economic science treatise entitled "Arthasastra" (profit treaty) gives some information about the use of an extremely detailed record. This treatise was written by Kautilya, Prime Minister in the reign of Chandragupta Maurya (313–289 BC).

Another very important civilization, the Incans, also used censuses. They used a statistics system called "quipos". Each quipo was both an instrument and a registry of information. Formed from a series of cords, the colors, combinations and knots on the cords had precise meanings. The quipos were passed to specially initiated guards that gathered together all of these statistics.

In Europe, the ancient Greeks and Romans also practiced censuses. Aristotle reported that the Greeks donated a measure of wheat per birth and a measure of barley per death to the goddess Athéna. In Rome, the first census was performed at the behest of King Servius Tullius (578–534 BC) in order to monitor revenues, and consequently raise taxes.

Later, depending on the country, censuses were practiced with different frequencies and on different scales.

In 786, Charlemagne ordered a count of all of his subjects over twelve years old; population counts were also initiated in Italy in the twelfth century; many cities performed censuses of their inhabitants in the fifteenth century, including Nuremberg (in 1449) and Strasbourg (in 1470). In the sixteenth century, France initiated marital status registers.

The seventeenth century saw the development of three different schools of thought: a German school associated with **descriptive statistics**, a French school associated with census ideology and methodology, and an English school that led to modern statistics.

In the history of censuses in Europe, there is a country that occupies a special place. In 1665, Sweden initiated registers of parishioners that were maintained by pastors; in 1668, a decree made it obligatory to be counted in these registers, and instead of being religious, the registers became administrative. 1736 saw the appearance of another decree, stating that the governor of each province had to report any changes in the population of the province to parliament. Swedish population statistics were officially recognized on the 3rd of February 1748 due to creation of the "Tabellverket" (administrative tables). The first summary for all Swedish provinces, realized in 1749, can be considered to be the first proper census in Europe, and the 11th of November 1756 marked the creation of the "Tabellkommissionen," the first official Division of Statistics. Since 1762, these tables of figures have been maintained by the Academy of Sciences.

Initially, the Swedish censuses were organized annually (1749–1751), and then every three years (1754–1772), but since 1775 they have been conducted every five years.

At the end of the eighteenth century an official institute for censuses was created in France (in 1791). In 1787 the principle of census has been registered in the Constitutional Charter of the USA (C. C. USA).

See also **official statistics**.

FURTHER READING
▶ **Data collection**
▶ **Demography**
▶ **Official statistics**
▶ **Population**
▶ **Sample**
▶ **Survey**

REFERENCES
Hecht, J.: L'idée du dénombrement jusqu'à la Révolution. Pour une histoire de la

statistique, tome 1, pp. 21–82 . Economica/INSEE (1978)

Central Limit Theorem

The central limit theorem is a fundamental theorem of statistics. In its simplest form, it prescribes that the sum of a sufficiently large number of independent identically distributed random variables approximately follows a **normal distribution**.

HISTORY

The central limit theorem was first established within the framework of **binomial distribution** by **Moivre, Abraham de** (1733). **Laplace, Pierre Simon de** (1810) formulated the proof of the theorem. **Poisson, Siméon Denis** (1824) also worked on this theorem, and **Chebyshev, Pafnutii Lvovich** (1890–1891) gave a rigorous demonstration of it in the middle of the nineteenth century.

At the beginning of the twentieth century, the Russian mathematician Liapounov, Aleksandr Mikhailovich (1901) created the generally recognized form of the central limit theorem by introducing its characteristic functions. **Markov, Andrei Andreevich** (1908) also worked on it and was the first to generalize the theorem to the case of independent variables.

According to Le Cam, L. (1986), the qualifier "central" was given to it by George Polyà (1920) due to the essential role that it plays in probability theory.

MATHEMATICAL ASPECTS

Let $X_1, X_2, \ldots, X_n$ be n independent random variables that are identically distributed (with any distribution) with a **mean** μ and a finite **variance** σ^2.

We define the sum $S_n = X_1 + X_2 + \ldots + X_n$ and we establish the ratio:

$$\frac{S_n - n \cdot \mu}{\sigma \cdot \sqrt{n}},$$

where $n \cdot \mu$ and $\sigma \cdot \sqrt{n}$ represent the mean and the **standard deviation** of S_n, respectively. The central limit theorem establishes that the distribution of this ratio tends to the standard normal distribution when n tends to infinity. This means that:

$$P\left(\frac{S_n - n \cdot \mu}{\sigma \sqrt{n}} \leq x\right) \xrightarrow[n \to +\infty]{} \Phi(x)$$

where $\Phi(x)$ is the **distribution function** of the standard **normal distribution**, expressed by:

$$\Phi(x) = \int_{-\infty}^{x} \frac{1}{\sqrt{2\pi}} \exp\left(-\frac{x^2}{2}\right) dx,$$
$$-\infty < x < \infty.$$

DOMAINS AND LIMITATIONS

The central limit theorem provides a simple method of approximately calculating probabilities related to the sums of random variables.

Besides its interest in relation to the **sampling** theorem, where the sums and the means play an important role, the central limit theorem is used to approximate **normal distributions** derived from summing identical distributions. We can for example, with the help of the central limit theorem, use the **normal distribution** to approximate the **binomial distribution**, the **Poisson distribution**, the **gamma distribution**, the **chi-square distribution**, the **Student distribution**, the **hypergeometric distribution**, the **Fisher distribution** and the **lognormal distribution**.

EXAMPLES

In a large batch of electrical items, the **probability** of choosing a defective item equals $p = \frac{1}{8}$. What is the probability that 4 defective items are chosen when 25 items are selected?

Let X the **dichotomous** variable be the result of a trial:

$$X = \begin{cases} 1 & \text{if the selected item is defective;} \\ 0 & \text{if it is not.} \end{cases}$$

The **random variable** X follows a **Bernoulli distribution** with **parameter** p. Consequently, the sum $S_n = X_1 + X_2 + \ldots + X_n$ follows a **binomial distribution** with a **mean**, np and a **variance** $np(1-p)$, which, following the central limit theorem, can be approximated by a **normal distribution** with a mean $\mu = np$ and a variance $\sigma^2 = np(1-p)$. We evaluate these values.

$$\mu = n \cdot p = 25 \cdot \tfrac{1}{8} = 3.125$$

$$\sigma^2 = n \cdot p(1-p) = 25 \cdot \tfrac{1}{8} \cdot \left(1 - \tfrac{1}{8}\right)$$

$$= 2.734 \,.$$

We then calculate $P(S_n > 4)$ in two different ways:

1. With the binomial distribution:
   From the **binomial table**, the probability of $P(S_n \leq 4) = 0.8047$. The probability of $P(S_n > 4)$ is then:

   $$P(S_n > 4) = 1 - P(S_n \leq 4) = 0.1953 \,,$$

2. With the normal approximation (obtained from the central limit theorem):
   In order to account for the discrete character of the random variable S_n, we must make a continuity correction; that is, we calculate the probability that S_n is greater then $4 + \frac{1}{2} = 4.5$

We have:

$$z = \frac{S_n - n \cdot p}{\sqrt{n \cdot p \, (1-p)}}$$

$$= \frac{4.5 - 3.125}{1.654}$$

$$= 0.832 \,.$$

From the **normal table**, we obtain the probability:

$$P(Z > z) = P(Z > 0.832)$$
$$= 1 - P(Z \leq 0.832)$$
$$= 1 - 0.7967 = 0.2033 \,.$$

FURTHER READING

▶ **Binomial distribution**
▶ **Chi-square distribution**
▶ **Convergence**
▶ **Convergence theorem**
▶ **Fisher distribution**
▶ **Gamma distribution**
▶ **Hypergeometric distribution**
▶ **Law of large numbers**
▶ **Lognormal distribution**
▶ **Normal distribution**
▶ **Poisson distribution**
▶ **Probability**
▶ **Probability distribution**
▶ **Student distribution**

REFERENCE

Laplace, P.S. de: Mémoire sur les approximations des formules qui sont fonctions de très grands nombres et sur leur application aux probabilités. Mémoires de l'Académie Royale des Sciences de Paris, 10. Reproduced in: Œuvres de Laplace **12**, 301–347 (1810)

Le Cam, L.: The Central Limit Theorem around 1935. Stat. Sci. **1**, 78–96 (1986)

Liapounov, A.M.: Sur une proposition de la théorie des probabilités. Bulletin de l'Academie Imperiale des Sciences de St.-Petersbourg **8**, 1–24 (1900)

Markov, A.A.: Extension des théorèmes limites du calcul des probabilités aux sommes des quantités liées en chaîne. Mem. Acad. Sci. St. Petersburg **8**, 365–397 (1908)

Moivre, A. de: Approximatio ad summam terminorum binomii $(a + b)^n$, in seriem expansi. Supplementum II to Miscellanae Analytica, pp. 1–7 (1733). Photographically reprinted in a rare pamphlet on Moivre and some of his discoveries. Published by Archibald, R.C. Isis **8**, 671–683 (1926)

Poisson, S.D.: Sur la probabilité des résultats moyens des observations. Connaissance des temps pour l'an 1827, pp. 273–302 (1824)

Polyà, G.: Ueber den zentralen Grenzwertsatz der Wahrscheinlichkeitsrechnung und das Momentproblem. Mathematische Zeitschrift **8**, 171–181 (1920)

Tchebychev, P.L. (1890–1891). Sur deux théorèmes relatifs aux probabilités. Acta Math. **14**, 305–315

Chebyshev, Pafnutii Lvovich

Chhebyshev, Pafnutii Lvovich (1821–1894) began studying at Moscow University in 1837, where he was influenced by Zernov, Nikolai Efimovich (the first Russian to get a doctorate in mathematical sciences) and Brashman, Nikolai Dmitrievich. After gaining his degree he could not find any teaching work in Moscow, so he went to St. Petersbourg where he organized conferences on algebra and probability theory. In 1859, he took the probability course given by Buni-akovsky, Viktor Yakovlevich at St. Petersbourg University.

His name lives on through the **Chebyshev inequality** (also known as the Bienaymé–Chebyshev inequality), which he proved. This was published in French just after **Bienaymé, Irénée-Jules** had an article published on the same topic in the *Journal de Mathématiques Pures et Appliquées* (also called the Journal of Liouville).

He initiated rigorous work into establishing a general version of the **central limit theorem** and is considered to be the founder of the mathematical school of St. Petersbourg.

Some principal works and articles of Chebyshev, Pafnutii Lvovich:

1845   An Essay on Elementary Analysis of the Theory of Probabilities (thesis) Crelle's Journal.

1867   Preuve de l'inégalité de Tchebychev. J. Math. Pure. Appl., 12, 177–184.

FURTHER READING

▶ **Central limit theorem**

REFERENCES

Heyde, C.E., Seneta, E.: I.J. Bienaymé. Statistical Theory Anticipated. Springer, Berlin Heidelberg New York (1977)

Chebyshev's Inequality

See **law of large numbers**.

Chi-Square Distance

Consider a **frequency table** with n rows and p columns, it is possible to calculate row profiles and column profiles. Let us then plot the n or p points from each profile. We can define the **distances** between these points. The Euclidean distance between the components of the profiles, on which a weighting is defined (each term has a weight that is the inverse of its **frequency**), is called the chi-square distance. The name of the distance is derived from the fact that the mathematical expression defining the distance is identical to that encountered in the elaboration of the **chi square goodness of fit test**.

MATHEMATICAL ASPECTS

Let (f_{ij}), be the **frequency** of the ith row and jth column in a frequency table with n rows an p columns. The chi-square distance between two rows i and i' is given by the formula:

$$d(i, i') = \sqrt{\sum_{j=1}^{p} \left(\frac{f_{ij}}{f_{i.}} - \frac{f_{i'j}}{f_{i'.}}\right)^2 \cdot \frac{1}{f_{.j}}},$$

where

$f_{i.}$ is the sum of the components of the ith row;

$f_{.j}$ is the sum of the components of the jth column;

$[\frac{f_{ij}}{f_{i.}}]$ is the ith row profile for $j = 1, 2, \ldots, p$.

Likewise, the distance between two columns j and j' is given by:

$$d(j, j') = \sqrt{\sum_{i=1}^{n} \left(\frac{f_{ij}}{f_{.j}} - \frac{f_{ij'}}{f_{.j'}}\right)^2 \cdot \frac{1}{f_{i.}}},$$

where $[\frac{f_{ij}}{f_{.j}}]$ is the jth column profile for $j = 1, \ldots, n$.

DOMAINS AND LIMITATIONS

The chi-square distance incorporates a weight that is inversely proportional to the total of each row (or column), which increases the importance of small deviations in the rows (or columns) which have a small sum with respect to those with more important sum package.

The chi-square distance has the property of distributional equivalence, meaning that it ensures that the **distances** between rows and columns are invariant when two columns (or two rows) with identical profiles are aggregated.

EXAMPLES

Consider a **contingency table** charting how satisfied employees working for three different businesses are. Let us establish a **distance table** using the chi-square distance.

Values for the studied **variable** X can fall into one of three **categories**:

- X_1: high satisfaction;
- X_2: medium satisfaction;
- X_3: low satisfaction.

The **observations** collected from **samples** of individuals from the three businesses are given below:

	Business 1	Business 2	Business 3	Total
X_1	20	55	30	105
X_2	18	40	15	73
X_3	12	5	5	22
Total	50	100	50	200

The relative **frequency table** is obtained by dividing all of the elements of the table by 200, the total number of observations:

	Business 1	Business 2	Business 3	Total
X_1	0.1	0.275	0.15	0.525
X_2	0.09	0.2	0.075	0.365
X_3	0.06	0.025	0.025	0.11
Total	0.25	0.5	0.25	1

We can calculate the difference in employee satisfaction between the the 3 enterprises. The column profile matrix is given below:

	Business 1	Business 2	Business 3	Total
X_1	0.4	0.55	0.6	1.55
X_2	0.36	0.4	0.3	1.06
X_3	0.24	0.05	0.1	0.39
Total	1	1	1	3

This allows us to calculate the **distances** between the different columns:

$$d^2(1, 2) = \frac{1}{0.525} \cdot (0.4 - 0.55)^2$$
$$+ \frac{1}{0.365} \cdot (0.36 - 0.4)^2$$
$$+ \frac{1}{0.11} \cdot (0.24 - 0.05)^2$$
$$= 0.375423$$
$$d(1, 2) = 0.613$$

We can calculate $d(1, 3)$ and $d(2, 3)$ in a similar way. The distances obtained are summarized in the following **distance table**:

	Business 1	Business 2	Business 3
Business 1	0	0.613	0.514
Business 2	0.613	0	0.234
Business 3	0.514	0.234	0

We can also calculate the **distances** between the rows, in other words the difference in employee satisfaction; to do this we need the line profile table:

	Business 1	Business 2	Business 3	Total
X_1	0.19	0.524	0.286	1
X_2	0.246	0.548	0.206	1
X_3	0.546	0.227	0.227	1
Total	0.982	1.299	0.719	3

This allows us to calculate the **distances** between the different rows:

$$d^2(1, 2) = \frac{1}{0.25} \cdot (0.19 - 0.246)^2$$
$$+ \frac{1}{0.5} \cdot (0.524 - 0.548)^2$$
$$+ \frac{1}{0.25} \cdot (0.286 - 0.206)^2$$
$$= 0.039296$$
$$d(1, 2) = 0.198$$

We can calculate $d(1, 3)$ and $d(2, 3)$ in a similar way. The differences between the degrees of employee satisfaction are finally summarized in the following **distance table**:

	X_1	X_2	X_3
X_1	0	0.198	0.835
X_2	0.198	0	0.754
X_3	0.835	0.754	0

FURTHER READING
▶ **Contingency table**
▶ **Distance**
▶ **Distance table**
▶ **Frequency**

Chi-square Distribution

A **random variable** X follows a chi-square distribution with n **degrees of freedom** if its **density function** is:

$$f(x) = \frac{x^{\frac{n}{2}-1} \exp\left(-\frac{x}{2}\right)}{2^{\frac{n}{2}} \Gamma\left(\frac{n}{2}\right)}, \quad x \geq 0,$$

where Γ is the gamma function (see **Gamma distribution**).

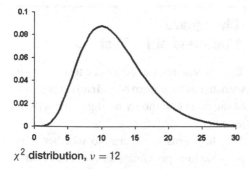

χ^2 distribution, $\nu = 12$

The chi-square distribution is a **continuous probability distribution**.

HISTORY

According to Sheynin (1977), the chi-square distribution was discovered by Ernst Karl Abbe in 1863. Maxwell obtained it for three degrees of freedom a few years before (1860), and Boltzman discovered the general case in 1881.

However, according to Lancaster (1966), Bienaymé obtained the chi-square distribution in 1838 as the limit of the discrete **random variable**

$$\sum_{i=1}^{k} \frac{(n_i - np_i)^2}{np_i},$$

if $(N_1, N_2, \ldots, N_k)$ follow a joint multinomial distribution of **parameters** $n, p_1, p_2, \ldots, p_k$.

Ellis demonstrated in 1844 that the sum of k **random variables** distributed according to a chi-square distribution with two **degrees of freedom** follows a chi-square distribution with $2k$ degrees of freedom. The general result was demonstrated in 1852 by Bienaymé.

The works of **Pearson, Karl** are very important in this field. In 1900 he used the chi-square distribution to approximate the chi-square **statistic** used in different tests based on **contingency tables**.

MATHEMATICAL ASPECTS

The chi-square distribution appears in the theory of **random variables** distributed according to a **normal distribution**. In this, it is the distribution of the sum of squares of normal, centered and reduced random variables (with a **mean** equal to 0 and a **variance** equal to 1).

Consider $Z_1, Z_2, \ldots, Z_n$, n independent, standard normal **random variables**. Their sum of squares:

$$X = Z_1^2 + Z_2^2 + \ldots + Z_n^2 = \sum_{i=1}^{n} Z_i^2$$

is a random variable distributed according to a chi-square distribution with n **degrees of freedom**.

The **expected value** of the chi-square distribution is given by:

$$E[X] = n.$$

The **variance** is equal to:

$$\mathrm{Var}(X) = 2n.$$

The chi-square distribution is related to other **continuous probability distributions**:

- The chi-square distribution is a particular case of the **gamma distribution**.
- If two **random variables** X_1 and X_2 follow a chi-square distribution with, respectively, n_1 and n_2 **degrees of freedom**, then the random variable

$$Y = \frac{X_1/n_1}{X_2/n_2}$$

follows a **Fisher distribution** with n_1 and n_2 degrees of freedom.

- When the number of **degrees of freedom** n tends towards infinity, the chi-square distribution tends (relatively slowly) towards a **normal distribution**.

DOMAINS AND LIMITATIONS

The chi-square distribution is used in many approaches to **hypothesis testing**, the most important being the **goodness of fit test** which involves comparing the observed **frequencies** and the hypothetical frequencies of specific classes.

It is also used for comparisons between the observed **variance** and the hypothetical variance of normally distributed **samples**, and to test the **independence** of two **variables**.

FURTHER READING

▶ **Chi-square goodness of fit test**

▶ **Chi-square table**

▶ **Chi-square test**

▶ **Continuous probability distribution**

▶ **Fisher distribution**

▶ **Gamma distribution**

▶ **Normal distribution**

REFERENCES

Lancaster, H.O.: Forerunners of the Pearson chi-square. Aust. J. Stat. **8**, 117–126 (1966)

Pearson, K.: On the criterion, that a given system of deviations from the probable in the case of a correlated system of variables is such that it can be reasonably supposed to have arisen from random sampling. In: Karl Pearson's Early Statistical Papers. Cambridge University Press, pp. 339–357. First published in 1900 in Philos. Mag. (5th Ser) **50**, 157–175 (1948)

Sheynin, O.B.: On the history of some statistical laws of distribution. In: Kendall, M., Plackett, R.L. (eds.) Studies in the History of Statistics and Probability, vol. II. Griffin, London (1977)

Chi-square Goodness of Fit Test

The chi-square goodness of fit test is, along with the **Kolmogorov–Smirnov test**, one of the most commonly used goodness of fit tests.

This test aims to determine whether it is possible to approximate an observed distribution by a particular **probability distribution (normal distribution, Poisson distribution**, etc).

HISTORY

The chi-square goodness of fit test is the oldest and most well-known of the goodness of fit tests. It was first presented in 1900 by **Pearson, Karl**.

MATHEMATICAL ASPECTS

Let $X_1, \ldots, X_n$ be a **sample** of n observations. The steps used to perform the chi-square goodness of fit test are then as follows:

1. State the hypothesis. The **null hypothesis** will take the following form:

$$H_0: \quad F = F_0 \,,$$

where F_0 is the presumed **distribution function** of the distribution.

2. Distribute the observations in k disjoint classes:

$$[a_{i-1}, a_i] \,.$$

We denote the number of observations contained in the ith class, $i = 1, \ldots, k$, by n_i.

3. Calculate the theoretical probabilities for every class on the base of the presumed distribution function F_0:

$$p_i = F_0(a_i) - F_0(a_{i-1}) \,, \quad i = 1, \ldots, k \,.$$

4. Obtain the expected frequencies for every class

$$e_i = n \cdot p_i, \quad i = 1, \ldots, k,$$

where n is the size of the **sample**.

5. Calculate the χ^2 (chi-square) statistic:

$$\chi^2 = \sum_{i=1}^{k} \frac{(n_i - e_i)^2}{e_i}.$$

If H_0 is true, the χ^2 statistic follows a **chi-square distribution** with v degrees of freedom, where:

$$v = \left(k - 1 - \begin{array}{c} \text{number of estimated} \\ \text{parameters} \end{array} \right).$$

For example, when testing the goodness of fit to a **normal distribution**, the number of degrees of freedom equals:

- $k - 1$ if the **mean** μ and the **standard deviation** σ of the **population** are known;
- $k - 2$ if one out of μ or σ is unknown and will be estimated in order to proceed with the test;
- $k - 3$ if both parameters μ and σ are unknown and both are estimated from the corresponding values of the sample.

6. Reject H_0 if the deviation between the observed and estimated frequencies is big; that is:

$$\text{if } \chi^2 > \chi^2_{v,\alpha},$$

where $\chi^2_{v,\alpha}$ is the **value** given in the **chi-square table** for a particular **significance level** α.

DOMAINS AND LIMITATIONS

To apply the chi-square goodness of fit test, it is important that n is big enough and that the estimated frequencies, e_i, are not too small. We normally state that the estimated frequencies must be greater then 5, except for extreme classes, where they can be smaller then 5 but greater then 1. If this constraint is not satisfied, we must regroup the classes in order to satisfy this rule.

EXAMPLES
Goodness of Fit to the Binomial Distribution

We throw a coin four times and count the number of times that "heads" appears. This **experiment** is performed 160 times. The observed frequencies are as follows:

Number of "heads"	Number of experiments
x_i	(n_i)
0	17
1	52
2	54
3	31
4	6
Total	160

1. If the **experiment** was performed correctly and the coin is not forged, the distribution of the number of "heads" obtained should follow the **binomial distribution**. We then state a **null hypothesis** that the observed distribution can be approximated by the binomial distribution, and we will proceed with a **goodness of fit test** in order to determine whether this **hypothesis** can be accepted or not.

2. In this example, the different number of "heads" that can be obtained per experiment (0, 1, 2, 3 and 4) are each considered to be a class.

3. The **random variable** X (number of "heads" obtained after four throws of a coin) follows a binomial distribution if

$$P(X = x) = C_n^x \cdot p^x \cdot q^{n-x},$$

where:

n is the number of independent trials $= 4$;

p is the **probability** of a success ("heads") $= 0.5$;

q is the probability of a failure ("tails") $= 0.5$;

C_n^x is the number of combinations of x objects from n.

We then have the following theoretical probabilities for four throws:

$$P(X = 0) = \frac{1}{16}$$
$$P(X = 1) = \frac{4}{16}$$
$$P(X = 2) = \frac{6}{16}$$
$$P(X = 3) = \frac{4}{16}$$
$$P(X = 4) = \frac{1}{16}$$

4. After the experiment has been performed 160 times, the expected number of heads for each possible **value** of X is given by:

$$e_i = 160 \cdot P(X = x_i).$$

We obtain the following table:

Number of "heads" x_i	Observed frequency (n_i)	Expected frequency (e_i)
0	17	10
1	52	40
2	54	60
3	31	40
4	6	10
Total	160	160

5. The χ^2 (chi-square) statistic is then:

$$\chi^2 = \sum_{i=1}^{k} \frac{(n_i - e_i)^2}{e_i},$$

where k is the number of possible values of X.

$$\chi^2 = \frac{(17 - 10)^2}{10} + \ldots + \frac{(6 - 10)^2}{10}$$
$$= 12.725.$$

6. Choosing a **significance level** α of 5%, we find that the value of $\chi^2_{\nu,\alpha}$ for $k - 1 = 4$ degrees of freedom is:

$$\chi^2_{4,0.05} = 9.488.$$

Since the calculated value of χ^2 is greater than the value obtained from the table, we reject the **null hypothesis** and conclude that the binomial distribution does not give a good approximation to our observed distribution. We can then conclude that the coins were probably forged, or that they were not correctly thrown.

Goodness of Fit to the Normal Distribution

The diameters of cables produced by a factory were studied.

A **frequency table** of the observed distribution of diameters is given below:

Cable diameter (in mm)	Observed frequency n_i
19.70–19.80	5
19.80–19.90	12
19.90–20.00	35
20.00–20.10	42
20.10–20.20	28
20.20–20.30	14
20.30–20.40	4
Total	140

1. We perform a **goodness of fit test** for a **normal distribution**. The null hypothesis is therefore that the observed distribution can be approximated by a normal distribution.
2. The previous table shows the observed diameters divided up into classes.
3. If the random variable X (cable diameter) follows the normal distribution, the random variable

$$Z = \frac{X - \mu}{\sigma}$$

follows a standard normal distribution. The **mean** μ and the **standard deviation** σ of the **population** are unknown and are estimated using the mean $\bar{x}$ and the standard deviation S of the **sample**:

$$\bar{x} = \frac{\sum\limits_{i=1}^{7} \delta_i \cdot n_i}{n} = \frac{2806.40}{140} = 20.05,$$

$$S = \sqrt{\frac{\sum\limits_{i=1}^{7} n_i \cdot (\delta_i - \bar{x})^2}{n-1}} = \sqrt{\frac{2.477}{139}}$$

$$= 0.134$$

where the δ_i are the centers of the classes (in this example, the mean diameter of a class; for $i = 1$: $\delta_1 = 19.75$) and n is the total number of observations.

We can then calculate the theoretical probabilities associated with each class. The detailed calculations for the first two classes are:

$$p_1 = P(X \leq 19.8)$$

$$= P(Z \leq \frac{19.8 - \bar{x}}{S})$$

$$= P(Z \leq -1.835)$$

$$= 1 - P(Z \leq 1.835)$$

$$p_2 = P(19.8 \leq X \leq 19.9)$$

$$= P\left(\frac{19.8 - \bar{x}}{S} \leq Z \leq \frac{19.9 - \bar{x}}{S}\right)$$

$$= P(-1.835 \leq Z \leq -1.089)$$

$$= P(Z \leq 1.835) - P(Z \leq 1.089)$$

These probabilities can be found by consulting the **normal table**. We get:

$$p_1 = P(X \leq 19.8) = 0.03325$$

$$p_2 = P(19.8 \leq X \leq 19.9) = 0.10476$$

$$p_3 = P(19.9 \leq X \leq 20.0) = 0.22767$$

$$p_4 = P(20.0 \leq X \leq 20.1) = 0.29083$$

$$p_5 = P(20.1 \leq X \leq 20.2) = 0.21825$$

$$p_6 = P(20.2 \leq X \leq 20.3) = 0.09622$$

$$p_7 = P(X > 20.3) = 0.02902$$

4. The expected frequencies for the classes are then given by:

$$e_i = n \cdot p_i,$$

which yields the following table:

Cable diameter (in mm)	Observed frequency n_i	Expected frequency e_i
19.70–19.80	5	4.655
19.80–19.90	12	14.666
19.90–20.00	35	31.874
20.00–20.10	42	40.716
20.10–20.20	28	30.555
20.20–20.30	14	13.471
20.30–20.40	4	4.063
Total	140	140

5. The χ^2 (chi-square) statistic is then:

$$\chi^2 = \sum_{i=1}^{k} \frac{(n_i - e_i)^2}{e_i},$$

C

where $k = 7$ is the number of classes.

$$\chi^2 = \frac{(5 - 4.655)^2}{4.655}$$
$$+ \frac{(12 - 14.666)^2}{14.666}$$
$$+ \ldots + \frac{(4 - 4.063)^2}{4.063}$$
$$= 1.0927 .$$

6. Choosing a significance level $\alpha = 5\%$, we find that the value of $\chi^2_{v,\alpha}$ with $k - 3 = 7 - 3 = 4$ degrees of freedom in the chi-square table is:

$$\chi^2_{4,0.05} = 9.49 .$$

Since the value calculated from χ^2 is smaller then the value obtained from the chi-square table, we do not reject the null hypothesis and we conclude that the difference between the observed distribution and the normal distribution is not significant at a **significance level** of 5%.

FURTHER READING
▶ **Chi-square distribution**
▶ **Chi-square table**
▶ **Goodness of fit test**
▶ **Hypothesis testing**
▶ **Kolmogorov–Smirnov test**

REFERENCE

Pearson, K.: On the criterion, that a given system of deviations from the probable in the case of a correlated system of variables is such that it can be reasonably supposed to have arisen from random sampling. In: Karl Pearson's Early Statistical Papers. Cambridge University Press, pp. 339–357. First published in 1900 in Philos. Mag. (5th Ser) **50**, 157–175 (1948)

Chi-Square Table

The chi-square table gives the values obtained from the **distribution function** of a **random variable** that follows a **chi-square distribution**.

HISTORY

One of the first chi-square tables was published in 1902, by Elderton. It contains **distribution function** values that are given to six decimal places.

In 1922, **Pearson, Karl** reported a table of values for the incomplete gamma function, down to seven decimals.

MATHEMATICAL ASPECTS

Let X be a **random variable** that follows a **chi-square distribution** with v degrees of freedom. The **density function** of the random variable X is given by:

$$f(t) = \frac{t^{\frac{v}{2}-1} \exp\left(-\frac{t}{2}\right)}{2^{\frac{v}{2}} \Gamma\left(\frac{v}{2}\right)}, \quad t \geq 0,$$

where Γ represents the gamma function (see **gamma distribution**).

The **distribution function** of the random variable X is defined by:

$$F(x) = P(X \leq x) = \int_0^x f(t) \, dt .$$

The chi-square table gives the values of the distribution function $F(x)$ for different values of v.

We often use the chi-square table in the opposite way, to find the value of x that corresponds to a given **probability**.

We generally denote as $\chi^2_{v,\alpha}$ the value of the random variable X for which

$$P(X \leq \chi^2_{v,\alpha}) = 1 - \alpha .$$

Note: the notation χ^2 is read "chi-square".

EXAMPLES

See Appendix F.

The chi-square table allows us, for a given number of degrees of freedom v, to determine:

1. The **value** of the **distribution function** $F(x)$, given x.
2. The value of $\chi^2_{v,\alpha}$, given the **probability** $P(X \leq \chi^2_{v,\alpha})$.

FURTHER READING

▶ **Chi-square distribution**
▶ **Chi-square goodness of fit test**
▶ **Chi-square test**
▶ **Chi-square test of independence**
▶ **Statistical table**

REFERENCES

Elderton, W.P.: Tables for testing the goodness of fit of theory to observation. Biometrika **1**, 155–163 (1902)

Pearson, K.: Tables of the Incomplete Γ-function. H.M. Stationery Office (Cambridge University Press, Cambridge since 1934), London (1922)

Chi-Square Test

There are a number of chi-square tests, all of which involve comparing the test results to the values from the **chi-square distribution**. The most well-known of these tests are introduced below:

- The **chi-square test of independence** is used to determine whether two qualitative categorical variables associated with a **sample** are independent.
- The **chi-square goodness of fit test** is used to determine whether the distribution observed for a sample can be approximated by a theoretical distribution. We

might want to know, for example, whether the distribution observed for the sample corresponds to a particular **probability distribution** (**normal distribution**, **Poisson distribution**, etc).

- The chi-square test for an unknown **variance** is used when we want to test whether this variance takes a particular constant **value**.
- The chi-square test is used to test for homogeneity of the variances calculated for many samples drawn from a normally distributed **population**.

HISTORY

In 1937, Bartlett, M.S. proposed a method of testing the homogeneity of the **variance** for many samples drawn from a normally distributed **population**.

See also **chi-square test of independence** and **chi-square goodness of fit test**.

MATHEMATICAL ASPECTS

The mathematical aspects of the **chi-square test of independence** and those of the **chi-square goodness of fit test** are dealt with in their corresponding entries.

The chi-square test used to check whether an unknown **variance** takes a particular constant **value** is the following:

Let $(x_1, \ldots, x_n)$ be a random **sample** coming from a normally distributed **population** of unknown **mean** μ and of unknown **variance** σ^2.

We have good reason to believe that the variance of the population equals a presumed value σ_0^2. The hypotheses for each case are described below.

A: Two-sided case:

$$H_0: \quad \sigma^2 = \sigma_0^2$$
$$H_1: \quad \sigma^2 \neq \sigma_0^2$$

B: One-sided test:

$$H_0: \quad \sigma^2 \leq \sigma_0^2$$
$$H_1: \quad \sigma^2 > \sigma_0^2$$

C: One-sided test:

$$H_0: \quad \sigma^2 \geq \sigma_0^2$$
$$H_1: \quad \sigma^2 < \sigma_0^2$$

We then determine the **statistic** of the given chi-square test using:

$$\chi^2 = \frac{\sum_{i=1}^{n}(x_i - \bar{x})^2}{\sigma_0^2} .$$

This statistic is, under H_0, chi-square distributed with $n - 1$ degrees of freedom. In other words, we look for the value of $\chi_{n-1,\alpha}^2$ in the **chi-square table**, and we then compare that value to the calculated value χ^2. The decision rules depend on the case, and are as follows.

Case A

If $\chi^2 \geq \chi_{n-1,\alpha_1}^2$ or if $\chi^2 \leq \chi_{n-1,1-\alpha_2}^2$ we reject the **null hypothesis** H_0 for the **alternative hypothesis** H_1, where we have split the **significance level** α into α_1 and α_2 such that $\alpha_1 + \alpha_2 = \alpha$. Otherwise we do not reject the null hypothesis H_0.

Case B

If $\chi^2 < \chi_{n-1,\alpha}^2$ we do not reject the null hypothesis H_0. If $\chi^2 \geq \chi_{n-1,\alpha}^2$ we reject the null hypothesis H_0 for the alternative hypothesis H_1.

Case C

If $\chi^2 > \chi_{n-1,1-\alpha}^2$ we do not reject the null hypothesis H_0. If $\chi^2 \leq \chi_{n-1,1-\alpha}^2$ we reject the null hypothesis H_0 for the alternative hypothesis H_1.

Other chi-square tests are proposed in the work of Ostle, B. (1963).

DOMAINS AND LIMITATIONS

χ^2 (chi-square) **statistic** must be calculated using absolute frequencies and not relative ones.

Note that the chi-square test can be unreliable for small samples, especially when some of the estimated frequencies are small (< 5). This issue can often be resolved by grouping categories together, if such grouping is sensible.

EXAMPLES

Consider a batch of items produced by a machine. They can be divided up into classes depending on their diameters (in mm), as in the following table:

Diameter (mm) x_i	Number of items n_i
59.5	2
59.6	6
59.7	7
59.8	15
59.9	17
60.0	25
60.1	15
60.2	7
60.3	3
60.4	1
60.5	2
Total	$N = 100$

We have a random **sample** drawn from a normally distributed **population** where the **mean** and **variance** are not known. The vendor of these items would like the variance σ^2 to be smaller than or equal to 0.05. We test the following hypotheses:

null hypothesis H_0 : $\sigma^2 \leq 0.05$

alternative hypothesis H_1 : $\sigma^2 > 0.05$.

In this case we use the one-tailed hypothesis test.

We start by calculating the mean for the sample:

$$\bar{x} = \frac{\sum_{i=1}^{11} n_i \cdot x_i}{N} = \frac{5995}{100} = 59.95 .$$

We can then calculate the χ^2 **statistics**:

$$\chi^2 = \frac{\sum_{i=1}^{11} n_i \cdot (x_i - \bar{x})^2}{\sigma_0^2}$$

$$= \frac{2 \cdot (-0.45)^2 + \ldots + 2 \cdot (0.55)^2}{0.05}$$

$$= \frac{3.97}{0.05} = 79.4 .$$

Using a **significance level** of $\alpha = 5\%$, we then find the **value** of $\chi^2_{99,0.05}$ ($= 123.2$) in the **chi-square table**.
As $\chi^2 = 79.4 < \chi^2_{99,0.05}$, we do not reject the null hypothesis, which means that the vendor should be happy to sell these items since they are not significantly different in diameter.

FURTHER READING
▶ **Chi-square distribution**
▶ **Chi-square goodness of fit test**
▶ **Chi-square table**
▶ **Chi-square test of independence**

REFERENCES
Bartlett, M.S.: Some examples of statistical methods of research in agriculture and applied biology. J. Roy. Stat. Soc. (Suppl.) **4**, 137–183 (1937)

Ostle, B.: Statistics in Research: Basic Concepts and Techniques for Research Workers. Iowa State College Press, Ames, IA (1954)

Chi-square Test of Independence

The chi-square test of independence aims to determine whether two variables associated with a **sample** are independent or not. The variables studied are categorical qualitative variables.

The chi-square independence test is performed using a **contingency table**.

HISTORY
The first contingency tables were used only for enumeration. However, encouraged by the work of **Quetelet, Adolphe** (1849), statisticians began to take an interest in the associations between the variables used in the tables. For example, **Pearson, Karl** (1900) performed fundamental work on contingency tables.
Yule, George Udny (1900) proposed a somewhat different approach to the study of contingency tables to Pearson's, which lead to a disagreement between them. Pearson also argued with **Fisher, Ronald Aylmer** about the number of degrees of freedom to use in the chi-square test of independence. Everyone used different numbers until Fisher, R.A. (1922) was eventually proved to be correct.

MATHEMATICAL ASPECTS
Consider two qualitative categorical variables X and Y. We have a **sample** containing n observations of these variables.

These observations can be presented in a **contingency table**.

We denote the observed **frequency** of the **category** i of the variable X and the category j of the variable Y as n_{ij}.

		Categories of variable Y			
		Y_1	...	Y_c	Total
Categories	X_1	n_{11}	...	n_{1c}	$n_{1.}$
of	...	...	...	...	...
variable X	X_r	n_{r1}	...	n_{rc}	$n_{r.}$
	Total	$n_{.1}$	...	$n_{.c}$	$n_{..}$

The hypotheses to be tested are:

Null hyp. H_0:   the two variables are independent,

Alternative hyp. H_1 : the two variables are not independent.

Steps Involved in the Test

1. Compute the expected frequencies, denoted by e_{ij}, for each case in the **contingency table** under the **independence hypothesis**:

$$e_{ij} = \frac{n_{i.} \cdot n_{.j}}{n_{..}},$$

$$n_{i.} = \sum_{k=1}^{c} n_{ik} \text{ and } n_{.j} = \sum_{k=1}^{r} n_{kj},$$

where c represents the number of columns (or number of categories of variable X in the contingency table) and r the number of rows (or the number of categories of variable Y).

2. Calculate the **value** of the χ^2 (chi-square) statistic, which is really a measure of the deviation of the observed frequencies n_{ij} from the expected frequencies e_{ij}:

$$\chi^2 = \sum_{i=1}^{c} \sum_{j=1}^{r} \frac{(n_{ij} - e_{ij})^2}{e_{ij}}.$$

3. Choose the **significance level** α to be used in the test and compare the calculated **value** of χ^2 with the value obtained from the **chi-square table**, $\chi^2_{v,\alpha}$. The number of degrees of freedom correspond to the number of cases in the table that can take arbitrary values; the values taken by the other cases are imposed on them by the row and column totals. So, the number of degrees of freedom is given by:

$$v = (r-1)(c-1).$$

4. If the calculated χ^2 is smaller then the $\chi^2_{v,\alpha}$ from the table, we do not reject the null hypothesis. The two variables can be considered to be independent.

However, if the calculated χ^2 is greater then the $\chi^2_{v,\alpha}$ from the table, we reject the null hypothesis for the alternative hypothesis. We can then conclude that the two variables are not independent.

DOMAINS AND LIMITATIONS

Certain conditions must be fulfilled in order to be able to apply the chi-square test of independence:

1. The **sample**, which contains n observations, must be a random sample;
2. Each individual observation can only appear in one **category** for each **variable**. In other words, each individual observation can only appear in one line and one column of the **contingency table**.

Note that the chi-square test of independence is not very reliable for small samples, especially when the estimated frequencies are small (< 5). To avoid this issue we can group categories together, but only when this groups obtained are sensible.

EXAMPLES

We want to determine whether the proportion of smokers is independent of gender. The two variables to be studied are categorical and qualitative and contain two categories each:

- Variable "gender:" M or F;
- Variable "smoking status:" "smokes" or "does not smoke."

The hypotheses are then:

H_0: chance of being a smoker is independent of gender

H_1: chance of being a smoker is not independent of gender.

The **contingency table** obtained from a **sample** of 100 individuals ($n = 100$) is shown below:

		Smoking status		
		"smokes"	"does not smoke"	Total
Gender	M	21	44	65
	F	10	25	35
	Total	31	69	100

We now denote the observed frequencies as n_{ij} ($i = 1, 2, j = 1, 2$).

We then estimate all of the frequencies in the table based on the **hypothesis** that the two variables are independent of each other. We denote these estimated frequencies by e_{ij}:

$$e_{ij} = \frac{n_{i.} \cdot n_{.j}}{n_{..}}.$$

We therefore obtain:

$$e_{11} = \frac{65 \cdot 31}{100} = 20.15$$

$$e_{12} = \frac{65 \cdot 69}{100} = 44.85$$

$$e_{21} = \frac{35 \cdot 31}{100} = 10.85$$

$$e_{12} = \frac{35 \cdot 69}{100} = 24.15.$$

The estimated **frequency table** is given below:

		Smoking status		
		"smokes"	"does not smoke"	Total
Gender	M	20.15	44.85	65
	F	10.85	24.15	35
	Total	31	69	100

If the **null hypothesis** H_0 is true, the statistic

$$\chi^2 = \sum_{i=1}^{2} \sum_{j=1}^{2} \frac{(n_{ij} - e_{ij})^2}{e_{ij}}$$

is chi-square-distributed with $(r-1)(c-1) = (2-1)(2-1) = 1$ degree of freedom and

$$\chi^2 = 0.036 + 0.016 + 0.066 + 0.030$$
$$= 0.148.$$

If a **significance level** of 5% is selected, the **value** of $\chi^2_{1, 0.05}$ is 3.84, from the **chi-square table**.

Since the calculated value of χ^2 is smaller then the value found in the chi-square table, we do not reject the null hypothesis and we conclude that the two variables studied are independent.

FURTHER READING

▶ **Chi-square distribution**
▶ **Chi-square table**
▶ **Contingency table**
▶ **Test of independence**

REFERENCE

Fisher, R.A.: On the interpretation of χ^2 from contingency tables, and the calculation of P. J. Roy. Stat. Soc. Ser. A **85**, 87–94 (1922)

Pearson, K.: On the criterion, that a given system of deviations from the probable in the case of a correlated system of variables is such that it can be reasonably supposed to have arisen from random sampling. In: Karl Pearson's Early Statistical Papers. Cambridge University Press, pp. 339–357. First published in 1900 in Philos. Mag. (5th Ser) **50**, 157–175 (1948)

Quetelet, A.: Letters addressed to H.R.H. the Grand Duke of Saxe Coburg and Gotha, on the Theory of Probabilities as Applied to the Moral and Political Sciences. (French translation by Downs, Olinthus Gregory). Charles & Edwin Layton, London (1849)

Yule, G.U.: On the association of attributes in statistics: with illustration from the material of the childhood society. Philos. Trans. Roy. Soc. Lond. Ser. A **194**, 257–319 (1900)

Classification

Classification is the grouping together of similar objects. If each object is characterized by p **variables**, classification can be performed according to rational criteria. Depending on the criteria used, an object could potentially belong to several classes.

HISTORY

Classifying the residents of a locality or a country according to their sex and other physical characteristics is an activity that dates back to ancient times. The Hindus, the ancient Greeks and the Romans all developed multiple typologies for human beings. The oldest comes from Galen (129–199 A.D.).

Later on, the concept of classification spread to the fields of biology and zoology; the works of Linné (1707–1778) should be mentioned in this regard.

The first ideas regarding actual methods of **cluster analysis** are attributed to Adanson (eighteenth century). Zubin (1938), Tryon (1939) and Thorndike (1953) also attempted to develop some methods, but the true development of classification methods coincides with the advent of the computer.

MATHEMATICAL ASPECTS

Classification methods can be divided into two large **categories**, one based on **probabilities** and the other not.

The first category contains, for example, discriminating analysis. The second category can be further subdivided into two groups. The first group contains what are known as optimal classification methods. In the second group, we can distinguish between several subtypes of classification method:

- Partition methods that consist of distributing n objects among g groups in such a way that each object exclusively belongs to just one group. The number of groups g is fixed beforehand, and the partition applied most closely satisfies the classification criteria.

- Partition methods incorporating infringing classes, where an object is allowed to belong to several groups simultaneously.

- Hierarchical classification methods, where the structure of the **data** at different levels of classification is taken into account. We can then take account for the relationships that exist between the different groups created during the partition. Two different kinds of hierarchical techniques exist: agglomerating techniques and dividing techniques.

Agglomerating methods start with separated objects, meaning that the n objects are initially distributed into n groups. Two groups are agglomerated in each subsequent step until there is only one group left.

In contrast, dividing methods start with all of the objects grouped together, meaning that all of the n objects are in one single group to start with. New groups are created at each subsequent step until there are n groups.

• Geometric classification, in which the objects are depicted on a **scatter plot** and then grouped according to position on the plot. In a **graphical representation**, the proximities of the objects to each other in the graphic correspond to the similarities between the objects.

The first three types of classification are generally grouped together under the term **cluster analysis**. What they have in common is the fact that the objects to be classified must present a certain amount of structure that allows us to measure the degree of similarity between the objects.

Each type of classification contains a multitude of methods that allow us to create classes of similar objects.

DOMAINS AND LIMITATIONS

Classification can be used in two cases:
• Description cases;
• Prediction cases.

In the first case, the classification is done on the basis of some generally accepted standard characteristics. For example, professions can be classified into freelance, managers, workers, and so on, and one can calculate average salaries, average frequency of health problems, and so on, for each class.

In the second case, classification will lead to a prediction and then to an action. For example, if the foxes in a particular region exhibit apathetic behavior and excessive salivation, we can conclude that there is a new rabies epidemic. This should then prompt a vaccine campaign.

FURTHER READING
▶ **Cluster analysis**
▶ **Complete linkage method**
▶ **Data analysis**

REFERENCES

Everitt, B.S.: Cluster Analysis. Halstead, London (1974)

Gordon, A.D.: Classification. Methods for the Exploratory Analysis of Multivariate Data. Chapman & Hall, London (1981)

Kaufman, L., Rousseeuw, P.J.: Finding Groups in Data: An Introduction to Cluster Analysis. Wiley, New York (1990)

Thorndike, R.L.: Who belongs in a family? Psychometrika **18**, 267–276 (1953)

Tryon, R.C.: Cluster Analysis. McGraw-Hill, New York (1939)

Zubin, J.: A technique for measuring like-mindedness. J. Abnorm. Social Psychol. **33**, 508–516 (1938)

Cluster Analysis

Clustering is the partitioning of a data set into subsets or clusters, so that the degree of association is strong between members of the same cluster and weak between members of different clusters according to some defined distance measure.

Several methods of performing **cluster analysis** exist:

- Partitional clustering
- Hierarchical clustering.

HISTORY

See **classification** and **data analysis**.

MATHEMATICAL ASPECTS

To carry out **cluster analysis** on a set of n objects, we need to define a **distance** between the objects (or more generally a measure of the similarity between the objects) that need to be classified. The existence of some kind of structure within the set of objects is assumed.

To carry out a hierarchical classification of a set E of objects $\{x_1, x_2, \ldots, x_n\}$, it is necessary to define a **distance** associated with E that can be used to obtain a **distance table** between the objects of E. Similarly, a distance must also be defined for any subsets of E.

One approach to hierarchical clustering is to use the agglomerating method. It can be summarized in the following algorithm:

1. Locate the pair of objects (x_i, x_j) which have the smallest **distance** between each other.

2. Aggregate the pair of objects (x_i, x_j) into a single element α and re-establish a new **distance table**. This is achieved by suppressing the lines and columns associated with x_i and x_j and replacing them with a line and a column associated with α. The new distance table will have a line and a column less than the previous table.

3. Repeat these two operations until the desired number of classes are obtained or until all of the objects are gathered into the same class.

Note that the distance between the group formed from aggregated elements and the other elements can be defined in different ways, leading to different methods. Examples include the single link method and the **complete linkage method**.

The single link method is a hierarchical classification method that uses the Euclidean **distance** to establish a **distance table,** and the distance between two classes is given by the Euclidean distance between the two closest elements (the minimum distance).

In the **complete linkage method**, the **distance** between two classes is given by the Euclidean distance between the two elements furthest away (the maximum distance).

Given that the only difference between these two methods is that the distance between two classes is either the minimum and the maximum distance, only the single link method will be considered here.

For a set $E = \{X_1, X_2, \ldots, X_n\}$, the **distance table** for the elements of E is then established.

Since this table is symmetric and null along its diagonal, only one half of the table is considered:

$$d(X_1, X_2) \quad d(X_1, X_3) \quad \ldots \quad d(X_1, X_n)$$
$$d(X_2, X_3) \quad \ldots \quad d(X_2, X_n)$$
$$\ddots \quad \ldots$$
$$d(X_{n-1}, X_n)$$

where $d(X_i, X_j)$ is the Euclidean distance between X_i and X_j for $i < j$, where the **values** of i and j are between 1 and n.

The **algorithm** for the single link method is as follows:

- Search for the minimum $d(X_i, X_j)$ for $i < j$;

- The elements X_i and X_j are aggregated into a new group $C_k = X_i \cup X_j$;
- The set E is then partitioned into

  $\{X_1\}, \ldots, \{X_{i-1}\}, \{X_i, X_j\}, \{X_{i+1}\}, \ldots,$
  $\{X_{j-1}\}, \{X_{j+1}\}, \ldots, \{X_n\}$;

- The **distance table** is then recreated without the lines and columns associated with X_i and X_j, and with a line and a column representing the **distances** between X_m and $C_k, m = 1, 2, \ldots, n\ m \neq i$ and $m \neq j$, given by:

$$d(C_k, X_m) = \min\{d(X_i, X_m); d(X_j, X_m)\}.$$

The algorithm is repeated until the desired number of groups is attained or until there is only one group containing all of the elements.

In the general case, the **distance** between two groups is given by:

$$d(C_k, C_m) = \min\{d(X_i, X_j) \text{ with } X_i$$
$$\text{belonging to } C_k \text{ and } X_j \text{ to } C_m\},$$

The formula quoted previously applies to the particular case when the groups are composed, respectively, of two elements and one single element.

This series of agglomerations can be represented by a **dendrogram**, where the abscissa shows the distance separating the objects. Note that we could find more than one pair when we search for the pair of closest elements. In this case, the pair that is selected for aggregation in the first step does not influence later steps (provided the algorithm does not finish at this step), because the other pair of closest elements will be aggregated in the following step. The aggregation order is not shown on the dendrogram because it reports the distance that separates two grouped objects.

DOMAINS AND LIMITATIONS

The choice of the **distance** between the group formed of aggregated elements and the other elements can be operated in several ways, according to the method that is used, as for example in the single link method and in the **complete linkage method**.

EXAMPLES

Let us illustrate how the **single link method** of cluster analysis can be applied to the examination grades obtained by five students each studying four courses: English, French, maths and physics.

We want to divide these five students into two groups using the single link method.

The grades obtained in the examinations, which range from 1 to 6, are summarized in the following table:

	English	French	Maths	Physics
Alain	5.0	3.5	4.0	4.5
Jean	5.5	4.0	5.0	4.5
Marc	4.5	4.5	4.0	3.5
Paul	4.0	5.5	3.5	4.0
Pierre	4.0	4.5	3.0	3.5

We then work out the Euclidian **distances** between the students and use them to create a **distance table**:

	Alain	Jean	Marc	Paul	Pierre
Alain	0	1.22	1.5	2.35	2
Jean	1.22	0	1.8	2.65	2.74
Marc	1.5	1.8	0	1.32	1.12
Paul	2.35	2.65	1.32	0	1.22
Pierre	2	2.74	1.12	1.22	0

By only considering the upper part of this symmetric table, we obtain:

	Jean	Marc	Paul	Pierre
Alain	1.22	1.5	2.35	2
Jean		1.8	2.65	2.74
Marc			1.32	1.12
Paul				1.22

The minimum **distance** is 1.12, between Marc and Pierre; we therefore form the first group from these two students. We then calculate the new distances.

For example, we calculate the new distance between Marc and Pierre on one side and Alain on the other by taking the minimum distance between Marc and Alain and the minimum distance between Pierre and Alain:

$d(\{Marc,Pierre\},Alain)$

$= \min\{d(Marc,Alain);d(Pierre,Alain)\}$

$= \min\{1.5; 2\} = 1.5$,

also

$d(\{Marc,Pierre\},Jean)$

$= \min\{d(Marc,Jean);d(Pierre,Jean)\}$

$= \min\{1.8; 2.74\} = 1.8$,

and

$d(\{Marc,Pierre\},Paul)$

$= \min\{d(Marc,Paul);d(Pierre,Paul)\}$

$= \min\{1.32; 1.22\} = 1.22$.

The new **distance table** takes the following form:

	Jean	Marc and Pierre	Paul
Alain	1.22	1.5	2.35
Jean		1.8	2.65
Marc and Pierre			1.22

The minimum distance is now 1.22, between Alain and Jean and also between the group

of Marc and Pierre on the one side and Paul on the other side (in other words, two pairs exhibit the minimum distance); let us choose to regroup Alain and Jean first. The other pair will be aggregated in the next step. We rebuild the **distance table** and obtain:

$d(\{Alain,Jean\}, \{Marc,Pierre\})$

$= \min\{d(Alain,\{Marc,Pierre\}),$

$d(Jean,\{Marc,Pierre\})\}$

$= \min\{1.5; 1.8\} = 1.5$

as well as:

$d(\{Alain,Jean\},Paul)$

$= \min\{d(Alain,Paul);d(Jean,Paul)\}$

$= \min\{2.35; 2.65\} = 2.35$.

This gives the following **distance table**:

	Marc and Pierre	Paul
Alain and Jean	1.5	2.35
Marc and Pierre		1.22

Notice that Paul must now be integrated in the group formed from Marc and Pierre, and the new **distance** will be:

$d(\{(Marc,Pierre),Paul\},\{Alain,Jean\})$

$= \min\{d(\{Marc,Pierre\},\{Alain,Jean\}),$

$d(Paul,\{Alain,Jean\})\}$

$= \min\{1.5; 2.35\} = 1.5$

which gives the following **distance table**:

	Alain and Jean
Marc, Pierre and Paul	1.5

We finally obtain two groups: {Alain and Jean} and {Marc, Pierre and Paul} which are separated by a **distance** of 1.5.

The following **dendrogram** illustrates the successive aggregations:

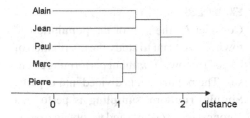

FURTHER READING
- ▶ **Classification**
- ▶ **Complete linkage method**
- ▶ **Dendrogram**
- ▶ **Distance**
- ▶ **Distance table**

REFERENCES

Celeux, G., Diday, E., Govaert, G., Lecheval-lier, Y., Ralambondrainy, H.: Classification automatique des données—aspects statistiques et informatiques. Dunod, Paris (1989)

Everitt, B.S.: Cluster Analysis. Halstead, London (1974)

Gordon, A.D.: Classification. Methods for the Exploratory Analysis of Multivariate Data. Chapman & Hall, London (1981)

Jambu, M., Lebeaux, M.O.: Classification automatique pour l'analyse de données. Dunod, Paris (1978)

Kaufman, L., Rousseeuw, P.J.: Finding Groups in Data: An Introduction to Cluster Analysis. Wiley, New York (1990)

Lerman, L.C.: Classification et analyse ordinale des données. Dunod, Paris (1981)

Tomassone, R., Daudin, J.J., Danzart, M., Masson, J.P.: Discrimination et classement. Masson, Paris (1988)

Cluster Sampling

In cluster **sampling**, the first step is to divide the **population** into subsets called clusters. Each cluster consists of individuals that are supposed to be representative of the population.

Cluster **sampling** then involves choosing a random **sample** of clusters and then observing all of the individuals that belong to each of them.

HISTORY
See **sampling**.

MATHEMATICAL ASPECTS

Cluster sampling is the process of randomly extracting representative sets (known as clusters) from a larger population of units and then applying a questionnaire to all of the units in the clusters. The clusters often consist of geographical units, like city districts. In this case, the method involves dividing a city into districts, and then selecting the districts to be included in the sample. Finally, all of the people or households in the chosen district are questioned.

There are two principal reasons to perform cluster sampling. In many inquiries, there is no complete and reliable list of the population units on which to base the sampling, or it may be that it is too expensive to create such a list. For example, in many countries, including industrialized ones, it is rare to have complete and up-to-date lists of all of the members of the population, households or rural estates. In this situation, sampling can be achieved in a geographical manner: each urban region is divided up into districts and each rural region into rural estates. The districts and the agricultural areas are considered to be clusters and we use the complete list of clusters because we do not have a complete and up-to-date list of all population units. Therefore, we sample a requisite number of clusters from the list and then question all of the units in the selected cluster.

DOMAINS AND LIMITATIONS

The advantage of cluster **sampling** is that it is not necessary to have a complete, up-to-date list of all of the units of the **population** to perform analysis.

For example, in many countries, there are no updated lists of people or housing. The costs of creating such lists are often prohibitive.

It is therefore easier to analyze subsets of the **population** (known as clusters).

In general, cluster **sampling** provides **estimations** that are not as precise as **simple random sampling**, but this drop in accuracy is easily offset by the far lower cost of cluster sampling.

In order to perform cluster **sampling** as efficiently as possible:

- The clusters should not be too big, and there should be a large enough number of clusters,
- cluster sizes should be as uniform as possible;
- The individuals belonging to each cluster must be as heterogenous as possible with respect to the parameter being observed.

Another reason to use cluster sampling is cost. Even when a complete and up-to-date list of all population units exists, it may be preferable to use cluster sampling from an economic point of view, since it is completed faster, involves fewer workers and minimizes transport costs. It is therefore more appropriate to use cluster sampling if the money saved by doing so is far more signif-

icant than the increase in sampling variance that will result.

EXAMPLES

Consider N, the size of the **population** of town X. We want to study the distribution of "Age" for town X without performing a census. The population is divided into G parts. **Simple random sampling** is performed amongst these G parts and we obtain g parts. The final **sample** will be composed of all the individuals in the g selected parts.

FURTHER READING

▶ **Sampling**
▶ **Simple random sampling**

REFERENCES

Hansen, M.H., Hurwitz, W.N., Madow, M.G.: Sample Survey Methods and Theory. Vol. I. Methods and Applications. Vol. II. Theory. Chapman & Hall, London (1953)

Coefficient of Determination

The coefficient of determination, denoted R^2, is the quotient of the explained variation (sum of squares due to regression) to the total variation (total sum of squares total SS (TSS)) in a **model** of simple or **multiple linear regression**:

$$R^2 = \frac{\text{Explained variation}}{\text{Total variation}} .$$

It equals the square of the **correlation coefficient**, and it can take values between 0 and 1. It is often expressed as a **percentage**.

HISTORY

See **correlation coefficient**.

MATHEMATICAL ASPECTS

Consider the following **model** for multiple regression:

$$Y_i = \beta_0 + \beta_1 X_{i1} + \cdots + \beta_p X_{ip} + \varepsilon_i$$

for $i = 1, \ldots, n$, where

Y_i are the dependent variables,

X_{ij} $(i = 1, \ldots, n, j = 1, \ldots, p)$ are the independent variables,

ε_i are the random nonobservable **error** terms,

β_j $(j = 1, \ldots, p)$ are the parameters to be estimated.

Estimating the parameters $\beta_0, \beta_1, \ldots, \beta_p$ yields the estimation

$$\hat{Y}_i = \hat{\beta}_0 + \hat{\beta}_1 X_{i1} + \cdots + \hat{\beta}_p X_{ip}.$$

The coefficient of determination allows us to measure the quality of fit of the regression equation to the measured values.

To determine the quality of the fit of the **regression** equation, consider the gap between the observed **value** and the estimated value for each **observation** of the **sample**. This gap (or **residual**) can also be expressed in the following way:

$$\sum_{i=1}^{n}(Y_i - \bar{Y})^2 = \sum_{i=1}^{n}(Y_i - \hat{Y}_i)^2$$

$$+ \sum_{i=1}^{n}(\hat{Y}_i - \bar{Y})^2$$

$$TSS = RSS + REGSS$$

where

TSS   is the total sum of squares,
RSS   the residual sum of squares and
REGSS the sum of the squares of the regression.

These concepts and the relationships between them are presented in the following graph:

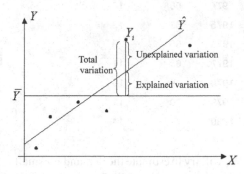

Using these concepts, we can define R^2, which is the determination coefficient. It measures the proportion of variation in **variable** Y, which is described by the **regression** equation as:

$$R^2 = \frac{REGSS}{TSS} = \frac{\displaystyle\sum_{i=1}^{n}(\hat{Y}_i - \bar{Y})^2}{\displaystyle\sum_{i=1}^{n}(Y_i - \bar{Y})^2}$$

If the **regression** function is to be used to make predictions about subsequent observations, it is preferable to have a high **value** of R^2, because the higher the value of R^2, the smaller the unexplained variation.

EXAMPLES

The following table gives values for the Gross National Product (GNP) and the demand for domestic products covering the 1969–1980 period for a particular country.

Year	GNP X	Demand for domestic products Y
1969	50	6
1970	52	8
1971	55	9
1972	59	10

Year	GNP X	Demand for domestic products Y
1973	57	8
1974	58	10
1975	62	12
1976	65	9
1977	68	11
1978	69	10
1979	70	11
1980	72	14

We will try to estimate the demand for small goods as a function of GNP according to the **model**

$$Y_i = a + b \cdot X_i + \varepsilon_i, \quad i = 1, \ldots, 12.$$

Estimating the **parameters** a and b by the **least squares** method yields the following **estimators**:

$$\hat{b} = \frac{\sum_{i=1}^{12}(X_i - \bar{X})(Y_i - \bar{Y})}{\sum_{i=1}^{12}(X_i - \bar{X})^2} = 0.226,$$

$$\hat{a} = \bar{Y} - \hat{b} \cdot \bar{X} = -4.047.$$

The estimated line is written

$$\hat{Y} = -4.047 + 0.226 \cdot X.$$

The quality of the fit of the measured points to the regression line is given by the determination coefficient:

$$R^2 = \frac{\text{REGSS}}{\text{TSS}} = \frac{\sum_{i=1}^{12}(\hat{Y}_i - \bar{Y})^2}{\sum_{i=1}^{12}(Y_i - \bar{Y})^2}.$$

We can calculate the **mean** using:

$$\bar{Y} = \frac{\sum_{i=1}^{12} Y_i}{n} = 9.833.$$

Y_i	$\hat{Y}_i$	$(Y_i - \bar{Y})^2$	$(\hat{Y}_i - \bar{Y})^2$
6	7.260	14.694	6.622
8	7.712	3.361	4.500
9	8.390	0.694	2.083
10	9.294	0.028	0.291
8	8.842	3.361	0.983
10	9.068	0.028	0.586
12	9.972	4.694	0.019
9	10.650	0.694	0.667
11	11.328	1.361	2.234
10	11.554	0.028	2.961
11	11.780	1.361	3.789
14	12.232	17.361	4.754
Total		47.667	30.489

We therefore obtain:

$$R^2 = \frac{30.489}{47.667} = 0.6396$$

or, in percent:

$$R^2 = 63.96\%.$$

We can therefore conclude that, according to the **model** chosen, 63.96% of the variation in the demand for small goods is explained by the variation in the GNP.

Obviously the **value** of R^2 cannot exceed 100%. While 63.96% is relatively high, it is not close enough to 100% to rule out trying to modify the **model** further.

This analysis also shows that other variables apart from the GNP should be taken into account when determining the function corresponding to the demand for small goods, since the GNP only partially explains the variation.

FURTHER READING
▶ **Correlation coefficient**
▶ **Multiple linear regression**
▶ **Regression analysis**
▶ **Simple linear regression**

Coefficient of Kurtosis

The coefficient of kurtosis is used to measure the peakness or flatness of a curve. It is based on the **moments** of the distribution. This coefficient is one of the **measures of kurtosis**.

HISTORY
See **coefficient of skewness**.

MATHEMATICAL ASPECTS
The coefficient of kurtosis (β_2) is based on the centered fourth order **moment** of a **distribution** which is equal to:

$$\mu_4 = E\left[(X - \mu)^4\right].$$

In order to obtain a coefficient of kurtosis that is independent of the units of measurement, the fourth-order **moment** is divided by the **standard deviation** of the **population** σ raised to the fourth power. The coefficient of kurtosis then becomes equal to:

$$\beta_2 = \frac{\mu_4}{\sigma^4}.$$

For a **sample** $(x_1, x_2, \ldots, x_n)$, the **estimator** of this coefficient is denoted by b_2. It is equal to:

$$b_2 = \frac{m_4}{S^4},$$

where m_4 is the centered fourth-order **moment** of the sample, given by:

$$m_4 = \frac{1}{n} \cdot \sum_{i=1}^{n} (x_i - \bar{x})^4,$$

where $\bar{x}$ is the **arithmetic mean**, n is the total number of **observations** and S^4 is the **standard deviation** of the sample raised to the fourth power.

For the case where a **random variable** X takes **values** x_i with **frequencies** f_i, $i = 1, 2, \ldots, h$, the centered fourth-order **moment** of the **sample** is given by the formula:

$$m_4 = \frac{1}{n} \cdot \sum_{i=1}^{h} f_i \cdot (x_i - \bar{x})^4.$$

DOMAINS AND LIMITATIONS
For a **normal distribution**, the coefficient of kurtosis is equal to 3. Therefore a curve will be called platikurtic (meaning flatter than the normal distribution) if it has a kurtosis coefficient smaller than 3. It will be leptokurtic (meaning sharper than the normal distribution) if β_2 is greater than 3.

Let us now prove that the coefficient of kurtosis is equal to 3 for the **normal distribution**. We know that $\beta_2 = \frac{\mu_4}{\sigma^4}$, meaning that the centered fourth-order **moment** is divided by the **standard deviation** raised to the fourth power. It can be proved that the centered sth order moment, denoted μ_s, satisfies the following relation for a **normal distribution**:

$$\mu_s = (s - 1)\sigma^2 \cdot \mu_{s-2}.$$

This formula is a recursive formula which expresses higher order **moments** as a function of lower order moments.

Given that $\mu_0 = 1$ (the zero-order **moment** of any **random variable** is equal to 1, since it is the **expected value** of this **variable** raised to the power zero) and that $\mu_1 = 0$ (the centered first-order **moment** is zero for any random variable), we have:

$$\mu_2 = \sigma^2$$

$$\mu_3 = 0$$

$$\mu_4 = 3\sigma^4$$

etc. .

The coefficient of kurtosis is then equal to:

$$\beta_2 = \frac{\mu_4}{\sigma^4} = \frac{3\sigma^4}{\sigma^4} = 3.$$

EXAMPLES

We want to calculate the kurtosis of the distribution of daily turnover for 75 bakeries. Let us calculate the coefficient of kurtosis β_2 using the following **data**:

Table categorizing the daily turnovers of 75 bakeries

Turnover	Frequencies
215–235	4
235–255	6
255–275	13
275–295	22
295–315	15
315–335	6
335–355	5
355–375	4

The fourth-order **moment** of the **sample** is given by:

$$m_4 = \frac{1}{n} \sum_{i=1}^{h} f_i(x_i - \bar{x})^4,$$

where $n = 75$, $\bar{x} = 290.60$ and x_i is the center of class **interval** i. We can summarize the calculations in the following table:

x_i	$x_i - \bar{x}$	f_i	$f_i(x_i - \bar{x})^4$
225	−65.60	4	74075629.16
245	−45.60	6	25942428.06
265	−25.60	13	5583457.48
285	−5.60	22	21635.89
305	14.40	15	644972.54
325	34.40	6	8402045.34
345	54.40	5	43789058.05
365	74.40	4	122560841.32
			281020067.84

Since $S = 33.88$, the coefficient of kurtosis is equal to:

$$\beta_2 = \frac{\frac{1}{75}(281020067.84)}{(33.88)^4} = 2.84.$$

Since β_2 is smaller than 3, we can conclude that the distribution of the daily turnover in 75 bakeries is platikurtic, meaning that it is flatter than the **normal distribution**.

FURTHER READING
▶ **Measure of kurtosis**
▶ **Measure of shape**

REFERENCES
See **coefficient of skewness β_1 de Pearson**.

Coefficient of Skewness

The coefficient of skewness measures the skewness of a distribution. It is based on the notion of the **moment** of the distribution. This coefficient is one of the **measures of skewness**.

HISTORY
Between the end of the nineteenth century and the beginning of the twentieth century, **Pearson, Karl** studied large sets of **data** which sometimes deviated significantly from normality and exhibited considerable skewness.

He first used the following coefficient as a **measure of skewness**:

$$\text{skewness} = \frac{\bar{x} - \text{mode}}{S},$$

where $\bar{x}$ represents the **arithmetic mean** and S the **standard deviation**.

This measure is equal to zero if the data are distributed symmetrically.

He discovered empirically that for a moderately asymmetric distribution (the gamma distribution):

$$M_o - \bar{x} \approx 3 \cdot (M_d - \bar{x}),$$

where M_o and M_d denote the mode and the median of data set. By substituting this expression into the previous coefficient, the following alternative formula is obtained:

$$\text{skewness} = \frac{3 \cdot (\bar{x} - M_d)}{S}.$$

Following this, **Pearson, K.** (1894,1895) introduced a coefficient of skewness, known as the β_1 coefficient, based on calculations of the centered **moments**. This coefficient is more difficult to calculate but it is more descriptive and better adapted to large numbers of **observations**.

Pearson, K. also created the **coefficient of kurtosis** (β_2), which is used to measure the oblateness of a curve. This coefficient is also based on the **moments** of the distribution being studied.

Tables giving the limit values of the coefficients β_1 and β_2 can be found in the works of Pearson and Hartley (1966, 1972). If the sample estimates a fall outside the limit for β_1, β_2, we conclude that the population is significantly curved or skewed.

MATHEMATICAL ASPECTS

The skewness coefficient is based on the centered third-order **moment** of the distribution in question, which is equal to:

$$\mu_3 = E\left[(X - \mu)^3\right].$$

To obtain a coefficient of skewness that is independent of the measuring unit, the third-order **moment** is divided by the **standard deviation** of the **population** σ raised to the

third power. The coefficient obtained, designated by $\sqrt{\beta_1}$, is equal to:

$$\sqrt{\beta_1} = \frac{\mu_3}{\sigma^3}.$$

The **estimator** of this coefficient, calculated for a **sample** $(x_1, x_2, \ldots x_n)$, is denoted by $\sqrt{b_1}$. It is equal to:

$$\sqrt{b_1} = \frac{m_3}{S^3},$$

where m_3 is the centered third-order **moment** of the sample, given by:

$$m_3 = \frac{1}{n} \cdot \sum_{i=1}^{n}(x_i - \bar{x})^3.$$

Here $\bar{x}$ is the **arithmetic mean**, n is the total number of **observations** and S^3 is the **standard deviation** raised to the third power.

For the case where a **random variable** X takes **values** x_i with **frequencies** f_i, $i = 1, 2, \ldots, h$, the centered third-order **moment** of the **sample** is given by the formula:

$$m_3 = \frac{1}{n} \cdot \sum_{i=1}^{h} f_i \cdot (x_i - \bar{x})^3.$$

If the coefficient is positive, the distribution spreads to the the right. If it is negative, the distribution expands to the left. If it is close to zero, the distribution is approximately symmetric.

If the **sample** is taken from a normal **population**, the **statistic** $\sqrt{b_1}$ roughly follows a **normal distribution** with a **mean** of 0 and a **standard deviation** of $\sqrt{\frac{6}{n}}$. If the **size of the sample** n is bigger than 150, the **normal table** can be used to test the skewness **hypothesis**.

DOMAINS AND LIMITATIONS

This coefficient (similar to the other **measures of skewness**) is only of interest if it can be used to compare the shapes of two or more distributions.

EXAMPLES

Suppose that we want to compare the shapes of the daily turnover distributions obtained for 75 bakeries for two different years. We then calculate the skewness coefficient in both cases.

The **data** are categorized in the table below:

Turnover	Frequencies for year 1	Frequencies for year 2
215–235	4	25
235–255	6	15
255–275	13	9
275–295	22	8
295–315	15	6
315–335	6	5
335–355	5	4
355–375	4	3

The third-order **moment** of the **sample** is given by:

$$m_3 = \frac{1}{n} \cdot \sum_{i=1}^{h} f_i \cdot (x_i - \bar{x})^3.$$

For year 1, $n = 75$, $\bar{x} = 290.60$ and x_i is the center of each class **interval** i. The calculations are summarized in the following table:

x_i	$x_i - \bar{x}$	f_i	$f_i(x_i - \bar{x})^3$
225	−65.60	4	−1129201.664
245	−45.60	6	−568912.896
265	−25.60	13	−218103.808
285	−5.60	22	−3863.652
305	14.40	15	44789.760
325	34.40	6	244245.504
345	54.40	5	804945.920
365	74.40	4	1647323.136
			821222.410

Since $S = 33.88$, the coefficient of skewness is equal to:

$$\sqrt{b_1} = \frac{\frac{1}{75}(821222.41)}{(33.88)^3} = \frac{10949.632}{38889.307} = 0.282.$$

For year 2, $n = 75$ and $\bar{x} = 265.27$. The calculations are summarized in the following table:

x_i	$x_i - \bar{x}$	f_i	$f_i(x_i - \bar{x})^3$
225	−40.27	25	−1632213.81
245	−20.27	15	−124864.28
265	−0.27	9	−0.17
285	19.73	8	61473.98
305	39.73	6	376371.09
325	59.73	5	1065663.90
345	79.73	4	2027588.19
365	99.73	3	2976063.94
			4750082.83

Since $S = 42.01$, the coefficient of skewness is equal to:

$$\sqrt{b_1} = \frac{\frac{1}{75}(4750082.83)}{(42.01)^3} = \frac{63334.438}{74140.933} = 0.854.$$

The coefficient of skewness for year 1 is close to zero ($\sqrt{b_1} = 0.282$), so the daily turnover distribution for the 75 bakeries for year 1 is very close to being a symmetrical distribution. For year 2, the skewness coefficient is higher; this means that the distribution spreads towards the right.

FURTHER READING

▶ **Measure of shape**
▶ **Measure of skewness**

REFERENCES

Pearson, E.S., Hartley, H.O.: Biometrika Tables for Statisticians, vols. I and II. Cambridge University Press, Cambridge (1966,1972)

Pearson, K.: Contributions to the mathematical theory of evolution. I. In: Karl Pearson's Early Statistical Papers. Cambridge University Press, Cambridge, pp. 1–40 (1948). First published in 1894 as: On the dissection of asymmetrical frequency curves. Philos. Trans. Roy. Soc. Lond. Ser. A **185**, 71–110

Pearson, K.: Contributions to the mathematical theory of evolution. II: Skew variation in homogeneous material. In: Karl Pearson's Early Statistical Papers. Cambridge University Press, Cambridge, pp. 41–112 (1948). First published in 1895 in Philos. Trans. Roy. Soc. Lond. Ser. A **186**, 343–414

Coefficient of Variation

The coefficient of variation is a **measure of relative dispersion**. It describes the **standard deviation** as a percentage of the **arithmetic mean**.

This coefficient can be used to compare the dispersions of **quantitative variables** that are not expressed in the same units (for example, when comparing the salaries in different countries, given in different currencies), or the dispersions of variables that have very different **means**.

MATHEMATICAL ASPECTS

The coefficient of variation CV is defined as the ratio of the **standard deviation** to the **arithmetic mean** for a set of **observations**;

in other words:

$$CV = \frac{S}{\bar{x}} \cdot 100$$

for a **sample**, where:

S is the standard deviation of the sample, and

$\bar{x}$ is the arithmetic mean of the sample,

or:

$$CV = \frac{\sigma}{\mu} \cdot 100$$

for a **population**, where

σ is the standard deviation of the population, and

μ is the mean of the population.

This coefficient is independent of the unit of measurement used for the variable.

EXAMPLES

Let us study the salary distributions for two companies from two different countries.

According to a **survey**, the **arithmetic means** and the **standard deviations** of the salaries are as follows:

Company A:

$$\bar{x} = 2500\,\text{CHF},$$
$$S = 200\,\text{CHF},$$
$$CV = \frac{200}{2500} \cdot 100 = 8\%.$$

Company B:

$$\bar{x} = 1000\,\text{CHF},$$
$$S = 50\,\text{CHF},$$
$$CV = \frac{50}{1000} \cdot 100 = 5\%.$$

The **standard deviation** represents 8% of the **arithmetic mean** for company A, and 5% for company B. The salary distribution is a bit more homogeneous for company B than for company A.

FURTHER READING
▶ **Arithmetic mean**
▶ **Measure of dispersion**
▶ **Standard deviation**

REFERENCES
Johnson, N.L., Leone, F.C.: Statistics and experimental design in engineering and the physical sciences. Wiley, New York (1964)

Collinearity

Variables are known to be mathematically collinear if one of them is a linear combination of the other variables. They are known as statistically collinear if one of them is approximately a linear combination of other variables. In the case of a regression model where the explanatory variables are strongly correlated to each other, we say that there is collinearity (or multicollinearity) between the explanatory variables. In the first case, it is simply impossible to define least squares estimators, and in the second case, these estimators can exhibit considerable variance.

HISTORY
The term "collinearity" was first used in mathematics at the beginning of the twentieth century, due to the rediscovery of the theorem of Pappus of Alexandria (a third-century mathematician). Let A, B, C be three points on a line and A', B', C' be three points on a different line. If we relate the pairs using AB'.A'B, CA'.AC' and BC'.B'C, their intersections will occur in a line; in other words, the three intersection points will be collinear.

MATHEMATICAL ASPECTS
In the case of a matrix of explanatory variables $\mathbf{X}$, collinearity means that one of the columns of $\mathbf{X}$ is (approximately) a linear combination of the other columns. This implies that $\mathbf{X'X}$ is almost singular. Consequently, the **estimator** obtained by the least squares method $\widehat{\beta} = \left(\mathbf{X'X}\right)^{-1}\mathbf{X'Y}$ is obtained by inverting an almost singular **matrix**, which causes its components to become unstable. The **ridge regression** technique was created in order to deal with these collinearity problems.

A collinear relation between more than two variables will not always be the result of observing the pairwise correlations between the variables. A better indication of the presence of a collinearity problem is provided by variance inflation factors, VIF. The variance inflation factor of an explanatory variable X_j is defined by:

$$VIF_j = \frac{1}{1 - R_j^2},$$

where R_j^2 is the **coefficient of determination** of the model

$$X_j = \beta_0 + \sum_{k \neq j} \beta_k X_k + \varepsilon.$$

The coefficient VIF takes values of between 1 and ∞. If the X_j are mathematically collinear with other variables, we get $R_j^2 = 1$ and $VIF_j = \infty$. On the other hand, if the X_j are reciprocally independent, we have $R_j^2 = 0$ and $VIF_j = 1$. In practice, we consider that there is a real problem with collinearity when VIF_j is greater then 100, which corresponds to a R_j^2 that is greater then 0.99.

DOMAINS AND LIMITATIONS
Inverting a singular matrix, similar to inverting 0, is not a valid operation. Using the

same principle, inverting an almost singular matrix is similar to inverting a very small number. Some of the elements of the matrix must therefore be very big. Consequently, when the explanatory variables are collinear, some elements of the matrix $(\mathbf{X}'\mathbf{X})^{-1}$ of $\widehat{\beta}$ will probably be very large. This is why collinearity leads to unstable regression estimators. Aside from this problem, collinearity also results in a calculation problem; it is difficult to precisely calculate the inverse of an almost singular matrix.

EXAMPLES

Thirteen portions of cement are examined in the following example. Each portion contains four ingredients, as described in the table. The goal of the experiment is to determine how the quantities X_1, X_2, X_3 and X_4, corresponding to the quantities of these four ingredients, affect the quantity Y of heat given out as the cement hardens.

Y_i quantity of heat given out during the hardening of the ith portion (in joules);

X_{i1} quantity of ingredient 1 (tricalcium aluminate) in the ith portion;

X_{i2} quantity of ingredient 2 (tricalcium silicate) in the ith portion;

X_{i3} quantity of the ingredient 3 (tetracalcium aluminoferrite) in the ith portion;

X_{i4} quantity of ingredient 4 (dicalcium silicate) in the ith portion.

Table: Heat given out by the cement portions during hardening

Por-tion	Ingre-dient	Ingre-dient	Ingre-dient	Ingre-dient	Heat
i	1 X_1	2 X_2	3 X_3	4 X_4	Y
1	7	26	6	60	78.5
2	1	29	12	52	74.3
3	11	56	8	20	104.3
4	11	31	8	47	87.6
5	7	54	6	33	95.9
6	11	55	9	22	109.2
7	3	71	17	6	102.7
8	1	31	22	44	72.5
9	2	54	18	22	93.1
10	21	48	4	26	115.9
11	1	40	23	34	83.9
12	11	66	9	12	113.3
13	10	68	8	12	109.4

Source: Birkes & Dodge (1993)

We start with a **simple linear regression**. The model used for the linear regression is:

$$Y = \beta_0 + \beta_1 X_1 + \beta_2 X_2 + \beta_3 X_3 + \beta_4 X_4 + \varepsilon.$$

We obtain the following results:

Variable	$\widehat{\beta}_i$	S.d.	t_c
Constant	62.41	70.07	0.89
X_1	1.5511	0.7448	2.08
X_2	0.5102	0.7238	0.70
X_3	0.1019	0.7547	0.14
X_4	−0.1441	0.7091	−0.20

We can see that only coefficient X_1 is significantly greater then zero; in the other cases $t_c < t_{(\alpha/2, n-2)}$ (value taken from the Student table) for a **significance level** of $\alpha = 0.1$. Moreover, the standard deviations of the estimated coefficients $\widehat{\beta}_2$, $\widehat{\beta}_3$ and $\widehat{\beta}_4$ are greater then the coefficients themselves. When the variables are strongly correlated, it is known that the effect of one can mask the effect of another. Because of this, the coefficients can appear to be insignificantly different from zero.

To verify the presence of *multicollinearity* for a couple of variables, we calculate the correlation matrix.

	X_1	X_2	X_3	X_4
X_1	1	0.229	−0.824	0.245
X_2	0.229	1	−0.139	−0.973
X_3	−0.824	−0.139	1	0.030
X_4	0.245	−0.973	0.030	1

We note that a strong negative correlation (−0.973) exists between X_2 and X_4. Looking at the data, we can see the reason for that. Aside from portions 1 to 5, the total quantity of silicates ($X_2 + X_4$) is almost constant across the portions, and is approximately 77; therefore, X_4 is approximately $77 - X_2$. this situation does not allow us to distinguish between the individual effects of X_2 and those of X_4. For example, we see that the four largest values of X_4 (60, 52, 47 and 44) correspond to values of Y smaller then the mean heat 95.4. Therefore, at first sight it seems that large quantities of ingredient 4 will lead to small amounts of heat. However, we also note that the four largest values of X_4 correspond to the four smallest values of X_2, giving a negative correlation between X_2 and X_4. This suggests that ingredient 4 taken alone has a small effect on variable Y, and that the small quantities of 2 taken alone can explain the small amount of heat emitted. Hence, the linear dependence between two explanatory variables (X_2 and X_4) makes it more complicated to see the effect of each variable alone on the response variable Y.

FURTHER READING
▶ **Multiple linear regression**
▶ **Ridge regression**

REFERENCES
Birkes, D., Dodge, Y.: Alternative Methods of Regression. Wiley, New York (1993)

Bock, R.D.: Multivariate Statistical Methods in Behavioral Research. McGraw-Hill, New York (1975)

Combination

A combination is an un-ordered collection of unique elements or objects.

A k-combination is a subset with k elements. The number of k-combinations from a set of n elements is the number of **arrangements**.

HISTORY
See **combinatory analysis**.

MATHEMATICAL ASPECTS
1. *Combination without repetition*

   Combination without repetition describe the situation where each object drawn is not placed back for the next drawing. Each object can therefore only occur once in each group.

   The number of combination without repetition of k objects among n is given by:

   $$C_n^k = \binom{n}{k} = \frac{n!}{k! \cdot (n-k)!} \, .$$

2. *Combination with repetitions*

   Combination with repetitions (or with remittance) are used when each drawn object is placed back for the next drawing. Each object can then occur r times in each group, $r = 0, 1, \ldots, k$.

The number of combinations with repetitions of k objects among n is given by:

$$K_n^k = \binom{n+k-1}{k} = \frac{(n+k-1)!}{k! \cdot (n-1)!}.$$

EXAMPLES

1. *Combinations without repetition*
   Consider the situation where we must choose a committee of three people from an assembly of eight people. How many different committees could potentially be picked from this assembly, if each person can only be selected once in each group? Here we need to calculate the number of possible combinations of three people from eight:

$$C_n^k = \frac{n!}{k! \cdot (n-k)!} = \frac{8!}{3! \cdot (8-3)!}$$

$$= \frac{40320}{6 \cdot 120} = 56$$

Therefore, it is possible to form 56 different committees containing three people from an assembly of eight people.

2. *Combinations with repetitions* Consider an urn containing six numbered balls. We carry out four successive drawings, and place the drawn ball back into the urn after each drawing. How many different combinations could occur from this drawing? In this case we want to find the number of combinations with repetition (because each drawn ball is placed back in the urn before the next drawing). We obtain

$$K_n^k = \frac{(n+k-1)!}{k! \cdot (n-1)!} = \frac{9!}{4! \cdot (6-1)!}$$

$$= \frac{362880}{24 \cdot 120} = 126$$

different combinations.

FURTHER READING
▶ **Arrangement**
▶ **Binomial distribution**
▶ **Combinatory analysis**
▶ **Permutation**

Combinatory Analysis

Combinatory analysis refers to a group of techniques that can be used to determine the number of elements in a particular set without having to count them one-by-one.

The elements in question could be the results from a scientific **experiment** or the different potential outcomes of a random **event**.

Three particular concepts are important in combinatory analysis:

- Permutations;
- Combinations;
- Arrangements.

HISTORY
Combinatory analysis has interested mathematicians for centuries. According to Takacs (1982), such analysis dates back to ancient Greece. However, the Hindus, the Persians (including the poet and mathematician Khayyâm, Omar) and (especially) the Chinese also studied such problems. A 3000 year-old Chinese book "I Ching" describes the possible **arrangements** of a set of n elements, where $n \leq 6$. In 1303, Chu, Shih-chieh published a work entitled "Ssu Yuan Yü Chien" (Precious mirror of the four elements). The cover of the book depicts a triangle that shows the combinations of k elements taken from a set of size n where $0 \leq k \leq n$. This arithmetic triangle was also explored by several European mathematicians such as Stifel, Tartaglia and Hérigone, and especially Pascal, who wrote the "Trai-

té du triangle arithmétique" (Treatise of the arithmetic triangle) in 1654 (although it was not published until after his death in 1665). Another document on **combinations** was published in 1617 by Puteanus, Erycius called "Erycii Puteani Pretatis Thaumata in Bernardi Bauhusii è Societate Jesu Proteum Parthenium". However, combinatory analysis only revealed its true power with the works of Fermat (1601–1665) and Pascal (1623–1662). The term "combinatory analysis" was introduced by Leibniz (1646–1716) in 1666. In his work "Dissertatio de Arte Combinatoria," he systematically studied problems related to **arrangements, permutations** and **combinations**.

Other works in this field should be mentioned here, such as those of Wallis, J. (1616–1703), reported in "The Doctrine of Permutations and Combinations" (an essential and fundamental part of the "Doctrines of Chances"), or those of **Bernoulli,** J., **Moivre, A. de,** Cardano, G. (1501–1576), and Galileo (1564–1642).

In the second half of the nineteenth century, Cayley (1829–1895) solved some problems related to this type of analysis via graphics that he called "trees". Finally, we should also mention the important work of MacMahon (1854–1929), "Combinatory Analysis" (1915–1916).

EXAMPLES
See **arrangement, combination** and **permutation**.

FURTHER READING
► **Arithmetic triangle**
► **Arrangement**
► **Binomial**
► **Binomial distribution**
► **Combination**
► **Permutation**

REFERENCES

MacMahon, P.A.: Combinatory Analysis, vols. I and II. Cambridge University Press, Cambridge (1915–1916)

Stigler, S.: The History of Statistics, the Measurement of Uncertainty Before 1900. Belknap, London (1986)

Takács, L.: Combinatorics. In: Kotz, S., Johnson, N.L. (eds.) Encyclopedia of Statistical Sciences, vol. 2. Wiley, New York (1982)

Compatibility

Two **events** are said to be compatible if the occurrence of the first event does not prevent the occurrence of the second (or in other words, if the **intersection** between the two events is not null):

$$P(A \cap B) \neq 0 .$$

We can represent two compatible **events** A and B schematically in the following way:

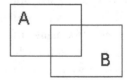

Two **events** A and B are incompatible (or mutually exclusive) if the occurrence of A prevents the occurrence of B, or vice versa. We can represent this in the following way:

This means that the **probability** that these two events happen at the same time is zero:

$$A \cap B = \phi \longrightarrow P(A \cap B) = 0.$$

MATHEMATICAL ASPECTS

If two **events** A and B are compatible, the **probability** that at least one of the events occurs can be obtained using the following formula:

$$P(A \cup B) = P(A) + P(B) - P(A \cap B).$$

On the other hand, if the two **events** A and B are incompatible, the **probability** that the two events A and B happen at the same time is zero:

$$P(A \cap B) = 0.$$

The **probability** that at least one of the events occurs can be obtained by simply adding the individual probabilities of A and B:

$$P(A \cup B) = P(A) + P(B).$$

EXAMPLES

Consider a **random experiment** that involves drawing a card from a pack of 52 cards. We are interested in the three following **events**:

$$A = \text{"draw a heart"}$$
$$B = \text{"draw a queen"}$$
$$C = \text{"draw a club"}.$$

The **probabilities** associated with each of these **events** are:

$$P(A) = \frac{13}{52}$$
$$P(B) = \frac{4}{52}$$
$$P(C) = \frac{13}{52}.$$

The **events** A and B are compatible, because it is possible to draw both a heart and a queen at the same time (the queen of hearts). Therefore, the intersection between A and B is the queen of hearts. The probability of this event is given by:

$$P(A \cap B) = \frac{1}{52}.$$

The **probability** of the union of the two **events** A and B (drawing either a heart or a queen) is then equal to:

$$P(A \cup B) = P(A) + P(B) - P(A \cap B)$$
$$= \frac{13}{52} + \frac{4}{52} - \frac{1}{52}$$
$$= \frac{4}{13}.$$

On the other hand, the **events** A and C are incompatible, because a card cannot be both a heart and a club! The **intersection** between A and C is an empty set.

$$A \cap C = \phi.$$

The **probability** of the union of the two events A and C (drawing a heart or a club) is simply given by the sum of the probabilities of each event:

$$P(A \cup C) = P(A) + P(C)$$
$$= \frac{13}{52} + \frac{13}{52}$$
$$= \frac{1}{2}.$$

FURTHER READING
Event
Independence
Probability

Complementary

Consider the **sample space** Ω for a **random experiment**.

For any **event** A, an element of Ω, we can determine a new event B that contains all of the elements of the **sample space** Ω that are not included in A.

This event B is called the complement of A with respect to Ω and is obtained by the negation of A.

MATHEMATICAL ASPECTS

Consider an **event** A, which is an element of the **sample space** Ω.

The compliment of A with respect to Ω is denoted $\bar{A}$. It is given by the negation of A:

$$\bar{A} = \Omega - A$$
$$= \{w \in \Omega; \ w \notin A\} \ .$$

EXAMPLES

Consider a **random experiment** that consists of flipping a coin three times.

The **sample space** of this experiment is

$$\Omega = \{TTT, TTH, THT, THH,$$
$$HTT, HTH, HHT, HHH\} \ .$$

Consider the **event**

$$A = \text{"Heads } (H) \text{ occurs twice"}$$
$$= \{THH, HTH, HHT\} \ .$$

The compliment of A with respect to Ω is equal to

$$\bar{A} = \{TTT, TTH, THT, HTT, HHH\}$$
$$= \text{"Heads } (H) \text{ does not occur twice"} \ .$$

FURTHER READING
► **Event**
► **Random experiment**
► **Sample space**

Complete Linkage Method

The complete linkage method is a hierarchical **classification** method where the distance between two classes is defined as the greatest distance that could be obtained if we select one element from each class and measure the distance between these elements. In other words, it is the distance between the most distant elements from each class.

For example, the **distance** used to construct the **distance table** is the Euclidian distance. Using the complete linkage method, the distance between two classes is given by the Euclidian distance between the most distant elements (the maximum distance).:

MATHEMATICAL ASPECTS
See **cluster analysis**.

FURTHER READING
► **Classification**
► **Cluster analysis**
► **Dendrogram**
► **Distance**
► **Distance table**

Completely Randomized Design

A completely randomized design is a type of experimental design where the experimental units are randomly assigned to the different treatments.

It is used when the experimental units are believed to be "uniform;" that is, when there is no uncontrolled **factor** in the **experiment**.

HISTORY
See **design of experiments**.

EXAMPLES

We want to test five different drugs based on aspirin. To do this, we randomly distribute the five types of drug to 40 patients. Denoting the five drugs by A, B, C, D and E, we obtain the following random distribution:

A is attributed to 10 people;

B is attributed to 12 people;

C is attributed to 4 people;

D is attributed to 7 people;

E is attributed to 7 people.

We have then a completely randomized design where the treatments (drugs) are randomly attributed to the experimental units (patients), and each patient receives only one treatment. We also assume that the patients are "uniform:" that there are no differences between them. Moreover, we assume that there is no uncontrolled **factor** that intervenes during the treatment.

In this example, the completely randomized design is a **factorial experiment** that uses only one factor: the aspirin. The five types of aspirin are different levels of the factor.

FURTHER READING
▶ **Design of experiments**
▶ **Experiment**

Composite Index Number

Composite **index numbers** allow us to measure, with a single number, the relative variations within a group of **variables** upon moving from one situation to another.

The consumer price index, the wholesale price index, the employment index and the Dow-Jones index are all examples of composite index numbers.

The aim of using composite index numbers is to summarize all of the simple index numbers contained in a complex number (a value formed from a set of simple values) in just one index.

The most commonly used composite index numbers are:
- The **Laspeyres index**;
- The **Paasche index**;
- The **Fisher index**.

HISTORY
See **index number**.

MATHEMATICAL ASPECTS
There are several methods of creating composite index numbers.

To illustrate these methods, let us use a scenario where a price index is determined for the current period n with respect to a reference period 0.

1. *Index number of the arithmetic means* (the sum method):

$$I_{n/0} = \frac{\sum P_n}{\sum P_0} \cdot 100 \,,$$

where $\sum P_n$ is the sum of the prices of the items at the current period, and $\sum P_0$ is the sum of the prices of the items at the reference period.

2. *Arithmetic mean of simple index numbers:*

$$I_{n/0} = \frac{1}{N} \cdot \sum \left(\frac{P_n}{P_0} \right) \cdot 100 \,,$$

where N is the number of goods considered and $\frac{P_n}{P_0}$ is the **simple index number** of each item.

In these two methods, each item has the same importance. This is a situation which often does not correspond to reality.

3. *Index number of weighted arithmetic means* (the weighted sum method):
The general formula for an index number calculated by the weighted sum method is as follows:

$$I_{n/0} = \frac{\sum P_n \cdot Q}{\sum P_0 \cdot Q} \cdot 100 \,.$$

Choosing the quantity Q for each item considered could prove problematic here: Q must be the same for both the numerator and the denominator when calculating a price index.

In the **Laspeyres index**, the value of Q corresponding to the reference year is used. In the **Paasche index**, the value of Q for the current year is used. Other statisticians have proposed using the value of Q for a given year.

EXAMPLES

Consider the following table indicating the (fictitious) prices of three consumer goods in the reference year (1970) and their current prices.

Goods	Price (francs)	
	1970 (P_0)	Now (P_n)
Milk	0.20	1.20
Bread	0.15	1.10
Butter	0.50	2.00

Using these numbers, we now examine the three main methods of constructing composite **index numbers**.

1. *Index number of arithmetic means* (the sum method):

$$I_{n/0} = \frac{\sum P_n}{\sum P_0} \cdot 100$$
$$= \frac{4.30}{0.85} \cdot 100 = 505.9 \,.$$

According to this method, the price index has increased by 405.9% (505.9 − 100) between the reference year and the current year.

2. *Arithmetic mean of simple index numbers:*

$$I_{n/0} = \frac{1}{N} \cdot \sum \left(\frac{P_n}{P_0}\right) \cdot 100$$
$$= \frac{1}{3} \cdot \left(\frac{1.20}{0.20} + \frac{1.10}{0.15} + \frac{2.00}{0.50}\right) \cdot 100$$
$$= 577.8 \,.$$

This method gives a slightly different result from the previous one, since we obtain an increase of 477.8% (577.8 − 100) in the price index between the reference year and the current year.

3. *Index number of weighted arithmetic means* (the weighted sum method):

$$I_{n/0} = \frac{\sum P_n \cdot Q}{\sum P_0 \cdot Q} \cdot 100 \,.$$

This method is used in conjunction with the **Laspeyres index** or the **Paasche index**.

FURTHER READING
▶ **Fisher index**
▶ **Index number**
▶ **Laspeyres index**
▶ **Paasche index**
▶ **Simple index number**

Conditional Probability

The probability of an event given that another event is known to have occured.

The conditional probability is denoted $P(A|B)$, which is read as the "probability of A conditioned by B."

HISTORY

The concept of **independence** dominated probability theory until the beginning of the twentieth century. In 1933, Kolmogorov, Andrei Nikolaievich introduced the concept of conditional probability; this concept now plays an essential role in theoretical and applied probability and statistics.

MATHEMATICAL ASPECTS

Consider a **random experiment** for which we know the **sample space** Ω. Consider two events A and B from this space. The **probability** of A, $P(A)$ depends on the set of possible events in the **experiment** (Ω).

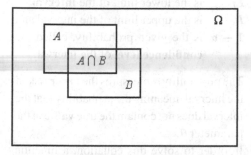

Now consider that we have supplementary information concerning the experiment: that the event B has occurred. The probability of the event A occurring will then be a function of the space B rather than a function of Ω. The probability of A conditioned by B is calculated as follows:

$$P(A|B) = \frac{P(A \cap B)}{P(B)}.$$

If A and B are two incompatible events, the intersection between A and B is an empty space. We will then have $P(A|B) = P(B|A) = 0$.

DOMAINS AND LIMITATIONS

The concept of conditional probability is one of the most important ones in probability the-

ory. This importance is mainly due to the following points:

1. We are often interested in calculating the probability of an **event** when some information about the result is already known. In this case, the probability required is a conditional probability.
2. Even when partial information on the result is not known, conditional probabilities can be useful when calculating the probabilities required.

EXAMPLES

Consider a group of 100 cars distributed according to two criteria, comfort and speed. We will make the following distinctions:

$$\text{a car can be } \begin{cases} \text{fast} \\ \text{slow} \end{cases},$$

$$\text{a car can be } \begin{cases} \text{comfortable} \\ \text{uncomfortable} \end{cases}.$$

A partition of the 100 cars based on these criteria is provided in the following table:

	fast	slow	total
comfortable	40	10	50
uncomfortable	20	30	50
total	60	40	100

Consider the following events:

$A =$ "a fast car is chosen"

and $B =$ "a comfortable car is chosen."

The probability of these two events are:

$$P(A) = 0.6,$$
$$P(B) = 0.5.$$

The probability of choosing a fast car is then of 0.6, and that of choosing a comfortable car is 0.5.

Now imagine that we are given supplementary information: a fast car was chosen. What, then, is the probability that this car is also comfortable?

We calculate the probability of B knowing that A has occurred, or the conditional probability of B depending on A:

We find in the table that

$$P(A \cap B) = 0.4 \,.$$

$$\Rightarrow P(B|A) = \frac{P(A \cap B)}{P(A)} = \frac{0.4}{0.6} = 0.667 \,.$$

The probability that the car is comfortable, given that we know that it is fast, is therefore $\frac{2}{3}$.

FURTHER READING
▶ **Event**
▶ **Probability**
▶ **Random experiment**
▶ **Sample space**

REFERENCES

Kolmogorov, A.N.: Grundbegriffe der Wahrscheinlichkeitsrechnung. Springer, Berlin Heidelberg New York (1933)

Kolmogorov, A.N.: Foundations of the Theory of Probability. Chelsea Publishing Company, New York (1956)

Confidence Interval

A confidence interval is any **interval** constructed around an **estimator** that has a particular **probability** of containing the true **value** of the corresponding **parameter** of a **population**.

HISTORY

According to Desrosières, A. (1988), Bowley, A.L. was one of the first to be interested in the concept of the confidence interval. Bowley presented his first confidence interval calculations to the Royal Statistical Society in 1906.

MATHEMATICAL ASPECTS

In order to construct a confidence interval that contains the true **value** of the **parameter** θ with a given **probability**, an equation of the following form must be solved:

$$P(L_i \le \theta \le L_s) = 1 - \alpha \,,$$

where

θ is the parameter to be estimated,
L_i is the lower limit of the **interval**,
L_s is the upper limit of the **interval** and
$1 - \alpha$ is the given probability, called the **confidence level** of the interval.

The probability α measures the **error** risk of the interval, meaning the probability that the interval does not contain the true **value** of the parameter θ.

In order to solve this equation, a function $f(t, \theta)$ must be defined where t is an **estimator** of θ, for which the **probability distribution** is known.

Defining this **interval** for $f(t, \theta)$ involves writing the equation:

$$P(k_1 \le f(t, \theta) \le k_2) = 1 - \alpha \,,$$

where the constants k_1 and k_2 are given by the **probability distribution** of the function $f(t, \theta)$. Generally, the error risk α is divided into two equal parts at $\frac{\alpha}{2}$ distributed on each side of the distribution of $f(t, \theta)$. If, for example, the function $f(t, \theta)$ follows a centered and reduced **normal distribution**, the constants k_1 and k_2 will be symmetric and can be represented by $-z_{\frac{\alpha}{2}}$ and $+z_{\frac{\alpha}{2}}$, as shown in the following figure.

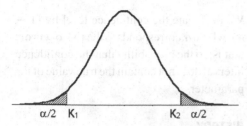

α/2 K₁   K₂ α/2

Once the constants k_1 and k_2 have been found, the **parameter** θ must be isolated in the equation given above. The confidence interval θ is found in this way for the **confidence level** $1 - \alpha$.

DOMAINS AND LIMITATIONS

One should be very careful when interpreting a confidence interval. If, for a **confidence level** of 95%, we find a confidence interval for a **mean** of μ where the lower and upper limits are k_1 and k_2 respectively, we can conclude the following (for example):
"On the basis of the studied **sample**, we can affirm that it is probable that the **mean** of the **population** can be found in the **interval** established."
It would not be correct to conclude that there is a 95% chance of finding the **mean** of the **population** in the **interval**. Indeed, since μ and the limits k_1 and k_2 of the interval are constants, the interval may or may not contain μ. However, if the statistician has the ability to repeat the **experiment** (which consists of drawing a **sample** from the population) several times, 95% of the intervals obtained will contain the true **value** of μ.

EXAMPLES

A business that fabricates lightbulbs wants to test the average lifespan of its lightbulbs. The distribution of the **random variable** X, which represents the life span in hours, is a **normal distribution** with **mean** μ and **standard deviation** $\sigma = 30$.

In order to estimate μ, the business burns out $n = 25$ lightbulbs.

It obtains an average lifespan of $\bar{x} = 860$ hours. It wants to establish a confidence interval around the **estimator** $\bar{x}$ at a **confidence level** of 0.95. Therefore, the first step is to obtain a function $f(t, \theta) = f(\bar{x}, \mu)$ for the known distribution. Here we use:

$$f(t, \theta) = f(\bar{x}, \mu) = \frac{\bar{x} - \mu}{\frac{\sigma}{\sqrt{n}}},$$

which follows a centered and reduced **normal distribution**. The equation $P(k_1 \le f(t, \theta) \le k_2) = 1 - \alpha$ becomes:

$$P\left(-z_{0.025} \le \frac{\bar{x} - \mu}{\frac{\sigma}{\sqrt{n}}} \le z_{0.025}\right) = 0.95,$$

because the **error** risk α has been divided into two equal parts at $\frac{\alpha}{2} = 0.025$.

The table for the centered and reduced **normal distribution**, the **normal table**, gives $z_{0.025} = 1.96$. Therefore:

$$P\left(-1.96 \le \frac{\bar{x} - \mu}{\frac{\sigma}{\sqrt{n}}} \le 1.96\right) = 0.95.$$

To obtain the confidence interval for μ at a **confidence level** of 0.95, μ must be isolated in the equation above:

$$P\left(-1.96 \le \frac{\bar{x} - \mu}{\frac{\sigma}{\sqrt{n}}} \le 1.96\right) = 0.95$$

$$P\left(-1.96\frac{\sigma}{\sqrt{n}} \le \bar{x} - \mu \le 1.96\frac{\sigma}{\sqrt{n}}\right) = 0.95$$

$$P\left(\bar{x} - 1.96\frac{\sigma}{\sqrt{n}} \le \mu \le \bar{x} + 1.96\frac{\sigma}{\sqrt{n}}\right) = 0.95.$$

By replacing $\bar{x}$, σ and n with their respective **values**, we obtain:

$$P\left(860 - 1.96\frac{30}{\sqrt{25}} \leq \mu \leq 860\right.$$

$$\left. + 1.96\frac{30}{\sqrt{25}}\right) = 0.95$$

$$P(848.24 \leq \mu \leq 871.76) = 0.95.$$

The confidence interval for μ at the confidence level 0.95 is therefore:

$$[848.24, 871.76].$$

This means that we can affirm with a **probability** of 95% that this **interval** contains the true **value** of the **parameter** μ that corresponds to the average lifespan of the lightbulbs.

FURTHER READING
▶ **Confidence level**
▶ **Estimation**
▶ **Estimator**

REFERENCES

Bowley, A.L.: Presidential address to the economic section of the British Association. J. Roy. Stat. Soc. **69**, 540–558 (1906)

Desrosières, A. La partie pour le tout: comment généraliser? La préhistoire de la contrainte de représentativité. Estimation et sondages. Cinq contributions à l'histoire de la statistique. Economica, Paris (1988)

Confidence Level

The confidence level is the **probability** that the **confidence interval** constructed around an **estimator** contains the true **value** of the corresponding **parameter** of the **population**.

We designate the confidence level by $(1 - \alpha)$, where α corresponds to the risk of **error**; that is, to the probability that the confidence interval does not contain the true value of the parameter.

HISTORY
The first example of a confidence interval appears in the work of Laplace (1812). According to Desrosières, A. (1988), Bowley, A.L. was one of the first to become interested in the concept of the confidence interval.

See **hypothesis testing**.

MATHEMATICAL ASPECTS
Let θ be a parameter associated with a population. θ is to be estimated and T is its estimator from a random sample. We evaluate the precision of T as the estimator of θ by constructing a **confidence interval** around the estimator, which is often interpreted as an error margin.

In order to construct this confidence interval, we generally proceed in the following manner. From the distribution of the estimator T, we determine an interval that is likely to contain the true value of the parameter. Let us denote this interval by $(T - \varepsilon, T + \varepsilon)$ and the probability of true value of the paramter being in this interval as $(1 - \alpha)$. We can then say that the error margin ε is related to α by the probability:

$$P(T - \varepsilon \leq \theta \leq T + \varepsilon) = 1 - \alpha.$$

The level of probability associated with an interval of estimation is called the confidence level or the confidence degree.

The interval $T - \varepsilon \leq \theta \leq T + \varepsilon$ is called the confidence interval for θ at the confidence level $1 - \alpha$. Let us use, for example, $\alpha = 5\%$, which will give the confidence interval of the parameter θ to a probability level of 95%. This means that, if we use T as an estimator of θ, then the interval indicated will on average contain the true value of the parameter 95 times out of 100 samplings, and it will not contain it 5 times.

The quantity ε of the confidence interval corresponds to half od the length of the interval. This parameter therefore gives us an idea of the error margin for the estimator. For a given confidence level $1 - \alpha$, the smaller the confidence interval, more efficient the estimator.

DOMAINS AND LIMITATIONS

The most commonly used confidence levels are 90%, 95% and 99%. However, if necessary, other levels can be used instead.

Although we would like to use the highest confidence level in order to maximize the **probability** that the confidence interval contains the true value of the **parameter**, but if we increase the confidence level, the **interval** increases as well. Therefore, what we gain in terms of confidence is lost in terms of precision, so have to find a compromise.

EXAMPLES

A company that produces ligthtbulbs wants to study the mean lifetime of its bulbs. The distribution of the **random variable** X that represents the lifetime in hours is a **normal distribution** of **mean** μ and **standard deviation** $\sigma = 30$.

In order to estimate μ, the company burns out a random **sample** of $n = 25$ bulbs.

The company obtains an average bulb lifetime of $\bar{x} = 860$ hours. It then wants to construct a 95% **confidence interval** for μ around the **estimator** $\bar{x}$.

The standard deviation σ of the population is known; the value of ε is $z_{\alpha/2} \cdot \sigma_{\bar{X}}$. The value of $z_{\alpha/2}$ is obtained from the normal table, and it depends on the probability attributed to the parameter α. We then deduce the confidence interval of the estimator of μ at the probability level $1 - \alpha$:

$$\bar{X} - z_{\alpha/2}\sigma_{\bar{X}} \le \mu \le \bar{X} + z_{\alpha/2}\sigma_{\bar{X}}.$$

From the **hypothesis** that the bulb lifetime X follows a normal distribution with mean μ and standard deviation $\sigma = 30$, we deduce that the expression

$$\frac{\bar{x} - \mu}{\frac{\sigma}{\sqrt{n}}}$$

follows a standard normal distribution. From this we obtain:

$$P\left[-z_{0.025} \le \frac{\bar{x} - \mu}{\frac{\sigma}{\sqrt{n}}} \le z_{0.025}\right] = 0.95,$$

where the risk of **error** α is divided into two parts that both equal $\frac{\alpha}{2} = 0.025$.

The table of the standard normal distribution (the **normal table**) gives $z_{0.025} = 1.96$. We then have:

$$P\left(-1.96 \le \frac{\bar{x} - \mu}{\frac{\sigma}{\sqrt{n}}} \le 1.96\right) = 0.95.$$

To get the confidence interval for μ, we must isolate μ in the following equation via the following transformations:

$$P\left(-1.96\frac{\sigma}{\sqrt{n}} \le \bar{x} - \mu \le 1.96\frac{\sigma}{\sqrt{n}}\right)$$
$$= 0.95,$$

$$P\left(\bar{x} - 1.96\frac{\sigma}{\sqrt{n}} \le \mu \le \bar{x} + 1.96\frac{\sigma}{\sqrt{n}}\right)$$
$$= 0.95.$$

Substituting $\bar{x}$, σ and n for their respective values, we evaluate the confidence interval:

$$860 - 1.96 \cdot \frac{30}{\sqrt{25}}$$
$$\leq \mu \leq 860 + 1.96 \cdot \frac{30}{\sqrt{25}}$$
$$848.24 \leq \mu \leq 871.76.$$

We can affirm with a confidence level of 95% that this **interval** contains the true value of the **parameter** μ, which corresponds to the mean bulb lifetime.

FURTHER READING

▶ **Confidence interval**
▶ **Estimation**
▶ **Estimator**
▶ **Hypothesis testing**

Contingency Table

A contingency table is a crossed table containing various attributes of a population or an observed sample. Contingency table analysis consists of discovering and studying the relations (if they exist) between these attributes.

A contingency table can be a two-dimensional table with r lines and c columns relating to two qualitative categorical variables possessing, respectively, r and c categories. It can also be multidimensional when the number of qualitative variables is greater then two: if, for example, the elements of a **population** or a **sample** are characterized by three attributes, the associated contingency table has the dimensions $I \times J \times K$, where I represents the number of categories defining the first attribute, J the number of categories of the second attribute and K the number of the categories of the third attribute.

HISTORY

The term "contingency," used in relation to a crossed table of categorical data, seems to have originated with **Pearson, Karl** (1904), who used he term "contingency" to mean a measure of the total **deviation** relative to the **independence**.

See also **chi-square test of independence**.

MATHEMATICAL ASPECTS

If we consider a two-dimensional table, containing entries for two qualitative categorical variables X and Y that have, respectively, r and c categories, the contingency table is:

		Categories of the variable Y		
		Y_1	$\ldots$ Y_c	Total
Categories	X_1	n_{11}	$\ldots$ n_{1c}	$n_{1.}$
of the	$\ldots$	$\ldots$	$\ldots$ $\ldots$	$\ldots$
variable X	X_r	n_{r1}	$\ldots$ n_{rc}	$n_{r.}$
	Total	$n_{.1}$	$\ldots$ $n_{.c}$	$n_{..}$

where

n_{ij} represents the observed **frequency** for category i of **variable** X and category j of variable Y;

$n_{i.}$ represents the sum of the frequencies observed for category i of variable X,

$n_{.j}$ represents the sum of the observed frequencies for category j of variable Y, and

$n_{..}$ indicates the total number of observations.

In the case of a multidimensional table, the elements of the table are denoted by n_{ijk}, representing the observed frequency for category i of variable X, category j of variable Y and category k of variable Z.

DOMAINS AND LIMITATIONS

The **independence** of two categorical qualitative variables represented in the contingency table can be assessed by performing a **chi-square test of independence**.

FURTHER READING
▶ **Chi-square test of independence**
▶ **Frequency distribution**

REFERENCES
Fienberg, S.E.: The Analysis of Cross-Classified Categorical Data, 2nd edn. MIT Press, Cambridge, MA (1980)

Pearson, K.: On the theory of contingency and its relation to association and normal correlation. Drapers' Company Research Memoirs, Biometric Ser. I., pp. 1–35 (1904)

Continuous Distribution Function

The **distribution function** of a continuous **random variable** is defined to be the **probability** that the random variable takes a **value** less than or equal to a real number.

HISTORY
See **probability**.

MATHEMATICAL ASPECTS
The function defined by

$$F(b) = P(X \leq b) = \int_{-\infty}^{b} f(x)\, dx.$$

is called the **distribution function** of a continuous **random variable**.

In other words, the **density function** f is the derivative of the continuous distribution function.

Properties of the Continuous Distribution Function
1. $F(x)$ is a continually increasing function for all x;
2. F takes its **values** in the **interval** $[0, 1]$;
3. $\lim_{b \to -\infty} F(b) = 0$;
4. $\lim_{b \to \infty} F(b) = 1$;
5. $F(x)$ is a continuous and differentiable function.

This distribution function can be graphically represented on a system of axes. The different **values** of the **random variable** X are plotted on the abscissa and the corresponding values of $F(x)$ on the ordinate.

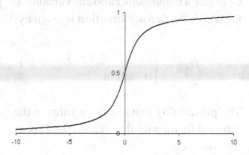

DOMAINS AND LIMITATIONS
The **probability** that the continuous **random variable** X takes a **value** in the **interval** $]a, b]$ for all $a < b$, meaning that $P(a < X \leq b)$, is equal to $F(b) - F(a)$, where F is the distribution function of the random variable X.

Demonstration: The **event** $\{X \leq b\}$ can be written as the union of two mutually exclusive events: $\{X \leq a\}$ and $\{a < X \leq b\}$:

$$\{X \leq b\} = \{X \leq a\} \cup \{a < X \leq b\}.$$

By finding the **probability** on each side of the equation, we obtain:

$$P(X \leq b) = P(\{X \leq a\} \cup \{a < X \leq b\})$$
$$= P(X \leq a) + P(a < X \leq b).$$

The sum of probabilities result from the fact that the two events are exclusive.

By subtracting $P(X \leq a)$ on each side, we have:

$$P(a < X \leq b) = P(X \leq b) - P(X \leq a).$$

Finally, from the definition of the **distribution function**, we obtain:

$$P(a < X \leq b) = F(b) - F(a).$$

EXAMPLES

Consider a continuous **random variable** X for which the **density function** is given by:

$$f(x) = \begin{cases} 1 & \text{if } 0 < x < 1 \\ 0 & \text{if not} \end{cases}.$$

The **probability** that X takes a **value** in the **interval** $[a, b]$, with $0 < a$ and $b < 1$, is as follows:

$$\begin{aligned} P(a \leq X \leq b) &= \int_a^b f(x)\,dx \\ &= \int_a^b 1\,dx \\ &= x \Big|_a^b \\ &= b - a. \end{aligned}$$

Therefore, for $0 < x < 1$ the distribution function is:

$$\begin{aligned} F(x) &= P(X \leq x) \\ &= P(0 \leq X \leq x) \\ &= x. \end{aligned}$$

This function is presented in the following figure:

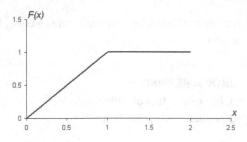

FURTHER READING

▶ **Density function**
▶ **Probability**
▶ **Random experiment**
▶ **Random variable**
▶ **Value**

Continuous Probability Distribution

Every **random variable** has a corresponding **frequency distribution**. For a continuous random variable, this distribution is continuous too.

A continuous probability distribution is a **model** that represents the **frequency distribution** of a continuous **variable** in the best way.

MATHEMATICAL ASPECTS

The **probability distribution** of a continuous **random variable** X is given by its **density function** $f(x)$ or its **distribution function** $F(x)$.

It can generally be characterized by its **expected value**:

$$E[X] = \int_D x \cdot f(x)\,dx = \mu$$

and its **variance**:

$$\begin{aligned} \text{Var}(X) &= \int_D (x - \mu)^2 \cdot f(x)\,dx \\ &= E\left[(X - \mu)^2\right] \\ &= E[X^2] - E[X]^2, \end{aligned}$$

where D represents the **interval** covering the range of **values** that X can take.

One essential property of a continuous **random variable** is that the **probability** that it will take a specific numerical **value** is zero, whereas the probability that it will take a value over an **interval** (finite or infinite) is usually nonzero.

DOMAINS AND LIMITATIONS

The most famous continuous probability distribution is the **normal distribution**. Continuous probability distributions are often used to approximate **discrete probability distributions**. They are used in model construction just as much as they are used when applying statistical techniques.

FURTHER READING

▶ **Beta distribution**
▶ **Cauchy distribution**
▶ **Chi-square distribution**
▶ **Continuous distribution function**
▶ **Density function**
▶ **Discrete probability distribution**
▶ **Expected value**
▶ **Exponential distribution**
▶ **Fisher distribution**
▶ **Gamma distribution**
▶ **Laplace distribution**
▶ **Lognormal distribution**
▶ **Normal distribution**
▶ **Probability**
▶ **Probability distribution**
▶ **Random variable**
▶ **Student distribution**
▶ **Uniform distribution**
▶ **Variance of a random variable**

REFERENCES

Johnson, N.L., Kotz, S.: Distributions in Statistics: Continuous Univariate Distri-butions, vols. 1 and 2. Wiley, New York (1970)

Contrast

C

In analysis of variance a contrast is a linear combination of the **observations** or factor levels or treatments in a **factorial experiment**, where the sum of the coefficients is zero.

HISTORY

According to Scheffé, H. (1953), Tukey, J.W. (1949 and 1951) was the first to propose a method of simultaneously **estimating** all of the contrasts.

MATHEMATICAL ASPECTS

Consider $T_1, T_2, \ldots, T_k$, which are the sums of $n_1, n_2, \ldots, n_k$ **observations**. The linear function

$$c_j = c_{1j} \cdot T_1 + c_{2j} \cdot T_2 + \cdots + c_{kj} \cdot T_k$$

is a contrast if and only if

$$\sum_{i=1}^{k} n_i \cdot c_{ij} = 0 .$$

If each $n_i = n$, meaning that if T_i is the sum of the same number of **observations**, the condition is reduced to:

$$\sum_{i=1}^{k} c_{ij} = 0 .$$

DOMAINS AND LIMITATIONS

In most **experiments** involving several **treatments**, it is interesting for the experimenter to make comparisons between the different treatments. The statistician uses contrasts to carry out this type of comparison.

EXAMPLES

When an **analysis of variance** is carried out for a three-**level factor**, some contrasts of interest are:

$$c_1 = T_1 - T_2$$
$$c_2 = T_1 - T_3$$
$$c_3 = T_2 - T_3$$
$$c_4 = T_1 - 2 \cdot T_2 + T_3 .$$

FURTHER READING

▶ **Analysis of variance**

▶ **Experiment**

▶ **Factorial experiment**

REFERENCES

Ostle, B.: Statistics in Research: Basic Concepts and Techniques for Research Workers. Iowa State College Press, Ames, IA (1954)

Scheffé, H.: A method for judging all contrasts in the analysis of variance. Biometrika **40**, 87–104 (1953)

Tukey, J.W.: Comparing individual means in the analysis of variance. Biometrics **5**, 99–114 (1949)

Tukey, J.W.: Quick and Dirty Methods in Statistics. Part II: Simple Analyses for Standard Designs. Quality Control Conference Papers 1951. American Society for Quality Control, New York, pp. 189–197 (1951)

Convergence

In statistics, the term "convergence" is related to **probability** theory. This statistical convergence is often termed stochastic convergence in order to distinguish it from classical convergence.

MATHEMATICAL ASPECTS

Different types of stochastic convergence can be defined. Let $\{x_n\}_{n \in N}$ be a set of random variables. The most important types of stochastic convergence are:

1. $\{X_n\}_{n \in \mathbb{N}}$ converges *in distribution* to a random variable X if

$$\lim_{n \to \infty} F_{X_n}(z) = F_X(z) \quad \forall z ,$$

where F_{X_n} and F_X are the distribution functions of X_n and X, *respectively*.

This convergence is simply the point convergence (well-known in mathematics) of the set of the distribution functions of the X_n.

2. $\{X_n\}_{n \in \mathbb{N}}$ converges *in probability* to a random variable X if:

$$\lim_{n \to \infty} P\left(|X_n - X| > \varepsilon\right) = 0 ,$$

$$\text{for every } \varepsilon > 0 .$$

3. $\{X_n\}_{n \in \mathbb{N}}$ exhibits *almost sure* convergence to a random variable X if:

$$P\left(\left\{w|\lim_{n \to \infty} X_n(w) = X(w)\right\}\right) = 1 .$$

4. Suppose that all elements of X_n have a finite **expectancy**. The set $\{X_n\}_{n \in \mathbb{N}}$ converges *in mean square* to X if:

$$\lim_{n \to \infty} E\left[(X_n - X)^2\right] = 0 .$$

Note that:

- Almost sure convergence and mean square convergence both imply a convergence in probability;
- Convergence in probability (weak convergence) implies convergence in distribution.

EXAMPLES

Let X_i be independent random variables uniformly distributed over [0, 1]. We define the following set of random variables from X_i:

$$Z_n = n \cdot \min_{i=1,\dots,n} X_i .$$

We can show that the set $\{Z_n\}_{n\in\mathbb{N}}$ converges in distribution to an **exponential distribution** Z with a parameter of 1 as follows:

$$1 - F_{Z_n}(t) = P(Z_n > t)$$

$$= P\left(\min_{i=1,\dots,n} X_i > \frac{t}{n}\right)$$

$$= P\left(X_1 > \frac{t}{n} \text{ and } X_2 > \frac{t}{n} \text{ and } \dots X_n > \frac{t}{n}\right)$$

$$\overset{\text{ind.}}{=} \prod_{i=1}^{n} P\left(X_i > \frac{t}{n}\right) = \left(1 - \frac{t}{n}\right)^n .$$

Now, for $\lim_{n\to\infty}$, $\left(1 - \frac{t}{n}\right)^n = \exp(-t)$ since:

$$\lim_{n\to\infty} F_{Z_n} = \lim_{n\to\infty} P(Z_n \le t)$$

$$= 1 - \exp(-t) = F_Z .$$

Finally, let S_n be the number of successes obtained during n Bernoulli trials with a probability of success p. **Bernoulli's theorem** tells us that $\frac{S_n}{n}$ converges in probability to a "random" variable that takes the value p with probability 1.

FURTHER READING

▶ **Bernoulli's theorem**
▶ **Central limit theorem**
▶ **Convergence theorem**
▶ **De Moivre–Laplace Theorem**
▶ **Law of large numbers**
▶ **Probability**
▶ **Random variable**
▶ **Stochastic process**

REFERENCES

Le Cam, I.M., Yang, C.L.: Asymptotics in Statistics: Some Basic Concepts. Springer, Berlin Heidelberg New York (1990)

Staudte, R.G., Sheater, S.J.: Robust Estimation and Testing. Wiley, New York (1990)

Convergence Theorem

The convergence theorem leads to the most important theoretical results in probability theory. Among them, we find the **law of large numbers** and the **central limit theorem**.

EXAMPLES

The **central limit theorem** and the **law of large numbers** are both convergence theorems.

The law of large numbers states that the **mean** of a sum of identically distributed random variables converges to their common **mathematical expectation**.

On the other hand, the central limit theorem states that the distribution of the sum of a sufficiently large number of random variables tends to approximate the **normal distribution**.

FURTHER READING

▶ **Central limit theorem**
▶ **Law of large numbers**

Correlation Coefficient

The simple correlation coefficient is a measure of the strength of the linear relation between two **random variables**.

The correlation coefficient can take **values** that occur in the interval $[-1; 1]$. The two extreme values of this interval represent a perfectly linear relation between the **variables**, "positive" in the first case and "negative" in the other. The value 0 (zero) implies the absence of a linear relation.

The correlation coefficient presented here is also called the Bravais–Pearson correlation coefficient.

HISTORY

The concept of correlation originated in the 1880s with the works of **Galton, F.**. In his autobiography *Memories of My Life* (1890), he writes that he thought of this concept during a walk in the grounds of Naworth Castle, when a rain shower forced him to find shelter. According to Stigler, S.M. (1989), Porter, T.M. (1986) was carrying out historical research when he found a forgotten article written by Galton in 1890 in The North American Review, under the title "Kinship and correlation". In this article, which he published right after its discovery, Galton (1908) explained the nature and the importance of the concept of correlation.

This discovery was related to previous works of the mathematician, notably those on heredity and linear regression. Galton had been interested in this field of study since 1860. He published a work entitled "Natural inheritance" (1889), which was the starting point for his thoughts on correlation.

In 1888, in an article sent to the Royal Statistical Society entitled "Co-relations and their measurement chiefly from anthropometric data," Galton used the term "correlation" for the first time, although he was still alternating between the terms "co-relation" and "correlation" and he spoke of a "co-relation index." On the other hand, he invoke the concept of a negative correlation. According to Stigler (1989), Galton only appeared to suggest that correlation was a positive relationship.

Pearson, Karl wrote in 1920 that correlation had been discovered by Galton, whose work "Natural inheritance" (1889) pushed him to study this concept too, along with two other researchers, Weldon and Edgeworth. Pearson and Edgeworth then developed the theory of correlation.

Weldon thought the correlation coefficient should be called the "Galton function." However, Edgeworth replaced Galton's term "co-relation index" and Weldon's term "Galton function" by the term "correlation coefficient."

According to Mudholkar (1982), **Pearson, K.** systemized the analysis of correlation and established a theory of correlation for three **variables**. Researchers from University College, most notably his assistant Yule, G.U., were also interested in developing multiple correlation.

Spearman published the first study on rank correlation in 1904.

Among the works that were carried out in this field, it is worth highlighting those of Yule, who in an article entitled "Why do we sometimes get non-sense-correlation between time-series" (1926) discussed the problem of correlation analysis interpretation. Finally, correlation robustness was investigated by Mosteller and Tukey (1977).

MATHEMATICAL ASPECTS
Simple Linear Correlation Coefficient

Simple linear correlation is the term used to describe a linear dependence between two **quantitative variables** X and Y (see **simple linear regression**).

If X and Y are **random variables** that follow an unknown **joint distribution**, then the simple linear correlation coefficient is equal to the **covariance** between X and Y divided by the product of their **standard deviations**:

$$\rho = \frac{\text{Cov}(X, Y)}{\sigma_X \sigma_Y}.$$

Here $\text{Cov}(X, Y)$ is the measured **covariance** between X and Y; σ_X and σ_Y are the respective **standard deviations** of X and Y.

Given a sample of size n, (X_1, Y_1), $(X_2, Y_2), \ldots, (X_n, Y_n)$ from the joint distribution, the quantity

$$r = \frac{\sum_{i=1}^{n}(X_i - \bar{X})(Y_i - \bar{Y})}{\sqrt{\sum_{i=1}^{n}(X_i - \bar{X})^2 \sum_{i=1}^{n}(Y_i - \bar{Y})^2}}$$

is an **estimation** of ρ; it is the **sampling** correlation.

If we denote $(X_i - \bar{X})$ by x_i and $(Y_i - \bar{Y})$ by y_i, we can write this equation as:

$$r = \frac{\sum_{i=1}^{n} x_i y_i}{\sqrt{\left(\sum_{i=1}^{n} x_i^2\right)\left(\sum_{i=1}^{n} y_i^2\right)}}.$$

Test of Hypothesis

To test the **null hypothesis**

$$H_0: \rho = 0$$

against the **alternative hypothesis**

$$H_1: \rho \neq 0,$$

we calculate the **statistic** t:

$$t = \frac{r}{S_r},$$

where S_r is the **standard deviation** of the estimator r:

$$S_r = \sqrt{\frac{1 - r^2}{n - 2}}.$$

Under H_0, the **statistic** t follows a **Student distribution** with $n - 2$ **degrees of freedom**. For a given significance level α, H_0 is rejected if $|t| \geq t_{\frac{\alpha}{2}, n-2}$; the **value** of $t_{\frac{\alpha}{2}, n-2}$ is the **critical value** of the test given in the **Student table**.

Multiple Correlation Coefficient

Known as the coefficient of determination denoted by R^2, determines whether the hyperplane estimated from a **multiple linear regression** is correctly adjusted to the data points.

The **value** of the multiple **determination coefficient** R^2 is equal to:

$$R^2 = \frac{\text{Explained variation}}{\text{Total variation}}$$
$$= \frac{\sum_{i=1}^{n}(\hat{Y}_i - \bar{Y})^2}{\sum_{i=1}^{n}(Y_i - \bar{Y})^2}.$$

It corresponds to the square of the multiple correlation coefficient. Notice that

$$0 \leq R^2 \leq 1.$$

In the case of **simple linear regression**, the following relation can be derived:

$$r = \text{sign}(\hat{\beta}_1)\sqrt{R^2},$$

where $\hat{\beta}_1$ is the **estimator** of the **regression** coefficient β_1, and it is given by:

$$\hat{\beta}_1 = \frac{\sum_{i=1}^{n}(X_i - \bar{X})(Y_i - \bar{Y})}{\sum_{i=1}^{n}(X_i - \bar{X})^2}.$$

DOMAINS AND LIMITATIONS

If there is a linear relation between two variables, the correlation coefficient is equal to 1 or -1.

A positive relation ($+$) means that the two variables vary in the same direction. If the individuals obtain high scores in the first variable (for example the **independent variable**), they will have a tendency to obtain high scores in the second variable (the **dependant variable**). The opposite is also true.

A negative relation ($-$) means that the individuals that obtain high scores in the first **variable** will have a tendency to obtain low scores in the second one, and vice versa.

Note that if the **variables** are independent the correlation coefficient is equal to zero. The reciprocal conclusion is not necessarily true. The fact that two or more variables are related in a statistical way is not sufficient to conclude that a cause and effect relation exists. The existence of a statistical correlation is not a proof of causality.

Statistics provides numerous correlation coefficients. The choice of which to use for a particular set of **data** depends on different factors, such as:

- The type of scale used to express the variable;
- The nature of the underlying distribution (continuous or discrete);
- The characteristics of the distribution of the scores (linear or nonlinear).

EXAMPLES

The **data** for two **variables** X and Y are shown in the table below:

No of order	X	Y	$x = X - \bar{X}$	$y = Y - \bar{Y}$	xy	x^2	y^2
1	174	64	−1.5	−1.3	1.95	2.25	1.69
2	175	59	−0.5	−6.3	3.14	0.25	36.69
3	180	64	4.5	−1.3	−5.85	20.25	1.69
4	168	62	−7.5	−3.3	24.75	56.25	10.89
5	175	51	−0.5	−14.3	7.15	0.25	204.49
6	170	60	−5.5	−5.3	29.15	30.25	28.09
7	170	68	−5.5	2.7	−14.85	30.25	7.29
8	178	63	2.5	−2.3	−5.75	6.25	5.29
9	187	92	11.5	26.7	307.05	132.25	712.89
10	178	70	2.5	4.7	11.75	6.25	22.09
Total	1755	653			358.5	284.5	1034.1

$$\bar{X} = 175.5 \quad \text{and} \quad \bar{Y} = 65.3.$$

We now perform the necessary calculations to obtain the correlation coefficient between the two variables. Applying the formula gives:

$$r = \frac{\sum_{i=1}^{n} x_i y_i}{\sqrt{\left(\sum_{i=1}^{n} x_i^2\right)\left(\sum_{i=1}^{n} y_i^2\right)}}$$

$$= \frac{358.5}{\sqrt{284.5 \cdot 1034.1}} = 0.66.$$

Test of Hypothesis

We can calculate the estimated **standard deviation** of r:

$$S_r = \sqrt{\frac{1 - r^2}{n - 2}} = \sqrt{\frac{0.56}{10 - 2}} = 0.266.$$

Calculating the **statistic** t gives:

$$t = \frac{r - 0}{S_r} = \frac{0.66}{0.266} = 2.485 \, .$$

If we choose a significance level α of 5%, the **value** from the **Student table**, $t_{0.025,8}$, is equal to 2.306.

Since $|t| = 2.485 > t_{0.025,8} = 2.306$, the **null hypothesis**

$$H_0 : \rho = 0$$

is rejected.

FURTHER READING

▶ **Coefficient of determination**
▶ **Covariance**
▶ **Dependence**
▶ **Kendall rank correlation coefficient**
▶ **Multiple linear regression**
▶ **Regression analysis**
▶ **Simple linear regression**
▶ **Spearman rank correlation coefficient**

REFERENCES

Galton, F.: Co-relations and their measurement, chiefly from anthropological data. Proc. Roy. Soc. Lond. **45**, 135–145 (1888)

Galton, F.: Natural Inheritance. Macmillan, London (1889)

Galton, F.: Kinship and correlation. North Am. Rev. **150**, 419–431 (1890)

Galton, F.: Memories of My Life. Methuen, London (1908)

Mosteller, F., Tukey, J.W.: Data Analysis and Regression: A Second Course in Statistics. Addison-Wesley, Reading, MA (1977)

Mudholkar, G.S.: Multiple correlation coefficient. In: Kotz, S., Johnson, N.L. (eds.) Encyclopedia of Statistical Sciences, vol. 5. Wiley, New York (1982)

Pearson, K.: Studies in the history of statistics and probability. Biometrika **13**, 25–45 (1920). Reprinted in: Pearson, E.S., Kendall, M.G. (eds.) Studies in the History of Statistics and Probability, vol. I. Griffin, London

Porter, T.M.: The Rise of Statistical Thinking, 1820–1900. Princeton University Press, Princeton, NJ (1986)

Stigler, S.: Francis Galton's account of the invention of correlation. Stat. Sci. **4**, 73–79 (1989)

Yule, G.U.: On the theory of correlation. J. Roy. Stat. Soc. **60**, 812–854 (1897)

Yule, G.U. (1926) Why do we sometimes get nonsense-correlations between time-series? A study in sampling and the nature of time-series. J. Roy. Stat. Soc. (**2**) 89, 1–64

Correspondence Analysis

Correspondence analysis is a **data analysis** technique that is used to describe **contingency tables** (or crossed tables). This analysis takes the form of a **graphical representation** of the associations and the "correspondence" between rows and columns.

HISTORY

The theoretical principles of correspondence analysis date back to the works of Hartley, H.O. (1935) (published under his original name Hirschfeld) and of **Fisher, R.A.** (1940) on **contingency tables**. They were first presented in the framework of inferential statistics.

The term "correspondence analysis" first appeared in the autumn of 1962, and the first presentation of this method that referred to this term was given by Benzécri, J.P. in the winter of 1963. In 1976 the works of Benzécri, J.P., which retraced twelve years of his laboratory work, were published, and since then the algebraic and geometrical properties of this descriptive analytical tool have become more widely known and used.

MATHEMATICAL ASPECTS

Consider a contingency table relating to two **categorial qualitative variables** X and Y that have, respectively, r and c **categories**:

	Y_1	Y_2	...	Y_c	Total
X_1	n_{11}	n_{12}	...	n_{1c}	$n_{1.}$
X_2	n_{21}	n_{22}	...	n_{2c}	$n_{2.}$
...	...	...	...	...	...
X_r	n_{r1}	n_{r2}	...	n_{rc}	$n_{r.}$
Total	$n_{.1}$	$n_{.2}$	...	$n_{.c}$	$n_{..}$

where

n_{ij} represents the **frequency** that category i of **variable** X and category j of variable Y is observed,

$n_{i.}$ represents the sum of the observed frequencies for category i of variable X,

$n_{.j}$ represents the sum of the observed frequencies for category j of variable Y,

$n_{..}$ represents the total number of **observations**.

We will assume that $r \geq c$; if not we take the **transpose** of the initial table and use this transpose as the new **contingency table**. The correspondence analysis of a contingency table with more lines than columns, is performed as follows:

1. Tables of row profiles X_I and column profiles X_J are constructed..

   For a fixed line (column), the line (column) profile is the line (column) obtained by dividing each element in this row (column) by the sum of the elements in the line (column).

The line profile of row i is obtained by dividing each term of row i by $n_{i.}$, which is the sum of the observed **frequencies** in the row.

The table of row profiles is constructed by replacing each row of the **contingency table** with its profile:

	Y_1	Y_2	...	Y_c	Total
X_1	$\frac{n_{11}}{n_{1.}}$	$\frac{n_{12}}{n_{1.}}$	...	$\frac{n_{1c}}{n_{1.}}$	1
X_2	$\frac{n_{21}}{n_{2.}}$	$\frac{n_{22}}{n_{2.}}$	...	$\frac{n_{2c}}{n_{2.}}$	1
...	...	...	...	...	...
X_r	$\frac{n_{r1}}{n_{r.}}$	$\frac{n_{r2}}{n_{r.}}$	...	$\frac{n_{rc}}{n_{r.}}$	1
Total	$n'_{.1}$	$n'_{.2}$	...	$n'_{.c}$	r

It is also common to multiply each element of the table by 100 in order to convert the terms into percentages and to make the sum of terms in each row 100%.

The column profile matrix is constructed in a similar way, but this time each column of the **contingency table** is replaced with its profile: the column profile of column j is obtained by dividing each term of column j by $n_{.j}$, which is the sum of **frequencies** observed for the category corresponding to this column.

	Y_1	Y_2	...	Y_c	Total
X_1	$\frac{n_{11}}{n_{.1}}$	$\frac{n_{12}}{n_{.2}}$	...	$\frac{n_{1c}}{n_{.c}}$	$n'_{1.}$
X_2	$\frac{n_{21}}{n_{.1}}$	$\frac{n_{22}}{n_{.2}}$	...	$\frac{n_{2c}}{n_{.c}}$	$n'_{2.}$
...	...	...	...	...	...
X_r	$\frac{n_{r1}}{n_{.1}}$	$\frac{n_{r2}}{n_{.2}}$	...	$\frac{n_{rc}}{n_{.c}}$	$n'_{r.}$
Total	1	1	...	1	c

The tables of row profiles and column profiles correspond to a transformation of the **contingency table** that is used to make the rows and columns comparable.

2. Determine the **inertia matrix** V.
   This is done in the following way:
   - The weighted mean of the r column coordinates is calculated:

   $$g_j = \sum_{i=1}^{r} \frac{n_{i.}}{n_{..}} \cdot \frac{n_{ij}}{n_{i.}} = \frac{n_{.j}}{n_{..}}, j = 1, \ldots, c.$$

   - The c obtained values g_j are written r times in the rows of a matrix G;
   - The diagonal matrix D_I is constructed with diagonal elements of $\frac{n_{i.}}{n_{..}}$;
   - Finally, the **inertia matrix** V is calculated using the following formula:

   $$V = (X_I - G)' \cdot D_I \cdot (X_I - G).$$

3. Using the **matrix** M, which consists of $\frac{n_{..}}{n_{.j}}$ terms on its diagonal and zero terms elsewhere, we determine the matrix C:

   $$C = \sqrt{M} \cdot V \cdot \sqrt{M}.$$

4. Find the **eigenvalues** (denoted k_l) and the **eigenvectors** (denoted v_l) of this matrix C.
   The c **eigenvalues** $k_c, k_{c-1}, \ldots, k_1$ (written in decreasing order) are the **inertia**. The corresponding **eigenvectors** are called the **factorial axes** (or axes of inertia).
   For each **eigenvalue**, we calculate the corresponding inertia explained by the **factorial axis**. For example, the first factorial axis explains:

   $$\frac{100 \cdot k_1}{\sum_{l=1}^{c} k_l} \text{ (in \%) of \textbf{inertia}.}$$

   In the same way, the two first **factorial axes** explain:

   $$\frac{100 \cdot (k_1 + k_2)}{\sum_{l=1}^{c} k_l} \text{ (in \%) of \textbf{inertia}.}$$

If we want to know, for example, the number of **eigenvalues** and therefore the factorial axes that explain at least 3/4 of the total inertia, we sum the explained inertia from each of the eigenvalues until we obtain 75%.
We then calculate the main axes of inertia from these factorial axes.

5. The main axes of **inertia**, denoted u_l, are then given by:

   $$u_l = \sqrt{M^{-1}} \cdot v_l,$$

   meaning that its jth component is:

   $$u_{jl} = \sqrt{\frac{n_{.j}}{n_{..}}} \cdot v_{jl}.$$

6. We then calculate the main components, denoted by y_k, which are the orthogonal projections of the row coordinates on the main axes of inertia: the ith **coordinate** of the lth main component takes the following **value**:

   $$y_{il} = x_i \cdot M \cdot u_l,$$

   meaning that

   $$y_{il} = \sum_{j=1}^{c} \frac{n_{ij}}{n_{i.}} \sqrt{\frac{n_{..}}{n_{.j}}} \cdot v_{jl}$$

   is the **coordinate** of row i on the lth axis.

7. After the main components y_l (of the row coordinates) have been calculated, we determine the main components of the column coordinates (denoted by z_l) using the y_l, thanks to the transaction formulae:

   $$z_{jl} = \frac{1}{\sqrt{k_l}} \sum_{i=1}^{r} \frac{n_{ij}}{n_{.j}} \cdot y_{il},$$

   $$\text{for } j = 1, 2, \ldots, c;$$

   $$y_{il} = \frac{1}{\sqrt{k_l}} \sum_{i=1}^{c} \frac{n_{ij}}{n_{i.}} \cdot z_{jl},$$

   $$\text{for } i = 1, 2, \ldots, r.$$

z_{jl} is the **coordinate** of the column j on the lth **factorial axis**.

8. The simultaneous representation of the row coordinates and the column coordinates on a **scatter plot** with two **factorial axes**, gives a signification to the axis depending on the points it is related to.

The quality of the representation is related to the proportion of the inertia explained by the two main axes used. The closer the explained **inertia** is to 1, the better the quality.

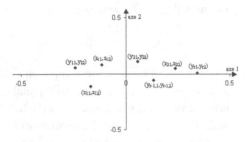

9. It can be useful to insert additional point-rows or column coordinates (illustrative **variables**) along with the active variables used for the correspondence analysis. Consider an extra row-coordinate:

$$(n_{s1}, n_{s2}, n_{s3}, \ldots, n_{sc})$$

of profile-row

$$\left(\frac{n_{s1}}{n_{s.}}, \frac{n_{s2}}{n_{s.}}, \ldots, \frac{n_{sc}}{n_{s.}} \right).$$

The lth main component of the extra row-dot is given by the second formula of point 7):

$$y_{il} = \frac{1}{\sqrt{k_l}} \sum_{j=1}^{c} \frac{n_{sj}}{n_{s.}} \cdot z_{jl},$$

$$\text{for } i = 1, 2, \ldots, r.$$

We proceed the same way for a column coordinate, but apply the first formula of point 7 instead.

DOMAINS AND LIMITATIONS

When simultaneously representing the row coordinate and the column coordinate, it is not advisable to interpret eventual proximities crossed between lines and columns because the two points are not in the same initial space. On the other hand it is interesting to interpret the position of a row coordinate by comparing it to the set of column coordinate (and vice versa).

EXAMPLES

Consider the example of a company that wants to find out how healthy its staff. One of the subjects covered in the questionnaire concerns the number of medical visits (dentists included) per year. Three possible answers are proposed:

- Between 0 and 6 visits;
- Between 7 and 12 visits;
- More than 12 visits per year.

The staff questioned are distributed into five **categories**:

- Managerial staff members over 40;
- Managerial staff members under 40;
- Employees over 40;
- Employees under 40;
- Office personnel.

The results are reported in the following **contingency table**, to which margins have been added:

	Number of visits per year			
	0 to 6	7 to 12	> 12	Total
Managerial staff members > 40	5	7	3	15
Managerial staff members < 40	5	5	2	12

	Number of visits per year			
	0 to 6	7 to 12	> 12	Total
Employees > 40	2	12	6	20
Employees < 40	24	18	12	54
Office personnel	12	8	4	24
Total	48	50	27	125

Using these **data** we will describe the eight main steps that should be followed to obtain a **graphical representation** of employee health via correspondence analysis.

1. We first determine the table of line profiles matrix X_I, by dividing each element by the sum of the elements of the line in which it is located:

$$\begin{array}{cccc} 0.333 & 0.467 & 0.2 & 1 \\ 0.417 & 0.417 & 0.167 & 1 \\ 0.1 & 0.6 & 0.3 & 1 \\ 0.444 & 0.333 & 0.222 & 1 \\ 0.5 & 0.333 & 0.167 & 1 \\ 1.794 & 2.15 & 1.056 & 5 \end{array}$$

and the table of column profiles by dividing each element by the sum of the elements in the corresponding column:

$$\begin{array}{cccc} 0.104 & 0.14 & 0.111 & 0.355 \\ 0.104 & 0.1 & 0.074 & 0.278 \\ 0.042 & 0.24 & 0.222 & 0.504 \\ 0.5 & 0.36 & 0.444 & 1.304 \\ 0.25 & 0.16 & 0.148 & 0.558 \\ 1 & 1 & 1 & 3 \end{array}$$

2. We then calculate the matrix of inertia V for the **frequencies** (given by $\frac{n_{i.}}{n_{..}}$ for values of i from 1 to 5) by proceeding in the following way:

- We start by calculating the weighted **mean** of the five line-dots:

$$g_j = \sum_{i=1}^{5} \frac{n_{i.}}{n_{..}} \cdot \frac{n_{ij}}{n_{i.}} = \frac{n_{.j}}{n_{..}}$$

$$\text{for } j = 1, 2 \text{ and } 3;$$

- We then write these three values five times in the **matrix** G, as shown below:

$$G = \begin{bmatrix} 0.384 & 0.400 & 0.216 \\ 0.384 & 0.400 & 0.216 \\ 0.384 & 0.400 & 0.216 \\ 0.384 & 0.400 & 0.216 \\ 0.384 & 0.400 & 0.216 \end{bmatrix}.$$

- We then construct a diagonal **matrix** D_I containing the $\frac{n_{i.}}{n_{..}}$;
- Finally, we calculate the **matrix of inertia** V, as given by the following formula:

$$V = (X_I - G)' \cdot D_I \cdot (X_I - G),$$

which, in this case, gives:

$$V = \begin{bmatrix} 0.0175 & -0.0127 & -0.0048 \\ -0.0127 & 0.0097 & 0.0029 \\ -0.0048 & 0.0029 & 0.0019 \end{bmatrix}.$$

3. We define a third-order square **matrix** M that contains $\frac{n_{..}}{n_{.j}}$ terms on its diagonal and zero terms everywhere else:

$$M = \begin{bmatrix} 2.604 & 0 & 0 \\ 0 & 2.5 & 0 \\ 0 & 0 & 4.630 \end{bmatrix}.$$

The square root of M, denoted $\sqrt{M}$, is obtained by taking the square root of each diagonal element of M. Using this new **matrix**, we determine

$$C = \sqrt{M} \cdot V \cdot \sqrt{M}$$

$$C = \begin{bmatrix} 0.0455 & -0.0323 & -0.0167 \\ -0.0323 & 0.0243 & 0.0100 \\ -0.0167 & 0.0100 & 0.0087 \end{bmatrix}.$$

C

4. The **eigenvalues** of C are obtained by diagonalizing the **matrix**. Arranging these values in decreasing order gives:

$$k_1 = 0.0746$$
$$k_2 = 0.0039$$
$$k_3 = 0.$$

The explained inertia is determined for each of these values. For example:

$$k_1 : \frac{0.0746}{0.0746 + 0.0039} = 95.03\%,$$

$$k_2 : \frac{0.0039}{0.0746 + 0.0039} = 4.97\%.$$

For the last one, k_3, the explained inertia is zero.

The first two **factorial axes**, associated with the **eigenvalues** k_1 and k_2, explain all of the inertia. Since the third eigenvalue is zero it is not necessary to calculate the **eigenvector** that is associated with it. We focus on calculating the first two normalized eigenvectors then:

$$v_1 = \begin{bmatrix} 0.7807 \\ -0.5576 \\ -0.2821 \end{bmatrix} \quad \text{and}$$

$$v_2 = \begin{bmatrix} 0.0807 \\ 0.5377 \\ -0.8393 \end{bmatrix}.$$

5. We then calculate the main **axes** of inertia by:

$$u_i = \sqrt{M^{-1}} \cdot v_i, \quad \text{for} \quad i = 1 \text{ and } 2,$$

where $\sqrt{M^{-1}}$ is obtained by inverting the diagonal elements of $\sqrt{M}$.

We find:

$$u_1 = \begin{bmatrix} 0.4838 \\ -0.3527 \\ -0.1311 \end{bmatrix} \quad \text{and}$$

$$u_2 = \begin{bmatrix} 0.0500 \\ 0.3401 \\ -0.3901 \end{bmatrix}.$$

6. We then calculate the main components by projecting the rows onto the main axes of inertia. Constructing an auxiliary **matrix** U formed from the two **vectors** u_1 and u_2, we define:

$$Y = X_I \cdot M \cdot U$$
$$= \begin{bmatrix} -0.1129 & 0.0790 \\ 0.0564 & 0.1075 \\ -0.5851 & 0.0186 \\ 0.1311 & -0.0600 \\ 0.2349 & 0.0475 \end{bmatrix}.$$

We can see the coordinates of the five rows written horizontally in this **matrix** Y; for example, the first column indicates the components related to the first **factorial axis** and the second indicates the components related to the second axis.

7. We then use the coordinates of the rows to find those of the columns via the following transition formulae written as matrices:

$$Z = K \cdot Y' \cdot X_J,$$

where:

K is the second-order diagonal **matrix** that has $1/\sqrt{k_i}$ terms on its diagonal and zero terms elsewhere;

Y' is the transpose of the matrix containing the coordinates of the rows, and;

X_J is the column profile matrix.

We obtain the matrix:

$$Z = \begin{bmatrix} 0.3442 & -0.2409 & -0.1658 \\ 0.0081 & 0.0531 & -0.1128 \end{bmatrix},$$

where each column contains the components of one of the three column coordinates; for example, the first line corresponds to each coordinate on the first **factorial axis** and the second line to each coordinate on the second axis.

We can verify the transition formula that gives Y from Z:

$$Y = X_1 \cdot Z' \cdot K$$

using the same notation as seen previously.

8. We can now represent the five **categories** of people questioned and the three categories of answers proposed on the same factorial plot:

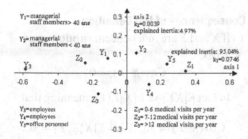

We can study this factorial plot at three different levels of analysis, depending:

- The set of **categories** for the people questioned;
- The set of modalities for the medical visits;
- Both at the same time.

In the first case, close proximity between two rows (between two **categories** of personnel) signifies similar medical visit profiles. On the factorial plot, this is the case for the

employees under 40 (Y_4) and the office personnel (Y_5). We can verify from the table of line profiles that the **percentages** for these two categories are indeed very similar.

Similarly, the proximity between two columns (representing two **categories** related to the number of medical visits) indicates similar distributions of people within the business for these categories. This can be seen for the modalities Z_2 (from 7 to 12 medical visits per year) and Z_3 (more than 12 visits per year).

If we consider the rows and the columns simultaneously (and not separately as we did previously), it becomes possible to identify similarities between categories for certain modalities. For example, the employees under 40 (Y_4) and the office personnel (Y_5) seem to have the same behavior towards health: high proportions of them (0.44 and 0.5 respectively) go to the doctor less than 6 times per year (Z_1).

In conclusion, axis 1 is confronted on one side with the categories indicating an average or high number of visits (Z_2 and Z_3)—the employees or the managerial staff members over 40 (Y_1 and Y_3)—and on the other side with the modalities associated with low numbers of visits (Z_1): Y_2, Y_4 and Y_5. The first factor can then be interpreted as the importance of medical control according to age.

FURTHER READING

- ▶ **Contingency table**
- ▶ **Data analysis**
- ▶ **Eigenvalue**
- ▶ **Eigenvector**
- ▶ **Factorial axis**
- ▶ **Graphical representation**
- ▶ **Inertia matrix**
- ▶ **Matrix**
- ▶ **Scatterplot**

REFERENCES

Benzécri, J.P.: L'Analyse des données. Vol. 1: La Taxinomie. Vol. 2: L'Analyse factorielle des correspondances. Dunod, Paris (1976)

Benzécri, J.P.: Histoire et préhistoire de l'analyse des données, les cahiers de l'analyse des données 1, no. 1–4. Dunod, Paris (1976)

Fisher, R.A.: The precision of discriminant functions. Ann. Eugen. (London) **10**, 422–429 (1940)

Greenacre, M.: Theory and Applications of Correspondence Analysis. Academic, London (1984)

Hirschfeld, H.O.: A connection between correlation and contingency. Proc. Camb. Philos. Soc. **31**, 520–524 (1935)

Lebart, L., Salem, A.: Analyse Statistique des Données Textuelles. Dunod, Paris (1988)

Saporta, G.: Probabilité, analyse de données et statistiques. Technip, Paris (1990)

Covariance

The covariance between two **random variables** X and Y is the measure of how much two random variables vary together.

If X and Y are independent **random variables**, the covariance of X and Y is zero. The converse, however, is not true.

MATHEMATICAL ASPECTS

Consider X and Y, two **random variables** defined in the same **sample space** Ω. The covariance of X and Y, denoted by Cov(X, Y), is defined by

$$\text{Cov}(X, Y) = E[(X - E[X])(Y - E[Y])],$$

where $E[.]$ is the **expected value**.

Developing the right side of the equation gives:

$$\begin{aligned}
\text{Cov}(X, Y) &= E[XY - E[X]Y - XE[Y] \\
&\quad + E[X]E[Y]] \\
&= E[XY] - E[X]E[Y] \\
&\quad - E[X]E[Y] + E[X]E[Y] \\
&= E[XY] - E[X]E[Y].
\end{aligned}$$

Properties of Covariance

Consider X, Y and Z, which are **random variables** defined in the same **sample space** Ω, and a, b, c and d, which are constants. We find that:

1. $\text{Cov}(X, Y) = \text{Cov}(Y, X)$
2. $\text{Cov}(X, c) = 0$
3. $\text{Cov}(aX + bY, Z) = a\,\text{Cov}(X, Z) + b\text{Cov}(Y, Z)$
4. $\text{Cov}(X, cY + dZ) = c\,\text{Cov}(X, Y) + d\text{Cov}(X, Z)$
5. $\text{Cov}(aX + b, cY + d) = ac\,\text{Cov}(X, Y)$.

Consequences of the Definition

1. If X and Y are independent **random variables**,

$$\text{Cov}(X, Y) = 0.$$

In fact $E[XY] = E[X]E[Y]$, meaning that:

$$\text{Cov}(X, Y) = E[XY] - E[X]E[Y] = 0.$$

The reverse is not generally true: $\text{Cov}(X, Y) = 0$ does not necessarily imply that X and Y are independent.

2. $\text{Cov}(X, X) = \text{Var}(X)$
   where $\text{Var}(X)$ represents the **variance** of X.
   In fact:

$$\begin{aligned}
\text{Cov}(X, X) &= E[XX] - E[X]E[X] \\
&= E[X^2] - (E[X])^2 \\
&= \text{Var}(X).
\end{aligned}$$

DOMAINS AND LIMITATIONS

Consider two **random variables** X and Y, and their sum $X + Y$. We then have:

$$E[X + Y] = E[X] + E[Y] \quad \text{and}$$
$$\text{Var}(X + Y) = \text{Var}(X) + \text{Var}(Y)$$
$$+ 2\text{Cov}(X, Y).$$

We now show these results for discrete variables. If $P_{ji} = P(X = x_i, Y = y_j)$ we have:

$$E[X + Y] = \sum_i \sum_j (x_i + y_j)P_{ji}.$$

$$= \sum_i \sum_j x_i P_{ji} + \sum_i \sum_j y_j P_{ji}$$

$$= \sum_i x_i \left(\sum_j P_{ji} \right)$$

$$+ \sum_j y_j \left(\sum_i P_{ji} \right)$$

$$= \sum_i x_i P_i + \sum_j y_j P_j$$

$$= E[X] + E[Y].$$

Moreover:

$$\text{Var}(X + Y) = E[(X + Y)^2] - (E[X + Y])^2$$
$$= E[X^2] + 2E[XY] + E[Y^2]$$
$$- (E[X + Y])^2$$
$$= E[X^2] + 2E[XY] + E[Y^2]$$
$$- (E[X] + E[Y])^2$$
$$= E[X^2] + 2E[XY] + E[Y^2]$$
$$- (E[X])^2 - 2E[X]E[Y] - (E[Y])^2$$
$$= \text{Var}(X) + \text{Var}(Y) + 2(E[XY]$$
$$- E[X]E[Y])$$
$$= \text{Var}(X) + \text{Var}(Y) + 2\text{Cov}(X, Y).$$

These results can be generalized for n **random variables** $X_1, X_2, \ldots, X_n$, with x_i

having an **expected value** equal to $E(X_i)$ and a **variance** equal to $\text{Var}(X_i)$. We then have:

$$E[X_1 + X_2 + \cdots + X_n]$$
$$= E[X_1] + E[X_2] + \cdots + E[X_n]$$
$$= \sum_{i=1}^{n} E[X_i].$$

$$\text{Var}(X_1 + X_2 + \cdots + X_n)$$
$$= \text{Var}(X_1) + \text{Var}(X_2) + \cdots + \text{Var}(X_n)$$
$$+ 2[\text{Cov}(X_1, X_2) + \cdots + \text{Cov}(X_1, X_n)$$
$$+ \text{Cov}(X_2, X_3) + \cdots + \text{Cov}(X_2, X_n)$$
$$+ \cdots + \text{Cov}(X_{n-1}, X_n)]$$
$$= \sum_{i=1}^{n} \text{Var}(X_i) + 2 \sum_{i=1}^{n-1} \sum_{j>i} \text{Cov}(X_i, X_j).$$

EXAMPLES

Consider two psychological tests carried out in succession. Each subject receives a grade X of between 0 and 3 for the first test and a grade Y of between 0 and 2 for the second test. Given that the **probabilities** of X being equal to 0, 1, 2 and 3 are respectively 0.16, 0.3, 0.41 and 0.13, and that the probabilities of Y being equal to 0, 1 and 2 are respectively 0.55, 0.32 and 0.13, we have:

$$E[X] = 0 \cdot 0.16 + 1 \cdot 0.3 + 2 \cdot 0.41$$
$$+ 3 \cdot 0.13$$
$$= 1.51$$
$$E[Y] = 0 \cdot 0.55 + 1 \cdot 0.32 + 2 \cdot 0.13$$
$$= 0.58$$
$$E[XY] = 0 \cdot 0 \cdot 0.16 \cdot 0.55$$
$$+ 0 \cdot 1 \cdot 0.16 \cdot 0.32 + \ldots$$
$$+ 3 \cdot 2 \cdot 0.13 \cdot 0.13$$
$$= 0.88.$$

We can then calculate the covariance of X and Y:

$$\text{Cov}(X, Y) = E[XY] - E[X]E[Y]$$
$$= 0.88 - (1.51 \cdot 0.58)$$
$$= 0.00428.$$

FURTHER READING
▸ **Correlation coefficient**
▸ **Expected value**
▸ **Random variable**
▸ **Variance of a random variable**

Covariance Analysis

Covariance analysis is a method used to estimate and test the effects of **treatments**. It checks whether there is a significant difference between the **means** of several treatments by taking into account the observed **values** of the **variable** before the treatment. **Covariance analysis** is a precise way of performing **treatment** comparisons because it involves adjusting the response variable Y to a concomitant variable X which corresponds to the **values** observed before the treatment.

HISTORY
Covariance analysis dates back to 1930. It was first developed by **Fisher, R.A.** (1932). After that, other authors applied covariance analysis to agricultural and medical problems. For example, Bartlett, M.S. (1937) applied covariance analysis to his studies on cotton cultivation in Egypt and on milk yields from cows in winter.
Delurry, D.B. (1948) used covariance analysis to compare the effects of different medications (atropine, quinidine, atrophine) on rat muscles.

MATHEMATICAL ASPECTS
We consider here a covariance analysis of a **completely randomized design** implying one **factor**.
The linear **model** that we will consider is the following:

$$Y_{ij} = \mu + \tau_i + \beta X_{ij} + \varepsilon_{ij},$$
$$i = 1, 2, \ldots, t, \quad j = 1, 2, \ldots, n_i$$

where

Y_{ij} represents **observation** j, receiving **treatment** i,

μ is the general **mean** common to all treatments,

τ_i is the actual effect of treatment i on th observation,

X_{ij} is the **value** of the concomitant variable, and

ε_{ij} is the experimental error in observation Y_{ij}.

Calculations
In order to calculate the F ratio that will help us to determine whether there is a significant difference between **treatments**, we need to work out sums of squares and sums of products. Therefore, if $\bar{X}_{i.}$ and $\bar{Y}_{i.}$ are respectively the **means** of the X values and the Y values for treatment i, and if $\bar{X}_{..}$ and $\bar{Y}_{..}$ are respectively the means of all the **values** of X and Y, we obtain the formulae given below.
1. The total sum of squares for X:

$$S_{XX} = \sum_{i=1}^{t} \sum_{j=1}^{n_i} (X_{ij} - \bar{X}_{..})^2.$$

2. The total sum of squares for Y (S_{YY}) is calculated in the same way as S_{XX}, but X is replaced by Y.

3. The total sum of products of X and Y:

$$S_{XY} = \sum_{i=1}^{t} \sum_{j=1}^{n_i} (X_{ij} - \bar{X}_{..})(Y_{ij} - \bar{Y}_{..}).$$

4. The sum of squares of the treatments for X:

$$T_{XX} = \sum_{i=1}^{t} \sum_{j=1}^{n_i} (\bar{X}_{i.} - \bar{X}_{..})^2.$$

5. The sum of squares of the treatments for Y (T_{YY})

is calculated in the same way as T_{XX}, but X is replaced by Y.

6. The sum of the products of the treatments of X and Y:

$$T_{XY} = \sum_{i=1}^{t} \sum_{j=1}^{n_i} (\bar{X}_{i.} - \bar{X}_{..})(\bar{Y}_{i.} - \bar{Y}_{..}).$$

7. The sum of squares of the **errors** for X:

$$E_{XX} = \sum_{i=1}^{t} \sum_{j=1}^{n_i} (X_{ij} - \bar{X}_{i.})^2.$$

8. The sum of the squares of the **errors** for Y: is calculated in the same way as E_{XX}, but X is replaced by Y.

9. The sum of products of the **errors** X and Y:

$$E_{XY} = \sum_{i=1}^{t} \sum_{j=1}^{n_i} (X_{ij} - \bar{X}_{i.})(Y_{ij} - \bar{Y}_{i.}).$$

Substituting in appropriate values and calculating these formulae corresponds to an **analysis of variance** for each of X, Y and XY. The **degrees of freedom** associated with these different formulae are as follows:

1. For the total sum: $\sum_{i=1}^{t} n_i - 1.$

2. For the sum of the **treatments**: $t - 1$.

3. For the sum of the **errors**: $\sum_{i=1}^{t} n_i - t.$

Adjustment of the **variable** Y to the concomitant variable X yields two new sums of squares:

1. The adjusted total sum of squares:

$$SS_{tot} = S_{YY} - \frac{S_{XY}^2}{S_{XX}}.$$

2. The adjusted sum of the squares of the errors:

$$SS_{err} = E_{YY} - \frac{E_{XY}^2}{E_{XX}}.$$

The new **degrees of freedom** for these two sums are:

1. $\sum_{i=1}^{t} n_i - 2$;
2. $\sum_{i=1}^{t} n_i - t - 1$, where a **degree of freedom** is subtracted due to the adjustment. The third adjusted sum of squares, the adjusted sum of the squares of the **treatments**, is given by:

$$SS_{tr} = SS_{tot} - SS_{err}.$$

This has the same number of **degrees of freedom** $t - 1$ as before.

Covariance Analysis Table

We now have all of the elements needed to establish the covariance analysis table. The sums of squares divided by the number of **degrees of freedom** gives the **means** of the squares.

Source of variation	Degrees of freedom	Sum of squares and of products		
		$\sum_{i=1}^{t} x_i^2$	$\sum_{i=1}^{t} x_i y_i$	$\sum_{i=1}^{t} y_i^2$
Treatments	$t - 1$	T_{XX}	T_{XY}	T_{YY}
Errors	$\sum_{i=1}^{t} n_i - t$	E_{XX}	E_{XY}	E_{YY}
Total	$\sum_{i=1}^{t} n_i - 1$	S_{XX}	S_{XY}	S_{YY}

Note: the numbers in the $\sum x_i^2$ and $\sum y_i^2$ columns cannot be negative; on the other hand, the numbers in the $\sum_{i=1}^{t} x_i y_i$ column can be negative.

Adjustment

Source of variation	Degrees of freedom	Sum of squares	Mean of squares
Treatments	$t-1$	SS_{tr}	MC_{tr}
Errors	$\sum_{i=1}^{t} n_i - t - 1$	SS_{err}	MC_{err}
Total	$\sum_{i=1}^{t} n_i - 2$	TSS	

F Ratio: Testing the Treatments

The *F* **ratio,** used to test the **null hypothesis** that there is a significant difference between the **means** of the **treatments** once adjusted to the variable X, is given by:

$$F = \frac{MC_{tr}}{MC_{err}}.$$

The ratio follows a **Fisher distribution** with $t-1$ and $\sum_{i=1}^{t} n_i - t - 1$ **degrees of freedom**. The **null hypothesis**

$$H_0: \tau_1 = \tau_2 = \ldots = \tau_t$$

will be rejected at the significance level α if the *F* ratio is superior or equal to the **value** of the **Fisher table**, in other words if:

$$F \geq F_{t-1, \sum_{i=1}^{t} n_i - t - 1, \alpha}.$$

It is clear that we assume that the β coefficient is different from zero when performing covariance analysis. If this is not the case, a simple **analysis of variance** is sufficient.

Test Concerning the β Slope

So, we would like to know whether there is a significant effect of the concomitant vari-able before the application of the treatment. To test this **hypothesis**, we will assume the **null hypothesis** to be

$$H_0: \beta = 0$$

and the **alternative hypothesis** to be

$$H_1: \beta \neq 0.$$

The *F* ratio can be established:

$$F = \frac{E_{XY}^2 / E_{XX}}{MC_{err}}.$$

It follows a **Fisher distribution** with 1 and $\sum_{i=1}^{t} n_i - t - 1$ **degrees of freedom**. The **null hypothesis** will be rejected at the significance level α if the *F* ratio is superior to or equal to the value in the **Fisher table**, in other words if:

$$F \geq F_{1, \sum_{i=1}^{t} n_i - t - 1, \alpha}.$$

DOMAINS AND LIMITATIONS

The basic **hypotheses** that need to be constructed before initiating a covariance analysis are the same as those used for an **analysis of variance** or a **regression analysis**. These are the hypotheses of normality, homogeneity (homoscedasticity), **variances** and independence.

In covariance analysis, as in an analysis of variance, the **null hypothesis** stipulates that the independent samples come from different **populations** that have identical **means**. Moreover, since there are always conditions associated with any statistical technique, those that apply to covariance analysis are as follows:

1. The **population** distributions must be approximately normal, if not completely normal.

2. The populations from which the **samples** are taken must have the same **variance** σ^2, meaning:

$$\sigma_1^2 = \sigma_2^2 = \ldots = \sigma_k^2,$$

where k is the number of populations to be compared.

3. The samples must be chosen randomly and all of the samples must be independent.

We must also add a basic **hypothesis** specific to covariance analysis, which is that the **treatments** that were carried out must not influence the **values** of the concomitant **variable** X.

EXAMPLES

Consider an **experiment** consisting of comparing the effects of three different diets on a **population** of cows.

The **data** are presented in the form of a table, which includes the three different diets, each of which was administered to five porch. The initial weights are denoted by the concomitant variable X (in kg), and the gains in weight (after **treatment**) are denoted by Y:

Diets

	1		2		3
X	Y	X	Y	X	Y
32	167	26	182	36	158
29	172	33	171	34	191
22	132	22	173	37	140
23	158	28	163	37	192
35	169	22	182	32	162

We first calculate the various sums of squares and products:

1. The total sum of squares for X:

$$S_{XX} = \sum_{i=1}^{3}\sum_{j=1}^{5}(X_{ij} - \bar{X}_{..})^2$$

$$= (32 - 29.8667)^2 + \ldots$$
$$+ (32 - 29.8667)^2$$
$$= 453.73.$$

2. The total sum of squares for Y:

$$S_{YY} = \sum_{i=1}^{3}\sum_{j=1}^{5}(Y_{ij} - \bar{Y}_{..})^2$$

$$= (167 - 167.4667)^2 + \ldots$$
$$+ (162 - 167.4667)^2$$
$$= 3885.73.$$

3. The total sum of the products of X and Y:

$$S_{XY} = \sum_{i=1}^{3}\sum_{j=1}^{5}(X_{ij} - \bar{X}_{..})(Y_{ij} - \bar{Y}_{..})$$

$$= (32 - 29.8667)$$
$$\cdot (167 - 167.4667) + \ldots$$
$$+ (32 - 29.8667)$$
$$\cdot (162 - 167.4667)$$
$$= 158.93.$$

4. The sum of the squares of the treatments for X:

$$T_{XX} = \sum_{i=1}^{3}\sum_{j=1}^{5}(\bar{X}_{i.} - \bar{X}_{..})^2$$

$$= 5(28.2 - 29.8667)^2$$
$$+ 5(26.2 - 29.8667)^2$$
$$+ 5(35.2 - 29.8667)^2$$
$$= 223.33.$$

5. The sum of the squares of the treatments for Y:

$$T_{YY} = \sum_{i=1}^{3}\sum_{j=1}^{5}(\bar{Y}_{i.} - \bar{Y}_{..})^2$$

$$= 5(159.6 - 167.4667)^2$$
$$+ 5(174.2 - 167.4667)^2$$
$$+ 5(168.6 - 167.4667)^2$$
$$= 542.53.$$

6. The sum of the products of the treatments of X and Y:

$$T_{XY} = \sum_{i=1}^{3}\sum_{j=1}^{5}(\bar{X}_{i.} - \bar{X}_{..})(\bar{Y}_{i.} - \bar{Y}_{..})$$

$$= 5(28.2 - 29.8667)$$
$$\cdot (159.6 - 167.4667) + \ldots$$
$$+ 5(35.2 - 29.8667)$$
$$\cdot (168.6 - 167.4667)$$
$$= -27.67.$$

7. The sum of the squares of the errors for X:

$$E_{XX} = \sum_{i=1}^{3}\sum_{j=1}^{5}(X_{ij} - \bar{X}_{i.})^2$$

$$= (32 - 28.2)^2 + \ldots + (32 - 35.2)^2$$
$$= 230.40.$$

8. The sum of the squares of the errors for Y:

$$E_{YY} = \sum_{i=1}^{3}\sum_{j=1}^{5}(Y_{ij} - \bar{Y}_{i.})^2$$

$$= (167 - 159.6)^2 + \ldots$$
$$+ (162 - 168.6)^2$$
$$= 3343.20.$$

9. The sum of the products of the errors of X and Y:

$$E_{XY} = \sum_{i=1}^{3}\sum_{j=1}^{5}(X_{ij} - \bar{X}_{i.})(Y_{ij} - \bar{Y}_{i.})$$

$$= (32 - 28.2)(167 - 159.6) + \ldots$$
$$+ (32 - 35.2)(162 - 168.6)$$
$$= 186.60.$$

The **degrees of freedom** associated with these different calculations are as follows:
1. For the total sums:

$$\sum_{i=1}^{3} n_i - 1 = 15 - 1 = 14.$$

2. For the sums of **treatments**:

$$t - 1 = 3 - 1 = 2.$$

3. For the sums of **errors**:

$$\sum_{i=1}^{3} n_i - t = 15 - 3 = 12.$$

Adjusting the **variable** Y to the concomitant variable X yields two new sums of squares:
1. The total adjusted sum of squares:

$$SS_{\text{tot}} = S_{YY} - \frac{S_{XY}^2}{S_{XX}}$$

$$= 3885.73 - \frac{158.93^2}{453.73}$$

$$= 3830.06.$$

2. The adjusted sum of the squares of the errors:

$$SS_{\text{err}} = E_{YY} - \frac{E_{XY}^2}{E_{XX}}$$

$$= 3343.20 - \frac{186.60^2}{230.40}$$

$$= 3192.07.$$

The new **degrees of freedom** for these two sums are:
1. $\sum_{i=1}^{3} n_i - 2 = 15 - 2 = 13$;
2. $\sum_{i=1}^{3} n_i - t - 1 = 15 - 3 - 1 = 11$.
The adjusted sum of the squares of the **treatments** is given by:

$$SS_{\text{tr}} = SS_{\text{tot}} - SS_{\text{err}}$$

$$= 3830.06 - 3192.07$$

$$= 637.99.$$

This has the same number of **degrees of freedom** as before:

$$t - 1 = 3 - 1 = 2.$$

We now have all of the elements required in order to establish a covariance analysis table. The sums of squares divided by the **degrees of freedom** gives the **means** of the squares.

Source of variation	Degrees of freedom	$\sum_{i=1}^{3} x_i^2$	$\sum_{i=1}^{3} x_i y_i$	$\sum_{i=1}^{3} y_i^2$
Treatments	2	223.33	−27.67	543.53
Errors	12	230.40	186.60	3343.20
Total	14	453.73	158.93	3885.73

Adjustment

Source of variation	Degrees of freedom	Sum of squares	Mean of squares
Treatments	2	637.99	318.995
Errors	11	3192.07	290.188
Total	13	3830.06	

The F **ratio**, which is used to test the **null hypothesis** that there is no significant difference between the **means** of the **treatments** once adjusted to the variable Y, is given by:

$$F = \frac{MC_{\text{tr}}}{MC_{\text{err}}} = \frac{318.995}{290.188} = 1.099.$$

If we choose a significance level of $\alpha = 0.05$, the value of F in the **Fisher table** is equal to:

$$F_{2,11,0.05} = 3.98.$$

Since $F < F_{2,11,0.05}$, we cannot reject the **null hypothesis** and so we conclude that there is no significant difference between the responses to the three diets once the **variable** Y is adjusted to the initial weight X.

FURTHER READING

▶ **Analysis of variance**
▶ **Design of experiments**
▶ **Missing data analysis**
▶ **Regression analysis**

C

REFERENCES

Bartlett, M.S.: Some examples of statistical methods of research in agriculture and applied biology. J. Roy. Stat. Soc. (Suppl.) 4, 137–183 (1937)

DeLury, D.B.: The analysis of covariance. Biometrics 4, 153–170 (1948)

Fisher, R.A.: Statistical Methods for Research Workers. Oliver & Boyd, Edinburgh (1925)

Huitema, B.E.: The Analysis of Covariance and Alternatives. Wiley, New York (1980)

Wildt, A.R., Ahtola, O.: Analysis of Covariance (Sage University Papers Series on Quantitative Applications in the Social Sciences, Paper 12). Sage, Thousand Oaks, CA (1978)

Covariation

It is often interesting, particularly in economics, to compare two **time series**.

Since we wish to measure the level of **dependence** between two **variables**, this is somewhat reminiscent of the concept of correlation. However, in this case, since the **time series** are bound by a third variable, time, finding the **correlation coefficient** would only lead to an artificial relation.

Indeed, if two **time series** are considered, x_t and y_t, which represent completely independent phenomena and are linear functions

of time:
$$x_t = a \cdot t + b,$$
$$y_t = c \cdot t + d,$$

where a, b, c and d are constants, it is always possible to eliminate the time factor t between the two equations and to obtain a functional relation of the type $y = e \cdot x + f$. This relation states that there is a linear dependence between the two time series, which is not the case.

Therefore, measuring the correlation between the evolutions of two phenomena over time does not imply the existence of a link between them. The term covariation is therefore used instead of correlation, and this dependence is measured using a covariation coefficient. We can distinguish between:
- The linear covariation coefficient;
- The tendency covariation coefficient.

HISTORY
See **correlation coefficient** and **time series**.

MATHEMATICAL ASPECTS
In order to compare two **time series** y_t and x_t, the first step is to attempt to represent them on the same graphic.

However, visual comparison is generally difficult. The following change of **variables** is performed:

$$Y_t = \frac{y_t - \bar{y}}{S_y} \quad \text{and} \quad X_t = \frac{x_t - \bar{x}}{S_x},$$

which are the centered and reduced variables where S_y and S_x are the **standard deviations** of the respective **time series**.

We can distinguish between the following covariation coefficients:
- *The linear covariation coefficient*
  The form of this expression is identical to the one for the **correlation coefficient** r,

but here the calculations do not have the same grasp because the goal is to detect the eventual existence of relation between variations that are themselves related to time and to measure the order of magnitude

$$C = \frac{\sum_{t=1}^{n}(x_t - \bar{x}) \cdot (y_t - \bar{y})}{\sqrt{\sum_{t=1}^{n}(x_t - \bar{x})^2 \cdot \sum_{t=1}^{n}(y_t - \bar{y})^2}}.$$

This yields values of between -1 and $+1$. If it is close to ± 1, there is a linear **relation** between the time evolutions of the two **variables**.

Notice that:

$$C = \frac{\sum_{t=1}^{n} X_t \cdot Y_t}{n}.$$

Here n is the number of **observations**, while Y_t and X_t are the centered and reduced series obtained by a change of **variable**, respectively.
- *The tendency covariation coefficient*
  The influence exerted by the **means** is eliminated by calculating:

$$K = \frac{\sum_{t=1}^{n}(x_t - T_{x_t}) \cdot (y_t - T_{y_t})}{\sqrt{\sum_{t=1}^{n}(x_t - T_{x_t})^2 \cdot \sum_{t=1}^{n}(y_t - T_{y_t})^2}}.$$

The means $\bar{x}$ and $\bar{y}$ have simply been replaced with the **values** of the **secular trends** T_{x_t} and T_{y_t} of each **time series**.

The tendency covariation coefficient K also takes values between -1 to $+1$, and the closer it gets to ± 1, the stronger the covariation between the time series.

DOMAINS AND LIMITATIONS

There are many examples of the need to compare two **time series** in economics: for example, when comparing the evolution of the price of a product to the evolution of the quantity of the product on the market, or the evolution of the national revenue to the evolution of real estate transactions. It is important to know whether there is some kind of dependence between the two phenomena that evolve over time: this is the goal of measuring the covariation.

Visually comparing two **time series** is an important operation, but this is often a difficult task because:

- The **data** undergoing comparison may come from very different domains and present very different orders of magnitude, so it is preferable to study the deviations from the **mean**.

- The peaks and troughs of two time series may have very different amplitudes; it is then preferable to homogenize the dispersions by linking the variations back to the **standard deviation** of the time series.

Visual comparison is simplified if we consider the centered and reduced **variables** obtained via the following variable changes:

$$Y_t = \frac{y_t - \bar{y}}{S_y} \quad \text{and} \quad X_t = \frac{x_t - \bar{x}}{S_x}.$$

Also, in a similar way to the **correlation coefficient**, nonlinear relations can exist between two variables that give a C **value** that is close to zero.

It is therefore important to be cautious during interpretation.

The tendency covariation coefficient is preferentially used when the relation between the time series is not linear.

EXAMPLES

Let us study the covariation between two **time series**.

The **variable** x_t represents the annual production of an agricultural product; the variable y_t is its average annual price per unit in constant euros.

t	x_t	y_t
1	320	5.3
2	660	3.2
3	300	2.2
4	190	3.4
5	320	2.7
6	240	3.5
7	360	2.0
8	170	2.5

$$\sum_{t=1}^{8} x_t = 2560, \qquad \sum_{t=1}^{8} y_t = 24.8,$$

giving $\bar{x} = 320$ and $\bar{y} = 3.1$.

$x_t - \bar{x}$	$(x_t - \bar{x})^2$	$y_t - \bar{y}$	$(y_t - \bar{y})^2$
0	0	2.2	4.84
340	115600	0.1	0.01
−20	400	−0.9	0.81
−130	16900	0.3	0.09
0	0	−0.4	0.16
−80	6400	0.4	0.16
40	1600	−1.1	1.21
−150	22500	0.6	0.36

$$\sum_{t=1}^{8}(x_t - \bar{x})^2 = 163400,$$

$$\sum_{t=1}^{8}(y_t - \bar{y})^2 = 7.64,$$

giving $\sigma_x = 142.9$ and $\sigma_y = 0.98$.

The centered and reduced **values** X_t and Y_t are then calculated.

X_t	Y_t	$X_t \cdot Y_t$	$X_t \cdot Y_{t-1}$
0.00	2.25	0.00	5.36
2.38	0.10	0.24	−0.01
−0.14	−0.92	0.13	0.84
−0.91	0.31	−0.28	0.00
0.00	−0.41	0.00	0.23
−0.56	0.41	−0.23	0.11
0.28	−1.13	−0.32	1.18
−1.05	−0.61	0.64	

The linear covariation coefficient is then calculated:

$$C = \frac{\sum_{t=1}^{8} X_t \cdot Y_t}{8} = \frac{0.18}{8} = 0.0225 \, .$$

If the **observations** X_t are compared with Y_{t-1}, meaning the production this year is compared with that of the previous year, we obtain:

$$C = \frac{\sum_{t=2}^{8} X_t \cdot Y_{t-1}}{8} = \frac{7.71}{8} = 0.964 \, .$$

The linear covariation coefficient for a shift of one year is very strong. There is a strong (positive) covariation with a shift of one year between the two **variables**.

FURTHER READING
▶ **Correlation coefficient**
▶ **Moving average**
▶ **Secular trend**
▶ **Standard deviation**
▶ **Time series**

REFERENCES
Kendall, M.G.: Time Series. Griffin, London (1973)

Py, B.: Statistique Déscriptive. Economica, Paris (1987)

Cox, David R.

Cox, David R. was born in 1924 in Birmingham in England. He studied mathematics at the University of Cambridge and obtained his doctorate in applied mathematics at the University of Leeds in 1949. From 1966 to 1988, he was a professor of statistics at Imperial College London, and then from 1988 to 1994 he taught at Nuffield College, Oxford.

Cox, David is an eminent statistician. He was knighted by Queen Elizabeth II in 1982 in gratitude for his contributions to statistical science, and has been named Doctor Honoris Causa by many universities in England and elsewhere. From 1981 to 1983 he was President of the Royal Statistical Society; he was also President of the Bernoulli Society from 1972 to 1983 and President of the International Statistical Institute from 1995 to 1997.

Due to the variety of subjects that he has studied and developed, Professor Cox has had a profound impact in his field. He was named Doctor Honoris Causa of the University of Neuchâtel in 1992.

Cox, Sir David is the author and the coauthor of more then 250 articles and 16 books, and between 1966 and 1991 he was the editor of Biometrika.

Some principal works and articles of Cox, Sir David:

1964 (and Box, G.E.P.) An analysis of transformations (with discussion) J. Roy. Stat. Soc. Ser. B 26, 211–243.

1970 The Analysis of Binary Data. Methuen, London.

1973 (and Hinkley, D.V.) Theoretical Statistics. Chapman & Hall, London.

1974 (and Atkinson, A.C.) Planning experiments for discriminating between models. J. Roy. Stat. Soc. Ser. B 36, 321–348.

1978 (and Hinkley, D.V.) Problems and Solutions in Theoretical Statistics. Chapman & Hall, London.

1981 (and Snell, E.J.) Applied Statistics: Principles and Examples. Chapman & Hall.

1981 Theory and general principles in statistics. J. Roy. Stat. Soc. Ser. A 144, 289–297.

2000 (and Reid, N.) Theory of Design Experiments. Chapman & Hall, London.

Cox, Mary Gertrude

Cox, Gertrude Mary was born in 1900, near Dayton, Iowa, USA. Her ambition was to help people, and so she initially studied a social sciences course for two years. Then she worked in orphanage for young boys in Montana for two years. In order to become a director of the orphanage, she decided to continue her education at Iowa State College. She graduated from Iowa State College in 1929. To pay for her studies, Cox, Gertrude worked with **Snedecor, George Waddel**, her professor, in his statistical laboratory, which led to her becoming interested in statistics. After graduating, she started studying for a doctorate in psychology. In 1933, before finishing her doctorate, Snedecor, George, then the director of the Iowa State Statistical Laboratory, convinced her to become his assistant, which

she agreed to, although she remained in the field of psychology because she worked on the evaluation of statistical test in psychology and the analysis of psychological data.

On 1st November 1940, she became the director of the Department of Experimental Statistics of the State of North Carolina.

In 1945, the General Education Board gave her permission to create an institute of statistics at the University of North Carolina, with a department of mathematical statistics at Chapel Hill.

She was a founder member of the Journal of the International Biometric Society in 1947, and she was a director of it from 1947 to 1955 and president of it from 1968 to 1969. She was president of the American Statistical Association (ASA) in 1956. She died in 1978.

Principal work of Cox, M. Gertrude:

1957 (and Cochran, W.) Experimental Designs, 2nd edn. Wiley, New York

C_p Criterion

The C_p citerion is a model selection citerion in linear regression. For a linear regression model with p parameters including any constant term, in the model, a rule of thumb is to select a model in which the value of C_p is close to the number of terms in the model.

HISTORY

Introduced by Mallows, Colin L. in 1964, the model selection criterion C_p has been used ever since as a criterion for evaluating the goodness of fit of a regression model.

MATHEMATICAL ASPECTS

Let $Y_i = \beta_0 + \sum_{j=1}^{p-1} X_{ji}\beta_j + \varepsilon_i$, $i = 1,\ldots,n$ be a **multiple linear regression** model. Denote the **mean square error** as $MSE(\widehat{y}_i)$. The criterion introduced in this section can be used to choose the model with the minimal sum of mean square errors:

$$
\sum MSE(\widehat{y}_i) = \sum \left(E\left((\widehat{y}_i - y_i)^2 \right) \right.
$$
$$
\left. - \sigma^2 (1 - 2h_{ii}) \right)
$$
$$
= E\left(\sum (\widehat{y}_i - y_i)^2 \right)
$$
$$
- \sigma^2 \sum (1 - 2h_{ii})
$$
$$
= E(\mathrm{RSS}) - \sigma^2(n - 2p),
$$

where

n is the number of observations,
p is the number of estimated parameters,
h_{ii} are the diagonal elements of the **hat matrix**, and
$\widehat{y}_i$ is the estimator of y_i.

Recall the following property of the h_{ii}:

$$
\sum h_{ii} = p.
$$

Define the coefficient

$$
\Gamma_p = \frac{\sum MSE(\widehat{y}_i)}{\sigma^2}
$$
$$
= \frac{E(\mathrm{RSS}) - \sigma^2 (n - 2p)}{\sigma^2}
$$
$$
= \frac{E(\mathrm{RSS})}{\sigma^2} - n + 2p.
$$

If the model is correct, we must have:

$$
E(\mathrm{RSS}) = (n - p)\sigma^2,
$$

which implies

$$
\Gamma_p = p.
$$

In practice, we estimate Γ_p by

$$
C_p = \frac{\mathrm{RSS}}{\widehat{\sigma}^2} - n + 2p,
$$

where $\widehat{\sigma}^2$ is an estimator of σ^2. Here we estimate σ^2 using the s^2 of the full model. For this full model, we actually obtain $C_p = p$, which is not an interesting result. However, for all of the other models, where we use only a *subset* of the explanatory variables, the coefficient C_p can have values that are different from p. From the models that incorporate only a subset of the explanatory variables, we then choose those for which the value of C_p is the closest to p.

If we have k explanatory variables, we can also define the coefficient C_p for a model that incorporates a subset $X_1,\ldots,X_{p-1}, p \leq k$ of the k explanatory variables in the following manner:

$$
C_p = \frac{(n - k)\,\mathrm{RSS}\left(X_1,\ldots,X_{p-1} \right)}{\mathrm{RSS}\left(X_1,\ldots,X_k \right)}
$$
$$
- n + 2p,
$$

where $\mathrm{RSS}\left(X_1,\ldots,X_{p-1} \right)$ is the sum of the squares of the residuals related to the model with $p - 1$ explanatory variables, and $\mathrm{RSS}(X_1,\ldots,X_k)$ is the sum of the squares of the residuals related to the model with all k explanatory variables. Out of the two models that incorporate $p - 1$ explanatory variables, we choose, according to the criterion, the one for which the value of the coefficient C_p is the closest to p.

DOMAINS AND LIMITATIONS

The C_p criterion is used when selecting variables. When used with the R^2 criterion, this criterion can tell us about the goodness of fit of the chosen model. The underlying idea of such a procedure is the following: instead of trying to explain one variable using all of the available explanatory variables, it is sometimes possible to determine an underlying model with a subset of these variables, and the explanatory power of this model is almost the same as that of the model containing all of the explanatory variables. Another reason for this is that the collinearity of the explanatory variables tends to decrease estimator precision, and so it can be useful to delete some superficial variables.

EXAMPLES

Consider some data related to the Chicago fires of 1975. We denote the variable corresponding to the logarithm of the number of fires per 1000 households per district i of Chicago in 1975 by Y, and the variables corresponding to the proportion of households constructed before 1940, to the number of thefts and to the median revenue per district i by X_1, X_2 and X_3, respectively.

Since this set contains three explanatory variables, we have $2^3 = 8$ possible models. We divide the eight possible equations into four sets:

1. Set A contains the only equation without explanatory variables:

$$Y = \beta_0 + \varepsilon.$$

2. Set B contains three equations with one explanatory variable:

$$Y = \beta_0 + \beta_1 X_1 + \varepsilon$$
$$Y = \beta_0 + \beta_2 X_2 + \varepsilon$$
$$Y = \beta_0 + \beta_3 X_3 + \varepsilon.$$

3. Set C contains three equations with two explanatory variables:

$$Y = \beta_0 + \beta_1 X_1 + \beta_2 X_2 + \varepsilon$$
$$Y = \beta_0 + \beta_1 X_1 + \beta_3 X_3 + \varepsilon$$
$$Y = \beta_0 + \beta_2 X_2 + \beta_3 X_3 + \varepsilon.$$

4. Set D contains one equation with three explanatory variables:

$$Y = \beta_0 + \beta_1 X_1 + \beta_2 X_2 + \beta_3 X_3 + \varepsilon.$$

C

District	X_1	X_2	X_3	Y
i	x_{i1}	x_{i2}	x_{i3}	y_i
1	0.604	29	11.744	1.825
2	0.765	44	9.323	2.251
3	0.735	36	9.948	2.351
4	0.669	37	10.656	2.041
5	0.814	53	9.730	2.152
6	0.526	68	8.231	3.529
7	0.426	75	21.480	2.398
8	0.785	18	11.104	1.932
9	0.901	31	10.694	1.988
10	0.898	25	9.631	2.715
11	0.827	34	7.995	3.371
12	0.402	14	13.722	0.788
13	0.279	11	16.250	1.740
14	0.077	11	13.686	0.693
15	0.638	22	12.405	0.916
16	0.512	17	12.198	1.099
17	0.851	27	11.600	1.686
18	0.444	9	12.765	0.788
19	0.842	29	11.084	1.974
20	0.898	30	10.510	2.715
21	0.727	40	9.784	2.803
22	0.729	32	7.342	2.912
23	0.631	41	6.565	3.589
24	0.830	147	7.459	3.681
25	0.783	22	8.014	2.918
26	0.790	29	8.177	3.148
27	0.480	46	8.212	2.501
28	0.715	23	11.230	1.723
29	0.731	4	8.330	3.082
30	0.650	31	5.583	3.073
31	0.754	39	8.564	2.197
32	0.208	15	12.102	1.281
33	0.618	32	11.876	1.609
34	0.781	27	9.742	3.353

District	X_1	X_2	X_3	Y
i	x_{i1}	x_{i2}	x_{i3}	y_i
35	0.686	32	7.520	2.856
36	0.734	34	7.388	2.425
37	0.020	17	13.842	1.224
38	0.570	46	11.040	2.477
39	0.559	42	10.332	2.351
40	0.675	43	10.908	2.370
41	0.580	34	11.156	2.380
42	0.152	19	13.323	1.569
43	0.408	25	12.960	2.342
44	0.578	28	11.260	2.747
45	0.114	3	10.080	1.946
46	0.492	23	11.428	1.960
47	0.466	27	13.731	1.589

Source: Andrews and Herzberg (1985)

We denote the resulting models in the following way: **1** for X_1, **2** for X_2, **3** for X_3, **12** for X_1 and X_2, **13** for X_1 and X_3, **23** for X_2 and X_3, and **123** for the full model. The following table represents the results obtained for the C_p criterion for each model:

Model	C_p
1	32.356
2	29.937
3	18.354
12	18.936
13	14.817
23	3.681
123	4.000

If we consider a model from group B (containing one explanatory variable), the number of estimated parameters is $p = 2$ and none of the C_p values for the three models approaches 2. If we now consider a model from group C (with two explanatory variables), p equals 3 and the C_p of model **23** approaches this. Finally, for the complete model we find that $C_p = 4$, which is also the number of estimated parameters, but this is not an interesting result as previously explained. Therefore, the most reasonable choice for the model appears to be:

$$Y = \beta_0 + \beta_2 X_2 + \beta_3 X_3 + \varepsilon.$$

FURTHER READING
▶ **Coefficient of determination**
▶ **Collinearity**
▶ **Hat matrix**
▶ **Mean squared error**
▶ **Regression analysis**

REFERENCES
Andrews D.F., Hertzberg, A.M.: Data: A Collection of Problems from Many Fields for Students and Research Workers. Springer, Berlin Heidelberg New York (1985)

Mallows, C.L.: Choosing variables in a linear regression: A graphical aid. Presented at the Central Regional Meeting of the Institute of Mathematical Statistics, Manhattan, KS, 7–9 May 1964 (1964)

Mallows, C.L.: Some comments on Cp. Technometrics **15**, 661–675 (1973)

Cramér, Harald

Cramér, Harald (1893–1985) entered the University of Stockholm in 1912 in order to study chemistry and mathematics; he became a student of Leffler, Mittag and Riesz, Marcel. In 1919, Cramér was named assistant professor at the University of Stockholm. At the same time, he worked as an actuary for an insurance company, Svenska Life Assurance, which allowed him to study probability and statistics.

His main work in actuarial mathematics is *Collective Risk Theory*. In 1929, he was asked to create a new department in Stockholm, and he became the first Swedish professor of actuarial and statistical mathematics. At the end of the Second World War he wrote his principal work *Mathematical Methods of Statistics*, which was published for the first time in 1945 and was recently (in 1999) republished.
Some principal works and articles of Cramér, Harald

1946 Mathematical Methods of Statistics. Princeton University Press, Princeton, NJ.

1946 Collective Risk Theory: A Survey of the Theory from the Point of View of the Theory of Stochastic Processes. Skandia Jubilee Volume, Stockholm.

Criterion Of Total Mean Squared Error

The criterion of total mean squared error is a way of comparing estimations of the parameters of a biased or unbiased model.

MATHEMATICAL ASPECTS
Let

$$\widehat{\boldsymbol{\beta}} = \left(\widehat{\beta}_1, \ldots, \widehat{\beta}_{p-1}\right)$$

be a vector of estimators for the parameters of a regression model. We define the total mean square error, *TMSE*, of the vector $\widehat{\boldsymbol{\beta}}$ of estimators as being the sum of the mean squared errors (*MSE*) of its components.
We recall that

$$MSE\left(\widehat{\beta}_j\right) = E\left(\left(\widehat{\beta}_j - \beta_j\right)^2\right)$$
$$= V\left(\widehat{\beta}_j\right) + \left(E\left(\widehat{\beta}_j\right) - \beta_j\right)^2,$$

where $E(.)$ and $V(.)$ are the usual symbols used for the **expected value** and the **variance**. We define the total mean squared error as:

$$TMSE(\widehat{\beta}_j) = \sum_{j=1}^{p-1} MSE\left(\widehat{\beta}_j\right)$$
$$= \sum_{j=1}^{p-1} E\left(\left(\widehat{\beta}_j - \beta_j\right)^2\right)$$
$$= \sum_{j=1}^{p-1} \left[\mathrm{Var}\left(\widehat{\beta}_j\right) + \left(E\left(\widehat{\beta}_j\right) - \beta_j\right)^2 \right]$$
$$= (p-1)\sigma^2 \cdot \mathrm{Trace}\,(\mathbf{V})$$
$$+ \sum_{j=1}^{p-1} \left(E\left(\widehat{\beta}_j\right) - \beta_j\right)^2.$$

where $\mathbf{V}$ is the **variance-covariance matrix** of $\widehat{\boldsymbol{\beta}}$.

DOMAINS AND LIMITATIONS
Unfortunately, when we want to calculate the total mean squared error

$$TMSE\left(\widehat{\boldsymbol{\beta}}\right) = \sum_{j=1}^{p-1} E\left(\left(\widehat{\beta}_j - \beta_j\right)^2\right)$$

of a vector of estimators

$$\widehat{\boldsymbol{\beta}} = \left(\widehat{\beta}_1, \ldots, \widehat{\beta}_{p-1}\right)$$

for the parameters of the model

$$Y_i = \beta_0 + \beta_1 X_{i1}^s + \ldots + \beta_{p-1} X_{ip-1}^s + \varepsilon_i,$$

we need to know the values of the model parameters β_j, which are obviously unknown for all real data. The notation X_j^s means that the data for the jth explanatory variable were standardized (see **standardized data**).
Therefore, in order to estimate these *TMSE* we generate data from a structural similarly

model. Using a generator of pseudo-random numbers, we can simulate all of the data for the artificial model, analyze it with different models of regression (such as simple regression or **ridge regression**), and calculate what we call the *total squared error*, *TSE*, of the vectors of the estimators $\widehat{\beta}$ related to each method:

$$TSE(\widehat{\beta}) = \sum_{j=1}^{p-1} (\widehat{\beta}_j - \beta_j)^2 .$$

We repeat this operation 100 times, ensuring that the 100 data sets are pseudo-independent. For each model, the average of 100 *TSE* gives a good estimation of the *TMSE*. Note that some statisticians prefer the model obtained by selecting variables due to its simplicity. On the other hand, other statisticians prefer the ridge method because it uses all of the available information.

EXAMPLES

We can generally compare the following regression methods: **linear regression** by mean squares, ridge regression, or the variable selection method.

In the following example, thirteen portions of cement have been examined. Each portion is composed of four ingredients, given in the table. The aim is to determine how the quantities x_{i1}, x_{i2}, x_{i3} and x_{i4} of these four ingredients influence the quantity y_i, the heat given out due to the hardening of the cement. Heat given out by the cement

Portion	Ingredient				Heat
	1	2	3	4	
i	x_{i1}	x_{i2}	x_{i3}	x_{i4}	y_i
1	7	26	6	60	78.5
2	1	29	15	52	74.3
3	11	56	8	20	104.3

Portion	Ingredient				Heat
	1	2	3	4	
i	x_{i1}	x_{i2}	x_{i3}	x_{i4}	y_i
4	11	31	8	47	87.6
5	7	52	6	33	95.9
6	11	55	9	22	109.2
7	3	71	17	6	102.7
8	1	31	22	44	72.5
9	2	54	18	22	93.1
10	21	47	4	26	115.9
11	1	40	23	34	83.9
12	11	66	9	12	113.3
13	10	68	8	12	109.4

y_i quantity of heat given out due to the hardening of the ith portion (in joules);

x_{i1} quantity of ingredient 1 (tricalcium aluminate) in the ith portion;

x_{i2} quantity of ingredient 2 (tricalcium silicate) in the ith portion;

x_{i3} quantity of ingredient 3 (tetracalcium alumino-ferrite) in the ith portion;

x_{i4} quantity of ingredient 4 (dicalcium silicate) in the ith portion.

In this paragraph we will compare the estimators obtained by **least squares** (LS) regression with those obtained by ridge regression (R) and those obtained with the variable selection method (SV) via the total mean squared error *TMSE*. The three estimation vectors obtained from each method are:

$$\widehat{Y}_{LS} = 95.4 + 9.12X_1{}^s + 7.94X_2{}^s$$
$$+ 0.65X_3{}^s - 2.41X_4{}^s$$
$$\widehat{Y}_R = 95.4 + 7.64X_1{}^s + 4.67X_2{}^s$$
$$- 0.91X_3{}^s - 5.84X_4{}^s$$
$$\widehat{Y}_{SV} = 95.4 + 8.64X_1{}^s + 10.3X_2{}^s$$

We note here that all of the estimations were obtained from standardized explanatory variables. We compare these three estimations using the total mean squared errors of the three vectors:

$$\hat{\boldsymbol{\beta}}_{LS} = \left(\hat{\beta}_{LS_1}, \hat{\beta}_{LS_2}, \hat{\beta}_{LS_3}, \hat{\beta}_{LS_4} \right)'$$

$$\hat{\boldsymbol{\beta}}_{R} = \left(\hat{\beta}_{R_1}, \hat{\beta}_{R_2}, \hat{\beta}_{R_3}, \hat{\beta}_{R_4} \right)'$$

$$\hat{\boldsymbol{\beta}}_{SV} = \left(\hat{\beta}_{SV_1}, \hat{\beta}_{SV_2}, \hat{\beta}_{SV_3}, \hat{\beta}_{SV_4} \right)'.$$

Here the subscript LS corresponds to the method of least squares, R to the ridge method and SV to the variable selection method, respectively. For this latter method, the estimations for the coefficients of the unselected variables in the model are considered to be zero. In our case we have:

$$\hat{\boldsymbol{\beta}}_{LS} = (9.12, 7.94, 0.65, -2.41)'$$

$$\hat{\boldsymbol{\beta}}_{R} = (7.64, 4.67, -0.91, -5.84)'$$

$$\hat{\boldsymbol{\beta}}_{SV} = (8.64, 10.3, 0, 0)'.$$

We have chosen to approximate the underlying process that results in the cement data by the following least squares equation:

$$Y_{iMC} = 95.4 + 9.12X_{i1}^s + 7.94X_{i2}^s + 0.65X_{i3}^s - 2.41X_{i4}^s + \varepsilon_i.$$

The procedure consists of generating 13 random error terms $\varepsilon_1, \ldots, \varepsilon_{13}$ 100 times based on a normal distribution with mean 0 and standard deviation 2.446 (recall that 2.446 is the least squares estimator of σ for the cement data). We then calculate $Y_{1LS}, \ldots, Y_{13LS}$ using the X_{ij}^s values in the data table. In this way, we generate 100 $Y_{1LS}, \ldots, Y_{13LS}$ samples from 100 $\varepsilon_1, \ldots, \varepsilon_{13}$ samples.

The three methods are applied to each of these 100 $Y_{1LS}, \ldots, Y_{13LS}$ samples (always using the same values for X_{ij}^s), which yields 100 estimators $\hat{\boldsymbol{\beta}}_{LS}$, $\hat{\boldsymbol{\beta}}_{R}$ and $\hat{\boldsymbol{\beta}}_{SV}$. Note that

these three methods are applied to these 100 samples without any influence from the results from the equations

$$\widehat{Y}_{iLS} = 95.4 + 9.12X_{i1}^s + 7.94X_{i2}^s + 0.65X_{i3}^s - 2.41X_{i4}^s,$$

$$\widehat{Y}_{iR} = 95.4 + 7.64X_{i1}^s + 4.67X_{i2}^s - 0.91X_{i3}^s - 5.84X_{i4}^s \text{ and}$$

$$\widehat{Y}_{iSV} = 95.4 + 8.64X_{i1}^s + 10.3X_{i2}^s$$

obtained for the original sample. Despite the fact that the variable selection method has chosen the variables X_{i1} and X_{i2} in $\widehat{Y}_{iSV} = 95.4 + 8.64X_{i1}^s + 10.3X_{i2}^s$, it is possible that, for one of these 100 samples, the method has selected better X_{i2} and X_{i3} variables, or only X_{i3}, or all of the subset of the four available variables. In the same way, despite the fact that the equation $Y_{iR} = 95.4 + 7.64X_{i1}^s + 4.67xX_{i2}^s - 0.91X_{i3}^s - 5.84X_{i4}^s$ was obtained with $k = 0.157$, the value of k is recalculated for each of these 100 samples (for more details refer to the ridge regression example). From these 100 estimations of $\hat{\boldsymbol{\beta}}_{LS}$, $\hat{\boldsymbol{\beta}}_{SV}$ and $\hat{\boldsymbol{\beta}}_{R}$, we can calculate 100 ETCs for each method, which we label as:

$$ETC_{LS} = \left(\hat{\beta}_{LS_1} - 9.12 \right)^2$$
$$+ \left(\hat{\beta}_{LS_2} - 7.94 \right)^2$$
$$+ \left(\hat{\beta}_{LS_3} - 0.65 \right)^2$$
$$+ \left(\hat{\beta}_{LS_4} + 2.41 \right)^2$$

$$ETC_{R} = \left(\hat{\beta}_{R_1} - 9.12 \right)^2$$
$$+ \left(\hat{\beta}_{R_2} - 7.94 \right)^2$$
$$+ \left(\hat{\beta}_{R_3} - 0.65 \right)^2$$
$$+ \left(\beta_{R_4} + 2.41 \right)^2$$

$$ETC_{SV} = \left(\hat{\beta}_{SV_1} - 9.12 \right)^2$$
$$+ \left(\hat{\beta}_{SV_2} - 7.94 \right)^2$$
$$+ \left(\hat{\beta}_{SV_3} - 0.65 \right)^2$$
$$+ \left(\hat{\beta}_{SV_4} + 2.41 \right)^2 .$$

The means of the 100 values of ETC_{LS}, ETC_R and ETC_{SV} are the estimations of $TMSE_{LS}$, $TMSE_{SV}$ and $TMSE_R$: the $TMSE$ estimations for the three considered methods.

After this simulation was performed, the following estimations were obtained:

$$TMSE_{LS} = 270 ,$$
$$TMSE_R = 75 .$$
$$TMSE_{SV} = 166 .$$

These give the following differences:

$$TMSE_{LS} - TMSE_{SV} = 104 ,$$
$$TMSE_{LS} - TMSE_R = 195 ,$$
$$TMSE_{SV} - TMSE_R = 91 .$$

Since the standard deviations of 100 observed differences are respectively 350, 290 and 280, we can calculate the approximate 95% confidence intervals for the differences between the $TMSE$s of two methods

$$TMSE_{LS} - TMSE_{SV} = 104 \pm \frac{2 \cdot 350}{\sqrt{100}} ,$$
$$TMSE_{LS} - TMSE_R = 195 \pm \frac{2 \cdot 290}{\sqrt{100}} ,$$
$$TMSE_{SV} - TMSE_R = 91 \pm \frac{2 \cdot 280}{\sqrt{100}} .$$

We get

$$34 < TMSE_{LS} - TMSE_{SV} < 174 ,$$
$$137 < TMSE_{LS} - TMSE_R < 253 ,$$
$$35 < TMSE_{SV} - TMSE_R < 147 .$$

We can therefore conclude (at least for the particular model used to generate the simulated data, and taking into account our aim—to get a small $TMSE$), that the ridge method is the best of the methods considered, followed by the varible selection procedure.

FURTHER READING
▶ **Bias**
▶ **Expected value**
▶ **Hat matrix**
▶ **Mean squared error**
▶ **Ridge regression**
▶ **Standardized data**
▶ **Variance**
▶ **Variance–covariance matrix**
▶ **Weighted least-squares method**

REFERENCES
Box, G.E.P, Draper, N.R.: A basis for the selection of response surface design. J. Am. Stat. Assoc. **54**, 622–654 (1959)

Dodge, Y.: Analyse de régression appliquée. Dunod, Paris (1999)

Critical Value

In **hypothesis testing**, the critical value is the limit **value** at which we take the decision to reject the **null hypothesis** H_0, for a given **significance level**.

HISTORY
The concept of a critical value was introduced by **Neyman, Jerzy** and **Pearson, Egon Sharpe** in 1928.

MATHEMATICAL ASPECTS
The critical value depends on the type of the test used (**two-sided test** or **one-sided**

test on the right or the left), the **probability distribution** and the **significance level** α.

DOMAINS AND LIMITATIONS
The critical value is determined from the **probability distribution** of the **statistic** associated with the test. It is determined by consulting the **statistical table** corresponding to this probability distribution (**normal table**, **Student table**, **Fisher table**, **chi-square table**, etc).

EXAMPLES
A company produces steel cables. Using a **sample** of size $n = 100$, it wants to verify whether the diameters of the cables conform closely enough to the required diameter 0.9 cm in general.
The **standard deviation** σ of the **population** is known and equals 0.05 cm.
In this case, **hypothesis testing** involves a **two-sided test**. The hypotheses are the following:

> **null hypothesis** H_0: $\mu = 0.9$
> **alternative hypothesis** H_1: $\mu \neq 0.9$.

To a **significance level** of $\alpha = 5\%$, by looking at the **normal table** we find that the critical value $z_{\frac{\alpha}{2}}$ equals 1.96.

FURTHER READING
▶ **Confidence interval**
▶ **Hypothesis testing**
▶ **Significance level**
▶ **Statistical table**

REFERENCE
Neyman, J., Pearson, E.S.: On the use and interpretation of certain test criteria for purposes of statistical inference, Parts I and II. Biometrika **20**A, 175–240, 263–294 (1928)

Cyclical Fluctuation

Cyclical fluctuations is a term used to describe oscillations that occur over long periods about the **secular trend** line or curve of a **time series**.

HISTORY
See **time series**.

MATHEMATICAL ASPECTS
Consider Y_t, a **time series** given by its components; Y_t can be written as:
- $Y_t = T_t \cdot S_t \cdot C_t \cdot I_t$ (multiplicative **model**), or;
- $Y_t = T_t + S_t + C_t + I_t$ (additive model).

where

Y_t is the **data** at time t;
T_t is the **secular trend** at time t;
S_t is the **seasonal variation** at time t;
C_t is the cyclical fluctuation at time t, and;
I_t is the **irregular variation** at time t.

The first step when investigating a time series is always to determine the secular trend T_t, and then to determine the seasonal variation S_t. It is then possible to adjust the initial data of the **time series** Y_t according to these two components:
- $\dfrac{Y_t}{S_t \cdot T_t} = C_t \cdot I_t$ (multiplicative **model**);
- $Y_t - S_t - T_t = C_t + I_t$ (additive model).

To avoid cyclical fluctuations, a weighted **moving average** is established over a few months only. The use of moving averages allows use to smooth the **irregular variations** I_t by preserving the cyclical fluctuations C_t.

The choice of a weighted **moving average** allows us to give more weight to the central **values** compared to the extreme values, in order to reproduce cyclical fluctuations in a more accurate way. Therefore, large weights will be given to the central values and small weights to the extreme values.

For example, for a **moving average** considered for an interval of five months, the weights $-0.1, 0.3, 0.6, 0.3$ and -0.1 can be used; since their sum is 1, there will be no need for normalization.

If the **values** of $C_t \cdot I_t$ (resulting from the adjustments performed with respect to the **secular trend** and to the **seasonal variations**) are denoted by X_t, the value of the cyclical fluctuation for the month t is determined by:

$$C_t = -0.1 \cdot X_{t-2} + 0.3 \cdot X_{t-1} + 0.6 \cdot X_t + 0.3 \cdot X_{t+1} - 0.1 \cdot X_{t+2}.$$

DOMAINS AND LIMITATIONS

Estimating cyclical fluctuations allows us to:
- Determine the maxima or minima that a **time series** can attain.
- Perform short- or medium-term **forecasting**.
- Identify the cyclical components.

The limitations and advantages of the use of weighted **moving averages** when evaluating cyclical fluctuations are the following:
- Weighted moving averages can smooth a curve with cyclical fluctuations which still retaining most of the original fluctuation, because they preserve the amplitudes of the cycles in an accurate way.
- The use of an odd number of months to establish the moving average facilitates better centering of the **values** obtained.
- It is difficult to study the cyclical fluctuation of a **time series** because the cycles

usually vary in length and amplitude. This is due to the presence of a multitude of factors, where the effects of these factors can change from one cycle to the other. None of the models used to explain and predict such fluctuations have been found to be completely satisfactory.

EXAMPLES

Let us establish a **moving average** considered over five months of **data** adjusted according to the **secular trend** and **seasonal variations**.

Let us also use the weights $-0.1, 0.3, 0.6, 0.3, -0.1$; since their sum is equal to 1, there is no need for normalization.

Let X_i be the adjusted **values** of $C_i \cdot I_i$:

$$C_i = -0.1 \cdot X_{i-2} + 0.3 \cdot X_{i-1} + 0.6 \cdot X_i + 0.3 \cdot X_{i+1} - 0.1 \cdot X_{i+2}.$$

The table below shows the electrical power consumed by street lights every month in millions of kilowatt hours during the years 1952 and 1953. The **data** have been adjusted according to the **secular trend** and **seasonal variations**.

Year	Month	Data X_i	Moving average for 5 months C_i
1952	J	99.9	
	F	100.4	
	M	100.2	100.1
	A	99.0	99.1
	M	98.1	98.4
	J	99.0	98.7
	J	98.5	98.3
	A	97.8	98.3
	S	100.3	99.9
	O	101.1	101.3
	N	101.2	101.1
	D	100.4	100.7

Year	Month	Data X_i	Moving average for 5 months C_i
1953	J	100.6	100.3
	F	100.1	100.4
	M	100.5	100.1
	A	99.2	99.5
	M	98.9	98.7
	J	98.2	98.2
	J	98.4	98.5
	A	99.7	99.5
	S	100.4	100.5
	O	101.0	101.0
	N	101.1	
	D	101.1	

No significant cyclical effects appear in these data and non significant effects do not mean no effects. The beginning of an economic cycle is often sought, but this only appears every 20 years.

FURTHER READING
► **Forecasting**
► **Irregular variation**
► **Moving average**
► **Seasonal variation**
► **Secular trend**
► **Time series**

REFERENCE
Box, G.E.P., Jenkins, G.M.: Time Series Analysis: Forecasting and Control (Series in Time Series Analysis). Holden Day, San Francisco (1970)

Wold, H.O. (ed.): Bibliography on Time Series and Stochastic Processes. Oliver & Boyd, Edinburgh (1965)

Daniels, Henry E.

Daniels, Henry Ellis was born in London in October 1912. After graduating from Edinburgh University in 1933, he continued his studies at Cambridge University. After gaining his doctorate at Edinburgh University, he went back to Cambridge as lecturer in mathematics in 1947. In 1957, Daniels, Henry became the first professor of mathematics at Birmingham University, a post he held until 1978, up to his retirement. The research field that interested Daniels, Henry were inferential statistics, saddlepoint approximations in statistics, epidemiological modeling and statistical theory as applied to textile technology. When he was in Birmingham, he founded the annual meeting of statisticians in Gregynog Hall in Powys. These annual meetings gradually became one of the most well-received among English statisticians.

From 1974 to 1975 he was president of the Royal Statistical Society, which awarded him the Guy Medal in silver in 1957 and in gold in 1984. In 1980 he was elected a member of the Royal Society and in 1985 an honored member of the International Statistical Society. He died in 2000.

Some principal works and articles of Daniels, Henry E.:

1954 Saddlepoint approximations in statistics. Ann. Math. Stat. 25, 631–650.

1955 Discussion of a paper by Box, G.E.P. and Anderson, S.L.. J. Roy. Stat. Soc. Ser. B 17, 27–28.

1958 Discussion of paper by Cox, D.R.. J. Roy. Stat. Soc. Ser. B 20, 236–238.

1987 Tail probability approximations. Int. Stat. Rev. 55, 137–48.

Data

A datum (plural data) is the result of an **observation** made on a **population** or on a **sample**.

The word "datum" is Latin, and means "something given;" it is used in mathematics to denote an item of information (not necessarily numerical) from which a conclusion can be made.

Note that a number (or any other form of description) that does not necessarily contain any information should not be confused with a datum, which does contain information.

The data obtained from observations are related to the **variable** being studied. These data are quantitative, qualitative, discrete or continuous if the corresponding variable is quantitative, qualitative, discrete or continuous, respectively.

FURTHER READING
► **Binary data**
► **Categorical data**
► **Incomplete data**
► **Observation**
► **Population**
► **Sample**
► **Spatial data**
► **Standardized data**
► **Value**
► **Variable**

REFERENCES
Federer, W.T.: Data collection. In: Kotz, S., Johnson, N.L. (eds.) Encyclopedia of Statistical Sciences, vol. 2. Wiley, New York (1982)

Data Analysis

Often, one of the first steps performed in scientific research is to collect **data**. These data are generally organized into two-dimensional tables. This organization usually makes it easier to extract information about the data—in other words, to analyze them.

In its widest sense, data analysis can be considered to be the essence of **statistics** to which all other aspects of the subject are linked.

HISTORY
Since **data** analysis encompasses many different methods of statistical analysis, it is difficult to briefly overview the history of data analysis. Nevertheless, we can rapidly review the chronology of the most fundamental aspects of the subject:

- The first publication on exploratory data analysis dates back to 1970–1971, and was written by Tukey, J.W.. This was the first version of his work *Exploratory Data Analysis*, published in 1977.
- The theoretical principles of **correspondence analysis** are due to Hartley, H.O. (1935) (published under his original German name Hirschfeld) and to **Fisher, R.A.** (1940). However, the theory was largely developed in the 1970s by Benzécri, J.P.
- The first studies on **classification** were carried out in biology and in zoology. The oldest form of typology was conceived by Galen (129–199 A.D.). Numerical classification methods derive from the ideas of Adanson (in the eighteenth century), and were developed, amongst others, by Zubin (1938) and Thorndike (1953).

DOMAINS AND LIMITATIONS
The field of **data** analysis comprises many different statistical methods.

Types of data analysis can be classified in the following way:

1. *Exploratory data analysis*, which involves (as its name implies) exploring the data, via:
   - Representing the data graphically,
   - Data **transformation** (if required),
   - Detecting **outlier observations**,
   - Elaborating research **hypotheses** that were not envisaged at the start of the **experiment**,
   - **Robust estimation**.

2. *Initial data analysis*, which involves:
- Choosing the statistical methods to be applied to the data.
3. *Multivariate data analysis*, which includes:
- Diseriminant analysis,
- Data **transformation**, which reduces the number of dimensions and facilitates interpretation,
- Searching for structure.
4. *Specific forms of data analysis* that are suited to different analytical tasks; these forms include:
- **Correspondence analysis**,
- Multiple correspondence analysis,
- **Classification**.
5. *Confirmatory data analysis*, which involves evaluating and testing analytical results; this includes:
- **Parameter estimation**,
- **Hypotheses tests**,
- The generalization and the conclusion.

FURTHER READING
- ▶ **Classification**
- ▶ **Cluster analysis**
- ▶ **Correspondence analysis**
- ▶ **Data**
- ▶ **Exploratory data analysis**
- ▶ **Hypothesis testing**
- ▶ **Statistics**
- ▶ **Transformation**

REFERENCES

Benzécri, J.P.: L'Analyse des données. Vol. 1: La Taxinomie. Vol. 2: L'Analyse factorielle des correspondances. Dunod, Paris (1976)

Benzécri, J.P.: Histoire et préhistoire de l'analyse des données, les cahiers de l'analyse des données 1, no. 1–4. Dunod, Paris (1976)

Everitt, B.S.: Cluster Analysis. Halstead, London (1974)

Fisher, R.A.: The precision of discriminant functions. Ann. Eugen. (London) **10**, 422–429 (1940)

Gower, J.C.: Classification, geometry and data analysis. In: Bock, H.H. (ed.) Classification and Related Methods of Data Analysis. Elsevier, Amsterdam (1988)

Hirschfeld, H.O.: A connection between correlation and contingency. Proc. Camb. Philos. Soc. **31**, 520–524 (1935)

Thorndike, R.L.: Who belongs in a family? Psychometrika **18**, 267–276 (1953)

Tukey, J.W.: Explanatory Data Analysis, limited preliminary edition. Addison-Wesley, Reading, MA (1970–1971)

Tukey, J.W.: Exploratory Data Analysis. Addison-Wesley, Reading, MA (1977)

Zubin, J.: A technique for measuring likemindedness. J. Abnorm. Social Psychol. **33**, 508–516 (1938)

Data Collection

When collecting **data**, we need to consider several issues. First, it is necessary to define why we need to collect the data, and what the data will be used for. Second, we need to consider the type of data that should be collected: it is essential that the data collected are related to the goal of the study. Finally, we must consider how the data are to be collected. There are three approaches to data collection:

1. Register
2. **Sampling** and **census**
3. Experimental research

1. *Register*
One form of accessible data is a registered one. For example, there are registered data on births, deaths, marriages, daily temperatures, monthly rainfall and car sales.

There are no general statistical methods that ensure that valuable conclusions are drawn from these **data**. Each set of data should be considered on its own merits, and one should be careful when making **inference** on the **population**.

2. *Sampling and census*
Sampling methods can be divided into two categories: random methods and nonrandom methods.

Nonrandom methods involve constructing, by empirical means, a **sample** that resembles the **population** from which it is taken as much as possible. The most commonly used nonrandom method is **quota sampling**.

Random methods use a probabilistic procedure to derive a **sample** from the **population**. Given the fact that the probability that a given unit is selected in the sample is known, the error due to sampling can then be calculated. The main random methods are **simple random sampling**, **stratified sampling**, **systematic sampling** and **batch sampling**.

In a **census**, all of the objects in a **population** are observed, yielding data for the whole population. Clearly, this type of investigation is very costly when the population is very large. This is why censuses are not performed very often.

3. *Experimental research*
In an **experiment**, data collection is performed based upon a particular experimental design. These experimental designs are applied to **data** collection in all research fields.

HISTORY
It is likely that the oldest form of **data** collection dates back to population censuses performed in antiquity.

Antille and Ujvari (1991) mention the existence of a position for an official statistician in China during the Chow dynasty (111–211 B.C.).

The Roman author Tacitus says that Emperor Augustus ordered all of the soldiers, ships and wealth in the Empire to be counted.

Evidence for **censuses** can also be found in the Bible; Saint Luke reports that "Caesar Augustus ordered a decree prescribing a census of the whole world (...) and all went to be inscribed, each in his own town."

A form of statistics can also be found during this time. Its name betrays its administrative origin, because it comes from the Latin word "status:" the State.

FURTHER READING
▶ **Census**
▶ **Design of experiments**
▶ **Sampling**
▶ **Survey**

REFERENCES

Antille, G., Ujvari, A.: Pratique de la Statistique Inférentielle. PAN, Neuchâtel, Switzerland (1991)

Boursin, J.-L.: Les structures du hasard. Seuil, Paris (1986)

Decile

Deciles are **measures of position** calculated on a set of **data**.

The deciles are the **values** that separate a distribution into ten equal parts, where each part contains the same number of **observations**). The decile is a member of the wider family of **quantiles**.

The *x*th decile indicates the value where 10*x*% of the observations occur below this value and $(100 - 10x)\%$ of the observations occur above this value. For example, the eighth decile is the value where 80% of the observations fall below this and 20% occur above it.

The fifth decile represents the **median**.

There will therefore be nine deciles for a given distribution:

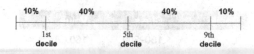

MATHEMATICAL ASPECTS

The process used to calculate deciles is similar to that used to calculate the **median** or **quartiles**.

When all of the raw **observations** are available, the process used to calculate deciles is as follows:

1. Organize the *n* **observations** into a **frequency distribution**
2. The deciles correspond to the **observations** for which the cumulative relative **frequencies** exceed 10%, 20%, 30%,…, 80%, 90%.

Some authors propose using the following formula to precisely determine the **values** of the different deciles:

Calculating the jth decile:

Take *i* to be the integer part of $j \cdot \frac{n+1}{10}$ and *k* the fractional part of $j \cdot \frac{n+1}{10}$.

Take x_i and x_{i+1} to be the **values** of the **observations** at the *i*th and $(i+1)$th posi-

tions (when the observations are arranged in increasing order).

The *j*th decile is then equal to:

$$D_j = x_i + k \cdot (x_{i+1} - x_i).$$

When the **observations** are grouped into classes, the deciles are determined in the following way:

1. Determine the class containing the desired decile located:

   • *First decile*: the first class for which the cumulative relative **frequency** exceeds 10%.

   • *Second decile*: the first class for which the cumulative relative **frequency** exceeds 20%.

   …

   • *Ninth decile*: the first class for which the cumulative relative **frequency** exceeds 90%.

2. Calculate the **value** of the decile based on the **hypothesis** that the **observations** are uniformly distributed in each class:

$$\text{decile} = L_1 + \left[\frac{(N \cdot q) - \sum f_{\text{inf}}}{f_{\text{decile}}} \right] \cdot c.$$

where

$L_1 =$	lower limit of the class of the decile
$N =$	total number of **observations**
$q =$	1/10 for the first decile
$q =$	2/10 for the second decile
	…
$q =$	9/10 for the ninth decile
$\sum f_{\text{inf}} =$	sum of the **frequencies** lower than the class of the decile
$f_{\text{decile}} =$	frequency of the class of the decile
$c =$	size of the **interval** of the class of the decile

DOMAINS AND LIMITATIONS

The calculation of deciles only has meaning for a **quantitative variable** that can take **values** over a given **interval**.

The number of **observations** needs to be relatively high, because the calculation of deciles involves dividing the set of observations into ten parts.

Deciles are relatively frequently used in practice; for example, when interpreting the distribution of revenue in a city or state.

EXAMPLES

Consider an example where the deciles are calculated for a **frequency distribution** of a continuous **variable** where the **observations** are grouped into classes.

The following **frequency table** represents the profits (in thousands of euros) of 500 bakeries:

Profit (in thousands of euros)	Frequencies	Cumulated frequency	Relative cumulated frequency
100–150	40	40	0.08
150–200	50	90	0.18
200–250	60	150	0.30
250–300	70	220	0.44
300–350	100	320	0.62
350–400	80	400	0.80
400–450	60	460	0.92
450–500	40	500	1.00
Total	500		

The first decile is in the class 150–200 (this is where the cumulative relative **frequency** exceeds 10%).

By assuming that the **observations** are distributed uniformly in each class, we obtain the following **value** for the first decile:

$$\text{1st decile} = 150 + \left[\frac{(500 \cdot \frac{1}{10}) - 40}{50} \right] \cdot 50$$

$$= 160 \,.$$

The second decile falls in the 200–250 class. The value of the second decile is equal to

$$\text{2nd decile} = 200 + \left[\frac{(500 \cdot \frac{2}{10}) - 90}{60} \right] \cdot 50$$

$$= 208.33 \,.$$

We can calculate the other deciles in the same way, which yields:

Decile	Class	Value
1	150–200	160.00
2	200–250	208.33
3	200–250	250.00
4	250–300	285.71
5	300–350	315.00
6	300–350	340.00
7	350–400	368.75
8	350–400	400.00
9	400–450	441.67

We can conclude, for example, that 10% of the 500 bakeries make a profit of between 100000 and 160000 euros, 50% make a profit that is lower than 315000 euros, and 10% make a profit of between 441670 and 500000 euros.

FURTHER READING

► **Measure of location**
► **Median**
► **Percentile**
► **Quantile**
► **Quartile**

Degree of Freedom

The number of degrees of freedom is a **parameter** from the **chi-square distribution**. It is also a parameter used in other **probability distributions** related to the chi-square distribution, such as the **Student distribution** and the **Fisher distribution**. In another context, the number of degrees of freedom refers to the number of linearly independent terms involved when calculating the sum of squares based on n independent **observations**.

HISTORY
The term "degree of freedom" was introduced by **Fisher, R.A.** in 1925.

MATHEMATICAL ASPECTS
Let $Y_1, Y_2, \ldots, Y_n$ be a random sample of size n taken from a population with an unknown mean $\bar{Y}$. The sum of the **deviations** of n **observations** with respect to their **arithmetic mean** is always equal to zero:

$$\sum_{i=1}^{n}(Y_i - \bar{Y}) = 0 .$$

This requirement is a constraint on each deviation $Y_i - \bar{Y}$ used when calculating the **variance**:

$$S^2 = \frac{\sum_{i=1}^{n}(Y_i - \bar{Y})^2}{n - 1} .$$

This constraint implies that $n - 1$ deviations completely determine the nth deviation. The n deviations (and also the sum of their squares and the variance in the S^2 of the **sample**) therefore have $n-1$ degrees of freedom.

FURTHER READING
▶ **Chi-square distribution**
▶ **Fisher distribution**
▶ **Parameter**
▶ **Student distribution**
▶ **Variance**

REFERENCES
Fisher, R.A.: Applications of "Student's" distribution. Metron **5**, 90–104 (1925)

Deming, W. Edwards

Deming, W. Edwards was born in 1900, in Sioux City, in Iowa. He studied science at Wyoming University, graduating in 1921, and at Colorado University, where he completed his Master degree in mathematics and physics in 1924. Then he went to Yale University, where he received his doctorate in physics in 1928. After his doctorate, he worked for ten years in the Laboratory of the Ministry for Agriculture. In 1939, Deming moved to the Census Bureau in Washington. There he used his theoretical knowledge to initiate the first censuses to be performed by sampling. These censuses used techniques that later provided an example for similar censuses performed around the rest of the world. He left the Census Bureau in 1946 to become consultant in statistical studies and a professor of statistics at New York University. He died in Washington D.C. in 1993.

For a long time (up to 1980), the theories of Deming, W.E. were ignored by American companies. In 1947, Deming went to Tokyo as a consultant to apply his techniques of sampling. From 1950 onwards, Japanese industry adopted the management theories of Deming. Within ten years many Japanese products were being exported to America because they were better and less expensive than equivalent products manufactured in the United States.

Some principal works and articles of Deming, W. Edwards:

1938 Statistical Adjustment of Data. Wiley, New York .

1950 Some Theory of Sampling. Wiley, New York .

1960 Sample Design in Business Research. Wiley, New York .

Demography

Demography is the study of human **populations**. It involves analyzing phenomena such as births, deaths, migration, marriages, divorces, fertility rate, mortality rate and age pyramids.

These phenomena are treated from both biological and socio-economic points of view. The methods used in demography are advanced mathematics and statistics as well as many fields of social science.

HISTORY

The first demographic studies date back to the middle of the seventeenth century. In this period, Graunt, John, an English scientist, analyzed the only population data available: a list of deaths in the London area, classified according to their cause. In 1662 he published a study in which he tried to evaluate the average size of a family, the importance of migrational movement, and other elements related to the structure of the population. In collaboration with Petty, Sir William, Graunt proposed that more serious studies of

the population should be made and that a center where statistical **data** would be gathered should be created.

During the eighteenth century the same types of analyses were performed and progressively improved, but it wasn't until the nineteenth century that several European countries as well as the United States undertook national **censuses** and established the regular records of births, marriages and deaths.

These reports shows that different regions offered different chances of survival for their inhabitants. These conclusions ultimately resulted in improved working and hygiene conditions.

Using the demographic data gathered, predictions became possible, and the first demographic journals and reviews appeared around the end of the century. These included "Demography" in the United States, "Population" in France and "Population Studies" in Great Britain.

In the middle of the twentieth century demographic studies began to focus on the global population, as the demographic problems of the Third World became an important issue.

FURTHER READING
► **Census**
► **Population**

REFERENCES

Cox, P.R.: Demography, 5th edn. Cambridge University Press, Cambridge (1976)

Hauser, P.M., Duncan, O.D.: The Study of Population: An Inventory and Appraisal. University of Chicago Press, Chicago, IL (1959)

Dendrogram

A dendrogram is a graphical representation of different aggregations made during a **cluster analysis**. It consists of knots that correspond to groups and branches that represent the associations made at each step. The structure of the dendrogram is determined by the order in which the aggregations are made. If a **scale** is added to the dendrogram it is possible to represent the **distances** over which the aggregations took place.

HISTORY

In a more general sense, a dendrogram (from the Greek "dendron", meaning tree) is a tree diagram that illustrates the relations that exist between the members of a set.

The first examples of dendrograms were the phylogenetic trees used by systematic specialists. The term "dendrogram" seems to have been used for the first time in the work of Mayr et al. (1953).

MATHEMATICAL ASPECTS

During **cluster analysis** on a set of objects, aggregations are achieved with the help of a **distance table** and the chosen linkage method (**single linkage method**, **complete linkage method**, etc.).

Each extreme point of the dendrogram represents a different class produced by automatic classification.

The dendrogram is then constructed by representing each group by a knot placed at a particular position with respect to the horizontal **scale**, where the position depends upon the **distance** over which the aggregate is formed.

The objects are placed at the zero level of the scale. To stop the branches connecting the knots from becoming entangled, the procedure of drawing a dendrogram is performed in a systematic way.

- The first two grouped elements are placed, one on top of the other, at the zero level of the **scale**. A horizontal line is then drawn next to each element, from the zero level to the aggregation **distance** for each element. The resulting class is then represented by a vertical line that connects the ends of the lines for the elements (forming a "knot"). The middle of this vertical line provides the starting point for a new horizontal line for this class when it is aggregated with another class. The dendrogram therefore consists of branches constructed from horizontal and vertical lines.

- Two special cases can occur during the next step:

  1. Two classes that each consist of one element are aggregated; these classes are placed at the zero level and the resulting knot is located at the aggregation distance, as described previously;

  2. An element is aggregated to a class formed previously. The element is aggregated from the zero level, and is inserted between existing classes if necessary). The knot representing this aggregation is placed at the (horizontal) **distance** corresponding to the aggregation distance between the class and the element.

This procedure continues until the desired configuration is obtained. If there are still classes consisting of a single element at the end of the aggregation process, they are simply added from the zero level of the dendrogram. When the **classification** has been completed, each class will corresponds to a particular part of the dendrogram.

DOMAINS AND LIMITATIONS

The dendrogram describes the ordered path of the set of operations performed during **cluster analysis**. It illustrates this type of **classification** in a very precise manner.

This strictly defined approach to constructing a dendrogram is sometimes modified due to circumstances. For example, as we will see in one of the examples below, the aggregation distances of two or more successive steps may be the same, and so the procedure must then be changed to make sure that the branches of the dendrogram do not get entangled.

In the case that we have described above, the dendrogram is drawn horizontally. Obviously, it is also possible to plot a dendrogram from top to bottom, or from bottom to top. The branches can even consist of diagonal lines.

EXAMPLES

We will perform a **cluster analysis** using the **single link method** and establish the corresponding dendrogram, illustrating the different aggregations that may occur.

Consider the grades obtained by five students during examinations for four different courses: English, French, maths and physics.

We would like to separate these five students into two groups using the **single link method**, in order to test a new teaching method.

These grades, which can take values from 1 to 6, are summarized in the following table:

	English	French	Maths	Physics
Alain	5.0	3.5	4.0	4.5
Jean	5.5	4.0	5.0	4.5
Marc	4.5	4.5	4.0	3.5
Paul	4.0	5.5	3.5	4.0
Pierre	4.0	4.5	3.0	3.5

The **distance table** is obtained by calculating the Euclidean **distances** between the students and adding them to the following table:

	Alain	Jean	Marc	Paul	Pierre
Alain	0	1.22	1.5	2.35	2
Jean	1.22	0	1.8	2.65	2.74
Marc	1.5	1.8	0	1.32	1.12
Paul	2.35	2.65	1.32	0	1.22
Pierre	2	2.74	1.12	1.22	0

Using the **single link method**, based on the minimum **distance** between the objects in two classes, the following partitions are obtained at each step:

First step: Marc and Pierre are grouped at an aggregation distance of 1.12: {Alain}, {Jean}, {Marc,Pierre}, {Paul} *Second step*: We revise the distance table. The new **distance table** is given below:

	Jean	Marc and Pierre	Paul
Alain	1.22	1.5	2.35
Jean		1.8	2.65
Marc and Pierre			1.22

Alain and Jean are then grouped at a distance of 1.22, and we obtain the following partition: {Alain, Jean}, {Marc, Pierre}, {Paul} *Third step*: The new distance table is given below:

	Marc and Pierre	Paul
Alain and Jean	1.5	2.35
Marc and Pierre		1.22

Paul is then grouped with Marc and Pierre at a distance of 1.22, yielding the following partition: {Alain, Jean}, {Marc, Pierre, Paul} *Fourth step*: The two groups are aggregated at a distance of 1.5.

This gives the following **dendrogram**:

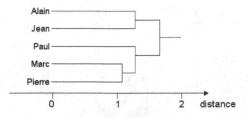

Notice that Paul must be positioned before the group of Marc and Pierre in order to avoid entangling the branches of the diagram.

FURTHER READING
- ► **Classification**
- ► **Cluster analysis**
- ► **Complete linkage method**
- ► **Distance**
- ► **Distance table**

REFERENCES
Mayr, E., Linsley, E.G., Usinger, R.L.: Methods and Principles of Systematic Zoology. McGraw-Hill, New York (1953)

Sneath, P.H.A., Sokal, R.R.: Numerical Taxonomy: The Principles and Practice of Numerical Classification (A Series of Books in Biology). W.H. Freeman, San Francisco, CA (1973)

Sokal, R.R., Sneath, P.H.: Principles of Numerical Taxonomy. Freeman, San Francisco (1963)

Density Function

The density function of a continuous **random variable** allows us to determine the **probability** that a random variable X takes values in a given **interval**.

HISTORY
See **probability**.

MATHEMATICAL ASPECTS
Consider $P(a \leq X \leq b)$, the **probability** that a continuous **random variable** X takes a **value** in the **interval** $[a, b]$. This probability is defined by:

$$P(a \leq X \leq b) = \int_a^b f(x)\,dx.$$

where $f(x)$ is the density function of the random variable X.

The density function is graphically represented on an axis system. The different **values** of the **random variable** X are placed on the abscissa, and those taken by the function f are placed on the **ordinate**.

The graph of the function f does not allow us to determine the **probability** for one particular point, but instead to visualize the probability for an **interval** on a surface.

The total surface under the curve corresponds to a value of 1:

$$\int f(x)\,dx = 1.$$

The variable X is a continuous random variable if there is a non-negative function f that is defined for real numbers and for which the following property holds for every interval $[a, b]$:

$$P(a \leq X \leq b) = \int_a^b f(x)\,dx,$$

Here $P(a \leq X \leq b)$ is the **probability function**. Therefore, the **probability** that the continuous **random variable** X takes a **value** in the **interval** $[a, b]$ can be obtained by integrating the probability function over $[a, b]$. There is also the following condition on f:

$$\int_{-\infty}^{\infty} f(x)\,dx = P(\infty \leq X \leq \infty) = 1.$$

We can represent the density function graphically on the system of axes. The different values of the random variable X are placed on the abscissa, and those of the function f on the ordinate.

DOMAINS AND LIMITATIONS
If $a = b$ in $P(a \leq X \leq b) = \int_a^b f(x)\,dx$, then

$$P(X = a) = \int_a^a f(x)\,dx = 0.$$

This means that the **probability** that a continuous **random variable** takes an isolated fixed **value** is always zero.

It therefore does not matter whether we include or exclude the boundaries of the **interval** when calculating the **probability** associated with an interval.

EXAMPLES
Consider X, a continuous **random variable**, for which the density function is

$$f(x) = \begin{cases} \frac{3}{4}(-x^2 + 2x), & \text{if } 0 < x < 2 \\ 0, & \text{if not} \end{cases}.$$

We can calculate the **probability** of X being higher than 1. We obtain:

$$P(X > 1) = \int_1^\infty f(x)\,dx,$$

which can be divided into:

$$P(X > 1) = \int_1^2 \frac{3}{4}(-x^2 + 2x)\,dx$$
$$+ \int_2^\infty 0\,dx$$
$$= \frac{3}{4}\left(-\frac{x^3}{3} + x^2 \Big|_1^2 \right)$$
$$= \frac{1}{2}.$$

The density function and the region where $P(X > 1)$ can be represented graphically:

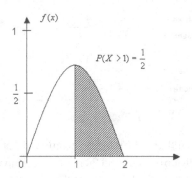

The shaded surface represents the region where $P(X > 1)$.

FURTHER READING
▶ **Continuous distribution function**
▶ **Continuous probability distribution**
▶ **Joint density function**
▶ **Probability**

Dependence

The concept of dependence can have two meanings. The first concerns the **events** of a **random experiment**. It is said that two events are dependent if the occurrence of one depends on the occurrence of the other.

The second meaning of the word concerns the **relation**, generally a functional relation, that can exist between **random variables**. This dependence can be measured, and in most cases these measurements require random varible **covariance**.

The most commonly used measure of dependence is the **correlation coefficient**. The **null hypothesis**, which states that the two **random variables** X and Y are independent, can be tested against the **alternative hypothesis** of dependence. Tests that are used for this purpose include the **chi-square test of independence**, as well as tests based on the **Spearman rank corre-**

lation coefficient and the **Kendall rank correlation coefficient**.

FURTHER READING
► **Chi-square test of independence**
► **Correlation coefficient**
► **Covariance**
► **Kendall rank correlation coefficient**
► **Spearman rank correlation coefficient**
► **Test of independence**

Dependent Variable

The dependent variable (or response variable) in a regression **model** is the **variable** that is considered to vary depending on the other variables called independent variables incorporated into the analysis.

FURTHER READING
► **Independent variable**
► **Multiple linear regression**
► **Regression analysis**
► **Simple linear regression**

Descriptive Epidemiology

Descriptive epidemiology involves describing the frequencies and patterns of illnesses among the population.
It is based on the collection of health-related information (in mortality tables, morbidity registers, illnesses that must be declared ...) and on information that may have an impact on the health of the population (data on atmospheric pollution, risk behavior ...), which are used to obtain a statistical picture of the general health of the population.

HISTORY
See **epidemiology**.

MATHEMATICAL ASPECTS
See **cause and effect in epidemiology, odds and odds ratio, relative risk, attributable risk, avoidable risk, incidence rate, prevalence rate**.

DOMAINS AND LIMITATIONS
Despite the fact that descriptive epidemiology only yields "elementary" information, it is still vitally important:

● It allows us to study the scales and the patterns of health phenomenona (by studying their **prevalence** and their **incidence**), and it facilitates epidemiological surveillance,
● It aids decision-making related to the planning and administration of health establishments and programs,
● It can lead to hypotheses on the risk factors of an illness (example: the increased rate of incidence of skin cancers in the south of France, believed to be related to the increased intensity of the sun's rays in the south of France compared to the north.).

Descriptive epidemiology cannot be used to relate a cause to an effect; it is not a predictive tool.

FURTHER READING
See **epidemiology**.

REFERENCES
See **epidemiology**.

Descriptive Statistics

The information gathered in a study can often take different forms, such as frequency data (for example, the number of votes cast for a candidate in elections) and scale data.

These data are often initially arranged or organized in such a way that they are difficult to read and interpret.

Descriptive **statistics** offers us some procedures that allow us to represent data in a readable and worthwhile form.

Some of these procedures allow us to obtain a **graphical representation** of the data, for example in the following forms:

- Histogram,
- Bar chart,
- Pie chart,
- Stem and leaf,
- Box plot, etc.

while others allow us to obtain a set of parameters that summarize important properties of the basic data:

- Mean,
- Standard deviation,
- Correlation coefficient,
- Index number, etc.

The descriptive statistics also encompasses methods of **data analysis**, such as the **correspondence analysis**, that consist of graphically representing the associations between the rows and the columns of a **contingency table**.

HISTORY

The first known form of descriptive **statistics** was the **census**. Censuses were ordered by the rulers of ancient civilizations, who wanted to count their subjects and to monitor their professions and goods.

See **statistics** and **graphical representation**.

FURTHER READING

▶ Box plot

▶ **Census**
▶ **Correspondence analysis**
▶ **Data analysis**
▶ **Dendrogram**
▶ **Graphical representation**
▶ **Histogram**
▶ **Index number**
▶ **Line chart**
▶ **Pie chart**
▶ **Stem-and-leaf diagram**

Design of Experiments

Designing an experiment is like programming the **experiment** in some ways. Each **factor** involved in the experiment can take a certain number of different values (called levels), and the experimental design employed specifies the levels of the one or many factors (or combinations of factors) used in the experiment.

HISTORY

Experimental designs were first used in the 1920s, mostly in the agricultural domain. Sir **Fisher, Ronald Aylmer** was the first to use mathematical statistics when designing experiments. In 1926 he wrote a paper outlining the principles of experimental design in non-mathematical terms.

Federer, W.T. and Balaam, L.N. (1973) provided a very detailed bibliography of literature related to experimental design before 1969, incorporating 8000 references.

DOMAINS AND LIMITATIONS

The goal of the experimental design is to find with the most efficient and economic methods that allow us to reach solid and adequate conclusions on the results from the experiment.

The most frequently applied experimental designs are the **completely randomized design**, the **randomized block design** and the **Latin square design**.

Each design implies a different mathematical analysis to those used for the other designs, since the designs really correspond to different mathematical **models**. Examples of these types of analysis include **variance analysis**, **covariance analysis** and **regression analysis**.

FURTHER READING
► **Analysis of variance**
► **Experiment**
► **Model**
► **Optimal design**
► **Regression analysis**

REFERENCES
Federer, W.T., Balaam, L.N.: Bibliography on Experiment and Treatment Design Pre-1968. Hafner, New York (1973)

Fisher, R.A.: The arrangement of field experiments. J. Ministry Agric. **33**, 503–513 (1926)

Determinant

Any square **matrix** A of order n has a special number associated with it, known as the determinant of matrix A.

MATHEMATICAL ASPECTS
Consider $A = (a_{ij})$, a square **matrix** of order n. The determinant of A, denoted $\det(A)$ or $|A|$, is given by the following sum:

$$\det(A) = \sum (\pm) a_{1i} \cdot a_{2j} \cdot \ldots \cdot a_{nr},$$

where the sum is made over all of the **permutations** of the second index. The sign is positive if the number of **inversions** in $(i, j, \ldots, r)$ is even (an even permutation) and it is negative if the number is odd (an odd permutation).

The determinant can also be defined by "developing" along a line or a column. For example, we can develop the matrix along the first line:

$$\det(A) = \sum_{j=1}^{n} (-1)^{1+j} \cdot a_{1j} \cdot A_{1j},$$

where A_{1j} is the determinant of the square "submatrix" of order $n-1$ obtained from A by erasing the first line and the jth column. A_{1j} is called the cofactor of element a_{1j}.

Since we can arbitrarily choose the line or column, we can write:

$$\det(A) = \sum_{j=1}^{n} (-1)^{i+j} \cdot a_{ij} \cdot A_{ij}$$

for a fixed i (developing along the ith line), or

$$\det(A) = \sum_{i=1}^{n} (-1)^{i+j} \cdot a_{ij} \cdot A_{ij}$$

for a fixed j (developing along the jth column).

This second way of defining the determinant is recursive, because the determinant of a square matrix of order n is calculated using the determinants of the matrices of order $n-1$.

Properties of the Determinant
• The determinant of a square **matrix** A is equal to the determinant of its **transposed** matrix A':

$$\det(A) = \det(A').$$

D

- If two lines (or two columns) are exchanged in matrix A, the sign of the determinant changes.
- If all of the elements of a line (or a column) are zero, the determinant is also zero.
- If a multiple of a line from matrix A is added to another line from A, the determinant remains the same. This means that if two lines from A are identical, it is easy to obtain a line where all of the elements are zero simply by subtracting one line from the other; because of the previous property, the determinant of A is zero. The same situation occurs when one line is a multiple of another.
- The determinant of a matrix that only has zeros under (or over) the diagonal is equal to the product of the elements on the diagonal (such a matrix is denoted "triangular").
- Consider A and B, two square matrices of order n. The determinant of the product of the two matrices is equal to the product of the determinants of the two matrices:

$$\det(A \cdot B) = \det(A) \cdot \det(B).$$

EXAMPLES

1) Consider A, a square **matrix** of order 2:

$$A = \begin{bmatrix} a & b \\ c & d \end{bmatrix}$$

$$\det(A) = a \cdot d - b \cdot c.$$

2) Consider B, a square matrix of order 3:

$$B = \begin{bmatrix} 3 & 2 & 1 \\ 4 & -1 & 0 \\ 2 & -2 & 0 \end{bmatrix}.$$

By developing along the last column:

$$\det(B) = 1 \cdot \begin{vmatrix} 4 & -1 \\ 2 & -2 \end{vmatrix} + 0 \cdot \begin{vmatrix} 3 & 2 \\ 2 & -2 \end{vmatrix}$$

$$+ 0 \cdot \begin{vmatrix} 3 & 2 \\ 4 & -1 \end{vmatrix}$$

$$= 1 \cdot \{4 \cdot (-2) - (-1) \cdot 2\}$$

$$= -8 + 2$$

$$= -6.$$

FURTHER READING
▶ **Inversion**
▶ **Matrix**
▶ **Permutation**

Deviation

The concept of deviation describes the difference between an observed **value** and a fixed value from the set of possible values of a **quantitative variable**.

This fixed **value** is often the **arithmetic mean** of the set of values or the **median**.

MATHEMATICAL ASPECTS

Consider the set of **values** $x_1, x_2, \ldots, x_n$. The **arithmetic mean** of these values is denoted by $\bar{x}$. The deviation of a given value x_i with respect to the arithmetic mean is equal to:

$$\text{deviation} = (x_i - \bar{x}).$$

In a similar way, if the **median** of these same **values** is denoted by m, the deviation of a value x_i with respect to the median is equal to:

$$\text{deviation} = (x_i - m).$$

FURTHER READING
▶ **Mean absolute deviation**
▶ **Standard deviation**
▶ **Variance**

Dichotomous Variable

A **variable** is called dichotomous if it can take only tow values.

The simplest example is that of the **qualitative categorical variable** "gender," which can take two values, "male" and "female". Note that quantitative variables can always be reduced and dichotomized. The variable "revenue" can, for example, be reduced to two categories: "low revenue" and "high revenue".

FURTHER READING
▶ **Binary data**
▶ **Category**
▶ **Qualitative categorical variable**
▶ **Variable**

Dichotomy

Dichotomy is the division of the individuals of a **population** or a **sample** into two groups, as a function of predetermined criteria.

A **variable** is called **dichotomous** when it can only take two **values**. Such data are called **binary data**.

FURTHER READING
▶ **Binary data**
▶ **Dichotomous variable**

Discrete Distribution Function

The distribution function of a discrete **random variable** is defined for all real numbers as the **probability** that the random variable takes a **value** less than or equal to this real number.

HISTORY
See **probability**.

MATHEMATICAL ASPECTS
The function F, defined by

$$F(b) = P(X \leq b).$$

is called the distribution function of the discrete **random variable** X.

Given a real number b, the distribution function therefore corresponds to the **probability** that X is less than or equal to b.

The distribution function can be graphically represented on a system of axes. The different **values** of the discrete **random variable** X are displayed on the abscissa and the cumulative **probabilities** corresponding to the different values of X, $F(x)$, are shown on the ordinate.

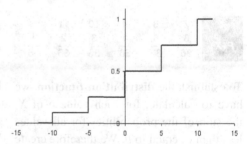

In the case where the possible **values** of the discrete **random variable** are $b_1, b_2, \ldots$ with $b_1 < b_2 < \ldots$, the discrete **distribution function** is a step function. $F(b_i)$ is constant over the **interval** $[b_i, b_{i+1}[$.

Properties of the Discrete Distribution Function

1. F is a nondecreasing function; in other words, if $a < b$, then $F(a) \leq F(b)$

2. F takes **values** from the **interval** $[0, 1]$.

3. $\lim_{b \to -\infty} F(b) = 0.$

4. $\lim_{b \to \infty} F(b) = 1.$

EXAMPLES

Consider a **random experiment** that consists of simultaneously throwing two dice. Consider the discrete **random variable** X, corresponding to the total score from the two dice.

Let us search for the **probability** of the **event** $\{X \leq 7\}$, which is by definition the **value** of the **distribution function** for $x = 7$.

The discrete **random variable** X takes its **values** from the set $E = \{2, 3, \ldots, 12\}$. The **probabilities** associated with each value of X are given by the following table:

X	2	3	4	5	6
P(X)	$\frac{1}{36}$	$\frac{2}{36}$	$\frac{3}{36}$	$\frac{4}{36}$	$\frac{5}{36}$

X	7	8	9	10	11	12
P(X)	$\frac{6}{36}$	$\frac{5}{36}$	$\frac{4}{36}$	$\frac{3}{36}$	$\frac{2}{36}$	$\frac{1}{36}$

To establish the **distribution function**, we have to calculate, for each value b of X, the sum of the probabilities for all values less than or equal to b. We therefore create a new table containing the cumulative probabilities:

b	2	3	4	5	6	7
P(X ≤ b)	$\frac{1}{36}$	$\frac{3}{36}$	$\frac{6}{36}$	$\frac{10}{36}$	$\frac{15}{36}$	$\frac{21}{36}$

b	8	9	10	11	12
P(X ≤ b)	$\frac{26}{36}$	$\frac{30}{36}$	$\frac{33}{36}$	$\frac{35}{36}$	$\frac{36}{36} = 1$

The **probability** of the **event** $\{X \leq 7\}$ is therefore equal to $\frac{21}{36}$.

We can represent the **distribution function** of the discrete **random variable** X as follows:

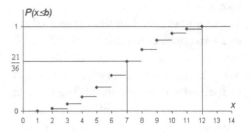

FURTHER READING
► **Probability**
► **Probability function**
► **Random experiment**
► **Value**

Discrete Probability Distribution

If each possible **value** of a discrete **random variable** is associated with a certain **probability**, we can obtain the discrete probability distribution of this random variable.

MATHEMATICAL ASPECTS

The probability distribution of a discrete **random variable** X is given by its **probability function** $P(x)$ or its **distribution function** $F(x)$.

It can generally be characterized by its **expected value**:

$$E[X] = \sum_D x \cdot P(X = x),$$

and its **variance**:

$$\mathrm{Var}(X) = \sum_D (x - E[X])^2 \cdot P(X = x),$$

where D represents the set from which X can take its **values**.

EXAMPLES

The discrete probability distributions that are most commonly used are the **Bernoulli distribution**, the **binomial distribution**, the **negative binomial distribution**, the **geometric distribution**, the **multinomial distribution**, the **hypergeometric distribution** and the **Poisson distribution**.

FURTHER READING
▶ **Bernoulli distribution**
▶ **Binomial distribution**
▶ **Continuous probability distribution**
▶ **Discrete distribution function**
▶ **Expected value**
▶ **Geometric distribution**
▶ **Hypergeometric distribution**
▶ **Joint probability distribution function**
▶ **Multinomial distribution**
▶ **Negative binomial distribution**
▶ **Poisson distribution**
▶ **Probability**
▶ **Probability distribution**
▶ **Probability function**
▶ **Random variable**
▶ **Variance of a random variable**

REFERENCES
Johnson, N.L., Kotz, S.: Distributions in Statistics: Discrete Distributions. Wiley, New York (1969)

Discrete Uniform Distribution

The discrete uniform distribution is a **discrete probability distribution**. The corresponding **continuous probability distribution** is the (continuous) **uniform distribution**.

Consider n **events**, each of which have the same **probability** $P(X = x) = \frac{1}{n}$; the ran-

dom variable X follows a discrete uniform distribution and its **probability function** is:

$$P(X = x) = \begin{cases} \frac{1}{n}, & \text{if } x = x_1, x_2, \ldots, x_n \\ 0, & \text{if not} \end{cases}.$$

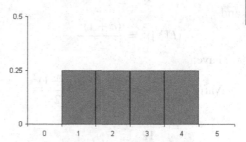

Discrete uniform distribution, $n = 4$

MATHEMATICAL ASPECTS
The **expected value** of the discrete uniform distribution over the set of first n natural numbers is, by definition, given by:

$$E[X] = \sum_{x=1}^{n} x \cdot P(X = x)$$
$$= \frac{1}{n} \cdot \sum_{x=1}^{n} x$$
$$= \frac{1}{n} \cdot (1 + 2 + \cdots + n)$$
$$= \frac{1}{n} \cdot \frac{n^2 + n}{2}$$
$$= \frac{n + 1}{2}.$$

The **variance** of this discrete uniform distribution is equal to:

$$\text{Var}(X) = E[X^2] - (E[X])^2.$$

Since

$$E[X^2] = \sum_{x=1}^{n} x^2 \cdot P(X = x)$$
$$= \frac{1}{n} \cdot \sum_{x=1}^{n} x^2$$

$$= \frac{1}{n} \cdot \left(1^2 + 2^2 + \cdots + n^2\right)$$

$$= \frac{1}{n} \cdot \frac{n(n+1)(2n+1)}{6}$$

$$= \frac{2n^2 + 3n + 1}{6},$$

and

$$(E[X])^2 = \frac{(n+1)^2}{2^2},$$

we have:

$$\text{Var}(X) = \frac{2n^2 + 3n + 1}{6} - \frac{(n+1)^2}{2^2}$$

$$= \frac{n^2 - 1}{12}.$$

DOMAINS AND LIMITATIONS

The discrete uniform distribution is often used to generate **random numbers** from any discrete or continuous **probability distribution**.

EXAMPLES

Consider the **random variable** X, the score obtained by throwing a die. If the die is not loaded, the **probability** of obtaining any particular score is equal to $\frac{1}{6}$. Therefore, we have:

$$P(X = 1) = \frac{1}{6}$$

$$P(X = 2) = \frac{1}{6}$$

$$\cdots$$

$$P(X = 6) = \frac{1}{6}.$$

The number of possible **events** is $n = 6$. We therefore have:

$$P(X = x) = \begin{cases} \frac{1}{6}, & \text{for } x = 1, 2, \ldots, 6 \\ 0, & \text{if not} \end{cases}.$$

In other words, the random variable X follows the discrete uniform distribution.

FURTHER READING

▶ **Discrete probability distribution**
▶ **Uniform distribution**

Dispersion

See **measure of dispersion**.

Distance

Distance is a numerical description of the spacing between two objects. A distance therefore corresponds to a real number:

- Zero, if both objects are the same;
- Strictly positive, if not.

MATHEMATICAL ASPECTS

Consider three objects X, Y and Z, and the distance between X and Y, $d(X, Y)$.
A distance has following properties:

- It is positive,

$$d(X, Y) > 0,$$

or zero if and only if the objects are the same

$$d(X, Y) = 0 \quad \Leftrightarrow \quad X = Y.$$

- It is symmetric, meaning that

$$d(X, Y) = d(Y, X).$$

- It verifies the following inequality, called the triangular inequality:

$$d(X, Z) \leq d(X, Y) + d(Y, Z).$$

This says that the distance from one object to another is smaller or equal to the distance obtained by passing through a third object.
Consider X and Y, which are two **vectors** with n components,

$$X = (x_i)' \text{ and } Y = (y_i)',$$
$$\text{for } i = 1, 2, \ldots, n.$$

A family of distances that is often used in such a case is given below:

$$d(X, Y) = \sqrt[p]{\sum_{i=1}^{n} |x_i - y_i|^p} \text{ with } p \geq 1.$$

These distances are called Minkowski distances.

The distance for $p = 1$ is called the absolute distance or the L_1 norm.

The distance for $p = 2$ is called the Euclidean distance, and it is defined by:

$$d(X, Y) = \sqrt{\sum_{i=1}^{n} (x_i - y_i)^2}.$$

The Euclidean distance is the most frequently used distance.

When the type of distance being used is not specified, it is generally the Euclidean distance that is being referred to.

There are also the weighted Minkowski distances, defined by:

$$d(X, Y) = \sqrt[p]{\sum_{i=1}^{n} w_i \cdot |x_i - y_i|^p} \text{ with } p \geq 1,$$

where $w_i, i = 1, \ldots, n$ represent the different weights, which sum to 1.

When $p = 1$ or 2, we obtain the weighted absolute and Euclidean distances respectively.

DOMAINS AND LIMITATIONS

For a set of r distinct objects $X_1, X_2, \ldots, X_r$, the distance between object i and object j for each pair (X_i, X_j) can be calculated. These distances are used to create a **distance table** or more generally a dissimilarity table containing terms that can be generalized to

$$d_{ij} = d(X_i, X_j), \quad \text{for } i, j = 1, 2, \ldots, r,$$

as used in methods of **cluster analysis**. Note that the term **measure of dissimilarity** is used if the triangular inequality is not satisfied.

EXAMPLES

Consider the example based on the grades obtained by five students in their English, French, maths and physics examinations.

These grades, which can take any value from 1 to 6, are summarized in the following table:

	English	French	Maths	Physics
Alain	5.0	3.5	4.0	4.5
Jean	3.5	4.0	5.0	5.5
Marc	4.5	4.5	4.0	3.5
Paul	6.0	5.5	5.5	5.0
Pierre	4.0	4.5	2.5	3.5

The Euclidean distance between Alain and Jean, denoted by d_2, is given by:

$$d_2(\text{Alain, Jean})$$
$$= \sqrt{(5.0 - 3.5)^2 + \cdots + (4.5 - 5.5)^2}$$
$$= \sqrt{2.25 + 0.25 + 1 + 1}$$
$$= \sqrt{4.5} = 2.12.$$

The absolute distance between Alain and Jean, denoted by d_1, is given by:

$$d_1(\text{Alain, Jean}) = |5.0 - 3.5| + \cdots$$
$$+ |4.5 - 5.5|$$
$$= 1.5 + 0.5 + 1 + 1$$
$$= 4.0.$$

Continuing the calculations, we obtain:

$$d_2(\text{Alain, Marc}) = \sqrt{2.25} = 1.5$$
$$d_2(\text{Alain, Paul}) = \sqrt{7.5} = 2.74$$

$d_2(\text{Alain, Pierre}) = \sqrt{5.25} = 2.29$

$d_2(\text{Jean, Marc}) = \sqrt{6.25} = 2.5$

$d_2(\text{Jean, Paul}) = \sqrt{9} = 3$

$d_2(\text{Jean, Pierre}) = \sqrt{10.75} = 3.28$

$d_2(\text{Marc, Paul}) = \sqrt{7.75} = 2.78$

$d_2(\text{Marc, Pierre}) = \sqrt{2.5} = 1.58$

$d_2(\text{Paul, Pierre}) = \sqrt{16.25} = 4.03$

$d_1(\text{Alain, Marc}) = 2.5$

$d_1(\text{Alain, Paul}) = 5.0$

$d_1(\text{Alain, Pierre}) = 4.5$

$d_1(\text{Jean, Marc}) = 4.5$

$d_1(\text{Jean, Paul}) = 5.0$

$d_1(\text{Jean, Pierre}) = 5.5$

$d_1(\text{Marc, Paul}) = 6.5$

$d_1(\text{Marc, Pierre}) = 2.0$

$d_1(\text{Paul, Pierre}) = 7.5$.

The Euclidean **distance table** is obtained by placing these results in the following table:

	Alain	Jean	Marc	Paul	Pierre
Alain	0	2.12	1.5	2.74	2.29
Jean	2.12	0	2.5	3	3.28
Marc	1.5	2.5	0	2.78	1.58
Paul	2.74	3	2.78	0	4.03
Pierre	2.29	3.28	1.58	4.03	0

Similarly, we can obtain the following absolute **distance table**:

	Alain	Jean	Marc	Paul	Pierre
Alain	0	4	2.5	5	4.5
Jean	4	0	4.5	5	5.5
Marc	2.5	4.5	0	6.5	2
Paul	5	5	6.5	0	7.5
Pierre	4.5	5.5	2	7.5	0

Note that the order of proximity of the individuals can vary depending on the chosen distance. For example, if the Euclidean distance is used, it is Alain who is closest to Marc ($d_2(\text{Alain, Marc}) = 1.5$), whereas Alain comes in second place if the absolute distance is used, ($d_1(\text{Alain, Marc}) = 2.5$), and it is Pierre who is closest to Marc, with a distance of 2 ($d_1(\text{Pierre.Marc}) = 2$).

FURTHER READING
▶ **Chi-square distance**
▶ **Cluster analysis**
▶ **Distance table**
▶ **Mahalanobis distance**
▶ **Measure of dissimilarity**

REFERENCES
Abdi, H.: Distance. In: Salkind, N.J. (ed.) Encyclopedia of Measurement and Statistics. Sage, Thousand Oaks (2007)

Distance Table

For a set of r individuals, the distance table is a square **matrix** of order r, and the general term (i, j) equals the **distance** between the ith and the jth individual. Thanks to the properties of distance, the distance table is symmetrical and its diagonal is zero.

MATHEMATICAL ASPECTS

Suppose we have r individuals, denoted $X_1, X_2, \ldots, X_r$; the distance table constructed from these r individuals is a square **matrix** D of order r and the general term is

$$d_{ij} = d\left(X_i, X_j\right),$$

where $d\left(X_i, X_j\right)$ is the **distance** between the ith and the jth individual.

The properties of distance allow us to state that:

- The diagonal elements d_{ii} of D are zero because $d(X_i, X_i) = 0$ (for $i = 1, 2, \ldots, r$),
- D is symmetrical; that is, $d_{ij} = d_{ji}$ for each i and for values of j from 1 to r.

DOMAINS AND LIMITATIONS

Using a distance table provides a type of **cluster analysis**, where the aim is to identify the closest or the most similar individuals in order to be able to group them.

The distances entered into distance table are generally assumed to be Euclidian **distances**, unless a different type of distance has been specified.

EXAMPLES

Let us take the example of the grades obtained by five students in four examinations: English, French, mathematics and physics.

These grades, which can take any multiple of 0.5 from 1 to 6, are summarized in the following table:

	English	French	Maths	Physics
Alain	5.0	3.5	4.0	4.5
Jean	3.5	4.0	5.0	5.5
Marc	4.5	4.5	4.0	3.5
Paul	6.0	5.5	5.5	5.0
Pierre	4.0	4.5	2.5	3.5

We attempt to measure the **distances** that separate pairs of students. To do this, we compare two lines of the table (corresponding to two students) at a time: to be precise, each grade in the second line is subtracted from the corresponding grade in the first line.

For example, if we calculate the distance from Alain to Jean using the Euclidian distance, we get:

$$d(Alain.Jean) = ((5.0 - 3.5)^2 + (3.5 - 4.0)^2 + (4.0 - 5.0)^2 + (4.5 - 5.5)^2)^{\frac{1}{2}}$$
$$= \sqrt{2.25 + 0.25 + 1 + 1}$$
$$= \sqrt{4.5} = 2.12 .$$

In the same way, the distance between Alain and Marc equals:

$$d(Alain, Marc) = ((5.0 - 4.5)^2 + (3.5 - 4.5)^2 + (4.0 - 4.0)^2 + (4.5 - 3.5)^2)^{\frac{1}{2}}$$
$$= \sqrt{0.25 + 1 + 0 + 1}$$
$$= \sqrt{2.25} = 1.5 .$$

And so on,

$$d(Alain, Paul) = \sqrt{7.5} = 2.74$$
$$d(Alain, Pierre) = \sqrt{5.25} = 2.29$$
$$d(Jean, Marc) = \sqrt{6.25} = 2.5$$
$$d(Jean, Paul) = \sqrt{9} = 3$$
$$d(Jean, Pierre) = \sqrt{10.75} = 3.28$$
$$d(Marc, Paul) = \sqrt{7.75} = 2.78$$
$$d(Marc, Pierre) = \sqrt{2.5} = 1.58$$
$$d(Paul, Pierre) = \sqrt{16.25} = 4.03 .$$

By organizing these results into a table, we obtain the following distance table:

	Alain	Jean	Marc	Paul	Pierre
Alain	0	2.12	1.5	2.74	2.29
Jean	2.12	0	2.5	3	3.28
Marc	1.5	2.5	0	2.78	1.58
Paul	2.74	3	2.78	0	4.03
Pierre	2.29	3.28	1.58	4.03	0

We can see that the values along the diagonal equal zero and that the **matrix** is symmetrical. This means that the distance from each student to the same student is zero, and the distance that separates the first student from the second is identical to the distance that separates the second from the first. So, $d(X_i, X_i) = 0$ and $d(X_i, X_j) = d(X_j, X_i)$.

FURTHER READING
► **Cluster analysis**
► **Dendrogram**
► **Distance**
► **Matrix**

Distribution Function

The distribution function of a **random variable** is defined for each real number as the **probability** that the random variable takes a **value** less than or equal to this real number.

Depending on whether the **random variable** is discrete or continuous, we obtain a **discrete distribution function** or a **continuous distribution function**.

HISTORY
See **probability**.

MATHEMATICAL ASPECTS
We define the distribution function of a **random variable** X at value b (with $0 < b < \infty$) by:

$$F(b) = P(X \le b).$$

FURTHER READING
► **Continuous distribution function**
► **Discrete distribution function**
► **Probability**

► **Random variable**
► **Value**

Dot Plot

The dot plot is a type of **frequency** graph. When the **data** set is relatively small (up to about 25 data points), the data and their **frequencies** can be represented in a dot plot. This type of **graphical representation** is particularly useful for identifying **outliers**.

MATHEMATICAL ASPECTS
The dot plot is constructed in the following way. A **scale** is established that contains all of the **values** in the **data** set. Each datum is then individually marked on the graph as a dot above the corresponding value; if there are several data with the same value, the dots are aligned one on top of each other.

EXAMPLES
Consider the following **data**:

5, 2, 4, 5, 3, 2, 4, 3, 5, 2, 3, 4, 3, 17 .

To construct the dot plot, we establish a scale from 0 to 20.
We then represent the **data** with dots above each **value**. We obtain:

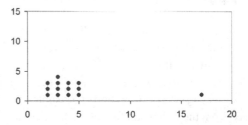

Note that this series includes an **outlier**, at 17.

FURTHER READING
▶ **Line chart**
▶ **Frequency distribution**
▶ **Histogram**
▶ **Graphical representation**

Durbin–Watson Test

The Durbin–Watson test introduces a **statistic** d that is used to test the autocorrelation of the residuals obtained from a linear regression model. This is a problem that often appears during the application of a linear model to a **time series**, when we want to test the independence of the residuals obtained in this way.

HISTORY
Durbin, J. and Watson, G.S. invented this test in 1950.

MATHEMATICAL ASPECTS
Consider the case of a **multiple linear regression** model containing $p - 1$ independent variables. The model is written:

$$Y_t = \beta_0 + \sum_{j=1}^{p-1} \beta_j X_{jt} + \varepsilon_t, \quad t = 1, \ldots, T,$$

where

Y_t is the **dependent variable**,

X_{jt}, with $j = 1, \ldots, p - 1$ are the independent variables

β_j, with $j = 1, \ldots, p - 1$ are the parameters to be estimated,

ε_t with $t = 1, \ldots, T$ is an unobservable random error term.

In the matrix form, the model is written as:

$$\mathbf{Y} = \mathbf{X}\boldsymbol{\beta} + \boldsymbol{\varepsilon},$$

where

$\mathbf{Y}$ is the vector $(n \times 1)$ of the observations related to the dependent variable (n observations),

$\boldsymbol{\beta}$ is a vector $(p \times 1)$ of the parameters to be estimated,

$\boldsymbol{\varepsilon}$ is a vector $(n \times 1)$ of the errors,

and $\mathbf{X} = \begin{pmatrix} 1 & X_{11} & \cdots & X_{1(p-1)} \\ \vdots & \vdots & & \vdots \\ 1 & X_{n1} & \cdots & X_{n(p-1)} \end{pmatrix}$ is the

matrix $(n \times p)$ of the design associated with the independent variables.

The residuals, obtained by the method of the **least squares**, are given by:

$$\mathbf{e} = \mathbf{Y} - \widehat{\mathbf{Y}} = \left[I - \mathbf{X} \left(\mathbf{X}'\mathbf{X} \right)^{-1} \mathbf{X}' \right] \mathbf{Y}.$$

The **statistic** d is defined as follows:

$$d = \frac{\sum_{t=2}^{T} \left(e_t - e_{t-1} \right)^2}{\sum_{t=1}^{T} e_t^2}.$$

The statistic d tests the hypothesis that the errors are independent against the alternative that they follow a first-order autoregressive process:

$$\varepsilon_t = \rho \varepsilon_{t-1} + u_t,$$

where $|\rho| < 1$ and the u_t are independent and are normally distributed with a **mean** of zero and a **variance** of σ^2. We call the term ρ the **autocorrelation**. In terms of ρ, the **hypothesis testing** is written as follows:

$$H_0: \quad \rho = 0$$
$$H_1: \quad \rho \neq 0.$$

DOMAINS AND LIMITATIONS
The construction of this statistic means that it can take values between 0 and 4. We have

$d = 2$ when $\widehat{\rho} = 0$. We denote the sample estimate of the observation of ρ by $\widehat{\rho}$. In order to test the hypothesis H_0, Durbin and Watson tabulated the critical values of d at a significance level of 5%; these critical values depend on the number of observations T and the number of explanatory variables $(p - 1)$. The table to test the positive autocorrelation at the α-level, the statistic test d is compared to lower and upper critical values (d_L, d_U), and

- reject H_0: $p = 0$ if $d < d_L$
- do not reject H_0 if $d > d_U$
- for $d_L < d < d_U$ the test is inconclusive.

Similarily to test negative autocorrelation at the α-level the statistic 4-d is compared to lower and upper critical values (d_L, d_U), and

- reject H_0: $p = 0$ if $H_4 - d < d_L$
- do not reject H_0 if $4 - d > d_U$
- for $d_L < 4 - d < d_U$ the test is nonconclusive.

Note that the model must contain a constant term, because established Durbin–Watson tables generally apply to models with a constant term. The variable to be explained should not be among the explanatory variables.

EXAMPLES

We consider an example of the number of highway accidents in the United States per million miles traveled by car between 1970 and 1984. The data for these years are given below:

4.9, 4.7, 4.5, 4.3, 3.6, 3.4,
3.3, 3.4, 3.4, 3.5, 3.5, 3.3,
2.9, 2.7, 2.7.

The following figure illustrates how the data are distributed with time. On the horizontal axis, zero corresponds to the year 1970, and the 15 to the year 1984.

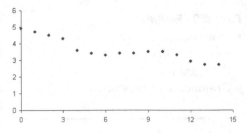

Number of accidents $= a + b \cdot t$,

with $t = 0, 1, \ldots, 14$.

The equation for the simple linear regression line obtained for these data is given below:

Number of accidents $= 4.59 - 0.141 \cdot t$.

We then obtain the following residuals:

0.31, 0.25,   0.19, 0.13, -0.43, -0.49,
-0.45, -0.21, -0.07, 0.17,   0.32,   0.26,
0.00, -0.06, 0.08.

The value of the statistic d is 0.53. In the Durbin–Watson table for a one-variable regression, we find the values:

$$d_{L,0.05} = 1.08$$

$$d_{U,0.05} = 1.36.$$

Therefore, we have the case where $0 < d < d_L$, so we reject the H_0 hypothesis in the favor of positive autocorrelation between the residuals of the model. The following figure illustrates the positive correlation between the residuals with time:

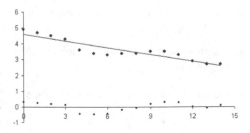

FURTHER READING
► **Autocorrelation**
► **Multiple linear regression**
► **Residual**
► **Simple linear regression**
► **Statistics**
► **Time series**

REFERENCES
Bourbonnais, R.: Econométrie, manuel et exercices corrigés, 2nd edn. Dunod, Paris (1998)

Durbin, J.: Alternative method to d-test. Biometrika **56**, 1–15 (1969)

Durbin, J., Watson, G.S.: Testing for serial correlation in least squares regression, I. Biometrika **37**, 409–428 (1950)

Durbin, J., Watson, G.S.: Testing for serial correlation in least squares regression, II. Biometrika **38**, 159–177 (1951)

Harvey, A.C.: The Econometric Analysis of Time Series. Philip Allan, Oxford (Wiley, New York) (1981)

D

E

Econometrics

Econometrics concerns the application of statistical methods to economic **data**. Economists use **statistics** to test their theories or to make forecasts.

Since economic data are not experimental in nature, and often contain some randomness, econometrics uses stochastic **models** rather than deterministic models.

HISTORY

The advantages of applying mathematics and **statistics** to the field of economics were quickly realized. In the second part of the seventeenth century, Petty, Sir William (1676) published an important article introducing the methodological foundations of econometrics.

According to Jaffé, W. (1968), Walras, Léon, professor at the University of Lausanne, is recognized as having originated general equilibrium economic theory, which provides the theoretical basis for modern econometrics.

On the 29th December 1930, in Cleveland, Ohio, a group of economists, statisticians and mathematicians founded an Econometrics Society to promote research into mathematical and statistical theories associated with the fields of economics. They created the bimonthly review "Econometrica,"

which appeared for the first time in January 1933.

The development of simultaneous equation **models** that could be used in **time series** for economic **forecasts** was the moment at which econometrics emerged as a distinct field, and it remains an important part of econometrics today.

FURTHER READING
▶ **Multiple linear regression**
▶ **Simple linear regression**
▶ **Time series**

REFERENCES

Jaffé, W.: Walras, Léon. In: Sills, D.L. (ed.) International Encyclopedia of the Social Sciences, vol. 16. Macmillan and Free Press, New York, pp. 447–453 (1968)

Petty, W.: Political arithmetic. In: William Petty, The Economic Writings . . . , vol 1. Kelley, New York, pp. 233–313 (1676)

Wonnacott, R.J., Wonnacott, T.H.: Econometrics. Wiley, New York (1970)

Edgeworth, Francis Y.

Edgeworth, Francis Ysidro was born in 1845 at Edgeworthstown (Ireland) and died in

1926. He contributed to various subjects, such as morality, economic sciences, probability and statistics. His early and important work on indices and the theory of utility were later followed by a statistical work that is now sometimes called "Bayesian".

The terms "module" and "fluctuation" originated with him.

He became the first editor of the Economic Journal in 1881.

Principal work of Edgeworth, Francis Ysidro:

1881   Mathematical Physics: An Essay on the Application of Mathematics to the Moral Sciences. Kegan Paul, London.

Efron, Bradley

Efron, Bradley was born in St. Paul, Minnesota in May 1938, to Efron, Esther and Efron, Miles, Jewish–Russian immigrants. He graduated in mathematics in 1960. During that year he arrived at Stanford, where he obtained his Ph.D. under the direction of Miller, Rupert and Solomon, Herb in the statistics department. He has taken up several positions at the University: Chair of Statistics, Associate Dean of Science, Chairman of the University Advisory Board, and Chair of the Faculty Senate. He is currently Professor of Statistics and Biostatistics at Stanford University.

He has been awarded many prizes, including the Ford Prize, the MacArthur Prize and the Wilks Medal for his research work into the use of computer applications in statistics, particularly regarding bootstrap and the jackknife techniques. He is a Member of the National Academy of Sciences and the American Academy of Arts and Science. He holds a fellowship in the IMS and the ASA. He also became President of the American Statistical Association in 2004.

The term "computer intensive statistical methods" originated with him and Diaconis, P. His research interests cover a wide range of statistical topics.

Some principal works and articles of Efron, Bradley:

1993   (with Tibshirani, R.J.) An Introduction to the Bootstrap. Chapman & Hall .

1983   Estimating the error rate of a prediction rule: Improvement on cross-validation. J. Am. Stat. Assoc. 78, 316-331.

1982   Jackknife, the Bootstrap and Other Resampling Plans. SIAM, Philadelphia, PA.

FURTHER READING
► **Bootstrap**
► **Resampling**

Eigenvalue

Let **A** be a square **matrix** of order n. If there is a **vector x** $\neq 0$ that gives a multiple $(k \cdot x)$ of itself when multiplied by **A**, then this vector is called an **eigenvector** and the multiplicative factor k is called an eigenvalue of the matrix **A**.

MATHEMATICAL ASPECTS
Let **A** be a square **matrix** of order n and **x** be a nonzero **eigenvector**. k is an eigenvalue of **A** if

$$\mathbf{A} \cdot \mathbf{x} = k \cdot \mathbf{x} .$$

that is, $(\mathbf{A} - k \cdot \mathbf{I}_n) \cdot \mathbf{x} = 0$, where $\mathbf{I}_n$ is the **identity matrix** of order n.

As $\mathbf{x}$ is nonzero, we can determine the possible values of k by finding the solutions of the equation $|\mathbf{A} - k \cdot \mathbf{I}_n| = 0$.

DOMAINS AND LIMITATIONS

For a square **matrix A** of order n, the **determinant** $|\mathbf{A} - k \cdot \mathbf{I}_n|$ is a polynomial of degree n in k that has at the most n not necessarily unique solutions. There will therefore be n eigenvalues at the most; some of these may be zeros.

EXAMPLES

Consider the following square **matrix** of order 3:

$$\mathbf{A} = \begin{bmatrix} 4 & 3 & 2 \\ 0 & 1 & 0 \\ -2 & 2 & 0 \end{bmatrix}$$

$$\mathbf{A} - k \cdot \mathbf{I}_3 = \begin{bmatrix} 4 & 3 & 2 \\ 0 & 1 & 0 \\ -2 & 2 & 0 \end{bmatrix}$$

$$- k \cdot \begin{bmatrix} 1 & 0 & 0 \\ 0 & 1 & 0 \\ 0 & 0 & 1 \end{bmatrix}$$

$$\mathbf{A} - k \cdot \mathbf{I}_3 = \begin{bmatrix} 4-k & 3 & 2 \\ 0 & 1-k & 0 \\ -2 & 2 & -k \end{bmatrix}.$$

The **determinant** of $\mathbf{A} - k \cdot \mathbf{I}_3$ is given by:

$$\begin{aligned} |\mathbf{A} - k \cdot \mathbf{I}_3| &= (1 - k) \cdot [(4 - k) \\ &\quad \cdot (-k) - 2 \cdot (-2)] \\ &= (1 - k) \cdot (k^2 - 4 \cdot k + 4) \\ &= (1 - k) \cdot (k - 2)^2. \end{aligned}$$

The eigenvalues are obtained by finding the solutions of the equations $(1 - k) \cdot (k - 2)^2 =$ 0. We find

$$k_1 = 1 \quad \text{and}$$
$$k_2 = k_3 = 2.$$

Therefore, two of the three eigenvalues are confounded.

FURTHER READING
▶ **Determinant**
▶ **Eigenvector**
▶ **Matrix**
▶ **Vector**

Eigenvector

Let $\mathbf{A}$ be a square **matrix** of order n. An "eigenvector" of $\mathbf{A}$ is a **vector x** containing n components (not all zeros) where the matrix product of $\mathbf{A}$ by $\mathbf{x}$ is a multiple of $\mathbf{x}$, for example $k \cdot \mathbf{x}$.

We say that the eigenvector $\mathbf{x}$ is associated with an **eigenvalue** k.

DOMAINS AND LIMITATIONS

Let $\mathbf{A}$ be a square **matrix** of order n; then the **vector x** (not zero) is an eigenvector of $\mathbf{A}$ if there is a number k such that

$$\mathbf{A} \cdot \mathbf{x} = k \cdot \mathbf{x}.$$

As $\mathbf{x}$ is nonzero, we determine the possible values of k by finding solutions to the equation $|\mathbf{A} - k \cdot \mathbf{I}_n| = 0$, where $\mathbf{I}_n$ is the **identity matrix** of order n.

We find the corresponding eigenvectors by resolving the system of equations defined by:

$$(\mathbf{A} - k \cdot \mathbf{I}_n) \cdot \mathbf{x} = 0.$$

Every nonzero multiple of an eigenvector can be considered to be an eigenvector. If c

E

is a nonzero constant, then

$$\mathbf{A} \cdot \mathbf{x} = k \cdot \mathbf{x}$$
$$c \cdot (\mathbf{A} \cdot \mathbf{x}) = c \cdot (k \cdot \mathbf{x})$$
$$\mathbf{A} \cdot (c \cdot \mathbf{x}) = k \cdot (c \cdot \mathbf{x}).$$

Therefore, $c \cdot \mathbf{x}$ is also an eigenvector if c is not zero.

It is possible to call an eigenvector, the vector of the **norm** 1 or unit vector on the axis of $c \cdot \mathbf{x}$, and the eigenvector will be on the **factorial axis**. This vector is unique (up to a sign).

EXAMPLES

We consider the following square **matrix** of order 3:

$$\mathbf{A} = \begin{bmatrix} 4 & 3 & 2 \\ 0 & 1 & 0 \\ -2 & 2 & 0 \end{bmatrix}.$$

The eigenvalues of this matrix $\mathbf{A}$ are obtained by making the **determinant** of $(\mathbf{A} - k \cdot \mathbf{I_3})$ equal to zero

$$\mathbf{A} - k \cdot \mathbf{I_3} = \begin{bmatrix} 4 & 3 & 2 \\ 0 & 1 & 0 \\ -2 & 2 & 0 \end{bmatrix}$$

$$- k \cdot \begin{bmatrix} 1 & 0 & 0 \\ 0 & 1 & 0 \\ 0 & 0 & 1 \end{bmatrix}$$

$$\mathbf{A} - k \cdot \mathbf{I_3} = \begin{bmatrix} 4-k & 3 & 2 \\ 0 & 1-k & 0 \\ -2 & 2 & -k \end{bmatrix}.$$

The determinant of $\mathbf{A} - k \cdot \mathbf{I_3}$ is given by:

$$|\mathbf{A} - k \cdot \mathbf{I_3}| = (1-k) \cdot [(4-k)$$
$$\cdot (-k) - 2 \cdot (-2)]$$
$$= (1-k) \cdot \left(k^2 - 4 \cdot k + 4\right)$$
$$= (1-k) \cdot (k-2)^2.$$

Hence, the eigenvalues are equal to :

$$k_1 = 1 \quad \text{and}$$
$$k_2 = k_3 = 2.$$

We now compute the eigenvectors associated with the eigenvalue k_1:

$$k_1 = 1,$$

$$\mathbf{A} - k_1 \cdot \mathbf{I_3} = \begin{bmatrix} 4-1 & 3 & 2 \\ 0 & 1-1 & 0 \\ -2 & 2 & -1 \end{bmatrix}$$

$$= \begin{bmatrix} 3 & 3 & 2 \\ 0 & 0 & 0 \\ -2 & 2 & -1 \end{bmatrix}.$$

Denoting the components of the eigenvector x as

$$\mathbf{x} = \begin{bmatrix} x_1 \\ x_2 \\ x_3 \end{bmatrix}.$$

we then need to resolve the following system of equations:

$$(\mathbf{A} - k_1 \cdot \mathbf{I_3}) \cdot \mathbf{x} = 0$$

$$\begin{bmatrix} 3 & 3 & 2 \\ 0 & 0 & 0 \\ -2 & 2 & -1 \end{bmatrix} \cdot \begin{bmatrix} x_1 \\ x_2 \\ x_3 \end{bmatrix} = \begin{bmatrix} 0 \\ 0 \\ 0 \end{bmatrix}.$$

that is

$$3x_1 + 3x_2 + 2x_3 = 0$$
$$-2x_1 + 2x_2 - x_3 = 0.$$

where the second equation has been eliminated because it is trivial.

The solution to this system of equations depending on x_2 is given by:

$$\mathbf{x} = \begin{bmatrix} 7 \cdot x_2 \\ x_2 \\ -12 \cdot x_2 \end{bmatrix}.$$

The vector $\mathbf{x}$ is an eigenvector of the matrix $\mathbf{A}$ for every nonzero **value** of x_2. To find the unit eigenvector, we let:

$$1 = \| \mathbf{x} \|^2 = (7 \cdot x_2)^2 + (x_2)^2 + (-12 \cdot x_2)^2$$
$$= (49 + 1 + 144) \cdot x_2^2 = 194 \cdot x_2^2.$$

from where $x_2^2 = \dfrac{1}{194}$,

and so $x_2 = \dfrac{1}{\sqrt{194}} = 0.0718$.

The unit eigenvector of the matrix $\mathbf{A}$ associated with the eigenvalue $k_1 = 1$ then equals:

$$\mathbf{x} = \begin{bmatrix} 0.50 \\ 0.07 \\ -0.86 \end{bmatrix}.$$

FURTHER READING
▶ **Determinant**
▶ **Eigenvalue**
▶ **Eigenvector**
▶ **Factorial axis**
▶ **Matrix**

REFERENCES

Dodge, Y.: Mathématiques de base pour économistes. Springer, Berlin Heidelberg New York (2002)

Meyer, C.D.: Matrix analysis and Applied Linear Algebra, SIAM (2000)

Epidemiology

Epidemiology (the word derives from the Greek words "epidemos" and "logos," meaning "study of epidemics") is the study of the frequency and distribution of illness, and the parameters and the risk factors that define the state of health of a population, in relation to time, space and groups of individuals.

HISTORY

Epidemiology originated in the seventeenth century, when Petty, William (1623–1687), an English doctor, economist and scientist, collected information on the population, which he described in the work *Political Arithmetic*. Graunt, John (1620–1694) proposed the first rigorous analysis of causes of death in 1662. **Quetelet, Adolphe** (1796–1874), Belgian astronomer and mathematician, is considered to be the foundor of modern population statistics, the mother discipline of epidemiology, statistics, econometrics and other quantitative disciplines describing social and biological characteristics of human life. With the work of Farr, William (1807–1883), epidemiology was recognized as a separate field from statistics, since he studied the causes of death and the way it varied with age, gender, season, place of residence and profession. The Scottish doctor Lind, James (1716–1794) proved that eating citrus fruit stopped scurvy, which marked an important step in the history of the epidemiology. Snow, John (1813–1858), a doctor in London, showed theat epidemics propagated by contagions, which was underlined by an in situ study that proved that drinking Thames water polluted by refuse from sewers contributes to the dissemination of cholera.

MATHEMATICAL ASPECTS

See **cause and effect in epidemiology, odds and odds ratio, relative risk, attributable risk, avoidable risk, incidence rate, prevalence rate**.

EXAMPLES

Let us take an example involving the mortality rates in two fictitious regions. Suppose that we want to analyze the annual mortality in two regions A and B, each populated with 10000 inhabitants. The mortality data are indicated in the following table:

	Region A	Region B
Deaths/Total population	500/ 10000	1000/ 10000
Total mortality per year	5%	10%
Annual mortality rate among those < 60 years old	5%	5%
Annual mortality rate among those > 60 years old	15%	15%
% of population < 60 years old	100%	50%

From these data, can we state that the inhabitants of region B encounter an increased chance of mortality than the inhabitants of region A? In other words, would we recommend that people leave region A to live in B? If we divide both populations by age, we note that, in spite, the different total mortality rates of regions A and B do not seem to be explained by the different age structures of A and B.

The previous example highlights the multitude of factors that should be considered during data analysis. The principal factors that we should consider during an epidemiological study include: age (as an absolute age and as the generation that the individual belongs to), gender and ethnicity, among others.

FURTHER READING
▶ **Attributable risk**
▶ **Avoidable risk**
▶ **Biostatistics**
▶ **Cause and effect in epidemiology**
▶ **Incidence rate**
▶ **Odds and odds ratio**
▶ **Prevalence rate**
▶ **Relative risk**

REFERENCES

Lilienfeld, A.M., Lilienfeld, D.E.: Foundations of Epidemiology, 2nd edn. Clarendon, Oxford (1980)

MacMahon, B., Pugh, T.F.: Epidemiology: Principles and Methods. Little Brown, Boston, MA (1970)

Morabia, A.: Epidemiologie Causale. Editions Médecine et Hygiène, Geneva (1996)

Rothmann, J.K.: Epidemiology. An Introduction. Oxford University Press (2002)

Error

The error is the difference between the estimated **value** and the true value (or reference value) of the concerned quantity.

HISTORY
The error distributions are the **probability** distributions which describe the error that appears during repeated measures of a same quantity under the same conditions. They were introduced in the second half of the eighteenth century to illustrate how the **arithmetic mean** can be used to obtain a good approximation of the reference **value** of the studied quantity.

In a letter to the President of the Royal Society, Simpson, T. (1756) suggested that the probability of obtaining a certain error during an **observation** can be described by a **discrete probability distribution**. He proposed the first two error distributions. Other discrete distributions were proposed and studied by Lagrange, J.L. (1776).

Simpson, T. (1757) also proposed continuous error distributions, as did **Laplace, P.S.** (1774 and 1781). However, the most important error distribution, the **normal distribution** was proposed by **Gauss, C.F.** (1809).

MATHEMATICAL ASPECTS

In the field of metrology, the following types of errors can be distinguished:

- The absolute error, which is the absolute value of the difference between the observed **value** and the reference value of a certain quantity;
- The relative error, which is the ratio between the absolute error and the value of the quantity itself. It characterizes the accuracy of a physical measure;
- The accidental error, which is not related to the measuring tool but to the experimenter himself;
- The experimental error, which is the error due to uncontrolled **variables**;
- The random error, which is the chance error resulting from a **combination** of errors due to the instrument and/or the user. The statistical properties of random errors and their **estimation** are studied using **probabilities**;
- The residual error, which corresponds to the difference between the estimated **value** and the observed value. The term **residual** is sometimes used for this error;
- The systematic error, the error that comes from consistent causes (for example the improper calibration of a measuring instrument), and which always happens in the same direction. The **bias** used in statistics is a particular example of this;
- The rounded error, which is the error that is created when a numerical **value** is replaced by a truncated value close to it. When performing calculations with

numbers that have n decimal places, the value of the rounded error is located between

$$-\frac{1}{2} \cdot 10^{-n} \quad \text{and} \quad \frac{1}{2} \cdot 10^{-n}.$$

During long calculation procedures, rounded errors can accumulate and produce very imprecise results. This type of error is often encountered when using a computer.

Every error that occurs in a **statistical** problem is a combination of the different types of errors listed above.

In **statistics**, and more specifically during **hypothesis testing**, there is also:

- **Type I error**, corresponding to the error committed when the **null hypothesis** is rejected when it is true;
- **Type II error**, the error committed when the null hypothesis is not rejected when it is false.

FURTHER READING

▶ **Analysis of residuals**
▶ **Bias**
▶ **Estimation**
▶ **Hypothesis testing**
▶ **Residual**
▶ **Type I error**
▶ **Type II error**

REFERENCES

Gauss, C.F.: Theoria Motus Corporum Coelestium. Werke, 7 (1809)

Lagrange, J.L.: Mémoire sur l'utilité de la méthode de prendre le milieu entre les résultats de plusieurs observations; dans lequel on examine les avantages de cette méthode par le calcul des probabilités; et où l'on résoud différents problèmes relatifs à cette matière. Misc. Taurinensia **5**, 167–232 (1776)

Laplace, P.S. de: Mémoire sur la probabilité des causes par les événements. Mem. Acad. Roy. Sci. (presented by various scientists) **6**, 621–656 (1774) (or Laplace, P.S. de (1891) Œuvres complètes, vol 8. Gauthier-Villars, Paris, pp. 27–65)

Laplace, P.S. de: Mémoire sur les probabilités. Mem. Acad. Roy. Sci. Paris, 227–332 (1781) (or Laplace, P.S. de (1891) Œuvres complètes, vol. 9. Gauthier-Villars, Paris, pp. 385–485.)

Simpson, T.: A letter to the Right Honorable George Earl of Macclesfield, President of the Royal Society, on the advantage of taking the mean of a number of observations in practical astronomy. Philos. Trans. Roy. Soc. Lond. **49**, 82–93 (1755)

Simpson, T.: Miscellaneous Tracts on Some Curious and Very Interesting Subjects in Mechanics, Physical-Astronomy and Speculative Mathematics. Nourse, London (1757)

Estimation

Estimation is the procedure that is used to determine the **value** of a particular **parameter** associated with a **population**. To estimate the parameter, a **sample** is drawn from the population and the value of the **estimator** for the unknown parameter is calculated.

Estimation is divided into two large **categories**: **point estimation** and **interval estimation**.

HISTORY

The concept of estimation dates back to the first works on mathematical **statistics**, notably by **Bernoulli, Jacques** (1713), **Laplace, P.S.** (1774) and **Bernoulli, Daniel** (1778).

The greatest advance in the theory of estimation, after the introduction of the **least squares** method, was probably the formulation of the **moments** method by **Pearson, K.** (1894, 1898). However, the foundations of the theory of estimation is due to **Fisher, R.A.**. In his first work of 1912, he introduced the **maximum likelihood** method. In 1922 he wrote a fundamental paper that clearly described what estimation really is for the first time.

Fisher, R.A. (1925) also introduced a set of definitions that were adopted to describe **estimators**. Terms such as "biased," "efficient" and "sufficient" estimators were introduced by him in his estimation theory.

DOMAINS AND LIMITATIONS

The following graph shows the relationship between sampling and estimation. "Sampling" is the process of obtaining a sample from a population, while "estimation" is the reverse process: from the sample to the population.

FURTHER READING
- ▶ **Confidence interval**
- ▶ **Estimator**
- ▶ **L_1 estimation**
- ▶ **Point estimation**
- ▶ **Robust estimation**
- ▶ **Sample**
- ▶ **Sampling**

REFERENCES

Armatte, M.: La construction des notions d'estimation et de vraisemblance chez Ronald A. Fisher. In: Mairesse, J. (ed.) Estimation et sondages. Economica, Paris (1988)

Bernoulli, D.: The most probable choice between several discrepant observations, translation and comments (1778). In: Pearson, E.S. and Kendall, M.G. (eds.) Studies in the History of Statistics and Probability. Griffin, London (1970)

Bernoulli, J.: Ars Conjectandi, Opus Posthumum. Accedit Tractatus de Seriebus infinitis, et Epistola Gallice scripta de ludo Pilae recticularis. Impensis Thurnisiorum, Fratrum, Basel (1713)

Fisher, R.A.: On an absolute criterium for fitting frequency curves. Mess. Math. **41**, 155–160 (1912)

Fisher, R.A.: On the mathematical foundations of theoretical statistics. Philos. Trans. Roy. Soc. Lond. Ser. A **222**, 309–368 (1922)

Fisher, R.A.: Theory of statistical estimation. Proc. Camb. Philos. Soc. **22**, 700–725 (1925)

Gauss, C.F.: Theoria Motus Corporum Coelestium. Werke, 7 (1809)

Laplace, P.S. de: Mémoire sur la probabilité des causes par les événements. Mem. Acad. Roy. Sci. (presented by various scientists) **6**, 621–656 (1774) (or Laplace, P.S. de (1891) Œuvres complètes, vol 8. Gauthier-Villars, Paris, pp. 27–65)

Laplace, P.S. de: Théorie analytique des probabilités. Courcier, Paris (1812)

Legendre, A.M.: Nouvelles méthodes pour la détermination des orbites des comètes. Courcier, Paris (1805)

Pearson, K.: Contributions to the mathematical theory of evolution. I. In: Karl Pearson's Early Statistical Papers. Cambridge University Press, Cambridge, pp. 1–40 (1948). First published in 1894 as: On the dissection of asymmetrical frequency curves. Philos. Trans. Roy. Soc. Lond. Ser. A **185**, 71–110

Pearson, K., Filon, L.N.G.: Mathematical contributions to the theory of evolution. IV: On the probable errors of frequency constants and on the influence of random selection on variation and correlation. Philos. Trans. Roy. Soc. Lond. Ser. A **191**, 229–311 (1898)

E

Estimator

Any statistical function of a **sample** used to estimate an unknown **parameter** of the **population** is called an estimator.

Any **value** obtained from this estimator is an **estimate** of the **parameter** of the **population**.

HISTORY

See **estimation**.

MATHEMATICAL ASPECTS

Consider θ, an unknown **parameter** defined for a **population**, and $(X_1, X_2, \ldots, X_n)$, a **sample** taken from this population. A statistical function (or simply a **statistic**) of this sample $g(X_1, \ldots, X_n)$ is used to estimate θ.

To distinguish the **statistic** $g(X_1, \ldots, X_n)$, which is a **random variable**, from the **value** that it takes in a particular case, it is common to call the statistic $g(X_1, \ldots, X_n)$ the estimator, and the **value** that is taken by it in a particular case is called an **estimate**.

An estimator should not be confused with a method of **estimation**.

A **parameter** of a **population** is usually denoted by a Greek letter, and its **estimator** by this same letter but with a circumflex accent (^) or a hat, to distinguish it from the parameter of the population. Sometimes the corresponding Roman letter is used instead.

DOMAINS AND LIMITATIONS

When calculating an estimator, such as the **arithmetic mean**, it is expected that the **value** of this estimator will approach the value of the true **parameter** (the **mean** of the **population** in this case) as the **sample size** increases.

It is therefore clear that a good estimator must be close to the true **parameter**. The closer the estimators cluster around the true parameters, the better the estimators.

In this sense, good estimators are strongly related to statistical **measures of dispersion** such as the **variance**.

In general estimators, should possess certain qualities, such as those described below.

1. *Estimator without bias:*

   An estimator $\hat{\theta}$ of an unknown **parameter** θ is said to be without bias if its **expected value** is equal to θ,

   $$E[\hat{\theta}] = \theta \,.$$

   The **mean** $\bar{x}$ and the **variance** S^2 of a **sample** are estimators without bias, respectively, of the mean μ and of the variance σ^2 of the **population**, with

   $$\bar{x} = \frac{1}{n} \cdot \sum_{i=1}^{n} x_i \quad \text{and}$$

   $$S^2 = \frac{1}{n-1} \cdot \sum_{i=1}^{n} (x_i - \bar{x})^2 \,.$$

2. *Efficient estimator:*

   An estimator $\hat{\theta}$ of a **parameter** θ is said to be efficient if the **variance** of $\hat{\theta}$ is smaller than the variance of any other estimator of θ.

   Therefore, for two estimators without bias of θ, one will be more efficient than the other if its **variance** is smaller.

3. *Consistent estimator:*

   An estimator $\hat{\theta}$ of a **parameter** θ is said to be consistent if the **probability** that it differs from θ decreases as the **sample size** increases.

   Therefore, an estimator is consistent if

   $$\lim_{n \to \infty} P(|\hat{\theta} - \theta| < \hat{\varepsilon}) = 1 \,,$$

however small the number $\hat{\varepsilon} > 0$ is.

EXAMPLES

Consider $(x_1, x_2, \ldots, x_n)$, a **sample** drawn from a **population** with a **mean** of μ and a **variance** of σ^2.

The estimator $\bar{x}$, calculated from the **sample** and used to estimate the **parameter** μ of the **population**, is an estimator without bias. Then:

$$E[\bar{x}] = E\left[\frac{1}{n} \sum_{i=1}^{n} x_i\right]$$

$$= \frac{1}{n} E\left[\sum_{i=1}^{n} x_i\right]$$

$$= \frac{1}{n} \sum_{i=1}^{n} E[x_i]$$

$$= \frac{1}{n} \sum_{i=1}^{n} \mu$$

$$= \frac{1}{n} n \cdot \mu$$

$$= \mu \,.$$

To estimate the **variance** σ^2 of the **population**, it is tempting to use the estimator S^2 calculated from the **sample** $(x_1, x_2, \ldots, x_n)$, which is defined here by:

$$S^2 = \frac{1}{n} \sum_{i=1}^{n} (x_i - \bar{x})^2 .$$

This estimator is a biased estimator σ^2 since:

$$\sum_{i=1}^{n} (x_i - \bar{x})^2 = \sum_{i=1}^{n} (x_i - \mu)^2 - n(\bar{x} - \mu)^2$$

in the following way:

$$\sum_{i=1}^{n} (x_i - \bar{x})^2$$

$$= \sum_{i=1}^{n} [(x_i - \mu) - (\bar{x} - \mu)]^2$$

$$= \sum_{i=1}^{n} \left[(x_i - \mu)^2 - 2(x_i - \mu)(\bar{x} - \mu) + (\bar{x} - \mu)^2 \right]$$

$$= \sum_{i=1}^{n} (x_i - \mu)^2 - 2(\bar{x} - \mu) \cdot \sum_{i=1}^{n} (x_i - \mu) + n(\bar{x} - \mu)^2$$

$$= \sum_{i=1}^{n} (x_i - \mu)^2 - 2(\bar{x} - \mu) \left(\sum_{i=1}^{n} x_i - n \cdot \mu \right) + n(\bar{x} - \mu)^2$$

$$= \sum_{i=1}^{n} (x_i - \mu)^2 - 2(\bar{x} - \mu)(n \cdot \bar{x} - n \cdot \mu) + n(\bar{x} - \mu)^2$$

$$= \sum_{i=1}^{n} (x_i - \mu)^2 - 2n(\bar{x} - \mu)(\bar{x} - \mu) + n(\bar{x} - \mu)^2$$

$$= \sum_{i=1}^{n} (x_i - \mu)^2 - 2n(\bar{x} - \mu)^2 + n(\bar{x} - \mu)^2$$

$$= \sum_{i=1}^{n} (x_i - \mu)^2 - n(\bar{x} - \mu)^2 .$$

Hence:

$$E[S^2] = E \left[\frac{1}{n} \sum_{i=1}^{n} (x_i - \bar{x})^2 \right]$$

$$= E \left[\frac{1}{n} \sum_{i=1}^{n} (x_i - \mu)^2 - (\bar{x} - \mu)^2 \right]$$

$$= \frac{1}{n} E \left[\sum_{i=1}^{n} (x_i - \mu)^2 \right] - E \left[(\bar{x} - \mu)^2 \right]$$

$$= \frac{1}{n} \sum_{i=1}^{n} E \left[(x_i - \mu)^2 \right] - E \left[(\bar{x} - \mu)^2 \right]$$

$$= \mathrm{Var}(x_i) - \mathrm{Var}(\bar{x})$$

$$= \sigma^2 - \frac{\sigma^2}{n}$$

$$= \frac{n-1}{n} \sigma^2 \neq \sigma^2 .$$

Therefore, to get an estimator of the **variance** σ^2 that is not biased, we avoid using S^2 and use the same value divided by $\frac{n-1}{n}$ instead:

$$S'^2 = \frac{n}{n-1} S^2$$

$$= \frac{n}{n-1} \cdot \frac{1}{n} \sum_{i=1}^{n} (x_i - \bar{x})^2$$

$$= \frac{1}{n-1} \sum_{i=1}^{n} (x_i - \bar{x})^2 .$$

FURTHER READING
▶ **Bias**
▶ **Estimation**
▶ **Sample**
▶ **Sampling**

REFERENCES
Hogg, R.V., Craig, A.T.: Introduction to Mathematical Statistics, 2nd edn. Macmillan, New York (1965)

Lehmann, E.L.: Theory of Point Estimation, 2nd edn. Wiley, New York (1983)

Event

In a **random experiment**, an event is a subset of the set the **sample space**. In other words, an event is a set of possible outcome in a random experiment.

MATHEMATICAL ASPECTS

Consider Ω, the **sample space** of a **random experiment**. Any subset of Ω is called an event.

Different Types of Events
- *Impossible events*

  An impossible event corresponds to an empty set. The **probability** of an impossible event is equal to 0:
  $$\text{if} \quad B = \phi, \quad P(B) = 0 .$$

- *Sure event*

  A sure event is an event that corresponds to the **sample space**. The **probability** of a sure event is equal to 1:
  $$P(\Omega) = 1 .$$

- *Simple event*

  A simple event is an event that only contains one outcome.

Operations on Events
- *Negation of an event*

  Consider the event E. The negation of this event is another event, called the opposite event of E, which is realized when E is not realized and vice versa. The opposite event of E is denoted by the set $\bar{E}$, the complement of E:
  $$\bar{E} = \Omega \backslash E = \{\omega \in \Omega; \omega \notin E\} .$$

- *Intersection of events*

  Consider the two events E and F. The intersection between these two events is another event that is realized when both events are realized simultaneously. It is called the "E and F" event, and it is denoted by $E \cap F$.
  $$E \cap F = \{\omega \in \Omega; \omega \in E \text{ and } \omega \in F\} .$$

  In a more general way, if $E_1, E_2, \ldots, E_r$ are r events, then
  $$\bigcap_{i=1}^{r} E_i = \{\omega \in \Omega; \omega \in E_i, \forall i = 1, \ldots, n\}$$

  is defined as the intersection of the r events, which is realized when all of the r events are realized simultaneously.

- *Union of events*

  Consider the two same events E and F. The union of these two events is another event, which is realized when at least one of the two events E or F is realized. It is sometimes called the "E or F" event, and it is denoted by $E \cup F$:
  $$E \cup F = \{\omega \in \Omega; \omega \in E \text{ or } \omega \in F\} .$$

Properties of Operations on the Events

Consider the three events E, F and G. The intersection and the union of these events satisfy the following properties:
- *Commutativity*:
  $$E \cap F = F \cap E ,$$
  $$E \cup F = F \cup E .$$

- *Associativity*:
  $$E \cap (F \cap G) = (E \cap F) \cap G ,$$
  $$E \cup (F \cup G) = (E \cup F) \cup G .$$

• *Idempotence*:

$$E \cap E = E,$$
$$E \cup E = E.$$

The distributivity of the intersection with respect to the union is:

$$E \cap (F \cup G) = (E \cap F) \cup (E \cap G).$$

The distributivity of the union with respect to the intersection is:

$$E \cup (F \cap G) = (E \cup F) \cap (E \cup G).$$

The main negation relations are:

$$\overline{E \cap F} = \bar{E} \cup \bar{F},$$
$$\overline{E \cup F} = \bar{E} \cap \bar{F}.$$

EXAMPLES

Consider a **random experiment** which involves simultaneously throwing a yellow die and a blue die.

The **sample space** of this **experiment** is formed from the set of pairs of possible scores from the two dice:

$$\Omega = \{(1, 1), (1, 2), (1, 3), \ldots, (2, 1),$$
$$(2, 2), \ldots, (6, 5), (6, 6)\}.$$

where, for example: $(1, 2) = $ "1″ on the yellow die and "2" on the blue die. The number of pairs (or simple events) that can be formed is given by the number of **arrangements** with repetition of two objects among six which is 36. For this **sample space**, it is possible to describe the following events, for example:

1. The sum of points is equal to six:

$$A = \{(1, 5), (2, 4), (3, 3), (4, 2), (5, 1)\}.$$

2. The sum of points is even:

$$B = \{(1, 1), (1, 3), (1, 5),$$
$$(2, 2), (2, 4), \ldots, (6, 4), (6, 6)\}.$$

3. The sum of points is less than 5:

$$C = \{(1, 1), (1, 2), (1, 3),$$
$$(2, 1), (2, 2), (3, 1)\}.$$

4. The sum of points is odd:

$$D = \{(1, 2), (1, 4), (1, 6),$$
$$(2, 1), (2, 3), \ldots, (6, 5)\} = \bar{B}.$$

5. The sum of points is even and less than 5:

$$E = \{(1, 1), (1, 3), (2, 2), (3, 1)\}$$
$$= B \cap C.$$

6. The blue die lands on "2":

$$F = \{(1, 2), (2, 2), (3, 2),$$
$$(4, 2), (5, 2), (6, 2)\}.$$

7. The blue die lands on "2" or the total score is less than 5:

$$G = \{(1, 2), (2, 2), (3, 2), (4, 2), (5, 2),$$
$$(6, 2), (1, 1), (1, 3), (2, 1), (3, 1)\}$$
$$= F \cup C.$$

8. A and D are incompatible.
9. A implies B.

FURTHER READING

▶ **Compatibility**
▶ **Conditional probability**
▶ **Independence**
▶ **Probability**
▶ **Random experiment**
▶ **Sample space**

Expected Value

The expected value of a **random variable** is the weighted **mean** of the **values** that the random variable can take, where the weight is the **probability** that a particular value is taken by the random variable.

HISTORY

The mathematical principle of the expected value first appeared in the work entitled *De Ratiociniis in Aleae Ludo*, published in 1657 by the Dutch scientist Huygens, Christiaan (1629–1695). His thoughts on the subject appear to have heavily influenced the later works of Pascal and Fermat on **probability**.

MATHEMATICAL ASPECTS

Depending on whether the **random variable** is discrete or continuous, we refer to either the expected value of a discrete random variable or the expected value of a continuous random variable.

Consider a discrete **random variable** X that has a **probability function** $p(X)$. The expected value of X, denoted by $E[X]$ or μ, is defined by

$$\mu = E[X] = \sum_{i=1}^{n} x_i p(x_i)$$

if X can take n **values**.

If the **random variable** X is continuous, the expected value becomes

$$\mu = E[X] = \int_D x f(x)\, dx,$$

if X takes **values** over the **interval** D, where $f(x)$ is the **density function** of X.

Properties of the Expected Value

Consider the two constants a and b, and the **random variable** X. We then have:

1. $E[aX + b] = aE[X] + b$;
2. $E[X + Y] = E[X] + E[Y]$;
3. $E[X - Y] = E[X] - E[Y]$;
4. If X and Y are independent, then:

$$E[X \cdot Y] = E[X] \cdot E[Y].$$

EXAMPLES

We will consider two examples, one concerning a discrete **random variable**, the other concerning a continuous random variable. Consider a game of chance where a die is thrown and the score noted. Suppose that you win a euro if it lands on an even number, two euros if it lands on a "1" or a "3," and lose three euros if it lands on a 5.

The **random variable** X considered here is the number of euros that are won or lost. The following table represents the different **values** of X and their respective probabilities:

X	−3	1	2
P(X)	1/6	3/6	2/6

The expected value is therefore equal to:

$$E[X] = \sum_{i=1}^{3} x_i p(x_i)$$
$$= -3 \cdot \tfrac{1}{6} + 1 \cdot \tfrac{3}{6} + 2 \cdot \tfrac{2}{6}$$
$$= \tfrac{2}{3}.$$

In other words, the player wins, on average, $\tfrac{2}{3}$ of a euro per throw.

Consider a continuous **random variable** X with the following **density function**:

$$f(x) = \begin{cases} 1 & \text{for } 0 < x < 1 \\ 0 & \text{elsewhere.} \end{cases}$$

The expected value of this **random variable** X is equal to:

$$E[X] = \int_0^1 x \cdot 1 \, dx$$
$$= \frac{x^2}{2} \Big|_0^1$$
$$= \frac{1}{2}.$$

FURTHER READING
► **Density function**
► **Probability function**
► **Weighted arithmetic mean**

Experiment

An experiment is an operation conducted under controlled conditions in order to discover a previously unknown effect, to test or establish a **hypothesis**, or to demonstrate a known law.

The goal of experimenting is to clarify the **relation** between the controllable conditions and the result of the experiment.

Experimental analysis is performed on **observations** which are affected not only by the controllable conditions, but also by uncontrolled conditions and measurement **errors**.

DOMAINS AND LIMITATIONS
When the possible results of the experiment can be described, and it is possible to attribute a **probability** of realization to each possible outcome, the experiment is called a **random experiment**. The set of all possible **outcome** of an experiment is called the **sample space**. A **factorial experiment** is when the experimenter organizes an experiment with two or more **factors**. The factors are the controllable conditions of the experiment. Experi-

mental errors come from uncontrolled conditions and from measurement **errors** that are present in any type of experiment.

An experimental design is established, depending on the aim of the experimentor or the possible resources that are available.

EXAMPLES
See **random experiment** and **factorial experiment**.

FURTHER READING
► **Design of experiments**
► **Factorial experiment**
► **Hypothesis testing**
► **Random experiment**

REFERENCES
Box, G.E.P., Hunter, W.G., Hunter, J.S.: Statistics for Experimenters. An Introduction to Design, Data Analysis, and Model Building. Wiley, New York (1978)

Rüegg, A.: Probabilités et statistique. Presses Polytechniques Romandes, Lausanne, Switzerland (1985)

Experimental Unit

An experimental unit is the smallest part of the experimental material to which we apply **treatment**.

The experimental design specifies the number of experimental units submitted to the different treatments.

FURTHER READING
► **Design of experiments**
► **Experiment**
► **Factor**
► **Treatment**

Exploratory Data Analysis

Exploratory data analysis is an approach to data analysis where the features and characteristics of the data are reviewed with an "open mind"; in other words, without attempting to apply any particular model to the data. It is often used upon first contact with the data, before any models have been chosen for the structural or stochastic components, and it is also used to look for deviations from common models.

HISTORY

Exploratory data analysis is a set of techniques that have been principally developed by **Tukey, John Wilder** since 1970. The philosophy behind this approach is to examine the data before applying a specific probability model. According to Tukey, J.W., exploratory data analysis is similar to detective work. In exploratory data analysis, these clues can be numerical and (very often) graphical. Indeed, Tukey introduced several new semigraphical data representation tools to help with exploratory data analysis, including the "box and whisker plot" (also known as the **box plot**) in 1972, and the **stem and leaf** diagram in 1977. This diagram is similar to the **histogram**, which dates from the eighteenth century.

MATHEMATICAL ASPECTS

A good way to summarize the essential information in a data with exploratory data analysis is provided by the *five-number summary*, which is presented in the form of a table as follows:

Median	
First quartile	Third quartile
Minimum	Maximum

Let n be the number of observations; we arrange the data in ascending order. We define:

$$\text{median rank} = \frac{n+1}{2}$$

$$\text{quartile rank} = \frac{\lfloor \text{median rank} \rfloor + 1}{2},$$

where $\lfloor x \rfloor$ is the value of x truncated down to the next smallest whole number.

In his book, **Tukey, J.W.** calls the first and the fourth quartiles "hinges".

DOMAINS AND LIMITATIONS

In the exploratory data analysis defined by Tukey, J.W., there are four principal topics. These are **graphical representation**, re-expression (which is simply the **transformation** of variables), residuals and resistance (which is synonymous to the concept of robustness). The resistance is a measure of sensitivity of the analysis or summary to "bad data". The need to study the resistance reflects the fact that even "good data rarely contains less then 5% error and so it is important to be able to protect the analysis against the adverse effects of error". Tukey's *resistant line* gives a robust fit to a set of points, which means that this line is not overly sensitive to any particular observation. The median is highly resistant, but the mean is not.

Graphical representations are used to analyze the behavior of the data, data fits, diagnostic measures and residuals. In this way we can spot any unexpected characteristics and recognizableir regularity in the data.

The development of the exploratory data analysis is closely associated with an emphasis on visual representation and the use of a variety of relatively new graphical techniques, such as the **stem and leaf** diagram and the **box plot**.

EXAMPLES

Income indices for Swiss cantons in 1993

Canton	Index	Canton	Index
Zurich	125.7	Schaffhouse	99.2
Bern	86.2	Appenzell Rh.-Ext.	84.3
Lucern	87.9	Appenzell Rh.-Int.	72.6
Uri	88.2	Saint-Gall	89.3
Schwytz	94.5	Grisons	92.4
Obwald	80.3	Argovie	98.0
Nidwald	108.9	Thurgovie	87.4
Glaris	101.4	Tessin	87.4
Zoug	170.2	Vaud	97.4
Fribourg	90.9	Valais	80.5
Soleure	88.3	Neuchatel	87.3
Basel-Stadt	124.2	Geneva	116.0
Basel-Land	105.1	Jura	75.1

The table provides the income indices of Swiss cantons per inhabitant (Switzerland = 100) in 1993. We can calculate the five-number summary statistics:

$$\text{median rank} = \frac{n+1}{2} = \frac{26+1}{2} = 13.5 \, .$$

Therefore, the median M_d is the mean of the thirteenth and the fourteenth observations:

$$M_d = \frac{89.3 + 90.9}{2} = 90.1 \, .$$

Then we calculate:

$$\text{quartile rank} = \frac{13+1}{2} = 7 \, .$$

The first quartile will be the seventh observation from the start of the data (when arranged in ascending order) and the third quartile will be the seventh observation from the end of the data:

$$\text{first quartile} = 87.3$$

$$\text{third quartile} = 101.4 \, .$$

The minimum and maximum are, respectively: 72.6 and 170.2.
This gives us the following five-number summary:

90.1	
87.3	101.4
72.6	170.2

Resistance (or Robustness) of the Median with Respect to the Mean

We consider the following numbers: 3, 3, 7, 7, 11, 11.
We first calculate the mean of these numbers:

$$\bar{x} = \frac{\sum_{i=1}^{n} x_i}{n}$$

$$= \frac{3+3+7+7+11+11}{6} = 7 \, .$$

Then we calculate the median M_d:

$$\text{median rank} = \frac{n+1}{2} = \frac{6+1}{2} = 3.5$$

$$M_d = \frac{7+7}{2} = 7 \, .$$

We note that, in this case, the mean and the median both equal 7.
Now suppose that we add another number: -1000. We therefore recalculate the mean and the median of the following numbers: $-1000, 3, 3, 3, 7, 11, 11$. The mean is:

$$\bar{x} = \frac{\sum_{i=1}^{n} x_i}{n}$$

$$= \frac{-1000+3+3+7+7+11+11}{7}$$

$$= -136.86 \, .$$

It is clear that the presence of even one "bad datum" can affect the mean to a large degree. The median is:

$$\text{median rank} = \frac{n+1}{2} = \frac{7+1}{2} = 4$$

$$M_d = 7 \, .$$

Unlike the mean, the median does not change, which tells us that the median is more resistant to extreme values ("bad data") than the mean.

FURTHER READING
▶ Box plot
▶ Graphical representation
▶ Residual
▶ Stem-and-leaf diagram
▶ Transformation

REFERENCES
Tukey, J.W.: Some graphical and semigraphical displays. In: Bancroft, T.A. (ed.) Statistical Papers in Honor of George W. Snedecor. Iowa State University Press, Ames, IA , pp. 293–316 (1972)

Tukey, J.W.: Exploratory Data Analysis. Addison-Wesley, Reading, MA (1977)

Exponential Distribution

A **random variable** X follows an exponential distribution with **parameter** θ if its **density function** is given by:

$$f(x) = \begin{cases} \theta \cdot e^{-\theta x} & \text{if } x \geq 0; \theta > 0 \\ 0 & \text{if not} \end{cases}.$$

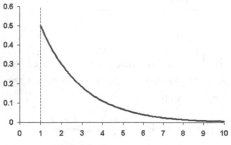

Exponential distribution, $\theta = 1$, $\sigma = 2$

The exponential distribution is also called the negative exponential.

The exponential distribution is a **continuous probability distribution**.

MATHEMATICAL ASPECTS
The **distribution function** of a **random variable** X that follows an exponential distribution with **parameter** θ is as follows:

$$F(x) = \int_0^x \theta \cdot e^{-\theta t} \, dt$$
$$= 1 - e^{-\theta x}.$$

The **expected value** is given by:

$$E[X] = \frac{1}{\theta}.$$

Since we have

$$E[X^2] = \frac{2}{\theta^2}.$$

the **variance** is equal to:

$$\text{Var}(X) = E[X^2] - (E[X])^2$$
$$= \frac{2}{\theta^2} - \frac{1}{\theta^2} = \frac{1}{\theta^2}.$$

The exponential distribution is the **continuous probability distribution** analog to the **geometric distribution**.

It is actually a particular case of the **gamma distribution**, where the **parameter** α of the gamma distribution is equal to 1.

It becomes a **chi-square distribution** with two **degrees of freedom** when the **parameter** θ of the exponential distribution is equal to $\frac{1}{2}$.

DOMAINS AND LIMITATIONS
The exponential distribution is used to describe random **events** in time. For example, lifespan is a characteristic that is frequently represented by an exponential **random variable**.

FURTHER READING

Extrapolation

Statistically speaking, an extrapolation is an **estimation** of the **dependent variable** for **values** of independent variable that are located outside of the set of **observations**.

Extrapolation is often used in **time series** when a **model** that is determined from the values $u_1, u_2, \ldots, u_n$ observed at times $t_1, t_2, \ldots, t_n$ (with $t_1 < t_2 < \ldots < t_n$) is used to predict the value of the variable at time t_{n+1}.

Generally, a **simple linear regression** function based on the **observations** (X_i, Y_i), $i = 1, 2, \ldots, n$ can be used to estimate **values** of Y for values of X that are located outside of the set $X_1, X_2, \ldots, X_n$. It is also possible to obtain extrapolations from a **multiple linear regression model**.

E

DOMAINS AND LIMITATIONS
Caution is required when using extrapolations because they are obtained based on the **hypothesis** that the **model** does not change for the **values** of the independent variable used for the extrapolation.

FURTHER READING

F

Factor

A factor is the term used to describe any controllable condition in an **experiment**. Each factor can take a predetermined number of **values**, called the **levels** of the factor. **Experiments** that use factors are called **factorial experiments**.

EXAMPLES

A sociologist wants to find out the opinions of the inhabitants of a city about the drug problem. He is interested in the influence of gender and income on the attitude to this issue.

The **levels** of the factor "gender" are: male (M) and female (F); the levels of the factor "income" correspond to low (L), average (A) and high (H), where the divisions between these levels are set by the experimenter.

The factorial space G of this **experiment** is composed of the following six **combinations** of the **levels** of the two factors:

$$G = \{(M,L); (M,A); (M,H); (F,L);$$
$$(F,A); (F,H)\}.$$

In this example, the factor "gender" is qualitative and takes nominal **values**; whereas the factor "income" is quantitative, and takes ordinal values.

FURTHER READING
▶ Analysis of variance
▶ Design of experiments
▶ Experiment
▶ Factorial experiment
▶ Treatment

REFERENCES
Raktoe, B.L. Hedayat, A., Federer, W.T.: Factorial Designs. Wiley, New York (1981)

Factor Level

The levels of a **factor** are the different values that it can take. The **experimental design** determines which levels, out of all of the possible combinations of such levels, are used in the **experiment**.

FURTHER READING
▶ Design of experiments
▶ Experiment
▶ Factor
▶ Treatment

Factorial Axis

Consider S, a square **matrix** of order n. If X_i is the **eigenvector** of S, associated with the **eigenvalue** k_i, we say that X_i is the ith factorial axis; more precisely it is the **axis** that carries the vector X_i.

We generally arrange eigenvalues in decreasing order, which explains the need to number the factorial axes.

HISTORY

See **correspondence analysis**.

DOMAINS AND LIMITATIONS

The concept of a factorial axis is used in correspondence analysis.

FURTHER READING

▶ **Correspondence analysis**
▶ **Eigenvalue**
▶ **Eigenvector**
▶ **Inertia**

REFERENCES

See **correspondence analysis**.

Factorial Experiment

A factorial experiment is an **experiment** in which all of the possible **treatments** that can be derived from two or more **factors**, where each factor has two or more **levels**, are studied, in a way such that the **main effects** and the **interactions** can be investigated.

The term "factorial experiment" describes an experiment where all of the different **factors** are combined in all possible ways, but it does not specify the experiment design used to perform such an experiment.

MATHEMATICAL ASPECTS

In a factorial experiment, a **model** equation can be used to relate the **dependent variable** to the **independent variable**.

Therefore, if the factorial experiment implicates two **factors**, the associated **model** is as follows:

$$Y_{ijk} = \mu + \alpha_i + \beta_j + (\alpha\beta)_{ij} + \varepsilon_{ijk},$$

$i = 1, 2, \ldots, a$ (**levels** of **factor** A),

$j = 1, 2, \ldots, b$ (levels of factor B),

$$k = 1, 2, \ldots, n_{ij} \begin{pmatrix} \text{number of } \mathbf{ex\text{-}} \\ \mathbf{perimental\ units} \\ \text{receiving the} \\ \mathbf{treatment}\ ij \end{pmatrix},$$

where:

μ is the general **mean** common to all of the treatments,

α_i is the effect of the ith level of factor A,

β_j is the effect of the jth level of factor B,

$(\alpha\beta)_{ij}$ is the effect of the interaction between ith level of A jth level of B, and

ε_{ijk} is the experimental **error** in the observation Y_{ijk}.

In general, μ is a constant and ϵ_{ijk} is a **random variable** distributed according to the **normal distribution** with a **mean** of zero and a **variance** of σ^2.

If the interaction term $(\alpha\beta)_{ij}$ is suppressed, an additive **model** is obtained, meaning that the effect of each **factor** is a constant that adds up to the general **mean** μ, and this does not depend on the **level** of the other factor. This is known as a model without **interactions**.

DOMAINS AND LIMITATIONS

In a factorial experiment where all of the **factors** have the same number of **levels**, the number of treatments employed in the experiment is usually given by the number of levels raised to a power equal to the number of factors. For example, for an experiment that employs three factors, each with two levels, the experiment is known as a 2^3 factorial experiment, which employs eight **treatments**.

The most commonly used notation for the **combinations** of **treatments** used in a factorial experiment can be described as follows. Use a capital letter to designate the **factor** and a numerical index to designate its **level**. Consider for example a factorial experiment with three factors: a factor A with two levels, a factor B with two levels, and a factor C with three levels. The 12 combinations will be:

$$A_1B_1C_1, \quad A_1B_1C_2, \quad A_1B_1C_3,$$
$$A_1B_2C_1, \quad A_1B_2C_2, \quad A_1B_2C_3,$$
$$A_2B_1C_1, \quad A_2B_1C_2, \quad A_2B_1C_3,$$
$$A_2B_2C_1, \quad A_2B_2C_2, \quad A_2B_2C_3.$$

Sometimes only the indices (written in the same order as the factors) are stated:

$$111, \quad 112, \quad 113,$$
$$121, \quad 122, \quad 123,$$
$$211, \quad 212, \quad 213,$$
$$221, \quad 222, \quad 223.$$

Double-factorial experiments, or more generally multiple-factorial experiments, are important for the following reasons:

- A double-factorial experiment uses resources more efficiently than two **experiments** that each employ a single **factor**. The first one takes less time and requires fewer **experimental units** to achieve a given **level** of precision.
- A double-factorial experiment allows the influence of one factor to be studied at each level of the other factor because the levels of both factors are varied. This leads to conclusions that are valid over a larger range of experimental conditions than if a series of one-factor designs is used.

- Finally, simultaneous research on two factors is necessary when **interactions** between the factors are present, meaning that the effect of a factor depends on the level of the other factor.

FURTHER READING
▶ **Design of experiments**
▶ **Experiment**

REFERENCES
Dodge, Y.: Principaux plans d'expériences. In: Aeschlimann, J., Bonjour, C., Stocker, E. (eds.) Méthodologies et techniques de plans d'expériences: 28ème cours de perfectionnement de l'Association vaudoise des chercheurs en physique, Saas-Fee, 2–8 March 1986 (1986)

Fisher Distribution

The **random variable** X follows a Fisher distribution if its **density function** takes the form:

$$f(x) = \frac{\Gamma\left(\frac{m+n}{2}\right)}{\Gamma\left(\frac{m}{2}\right)\Gamma\left(\frac{n}{2}\right)} \left(\frac{m}{n}\right)^{\frac{m}{2}} \cdot \frac{x^{\frac{m-2}{2}}}{\left(1+\frac{m}{n}x\right)^{\frac{m+n}{2}}},$$

where Γ is the gamma function (see **gamma distribution**) and m, n are the **degrees of freedom** ($m, n = 1, 2, \ldots$).

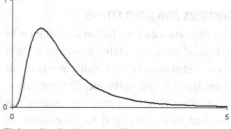

Fisher distribution, $m = 12$, $n = 8$

The Fisher distribution is a **continuous probability distribution**.

HISTORY

The Fisher distribution was discovered by **Fisher, R.A.** in 1925. The symbol F, used to denote the Fisher distribution, was introduced by Snedecor in 1934, in honor of Fisher, R.A..

MATHEMATICAL ASPECTS

If U and V are two independent **random variables** each following a **chi-square distribution**, with respectively m and n **degrees of freedom**, then the random variable:

$$F = \frac{U/m}{V/n}$$

follows a Fisher distribution with m and n **degrees of freedom**.

The **expected value** of a **random variable** F that follows a Fisher distribution is given by:

$$E[F] = \frac{n}{n-2} \quad \text{for } n > 2,$$

and the **variance** is equal to:

$$\text{Var}(F) = \frac{2n^2(m+n-2)}{m(n-2)^2(n-4)} \quad n > 4.$$

The Fisher distribution with 1 and v **degrees of freedom** is identical to the square of the **Student distribution** with v degrees of freedom.

DOMAINS AND LIMITATIONS

The importance of the Fisher distribution in statistical theory is related to its application to the distribution of the ratio of independent **estimators** of the **variance**. Currently, this distribution is most commonly used in the standard tests associated with **analysis of variance** and **regression** analysis.

FURTHER READING

► **Analysis of variance**
► **Chi-square distribution**
► **Continuous probability distribution**
► **Fisher table**
► **Fisher test**
► **Student distribution**

REFERENCES

Fisher, R.A.: Statistical Methods for Research Workers. Oliver & Boyd, Edinburgh (1925)

Snedecor, G.W.: Calculation and Interpretation of Analysis of Variance and Covariance. Collegiate, Ames, IA (1934)

Fisher Index

The Fisher **index** is a **composite index number** which allows us to study the increase in the cost of living (inflation). It is the **geometric mean** of two index numbers:

- The **Laspeyres index**, and
- The **Paasche index**.

$$\text{Fisher index} = \sqrt{\begin{array}{c}(\text{Laspeyres index}) \\ \times (\text{Paasche index})\end{array}} .$$

HISTORY

In 1922, the economist and mathematician Fisher, Irving established a model based on the Fisher index in order to circumvent some issues related to the use of the **index numbers** of Laspeyres and Paasche:

Since the **Laspeyres index** always uses the quantity of goods sold initially as a reference, it can overestimate any increase in the cost of living (because the consumption of more expensive goods may drop over time).

Conversely, since the **Paasche index** always uses the quantity of goods sold in the current period as a reference, it can underestimate the increase in the cost of living.

Because the Fisher index is the **geometric mean** of these two indices, it can be considered to be an ideal compromise between them.

MATHEMATICAL ASPECTS

Using the definitions of the **Laspeyres index** and the **Paasche index**, the Fisher index $F_{n/0}$ is calculated in the following way:

$$F_{n/0} = 100 \cdot \sqrt{\frac{\sum P_n Q_0 \cdot \sum P_n Q_n}{\sum P_0 Q_0 \cdot \sum P_0 Q_n}},$$

where P_n and Q_n represent the price of goods and the quantity of them sold in the current period n, while P_0 and Q_0 are the price and quantity sold of the same goods during reference period 0. Different values are obtained for different goods.

Notice that the Fisher index is reversible, meaning that:

$$F_{n/0} = \frac{1}{F_{0/n}}.$$

DOMAINS AND LIMITATIONS

Even though the Fisher index is a kind of "ideal compromise index number," it is rarely used in practice.

EXAMPLES

Consider the following fictitious table indicating the prices and the respective quantities sold of three consumer goods at the reference year 0 and at the current year n.

	Quantities (thousands)		Price (euros)	
Goods	1970	1988	1970	1988
	(Q_0)	(Q_n)	(P_0)	(P_n)
Milk	50.5	85.5	0.20	1.20
Bread	42.8	50.5	0.15	1.10
Butter	15.5	40.5	0.50	2.00

From the following table, we have:

$$\sum P_n Q_n = 239.15,$$

$$\sum P_0 Q_n = 44.925,$$

$$\sum P_n Q_0 = 138.68 \quad \text{and}$$

$$\sum P_0 Q_0 = 24.27.$$

Goods	$\sum P_n Q_n$	$\sum P_0 Q_n$	$\sum P_n Q_0$	$\sum P_0 Q_0$
Milk	102.60	17.100	60.60	10.10
Bread	55.55	7.575	47.08	6.42
Butter	81.00	20.250	31.00	7.75
Total	239.15	44.925	138.68	24.27

We can then find the **Paasche index**:

$$I_{n/0} = \frac{\sum P_n \cdot Q_n}{\sum P_0 \cdot Q_n} \cdot 100$$
$$= \frac{239.15}{44.925} \cdot 100 = 532.3,$$

and the **Laspeyres index**:

$$I_{n/0} = \frac{\sum P_n \cdot Q_0}{\sum P_0 \cdot Q_0} \cdot 100$$
$$= \frac{138.68}{24.27} \cdot 100 = 571.4.$$

The Fisher index is the square root of the product of the index numbers of Paasche and of Laspeyres (or the **geometric mean** of the two **index numbers**):

Fisher index $= \sqrt{532.3 \times 571.4} = 551.5$.

According to the Fisher index, the price of the goods considered has risen by 451.5% ($551.5 - 100$) between the reference year and the current year.

FURTHER READING
▶ **Composite index number**
▶ **Index number**
▶ **Laspeyres index**
▶ **Paasche index**
▶ **Simple index number**

REFERENCES

Fisher, I.: The Making of Index Numbers. Houghton Mifflin, Boston (1922)

Fisher, Irving

Fisher, Irving was born in New York on the 27th February 1867. He obtained his doctorate at Yale University in 1892. His mathematical approach to the theory of values and prices, first described in his thesis, resulted in him becoming known as one of the first American mathematical economists. He founded the Econometric Society in 1930 with Frisch, Ragnar and Roos, Charles F., and became its first president. Fisher, I. stayed at Yale University throughout his career. He began by teaching mathematics, then economics. After a stay in Europe, he was named Professor of Social Sciences in 1898, and died in New York in 1947.

Fisher was a prolific author: his list of publications contains more then 2000 titles. He was interested in both economic theory and scientific research. In 1920, he proposed econometrical and statistical methods for calculating indices. His "ideal index", which we now call the **Fisher index**, is the geometric mean of the **Laspeyres index** and the **Paasche index**.

Some principal works and articles of Fisher, Irving:

1892   Mathematical Investigations in the Theory of Value and Prices. Connecticut Academy of Arts and Science, New Haven, CT.

1906   The Nature of Capital and Income. Macmillan, New York.

1907   The Rate of Interest: Its Nature, Determination and Relation to Economic Phenomena. Macmillan, New York.

1910   Introduction to Economic Science. Macmillan, New York.

1911   The Purchasing Power of Money. Macmillan, New York.

1912   Elementary Principles of Economics. Macmillan, New York.

1921   The best form of index number. Am. Stat. Assoc. Quart., 17, 533–537.

1922   The Making of Index Numbers. Houghton Mifflin, Boston, MA.

1926   A statistical relation between unemployment and price changes. Int. Labour Rev., 13, 785–792.

1927   A statistical method for measuring "marginal utility" and testing the justice of a progressive income tax. In: Hollander, J.H. (ed) Economic Essays Contributed in Honor of John Bates Clark. Macmillan, New York.

1930   The Theory of Interest as Determined by Impatience to Spend Income and Opportunity to Invest It Macmillan, New York.

1932   Booms and Depressions. Adelphi, New York.

1933   The debt-deflation theory of great depressions. Econometrica, 1, 337–357.

1935   100% Money. Adelphi, New York.

FURTHER READING
▶ **Fisher index**
▶ **Index number**

Fisher, Ronald Aylmer

Born in 1890 in East Finchley, near London, Fisher, Ronald Aylmer studied mathematics at Harrow and at Gonville and Caius College, Cambridge. He began his statistical career in 1919 when he started work at the Institute of Agricultural Research at Rothamsted ("Rothamsted Experimental Station"). The time he spent there, which lasted until 1933, was very productive. In 1933, he moved to become Professor of Eugenics at University College London, and then from 1943 to 1957 he took over the Balfour Chair of Genetics at Cambridge. Even after he had retired, Fisher performed some research for the Mathematical Statistics Division of the Commonwealth Scientific and Industrial Research Organisation in Adelaide, Australia, where he died in 1962.

Fisher is recognized as being one of the founders of modern **statistics**. Well known in the field of genetics, he also contributed important work to the general theory of **maximum likelihood estimation**. His work on experimental design, required for the study of agricultural **experimental data**, is also worthy of mention, as is his development of the technique used for the **analysis of variance**.

Fisher originated the principles of **randomization**, randomized **blocks**, **Latin square designs** and factorial **arrangements**.

Some of the main works and articles of Fisher, R.A.:

1912   On an absolute criterion for fitting frequency curves. Mess. Math., 41, 155–160.

1918   The correlation between relatives on the supposition of Mendelian inheritance. Trans. Roy. Soc. Edinburgh, 52, 399–433.

1920   A mathematical examination of methods of determining the accuracy of an observation by the mean error and by the mean square error. Mon. Not. R. Astron. Soc., 80, 758–770.

1921   On the "probable error" of a coefficient of correlation deduced from a small sample. Metron, 1(4), 1–32.

1922   On the mathematical foundation of theoretical statistics. Phil. Trans. A, 222, 309–368.

1925   Theory of statistical estimation. Proc. Camb. Philos. Soc., 22, 700–725.

1925   Statistical Methods for Research Workers. Oliver & Boyd, London.

1934   Probability, likelihood and quantity of information in the logic of uncertain inference. Proc. Roy. Soc. A, 146, 1–8.

1935   The Design of Experiments. Oliver & Boyd, Edinburgh.

1971–1974 Bennett, J.H. (ed) Collected Papers of R.A. Fisher. Univ. Adelaide, Adelaide, Australia

FURTHER READING
► **Fisher distribution**
► **Fisher table**
► **Fisher test**

REFERENCES
Fisher-Box, J.: R.A. Fisher, the Life of a Scientist. Wiley, New York (1978)

Fisher Table

The Fisher table gives the values of the **distribution function** of a **random variable** that follows a **Fisher distribution**.

HISTORY
See **Fisher distribution**.

MATHEMATICAL ASPECTS
Let the **random variable** X follow a **Fisher distribution** with m and n degrees of freedom. The **density function** of the random variable X is given by:

$$f(t) = \frac{\Gamma\left(\frac{m+n}{2}\right)}{\Gamma\left(\frac{m}{2}\right)\Gamma\left(\frac{n}{2}\right)}\left(\frac{m}{n}\right)^{\frac{m}{2}}$$
$$\cdot \frac{t^{\frac{m-2}{2}}}{\left(1+\frac{m}{n}t\right)^{\frac{m+n}{2}}}, \quad t \geq 0,$$

where Γ represents the gamma function (see **gamma distribution**).

Given the degrees of freedom m and n, the Fisher table allows us to determine the **value** x for which the **probability** $P(X \leq x)$ equals a particular value.

The most commonly used Fisher tables give the values of x for

$$P(X \leq x) = 95\% \quad \text{and}$$
$$(X \leq x) = 97.5\%.$$

We generally use $(x =)\, F_{m,n,\alpha}$ to symbolize the value of the random variable X for which

$$P\left(X \leq F_{m,n,\alpha}\right) = 1 - \alpha.$$

DOMAINS AND LIMITATIONS
The Fisher table is used in hypothesis testing when it involves **statistics** distributed according to a **Fisher distribution**, and especially during **analysis of variance** and in **simple** and **multiple linear regression analysis**.

Merrington and Thompson (1943) provided Fisher tables for values up to five decimal places long, and Fisher and Yates (1938) produced tables up to two decimal places long.

EXAMPLES
If X follows a **Fisher distribution** with $m = 10$ and $n = 15$ degrees of freedom, the **value** of x that corresponds to a **probability** of 0.95 is 2.54:

$$P(X \leq 2.54) = 0.95.$$

In the same way, the value of x that corresponds to a probability of 0.975 is 3.06:

$$P(X \leq 3.06) = 0.975.$$

For an example of the use of the Fisher table, see **Fisher test**.

FURTHER READING
▶ **Analysis of variance**
▶ **Fisher distribution**
▶ **Multiple linear regression**
▶ **Simple linear regression**
▶ **Statistical table**

REFERENCES
Fisher, R.A., Yates, F.: Statistical Tables for Biological, Agricultural and Medical Research. Oliver and Boyd, Edinburgh and London (1963)

Merrington, M., Thompson, C.M.: Tables of percentage points of the inverted beta (F) distribution. Biometrika **33**, 73–88 (1943)

Fisher Test

A Fisher test is a hypothesis test on the observed **value** of a **statistic** of the form:

$$F = \frac{U}{m} \cdot \frac{n}{V},$$

where U and V are independent random variables that each follow a **chi-square distribution** with, respectively, m and n degrees of freedom.

HISTORY

See **Fisher distribution**.

MATHEMATICAL ASPECTS

We consider a linear **model** with one **factor** applied at t levels:

$$Y_{ij} = \mu + \tau_i + \varepsilon_{ij},$$

$$i = 1, 2, \ldots, t; \quad j = 1, 2, \ldots, n_i,$$

where

Y_{ij} represents the jth **observation** receiving the **treatment** i,

μ is the general **mean** common to all treatments,

τ_i is the real effect of the treatment i on the observation, and

ε_{ij} is the experimental **error** in the observation Y_{ij}.

The random variables ε_{ij} are independent and distributed normally with mean 0 and **variance** σ^2: $N(0, \sigma^2)$.

In this case, the **null hypothesis** that affirms that there is a significant difference between the t treatments, is as follows:

$$H_0: \quad \tau_1 = \tau_2 = \ldots = \tau_t.$$

The **alternative hypothesis** is as follows:

$H_1:$ the values of $\tau_i (i = 1, 2, \ldots, t)$ are not all identical.

To compare the variability of within treatment also called error or residuals with the variability between the treatments, we construct a ratio where the numerator is an **estimation** of the **variance** between treatment and the denominator an estimation of the variance within treatment or error. The Fisher test **statistic**, also called the F-ratio, is then given by:

$$F = \frac{\sum_{i=1}^{t} \frac{n_i \cdot (\bar{Y}_{i.} - \bar{Y}_{..})^2}{t - 1}}{\sum_{i=1}^{t} \sum_{j=1}^{n_i} \frac{(Y_{ij} - \bar{Y}_{i.})^2}{N - t}},$$

where

n_i is the number of observations in ith treatment $(i = 1, \ldots, t)$

$N = \sum_{i=1}^{t} n_i$ is the total number of observations,

$\bar{Y}_{i.} = \sum_{j=1}^{n_i} \frac{Y_{ij}}{n_i}$ is the mean of the ith treatment, for $i = 1, \ldots, t$, and

$\bar{Y}_{..} = \frac{1}{N} \sum_{i=1}^{t} \sum_{j=1}^{n_i} Y_{ij}$ is the global mean.

After choosing a **significance level** α for the test, we compare the calculated value F with the appropriate **critical value** in the **Fisher table** for $t - 1$ and $N - t$ degrees of freedom, denoted $F_{t-1, N-t, \alpha}$.

If $F \geq F_{t-1, N-t, \alpha}$, we reject the null hypothesis, which means that at least one of the treatments differs from the others.

If $F < F_{t-1, N-t, \alpha}$, we do not reject the null hypothesis; in other words, the treatments present no significant difference, which also means that the t samples derive from the same **population**.

DOMAINS AND LIMITATIONS

The Fisher test is most commonly used during the **analysis of variance**, in **covariance analysis**, and in **regression analysis**.

EXAMPLES

See **one-way analysis of variance** and **two-way analysis of variance**.

FURTHER READING

- ▶ **Analysis of variance**
- ▶ **Fisher distribution**
- ▶ **Fisher table**
- ▶ **Hypothesis testing**
- ▶ **Regression analysis**

Forecasting

Forecasting is the utilization of statistical data, economic theory and noneconomic conditions in order to form a reasonable opinion about future events.

Numerous methods are used to forecast the future. They are all based on a restricted number of basic hypotheses.

One such hypothesis affirms that the significant tendencies that have been observed historically will continue to hold in the future. Another one states that measurable fluctuations in a **variable** will be reproduced at regular intervals, so that these variations will become predictable.

The time series analysis performed in a forecast process is based on two hypotheses:

1. *Long-term forecast:*

   When we make long-term forecasts (more than five years into the future), analysis and extrapolation of the **secular tendency** become important.

   Long-term forecasts based on the secular tendency normally do not take cyclic fluctuations into account. Moreover, seasonal variations do not influence annual data and are not considered in a long-term forecast.

2. *Mean-term forecast:*

   In order to take the probable effect of cyclic fluctuations into account in the mechanism used to derive a mean-term forecast (one to five years ahead), we multiply the **value** of the projection of the secular tendency by the cyclic fluctuation, which gives the forecasted value. This, of course, assumes that the same data structures that produce the cyclic variations will hold in the future, and so the observed variations will continue to recur regularly. In reality, subsequent cycles have the tendency to vary enormously in terms of their periodicity, their amplitude and the models that they follow.

3. *Short-term forecast:*

   The **seasonal index** must be taken into account in a short-term forecast (a few months ahead), along with the projected values for the secular tendency and cyclic fluctuations.

   Generally, we do not try to forecast irregular variations.

MATHEMATICAL ASPECTS

Let Y be a **time series**. We can describe it using the following components:

- The **secular tendency** T;
- The **seasonal variations** S;
- The **cyclic fluctuations** C;
- The **irregular variations** I.

We distinguish between:

- The multiplicative **model** $Y = T \cdot S \cdot C \cdot I$;
- The additive model $Y = T + S + C + I$.

For a value of t that is greater than the time taken to perform an **observation**, we specify that:

Y_t is a forecast of the **value** Y at time t;

T_t is a projection of the secular tendency at time t;

C_t is the cyclic fluctuation at time t;

S_t is the forecast for the seasonal variation at time t.

In this case:

1. *For a long-term forecast*, the value of the time series can be estimated using:

$$Y_t = T_t.$$

2. *For a mid-term forecast*, the value of the time series can be estimated using:

$$Y_t = T_t \cdot C_t \text{ (multiplicative model)}$$

or $Y_t = T_t + C_t$ (additive model) .

3. *For a short-term forecast*, the value of the time series can be estimated using:

$$Y_t = T_t \cdot C_t \cdot S_t$$

(for the multiplicative model)

or $Y_t = T_t + C_t + S_t$

(for the additive model).

DOMAINS AND LIMITATIONS

Planning and making decisions are two activities that involve predicting the future, to some degree. In other words, the administrators that perform these tasks have to make forecasts.

It is worth noting that obtaining forecasts via time series analysis is worthwhile, even though the forecast usually turns out to be inaccurate. In reality, even when it is inaccurate, the forecast obtained in this way will probably be more accurate than another forecast based only on intuition. Therefore, despite its limitations and problems, time series analysis is a useful tool in the process of forecasting.

EXAMPLES

The concept of a "forecast" depends on the subject being investigated. For example:

- In meteorology:
  - A short-term forecast looks hours ahead,
  - A long-term forecast looks days ahead.
- In economics:
  - A short-term forecast looks months ahead;
  - A mid-term forecast looks one to five years ahead;
  - A long-term forecast looks more then five years ahead.

FURTHER READING

▶ **Cyclical fluctuation**
▶ **Irregular variation**
▶ **Seasonal index**
▶ **Seasonal variation**
▶ **Secular trend**
▶ **Time series**

Fractional Factorial Design

A fractional factorial experimental design is a **factorial experiment** in which only a fraction of the combinations of the **factor levels** possible is realized.

This type of design is used when an **experiment** contains a number of **factors** that are believed to be more important than the others, and/or there are a large number of factor **levels**. The design allows use to reduce the number of **experimental units** needed.

EXAMPLES

For a **factorial experiment** containing two **factors**, each with three **levels**, there are nine possible combinations. If four of these **observations** are suppressed, a fraction of the observations are left that can be schematized in the following way:

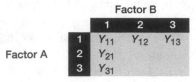

where Y_{ij} represents the **observation** taken for the ith level of factor A and the jth level of factor B.

FURTHER READING
▶ **Design of experiments**
▶ **Experiment**

REFERENCES
Yates, F.: The design and analysis of factorial experiments, Techn. comm. 35, Imperial Bureau of Soil Science, Harpenden (1937)

Frequency

The frequency or absolute frequency corresponds to the number of appearances of a particular **observation** or result in an **experiment**.

The absolute frequency can be distinguished from the relative frequency. The relative frequency of an **observation** is defined as the ratio of the number of appearances to the total number of observations.

For certain analyses, it is desirable to know the number of **observations** for which the **value** is, say, "less than or equal to" or "higher than" a given value. The number of values less than or equal to this given value is called the cumulative frequency.

The **percentage** of **observations** in which the **value** is "less than or equal to" or "more than" a given value may also be desired. The proportion of values less than or equal to this given value is called the cumulative relative frequency.

EXAMPLES
Consider, for example, the following 16 **observations**, which represent the heights (in cm) of a group of individuals:

174	169	172	174
171	179	174	176
177	161	174	172
177	168	171	172

Here, the frequency of observing 171 cm is 2; whereas the frequency of observing 174 cm is 4.

In this example, the relative frequency of 174 is $\frac{4}{16} = 0.25$ (or 25% of all of the **observations**). The absolute frequency is 4 and the total number of observations is 16.

FURTHER READING
▶ **Frequency curve**
▶ **Frequency distribution**
▶ **Frequency table**

Frequency Curve

If a **frequency polygon** is smoothed, a curve is obtained, called the frequency curve.

This smoothing can be performed if the number of **observations** in the **frequency distribution** becomes infinitely large and the **widths** of the classes become infinitely small.

The frequency curve corresponds to the limit shape of a **frequency polygon**.

HISTORY
Towards the end of the nineteenth century, the general tendency was to consider all

distributions to be **normal distributions**. Histograms that presented several **modes** were adjusted to form a set of normal frequency curves, and those that presented asymmetry were analyzed by transforming the data in such a way that the **histogram** resulting from this **transformation** could be compared to a normal curve.

In 1895 **Pearson, K.** proposed creating a set of various theoretical frequency curves in order to be able to obtain better approximations to the histograms.

FURTHER READING
- ▶ **Continuous probability distribution**
- ▶ **Frequency distribution**
- ▶ **Frequency polygon**
- ▶ **Normal distribution**

REFERENCES
Pearson, K.: Contributions to the mathematical theory of evolution. II: Skew variation in homogeneous material. In: Karl Pearson's Early Statistical Papers. Cambridge University Press, Cambridge, pp. 41–112 (1948). First published in 1895 in Philos. Trans. Roy. Soc. Lond. Ser. A **186**, 343–414

Frequency Distribution

The distribution of a **population** can be expressed in relative terms (for example as a **percentage**), in fractions or in absolute terms. In all of these cases, the result is called the **frequency** distribution.

We also obtain either a discrete frequency distribution or a continuous frequency distribution, depending on whether the **variable** being studied is discrete or continuous.

The number of appearances of a specific **value**, or the number of observations for a spe-

cific class, is called the **frequency** of that value or class.

The frequency distribution that is obtained in this manner can be represented either as a **frequency table** or in a graphical way, for example as a **histogram**, a **frequency polygon** or a **line chart**.

Any frequency distribution has a corresponding relative **frequency** distribution, which is the distribution of each **value** or each class with respect to the total number of **observations**.

F

HISTORY
See **graphical representation**.

DOMAINS AND LIMITATIONS
In practice, frequency distributions can vary considerably. For example:
- Some are symmetric; others are asymmetric:

Symmetric distribution

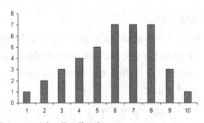

Asymmetric distribution

- Some only have one **mode** (unimodal distributions); others have several (plurimodal distributions):

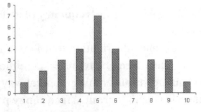

Unimodal distribution

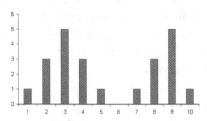

Bimodal distribution

This means that frequency distributions can be classified into four main types:
1. Unimodal symmetric distributions;
2. Unimodal asymmetric distributions;
3. J-shaped distributions;
4. U-shaped distributions.

During statistical studies, it is often necessary to compare two distributions. If both distributions are of the same type, it is possible to compare them by examining their main characteristics, such as the **measure of central tendency**, the **measure of dispersion** or the **measure of form**.

EXAMPLES

The following **data** represent the heights, measured in centimeters, observed for a **population** of 27 students from a junior high-school class:

169	177	178	181	173
172	175	171	175	172
173	174	176	170	172
173	172	171	176	176
175	168	167	166	170
173	169			

By ordering and regrouping these **observations** by **value**, we obtain the following **frequency** distribution:

Height	Frequency	Height	Frequency
166	1	174	1
167	1	175	3
168	1	176	3
169	2	177	1
170	2	178	1
171	2	179	0
172	4	180	0
173	4	181	1

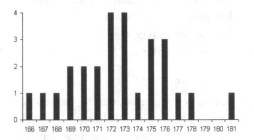

Here is another example, which concerns the **frequency** distribution of a continuous variable. After retrieving and classifying the personal incomes among the population of Neuchâtel (a state in Switzerland), for the period 1975–1976, we obtain the following **frequency table**:

Net revenue (in thousands of Swiss Francs)	Frequencies	Relative frequency
0–10	283	0.0047
10–20	13175	0.2176
20–50	40316	0.6661
50–80	5055	0.0835
80–120	1029	0.0170
120+	670	0.0111

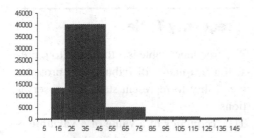

FURTHER READING
- ▶ **Frequency**
- ▶ **Frequency curve**
- ▶ **Frequency polygon**
- ▶ **Frequency table**
- ▶ **Histogram**
- ▶ **Interval**
- ▶ **Line chart**

Frequency Polygon

A frequency polygon is a **graphical representation** of a **frequency distribution**, which fits to the **histogram** of the frequency distribution. It is a type of frequency plot. It takes the form of a segmented line that joins the midpoint of the top of each rectangle in a histogram.

MATHEMATICAL ASPECTS
We construct a frequency polygon by plotting a point for each class of the **frequency distribution**. Each point corresponds to the midpoint of the class along the abscisse and the **frequency** of the class along the ordinate. Then we connect each point to its neighboring points by lines. The lines therefore effectively link the midpoint of the top of each rectangle in the histogram of the frequency distribution.

Normally we close the frequency polygon at the two ends of the distribution by choosing appropriate points on the horizontal axis and connecting them to the points for the first and last classes.

DOMAINS AND LIMITATIONS
A frequency polygon is usually constructed from data that has been grouped into intervals. In this case, it is preferable to employ intervals of the same width. If we fit a smooth curve to a frequency polygon, we get a **frequency curve**. This smoothing process involves regrouping the data such that the class width is as small as possible, and then replotting the distribution.
Clearly, we can only perform this type of smoothing if the number of units in the population is large enough to give a significant number of observations in each class after regrouping.

EXAMPLES
The **frequency table** below gives the annual mean precipitations for 69 cities in the USA:

Annual mean precipitation (in inches)	Frequency
0–10	4
10–20	9
20–30	5
30–40	24
40–50	21
50–60	4
60–70	2
Total	69

The frequency polygon is constructed via the **histogram** of the frequency distribution:

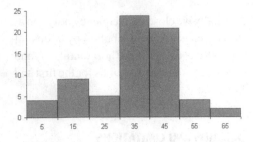

To construct the frequency polygon, we plot a point for each class. The position of this point along the abscisse is given by the midpoint of the class, and its position along the ordinate is given by the frequency of the class. In other words, each point occurs at the midpoint of the top of the rectangle for the class in the histogram. Each point is then joined to its nearest neighbors by lines. In order to close the polygon, a point is placed on the abscisse one class-width before the midpoint of the first class, and another point is placed on the abscisse one class-width after the midpoint of the last class. These points are then joined by lines to the points for the first and last classes, respectively. The frequency polygon we obtain from this procedure is as follows:

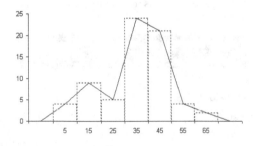

FURTHER READING
▶ **Frequency distribution**
▶ **Frequency table**
▶ **Graphical representation**
▶ **Histogram**
▶ **Ogive**

Frequency Table

The frequency table is a tool used to represent a **frequency distribution**. It provides the ability to represent statistical observations.

The **frequency** of a **value** of a **variable** is the number of times that this value appears in a population. A **frequency distribution** can then be defined as the list of the frequencies obtained for the different values taken by the **variable**.

MATHEMATICAL ASPECTS

Consider a set of n units described by a **variable** X that can take k values $x_1, x_2, \ldots, x_k$.
Let n_i be the number of units having the value x_i; n_i is then the **frequency** of value x_i.
The relative frequency of x_i is $f_i = \frac{n_i}{n}$.
Since the values are different and exhaustive, the sum of all of the frequencies n_i equals the total number n of units in the set, or the sum of the relative frequencies f_i equals unity; in other words:

$$\sum_{i=1}^{k} n_i = n \quad \text{and} \quad \sum_{i=1}^{k} f_i = 1.$$

The corresponding frequency table for this scenario is as follows:

Values of the variable X	Frequencies	Relative frequencies
x_1	n_1	f_1
x_2	n_2	f_2
...	...	...
x_k	n_k	f_k
Total	n	1

Sometimes the concept of cumulative frequency is important (see the second example for instance). This is the sum of all frequencies up to and including frequency f_i: $\sum_{j=1}^{i} f_j$.

DOMAINS AND LIMITATIONS

Constructing a frequency table is the simplest and the most commonly used approach to representing data. However, there are some rules that should be respected when creating such a table:

- The table must have a comprehensive and concise title, which mentions the units employed;
- The names of the lines and columns must be precise and short;
- The column totals should be provided;
- The source of the data given in the table must be indicated.

In general, the table must be comprehensible enough to understand without needing to read the accompanying text.

The number of classes chosen to represent the data depends on the size of the set studied. Using a large number of classes for a small set will result in irregular frequencies due to the small number of units per class. On the other hand, using small number of classes for a large set results in the loss of information about the structure of the data.

However, there is no general rule for determining the number of classes that should be used to construct a frequency table. For relatively small sets ($n \leq 200$), between 7 and 15 classes are recommended, although this rule is not absolute.

To simplify, it is frequent to use classes of the same width, and the class intervals should not overlap. In certain cases, for example if we have a variable that can take a large range of values, but a certain interval of values is expected to be far more frequent than the rest,

it can be useful to use an open class, such as "1000 or more."

EXAMPLES

Frequency distribution of Australian residents (in thousands) according to their matrimonial statuses, on the 30*th* June 1981 is given in the following table.

Matrimonial status	Frequency (in thousands)	Relative frequency
Single	6587	0.452
Married	6837	0.469
Divorced	403	0.028
Widowed	749	0.051
Total	14576	1.000

Source: ABS (1984) Pocket Year Book Australia. Australian Bureau of Statistics, Canberra, p. 11.

The data is presented in the form of a frequency table. The variable "Matrimonial status" can take four different values: single, married, divorced or widowed. The **frequency**, expressed in thousands, and the relative frequency are given for each value.

In the following example, we consider the case where observations are grouped into classes. This table represents the classification of 3014 people based on their incomes in 1955.

Income	Frequency		
	n_i	Relative	Relative Cumulative
Less than 1000	261	0.0866	0.0866
1000–1999	331	0.1098	0.1964
2000–2999	359	0.1191	0.3155
3000–3999	384	0.1274	0.4429
4000–4999	407	0.1350	0.5780

F

Income	Frequency		
	n_i	Relative	Relative Cumulative
5000–7499	703	0.2332	0.8112
7500–9999	277	0.0919	0.9031
10000 or more	292	0.0969	1.0000
Total	3014	1.0000	

From this table, we can see that 57.8% of these 3014 people have incomes that are smaller then 5000 dollars.

FURTHER READING
► **Frequency**
► **Frequency distribution**

Galton, Francis

Galton, Francis was born in 1822 near Birmingham, England, to a family of intellectuals: Darwin, Charles was his cousin, and his grandfather was a member of the Royal Society.

His interest in science first manifested itself in the fields of geography and meteorology. Elected as a member of the Royal Society in 1860, he began to become interested in genetics and statistical methods in 1864. Galton was close friends with **Pearson, K.**. Indeed, Pearson came to his financial aid when Galton founded the journal "Biometrika." Galton's "Eugenics Record Office" merged with Pearson's **biometry** laboratory at University College London and became known as the "Galton Laboratory."

He died in 1911, leaving more than 300 publications including 17 books, most notably on statistical methods related to **regression analysis** and the concept of **correlation**, both of which are attributed to him.

Some of the main works and articles of Galton, F.:

1869 Hereditary Genius: An Inquiry into its Laws and Consequences. Macmillan, London.

1877 Typical laws in heredity. Nature, 15, 492–495, 512–514, 532–533.

1889 Natural Inheritance. Macmillan, London.

1907 Probability, the Foundation of Eugenics. Henry Froude, London.

1908 Memories of my Life. Methuen, London.

1914–1930 Pearson, K. (ed) The Life, Letters and Labours of Francis Galton. Cambridge University Press, Cambridge

FURTHER READING
▶ **Correlation coefficient**
▶ **Regression analysis**

Gamma Distribution

A **random variable** X follows a gamma distribution with **parameter** α if its **density function** is given by

$$f(x) = \frac{\beta^{\alpha}}{\Gamma(\alpha)} x^{\alpha-1} \cdot e^{-\beta x},$$

$$\text{if } x > 0, \quad \alpha > 0, \quad \beta > 0,$$

where $\Gamma(\alpha)$ is the gamma function.

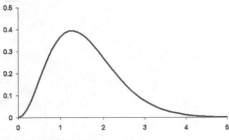

Gamma distribution, $\alpha = 2$, $\beta = 1$

The standard form of the gamma distribution is obtained by putting $\beta = 1$, which gives

$$f(x) = \frac{1}{\Gamma(\alpha)} x^{\alpha-1} \cdot e^{-x},$$

$$\text{if } x > 0, \ \alpha > 0.$$

The gamma function (Γ) appears frequently in statistical theory. It is defined by:

$$\Gamma(\alpha) = \int_0^\infty t^{\alpha-1} e^{-t} dt.$$

In **statistics**, we are only interested in **values** of $\alpha > 0$.

Integrating by parts, we obtain:

$$\Gamma(\alpha + 1) = \alpha \Gamma(\alpha).$$

Since

$$\Gamma(1) = 1,$$

we have

$\Gamma(\alpha+1) = \alpha!$ for every positive integer α.

HISTORY

According to Lancaster (1966), this **continuous probability distribution** was originated by **Laplace, P.S.** (1836).

MATHEMATICAL ASPECTS

The **expected value** of the gamma distribution is given by:

$$E[X] = \frac{1}{\beta} \cdot \frac{\Gamma(\alpha + 1)}{\Gamma(\alpha)} = \frac{\alpha}{\beta}.$$

Since

$$E[X^2] = \frac{1}{\beta^2} \cdot \frac{\Gamma(\alpha + 2)}{\Gamma(\alpha)} = \frac{\alpha(\alpha + 1)}{\beta^2},$$

the **variance** is equal to:

$$\begin{aligned}
\text{Var}(X) &= E[X^2] - (E[X])^2 \\
&= \frac{\alpha(\alpha + 1) - \alpha^2}{\beta^2} \\
&= \frac{\alpha}{\beta^2}.
\end{aligned}$$

The **chi-square distribution** is a particular case of the gamma distribution where $\alpha = \frac{n}{2}$ and $\beta = \frac{1}{2}$, and n is the number of **degrees of freedom** of the chi-square distribution. The gamma distribution with **parameter** $\alpha = 1$ gives the **exponential distribution**:

$$f(x) = \beta \cdot e^{-\beta x}.$$

FURTHER READING

▶ **Chi-square distribution**
▶ **Continuous probability distribution**
▶ **Exponential distribution**

REFERENCES

Lancaster, H.O.: Forerunners of the Pearson chi-square. Aust. J. Stat. **8**, 117–126 (1966)

Laplace, P.S. de: Théorie analytique des probabilités, suppl. to 3rd edn. Courcier, Paris (1836)

Gauss, Carl Friedrich

Gauss, Carl Friedrich was born in 1777 in Brunswick in Germany. He quickly became a renowned astronomer and mathematician and is still considered to be on a par with Archimedes and Newton as one of the greatest mathematicians of all time.

He obtained his doctorate in 1799, and then worked at the University of Helmsted. In 1807, he moved to Göttingen, where he became Laboratory Director. He spent the rest of his life in Göttingen, where he died in 1855.

His contributions to science, particularly physics, are of great importance. In **statistics**, his works concerned the theory of **estimation**, and the **least squares** method and the application of the **normal distribution** to problems related to measurement **errors** both originated with him.

Some of the main works and articles of Gauss, C.F.:

1803–1809 Disquisitiones de elementis ellipticis pallidis. Werke, 6, 1–24.

1809 Theoria motus corporum coelestium. Werke, 7 (1963 English translation by Davis, C.H., published by Dover, New York).

1816 Bestimmung der Genauigkeit der Beobachtungen. Werke, 4, 109–117.

1821, 1823 and 1826 Theoria combinationis observationum erroribus minimis obnoxiae, parts 1, 2 and suppl. Werke, 4, 1–108.

1823 Anwendungen der Wahrscheinlichkeitsrechnung auf eine Aufgabe der praktischen Geometrie. Werke, 9, 231–237.

1855 Méthode des Moindres Carrés: Mémoires sur la Combinaison des Observations. French translation of the work of Gauss, C.F. by Bertrand, J. (authorized by Gauss, C.F. himself). Mallet-Bachelier, Paris.

1957 Gauss' Work (1803–1826). On The Theory of Least Squares. English translation by Trotter, H.F. Technical

Report No. 5, Statistical Techniques Research Group, Princeton, NJ.

FURTHER READING
▶ **De Moivre Abraham**
▶ **Gauss–Markov theorem**
▶ **Normal distribution**

Gauss–Markov Theorem

The Gauss–Markov theorem postulates that when the error **probability distribution** is unknown in a linear **model**, then, amongst all of the linear unbiased estimators for the parameters of the linear **model**, the **estimator** obtained using the method of **least squares** is the one that minimizes the **variance**. The **mathematical expectation** of each error is assumed to be zero, and all of them have the same (unknown) variance.

HISTORY

Gauss, Carl Friedrich provided a proof of this theorem in the first part of his work "Theoria combinationis observationum erroribus minimis obnoxiae" (1821). Markov, Andrei Andreyevich rediscovered this theorem in 1900.

A version of the Gauss–Markov theorem written in modern notation, was provided by Graybill in 1976.

MATHEMATICAL ASPECTS

Consider the linear **model**

$$Y = X \cdot \beta + \varepsilon,$$

where

Y is the $n \times 1$ **vector** of the observations,

X is the $n \times p$ **matrix** of the independent variables that are considered fixed,

β is the $p \times 1$ vector of the unknown parameters, and

ε is the $n \times 1$ vector of the random errors.

If the error **probability distribution** is unknown but the following conditions are fulfilled:

1. The **mathematic expectation** $E[\varepsilon] = 0$,
2. The **variance** $\text{Var}(\varepsilon) = \sigma^2 \cdot \mathbf{I_n}$, where $\mathbf{I_n}$ is the **identity matrix**,
3. The matrix X has a full rank,

then the **estimator** of β,

$$\hat{\beta} = \left(X'X\right)^{-1} X'Y,$$

derived via the **least squares** method is the linear estimator without **bias** of β that has least variance.

FURTHER READING
▶ **Estimator**
▶ **Least squares**

REFERENCES
Gauss, C.F.: Theoria Combinationis Observationum Erroribus Minimis Obnoxiae, Parts 1, 2 and suppl. Werke **4**, 1–108 (1821, 1823, 1826)

Graybill, F.A.: Theory and Applications of the Linear Model. Duxbury, North Scituate, MA (Waldsworth and Brooks/Cole, Pacific Grove, CA) (1976)

Rao, C.R.: Linear Statistical Inference and Its Applications, 2nd edn. Wiley, New York (1973)

Generalized Inverse

The generalized inverse is analogous to the inverse of a nonsingular square **matrix**, but is used for a matrix of any dimension and rank. The generalized inverse is used in the resolution of systems of linear equations.

HISTORY
It appears that Fredholm (1903) was the first to consider the concept of a generalized inverse. Moore defined a single generalized inverse in his book *General Analysis* (1935), published after his death. However, his work was not used until the 1950s, when it experienced a surge in interest due to the application of the generalized inverse to problems related to **least squares**.

In 1955 Penrose, reusing and enlarging upon work published in 1951 by Bjerhammar, showed that the Moore's generalized inverse is a unique **matrix** G that satisfies the following four equations:

1. $\mathbf{A} = \mathbf{A} \cdot \mathbf{G} \cdot \mathbf{A}$;
2. $\mathbf{G} = \mathbf{G} \cdot \mathbf{A} \cdot \mathbf{G}$;
3. $(\mathbf{A} \cdot \mathbf{G})' = \mathbf{A} \cdot \mathbf{G}$;
4. $(\mathbf{G} \cdot \mathbf{A})' = \mathbf{G} \cdot \mathbf{A}$.

This unique generalized inverse is known as the Moore–Penrose inverse, and is denoted by $\mathbf{A}^+$.

MATHEMATICAL ASPECTS
The $n \times m$ **matrix** G is the generalized inverse of the $m \times n$ matrix $\mathbf{A}$ if

$$\mathbf{A} = \mathbf{A} \cdot \mathbf{G} \cdot \mathbf{A}.$$

The matrix $\mathbf{G}$ is unique if and only if $m = n$ and $\mathbf{A}$ is not singular.

The most common notation used for the generalized inverse of a **matrix** $\mathbf{A}$ is $\mathbf{A}^-$.

FURTHER READING
▶ **Inverse matrix**
▶ **Least squares**

REFERENCES
Bjerhammar, A.: Rectangular reciprocal matrices with special reference to geode-

tic calculations. Bull. Geod. **20**, 188–220 (1951)

Dodge, Y., Majumdar, D.: An algorithm for finding least square generalized inverses for classification models with arbitrary patterns. J. Stat. Comput. Simul. **9**, 1–17 (1979)

Fredholm, J.: Sur une classe d'équations fonctionnelles. Acta Math. **27**, 365–390 (1903)

Moore, E.H.: General Analysis, Part I. Mem. Am. Philos. Soc., Philadelphia, PA, pp. 147–209 (1935)

Penrose, R.: A generalized inverse for matrices. Proc. Camb. Philos. Soc. **51**, 406–413 (1955)

Rao, C.R.: A note on generalized inverse of a matrix with applications to problems in mathematical statistics. J. Roy. Stat. Soc. Ser. B **24**, 152–158 (1962)

Rao, C.R.: Calculus of generalized inverse of matrices. I. General theory. Sankhya **A29**, 317–342 (1967)

Rao, C.R., Mitra, S.K.: Generalized Inverses of Matrices and its Applications. Wiley, New York (1971)

Generalized Linear Regression

An extension of the linear regression model to settings more flexible in underlying assumptions that the linear regression requires. For example, instead of assuming that the errors should have equal variances, we could have the following forms:

- *Heteroscedasticity:* the errors observed in the model (also called residuals) can have different variances among the observa-

tions (or among the different groups of observations);

- *Autocorrelation:* there can be some **correlation** between the errors in the different observations.

MATHEMATICAL ASPECTS

We consider a general model of **multiple linear regression**:

$$Y_i = \beta_0 + \sum_{j=1}^{p-1} \beta_j X_{ij} + \varepsilon_i, \quad i = 1, \ldots, n,$$

where

Y_i is the **dependent variable**,

$X_{ji}, j = 1, \ldots, p - 1$ are the independent variables,

$\beta_j, j = 0, \ldots, p - 1$ are the parameters to be estimated, and

ε_i is the random nonobservable **error** term.

In matrix form, we write:

$$\mathbf{Y} = \mathbf{X}\boldsymbol{\beta} + \boldsymbol{\varepsilon},$$

where:

$\mathbf{Y}$ is the vector ($n \times 1$) of the observations related to the **dependent variable** (n observations),

$\boldsymbol{\beta}$ is the vector ($p \times 1$) of the parameters to be estimated

$\boldsymbol{\varepsilon}$ is the vector ($n \times 1$) of the errors,

and

$$\mathbf{X} = \begin{pmatrix} 1 & X_{11} & \cdots & X_{1(p-1)} \\ \vdots & \vdots & & \vdots \\ 1 & X_{n1} & \cdots & X_{n(p-1)} \end{pmatrix}$$

G

is the **matrix** ($n \times p$) related to the independent variables.

In an ordinary regression model, the hypotheses used for the errors ε_i are generally as follows:

- Each **expectation** equals zero (in other words, the observations are not biased),
- They are independent and are not correlated,
- their **variance** $V(\varepsilon_i)$ is constant and equals σ^2 (we suppose that the errors are of the same size).

If the chosen model is appropriate, the distribution of residuals must confirm these hypotheses (see **analysis of variance**).

If we find that this is not the case, or if we obtain supplementary information that indicates that the errors are correlated, then we can use generalized linear regression to obtain a more precise estimation for the parameters $\beta_0, \ldots, \beta_{p-1}$. We then introduce the variance–covariance **matrix** $V(\varepsilon)$ of (of dimension $n \times n$) of the errors, which is defined by the following equation for row i and column j:

- $V(\varepsilon_i)$ if $i = j$ (variance in ε_i),
- $\text{Cov}(\varepsilon_i, \varepsilon_j)$ (**covariance** between ε_i and ε_j) if $i \neq j$.

Often this matrix needs to be estimated, but we first consider the simplest case, where $V(\varepsilon)$ is known. We use the usual term for variance, σ^2: $V(\varepsilon) = \sigma^2 V$.

Estimation of the Vector β when V Is Known

By transforming our general model into an ordinary regression model, it is possible to estimate the parameters of the generalized model using:

$$\hat{\beta} = \left(X'V^{-1}X \right)^{-1} X'V^{-1}Y .$$

We can prove that:

1. The estimation is not biased; that is: $E(\hat{\beta}) = \beta$.
2. The following equality is verified:

$$V\left(\hat{\beta} \right) = \hat{\sigma}^2 \left(X'V^{-1}X \right)^{-1} ,$$

with $\hat{\sigma}^2 = \dfrac{1}{n-p} \varepsilon' V^{-1} \varepsilon$.

3. $\hat{\beta}$ is the best (meaning that it has the smallest variance) unbiased estimator of β; this result is known as the generalized Gauss–Markov theorem.

Estimation of the Vector β when V Is Unknown

When the variance–covariance matrix is not known, the model used is called generalized regression. The first step consists of expressing V as a function of a parameter θ, which allows us to estimate V using $\hat{V} = V(\hat{\theta})$. The estimations for the parameters are then obtained by substituting V with $\hat{V}$ in the formula given above:

$$\hat{\beta} = (X'\hat{V}^{-1}X)^{-1}X'\hat{V}^{-1}Y .$$

Transformation of V in Three Typical Cases

a) Heteroscedasticity

We assume that the observations have variance $\dfrac{\sigma^2}{w_i}$, where w_i is the weight, which can be different for each $i = 1, \ldots, n$. So:

$$V = \sigma^2 \begin{pmatrix} \frac{1}{w_1} & 0 & \cdots & 0 \\ 0 & \frac{1}{w_2} & \cdots & 0 \\ \vdots & \vdots & \ddots & \vdots \\ 0 & 0 & \cdots & \frac{1}{w_n} \end{pmatrix} .$$

By making

$$Y_w = WY$$
$$X_w = WX$$
$$\varepsilon_w = W\varepsilon ,$$

where $\mathbf{W}$ is the following square matrix of order n:

$$\mathbf{W} = \begin{pmatrix} \sqrt{w_1} & 0 & \cdots & 0 \\ 0 & \sqrt{w_2} & \cdots & 0 \\ \vdots & \vdots & \ddots & \vdots \\ 0 & 0 & \cdots & \sqrt{w_n} \end{pmatrix},$$

such that $\mathbf{W}'\mathbf{W} = \mathbf{V}^{-1}$, we obtain the equivalent model:

$$V(\varepsilon_w) = V(\mathbf{W}\varepsilon) = \mathbf{W}V(\varepsilon)\mathbf{W}'$$
$$= \sigma^2 \mathbf{W}\mathbf{V}\mathbf{W} = \sigma^2 \mathbf{I}_n,$$

where $\mathbf{I}_n$ is the identity matrix of dimension n. Since the variances of all of the errors are the same in this new model, we can apply the least squares method and we obtain the vector of the estimators:

$$\hat{\beta}_w = (\mathbf{X}'_w \mathbf{X}_w)^{-1} \mathbf{X}'_w \mathbf{Y}_w$$
$$= (\mathbf{X}'\mathbf{W}'\mathbf{W}\mathbf{X})^{-1}\mathbf{X}'\mathbf{W}'\mathbf{W}\mathbf{Y}$$
$$= (\mathbf{X}'\mathbf{V}^{-1}\mathbf{X})^{-1}\mathbf{X}'\mathbf{V}^{-1}\mathbf{Y},$$

so the vector of the estimated values for $\mathbf{Y} = \mathbf{W}^{-1}\mathbf{Y}_w$ is:

$$\hat{\mathbf{Y}} = \mathbf{W}^{-1}\hat{\mathbf{Y}}_w$$
$$= \mathbf{W}^{-1}\mathbf{W}\mathbf{X}(\mathbf{X}'\mathbf{V}^{-1}\mathbf{X})^{-1}\mathbf{X}'\mathbf{V}^{-1}\mathbf{Y}$$
$$= \mathbf{X}\hat{\beta}_w.$$

b) Heteroscedasticity by groups
We suppose that the observations fall into three groups (of size n_1, n_2, n_3) with respective variances σ_1^2, σ_2^2 and σ_3^2. So:

$$\mathbf{V} = \begin{pmatrix} \sigma_1^2 \mathbf{I}_{n_1} & \mathbf{O}_{n_2 \times n_1} & \mathbf{O}_{n_3 \times n_1} \\ \mathbf{O}_{n_1 \times n_2} & \sigma_2^2 \mathbf{I}_{n_2} & \mathbf{O}_{n_3 \times n_2} \\ \mathbf{O}_{n_1 \times n_3} & \mathbf{O}_{n_2 \times n_3} & \sigma_3^2 \mathbf{I}_{n_3} \end{pmatrix},$$

where $\mathbf{I}_{n_j}$ is the **identity matrix** of order n_j and $\mathbf{O}_{n_i \times n_j}$ is the null matrix of order $n_i \times n_j$.

The variances σ_1^2, σ_2^2, σ_3^2 can be estimated by performing ordinary regressions on the different groups of observations and by estimating σ^2 using S^2 (see **multiple linear regression**).

c) First-order autocorrelation
We assume that there is correlation between the errors in different observations. In particular, we assume first-order autocorrelation; that is the error ε_i for observation i depends on the error ε_{i-1} for the previous observation. We then have:

$$\varepsilon_i = \rho\varepsilon_{i-1} + u_i, \quad i = 2, \ldots, n,$$

where the usual hypotheses used for the errors of a regression model are applied to the u_i.
In this case, the variance–covariance matrix of errors can be expressed by:

$$\mathbf{V}(\varepsilon) = \sigma^2 \mathbf{V}$$
$$= \sigma^2 \frac{1}{1-\rho^2}$$
$$\cdot \begin{pmatrix} 1 & \rho & \rho^2 & \cdots & \rho^{n-1} \\ \rho & 1 & \rho & \cdots & \rho^{n-2} \\ \vdots & \rho & \ddots & \ddots & \vdots \\ \vdots & \vdots & \ddots & \ddots & \rho \\ \rho^{n-1} & \rho^{n-2} & \cdots & \rho & 1 \end{pmatrix}.$$

We can estimate $\mathbf{V}$ by replacing the following estimation of ρ in the matrix $\mathbf{V}$:

$$\hat{\rho} = \frac{\sum_{i=2}^n e_i e_{i-1}}{\sum_{i=2}^n e_{i-1}^2} \text{ (empirical correlation)},$$

where the e_i are the estimated error (residuals), with an ordinary regression.
We can test the first-order autocorrelation by performing a Durbin–Watson test on the residuals (as described under the entry for **serial correlation**).

EXAMPLES
Correction for heteroscedasticity when the data represent means

In order to test the efficiency of a new supplement on the growth of chickens, 40 chickens were divided into five groups (of size $n_1, n_2, \ldots, n_5$) and given the different doses of the supplement. The growth data obtained are summarized in the following table:

Group i	Number of chickens n_i	Mean weight Y_i	Mean dose X_i
1	12	1.7	5.8
2	8	1.9	6.4
3	6	1.2	4.8
4	9	2.3	6.9
5	5	1.8	6.2

Since the variables represent means, the variance of the error ε_i is $\frac{\sigma^2}{n_i}$.

The variance–covariance matrix of errors is then:

$$\mathbf{V}(\boldsymbol{\varepsilon}) = \sigma^2 \begin{pmatrix} \frac{1}{12} & 0 & 0 & 0 & 0 \\ 0 & \frac{1}{8} & 0 & 0 & 0 \\ 0 & 0 & \frac{1}{6} & 0 & 0 \\ 0 & 0 & 0 & \frac{1}{9} & 0 \\ 0 & 0 & 0 & 0 & \frac{1}{5} \end{pmatrix}$$

$$= \sigma^2 \mathbf{V}.$$

The **inverse matrix $\mathbf{V}^{-1}$** is then:

$$\mathbf{V}^{-1} = \begin{pmatrix} 12 & 0 & 0 & 0 & 0 \\ 0 & 8 & 0 & 0 & 0 \\ 0 & 0 & 6 & 0 & 0 \\ 0 & 0 & 0 & 9 & 0 \\ 0 & 0 & 0 & 0 & 5 \end{pmatrix},$$

and the estimations for the parameters β_0 and β_1 in the model

$$Y_i = \beta_0 + \beta_1 X_i + \varepsilon_i$$

are:

$$\hat{\beta}_0 = -1.23$$
$$\hat{\beta}_1 = 0.502.$$

This result is the same as that obtained using the **weighted least squares method**.

FURTHER READING
▶ **Analysis of variance**
▶ **Gauss–Markov theorem**
▶ **Model**
▶ **Multiple linear regression**
▶ **Nonlinear regression**
▶ **Regression analysis**

REFERENCES
Bishop, Y.M.M., Fienberg, S.E., Holland, P.W.: Discrete Multivariate Analysis: Theory and Practice. MIT Press, Cambridge, MA (1975)

Graybill, F.A.: Theory and Applications of the Linear Model. Duxbury, North Scituate, MA (Waldsworth and Brooks/Cole, Pacific Grove, CA) (1976)

Rao, C.R.: Linear Statistical Inference and Its Applications, 2nd edn. Wiley, New York (1973)

Generation of Random Numbers

Random numbers generation involves producing a series of numbers with no recognizable patterns or regularities, that is, appear random. These random numbers are often then used to perform **simulations** and to solve problems that are difficult or impossible to resolve analytically, using the **Monte Carlo method**.

HISTORY

The first works concerning random number generation were those of von Neumann, John (1951), who introduced the "middle four square" method. Lehmer (1951) introduced the simple congruence method, which is still used today.

Sowey published three articles giving a classified bibliography on random number generation and testing (Sowey, E.R. 1972, 1978 and 1986).

In 1996, Dodge, Yadolah proposed a natural method of generating random numbers from the digits after the decimal point in π. Some decimals of π presented in Appendix A.

MATHEMATICAL ASPECTS

There are several ways of generating **random numbers**. The first consists of pulling out notes that are numbered from 0 to 9 from an urn one by one, and then putting them back in the urn.

Certain physical systems can be used to generate **random numbers**: for example the roulette wheel. The numbers obtained can be listed in a random number table. These tables are useful if several **simulations** need to be performed for different **models** in order to test which one is the best. In this case, the same series of random numbers can be reused.

Some computers use electronic circuits or electromechanical means to provide the series of **random numbers**. Since these lists are not reproducible, von Neumann, John and Lehmer, D.H. proposed methods that are fully deterministic but give a series of numbers that have the appearance of being random. Such numbers are known as pseudorandom numbers.

These pseudorandom numbers are calculated from a predetermined algebraic formula. For example, the Lehmer method allows a series of pseudorandom numbers $x_0, x_1, \ldots, x_n, \ldots$ to be calculated from

$$x_i = a \cdot x_{i-1} \ (\text{modulo } m), \quad i = 1, 2, \ldots,$$

where $x_0 = b$ and a, b, m are given.

The choice of a, b, m influences the quality of the **sample** of pseudorandom numbers.

Independence between the pseudorandom numbers dictates that the length of the cycle, meaning the quantity of numbers generated before the same sequence is restarted, must be long enough. To obtain the longest possible cycle, the following criteria must be satisfied:

- a must not divide m,
- x_0 must not divide m,
- m must be large.

DOMAINS AND LIMITATIONS

If the **simulation** requires **observations** derived from nonuniform distributions, there are several techniques that can generate any law from a **uniform distribution**. Some of these methods are applicable to many distributions; others are more specific and are used for a particular distribution.

EXAMPLES

Let us use the Lehmer method to calculate the first pseudorandom numbers of the series created with $a = 33$, $b = 7$, and $m = 1000$:

$$x_0 = 7$$
$$x_1 = 7 \cdot 33 \ (\text{mod} 1000) = 231$$
$$x_2 = 231 \cdot 33 \ (\text{mod} 1000) = 623$$
$$x_3 = 623 \cdot 33 \ (\text{mod} 1000) = 559$$

etc.

A series of numbers r_i for which the distribution does not significantly differ from a uniform distribution is obtained by dividing x_i by m:

$$r_i = \frac{x_i}{m}, \quad i = 1, 2, \ldots$$

Therefore: $r_0 = 0.007$
$$r_1 = 0.231$$
$$r_2 = 0.623$$
$$r_3 = 0.559$$
etc.

FURTHER READING
▶ **Kendall Maurice George**
▶ **Monte Carlo method**
▶ **Random number**
▶ **Simulation**
▶ **Uniform distribution**

REFERENCES
Dodge, Y.: A natural random number generator, Int. Stat. Review, 64, 329–344 (1996)

Ripley, B.D.: Thoughts on pseudorandom number generators. J. Comput. Appl. Math. **31**, 153–163 (1990)

Genetic Statistics

In genetic statistics, genetic data analysis is performed using methods and concepts of classical **statistics** as well as those derived from the theory of stochastic processes. Genetics is the study of hereditary character and accidental variations; it is used to help explain transformism, in the practical domain, to the improvement of the units.

HISTORY
Genetics and statistics have been related for over 100 years. Sir **Galton, Francis** was very interested in studies of the human genetics, and particularly in eugenics, the study of methods that could be used improve the genetic quality of a human population, which obviously requires knowledge of heredity. Galton also invented regression and correlation coefficients in order to use them as statistical tools when investigating genetics. These methods were later developed further by **Pearson, Karl**. In 1900, the work of Mendel was rediscovered and classical mathematical and statistical tools for investigating Mendel genetics were put forward by Sir **Fisher, Ronald Aylmer**, Wright, Sewall and Haldane, J.B.S. in the period 1920–1950.

DOMAINS AND LIMITATIONS
Genetic statistics is used in domains such as biomathematics, bioinformatics, biology, **epidemiology** and genetics. Standard methods of **estimation** and hypothesis testing are used in order to estimate and test genetic parameters. The theory of stochastic processes is used to study the units of the evolution of a subject of a population, taking in account the random changes in the frequencies of genes.

EXAMPLES
The Hardy–Weinberg law is the central theoretical model used in population genetics. Formulated independently by the English mathematician Hardy, Godfrey H. and the German doctor Weinberg, W. (1908), the equations of this law show that when genetic variations first appear in some individuals in a population, they do not disappear upon hereditary transmission, but are instead maintained in future generations in the same proportions, conforming to Mendel laws.

The Hardy–Weinberg law allows us, under certain conditions, to calculate genotypic frequencies from allelic frequencies. By genotype, we mean the genetic properties of an individual that are received during hereditary transmission from the individual's parents. For example, identical twins have the same genotype. An allele is a possible version of a gene, constructed from a nucleotide chain. Therefore, a particular gene can occur in many different allelic forms in a given **population**. In an individual, if each gene is represented by two alleles that are composed of identical nucleotides, the individual is then homozygous for this gene, and if the alleles have different compositions, the individual is then heterozygous for this gene.

In the original version, the law tells that if a gene is controlled by two alleles A and B, that occur with frequencies p and $q = 1 - p$ respectively in the population at generation t, then the frequencies of the genotypes at generation $t+1$ are given by the following equation:

$$p^2 AA + 2pq AB + q^2 BB.$$

The basic hypothesis is that the size of the population N is big enough to minimize sampling variations; there also must be no selection, no mutation, no migration (no acquisition/loss of an allele), and successive generations must be discrete (no generation crossing).

To get an AA genotype, the individual must receive one allele of type A from both parents. If this process is random (so the hypothesis of **independence** holds), this event will occur with a probability of:

$$P(AA) = pp = p^2.$$

The same logic applies to the probability of obtaining the genotype BB:

$$P(BB) = qq = q^2.$$

Finally, for the genotype AB, two cases are possible: the individual received A from their father and B from their mother or vice versa, so:

$$P(AB) = pq + qp = 2pq.$$

Therefore, in an ideal population, the pHardy–Weinberg proportions are given by:

$$\begin{array}{ccc} AA & AB & BB \\ p^2 & 2pq & q^2. \end{array}$$

This situation can be generalized to a gene with many alleles $A_1, A_2, \ldots, A_k$. The frequency that homozygotes (A_iA_i) occur equals:

$$f(A_iA_i) = p_i^2, \quad i = 1, \ldots, k,$$

and the frequency that heterozygotes (A_iA_j) occur equals:

$$f(A_iA_j) = 2p_ip_j,$$
$$i \neq j, \quad i = 1, \ldots, k, \quad j = 1, \ldots, k.$$

The frequency p' of allele A in generation $t+1$ may also be of interest. By simple counting we have:

$$p' = \frac{2p^2N + 2pqN}{2N} = p^2 + pq$$
$$= p \cdot (p + q) = p \cdot (1 - q + q) = p.$$

The frequency of the allele A in generation $t+1$ is identical to the frequency of this allele in the previous generation and the initial generation.

FURTHER READING
▶ **Data**
▶ **Epidemiology**
▶ **Estimation**

▶ **Hypothesis testing**
▶ **Independence**
▶ **Parameter**
▶ **Population**
▶ **Statistics**
▶ **Stochastic process**

REFERENCES

Galton, F.: Typical laws of heredity. Proc. Roy. Inst. Great Britain 8, 282–301 (1877) (reprinted in: Stigler, S.M. (1986) The History of Statistics: The Measurement of Uncertainty Before 1900. Belknap Press, Cambridge, MA, p. 280)

Galton, F.: Co-relations and their measurement, chiefly from anthropological data. Proc. Roy. Soc. Lond. **45**, 135–145 (1888)

Hardy, G.H.: Mendelian proportions in a mixed population. Science **28**, 49–50 (1908)

Smith, C.A.B.: Statistics in human genetics. In: Kotz, S., Johnson, N.L. (eds.) Encyclopedia of Statistical Sciences, vol. 3. Wiley, New York (1983)

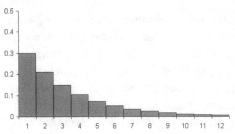

Geometric distribution, $p = 0.3$, $q = 0.7$

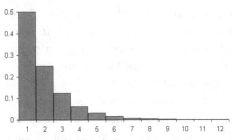

Geometric distribution, $p = 0.5$, $q = 0.5$

This distribution is called "geometric" because the successive terms of the **probability function** described above form a geometric progression with a ratio of q.

The geometric distribution is a particular case of the **negative binomial distribution** where $r = 1$, meaning that the process continues until the first success.

Geometric Distribution

Consider a process involving $k + 1$ **Bernoulli trials** with a **probability** of success p and a probability of failure q.

The **random variable** X, corresponding to the number of failures k that occur when the process is repeated until the first success, follows a geometric distribution with **parameter** p, and so we have:

$$P(X = k) = p \cdot q^k.$$

The geometric distribution is a **discrete probability distribution**.

MATHEMATICAL ASPECTS

The **expected value** of the geometric distribution is by definition equal to:

$$E[X] = \sum_{x=0}^{\infty} x \cdot P(X = x)$$

$$= \sum_{x=0}^{\infty} x \cdot p \cdot q^x$$

$$= \frac{q}{p}.$$

Indeed:

$$E[X] = \sum_{x=0}^{\infty} x \cdot q^{x-1} \cdot p$$

$$= p \cdot \sum_{x=0}^{\infty} x \cdot q^{x-1}$$

$$= 0 + p \left(1 + 2 \cdot q + 3 \cdot q^2 + \ldots \right).$$

Hence we have: $1 + 2 \cdot q + 3 \cdot q^2 + \ldots = \frac{1}{(1-q)^2}$
since:

$$q + q^2 + q^3 + \ldots = \frac{q}{1-q}$$

$$\left(q + q^2 + q^3 + \ldots \right)' = \left(\frac{q}{1-q} \right)'$$

$$1 + 2 \cdot q + 3 \cdot q^2 + \ldots = \frac{1}{(1-q)^2}.$$

The **variance** of the geometric distribution is by definition equal to:

$$\text{Var}(X) = E[X^2] - (E[X])^2$$

$$= \sum_{x=0}^{\infty} x^2 \cdot P(X = x) - \left(\frac{q}{p} \right)^2$$

$$= \sum_{x=0}^{\infty} x^2 \cdot p \cdot q^x - \left(\frac{q}{p} \right)^2$$

$$= p \cdot q(1 + 2^2 \cdot q + 3^2 q^2 + \ldots)$$

$$- \left(\frac{q}{p} \right)^2$$

$$= \frac{q}{p^2}.$$

Indeed, the sum $1 + 2^2 \cdot q + 3^2 q^2 + \ldots = \frac{1+q}{p^2}$,
since:

$$\left(q \left(1 + 2 \cdot q + 3 \cdot q^2 + \ldots \right) \right)'$$

$$= \left(\frac{q}{(1-q)^2} \right)'$$

$$1 + 2^2 \cdot q + 3^2 \cdot q^2 + \ldots$$

$$= \frac{1+q}{(1-q)^2} = \frac{1+q}{p^2}.$$

DOMAINS AND LIMITATIONS

The geometric distribution is used relatively frequently in meteorological models. It is also used in stochastic processes and the theory of waiting lines.

The geometric distribution can also be expressed in the following form:

$$P(X = k) = p \cdot q^{k-1},$$

where the **random variable** X represents the number of tasks required to attain the first success (including the success itself).

EXAMPLES

A fair die is thrown and we want to know the **probability** that a "six" will be thrown for the first time on the fourth throw.
We therefore have:

$p = \frac{1}{6}$ (**probability** that a six is thrown)

$q = \frac{5}{6}$ (**probability** that a six is not thrown)

$k = 3$ (since the fourth throw is a success, we have three failures)

Therefore, the probability is as follows:

$$P(X = 3) = \frac{1}{6} \cdot \left(\frac{5}{6} \right)^3 = 0.0965.$$

FURTHER READING
▶ **Bernoulli distribution**
▶ **Discrete probability distribution**
▶ **Negative binomial distribution**

Geometric Mean

The geometric **mean** is defined as the nth root of the product of n non-negative numbers.

HISTORY

According to Droesbeke, J.-J. and Tassi, Ph. (1990), the geometric mean and the **harmonic mean** were introduced by Jevons, William Stanley in 1874.

MATHEMATICAL ASPECTS

Let $x_1, x_2, \ldots, x_n$ be a set of n non-negative quantities, or of n observations related to a **quantitative variable** X. The geometric mean G of this set is:

$$G = \sqrt[n]{x_1 \cdot x_2 \cdot \ldots \cdot x_n}.$$

The geometric mean is sometimes expressed in a logarithmic form:

$$\ln(G) = \frac{\sum\limits_{i=1}^{n} \ln(x_i)}{n},$$

or

$$G = \exp\left[\frac{1}{n} \cdot \sum\limits_{i=1}^{n} \ln(x_i)\right].$$

We note that the logarithm of the geometric mean of a set of positive numbers is the **arithmetic mean** of the logarithms of these numbers (or the **weighted arithmetic mean** in the case of grouped observations).

DOMAINS AND LIMITATIONS

In practice, the geometric mean is mainly used to calculate the **mean** of a group of ratios, or particularly the mean of a group of indices.

Just like the **arithmetic mean**, the geometric mean takes every **observation** into account individually. However, it decreases the influence of outliers on the mean, which is why it is sometimes preferred to the arithmetic mean.

One important aspect of the geometric mean is that it only applies to positive numbers.

EXAMPLES

Here we have an example of how the geometric mean is used: in 1987, 1988 and metric mean is used: in 1987, 1988 and 1989, a businessman achieves profits of 10.000, 20.000 and 160.000 euros, respectively. Based on this information, we want to determine the mean rate at which these profits are increasing.

From 1987 to 1988, the profit doubled, and from 1988 to 1989 it increased eight-fold. If we simply calculate the **arithmetic mean** of these two numbers, we get

$$x = \frac{2+8}{2} = 5,$$

and we conclude that, on average, the profit increased five-fold annually. However, if we take the profit in 1987 (10000 euros) and multiply it by five annually, we obtain profits of 50000 in 1988 and 250000 euros in 1989. These two profits are much too large compared to the real profits.

On the other hand, if we calculate the mean using the geometric mean of these two increases instead of the arithmetic mean, we get:

$$G = \sqrt{2 \cdot 8} = \sqrt{16} = 4,$$

and we can correctly say that, on average, the profits quadrupled annually. Applying this mean rate of increase to the initial profit of 10000 euros in 1987, we obtain 40000 euros in 1988 and 160000 euros in 1989. Even if the profit in 1988 is too high, it is less so than the previous result, and the profit for 1989 is now correct.

To illustrate the use of the formula for the geometric mean in the logarithmic form, we will now find the mean of the following indices:

112, 99, 105, 96, 85, 100.

We calculate (using a table of logarithms):

$$
\begin{aligned}
\log(112) &= 2.0492 \\
\log(99) &= 1.9956 \\
\log(105) &= 2.0212 \\
\log(96) &= 1.9823 \\
\log(85) &= 1.9294 \\
\underline{\log(100)} &= \underline{2.0000} \\
&\ 11.9777
\end{aligned}
$$

We get then:

$$
\log(G) = \frac{11.9777}{6} = 1.9963 .
$$

Therefore $G = 99.15$.

FURTHER READING
▶ **Arithmetic mean**
▶ **Harmonic mean**
▶ **Mean**
▶ **Measure of central tendency**

REFERENCES
Droesbeke, J.J., Tassi, P.H.: Histoire de la Statistique. Editions "Que sais-je?" Presses universitaires de France, Paris (1990)

Jevons, W.S.: The Principles of Science: A Treatise on Logic and Scientific Methods (two volumes). Macmillan, London (1874)

Geometric Standard Deviation

The geometric standard deviation of a set of quantitative observations is a **measure of dispersion**. It corresponds to the deviation of observations around the **geometric mean**.

HISTORY
See L_1 **estimation**.

MATHEMATICAL ASPECTS
Let $x_1, x_2, \ldots, x_n$ be a set of n observations related to a **quantitative variable** X. The geometric standard deviation, denoted by σ_g, is calculated as follows:

$$
\log \sigma_g = \left[\frac{1}{n} \sum_{i=1}^{n} (\log x_i - \log G)^2 \right]^{\frac{1}{2}} .
$$

where $G = \sqrt[n]{x_1 \cdot x_2 \cdot \ldots \cdot x_n}$ is the geometric mean of the observations.

FURTHER READING
▶ L_1 **estimation**
▶ **Measure of dispersion**
▶ **Standard deviation**

Geostatistics

Geostatistics is the application of statistics to problems in geology and hydrology. Geostatistics naturally treat spatial data with known coordinates. Spatial hierarchical models follow the principle: model locally, analyze globally. Simple conditional models are constructed on all levels of the hierarchy (local modeling). The result is a joint model that can be very complex, but analysis is still possible (global analysis). Geostatistics encompasses the set used in the theory, as well as the techniques and statistical applications used to analyze and forecast the distribution of the values of the variable in space and (eventually) time.

HISTORY
The term geostatistics was first used by Hart (1954) in a geographical context, in order

G

to highlight the application of particular statistical techniques to observations covering a regional distribution. The first geostatistical concepts were formulated by Matheron (1963) at the Ecole des Mines de Paris in order to estimate the reserves of a mineral from spatially distributed data (observations).

DOMAINS AND LIMITATIONS

Geostatistics is used in a wide variety of domains, including forecasting (of precipitation, ground porosity, concentrations of heavy metals) and when treating images from satellites.

FURTHER READING
► **Spatial data**
► **Spatial statistics**

REFERENCES

Atteia, O., Dubois, J.-P., Webster, R.: Geostatistical analysis of soil contamination in the Swiss Jura. Environ. Pollut. **86**, 315–327 (1994)

Cressie, N.A.C.: Statistics for Spatial Data. Wiley, New York (1991)

Goovaerts, P.: Geostatistics for Natural Resources Evaluation. Oxford University Press, Oxford (1997)

Isaaks, E.H., Srivastava, R.M.: An Introduction to Applied Geostatistics. Oxford University Press, Oxford (1989)

Journel, A., Huijbregts, C.J.: Mining Geostatistics. Academic, New York (1981)

Matheron, G.: La théorie des variables regionalisées et ses applications. Masson, Paris (1965)

Oliver, M.A., Webster, R.: Kriging: a method of interpolation for geographical informa-

tion system. Int. J. Geogr. Inf. Syst. **4**(3), 313–332 (1990)

Pannatier, Y.: Variowin 2.2. Software for Spatial Data Analysis in 2D. Springer, Berlin Heidelberg New York (1996)

Gini, Corrado

Gini, Corrado (1884–1965) was born in Matta di Livenza, Italy. Although he graduated in law in 1905, he took courses in mathematics and biology during his studies. His subsequent inclinations towards both science and statistics led him to become a temporary professor of statistics at Cagliari University in 1909 and, in 1920 he acceded to the Chair of Statistics at the same university. In 1913, he began teaching at Padua University, and acceded to the Chair of Statistics at the University of Rome in 1927.

Between 1926 and 1932, he was also the President of the Central Institute of Statistics (ISTAT).

He founded two journals. The first one, Metron (founded in 1920), is an international journal of statistics, and the second one, Genus (founded in 1934), is the Journal of the Italian Committee for the Study of Population Problems.

The contributions of Gini to the field of statistics principally concern mean values and the variability, as well as associations between the random variables. He also contributed original works to economics, sociology, demography and biology.

Some principal works and articles of Gini, Corrado:

1910   Indici di concentrazione e di dependenza. Atti della III Riunione della

Societa Italiana per il Progresso delle Scienze, R12, 3–210.

1912 Variabilità e mutabilità. Studi economico-giuridici, Università di Cagliari, III 2a, R12, 211–382.

Gini Index

The Gini index is the most commonly used inequality index; it allows us to measure the degree of inequality in the distribution of incomes for a given population. Graphically, the Gini index is represented by the surface area between a line at 45° and the Lorenz curve (a graphical representation of the cumulative percentage of the total income versus the cumulative percentage of the population that receives that income, where income increases left to right). Dividing the surface area between the line at 45° and the Lorenz curve (area A in the following diagram) by the total surface area under the line ($A + B$) gives the Gini coefficient:

$$I_G = \frac{A}{A + B}.$$

The Gini coefficient is zero if there is no inequality in income distribution and a value of 1 if the income distribution is completely unequal, so $0 \le I_G \le 1$. The Gini index is simply the Gini coefficient multiplied by 100, in order to convert it into a percentage. Therefore, the closer the index is to 100%, the more unequal the income distribution is. In developed countries, values of the Gini index are around 40% (see the example).

HISTORY

The Gini coefficient was invented in 1912 by the Italian statistician and demographer **Gini, Corrado**.

MATHEMATICAL ASPECTS

The total surface area under the line at 45° ($A + B$) equals $\frac{1}{2}$, and so we can write:

$$I_G = \frac{A}{A + B} = \frac{A}{\frac{1}{2}} = 2A$$

$$= 2\left(\frac{1}{2} - B\right) \Rightarrow I_G = 1 - 2B \tag{1}$$

To calculate the value of the Gini coefficient, we therefore need to work out B (total surface under the Lorenz curve).

a) *Gini index: discrete case*

For a discrete Lorenz curve where the income distribution is arranged in increasing order ($x_1 \le \ldots \le x_n$), the Lorenz curve has the following profile:

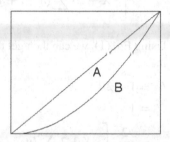

To calculate B, we need to divide the surface under the Lorenz curve into a series of polygons (where the first polygon is a triangle and others is are trapezia). The surface of the triangle is:

$$\frac{1}{2n} \frac{x_1}{X}.$$

The surface of the jth trapezium is:

$$\frac{1}{2n}\left[\sum_{i=1}^{j-1}\frac{x_i}{X} + \sum_{i=1}^{j}\frac{x_i}{X}\right]$$

$$= \frac{1}{2nX}\left[2 \cdot \sum_{i=1}^{j-1}x_i + x_j\right],$$

with $j = 2, \ldots, n$. We then have

$$B = \frac{1}{2n} \cdot \frac{x_1}{X}$$

$$+ \sum_{j=2}^{n} \frac{1}{2nX} \left[2 \cdot \sum_{i=1}^{j-1} x_i + x_j \right],$$

where:

$x_1, \ldots, x_n$ are the incomes;

X is the total revenue;

$\dfrac{x_1 + \ldots + x_n}{X}$ is the total cumulated revenue, in %.

After developing this equation, we get:

$$B = \frac{1}{2nX}$$

$$\cdot \left[-X + 2 \cdot \sum_{i=1}^{n} (n - i + 1)x_i \right]. \tag{2}$$

Using Eq. (1), we can then get the Gini index:

$$I_G = 1 - 2B$$

$$= 1$$

$$- \frac{2}{2nX} \left[-X + 2 \cdot \sum_{i=1}^{n} (n - i + 1)x_i \right]$$

$$= 1 + \frac{1}{n} - 2 \cdot \frac{\sum_{i=1}^{n} (n - i + 1) x_i}{nX}$$

$$= 1 + \frac{1}{n} - \frac{2}{n^2 \lambda (x)} \sum_{i=1}^{n} i x_{n-i+1},$$

where $\lambda (x) = \frac{1}{n} \sum_{i=1}^{n} x_i$ = mean income of the distribution.

The Gini coefficient can be reformulated in many ways. We could use, for example:

$$I_G = \sum_{i=1}^{n} \sum_{j=1}^{n} \frac{|x_i - x_j|}{2n^2 \lambda (x)}.$$

b) *Gini index: continuous case*

In this case, B corresponds to the integral from 0 to 1 of the Lorenz function. Therefore, we only need to insert this integral into Eq. (1):

$$I_G = 1 - 2B = 1 - 2 \int_0^1 L(P) \, dP,$$

where $P = F(y) = \int_0^y f(x) \, dx$ is the percentage of the population with incomes smaller then y. For example,

$$I_G = \int_0^1 \left(1 - (1 - P)^{\frac{\alpha-1}{\alpha}} \right) dP$$

$$= 1 - \frac{\alpha}{2\alpha - 1}$$

$$= 1 - 2 \left(1 - \frac{\alpha}{2\alpha - 1} \right).$$

DOMAINS AND LIMITATIONS

The Gini index (or coefficient) is the parameter most commonly used to measure the extent of inequality in an income distribution. It can also be used to measure the differences in unemployment between regions.

EXAMPLES

As an example of how to calculate the Gini coefficient (or index), let us take the discrete case.

We have: $I_G = 1 - 2B$

Recall that, in this case, we find surface B by calculating the sum of a series of polygons under the Lorenz curve.

The income data for a particular population in 1993–1994 are given in the following table, along with the areas associated the income intervals:

Population	Income	Surface Area
0	0	
94569	173100.4	0.000013733
37287	527244.8	0.000027323
186540	3242874.9	0.000726701
203530	4603346.0	0.00213263
266762	7329989.6	0.005465843
290524	9428392.6	0.010037284
297924	11153963.1	0.015437321
284686	12067808.2	0.02029755
251155	11898201.1	0.022956602
216164	11317647.7	0.023968454
182268	10455257.4	0.023539381
153903	9596182.1	0.022465371
127101	8560949.7	0.020488918
104788	7581237.4	0.018311098
85609	6621873.3	0.01597976
70108	5773641.3	0.013815408
57483	5021336.9	0.011848125
47385	4376126.7	0.010140353
39392	3835258.6	0.008701224
103116	11213361.3	0.024078927
71262	9463751.6	0.017876779
46387	7917846.5	0.012313061
50603	20181113.8	0.0114625049

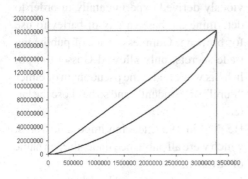

So, the surface B equals:

$$B = \frac{1}{2nX} \cdot \left[-X + 2 \cdot \sum_{i=1}^{n} (n - i + 1) \cdot x_i \right]$$
$$= 0.315246894 \,.$$

We then have:

$$I_G = 1 - 2B = 1 - 2 \cdot 0.315246894$$
$$= 0.369506212 \,.$$

FURTHER READING
▶ **Gini, Corrado**
▶ **Simple index number**
▶ **Uniform distribution**

REFERENCES
Greenwald, D. (ed.): Encyclopédie économique. Economica, Paris, pp. 164–165 (1984)

Flückiger, Y.: Analyse socio-économique des différences cantonales de chômage. Le chômage en Suisse: bilan et perspectives. Forum Helvet. **6**, 91–113 (1995)

Goodness of Fit Test

Performing a goodness of fit test on a **sample** allows us to determine whether the observed distribution corresponds to a particular **probability distribution** (such as the **normal distribution** or the **Poisson distribution**). This allows us to find out whether the observed sample was drawn from a **population** that follows this distribution.

HISTORY
See **chi-square goodness of fit test** and **Kolmogorov–Smirnov test**.

MATHEMATICAL ASPECTS
The goal of the goodness of fit test is to determine whether an observed **sample** was drawn from a **population** that follows a particular **probability distribution**.

The process used usually involves some **hypothesis testing**:

Null hypothesis H_0: $F = F_0$

Alternative hypothesis H_1: $F \neq F_0$

where F is an unknown **distribution function** of the underlying population and F_0 is the presumed distribution function.

The goodness of fit test involves comparing the empirical distribution with the presumed distribution. We reject the null hypothesis if the empirical distribution is not close enough to the presumed distribution. The precise rules that govern whether the null hypothesis is rejected or accepted depend on the type of the test used.

EXAMPLES

There are many goodness of fit tests, including the **chi-square goodness of fit test** and the **Kolmogorov–Smirnov test**.

FURTHER READING

▶ **Chi-square goodness of fit test**
▶ **Hypothesis testing**
▶ **Kolmogorov–Smirnov test**

Gosset, William Sealy

Gosset, William Sealy, better known by his pen name, "Student", was born in Canterbury, England in 1876. After studying mathematics and chemistry at New College, Oxford, he started working as a brewer for Guinness Breweries in Dublin in 1899. Guinness was a business that favored research, and made its laboratories available to its brewers and its chemists. Indeed, in 1900 it opened the "Guinness Research Laboratory," which had one of the greatest chemists around at that time, Horace Brown, as director. Work that was carried out in this laboratory was related to the quality and costs of the many varieties of barley and hop.

It was in this environment that Gosset developed his interest in **statistics**.

At Oxford, Gosset had studied mathematics, and so he was often called upon by his colleagues at Guinness when certain mathematical problems rose. That led to him studying the theory of **errors**, and he consulted **Pearson, K.**, whom he met in July 1905, on this topic.

The main difficulty encountered by Gosset during his work was small sample sizes. He therefore decided to attempt to find an appropriate method for handling the data from these small **samples**.

In 1906–1907, Gosset spent a year collaborating with **Pearson, K.**, in Pearson's laboratory in the University College London, attempting to develop methods related to the probable **error** of the **mean**.

In 1907, Gosset was made responsible for Guinness' Experimental Brewery, and he used the **Student table** that he had previously derived experimentally in order to determine the best variety of barley to use for brewing. Guinness, wary of publishing trade secrets, only allowed Gosset to publish his work under the pseudonym of either "Pupil" or "Student", and so he chose the latter.

He died in 1937, leaving important works, which were all published under the pen name of "Student".

Gosset's work remained largely ignored for many years, with few scientists using his table aside from the researchers at Guinness' breweries and those at Rothamsted Experimental Station (where **Fisher, R.A.** worked as a statistician).

Some of the main works and articles of Gosset, W.S.:

1907 On the error of counting with a haemacytometer. Biometrika 5, 351–360.

1908 The probable error of a mean. Biometrika 6, 1–25

1925 New tables for testing the significance of observations. Metron 5, 105–120.

1942 Pearson, E.S. and Wishart, J. (eds) Student's collected papers (foreword by McMullen, L.). Cambridge University Press, Cambridge (Issued by the Biometrika Office, University College London).

FURTHER READING
▶ Student distribution
▶ Student table
▶ Student test

REFERENCES

Fisher-Box, J.: Guinness, Gosset, Fisher, and small samples. Stat. Sci. **2**, 45–52 (1987)

McMullen, L., Pearson, E.S.: William Sealy Gosset, 1876–1937. In: Pearson, E.S., Kendall, M. (eds.) Studies in the History of Statistics and Probability, vol. I. Griffin, London (1970)

Plackett, R.L., Barnard, G.A. (eds.): Student: A Statistical Biography of William Sealy Gosset (Based on writings of E.S. Pearson). Clarendon, Oxford (1990)

Graeco-Latin Squares

See **Graeco-Latin square design**.

Graeco-Latin Square Design

A Graeco-Latin square design is a **design of experiment** in which the experimental units are grouped in three different ways. It is obtained by superposing two Latin squares of the same size.

If every Latin letter coincides exactly once with a Greek letter, the two Latin square designs are orthogonal. Together they form a Graeco-Latin square design.

In this design, each **treatment** (Latin letter) appears just once in each line, once in each column and once with each Greek letter.

HISTORY

The construction of a Graeco-Latin square, was originated by Euler, Leonhard (1782).

A book by **Fisher, Ronald Aylmer** and Yates, Frank (1963) gives Graeco-Latin tables of order 3 up to order 12 (not including the order of six). The book by Denes and Keedwell (1974) also contained comprehensive information on Graeco-Latin square designs. Dodge and Shah (1977) treat the case of the **estimation** when data is missing.

DOMAINS AND LIMITATIONS

Graeco-Latin square designs are used to reduce the effects of three sources of systematic error.

There are Graeco-Latin square designs for any size n except for $n = 1, 2$ and 6.

EXAMPLES

We want to test three different types of gasoline. To do this, we have three drivers and three vehicles. However, the test is too involved to perform in one day: we have to perform the experiment over three days. In this case, using drivers and vehicles over

G

three separate days could result in systematic errors. Therefore, in order to eliminate these errors we construct a Graeco-Latin square experimental design. In this case, each type of gasoline (**treatment**) will be tested just once with each driver, once with each vehicle and once each day.

In this Graeco-Latin square design, we represent the different types of gasoline by the Latin letters A, B and C and the different days by α, β and γ.

	Drivers		
	1	2	3
Vehicles 1	$A\alpha$	$B\beta$	$C\gamma$
2	$B\gamma$	$C\alpha$	$A\beta$
3	$C\beta$	$A\gamma$	$B\alpha$

An **analysis of variance** will then tell us whether, after eliminating the effects of the lines (vehicles), the columns (drivers) and the greek letters (days), there is a significant difference between the types of gasoline.

FURTHER READING
► **Analysis of variance**
► **Design of experiments**
► **Latin square designs**

REFERENCES

Dénes, J., Keedwell, A.D.: Latin squares and their applications. Academic, New York (1974)

Dodge, Y., Shah, K.R.: Estimation of parameters in Latin squares and Graeco-latin squares with missing observations. Commun. Stat. Theory **A6**(15), 1465–1472 (1977)

Euler, L.: Recherches sur une nouvelle espèce de carrés magiques. Verh. Zeeuw. Gen. Weten. Vlissengen **9**, 85–239 (1782)

Fisher, R.A., Yates, F.: Statistical Tables for Biological, Agricultural and Medical Research, 6th edn. Hafner (Macmillan), New York (1963)

Graphical Representation

Graphical representations encompass a wide variety of techniques that are used to clarify, interpret and analyze data by plotting points and drawing line segments, surfaces and other geometric forms or symbols.

The purpose of a graph is a rapid visualization of a data set. For instance, it should clearly illustrate the general behavior of the phenomenon investigated and highlight any important factors. It can be used, for example, as a means to translate or to complete a **frequency table**.

Therefore, graphical representation is a form of data representation.

HISTORY

The concept of plotting a point in coordinate space dates back to at least the ancient Greeks, but we had to wait until the work of Descartes, René for mathematicians to investigate this concept.

According to Royston, E. (1970), a German mathematician named Crome, A.W. was among the first to use graphical representation in **statistics**. He initially used it as a teaching tool.

In his works *Geographisch-statistische Darstellung der Staatskräfte* (1820) and *Ueber die Grösse und Bevölkerung der sämtlichen Europäischen Staaten* (1785), Crome employed different types of graphical representation, among them the **pie chart**.

Royston, E. (1970) also cites Playfair, W., whose work *The Commercial and Political Atlas* (1786) also referred to various graphical representations, especially the **line chart**. Playfair was very interested in the international trade balance, and illustrated his studies using different graphics such as the **histogram** and the pie chart.

The term histogram was used for the first time by **Pearson, Karl**, and Guerry (1833) appears to have been among the first to use the the line chart; in his *Essai sur la Statistique Morale de la France*, published in Paris in 1833, he described the frequencies of crimes by their characteristics; this study constituted one of the first uses of a frequency distribution.

Schmid, Calvin F. (1954) illustrates and describes the different types of graphical representations in his *Handbook of Graphic Presentation*.

MATHEMATICAL ASPECTS

The type of graphic employed depends upon the kind of **data** to be presented, the nature of the **variable** studied, and the goal of the study:

- A **quantitative graphic** is particularly useful for representing qualitative categorical variables. Such graphics include the **pie chart**, the **line chart** and the **pictogram**.
- A frequency graphic allows us to represent the (discrete or continuous) **frequency distribution** of a **quantitative variable**. Such graphics include the **his-togram** and the **stem and leaf** diagram.
- A cartesian graphic, employs a system of axes that are perpendicular to each other and intersect at a point known as the "origin." Such a coordinate system is termed "cartesian".

FURTHER READING
▶ **Frequency table**
▶ **Quantitative graph**

REFERENCES

Crome, A.F.W.: Ueber die Grösse und Bevölkerung der sämtlichen Europäischen Staaten. Weygand, Leipzig (1785)

Crome, A.F.W.: Geographisch-statistische Darstellung der Staatskräfte. Weygand, Leipzig (1820)

Fienberg, S.E.: Graphical method in statistics. Am. Stat. **33**, 165–178 (1979)

Guerry, A.M.: Essai sur la statistique morale de la France. Crochard, Paris (1833)

Playfair, W.: The Commercial and Political Atlas. Playfair, London (1786)

Royston, E.: A note on the history of the graphical presentation of data. In: Pearson, E.S., Kendall, M. (eds.) Studies in the History of Statistics and Probability, vol. I. Griffin, London (1970)

Schmid, C.F.: Handbook of Graphic Presentation. Ronald Press, New York (1954)

G

Hájek, Jaroslav

Hájek, Jaroslav was born in 1926 in Podebrady, Bohemia. A statistical engineer by profession, he obtained his doctorate in 1954. From 1954 to 1964, he worked as a researcher at the Institute of Mathematics of the Czechoslovakian Academy of Sciences. He then joined Charles University in Prague, where he was a professor from 1966 until his premature death in 1974. The principal works of Hájek, J. concern sampling probability theory and the rank test theory. In particular, he developed an asymptotic theory of the statistics of linear ranks. He was the first to apply the concept of invariance to the theory of rank testing.

Some principal works and articles of Hájek, Jaroslav:

1955 Some rank distributions and their applications. Cas. Pest. Mat. 80, 17–31 (in Czech); translation in (1960) Select. Transl. Math. Stat. Probab., 2, 41–61.

1958 Some contributions to the theory of probability sampling. Bull. Inst. Int. Stat., 36, 127–134.

1961 Some extensions of the Wald–Wolfowitz–Noether theorem. Ann. Math. Stat. 32, 506–523.

1964 Asymptotic theory of rejective sampling with varying probabilities from a finite population. Ann. Math. Stat. 35, 1419–1523.

1965 Extension of the Kolmogorov–Smirnov test to regression alternatives. In: Neyman, J. and LeCam, L. (eds) Bernoulli–Bayes–Laplace: Proceedings of an International Seminar, 1963. Springer, Berlin Heidelberg New York , pp. 45–60.

1981 Dupac, V. (ed) Sampling from a Finite Population. Marcel Dekker, New York.

Harmonic Mean

The harmonic **mean** of n observations is defined as n divided by the sum of the inverses of all of the observations.

HISTORY
See **geometric mean**.
The relationship between the harmonic mean, the geometric mean and the **arithmetic mean** is described by Mitrinovic, D.S. (1970).

MATHEMATICAL ASPECTS
Let $x_1, x_2, \ldots, x_n$ be n nonzero quantities, or n observations related to a **quantitative**

variable X. The harmonic mean H of these n quantities is calculated as follows:

$$H = \frac{n}{\sum_{i=1}^{n} \frac{1}{x_i}}.$$

If $\{x_i\}_{i=1,\dots,n}$ represents a finite series of positive numbers, we state that:

$$\min_i x_i \leq H \leq G \leq \bar{x} \leq \max_i x_i.$$

DOMAINS AND LIMITATIONS

The harmonic mean is rarely used in **statistics**. However, it can sometimes be useful, such as in the following cases:

- If a set of investments are invested at different interest rates, and they all give the same income, the unique rate at which all of the capital tied up in those investments must be invested to produce the same revenue as given by the set of investments is equal to the harmonic mean of the individual rates.

- Say we have a group of different materials, and each material can be purchased at a given price per amount of material (where the price per amount can be different for each material). We then buy a certain amount of each material, spending the same amount of money on each. In this case, the mean price per amount across all materials is given by the harmonic mean of the prices per amount for all of the materials.

- One property of the harmonic mean is that it is largely insensitive to outliers that have much larger values than the other data. For example, consider the following values: 1, 2, 3, 4, 5 and 100. Here the harmonic mean equals 2.62 and the **arithmetic mean** equals 19.17. However, the harmonic mean is much more sensitive

to outliers when they have much smaller values than the rest of the data. So, for the observations 1, 6, 6, 6, 6, 6, we get $H = 3.27$ whereas the arithmetic mean equals 5.17.

EXAMPLES

Three investments that each yield the same income have the following interest rates: 5%, 10% and 15%.

The harmonic mean gives the interest rate at which all of the capital would need to be invested in order to produce the same total income as the three separate investments:

$$H = \frac{3}{\left[\frac{1}{5} + \frac{1}{10} + \frac{1}{15}\right]} = \frac{3}{\frac{11}{30}} = 8.18\%.$$

We note that this result is different from the **arithmetic mean** of 10% $(5 + 10 + 15)/3$. A representative buys three lots of coffee, each of a different grade (quality), at 3, 2 and 1.5 euros per kg respectively. He buys 200 euros of each grade.

The mean price per kg of coffee is then obtained by dividing the total cost by the total quantity bought:

$$\text{mean price} = \frac{\text{total cost}}{\text{total quantity}}$$
$$= \frac{3 \cdot 200}{66.66 + 100 + 133.33} = 2.$$

This corresponds to the harmonic mean of the prices of the three different grades of coffee:

$$\text{mean price} = \frac{3}{\left[\frac{1}{3} + \frac{1}{2} + \frac{1}{1.5}\right]} = \frac{6}{3} = 2.$$

FURTHER READING

▶ **Arithmetic mean**
▶ **Geometric mean**

▶ **Mean**

▶ **Measure of central tendency**

REFERENCES
Mitrinovic, D.S. (with Vasic, P.M.): Analytic Inequalities. Springer, Berlin Heidelberg New York (1970)

Hat Matrix

The hat matrix is a **matrix** used in **regression analysis** and **analysis of variance**. It is defined as the matrix that converts values from the observed **variable** into estimations obtained with the **least squares** method. Therefore, when performing linear regression in the matrix form, if $\hat{\mathbf{Y}}$ is the **vector** formed from estimations calculated from the least squares parameters, and $\mathbf{Y}$ is a vector of observations related to the **dependent variable**, then $\hat{\mathbf{Y}}$ is given by vector $\mathbf{Y}$ multiplied by $\mathbf{H}$, that is, $\hat{\mathbf{Y}} = \mathbf{HY}$ converts to Y's into $\hat{\mathbf{Y}}$'s.

HISTORY
The hat matrix $\mathbf{H}$ was introduced by **Tukey, John Wilder** in 1972. An article by Hoaglin, D.C. and Welsch, R.E. (1978) gives the properties of the matrix $\mathbf{H}$ and also many examples of its application.

MATHEMATICAL ASPECTS
Consider the following linear regression **model**:
$$\mathbf{Y} = \mathbf{X} \cdot \boldsymbol{\beta} + \boldsymbol{\varepsilon}.$$
where

$\mathbf{Y}$ is an $(n \times 1)$ **vector** of observations on the **dependent variable**;

$\mathbf{X}$ is the $(n \times p)$ **matrix** of independent variables (there are p independent variables);

$\boldsymbol{\varepsilon}$ is the $(n \times 1)$ vector of **errors**, and;

$\boldsymbol{\beta}$ is the $(p \times 1)$ vector of parameters to be estimated.

The estimation $\hat{\boldsymbol{\beta}}$ of the vector $\boldsymbol{\beta}$ is given by
$$\hat{\boldsymbol{\beta}} = \left(\mathbf{X}' \cdot \mathbf{X}\right)^{-1} \cdot \mathbf{X}' \cdot \mathbf{Y}.$$

and we can calculate the estimated values $\hat{\mathbf{Y}}$ of $\mathbf{Y}$ if we know $\hat{\boldsymbol{\beta}}$:
$$\hat{\mathbf{Y}} = \mathbf{X} \cdot \hat{\boldsymbol{\beta}} = \mathbf{X} \cdot \left(\mathbf{X}' \cdot \mathbf{X}\right)^{-1} \cdot \mathbf{X}' \cdot \mathbf{Y}.$$

The matrix $\mathbf{H}$ is then defined by:
$$\mathbf{H} = \mathbf{X} \cdot \left(\mathbf{X}' \cdot \mathbf{X}\right)^{-1} \cdot \mathbf{X}'.$$

In particular, the diagonal element h_{ii} will be defined by:
$$h_{ii} = x_i \cdot \left(\mathbf{X}' \cdot \mathbf{X}\right)^{-1} \cdot x_i'.$$

where x_i is the ith line of $\mathbf{X}$.

DOMAINS AND LIMITATIONS
The matrix $\mathbf{H}$, which allows us to obtain n estimations of the **dependent variable** from n observations, is an idempotent symmetric square matrix of order n. The element (i, j) of this **matrix** measures the influence of the jth observation on the ith predicted **value**. In particular, the diagonal elements evaluate the effects of the observations on the corresponding estimations of the dependent variables. The value of each diagonal element of the matrix $\mathbf{H}$ ranges between 0 and 1.

Writing $\mathbf{H} = (h_{ij})$ for $i, j = 1, \ldots, n$, we have the relation:
$$h_{ii} = h_{ii}^2 + \sum_{\substack{i=1 \\ i \neq j}}^{n} \sum_{j=1}^{n} h_{ij}^2.$$

which is obtained based on the idempotent nature of $\mathbf{H}$; in other words $\mathbf{H} = \mathbf{H}^2$.

H

Therefore $\text{tr}(\mathbf{H}) = \sum_{i=1}^{n} h_{ii} = p =$ number of parameters to estimate.

The matrix $\mathbf{H}$ is used to determine leverage points in regression analysis.

EXAMPLES

Consider the following table where Y is a **dependent variable** related to the **independent variable** X:

X	Y
50	6
52	8
55	9
75	7
57	8
58	10

The **model** of **simple linear regression** is written in the following manner in the matrix form:

$$\mathbf{Y} = \mathbf{X} \cdot \boldsymbol{\beta} + \boldsymbol{\varepsilon} \,.$$

where

$$\mathbf{Y} = \begin{bmatrix} 6 \\ 8 \\ 9 \\ 7 \\ 8 \\ 10 \end{bmatrix}, \quad \mathbf{X} = \begin{bmatrix} 1 & 50 \\ 1 & 52 \\ 1 & 55 \\ 1 & 75 \\ 1 & 57 \\ 1 & 58 \end{bmatrix} \,.$$

$\boldsymbol{\varepsilon}$ is the (6×1) **vector** of errors, and $\boldsymbol{\beta}$ is the (2×1) **vector** of parameters.

We find the **matrix H** using the result:

$$\mathbf{H} = \mathbf{X} \cdot \left(\mathbf{X}' \cdot \mathbf{X} \right)^{-1} \cdot \mathbf{X}' \,.$$

By stepwise matrix calculations, we obtain:

$$(\mathbf{X}' \cdot \mathbf{X})^{-1} = \frac{1}{2393} \cdot \begin{bmatrix} 20467 & -347 \\ -347 & 6 \end{bmatrix}$$

$$= \begin{bmatrix} 8.5528 & -0.1450 \\ -0.1450 & 0.0025 \end{bmatrix} \,.$$

and finally:

$$\mathbf{H} = \mathbf{X} \cdot (\mathbf{X}' \cdot \mathbf{X})^{-1} \cdot \mathbf{X}'$$

$$= \begin{bmatrix} 0.32 & 0.28 & 0.22 & -0.17 & 0.18 & 0.16 \\ 0.28 & 0.25 & 0.21 & -0.08 & 0.18 & 0.16 \\ 0.22 & 0.21 & 0.19 & 0.04 & 0.17 & 0.17 \\ -0.17 & -0.08 & 0.04 & 0.90 & 0.13 & 0.17 \\ 0.18 & 0.18 & 0.17 & 0.13 & 0.17 & 0.17 \\ 0.16 & 0.16 & 0.17 & 0.17 & 0.17 & 0.17 \end{bmatrix} \,.$$

We remark, for example, that the weight of y_1 used during the **estimation** of $\hat{y}_1$ is 0.32. We then verify that the **trace** of $\mathbf{H}$ equals 2; in other words, it equals the number of parameters of the **model**.

$$\text{tr}(\mathbf{H}) = 0.32 + 0.25 + 0.19 + 0.90$$
$$+ 0.17 + 0.17$$
$$= 2 \,.$$

FURTHER READING

▶ **Leverage point**
▶ **Matrix**
▶ **Multiple linear regression**
▶ **Regression analysis**
▶ **Simple linear regression**

REFERENCES

Belsley, D.A., Kuh, E., Welsch, R.E.: Regression diagnostics. Wiley, New York pp. 16–19 (1980)

Hoaglin, D.C., Welsch, R.E.: The hat matrix in regression and ANOVA. Am. Stat. **32**, 17–22 (and correction at **32**, 146) (1978)

Tukey, J.W.: Some graphical and semigraphical displays. In: Bancroft, T.A. (ed.) Statistical Papers in Honor of George W. Snedecor. Iowa State University Press, Ames, IA , pp. 293–316 (1972)

Histogram

The histogram is a **graphical representation** of the distribution of **data** that has been grouped into classes. It consists of a series of rectangles, and is a type of frequency chart. Each data value is sorted and placed in an appropriate class **interval**. The number of data values within each class interval dictates the **frequency** (or relative frequency) of that class interval.

Each rectangle in the histogram represents a class of data. The width of the rectangle corresponds to the width of the class interval, and the surface of the rectangle represents the weight of the class.

HISTORY

The term histogram was used for the first time by **Pearson, Karl** in 1895.

Also see **graphical representation**.

MATHEMATICAL ASPECTS

The first step in the construction of a histogram consists of presenting the **data** in the form of a **frequency table**.

This requires that the class **intervals** are established and the data values are sorted and placed in the classes, which makes it possible to calculate the **frequencies** of the classes. The class intervals and frequencies are then added to the **frequency table**.

We then make use of the frequency table in order to construct the histogram. We divide the horizontal axis of the histogram into intervals, where the widths of these intervals correspond to those of the class intervals. We then draw the rectangles on the histogram. The width of each rectangle is the same as the width of the class that it corresponds to. The height of the rectangle is such that the surface area of the rectangle is equal to the relative frequency of the corresponding class. The sum of the surface areas of the rectangles must be equal to 1.

DOMAINS AND LIMITATIONS

Histograms are used to present a **data** set in a visual form that is easy to understand. They allow certain general characteristics (such as typical **values**, the range or the shape of the data) to be visualized and extracted.

Reviewing the shape of a histogram can allow us to detect the **probability model** followed by the data (**normal distribution**, **log-normal distribution**, ...).

It is also possible to detect unexpected behavior or abnormal **values** with a histogram.

This type of **graphical representation** is most frequently used in economics, but since it is an extremely simple way of visualizing a **data** set, it is used in many other fields too. We can also illustrate relative **frequency** in a histogram. In this case, the height of each rectangle equals the relative frequency of the corresponding class divided by the length of the class **interval**. In this case, if we sum the surface areas of all of the rectangles in the histogram, we obtain unity.

EXAMPLES

The following table gives raw **data** on the average annual precipitation in 69 cities of the USA. We will use these data to establish a **frequency table** and then a corresponding histogram.

Annual average precipitations in 69 cities of the USA (in inches)

Mob.	67.0	Chic.	34.4	St.L.	35.9
Sacr.	17.2	Loui.	43.1	At.C.	45.5
Wash.	38.9	Detr.	31.0	Char.	42.7
Boise	11.5	K.C.	37.0	Col.	37.0
Wich.	30.6	Conc.	36.2	Prov.	42.8
Bost.	42.5	N.Y.	40.2	Dall.	35.9
Jack.	49.2	Clev.	35.0	Norf.	44.7
Reno	7.2	Pitt.	36.2	Chey.	14.6

Buff.	36.1	Nash.	46.0	L.R.	48.5
Cinc.	39.0	Burl.	32.5	Hart.	43.4
Phil.	39.9	Milw.	29.1	Atl.	48.3
Memp.	49.1	Phoen.	7.0	Indi.	38.7
S.L.	15.2	Denv.	13.0	Port.	40.8
Char.	40.8	Miami	59.8	Dul.	30.2
Jun.	54.7	Peor.	35.1	G.Fls	15.0
S.F.	20.7	N.Or.	56.8	Albq	7.8
Jack.	54.5	S.S.M.	31.7	Ral.	42.5
Spok.	17.4	L.A.	14.0	Omaha	30.2
Ok.C.	31.4	Wilm.	40.2	Alb.	33.4
Col.	46.4	Hon.	22.9	Bism.	16.2
El Paso	7.8	D.Mn.	30.8	Port.	37.6
S-Tac	38.8	Balt.	41.8	S.Fls	24.7
S.J.	59.2	Mn-SP	25.9	Hstn.	48.2

Source: U.S. Census Bureau (1975) Statistical Abstract of the United States. U.S. Census Bureau, Washington, DC.

These **data** can be represented by the following **frequency table**:

Class	Frequency	Relative Frequency
0–10	4	0.058
10–20	9	0.130
20–30	5	0.072
30–40	24	0.348
40–50	21	0.304
50–60	4	0.058
60–70	2	0.030
Total	69	1.000

We can now construct the histogram:

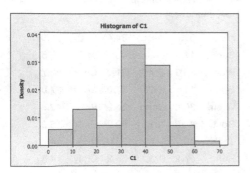

The horizontal axis is divided up into classes, and in this case the relative **frequencies** are given by the heights of the rectangles because the classes all have the same width.

FURTHER READING
▶ **Frequency distribution**
▶ **Frequency polygon**
▶ **Frequency table**
▶ **Graphical representation**
▶ **Ogive**
▶ **Stem-and-leaf diagram**

REFERENCES
Dodge, Y.: Some difficulties involving non-parametric estimation of a density function. J. Offic. Stat. **2**(2), 193–202 (1986)

Freedman, D., Pisani, R., Purves, R.: Statistics. Norton, New York (1978)

Pearson, K.: Contributions to the mathematical theory of evolution. II: Skew variation in homogeneous material. In: Karl Pearson's Early Statistical Papers. Cambridge University Press, Cambridge, pp. 41–112 (1948). First published in 1895 in Philos. Trans. Roy. Soc. Lond. Ser. A **186**, 343–414

Homogeneity Test

One issue that often needs to be considered when analyzing **categorical data** obtained for many groups is that of the homogeneity of the groups. In other words, we need to find out whether there are significant differences between these groups in relation to one or many qualitative categorical variables. A homogeneity test can show whether the differences are significant or not.

MATHEMATICAL ASPECTS

We consider the chi-square homogeneity test here, which is a specific type of **chi-square test**.

Let I be the number of groups considered and J the number of categories considered.

Let:

$n_{i.} = \sum_{j=1}^{J} n_{ij}$ correspond to the size of group i;

$n_{.j} = \sum_{i=1}^{I} n_{ij}$;

$n = \sum_{i=1}^{I} \sum_{j=1}^{J} n_{ij}$ be the total number of observations

n_{ij} be the empirical **frequency** (that is, the number of occurrences observed) corresponding to group i and category j;

m_{ij} is the theoretical frequency corresponding to group i and category j, which, assuming homogeneity among the groups, equals:

$$m_{ij} = \frac{n_{i.} \cdot n_{.j}}{n}.$$

If we represent the data in the form of a **contingency table** with I lines and J columns, we can calculate the $n_{i.}$ that contribute to the sum of the elements of line i and the $n_{.j}$ that contribute to the sum of all of the elements of column j.

We calculate

$$\chi_c^2 = \sum_{i=1}^{I} \sum_{j=1}^{J} \frac{\left(n_{ij} - m_{ij}\right)^2}{m_{ij}}.$$

and the chi-square homogeneity test is expressed in the following way: we reject the homogeneity hypothesis (at a significance level of 5%) if the value χ_c^2 is greater then the value of the χ^2 (chi-square) distribution with $(J - 1) \cdot (I - 1)$ degrees of freedom.

Note: We have assumed here that the same number of units are tested for each combination of group and category. However, we may want to test different numbers of units for different combinations. In this case, if we have a proportion p_{ij} of units for group i and category j, it is enough to replace m_{ij} by $n_{i.} \cdot p_{ij}$.

EXAMPLES

We consider a study performed in the pharmaceutical domain that concerns 100 people suffering from a particular illness. In order to examine the effect of a medical treatment, 100 people were chosen at random. Half of them were placed in a control group and received a placebo. The other patients received the medical treatment. Then the number of healthy people in each group was monitored for 24 hours following administration the treatment. The results are provided in the following table.

Observed frequency	Healthy for 24 hours	Not healthy	Total
Placebo	2	48	50
Treatment	9	41	50

The theoretical frequencies are obtained by assuming that the general state of health would have been the same for both groups if no treatment had been applied. In this case we obtain $m_{11} = m_{21} = \frac{50 \cdot 11}{100} = 5.5$ and $m_{12} = m_{22} = \frac{50 \cdot 89}{100} = 44.5$.

We calculate the value of χ^2 by comparing the theoretical frequencies with the observed frequencies:

$$\chi_c^2 = \frac{(2 - 5.5)^2}{5.5} + \frac{(48 - 44.5)^2}{44.5}$$
$$+ \frac{(9 - 5.5)^2}{5.5} + \frac{(41 - 44.5)^2}{44.5}$$
$$= 5.005.$$

If we then refer to the χ^2 distribution table for one degree of freedom, we obtain the value $\chi^2_{0.05} = 3.84$ for a significance level of 5%, which is smaller then the value we calculated, $\chi^2_c = 5.005$. We conclude that the groups were not homogeneous and the treatment is efficient.

FURTHER READING
▶ **Analysis of categorical data**
▶ **Analysis of variance**
▶ **Categorical data**
▶ **Chi-square distribution**
▶ **Chi-square test**
▶ **Frequency**

REFERENCE
See **analysis of categorical data**.

Hotelling, Harold

Hotelling, Harold (1895–1973) is considered to be one of the pioneers in the field of economical mathematics over the period 1920–1930. He introduced the T^2 multivariate test, principal components analysis and canonical correlation analysis.

He studied at University of Washington, where he obtained a B.A. in journalism in 1919. He then moved to Princeton University, obtaining a doctorate in mathematics from there in 1924. The began teaching at Stanford University that same year. His applications of mathematics to the social sciences initially concerned journalism and political science, and then he moved his focus to population and predictian.

In 1931, he moved to Colombia University, where he actively participated in the creation of its statistical department. During the Second World War, he performed statistical research for the military.

In 1946, he was hired by North Carolina University at Chapel Hill to create a statistics department there.

Some principal works and articles of Hotelling, Harold:

1933   Analysis of a complex of statistical variables with principal components. J. Educ. Psychol., 24, 417–441 and 498–520.

1936   Relation between two sets of variates. Biometrika 28, 321–377.

Huber, Peter J.

Huber, Peter J. was born at Wohlen (Switzerland) in 1934. He performed brilliantly during his studies and his doctorate in mathematics at the Federal Polytechnic School of Zurich, where he received the Silver Medal for the scientific quality of his thesis. He worked as Professor of Mathematical Statistics at the Federal Polytechnic School of Zurich. He then moved to the USA and worked at the most prestigious universities (Princeton, Yale, Berkeley) as an invited professor. In 1977 he was named Professor of Applied Mathematics at the Massachusetts Institute of Technology. He is member of the prestigious American Academy of Arts and Sciences, the Bernoulli Society and the National Science Foundation in the USA, in which foreign members are extremely rare. Since the publication of his article "Robust estimation of a location parameter" in 1964, he has been considered to be the founder of robust statistics.

Huber, Peter J. received the title of Docteur Honoris Causa from Neuchâtel University in 1994.

Some principal works and publications of Huber, Peter J.:

1964 Robust estimation of a location parameter. Ann. Math. Stat. 35, 73–101.

1968 Robust statistical procedures. SIAMCBMS-NSF Reg. Conf. Ser. Appl. Math.,

1981 Robust Statistics. Wiley, New York .

1995 Robustness: Where are we now? Student, Vol.1, 75–86.

Hypergeometric Distribution

The hypergeometric distribution describes the **probability** of success if a series of objects are drawn from a population (which contains some objects that represent failure while the others represent success), without replacement.

It is therefore used to describe a **random experiment** where there are only two possible results: "success" and "failure."

Consider a set of N **events** in which there are M "successes" and $N - M$ "failures." The **random variable** X, corresponding to the number of successes obtained if we draw n events without replacement follows a hypergeometric distribution with **parameters** N, M and n, denoted by $H(N, M, n)$.

The hypergeometric distribution is a **discrete probability distribution**.

The number of ways that n events can be drawn from N events is equal to:

$$C_N^n = \binom{N}{n} = \frac{n!}{N! \cdot (n - N)!} .$$

The number of elementary **events** depends on X and is:

$$C_M^x \cdot C_{N-M}^{n-x} .$$

which gives the following **probability function**:

$$P(X = x) = \frac{C_M^x \cdot C_{N-M}^{n-x}}{C_N^n} ,$$

$$\text{for } x = 0, 1, \ldots, n$$

(where $C_v^u = 0$ if $u < v$ by convention) .

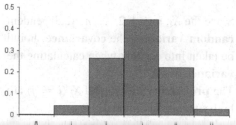

Hypergeometric distribution, $N = 12$, $M = 7$, $n - 5$

MATHEMATICAL ASPECTS

Consider the **random variable** $X = X_1 + X_2 + \ldots + X_n$, where:

$$X_i = \begin{cases} 1 & \text{if the } i\text{th drawing is a success} \\ 0 & \text{if the } i\text{th drawing is a failure} \end{cases} .$$

In this case, the **probability distribution** for X_i is:

X_i	0	1
$P(X_i)$	$\dfrac{N-M}{N}$	$\dfrac{M}{N}$

The **expected value** of X_i is therefore given by:

$$E[X_i] = \sum_{j=1}^{2} x_j P(x_i = x_j)$$

$$= 0 \cdot \frac{N-M}{N} + 1 \cdot \frac{M}{N}$$

$$= \frac{M}{N} .$$

Utilizing the fact that $X = X_1 + X_2 + \ldots + X_n$, we have:

$$E[X] = \sum_{i=1}^{n} E[X_i] = \sum_{i=1}^{n} \frac{M}{N} = n\frac{M}{N}.$$

The **variance** of X_i is, by definition:

$$\begin{aligned} \text{Var}(X_i) &= E[X_i^2] - (E[X_i])^2 \\ &= \frac{M}{N} - \left(\frac{M}{N}\right)^2 \\ &= \frac{M(N-M)}{N^2}. \end{aligned}$$

Since the X_i, $i = 1, 2, \ldots, n$ are dependent **random variables**, the **covariance** should be taken into account when calculating the **variance** of X.

The **probability** that X_i and X_j $(i \neq j)$ are both successes is equal to:

$$P(X_i = 1, X_j = 1) = \frac{M(M-1)}{N(N-1)}.$$

If we put $V = X_i \cdot X_j$, the **values** of V and the associated **probabilities** are:

V	0	1
P(V)	$1 - \dfrac{M(M-1)}{N(N-1)}$	$\dfrac{M(M-1)}{N(N-1)}$

The **expected value** of V is therefore:

$$\begin{aligned} E[V] &= 0 \cdot \left(1 - \frac{M(M-1)}{N(N-1)}\right) \\ &\quad + 1 \cdot \frac{M(M-1)}{N(N-1)}. \end{aligned}$$

The **covariance** of X_i and X_j is, by definition, equal to:

$$\begin{aligned} \text{Cov}(X_i, X_j) &= E[X_i \cdot X_j] - E[X_i] \cdot E[X_j] \\ &= \frac{M(M-1)}{N(N-1)} - \left(\frac{M}{N}\right)^2 \\ &= -\frac{M(N-M)}{N^2(N-1)} \\ &= -\frac{1}{N-1}\text{Var}(X_i). \end{aligned}$$

We can now calculate the **variance** of X:

$$\begin{aligned} \text{Var}(X) &= \sum_{i=1}^{n} \text{Var}(X_i) \\ &\quad + 2\sum_{j=1}^{n}\sum_{i<j} \text{Cov}(X_i, X_j) \\ &= \sum_{i=1}^{n} \text{Var}(X_i) \\ &\quad + n(n-1)\text{Cov}(X_i, X_j) \\ &= n\left[\text{Var}(X_i) - \frac{n-1}{N-1}\text{Var}(X_i)\right] \\ &= n\text{Var}(X_i)\frac{N-n}{N-1} \\ &= n\frac{M(N-M)}{N^2}\frac{N-n}{N-1} \\ &= \frac{N-n}{N-1}n\frac{M}{N}\left(1 - \frac{M}{N}\right). \end{aligned}$$

DOMAINS AND LIMITATIONS

The hypergeometric distribution is often used in quality control.

Suppose that a production line produces N products, which are then submitted to verification. A **sample** of size n is taken from this batch of products, and the number of defective products in this sample is noted. It is possible to to use this to obtain (by **inference**) information on the probable total number of defective products in the whole batch.

EXAMPLES

A box contains 30 fuses, and 12 of these are defective. If we take five fuses at random, the **probability** that none of them is defective is equal to:

$$\begin{aligned} P(X = 0) &= \frac{C_M^x C_{N-M}^{n-x}}{C_N^n} = \frac{C_{12}^0 C_{18}^5}{C_{30}^5} \\ &= 0.0601. \end{aligned}$$

FURTHER READING
- ▶ Bernoulli distribution
- ▶ Binomial distribution
- ▶ Discrete probability distribution

Hypothesis

A statistical hypothesis is an assertion regarding the distribution(s) of one or several **random variables**. It may concern the **parameters** of a given distribution or the **probability distribution** of a **population** under study.

The validity of the hypothesis is examined by performing **hypothesis testing** on **observations** collected for a **sample** of the studied **population**.

When performing **hypothesis testing** on the **probability distribution** of the **population** being studied, the hypothesis that the studied population follows a given probability distribution is called the **null hypothesis**. The hypothesis that affirms that the population does not follow a given probability distribution is called the **alternative hypothesis** (or opposite hypothesis).

If we perform **hypothesis testing** on the **parameters** of a **distribution**, the hypothesis that the studied parameter is equal to a given **value** is called the **null hypothesis**. The hypothesis that states that the value of the parameter is different to this given value is called the **alternative hypothesis**.

The **null hypothesis** is usually denoted by H_0 and the **alternative hypothesis** by H_1.

HISTORY

In **hypothesis testing**, the hypothesis that is to be tested is called the **null hypothesis**. We owe the term "null" to **Fisher, R.A.** (1935). Introducing this concept, he mentioned the well-known tea tasting problem, where a lady claimed to be able to recognize by taste whether the milk or the tea was poured into her cup first. The hypothesis to be tested was that the taste was absolutely not influenced by the order in which the tea was made.

Originally, the null hypothesis was usually taken to mean that a particular **treatment** has no effect, or that there was no difference between the effects of different treatments.

Nowadays the null hypothesis is mostly used to indicate the hypothesis has that to be tested, in contrast to the **alternative hypothesis**. Also see **hypothesis testing**.

EXAMPLES

Many problems involve repeating an **experiment** that has two possible results.

One example of this is the gender of a newborn child. In this case we are interested in the proportion of boys and girls in a given **population**. Consider p, the proportion of girls, which we would like to estimate from an observed **sample**. To determine whether the proportions of newborn boys and girls are the same, we make the statistical hypothesis that $p = \frac{1}{2}$.

FURTHER READING
- ▶ Alternative hypothesis
- ▶ Analysis of variance
- ▶ Hypothesis testing
- ▶ Null hypothesis

REFERENCES
Fisher, R.A.: The Design of Experiments. Oliver & Boyd, Edinburgh (1935)

Hypothesis Testing

Hypothesis testing is a procedure that allow, us to (depending on certain decision rules) confirm a starting **hypothesis**, called the **null hypothesis**, or to reject this null hypothesis in favor of the **alternative hypothesis**.

HISTORY

The theory behind hypothesis testing developed under study. The first steps were taken when works began to appear that discussed the significance (or insignificance) of a group of observations. Some examples of such works date from the eighteenth century, including those by Arbuthnott, J. (1710), Bernoulli, Daniel (1734) and **Laplace, Pierre Simon de** (1773). These works were seen more frequently in the nineteenth century, such as those by Gavarett (1840) and **Edgeworth, Francis Y.** (1885). The development of hypothesis testing occurred in parallel with the theory of **estimation**. Hypothesis testing seems to have been first elaborated by workers in the experimental sciences and the management domain. For example, the **Student test** was developed by **Gosset, William Sealy** during his time working for Guinness.

Neyman, Jerzy and **Pearson, Egon Sharpe** developed the mathematical theory of hypothesis testing, which they presented in an article published in 1928 in the review Biometrika. They were the first to recognize that the rational choice to be made during hypothesis testing had to be between the **null hypothesis** that we want to test and an **alternative hypothesis**. A second fundamental article on the theory of hypothesis testing was published in 1933 by the same mathematicians, where they also distinguished between a **type I error** and a **type II error**.

The works resulting from the collaboration between Neyman, J. and Pearson, E.S. are described in Pearson (1966) and in the biography of Neyman, published by Reid (1982).

MATHEMATICAL ASPECTS

Hypothesis testing of a **sample** generally involves the following steps:

1. Formulate the hypotheses:
   - The **null hypothesis** H_0,
   - The **alternative hypothesis** H_1.
2. Determine the **significance level** α of the test.
3. Determine the **probability distribution** that corresponds to the **sampling distribution**.
4. Calculate the **critical value** of the null hypothesis and deduce the **rejection region** or the **acceptance region**.
5. Establish the decision rules:
   - If the **statistics** observed in the sample are located in the acceptance region, we do not reject the null hypothesis H_0;
   - If the statistics observed on the sample are located in the rejection region, we reject the null hypothesis H_0 for the alternative hypothesis H_1.
6. Take the decision to accept or to reject the null hypothesis on the basis of the observed sample.

DOMAINS AND LIMITATIONS

The most frequent types of hypothesis testing are described below.

1. *Hypothesis testing of a sample*: We want to test whether the **value** of a **parameter** θ of the **population** is identical to a presumed value. The hypotheses will be as follows:

$$H_0: \quad \theta = \theta_0,$$
$$H_1: \quad \theta \neq \theta_0.$$

where θ_0 is the presumed value of the unknown parameter θ.

2. *Hypothesis testing on two samples*: The goal in this case is to find out whether two populations that are both described by a particular parameter are different. Let θ_1 and θ_2 be parameters that describe populations 1 and 2 respectively. We can then formulate the following hypotheses:

$$H_0: \quad \theta_1 = \theta_2 \,,$$
$$H_1: \quad \theta_1 \neq \theta_2 \,.$$

or

$$H_0: \quad \theta_1 - \theta_2 = 0 \,,$$
$$H_1: \quad \theta_1 - \theta_2 \neq 0 \,.$$

3. *Hypothesis testing of more than two samples*: As for a test performed on two samples, hypothesis testing is performed on more than two samples to determine whether these populations are different, based on comparing the same parameter from all of the populations being tested. In this case, we test the following hypotheses:

$$H_0: \quad \theta_1 = \theta_2 = \ldots = \theta_k \,,$$
$H_1: \quad$ The values of θ_i $(i = 1, 2, \ldots, k)$ are not all identical.

Here $\theta_1, \ldots, \theta_k$ are the unknown parameters of the populations and k is the number of populations to be compared.

In hypothesis testing theory, and in practice, we can distinguish between two types of tests: a parametric test and a nonparametric test.

Parametric Tests

A parametric test is a hypothesis test that presupposes a particular form for each of the distributions related to the underlying populations. This case applies, for example, when these populations follow a **normal distribution**.

The **Student test** is an example of a parametric test. This test compares the means of two normally distributed populations.

Nonparametric Test

A nonparametric test is a hypothesis test where it is not necessary to specify the parametric form of the distribution of the underlying population.

There are many examples of this type of test, including the **sign test**, the **Wilcoxon test**, the **signed Wilcoxon test**, the **Mann–Whitney test**, the **Kruskal–Wallis test**, and the **Kolmogorov–Smirnov test**.

H

EXAMPLES

For examples of parametric hypothesis testing, see **binomial test**, **Fisher test** or **Student test**. For examples of nonparametric hypothesis testing, see **Kolmogorov–Smirnov test**, **Kruskal–Wallis test**, **Mann–Whitney test**, **Wilcoxon test**, **signed Wilcoxon test** and **sign test**.

FURTHER READING

▶ **Acceptance region**
▶ **Alternative hypothesis**
▶ **Nonparametric test**
▶ **Null hypothesis**
▶ **One-sided test**
▶ **Parametric test**
▶ **Rejection region**
▶ **Sampling distribution**
▶ **Significance level**
▶ **Two-sided test**

REFERENCES

Arbuthnott, J.: An argument for Divine Providence, taken from the constant regularity observed in the births of both sexes. Philos. Trans. **27**, 186–190 (1710)

Bernoulli, D.: Quelle est la cause physique de l'inclinaison des planètes (...). Rec. Pièces Remport. Prix Acad. Roy. Sci. **3**, 95–122 (1734)

Edgeworth, F.Y.: Methods of Statistics. Jubilee Volume of the Royal Statistical Society, London (1885)

Gavarret, J.: Principes généraux de statistique médicale. Beché Jeune & Labé, Paris (1840)

Laplace, P.S. de: Mémoire sur l'inclinaison moyenne des orbites des comètes. Mém. Acad. Roy. Sci. Paris **7**, 503–524 (1773)

Lehmann, E.L.: Testing Statistical Hypotheses, 2nd edn. Wiley, New York (1986)

Neyman, J., Pearson, E.S.: On the use and interpretation of certain test criteria for purposes of statistical inference, Parts I and II. Biometrika **20**A, 175–240, 263–294 (1928)

Neyman, J., Pearson, E.S.: On the problem of the most efficient tests of statistical hypotheses. Philos. Trans. Roy. Soc. Lond. Ser. A **231**, 289–337 (1933)

Pearson, E.S.: The Neyman-Pearson story: 1926–34. In: David, F.N. (ed.) Research Papers in Statistics: Festschrift for J. Neyman. Wiley, New York (1966)

Reid, C.: Neyman—From Life. Springer, Berlin Heidelberg New York (1982)

Idempotent Matrix

See **matrix**.

Identity Matrix

See **matrix**.

Incidence

See **incidence rate**.

Incidence Rate

The **incidence** of an illness is defined as being the number of new cases of illness appearing during a determined period among the individuals of a **population**. This notion is similar to that of "stream."

The incidence rate is defined as being relative to the dimension of the population and the time; this value is expressed by relation to a number of individuals and to a duration. The incidence rate is the incidence I divided by the number of people at risk for the illness.

HISTORY

See **prevalence rate**.

MATHEMATICAL ASPECTS

Formally we have:

$$\text{incidence rate} = \frac{\text{risk}}{\text{duration of observation}}.$$

DOMAINS AND LIMITATIONS

An accurate incidence rate calculated in a **population** and during a given period will be interpreted as the index of existence of an epidemic in this population.

EXAMPLES

Let us calculate the **incidence** rate of breast cancer in a population of 150000 women observed during 2 years, according to the following table:

	Number of cases	Number of subjects	Incidence rate / 100000 / year
Group	[A]	[B]	$[\frac{A}{2B}]$
Non-exposed	40	35100	57.0
Passive smokers	140	55500	126.1
Active smokers	164	59400	138.0
Total	344	150000	114.7

The numerator of the incidence rate is the same as that of the **risk**: the number of new cases (or incident cases) of breast cancer during 2 years of observation among the preference group. The denominator of the incidence rate is obtained by multiplying the number of individuals in the preference group by the duration of observation (here 2 years). Normally, the rate is expressed as 1 year.

FURTHER READING
- ▶ **Attributable risk**
- ▶ **Avoidable risk**
- ▶ **Cause and effect in epidemiology**
- ▶ **Odds and odds ratio**
- ▶ **Prevalence rate**
- ▶ **Relative risk**
- ▶ **Risk**

REFERENCES
Lilienfeld, A.M., Lilienfeld, D.E.: Foundations of Epidemiology, 2nd edn. Clarendon, Oxford (1980)

MacMahon, B., Pugh, T.F.: Epidemiology: Principles and Methods. Little Brown, Boston, MA (1970)

Morabia, A.: Epidemiologie Causale. Editions Médecine et Hygiène, Geneva (1996)

Morabia, A.: L'Épidémiologie Clinique. Editions "Que sais-je?". Presses Universitaires de France, Paris (1996)

Incomplete Data

See **missing data**.

Independence

The notion of independence can have two meanings: the first concerns the **events** of a **random experiment**. Two events are said to be independent if the realization of the second event does not depend on the realization of the first.

The independence between two **events** A and B can be determined by **probabilities**: if the probability of event A when B is realized is identical to the probability of A when B is not realized, then A and B are called independent. The second meaning of the word independence concerns the **relation** that can exist between **random variables**. The independence of two random variables can be tested with the **Kendall test** and the **Spearman test**.

HISTORY
The independence notion was implicitly used long before a formal set of **probability** axioms was established. According to Maistrov, L.E. (1974), Cardano was already using the multiplication rule of probabilities $(P(A \cap B) = P(A) \cdot P(B))$. Maistrov, L.E. also mentions that the notions of independence and **dependence** between **events** were very familiar to Pascal, Blaise, de Fermat, Pierre, and Huygens, Christiaan.

MATHEMATICAL ASPECTS
Mathematically, we can say that **event** A is independent of event B if

$$P(A) = P(A \mid B) ,$$

where $P(A)$ is the **probability** of realizing event A and $P(A|B)$ is the **conditional probability** of event A as a function of the realization of B, meaning the probability that A happens knowing that B has happened.

Using the definition of **conditional probability**, the previous equation can be written as:

$$P(A) = \frac{P(A \cap B)}{P(B)}$$

or

$$P(A) \cdot P(B) = P(A \cap B) ,$$

which gives

$$P(B) = \frac{P(A \cap B)}{P(A)} = P(B \mid A) .$$

With this simple transformation, we have demonstrated that if A is independent of B, then B is also independent of A. We will say that **events** A and B are independent.

DOMAINS AND LIMITATIONS

If two **events** A and B are independent, then they are always compatible. The **probability** that both events A and B happen simultaneously can be determined by multiplying the individual probabilities of the two events:

$$P(A \cap B) = P(A) \cdot P(B) .$$

But if A and B are dependent, they can be compatible or incompatible. There is no absolute rule.

EXAMPLES

Consider a box containing four notes numbered from 1 to 4 from which we will make two successive drawings without putting the first note back in the box before drawing the second note.

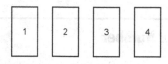

The **sample space** can be represented by the following 12 pairs:

$$\Omega = \{(1, 2), (1, 3), (1, 4), (2, 1), (2, 3),$$
$$(2, 4), (3, 1), (3, 2), (3, 4),$$
$$(4, 1), (4, 2), (4, 3)\} .$$

Consider the following two **events**:

$A =$ "pull out note No. 3 on the first drawing"
$= \{(3, 1), (3, 2), (3, 4)\}$
$B =$ "pull out note No. 3 on the second drawing"
$= \{(1, 3), (2, 3), (4, 3)\}$

Are these two **events** dependent or independent? If they are independent, the **probability** of **event** B must be identical to the probability of event A, regardless of the result of event A.

If A happens, then the following notes will remain in the box for the second drawing:

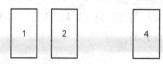

The **probability** that B happens is then null because note No. 3 is no longer in the box. On the other hand, if A is not fulfilled (for example, note No. 2 was drawn in the first drawing), the second drawing will have the following notes in the box:

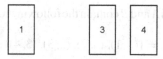

The **probability** that B will be fulfilled is $\frac{1}{3}$. We can therefore conclude that the probability of **event** B is dependent on event A or that events A and B are dependent.

We can verify our conclusion in the following way: if A and B are independent, the following equation must be verified:

$$P(A \cap B) = P(A) \cdot P(B) .$$

In our example, $P(A) = \frac{3}{12}$ and $P(B) = \frac{3}{12}$.

The intersection between A and B is null. Therefore we have

$$P(A \cap B) = 0 \neq P(A) \cdot \tfrac{1}{16} = P(B) \ .$$

If the equation is not confirmed, then **events** A and B are not independent. Therefore they are dependent.

Consider now a similar experiment, but by putting the drawn note back in the box before the second drawing.

The **sample space** of this new **random experiment** is composed of the following 16 pairs:

$$\Omega = \{(1, 1), (1, 2), (1, 3), (1, 4), (2, 1),$$
$$(2, 2), (2, 3), (2, 4), (3, 1), (3, 2),$$
$$(3, 3), (3, 4), (4, 1), (4, 2), (4, 3),$$
$$(4, 4)\} \ .$$

In these conditions, the second drawing will be done from a box that is identical to that of the first drawing:

Events A and B contain the following results:

$$A = \{(3, 1), (3, 2), (3, 3), (3, 4)\},$$
$$B = \{(1, 3), (2, 3), (3, 3), (4, 3)\}.$$

The **probability** of A, as well as of B, is therefore $\tfrac{4}{16}$. The intersection between A and B is given by the **event** $A \cap B = \{(3, 3)\}$; its probability is $\tfrac{1}{16}$. Therefore we have

$$P(A \cap B) = \tfrac{1}{16} = P(A) \cdot P(B) \ .$$

Since $P(A \cap B) = P(A) \cdot P(B)$, we can conclude that **event** B is independent of event A, or that the events A and B are independent.

FURTHER READING
► **Chi-square test of independence**
► **Compatibility**
► **Conditional probability**
► **Dependence**
► **Event**
► **Kendall rank correlation coefficient**
► **Probability**
► **Spearman rank correlation coefficient**
► **Test of independence**

REFERENCES
Maistrov, L.E.: Probability Theory—A History Sketch (transl. and edit. by Kotz, S.). Academic, New York (1974)

Independent Variable

In a regression model the variables which are considered to influence the dependent variables or to explain the variations of the latter, are all independent variables.

FURTHER READING
► **Dependent variable**
► **Multiple linear regression**
► **Regression analysis**
► **Simple linear regression**

Index Number

In its most general definition, an index number is a value representing the relative variation of a **variable** between two determined periods (or situations).

The **simple index numbers** should be distinguished from the **composite index numbers**.

- **Simple index numbers** describe the relative change of a single **variable**.
- **Composite index numbers** allow one to describe with a single number the comparison of the set of **values** that several **variables** take in a certain situation with respect to the set of values of the same variables in a reference situation.

The reference situation defines the basis of the index number. We can say, for example, that for reference year (basis year) 1980, a certain index number has a value of 120 in 1982.

HISTORY

The first studies using index numbers date back to the early 18th century.

In 1707, the Anglican Bishop Fleetwood undertook the study of the evolution of prices between the years 1440 and 1700. This study, whose results are presented in his work *Chronicon Preciosum* (1745), was done for a very specific purpose: during the founding of a college in 1440–1460, an essential clause was established stipulating that any admitted member had to leave the college if his fortune exceeded £5 per year. Fleetwood wanted to know if, taking into account the price evolution, such a promise could still be kept three centuries later.

He considered four products—wheat, meat, drink, and clothing—and studied the evolution of their prices. At the end of his work, he came to the conclusion that 5 pounds (£) in 1440–1460 had the same value as 30 pounds (£) in 1700.

Later, in 1738, Dutot, C. studied the diminution of the value of money in a work entitled *Réflexions politiques sur les finances et le commerce*. For this he studied the incomes of two kings, Louis XII and Louis XV. The two kings had the following incomes:

| Louis XII: | £ 7650000 in 1515 |
| Louis XV: | £100000000 in 1735 |

To determine which of the sovereigns had the highest disposable income, Dutot, C. noted the different prices, at the considered dates, of a certain number of goods: a chicken, a rabbit, a pigeon, the value of a day's work, etc.

Dutot's measurement was the following:

$$I = \frac{\sum P_1}{\sum P_0},$$

where $\sum P_1$ and $\sum P_0$ are the sums of the noted prices at the respective dates. Notice that this index number is not ponderated. Dutot came to the conclusion that the wealth of Louis XII was lower than that of Louis XV. The diminution of the value of money was also the object of a study by Carli, Gian Rinaldo. In 1764 the astronomy professor of Padoue studied the evolution of prices since the discovery of the Americas.

He took three products as references—grains, wine, and cooking oil—studied the prices on these products in the period 1500–1750.

Carli's measurement was the following:

$$I = \frac{1}{n} \cdot \sum \frac{P_1}{P_0}.$$

The obtained index number is an **arithmetic mean** of the relative prices for each observed product.

In 1798 a work entitled *An Account of Some Endeavours to Ascertain a Standard of Weight and Measure*, signed by Sir George Shuckburgh Evelyn, was published. This precursor to index numbers defines a price index number covering the years 1050 to 1800 by introducing the notions of reference year (1550) and relative prices. He also included the price of certain services in his **observations**.

It is Lowe, Joseph who should be seen, according to Kendall, M.G. (1977), as the true father of index numbers. His work, published in 1822, called *The present state of England*, treated many problems relative to the creation of index numbers. Among other things, he introduced the notion of price ponderation at different dates (P_1 and P_0) by the quantities observed at the first date (Q_0). The Lowe measurement is the following:

$$I = \frac{\sum (P_1 \cdot Q_0)}{\sum (P_0 \cdot Q_0)} .$$

Later on this index number would be known as the **Laspeyres index**.

In the second half of the 19th century, statisticians made many advances in this field.

In 1863, with '*A Serious Fall in the Value of Gold Ascertained and Its Social Effects Set Forth*, Jevons, W.S. studied the type of **mean** to be used during the creation of an index number. He recommended the use of the **geometrical mean**.

Between 1864 and 1880, three German scientists, Laspeyres, E. (1864, 1871), Drobisch, M.W. (1871), and Paasche, H. (1874), worked on the evolution of prices of material goods in Hamburg, without taking services into account. All three are nonponderated calculation methods.

The **Paasche index** is determined by the following formula:

$$I = \frac{\sum (P_1 \cdot Q_1)}{\sum (P_0 \cdot Q_1)} .$$

Palgrave, R.H.I (1886) proposed to ponderate the relative prices by the total **value** of the concerned good; this method leads to calculating the index number as follows:

$$I = \frac{\sum \left(P_1 \cdot Q_1 \cdot \frac{P_1}{P_0} \right)}{\sum (P_1 \cdot Q_1)} .$$

It was the works of statisticians such as I. Fisher who defined in 1922 an index number that carries his name and that constitutes the **geometric mean** of the **Laspeyres index** and the **Paasche index**:

$$I = \sqrt{\frac{\sum (P_1 \cdot Q_0) \sum (P_1 \cdot Q_1)}{\sum (P_0 \cdot Q_0) \sum (P_0 \cdot Q_1)}} ,$$

whereas in the same period, Marshall, A. and Edgeworth, F.W. proposed the following formula:

$$I = \frac{\sum (P_1 (Q_1 + Q_0))}{\sum (P_0 (Q_1 + Q_0))} .$$

Notice that with this formula it is the **mean** of the weights that is calculated.

Modern statistical studies still largely call upon the index numbers that have been defined here.

DOMAINS AND LIMITATIONS

The calculation of index numbers is a usual method of describing the changes in economic **variables**. Even if the most current index numbers essentially concern the price variable and the quantity variable, they can also be calculated for values, qualities, etc. Generally associated with the field of business and economy, index numbers are also widely used in other spheres of activity.

FURTHER READING
▶ **Composite index number**
▶ **Fisher index**
▶ **Laspeyres index**
▶ **Paasche index**
▶ **Simple index number**

REFERENCES

Carli, G.R.: Del valore etc. In: Opere Scelte di Carli, vol. I, p. 299. Custodi, Milan (1764)

Drobisch, M.W.: Über Mittelgrössen und die Anwendbarkeit derselben auf die Berechnung des Steigens und Sinkens des Geldwerts. Berichte über die Verhandlungen der König. Sachs. Ges. Wiss. Leipzig. Math-Phy. Klasse, 23, 25 (1871)

Dutot, C.: Réflexions politiques sur les finances, et le commerce. Vaillant and Nicolas Prevost, The Hague (1738)

Evelyn, Sir G.S.: An account of some endeavours to ascertain a standard of weight and measure. Philos. Trans. 113 (1798)

Fleetwood, W.: Chronicon Preciosum, 2nd edn, Fleetwood, London (1707)

Jevons, W.S.: A Serious Fall in the Value of Gold Ascertained and Its Social Effects Set Forth. Stanford, London (1863)

Kendall, M.G.: The early history of index numbers. In: Kendall, M., Plackett, R.L. (eds.) Studies in the History of Statistics and Probability, vol. II. Griffin, London (1977)

Laspeyres, E.: Hamburger Warenpreise 1850–1863 und die kalifornisch-australischen Geldentdeckung seit 1848. Jahrb. Natl. Stat. **3**, 81–118, 209–236 (1864)

Laspeyres, E.: Die Berechnung einer mittleren Waarenpreissteigerung. Jahrb. Natl. Stat. **16**, 296–314 (1871)

Lowe, J.: The Present State of England in Regard to Agriculture, Trade, and Finance. Kelley, London (1822)

Paasche, H.: Über die Preisentwicklung der letzten Jahre nach den Hamburger Borsen-notirungen. Jahrb. Natl. Stat. **23**, 168–178 (1874)

Palgrave, R.H.I.: Currency and standard of value in England, France and India etc. (1886) In: IUP (1969) British Parliamentary Papers: Session 1886: Third Report and Final Report of the Royal Commission on the Depression in Trade and Industry, with Minutes of Evidence and Appendices, Appendix B. Irish University Press, Shannon

Indicator

An indicator is a statistic (**official statistic**) whose objective is to give an indication about the state, behavior, and changing nature during some period of an economic or political phenomenon. An indicator must inform about the variations in the values as well as changes in the nature of the observed phenomena and must serve to instruct the decision makers.

At its most accurate, an indicator reflects the social or economic process and suggests the changes that should be made. It is an alarm signal that attracts attention and and serves as a call to action (pilotage). Many indicators are generally necessary for providing a global reflection of given phenomena and of the politics that one wishes to evaluate. A set of indicators provides information about the same subject and is called a system of indicators.

Not all statistics are indicators; indicators must:

1. Be quantifiable, which, strictly speaking, does not mean measurable.
2. Be unidimensional (avoid overlap with other indicators).
3. Cover the set of priority objectives.
4. Directly evaluate the performance of a policy.

5. Show ways to improve.
6. Refer to major aspects of the system and be changeable with changes in the political system.
7. Be verifiable.
8. Be relatable to each other.
9. Be as economic as possible.
10. Be readily available.
11. Be easily understood by the general public.
12. Be recognizable by everybody as valid and reliable.

DOMAINS AND LIMITATIONS
Construction of Indicators
Indicators, like all measures of social phenomena, cannot be exact and exhaustive. The construction of indicators involves philosophical considerations (judgement of values) and technical and scientific knowledge, but it is also based on logic and often reflects political preoccupations.

Collection and Choice of Indicators
To collect good indicators, an appropriate model of economic, social, and educational systems is needed. The model as generally adapted by the OCDE is a systemic model: in a given environment, we study a system's resources, process, and effects. But the choice of indicators united in a system of indicators also depends on the logistical feasibility of data collection. This explains the large difference among indicators. The number of indicators necessary for a good pilotage of an economic or social system is difficult to evaluate. If too numerous, the indicators lose importance and risk not playing an important role. But together they must represent the system to be evaluated. A too small number risks making the system invalid and unrepresentative.

Interpretation of Indicators
The interpretation of indicators is as important for the system as their choice, construction, and collection. Having the significance that the indicators can take for the politically responsible, for the financing and the success of certain projects, all precautions must be taken to assure the interpretative integrality, the quality, and the neutrality of the indicators.

Validation of Indicators
Before using these statistics as indicators of the performance, quality, and health of an economic, social, or educational system, they should be validated. If indicators' values are calculated scientifically using mathematical methods and instruments, their influence on the system to be described and their reliability as the foundation for a plan of action are not always scientifically proven. Moreover, when a set of indicators is used, it is still far from clear how these indicators combine their effects to produce the observed results. The methods for validating indicators have yet to be developed.

EXAMPLES
Economic Indicators
These include gross domestic product (GDP), national revenue, consumer price index, producer price index, unemployment rate, total employment, long-term and short-term interest rates, exports and imports of produced goods, and balance of payments.

Education Indicators
These include the cost of education, the actives in the system of teaching, the scholarization expectancy, the geographical disparities in access to the undergraduate level, the

general level of draftees, the level of education of young people following high school graduation, the benefits of an undergraduate degree with respect to finding employment, compulsory schooling of children from 2 to 5 years of age, and educational expenses for continuing education.

Science and Technology Indicators

These include Gross domestic expenditures on research and development (GDERD), per-capita GDERD, GDERD as a percentage of GDP, total personnel R&D expenditures, total university diplomas, percentage of GDERD financing by businesses, percentage of GDERD financed by the government, percentage of GDERD financed from abroad, and national patent applications.

FURTHER READING
▶ **Official statistics**

REFERENCES
OCDE: OCDE et les indicateurs internationaux de l'enseignement. Centre de la Recherche et l'innovation dans l'enseignement, OCDE, Paris (1992)

OCDE: Principaux indicateurs de la science et de la technologie. OCDE, Paris (1988)

Inertia

Inertia is a notion that is used largely in mechanics. The more inertia a body has around an axis, the more energy has to be used to put this body in rotation around this axis. The mass of the body and the (squared) distance from the body to the axis are used in the calculation of the inertia. The definition of **variance** can be recognized here if "mass" is replaced by "**frequency**" and "distance" by "deviation."

MATHEMATICAL ASPECTS
Suppose that we have a cloud of n points centered in a space of dimension p; the **inertia matrix** V can be constructed with the **coordinates** and **frequencies** of the points.

The **eigenvalues** $k_1, \ldots, k_p$ (written in decreasing order) of **matrix** V are the inertia explained by the p **factorial axes** of the cloud of points. Their sum, equal to the **trace** of V, is equal to the total inertia of the cloud. We will say, for example, that the first **factorial axis** (or inertia axis) explains:

$$100 \cdot \frac{k_1}{\sum\limits_{i=1}^{p} k_i} \text{ (as \%) of inertia .}$$

The first two inertia axes, for example, explain:

$$100 \cdot \frac{k_1 + k_2}{\sum\limits_{i=1}^{p} k_i} \text{ (in \%) of inertia .}$$

DOMAINS AND LIMITATIONS
The notion of inertia or, more specifically, the **inertia matrix** of a cloud of points is used, among other things, to perform a **correspondence analysis** between considered points.

FURTHER READING
▶ **Correspondence analysis**
▶ **Distance**
▶ **Eigenvalue**
▶ **Factorial axis**
▶ **Frequency**
▶ **Inertia matrix**

Inertia Matrix

Consider a cloud of points in a space of p dimensions. We suppose that this cloud is centered in the origin of the reference system; if not, it is enough to cause a change in variables by subtracting from each point the center of the cloud. Let us designate by $\mathbf{A}$ the **matrix** of the coordinates multiplied by the square root of the **frequency** associated to each point. Such a matrix will be, for n different points, of dimension $(n \times p)$. Suppose that p is smaller than or equal to n; otherwise we use the **transpose** of matrix $\mathbf{A}$ as a new matrix of coordinates.

We obtain the inertia matrix $\mathbf{V}$ by multiplying $\mathbf{A}$ by the transpose of $\mathbf{A}$, $\mathbf{A}'$. $\mathbf{V}$ is a symmetric matrix of order p.

The **trace** of this matrix $\mathbf{V}$, is exactly the total **inertia** (or total variance).

MATHEMATICAL ASPECTS

Let n be different points X_i of p components

$$X_i = (x_{i1}, x_{i2}, \ldots, x_{ip}) , \quad i = 1, 2, \ldots, n ,$$

each point X_i has one **frequency** f_i ($i = 1, 2, \ldots, n$).

If the points are not centered around the origin, then we should make the changing variables correspond to the shift to the origin of the center of the cloud:

$$X_i' = X_i - Y ,$$

where $\mathbf{Y} = (y_1, y_2, \ldots, y_p)$ with

$$y_j = \frac{1}{\sum_{i=1}^{n} f_i} \cdot \left(\sum_{i=1}^{n} f_i \cdot x_{ij} \right) .$$

The general term of inertia matrix V is then:

$$v_{ij} = \sum_{k=1}^{n} f_k \cdot (x_{ki} - y_i) \cdot (x_{kj} - y_j) .$$

In matrix form, $\mathbf{V} = \mathbf{A}' \cdot \mathbf{A}$, where the coefficients of matrix $\mathbf{A}$ are:

$$a_{ki} = \sqrt{f_k} \cdot (x_{ki} - y_i) .$$

The total **inertia** of the cloud of points is obtained by calculating the **trace** of the inertia matrix $\mathbf{V}$, which equals:

$$I_{\text{tot}} = \text{tr}(\mathbf{V}) = \sum_{i=1}^{p} v_{ii}$$

$$= \sum_{i=1}^{p} \sum_{k=1}^{n} f_k \cdot (x_{ki} - y_i)^2 .$$

DOMAINS AND LIMITATIONS

The inertia matrix of a cloud of points is used to perform a **correspondence analysis** between two considered points.

EXAMPLES

Consider the following points in the space of three dimensions:

$$(2; \quad 0; \quad 0.5)$$
$$(3; \quad 1; \quad 2)$$
$$(-2; -1; \quad -1)$$
$$(-1; -1; -0.5)$$
$$(-1; -1; -0.5)$$
$$(-1; -1; -0.5)$$

There are four different frequencies equalling 1, 1, 1, and 3.

The center of the cloud formed by its points is given by:

$$y_1 = \frac{2 + 3 - 2 + 3 \cdot (-1)}{6} = 0 ,$$

$$y_2 = \frac{1 - 1 + 3 \cdot (-1)}{6} = -0.5 ,$$

$$y_3 = \frac{0.5 + 2 - 1 + 3 \cdot (-0.5)}{6} = 0 .$$

The new coordinates of our four different points will be given by:

$(2; 0.5; 0.5)$ $(3; 1.5; 2)$ $(-2; -0.5; -1)$

$(-1; -0.5; -0.5)$ of frequency 3,

from which A, the matrix of the coordinates multiplied by the square root of the frequency, is calculated:

$$A = \begin{bmatrix} 2 & 0.5 & 0.5 \\ 3 & 1.5 & 2 \\ -2 & -0.5 & -1 \\ -\sqrt{3} & -\dfrac{\sqrt{3}}{2} & -\dfrac{\sqrt{3}}{2} \end{bmatrix}$$

and V, the inertia matrix:

$$V = A' \cdot A = \begin{bmatrix} 20 & 8 & 10.5 \\ 8 & 3.5 & 4.5 \\ 10.5 & 4.5 & 6 \end{bmatrix}.$$

The **inertia** of the cloud of points equals:

$$I_{tot} = \text{tr}(V) = 20 + 3.5 + 6 = 29.5.$$

FURTHER READING
► **Correspondence analysis**
► **Frequency**
► **Matrix**
► **Trace**
► **Transpose**
► **Variance**

REFERENCES

Lagarde, J. de: Initiation à l'analyse des données. Dunod, Paris (1983)

Inference

Inference is a form of reasoning by induction done on the basis of information collected on a **sample** from the perspective of generalizing the information to the **population** associated to this sample.

HISTORY

See **inferential statistics**.

FURTHER READING
► **Inferential statistics**

Inferential Statistics

Inferential or deductive statistics is complementary to **descriptive statistics** because the goal of most research is not to establish a certain number of indicators on a given sample but rather to estimate a certain number of parameters characterizing the population associated to the treated sample.

In inferential **statistics**, hypotheses are formulated in the parameters of a **population**. These hypotheses are then tested on the basis of observations made on a representative **sample** of the population.

HISTORY

The origins of inferential **statistics** coincide with those of probability theory and correspond to the works of **Bayes, Thomas** (1763), **de Moivre, Abraham** (1718), **Gauss, Carl Friedrich** (1809), and **de Laplace, Pierre Simon** (1812).

At that time, there were numerous researchers in the field of inferential statistics. We mention, for example, the works of **Galton, Francis** (1889) in relation to the **correlation** as well as the development of hypothesis testing due principally to **Pearson, Karl** (1900) and **Gosset, William**

Sealey, who was known as "Student" (1908). **Neyman, Jerzy, Pearson, Egon Sharpe** (1928), and **Fisher, Ronald Aylmer** (1956) also contributed in a fundamental way to the development of inferential statistics.

FURTHER READING
► **Bayes' theorem**
► **Descriptive statistics**
► **Estimation**
► **Hypothesis testing**
► **Maximum likelihood**
► **Optimization**
► **Probability**
► **Sampling**
► **Statistics**
► **Survey**

REFERENCES

Bayes, T.: An essay towards solving a problem in the doctrine of chances. Philos. Trans. Roy. Soc. Lond. **53**, 370–418 (1763). Published, by the instigation of Price, R., 2 years after his death. Republished with a biography by Barnard, George A. in 1958 and in Pearson, E.S., Kendall, M.G.: Studies in the History of Statistics and Probability. Griffin, London, pp. 131–153 (1970)

Fisher, R.A.: Statistical Methods and Scientific Inference. Oliver & Boyd, Edinburgh (1956)

Galton, F.: Natural Inheritance. Macmillan, London (1889)

Gauss, C.F.: Theoria Motus Corporum Coelestium. Werke, 7 (1809)

Gosset, S.W. "Student": The Probable Error of a Mean. Biometrika **6**, 1–25 (1908)

Laplace, P.S. de: Théorie analytique des probabilités. Courcier, Paris (1812)

Moivre, A. de: The Doctrine of Chances: or, A Method of Calculating the Probability of Events in Play. Pearson, London (1718)

Neyman, J., Pearson, E.S.: On the use and interpretation of certain test criteria for purposes of statistical inference, Parts I and II. Biometrika **20A**, 175–240, 263–294 (1928)

Pearson, K.: On the criterion, that a given system of deviations from the probable in the case of a correlated system of variables is such that it can be reasonably supposed to have arisen from random sampling. In: Karl Pearson's Early Statistical Papers. Cambridge University Press, pp. 339–357. First published in 1900 in Philos. Mag. (5th Ser) **50**, 157–175 (1948)

Inner Product

The inner product of x and y, denoted by $x' \cdot y$, is the sum of products, component by component, of two vectors.

Geometrically, the inner product of two vectors of dimension n corresponds, in the space of n dimensions, to the number obtained by multiplying the **norm** of one by those of the projection of another. This definition is equivalent to saying that the scalar product of two vectors is obtained by making the product of their norms and cosines of the angle determined by two vectors.

MATHEMATICAL ASPECTS

Let x and y be two vectors of dimension n, given by their components:

$$
x = \begin{bmatrix} x_1 \\ x_2 \\ \vdots \\ x_n \end{bmatrix} \quad \text{and} \quad y = \begin{bmatrix} y_1 \\ y_2 \\ \vdots \\ y_n \end{bmatrix}.
$$

The scalar product of $\mathbf{x}$ and $\mathbf{y}$, denoted by $\mathbf{x}' \cdot \mathbf{y}$, is defined by:

$$\mathbf{x}' \cdot \mathbf{y} = [x_1 \quad x_2 \quad \dots x_n] \cdot \begin{bmatrix} y_1 \\ y_2 \\ \vdots \\ y_n \end{bmatrix},$$

$$\mathbf{x}' \cdot \mathbf{y} = x_1 \cdot y_1 + x_2 \cdot y_2 + \dots + x_n \cdot y_n,$$

$$\mathbf{x}' \cdot \mathbf{y} = \sum_{i=1}^{n} x_i \cdot y_i.$$

The scalar product of a vector $\mathbf{x}$ with itself corresponds to its square **norm** (or length), denoted by $\| \mathbf{x} \|^2$:

$$\| \mathbf{x} \|^2 = \mathbf{x}' \cdot \mathbf{x} = \sum_{i=1}^{n} x_i^2.$$

With the help of the norm, we can calculate the scalar product

$$\mathbf{x}' \cdot \mathbf{y} = \| \mathbf{x} \| \cdot \| \mathbf{y} \| \cdot \cos \alpha,$$

where α is the angle between vectors $\mathbf{x}$ and $\mathbf{y}$.

The scalar product of two vectors has the following properties:

- It is commutative, that is, for two vectors of the same dimension $\mathbf{x}$ and $\mathbf{y}$, $\mathbf{x}' \cdot \mathbf{y} = \mathbf{y}' \cdot \mathbf{x}$. In other words, there is no difference if we project $\mathbf{x}$ onto $\mathbf{y}$ or vice versa.
- We can take the negative values if, for example, the projections of two vectors on which we made the projection have opposite directions.

EXAMPLES

We consider two vectors in the space of three dimensions, given by their components:

$$\mathbf{x} = \begin{bmatrix} 1 \\ 2 \\ 3 \end{bmatrix} \quad \text{and} \quad \mathbf{y} = \begin{bmatrix} 0 \\ 2 \\ -1 \end{bmatrix}.$$

The scalar product of $\mathbf{x}$ and $\mathbf{y}$, denoted by $\mathbf{x}' \cdot \mathbf{y}$, is defined by:

$$\mathbf{x}' \cdot \mathbf{y} = \begin{bmatrix} 1 \\ 2 \\ 3 \end{bmatrix}' \cdot \begin{bmatrix} 0 \\ 2 \\ -1 \end{bmatrix}$$

$$= [1 \ 2 \ 3] \cdot \begin{bmatrix} 0 \\ 2 \\ -1 \end{bmatrix},$$

$$\mathbf{x}' \cdot \mathbf{y} = 1 \cdot 0 + 2 \cdot 2 + 3 \cdot (-1) = 4 - 3,$$

$$\mathbf{x}' \cdot \mathbf{y} = 1.$$

The **norm** of $\mathbf{x}$, denoted by $\| \mathbf{x} \|$, is given by the square root of the scalar product by itself. Thus:

$$\mathbf{x}' \cdot \mathbf{x} = \begin{bmatrix} 1 \\ 2 \\ 3 \end{bmatrix}' \cdot \begin{bmatrix} 1 \\ 2 \\ 3 \end{bmatrix}$$

$$= [1 \ 2 \ 3] \cdot \begin{bmatrix} 1 \\ 2 \\ 3 \end{bmatrix},$$

$$\mathbf{x}' \cdot \mathbf{x} = 1^2 + 2^2 + 3^2 = 1 + 4 + 9,$$

$$\mathbf{x}' \cdot \mathbf{x} = 14,$$

from which $\| \mathbf{x} \| = \sqrt{14} \approx 3.74$.

Vector $\mathbf{x}$ has a length of 3.74 in the space of three dimensions.

In the same way, we can calculate the norm of vector $\mathbf{y}$. It is given by:

$$\| \mathbf{y} \|^2 = \mathbf{y}' \cdot \mathbf{y} = \sum_{i=1}^{3} y_i^2 = 2^2 + (-1)^2 = 5$$

and equals $\| \mathbf{y} \| = \sqrt{5}$.

Thus we can determine the angle α situated between the two vectors with the help of the relation:

$$\cos \alpha = \frac{\mathbf{x}' \cdot \mathbf{y}}{\| \mathbf{x} \| \cdot \| \mathbf{y} \|} = \frac{1}{\sqrt{14} \cdot \sqrt{5}}$$

$$\approx 0.1195.$$

This gives an approximative angle of $83.1°$ between vectors $\mathbf{x}$ and $\mathbf{y}$.

FURTHER READING
▸ **Matrix**
▸ **Transpose**
▸ **Vector**

Interaction

The term interaction is used in **design of experiment** methods, more precisely in **factorial experiments**, where a certain number of factors can be studied simultaneously. It appears when the effect on the **dependent variable** of a change in the **level** of one factor depends on levels of other factors.

HISTORY
The interaction concept in **design of experiment** methodology is due to **Fisher, R.A.** (1925, 1935).

MATHEMATICAL ASPECTS
The **model** of a two-way classification design with interaction is as follows:

$$Y_{ijk} = \mu + \alpha_i + \beta_j + (\alpha\beta)_{ij} + \varepsilon_{ijk},$$
$$i = 1, 2, \ldots, a,$$
$$j = 1, 2, \ldots, b, \quad \text{and}$$
$$k = 1, 2, \ldots, n_{ij},$$

where Y_{ijk} is the kth **observation** receiving **treatment** ij, μ is the general **mean** common to all treatments, α_i is the effect of the ith **level** of **factor** A, β_j is the effect of the jth level of factor B, $(\alpha\beta)_{ij}$ is the effect of the interaction between ith level of factor A and jth level of factor B, and ϵ_{ijk} is the experimental **error** of observation Y_{ijk}.

DOMAINS AND LIMITATIONS
The interactions between two **factors** are called first-order interactions, those between three factors are called second-order interactions, and so forth.

EXAMPLES
See **two-way analysis of variance**.

FURTHER READING
▸ **Analysis of variance**
▸ **Design of experiments**
▸ **Experiment**
▸ **Fisher distribution**
▸ **Fisher table**
▸ **Fisher test**
▸ **Two-way analysis of variance**

REFERENCES
Cox, D.R.: Interaction. Int. Stat. Rev. **52**, 1–25 (1984)

Fisher, R.A.: Statistical Methods for Research Workers. Oliver & Boyd, Edinburgh (1925)

Fisher, R.A.: The Design of Experiments. Oliver & Boyd, Edinburgh (1935)

Interquartile Range

The interquartile range is a **measure of dispersion** corresponding to the difference between the first and the third **quartiles**. It therefore corresponds to the **interval** that contains 50% of the most centered **observations** of the distribution.

MATHEMATICAL ASPECTS
Consider Q_1 and Q_3 the first and third **quartiles** of a distribution. The interquartile range is calculated as follows:

$$\text{Interquartile range} = Q_3 - Q_1.$$

DOMAINS AND LIMITATIONS

The interquartile range is a measure of variability that does not depend on the number of **observations**. Moreover, and at the opposite end of the **range**, this measure is less sensitive to **outliers**.

EXAMPLES

Consider the following set of ten **observations**:

$$0\ 1\ 1\ 2\ 2\ 2\ 3\ 3\ 4\ 5$$

The first **quartile** Q_1 is equal to 1 and the third quartile Q_3 to 3. The interquartile range is therefore equal to:

Interquartile range $= Q_3 - Q_1 = 3 - 1 = 2$.

We can therefore conclude that 50% of the **observations** are located in an **interval** of length 2, meaning the interval between 1 and 3.

FURTHER READING

▶ **Measure of dispersion**
▶ **Quartile**
▶ **Range**

Interval

An interval is determined by two boundaries, a and b. These are called the limit **values** of the interval.

Intervals can be closed, open, semiopen, and semiclosed.

– *Closed interval:*
  denoted $[a, b]$, represents the set of x with:

$$a \leq x \leq b.$$

– *Open interval:*
  denoted $]a, b[$, represents the set of x with:

$$a < x < b.$$

– *Interval that is semiopen on left or semi-closed interval on right:*
  – denoted $]a, b]$, represents the set of x with:

$$a < x \leq b.$$

– *Interval that is semiopen on right or semiclosed on left:*
  denoted $[a, b[$, represents the set of x with:

$$a \leq x < b.$$

FURTHER READING

▶ **Frequency distribution**
▶ **Histogram**
▶ **Ogive**

Inverse Matrix

For a square **matrix** $\mathbf{A}$ of order n, we call the inverse matrix of $\mathbf{A}$ the square matrix of order n such that the result of the matrix product of $\mathbf{A}$ and its inverse (like those of the inverse-multiplying $\mathbf{A}$) equals the **identity matrix** of order n. For any square matrix, the inverse matrix, if it exists, is unique. To have an invertible square matrix $\mathbf{A}$, its determinant should be other than zero.

MATHEMATICAL ASPECTS

Let $\mathbf{A}$ be a square **matrix** of order n of nonzero determinant. We call the inverse of $\mathbf{A}$ the unique square matrix of order n, denoted $\mathbf{A}^{-1}$, such as:

$$\mathbf{A} \cdot \mathbf{A}^{-1} = \mathbf{A}^{-1} \cdot \mathbf{A} = \mathbf{I_n},$$

where $\mathbf{I_n}$ is the identity matrix of order n. So we say that the matrix is non-singular. When $\det(\mathbf{A}) = 0$, that is, the matrix $\mathbf{A}$ is not invertible, we say that it is singular.

The condition that the determinant of **A** cannot be zero is explained by the properties of the determinant:

$$1 = \det(I_n) = \det\left(\mathbf{A} \cdot \mathbf{A}^{-1}\right)$$

$$= \det(\mathbf{A}) \cdot \det\left(\mathbf{A}^{-1}\right),$$

that is, $\det\left(\mathbf{A}^{-1}\right) = \frac{1}{\det(\mathbf{A})}$.

On the other hand, the expression $\mathbf{A}^{-1}$, from **A**, is obtained with the help of $\frac{1}{\det(\mathbf{A})}$ (which must be defined) and equals:

$$\mathbf{A}^{-1} = \frac{1}{\det(\mathbf{A})} \cdot \left[(-1)^{i+j} \cdot D_{ij}\right],$$

where D_{ij} is the determinant of order $n-1$, which we obtain by deleting in A' (the **transpose** of **A**) line i and column j. D_{ij} is called the cofactor of the element a_{ij}.

DOMAINS AND LIMITATIONS

The inverse matrix is principally used in linear regression in matrix form to determine the parameters estimates.

Generally, we can say that the notion of the inverse matrix allows to resolve a system of linear equations.

EXAMPLES

Let us treat the general case of a square matrix of order 2: Let $\mathbf{A} = \begin{bmatrix} a & b \\ c & d \end{bmatrix}$. The **determinant** of **A** is given by:

$$\det(\mathbf{A}) = a \cdot d - b \cdot c.$$

If the latter is not zero, we define:

$$\mathbf{A}^{-1} = \frac{1}{a \cdot d - b \cdot c} \cdot \begin{bmatrix} d & -b \\ -c & a \end{bmatrix}.$$

So we can verify that $\mathbf{A} \cdot \mathbf{A}^{-1} = \mathbf{A}^{-1} \cdot \mathbf{A} = \mathbf{I_2}$ by direct calculation.

Let **B** be a square matrix of order 3:

$$\mathbf{B} = \begin{bmatrix} 3 & 2 & 1 \\ 4 & -1 & 0 \\ 2 & -2 & 0 \end{bmatrix},$$

and let us determine the inverse matric, $\mathbf{B}^{-1}$, of **B** if it exists.

Let us calculate first the determinant of **B** by developing according to the last column:

$$\det(B) = 1 \cdot [4 \cdot (-2) - (-1) \cdot 2],$$

$$\det(B) = -8 + 2 = -6.$$

The latter being non zero, we can calculate $\mathbf{B}^{-1}$ by determining the D_{ij}, that is, the determinants of order 2 obtained by deleting the ith line and the jth column of the **transpose** $\mathbf{B}'$ of **B**:

$$\mathbf{B}' = \begin{bmatrix} 3 & 4 & 2 \\ 2 & -1 & -2 \\ 1 & 0 & 0 \end{bmatrix}$$

$$D_{11} = -1 \cdot 0 - (-2) \cdot 0 = 0,$$

$$D_{12} = 2 \cdot 0 - (-2) \cdot 1 = 2,$$

$$D_{13} = 2 \cdot 0 - (-1) \cdot 1 = 1,$$

$$\vdots$$

$$D_{33} = 3 \cdot (-1) - 4 \cdot 2 = -11.$$

Thus we obtain the inverse matrix $\mathbf{B}^{-1}$ by adjusting the correct sign to the element D_{ij} [that is, $(-)$ if the sum of i and j is even and $(+)$ if it is odd]:

$$B^{-1} = \frac{1}{-6} \cdot \begin{bmatrix} 0 & -2 & 1 \\ 0 & -2 & 4 \\ -6 & 10 & -11 \end{bmatrix}$$

$$= \frac{1}{6} \cdot \begin{bmatrix} 0 & 2 & -1 \\ 0 & 2 & -4 \\ 6 & -10 & 11 \end{bmatrix}.$$

FURTHER READING
▶ **Determinant**
▶ **Generalized inverse**
▶ **Matrix**
▶ **Multiple linear regression**
▶ **Transpose**

Inversion

Consider a **permutation** of the integers $1, 2, \ldots, n$. When in a pair of elements (not necessarily adjacent) of the permutation the first element is bigger than the second, it is called inversion. This way it is possible to count in a permutation the number of inversions, which is the number of pairs in which the big integer precedes the small one.

MATHEMATICAL ASPECTS

Consider a **permutation** of the integers $1, 2, \ldots, n$ given by $\sigma_1, \sigma_2, \ldots, \sigma_n$; the number of inversions is determined by counting the number of pairs (σ_i, σ_j) with $\sigma_i > \sigma_j$.
To make sure we do not forget any, we start by taking σ_1 and count the σ_i, $i > 1$, smaller than σ_1, and then we take σ_2 and count the elements that are inferior to it located to its right, proceeding like this until σ_{n-1}; the number of inversions of the considered permutation is then obtained.

DOMAINS AND LIMITATIONS

The number of inversions is used to calculate the **determinant** of a square **matrix**. The notation a_{ij} is used for the element of the matrix that is located at the intersection of the ith line and the jth column. By carrying out the product $a_{1i} \cdot a_{1j} \cdot \ldots \cdot a_{1r}$, for example, the parity of the number of inversions of the permutation $i\,j \ldots r$ determines the sign

attributed to the product for the calculation of the determinant.

For a **permutation** of the elements $1, 2, \ldots, n$, the number of inversions is always an integer located between 0 and $\frac{n \cdot (n-1)}{2}$.

EXAMPLES

Consider the integers $1, 2, 3, 4, 5$; a **permutation** of these five integers is given, for example, by:

$$1\ 4\ 5\ 2\ 3\,.$$

The number of inversions of this permutation can be determined by writing down the pairs where there is an inversion:

$$(4, 2)$$
$$(4, 3)$$
$$(5, 2)$$
$$(5, 3)\,,$$

which gives a number of inversions equal to 4.

For the **permutation** $1\ 2\ 3\ 4\ 5$ (identity permutation) we find that there is no inversion. For the permutation $5\ 4\ 3\ 2\ 1$ we obtain 10 inversions, because each pair is an inversion and therefore the number of inversions is equal to the number of **combinations** of two of five elements chosen without repetition, meaning:

$$C_5^2 = \frac{5!}{2! \cdot 3!} = 10\,.$$

FURTHER READING
▶ **Determinant**
▶ **Permutation**

Irregular Variation

Irregular variations or random variations constitute one of four components of a **time** series. They correspond to the movements that appear irregularly and generally during short periods.

Irregular variations do not follow a particular model and are not predictable.

In practice, all the components of time series that cannot be attributed to the influence of cyclic fluctuations or seasonal variations or those of the **secular tendency** are classed as irregular.

HISTORY

See **time series**.

MATHEMATICAL ASPECTS

Let Y_t be a **time series**; we can describe it with the help of its components:

- **Secular tendency** T_t
- Seasonal variations S_t
- Cyclic fluctuations C_t
- Irregular variations I_t

We distinguish:

- Multiplicative **model**: $Y_t = T_t \cdot S_t \cdot C_t \cdot I_t$.
- Additive model: $Y_t = T_t + S_t + C_t + I_t$.

When the secular tendency, the seasonal variations, and the cyclic fluctuations are determined, it is possible to adjust the initial data of the time series according to these three components. So we obtain the values of the irregular variations at each time t by the following expressions:

$$\frac{Y_t}{S_t \cdot T_t \cdot C_t} = I_t \text{ (multiplicative model)}$$

$$Y_t - S_t - T_t - C_t = I_t \text{ (additive model)}.$$

DOMAINS AND LIMITATIONS

The study of the components of **time series** shows that the irregular variations can be classed into two categories:

- The most numerous can be attributed to a large number of small causes, especially errors of measure that provoke variations of small amplitude.
- Those resulting from isolated accidental events of greater scale such as strikes, administrative decisions, financial booms, natural catastrophes, etc. In this case, to apply this method of the moving average we make a first data treatment to correct as well as we can the raw data.

Even if we normally suppose that the irregular variations do not produce durable variations, but only in a short time **interval**, we can imagine that they are strong enough to take the walk of cyclic or other fluctuations. In practice, we find that the irregular variations have a tendency to have a small amplitude and to follow a **normal distribution**. That means that there are frequently small deviations and rarely large ones.

FURTHER READING

▶ **Cyclical fluctuation**
▶ **Normal distribution**
▶ **Seasonal variation**
▶ **Secular trend**
▶ **Time series**

REFERENCES

Bowerman, B.L., O'Connel, R.T.: Time Series and Forecasting: An Applied Approach. Duxbury, Belmont, CA (1979)

Harnett, D.L., Murphy, J.L.: Introductory Statistical Analysis. Addison-Wesley, Reading, MA (1975)

Jackknife Method

During point estimations by **confidence interval**, it is often difficult to estimate the **bias** or **standard error** of the used **estimator**. The reasons are generally summarized in two points: (1) the lack of information about the form of the theoretical data distribution and (2) the complexity of the theoretical **model** if the hypothesis is considered to be too strong. The jackknife method permits a reduction of bials in numerical estimation of the standard error as well as a confidence. interval. It is usually made by generating artificial subsamples from the data of the complete **sample**.

HISTORY

The jackknife method was developed by Quenouille (1949, 1956) and **John Wilder Tukey** (1958). It is now the most widely used method thanks to computers, which can generate a large amount of data in a very short time.

The creation of the jackknife denomination is attributed to Tukey, J.W. (1958). The name of this method issued from a scout knife with many blades ready to be used in a variety of situations.

MATHEMATICAL ASPECTS

Let $x_1, \ldots, x_n$, be a number of independent observations of a **random variable**. We want to estimate a **parameter** θ of the **population**. The jackknife technique is the following: the **estimator** $\hat{\theta}$ of θ is calculated from the initial **sample**. The latter is separated into J groups with $m = \frac{n}{J}$ observations in each of them (to simplify the problem, we suppose that n is a multiple of J).

The estimator of θ by the jackknife method is the **mean** of the pseudovalues:

$$\hat{\bar{\theta}} = \frac{1}{J} \cdot \sum_{j=1}^{J} \hat{\theta}_j,$$

where $\hat{\theta}_j$ is called the *pseudovalue* and defined by:

$$\hat{\theta}_j = J \cdot \hat{\theta} - (J-1) \cdot \hat{\theta}_{(j)}$$

and the estimators $\hat{\theta}_{(j)}$ $(j = 1, \ldots, J)$ are calculated in the same way as $\hat{\theta}$, but with the jth group deleted.

The **estimation** of the **variance** is defined by:

$$S^2 = \text{Var}(\hat{\bar{\theta}}) = \sum_{j=1}^{J} \frac{(\hat{\theta}_j - \hat{\bar{\theta}})^2}{(J-1) \cdot J}.$$

The estimator of the **bias** is:

$$B(\hat{\hat{\theta}}) = (J - 1) \cdot (\hat{\theta}_{(.)} - \hat{\theta}),$$

where $\hat{\theta}_{(.)}$ is defined as follows:

$$\hat{\theta}_{(.)} = \frac{1}{J} \cdot \sum_{j=1}^{J} \hat{\theta}_{(j)}.$$

The **confidence interval** at the **significance level** α is given by:

$$\left[\hat{\hat{\theta}} - \frac{S \cdot t_{n-1,1-\frac{\alpha}{2}}}{\sqrt{n}}; \hat{\hat{\theta}} + \frac{S \cdot t_{n-1,1-\frac{\alpha}{2}}}{\sqrt{n}} \right].$$

One of the important properties of the jackknife method is the fact that we delete from the bias the term of order $\frac{1}{n}$.

Suppose that $a_1, a_2, \ldots$ are functions of θ; we can pose as a **mathematical expectation** of $\hat{\theta}$:

$$E[\hat{\theta}] = \theta + \frac{a_1}{n} + \frac{a_2}{n^2} + \ldots$$
$$= \theta + \frac{a_1}{m \cdot J} + \frac{a_2}{(m \cdot J)^2} + \ldots$$

and

$$E[\hat{\hat{\theta}}_j] = J \cdot E[\hat{\theta}] - (J-1) \cdot E[\hat{\theta}_{(j)}]$$

$$= J \cdot \left[\theta + \frac{a_1}{m \cdot J} + \frac{a_2}{(m \cdot J)^2} + \ldots \right]$$

$$- (J-1) \cdot \left[\theta + \frac{a_1}{m \cdot (J-1)} + \ldots \right]$$

$$= J \cdot \theta + \frac{a_1}{m} + \frac{a_2}{m^2 \cdot J} + \ldots - J \cdot \theta$$

$$+ \theta - \frac{a_1}{m} - \frac{a_2}{m^2 \cdot (J-1)} - \ldots$$

$$= \theta - a_2 \cdot \frac{1}{m^2 \cdot J \cdot (J-1)} + \ldots.$$

DOMAINS AND LIMITATIONS

The idea behind the jackknife method, which is relatively old, became modern thanks to informatics. The Jackknife generates samples from the original data and determine

the **confidence interval** at a chosen **significance level** of an unknown **statistic**.

In practical cases, the groups are composed only in an **observation** ($m = 1$). The estimators $\hat{\theta}_{(j)}$ are calculated by deleting the jth observation.

EXAMPLES

A large chain of stores wants to determine the evolution of its turnover between 1950 and 1960.

To obtain a correct **estimation**, we take a **sample** of 8 points of sale where we know the respective turnovers.

The data are summarized in the following table:

Point of sale	Turnover (mill $)	
	1950	1960
1	6.03	7.59
2	12.42	15.65
3	11.89	14.95
4	3.12	5.03
5	8.17	11.08
6	13.34	18.48
7	5.30	7.98
8	4.23	5.61

The total turnover of the 8 points of sale increased from 64.5 in 1950 to 86.37 mill \$ in 1960.

Thus the estimation of the increasing factor equals:

$$\hat{\theta} = \frac{86.37}{64.5} = 1.34.$$

We want to calculate the **bias** of this estimation as well as its **standard deviation**.

We calculate the $\hat{\theta}_{(j)}$ for j going from 1 to 8, that is, the factors of increase calculated without the data of the jth point of sale. We obtain:

$$\hat{\theta}_{(1)} = \frac{86.37 - 7.59}{64.5 - 6.03} = 1.347,$$

$$\hat{\theta}_{(2)} = 1.358 \qquad \hat{\theta}_{(6)} = 1.327$$

$$\hat{\theta}_{(3)} = 1.358 \qquad \hat{\theta}_{(7)} = 1.324$$

$$\hat{\theta}_{(4)} = 1.325 \qquad \hat{\theta}_{(8)} = 1.340$$

$$\hat{\theta}_{(5)} = 1.337.$$

So we can calculate $\hat{\theta}_{(.)}$:

$$\hat{\theta}_{(.)} = \frac{1}{J} \cdot \sum_{j=1}^{J} \hat{\theta}_{(j)} = \frac{1}{8} \cdot (10.716) = 1.339.$$

We calculate the pseudovalues $\hat{\theta}_j$. Recall that $\hat{\theta}_j$ is calculated as follows:

$$\hat{\theta}_j = J \cdot \hat{\theta} - (J - 1) \cdot \hat{\theta}_{(j)}.$$

We find:

$$\hat{\theta}_1 = 1.281 \qquad \hat{\theta}_5 = 1.356$$

$$\hat{\theta}_2 = 1.207 \qquad \hat{\theta}_6 = 1.423$$

$$\hat{\theta}_3 = 1.210 \qquad \hat{\theta}_7 = 1.443$$

$$\hat{\theta}_4 = 1.436 \qquad \hat{\theta}_8 = 1.333.$$

We calculate the **estimator** of θ by the jackknife method:

$$\hat{\bar{\theta}} = \frac{1}{J} \cdot \sum_{j=1}^{J} \hat{\theta}_j = \frac{1}{8} \cdot (10.689) = 1.336.$$

So we determine the bias:

$$B(\hat{\bar{\theta}}) = (J - 1) \cdot (\hat{\theta}_{(.)} - \hat{\theta})$$

$$= (8 - 1) \cdot (1.339 - 1.34)$$

$$= -0.0035.$$

The **standard error** is given by:

$$S = \sqrt{\text{Var}(\hat{\bar{\theta}})}$$

$$= \sqrt{\sum_{j=1}^{J} \frac{(\hat{\theta}_j - \hat{\bar{\theta}})^2}{(J - 1) \cdot J}}$$

$$= \sqrt{\frac{0.0649}{7 \cdot 8}} = 0.034.$$

So we can calculate the **confidence interval** of the increase factor between 1950 and 1960 of the turnover of the chain. For a **significance level** α of 5%, the critical **value** of the **Student table** for 7° of freedom is:

$$t_{7,0.975} = 2.365.$$

The confidence interval for the estimation of the increase factor is:

$$\left[\hat{\bar{\theta}} - \frac{S \cdot t_{7,0.975}}{\sqrt{n}}; \hat{\bar{\theta}} + \frac{S \cdot t_{7,0.975}}{\sqrt{n}} \right]$$

$$\left[1.34 - \frac{0.034 \cdot 2.365}{\sqrt{8}}, \right.$$

$$\left. 1.34 + \frac{0.034 \cdot 2.365}{\sqrt{8}} \right],$$

that is:

$$[1.312, 1.368].$$

So we can affirm that the mean increase in the turnover is approximately contained between 31% and 37% during the period 1950–1960.

FURTHER READING
► **Bias**
► **Confidence interval**
► **Estimation**
► **Estimator**
► **Point estimation**
► **Standard error**

REFERENCES
Efron, B.: The Jackknife, the Bootstrap and the other Resampling Plans. Society for Industrial and Applied Mathematics, Philadelphia (1982)

Quenouille, M.H.: Approximate tests of correlation in time series. J. Roy. Stat. Soc. Ser. B **11**, 68–84 (1949)

Quenouille, M.H.: Notes on bias in estimation. Biometrika **43**, 353–360 (1956)

Tukey, J.W.: Bias and confidence in not quite large samples. Ann. Math. Stat. **29**, 614 (1958)

Jeffreys Harold

Jeffreys, Harold was born in Fatfield, England in 1891. He studied mathematics, physics, physical chemistry, and geology at Armstrong College in Newcastle, where he obtained his degree in 1910. He then joined St. John's College at Cambridge University, where he became a Fellow in 1914 and stayed there for 75 years. He died there in 1989. He published his first paper (about photography) in 1910. History has preserved two of his works in particular:

1939 Theory of Probability. Oxford University Press, Oxford .

1946 (with Jeffreys, B.S.) Methods of Mathematical Physics. Cambridge University Press, Cambridge ; Macmillan, New York .

Joint Density Function

Suppose X_1, X_2 are continuous random variables defined on the same **sample space** and that

$$F(x_1, x_2) = P(X_1 \leq x_1, X_2 \leq x_2)$$
$$= \int_{-\infty}^{x_k} \int_{-\infty}^{x_1} f(t_1, t_2) \, dt_1, dt_2$$

for all x_1, x_k. Then $f(x_1, x_2)$ is the joint density function of (X_1, X_2) provided $f(x_1, x_2) \geq 0$ and that

$$\int_{-\infty}^{\infty} \int_{-\infty}^{\infty} f(x_1 x_2) \, dx_1 dx_2 = 1 \, .$$

Suppose $(X_1, \ldots, X_k)$ are continuous random variables defined on the same **sample space** and

$$F(x_1, \ldots, x_k) = P(X_1 \leq x_1, \ldots, X_k \leq x_k)$$
$$= \int_{-\infty}^{x_1} \int_{-\infty}^{x_k} f(t_1, \ldots, t_k) \, dt_1 \ldots dt_k$$

for all $x_1, \ldots, x_k$. Then $f(x_1, \ldots, x_k)$ is the joint density function of $(x_1, \ldots, x_k)$ provided $f(x_1, \ldots, x_k) \geq 0$ and

$$\int_{-\infty}^{\infty} \int_{-\infty}^{\infty} f(x_1, \ldots, x_k) \, dx_1 \ldots dx_k = 1 \, .$$

REFERENCES
Knight, K.: Mathematical statistics. Chapman & Hall/CRC, London (2000)

Joint Distribution Function

The joint distribution function of a **pair of random variables** is defined, as the **probability** that the first **random variable** takes a **value** inferior or equal to the first real number and that simultaneously the second variable takes a value inferior or equal to the second real number.

HISTORY
See **probability**.

MATHEMATICAL ASPECTS
Consider two **random variables** X and Y. The joint distribution function is defined as follows:

$$F_{xy}(a, b) = P(X \leq a, Y \leq b),$$
$$-\infty < a, b < \infty.$$

FURTHER READING
▶ **Distribution function**
▶ **Marginal distribution function**
▶ **Pair of random variables**

REFERENCES
Hogg, R.V., Craig, A.T.: Introduction to Mathematical Statistics, 2nd edn. Macmillan, New York (1965)

Ross, S.M.: Introduction to Probability Models, 8th edn. John Wiley, New York (2006)

Joint Probability Distribution Function

Suppose that X_1, X_2 are discrete random variables defined on the same **sample space**. Then the joint probability distribution function of $X = (X_1, X_2)$ is defined to be

$$f(x_1, x_2) = P(X_1 = x_1, X_2 = x_2).$$

The joint probability distribution function must satisfy the following conditions:

$$\sum_{x_1, x_2} f(x_1, x_2) = 1.$$

A multiple joint probability distribution function of $X = (X_1, \ldots, X_k)$ is defined to be
$$f(x_1, \ldots, x_k) = P(X_1 = x_1, \ldots, X_k = x_k)$$
satisfying

$$\sum_{(x_1, \ldots, x_k)} f(x_1, \ldots, x_k) = 1.$$

REFERENCES
Casella, G., Berger, R.L.: Statistical Inference. Duxbury, Pacific Grove (2001)

Kendall Maurice George

Kendall Maurice was born in 1907 in Kettering, Northamptonshire, England. He studied Mathematics at St. John's College, Cambridge. After graduation as a Mathematics Wrangler in 1929, he joined the British Civil Service in the Ministry of Agriculture. He was elected a Fellow of the Society in 1934. In 1937, he worked with G. Udny Yule in the revision of his standard statistical textbook, *Introduction to the Theory of Statistics.* He also work on the rank correlation coefficient which bears his name, **Kendall's tau**, which eventually led to a monograph on *Rank Correlation* in 1948.

In 1938 and 1939 he began work, along with Bernard Babington Smith, on the problem of **random number generation**, developing both one of the first early mechanical devices to produce random digits, and formulated a series of tests such as *frequency test, serial test and a poker test,* for statistical randomness in a given set of digits.

During the war he managed to produce volume one of the Advanced Theory of Statistics in 1943 and a second volume in 1946.

In 1957, he published *Multivariate Analysis* and in the same year he also developed, with W.R. Buckland, a *Dictionary of Statistical Terms.*

In 1953, he published The Analytics of Economic Time Series, and in 1961 he left the University of London and took a position as the Managing Director of a consulting company, Scientific Control Systems, and in the same year began a two-year term as President of the Royal Statistical Society.

In 1972, he became Director of the World Fertility Survey, a project sponsored by the International Statistical Institute and the United Nations. He continued this work until 1980, when illness forced him to retire. He was knighted in 1974 for his services to the theory of statistics, and received the Peace Medal of the United Nations in 1980 in recognition for his work on the World Fertility Survey. He was also elected a fellow of the British Academy and received the highest honor of the Royal Statistical Society, the Guy Medal in Gold. At the time of his death in 1983, he was Honorary President of the International Statistical Institute.

Some principal works and articles of Kendall, Maurice George:

1938 (with Babington Smith, B.) Randomness and Random Sampling Numbers. J. Roy. Stat. Soc. **101**:1, 147–166.

1979 (with Stuart, A.) Advanced theory of Statistics. Arnold.

1957 (with Buckland, W.R.) A Dictionary of Statistical Terms. International Statistical Institute, The Hague, Netherland.

1973 Time Series, Griffin, London.

FURTHER READING

▶ **Kendall rank correlation coefficient**

▶ **Random number generation**

Kendall Rank Correlation Coefficient

The Kendall rank correlation coefficient (Kendall τ) is a nonparametric measure of correlation.

HISTORY

This rank correlation coefficient was discussed as far back as the early 20th century by Fechner, G.T. (1897), Lipps, G.F. (1906), and Deuchler, G. (1914).

Kendall, M.G. (1938) not only rediscovered it independently but also studied it using a (nonparametric) approach. His 1970 monograph contains a complete detailed presentation of the theory as well as a biography.

MATHEMATICAL ASPECTS

Consider two random **variables** (X, Y) observed on a **sample** of size n with n pairs of **observations** (X_1, Y_1), (X_2, Y_2), ..., (X_n, Y_n). An indication of the correlation between X and Y can be obtained by ordering the **values** X_i in increasing order and by counting the number of corresponding values Y_i not satisfying this order.

Q will denote the number of **inversions** among the values of Y that are required to obtain the same (increasing) order as the values of X.

Since there are $\frac{n(n-1)}{2}$ distinct pairs that can be formed, $0 \leq Q \leq \frac{n(n-1)}{2}$; the value 0 is obtained when all the **values** Y_i are already in increasing order, and the value $\frac{n(n-1)}{2}$ is reached when all the values Y_i are in inverse order of X_i, each pair having to be switched to obtain the desired order.

The Kendall rank correlation coefficient, denoted by τ, is defined by:

$$\tau = 1 - \frac{4Q}{n(n-1)}.$$

If all the pairs are in increasing order, then:

$$\tau = 1 - \frac{4 \cdot 0}{n(n-1)} = 1.$$

If all the pairs are in reverse order, then:

$$\tau = 1 - \frac{4 \cdot \frac{1}{2} \cdot n(n-1)}{n(n-1)} = -1.$$

An equivalent definition of the Kendall rank coefficient can be given as follows: two **observations** are called concording if the two members of one observation are larger than the respective members of the other observation. For example, $(0.9, 1.1)$ and $(1.5, 2.4)$ are two concording observations because $0.9 < 1.5$ and $1.1 < 2.4$. Two observations are said to be discording if the two members of one observation are in opposite order to the respective members of the other observation. For example, $(0.8, 2.6)$ and $(1.3, 2.1)$ are two discording observations because $0.8 < 1.3$ and $2.6 > 2.1$.

Let N_c and N_d denote the total number of pairs of concording and discording **observations**, respectively.

Two pairs for which $X_i = X_j$ and $Y_i = Y_j$ are neither concording nor discording and are therefore not counted either in N_c or in N_d.

With this notation the Kendall rank coefficient is given by:

$$\tau = \frac{2(N_c - N_d)}{n(n-1)}.$$

Notice that when there are no pairs for which $X_i = X_j$ or $Y_i = Y_j$, the two formulations of τ are exactly the same. In the opposite situation, the **values** given by both formulas can be different.

Hypothesis Test

The Kendall rank correlation coefficient is often used as a statistical test to determine if there is a relation between two **random variables**. The test can be a **two-sided test** or a **one-sided test**. The **hypotheses** are:

A: Two-sided case:

  H_0: X and Y are mutually independent.

  H_1: There is either a positive or a negative correlation between X and Y.

There is a positive correlation when the large **values** of X tend to be associated with the large values of Y and the small values of X with the small values of Y. There is a negative correlation when the large values of X tend to be associated with the small values of Y and vice versa.

B: One-sided case:

  H_0: X and Y are mutually independent.

  H_1: There is a positive correlation between X and Y.

C: One-sided case:

  H_0: X and Y are mutually independent.

  H_1: There is a negative correlation between X and Y.

The statistical test is defined as follows:

$$T = N_c - N_d.$$

Decision Rules

The decision rules are different depending on the **hypotheses** that are made. That is why there are decision rules A, B, and C relative to the previous cases.

Decision rule A

Reject H_0 at the **significant level** α if

$$T > t_{n, 1-\frac{\alpha}{2}} \quad \text{or} \quad T < t_{n, \frac{\alpha}{2}},$$

where t is the **critical value** of the test given by the Kendall table; otherwise accept H_0.

Decision rule B

Reject H_0 at the **significant level** α if

$$T > t_{n, 1-\alpha}.$$

otherwise accept H_0.

Decision rule C Reject H_0 at the **significant level** α if

$$T < t_{n, \alpha}.$$

otherwise accept H_0.

It is also possible to use

$$\tau = 1 - \frac{4Q}{n(n-1)}$$

as a statistical test.

When X and Y are independently distributed in a **population**, the exact distribution of τ has an **expected value** of zero and a **variance** of:

$$\sigma_\tau^2 = \frac{2(2n+5)}{9n(n-1)}$$

and tends very quickly toward a normal distribution, the approximation being good enough for $n \geq 10$.

K

In this case, to test **independence** at a 5% level, for example, it is enough to verify if τ is located outside the bounds

$$\pm 1.96 \cdot \sigma_\tau$$

and to reject the **independence** hypothesis if that is the case.

DOMAINS AND LIMITATIONS

The Kendall rank correlation coefficient is used as a **hypothesis test** to study the dependence between two **random variables**. It can be considered as a **test of independence**. As a nonparametric correlation measurement, it can also be used with nominal or ordinal **data**.

A correlation measurement between two **variables** must satisfy the following points:
1. Its **values** are between -1 and $+1$.
2. There is a positive correlation between X and Y if the value of the correlation coefficient is positive; a perfect positive correlation corresponds to a value of $+1$.
3. There is a negative correlation between X and Y if the value of the correlation coefficient is negative; a perfect negative correlation corresponds to a value of -1.
4. There is a null correlation between X and Y when the correlation coefficient is close to zero; one can also say that X and Y are not correlated.

The Kendall rank correlation coefficient has the following advantages:
- The **data** can be nonnumerical **observations** as long as they can be classified according to a determined criterion.
- It is easy to calculate.
- The associated statistical test does not make a basic **hypothesis** based on the

shape of the distribution of the **population** from which the **samples** are taken.
The Kendall table gives the theoretical **values** of the **statistic** τ of the Kendall rank correlation coefficient used as a statistical test under the **independence hypothesis** of two **random variables**.
A Kendall table can be found in Kaarsemaker and van Wijngaarden (1953).
Here is a sample of the Kendall table for $n = 4, \ldots, 10$ and $\alpha = 0.01$ and 0.05:

n	$\alpha = 0.01$	$\alpha = 0.05$
4	6	4
5	8	6
6	11	9
7	15	11
8	18	14
9	22	16
10	25	19

EXAMPLES

In this example eight pairs of real twins take intelligence tests. The goal is to see if there is **independence** between the tests of the one who is born first and those of the one who is born second.
The data are given in the table below; the highest scores correspond to the best results.

Pair of twins	First born X_i	Second born Y_i
1	90	88
2	75	79
3	99	98
4	60	66
5	72	64
6	83	83
7	83	88
8	90	98

The pairs are then classified in increasing order for X, and the concording and discording pairs are determined. This gives:

Pair of twins (X_i, Y_i)	Concording pairs	Discording pairs
(60,66)	6	1
(72,64)	6	0
(75,79)	5	0
(83,83)	3	0
(83,88)	2	0
(90,88)	1	0
(90,98)	0	0
(99,98)	0	0
	$N_c = 23$	$N_d = 1$

The Kendall rank correlation coefficient is given by:

$$\tau = \frac{2(N_c - N_d)}{n(n-1)}$$
$$= \frac{2(23-1)}{8 \cdot 7}$$
$$= 0.7857.$$

Notice that since there are several **observations** for which $X_i = X_j$ or $Y_i = Y_j$, the value of the coefficient given by:

$$\tau = 1 - \frac{4Q}{n(n-1)} = 1 - \frac{4 \cdot 1}{56} = 0.9286$$

is different.

In both cases, we notice a positive correlation between the intelligence tests.

We will now carry out the **hypothesis test**:

H_0: There is **independence** between the intelligence tests of a pair of twins.

H_1: There is a positive correlation between the intelligence tests.

We chose a **significant level** of $\alpha = 0.05$. Since we are in case B, H_0 is rejected if

$$T > t_{8,0.95},$$

where $T = N_c - N_d$ and $t_{8,0.95}$ is the **value** of the Kendall table. Since $T = 22$ and $t_{8,0.95} = 14$, H_0 is rejected.

We can then conclude that there is a positive correlation between the results of the intelligence tests of a pair of twins.

FURTHER READING

▶ **Hypothesis testing**
▶ **Nonparametric test**
▶ **Test of independence**

REFERENCES

Deuchler, G.: Über die Methoden der Korrelationsrechnung in der Pädagogik und Psychologie. Zeitung für Pädagogische Psycholologie und Experimentelle Pädagogik, **15**, 114–131, 145–159, 229–242 (1914)

Fechner, G.T.: Kollcktivmasslchrc. W. Engelmann, Leipzig (1897)

Kaarsemaker, L., van Wijngaarden, A.: Tables for use in rank correlation. Stat. Neerland. **7**, 53 (1953)

Kendall, M.G.: A new measure of rank correlation. Biometrika **30**, 81–93 (1938)

Kendall, M.G.: Rank Correlation Methods. Griffin, London (1948)

Lipps, G.F.: Die Psychischen Massmethoden. F. Vieweg und Sohn, Braunschweig, Germany (1906)

Kiefer Jack Carl

Kiefer, Jack Carl was born in Cincinnati, Ohio in 1924. He entered the Massachusetts Institute of Technology in 1942, but after 1 year of studying engineering and economics he left to take on war-related work during World War II. His master's thesis,

Sequential Determination of the Maximum of a Function, was supervised by Harold Freeman. It has been the basis for his paper "Sequential minimax search for a maximum" which appeared in 1953 in the "Proceedings of the American Mathematical Society". In 1948 he went to the Department of Mathematical Statistics at Columbia University, where Abraham Wald was preeminent in a department that included Ted Anderson, Henry Scheffe, and Jack Wolfowitz. He wrote his doctoral thesis in decision theory under Wolfowitz and went to Cornell University in 1951 with Wolfowitz. In 1973 Kiefer was elected the first Horace White Professor at Cornell University, a position he held until 1979, when he retired and joined the faculty at the University of California at Berkeley. He died at the age of 57 in 1981.

Kiefer's research area was the design of experiments. Most of his 100 publications dealt with that topic. He also wrote papers on topics in mathematical statistics including decision theory, stochastic approximation, queuing theory, nonparametric inference, estimation, sequential analysis, and conditional inference.

Kiefer was a fellow of the Institute of Mathematical Statistics and the American Statistical Association and president of the Institute of Mathematical Statistics (1969–1970). He was elected to the American Academy of Arts and Sciences in 1972 and to the National Academy of Sciences (USA) in 1975.

Selected works and publications of Jack Carl Kiefer:

1987 Introduction to Statistical Inference. Springer, Berlin Heidelberg New York

FURTHER READING
▶ Design of experiments

REFERENCES
Kiefer, J.: Optimum experimental designs. J. Roy. Stat. Soc. Ser. B **21**, 272–319 (1959)

L.D. Brown, I. Olkin, J. Sacks and H.P. Wynn (eds.): Jack Karl Kiefer, Collected Papers. I: Statistical inference and probability (1951–1963). New York (1985)

L.D. Brown, I. Olkin, J. Sacks and H.P. Wynn (eds.): Jack Karl Kiefer, Collected Papers. II: Statistical inference and probability (1964–1984). New York (1985)

L.D. Brown, I. Olkin, J. Sacks and H.P. Wynn (eds.): Jack Karl Kiefer, Collected Papers. III: Design of experiments. New York (1985)

Kolmogorov, Andrei Nikolaevich

Born in Tambov, Russia in 1903, Kolmogorov, Andrei Nikolaevich is one of the founders of modern probability. In 1920, he entered Moscow State University and studied mathematics, history, and metallurgy. In 1925, he published his first article in probability on the inequalities of the partial sums of random variables, which became the principal reference in the field of stochastic processes. He received his doctorate in 1929 and published 18 articles on the law of large numbers as well as on intuitive logic. He was named professor at Moscow State University in 1931. In 1933, he published his monograph on probability theory.

In 1939 he was elected member of the Academy of Sciences of the USSR. He received the Lenin Prize in 1965 and the Order of Lenin on six different occasions, as well as the Lobachevsky Prize in 1987. He was elected member of many other foreign academies including the Romanian Academy of Sciences (1956), the Royal Statistical Society of London (1956), the Leopoldina Academy of Germany (1959), the American Academy of Arts and Sciences (1959), the London Mathematical Society (1959), the American Philosophical Society (1961), the Indian Institute of Statistics(1962), the Holland Academy of Sciences (1963), the Royal Society of London (1964), the National Academy of the United States (1967), and the Académie Française des Sciences (1968).

Selected principal works of Kolmogorov, Andrei Nikolaevich:

1933 Grundbegriffe der Wahrscheinlichkeitsrechnung. Springer, Berlin Heidelberg New York.

1933 Sulla determinazione empirica di una lege di distribuzione. Giornale dell'Instituto Italiano degli Attuari, 4, 83–91 (6.1).

1941 Local structure of turbulence in incompressible fluids with very high Reynolds number. Dan SSSR, 30, 229.

1941 Dissipation of energy in locally isotropic turbulence. Dokl. Akad. Nauk. SSSR, 32, 16–18.

1958 (with Uspenskii, V.A.) K opredeleniyu algoritma. (Toward the definition of an algorithm). Uspekhi Matematicheskikh Nauk 13(4):3–28, American Mathematical Society Translations Series 2(29):217–245, 1963.

1961 (with Fomin, S.V.) Measure, Lebesgue integrals and Hilbert space. Natascha Artin Brunswick and Alan Jeffrey. Academic, New York.

1963 On the representation of continuous functions of many variables by superposition of continuous functions of one variable and addition. Doklady Akademii Nauk SSR, 114, 953–956, 1957. English translation. Mathematical Society Transactions, 28, 55–59.

1965 Three approaches to the quantitative definition of information. Problems of Information Transmission, 1, 1–17. Translation of Problemy peredachi informatsii 1(1), 3–11 (1965).

1987 (with Uspenskii, V.A.) Algorithms and randomness. Teoria veroyatnostey i ee primeneniya (Probability theory and its applications), 3(32):389–412.

FURTHER READING
▶ **Kolmogorov–Smirnov test**

Kolmogorov–Smirnov Test

The Kolmogorov–Smirnov test is a nonparametric **goodness-of-fit test** and is used to determine wether two distributions differ, or whether an underlying probability distribution differes from a hypothesized distribution. It is used when we have two samples coming from two populations that can be different. Unlike the **Mann–Whitney test** and the **Wilcoxon test** where the goal is to detect the difference between two means or medians, the Kolmogorov–Smirnov test has the advantage of considering the distribution functions collectively. The Kolmogorov–Smirnov test can also be used as a **goodness-of-fit test**. In this case, we have only one random sample obtained from a **population** where the distribution function is specific and known.

HISTORY

The **goodness-of-fit test** for a **sample** was invented by Andrey Nikolaevich Kolmogorov (1933).

The Kolmogorov–Smirnov test for two samples was invented by Vladimir Ivanovich Smirnov (1939).

In Massey (1952) we find a Smirnov table for the Kolmogorov–Smirnov test for two samples, and in Miller (1956) we find a Kolmogorov table for the goodness-of-fit test.

MATHEMATICAL ASPECTS

Consider two independent random samples: $(X_1, X_2, \ldots, X_n)$, a sample of size n coming from a **population** 1, and $(Y_1, Y_2, \ldots, Y_m)$, a sample of dimension m coming from a population 2. We denote by, respectively, $F(x)$ and $G(x)$ their unknown **distribution functions**.

Hypotheses

The hypotheses to test are as follows:

A: Two-sided case:

> H_0: $F(x) = G(x)$ for each x
> H_1: $F(x) \neq G(x)$ or at least one **value** of x

B: One-sided case:

> H_0: $F(x) \leq G(x)$ for each x
> H_1: $F(x) > G(x)$ for at least one value of x

C: One-sided case:

> H_0: $F(x) \geq G(x)$ for each x
> H_1: $F(x) < G(x)$ for at least one value of x

In case A, we make the hypothesis that there is no difference between the distribution functions of these two populations. Both populations can then be seen as one population.

In case B, we make the hypothesis that the distribution function of population 1 is smaller than those of population 2. We sometimes say that, generally, X tends to be smaller than Y.

In case C, we make the hypothesis that X is greater than Y.

We denote by $H_1(x)$ the empirical distribution function of the **sample** $(X_1, X_2, \ldots, X_n)$ and by $H_2(x)$ the empirical distribution function of the sample $(Y_1, Y_2, \ldots, Y_m)$. The statistical test are defined as follows:

A: Two-tail case

The statistical test T_1 is defined as the greatest vertical **distance** between two empirical distribution functions:

$$T_1 = \sup_x |H_1(x) - H_2(x)| \, .$$

B: One-tail case

The statistical test T_2 is defined as the greatest vertical distance when $H_1(x)$ is greater than $H_2(x)$:

$$T_2 = \sup_x [H_1(x) - H_2(x)] \, .$$

C: One-tail case

The statistical test T_3 is defined as the greatest vertical distance when $H_2(x)$ is greater than $H_1(x)$:

$$T_3 = \sup_x [H_2(x) - H_1(x)] \, .$$

Decision Rule

We reject H_0 at the **significance level** α if the appropriate statistical test (T_1, T_2, or T_3) is greater than the **value** of the Smirnov table having for parameters n, m, and $1 - \alpha$, which we denote by $t_{n,m,1-\alpha}$, that is, if

$$T_1 (\text{or } T_2 \text{or } T_3) > t_{n,m,1-\alpha} .$$

If we want to test the goodness of fit of an unknown distribution function $F(x)$ of a random sample from a population with a specific and known distribution function $F_o(x)$, then the hypotheses will be the same as those for testing two samples, except that $F(x)$ and $G(x)$ are replaced by $F(x)$ and $F_o(x)$. If $H(x)$ is the empirical distribution function of a random sample, then the statistical tests T_1, T_2, and T_3 are defined as follows:

$$T_1 = \sup_{x} |F_o(x) - H(x)| ,$$

$$T_2 = \sup_{x} [F_o(x) - H(x)] ,$$

$$T_3 = \sup_{x} [H(x) - F_o(x)] .$$

The decision rule is as follows: reject H_0 at the significance level α if T_1 (or T_2 or T_3) is greater than the value of the Kolmogorov table having for parameters n and $1 - \alpha$, which we denote by $t_{n,1-\alpha}$, that is, if

$$T_1 (\text{or } T_2 \text{or } T_3) > t_{n,1-\alpha} .$$

DOMAINS AND LIMITATIONS

To perform the Kolmogorov–Smirnov test, the following must be observed:
1. Both samples must be taken randomly from their respective **populations**.
2. There must be mutual **independence** between two samples.
3. The measure scale must be at least ordinal.

4. To perform an exact test, the random variables must be continuous; otherwise the test is less precise.

EXAMPLES

The first example treats the Kolmogorov–Smirnov test for two samples and the second one for the **goodness-of-fit test**.

In a class, we count 25 pupils: 15 boys and 10 girls. We perform a test of mental calculations to see if the boys tend to be better than the girls in this domain.

The data are presented in the following table; the highest scores correspond to the results of the test.

Boys (X_i)		Girls (Y_i)	
19.8	17.5	17.7	14.1
12.3	17.9	7.1	23.6
10.6	21.1	21.0	11.1
11.3	16.4	10.7	20.3
13.3	7.7	8.6	15.7
14.0	15.2		
9.2	16.0		
15.6			

We test the **hypothesis** according to which the distributions of the results of the girls and those of the boys are identical. This means that the **population** from which the sample of X is taken has the same **distribution function** as the population from which the sample of Y is taken. Hence the **null hypothesis**:

$$H_0 : F(x) = G(x) \quad \text{for each } x .$$

If the two-tail case is applied here, we calculate:

$$T_1 = \sup_{x} |H_1(x) - H_2(x)| ,$$

where $H_1(x)$ and $H_2(x)$ are the empirical distribution functions of the samples

$(X_1, X_2, \ldots, X_{15})$ and $(Y_1, Y_2, \ldots, Y_{10})$, respectively. In the following table, we have classed the observations of two samples in increasing order to simplify the calculations of $H_1(x) - H_2(x)$.

X_i	Y_i	$H_1(x) - H_2(x)$
	7.1	$0 - 1/10 = -0.1$
7.7		$1/15 - 1/10 = -0.0333$
	8.6	$1/15 - 2/10 = -0.1333$
9.2		$2/15 - 2/10 = -0.0667$
10.6		$3/15 - 2/10 = 0$
	10.7	$3/15 - 3/10 = -0.1$
	11.1	$3/15 - 4/10 = -0.2$
11.3		$4/15 - 4/10 = -0.1333$
12.3		$5/15 - 4/10 = -0.0667$
13.3		$6/15 - 4/10 = 0$
14.0		$7/15 - 4/10 = 0.0667$
	14.1	$7/15 - 5/10 = -0.0333$
15.2		$8/15 - 5/10 = 0.0333$
15.6		$9/15 - 5/10 = 0.1$
	15.7	$9/15 - 6/10 = 0$
16.0		$10/15 - 6/10 = 0.0667$
16.4		$11/15 - 6/10 = 0.1333$
17.5		$12/15 - 6/10 = 0.2$
	17.7	$12/15 - 7/10 = 0.1$
17.9		$13/15 - 7/10 = 0.1667$
19.8		$14/15 - 7/10 = 0.2333$
	20.3	$14/15 - 8/10 = 0.1333$
	21.0	$14/15 - 9/10 = 0.0333$
21.1		$1 - 9/10 = 0.1$
	23.6	$1 - 1 = 0$

We have then:

$$T_1 = \sup_x |H_1(x) - H_2(x)|$$

$$= 0.2333.$$

The **value** of the Smirnov table for $n = 15$, $m = 10$, and $1 - \alpha = 0.95$ equals $t_{15,10,0.95} = 0.5$.

Thus $T_1 = 0.2333 < t_{15,10,0.95} = 0.5$, and H_0 cannot be rejected. This means that there is no significant difference in the level of mental calculations of girls and boys.

Consider the following random sample of dimension 10: $X_1 = 0.695, X_2 = 0.937, X_3 = 0.134, X_4 = 0.222, X_5 = 0.239, X_6 = 0.763, X_7 = 0.980, X_8 = 0.322, X_9 = 0.523, X_{10} = 0.578$.

We want to verify by the Kolmogorov–Smirnov test if this sample comes from a **uniform distribution**. The distribution function of the uniform distribution is given by:

$$F_o(x) = \begin{cases} 0 & \text{if } x < 0 \\ x & \text{if } 0 \le x < 1 \\ 1 & \text{otherwise} \end{cases}.$$

The **null hypothesis** H_0 is then as follows, where $F(x)$ is the unknown distribution function of the population associated to the sample:

$$H_0 : F(x) = F_o(x) \quad \text{for each } x.$$

If the two-tail case is applied, we calculate:

$$T_1 = \sup_x |F_o(x) - H(x)|,$$

where $H(x)$ is the empirical distribution function of the sample $(X_1, X_2, \ldots, X_{10})$. In the following table, we class the 10 observations in increasing order to simplify the calculation of $F_0(x) - H(x)$.

X_i	$F_o(x)$	$H(x)$	$F_o(x) - H(x)$
0.134	0.134	0.1	$0.134 - 0.1 = 0.034$
0.222	0.222	0.2	$0.222 - 0.2 = 0.022$
0.239	0.239	0.3	$0.239 - 0.3 = -0.061$
0.322	0.322	0.4	$0.322 - 0.4 = -0.078$
0.523	0.523	0.5	$0.523 - 0.5 = 0.023$
0.578	0.578	0.6	$0.578 - 0.6 = -0.022$
0.695	0.695	0.7	$0.695 - 0.7 = -0.005$
0.763	0.763	0.8	$0.763 - 0.8 = -0.037$
0.937	0.937	0.9	$0.937 - 0.9 = 0.037$
0.980	0.980	1.0	$0.980 - 1.0 = -0.020$

We obtain then:

$$T_1 = \sup_x |F_o(x) - H(x)| = 0.078 \,.$$

The value of the Kolmogorov table for $n = 10$ and $1 - \alpha = 0.95$ is $t_{10,0.95} = 0.409$.
If T_1 is smaller than $t_{10,0.95}$ $(0.078 < 0.409)$, then H_0 cannot be rejected. That means that the random sample could come from a uniformly distributed population.

FURTHER READING
▶ **Goodness of fit test**
▶ **Hypothesis testing**
▶ **Nonparametric test**

REFERENCE

Kolmogorov, A.N.: Sulla determinazione empirica di una legge di distribuzione. Giornale dell'Instituto Italiano degli Attuari **4**, 83–91 (6.1) (1933)

Massey, F.J.: Distribution table for the deviation between two sample cumulatives. Ann. Math. Stat. **23**, 435–441 (1952)

Miller, L.H.: Table of percentage points of Kolmogorov statistics. J. Am. Stat. Assoc. **31**, 111–121 (1956)

Smirnov, N.V.: Estimate of deviation between empirical distribution functions in two independent samples. (Russian). Bull. Moscow Univ. **2**(2), 3–16 (6.1, 6.2) (1939)

Smirnov, N.V.: Table for estimating the goodness of fit of empirical distributions. Ann. Math. Stat. **19**, 279–281 (6.1) (1948)

Kruskal-Wallis Table

The Kruskal–Wallis table gives the theoretical values of the **statistic** H of the **Kruskal–Wallis test** under the **hypothesis** that there is no difference among the k $(k \geq 2)$ populations that we want to compare.

HISTORY
See **Kruskal–Wallis test**.

MATHEMATICAL ASPECTS
Let k be the number of samples of probably different sizes $n_1, n_2, \ldots, n_k$. We designate by N the total number of observations:

$$N = \sum_{i=1}^{k} n_i \,.$$

We class the N observations in increasing order without taking into account which samples they belong to. We then give rank 1 to the smallest **value**, rank 2 to the next greatest value, and so on until rank N, which is given to the greatest value.
We denote by R_i the sum of the ranks given to the observations of sample i:

$$R_i = \sum_{j=1}^{n_i} R(X_{ij}) \,, \quad i = 1, 2, \ldots, k \,,$$

where X_{ij} represents observation j of sample i and $R(X_{ij})$ the corresponding rank. When many observations are identical and of the same rank, we give them a mean rank (see **Kruskal–Wallis test**). If there are no mean ranks, the statistical test is defined in the following way:

$$H = \left(\frac{12}{N(N+1)} \sum_{i=1}^{k} \frac{R_i^2}{n_i} \right) - 3(N+1) \,.$$

On what to do if there are mean ranks, see **Kruskal–Wallis test**.
The Kruskal–Wallis table gives the values of the **statistic** H of the Kruskal–Wallis test in the case of three samples, for different values of n_1, n_2, and n_3 (with $n_1, n_2, n_3 \leq 5$).

K

DOMAINS AND LIMITATIONS

The Kruskal–Wallis table is used for non-parametric tests that use ranks and particularly for tests with the same name.

When the number of samples i is greater than 3, we can make an approximation of the value of the Kruskal–Wallis table by the chi-square table with $k - 1$ degrees of freedom.

EXAMPLES

See Appendix D.

For an example of the use of the Kruskal–Wallis table, see **Kruskal–Wallis test**.

FURTHER READING

▶ **Chi-square table**
▶ **Kruskal-Wallis test**
▶ **Statistical table**

REFERENCES

Kruskal, W.H., Wallis, W.A.: Use of ranks in one-criterion variance analysis. J. Am. Stat. Assoc. **47**, 583–621 and errata, ibid. **48**, 907–911 (1952)

Kruskal-Wallis Test

The Kruskal–Wallis test is a **nonparametric test** that has as its goal to determine if all k populations are identical or if at least one of the populations tends to give observations that are different from those of other populations.

The test is used when we have k samples, with $k \geq 2$, coming from k populations that can be different.

HISTORY

The Kruskal–Wallis test was developed in 1952 by Kruskal, W.H. and Wallis, W.A.

MATHEMATICAL ASPECTS

The data are represented in k samples. We designate by n_i the dimension of the **sample** i, for $i = 1, \ldots, k$, and by N the total number of observations:

$$N = \sum_{i=1}^{k} n_i .$$

We class the N observations in increasing order without taking into account whether or not they belong to the same samples. We then give rank 1 to the smallest **value**, rank 2 to the next greatest value, and so on until N, which is given to the greatest value.

Let X_{ij} be the jth observation of sample i, and set $i = 1, \ldots, k$ and $j = 1, \ldots, n_i$; we then denote the rank given to X_{ij} by $R\left(X_{ij}\right)$.

If many observations have the same value, we give them a mean rank. The sum of the ranks given to the observations of sample i is denoted by R_i, and we have:

$$R_i = \sum_{j=1}^{n_i} R\left(X_{ij}\right), \quad i = 1, \ldots, k .$$

If there are no mean ranks (or if there is a limited number of them), then the statistical test is defined as follows:

$$H = \left(\frac{12}{N(N + 1)} \sum_{i=1}^{k} \frac{R_i^2}{n_i} \right) - 3\,(N + 1) .$$

If, on the contrary, there are many mean ranks, it is necessary to make a correction and to calculate:

$$\widetilde{H} = \frac{H}{1 - \dfrac{\sum_{i=1}^{g} \left(t_i^3 - t_i\right)}{N^3 - N}} ,$$

where g is the number of groups of mean ranks and t_i the dimension of ith such group.

Hypotheses

The goal of the Kruskal–Wallis test is to determine if all the populations are identical or if at least one of the populations tends to give observations different from other populations. The hypotheses are as follows:

H_0: There is no difference among the k populations.

H_1: At least one of the populations differs from the other populations.

Decision Rule

If there are 3 samples, each having a dimension smaller or equal to 5, and if there are no mean ranks (that is, if H is calculated), then we use the **Kruskal–Wallis table** to test H_0. The decision rule is the following: We reject the **null hypothesis** H_0 at the **significance level** α if T is greater than the value of the table with parameters n_i, $k - 1$, and $1 - \alpha$, denoted $h_{n_1,n_2,n_3,1-\alpha}$, and if there is no available exact table or if there are mean ranks, we can make an approximation of the value of the Kruskal–Wallis table by the distribution of the chi-square with $k - 1$ degrees of freedom (**chi-square distribution**), that is, if:

$$H > h_{n_1,n_2,n_3,1-\alpha} \quad \textbf{(Kruskal–Wallis table)}$$

or $H > \chi^2_{k-1,1-\alpha}$ **(chi-square table)**.

The corresponding decision rule is based on $\tilde{H} > \chi^2_{k-1,1-\alpha}$ (chi-quare table).

DOMAINS AND LIMITATIONS

The following rules should be respected to make the Kruskal–Wallis test:
1. All the samples must be random samples taken from their respective **populations**.
2. In addition to the **independence** inside each sample, there must be mutual independence among the different samples.
3. The scale of measure must be at least ordinal.

If the Kruskal–Wallis test makes us reject the **null hypothesis** H_0, we can use the **Wilcoxon test** for all the samples taken in pairs to determine which pairs of populations tend to be different.

EXAMPLES

We cook potatoes in 4 different oils. We want to verify if the quantity of fat absorbed by potatoes depends on the type of oil used. We conduct 5 different experiments with oil 1, 6 with oil 2, 4 with oil 3, and 5 with oil 4, and we obtain the following data:

Type of oil			
1	2	3	4
64	78	75	55
72	91	93	66
68	97	78	49
77	82	71	64
56	85		70
	77		

In this example, the number of samples equals 4 ($k = 4$) with the following respective dimensions:

$$n_1 = 5,$$
$$n_2 = 6,$$
$$n_3 = 4,$$
$$n_4 = 5.$$

The number of observations equals:

$$N = 5 + 6 + 4 + 5 = 20.$$

We class the observations in increasing order and give them a rank from 1 to 20 taking into account the mean ranks. We obtain the following table with the rank of the observations in parentheses and at the end of each sample the sum of the ranks given to the corresponding sample:

Type of oil			
1	2	3	4
64 (4.5)	78 (14.5)	75 (11)	55 (2)
72 (10)	91 (18)	93 (19)	66 (6)
68 (7)	97 (20)	78 (14.5)	49 (1)
77 (12.5)	82 (16)	71 (9)	64 (4.5)
56 (3)	85 (17)		70 (8)
	77 (12.5)		
$R_1 = 37$	$R_2 = 98$	$R_3 = 53.5$	$R_4 = 21.5$

We calculate:

$$H = \left(\frac{12}{N(N+1)} \sum_{i=1}^{k} \frac{R_i^2}{n_i} \right) - 3(N+1)$$

$$= \frac{12}{20(20+1)}$$

$$\cdot \left(\frac{37^2}{5} + \frac{98^2}{6} + \frac{53.5^2}{4} + \frac{21.5^2}{5} \right)$$

$$- 3(20+1)$$

$$= 13.64 \,.$$

If we make an adjustment to take into account the mean ranks, we get:

$$\tilde{H} = \frac{H}{1 - \frac{\sum_{i=1}^{8} (t_i^3 - t_i)}{N^3 - N}}$$

$$= \frac{13.64}{1 - \frac{8 - 2 + 8 - 2 + 8 - 2}{20^3 - 20}}$$

$$= 13.67 \,.$$

We see that the difference between H and $\tilde{H}$ is minimal.

The hypotheses are as follows:

H_0: There is no difference between the four oils.

H_1: At least one of the oils differs from the others.

The decision rule is as follows: Reject the **null hypothesis** H_0 at the **significance level** α if

$$H > \chi^2_{k-1, 1-\alpha} \,,$$

where $\chi^2_{k-1, 1-\alpha}$ is the **value** of the **chi-square table** at level $1 - \alpha$ and $k - 1 = 3$ degrees of freedom.

If we choose $\alpha = 0.05$, then the value of $\chi^2_{3, 0.95}$ is 7.81, and if H is greater than $\chi^2_{3, 0.95}$ (13.64 > 7.81), then we reject the hypothesis H_0.

If H_0 is rejected, we can use the procedure of multiple comparisons to see which pairs of oils are different.

FURTHER READING

▶ **Hypothesis testing**
▶ **Nonparametric test**
▶ **Wilcoxon test**

REFERENCE

Kruskal, W.H., Wallis, W.A.: Use of ranks in one-criterion variance analysis. J. Am. Stat. Assoc. **47**, 583–621 and errata, ibid. **48**, 907–911 (1952)

L_1 Estimation

The L_1 estimation method is an alternative to the **least-squares** method used in linear regression for the **estimation** of the **parameters** of a **model**. Instead of minimizing the **residual** sum of squares, the sum of the **residual** absolute values is minimized.

The L_1 **estimators** have the advantage of not being as sensitive to outliers as the **least-squares** method.

HISTORY

The use of the L_1 estimation method, in a very basic form, dates back to Galileo in 1632, who used it to determine the **distance** between the earth and a star that had recently been discovered.

Boscovich, R.J. (1757) proposed two criteria for adjusting a line:

- The respective sums of the positive and negative **residuals** must be equal, and the optimal line will pass by the centroid of the **observations**.
- The sum of the residuals in absolute value must be minimal.

Boscovich, R.J. proposed a geometrical method to solve the equations resulting from his criteria. The analytical procedure used for the **estimation** of the **parameters** for the two criteria of Boscovich, R.J. is due to **Laplace, P.S.** (1793). Edgeworth, F.Y. (1887) suppressed the first condition, saying that the sum of the **residuals** must be null, and developed a method that could be used to treat the case of **multiple linear regression**.

During the first half of the 20th century, statisticians were not very interested in linear regression based on L_1 estimation. The main reasons were:

- The numerical difficulties due to the absence of a closed formula.
- The absence of asymptotic theory for L_1 estimation in the regression **model**.
- The estimation by the **least-squares** method is satisfactory for small **samples**, and it is not certain that L_1 estimation is any better.

Starting in 1970, the field of L_1 estimation became the pole of research for a multitude of statisticians, and many works were published. We mention Basset, G. and Koenker, R. (1978), who developed the asymptotic theory of L_1 estimators. In 1987, a congress dedicated to these problems was held in Neuchâtel (Switzerland): The First International Conference on Statistical Data Analysis Based on the L_1-norm and Related Methods.

MATHEMATICAL ASPECTS

In the one-dimensional (1D) case, the L_1 **estimation** method consists in estimating the

parameter β_0 in the **model**

$$y_i = \beta_0 + \varepsilon_i \,,$$

so that the sum of the **errors** in absolute value is minimal:

$$\min_{\beta_0} \sum_{i=1}^{n} |\varepsilon_i| = \min_{\beta_0} \sum_{i=1}^{n} |y_i - \beta_0| \,.$$

This problem boils down to finding β_0 so that $\sum_{i=1}^{n} |y_i - \beta_0|$ is minimal. The value of β_0 that minimizes this sum is the **median** of the 1D **sample** $y_1, y_2, \ldots, y_n$.

In the **simple linear regression** model

$$y_i = \beta_0 + \beta_1 \cdot x_i + \varepsilon_i \,,$$

the **parameters** β_0 and β_1 estimated using the L_1 **estimation** method are usually calculated by an iterative **algorithm** that is used to minimize

$$\sum_{i=1}^{n} |y_i - \beta_0 - \beta_1 \cdot x_i| \,.$$

Using **linear programming** techniques it is possible today to calculate the L_1 **estimators** in a relatively easy way, especially in the **multiple linear regression** model.

DOMAINS AND LIMITATIONS

The problem of L_1 estimation lies in the absence of a general formula for the **estimation** of the **parameters**, unlike the situation with the **least-squares** method. For a set of 1D **data**, the **measure of central tendency** determined by L_1 estimation is the **median**, which is by definition the **value** that minimizes the sum of the absolute deviations.

FURTHER READING
▶ **Error**
▶ **Estimation**
▶ **Estimator**
▶ **Least squares**
▶ **Mean absolute deviation**
▶ **Parameter**
▶ **Residual**
▶ **Robust estimation**

REFERENCES
Basset, G., Koenker, R.: Asymptotic theory of least absolute error regression. J. Am. Stat. Assoc. **73**, 618–622 (1978)

Boscovich, R.J.: De Litteraria Expeditione per Pontificiam ditionem, et Synopsis amplioris Operis, ac habentur plura eius ex exemplaria etiam sensorum impressa. Bononiensi Scientiarium et Artium Instituto Atque Academia Commentarii, vol. IV, pp. 353–396 (1757)

Dodge, Y. (ed.): Statistical Data Analysis Based on the L_1-norm and Related Methods. Elsevier, Amsterdam (1987)

Edgeworth F.Y.: A new method of reducing observations relating to several quantities. Philos. Mag. (5th Ser) **24**, 222–223 (1887)

Lagrange Multiplier

The Lagrange multiplier method is a classical **optimization** method that allows to determine the local extremes of a function subject to certain constraints. It is named after the Italian-French mathematician and astronomer Joseph-Louis Lagrange.

MATHEMATICAL ASPECTS

Let $f(x, y)$ be the objective function to be maximized or minimized subject to

$g(x, y) = 0$. We form the auxiliary function, the Lagrangian:

$$F(x, y, \theta) = f(x, y) - \theta g(x, y),$$

where θ is an unknown **variable** called the Lagrange multiplier. We search the minima or the maxima of function F by equating to zero the partial derivatives with respect to x, y, and θ:

$$\frac{\partial F}{\partial x} = \frac{\partial f}{\partial x} - \theta \frac{\partial g}{\partial x} = 0$$

$$\frac{\partial F}{\partial y} = \frac{\partial f}{\partial y} - \theta \frac{\partial g}{\partial y} = 0$$

$$\frac{\partial F}{\partial \theta} = g(x, y) = 0.$$

These three equations can be solved for x, y, and θ. The solution provides the points where the constrained function has a minimum or a maximum. We have to decide if there is a maximum or a minimum.
We get a maximum if

$$\left[\frac{\partial^2 F}{\partial x^2} \right] \cdot \left[\frac{\partial^2 F}{\partial y^2} \right] - \left[\frac{\partial^2 F}{\partial x \partial y} \right]^2 > 0,$$

$$\frac{\partial^2 F}{\partial x^2} < 0 \quad \text{and} \quad \frac{\partial^2 F}{\partial y^2} < 0.$$

We get a minimum if

$$\left[\frac{\partial^2 F}{\partial x^2} \right] \cdot \left[\frac{\partial^2 F}{\partial y^2} \right] - \left[\frac{\partial^2 F}{\partial x \partial y} \right]^2 > 0,$$

$$\frac{\partial^2 F}{\partial x^2} > 0 \quad \text{and} \quad \frac{\partial^2 F}{\partial y^2} > 0.$$

When

$$\left[\frac{\partial^2 F}{\partial x^2} \right] \cdot \left[\frac{\partial^2 F}{\partial y^2} \right] - \left[\frac{\partial^2 F}{\partial x \partial y} \right]^2 \leq 0,$$

the test breaks down, and to determine if there is a minimum or a maximum, we have to examine the functions neighboring in (x, y).

The Lagrange multiplier method is extended to a function of n variables, $f(x_1, x_2, \ldots, x_n)$, that are subject to k constraints $g_j(x_1, x_2, \ldots, x_n) = 0, j = 1, 2, \ldots, k$ with $k \leq n$. Similarly, in the bivariate case, we define the Lagrangian:

$$F = f(x_1, x_2, \ldots, x_n)$$
$$- \sum_{j=1}^{k} \theta_j \cdot g_j(x_1, x_2, \ldots, x_n),$$

with the parameters $\theta_1, \ldots, \theta_k$ being the Lagrange multipliers. We establish the partial derivatives that lead to $n + k$ equations on $n + k$ unknown parameters. These lead to the coordinates of the function extremes (minimum or maximum), if they exist.

DOMAINS AND LIMITATIONS

It is not possible to apply the Lagrange multiplier method when the functions are not differentiable or when the imposed constraints are inequalities. In this case we can use other **optimization** methods properly chosen from **linear programming**.

EXAMPLES

Let

$$f(x, y) = 5x^2 + 6y^2 - x \cdot y$$

be the objective function and

$$g(x, y) = x + 2y - 24 = 0$$

the imposed constraint. We form the Lagrangian

$$F(x, y, \theta) = f(x, y) - \theta g(x, y)$$
$$= 5x^2 + 6y^2 - x \cdot y$$
$$- \theta \cdot (x + 2y - 24),$$

and we set to zero the partial derivatives

$$\frac{\partial F}{\partial x} = 10x - y - \theta = 0, \qquad (1)$$

$$\frac{\partial F}{\partial y} = 12y - x - 2\theta = 0, \qquad (2)$$

$$\frac{\partial F}{\partial \theta} = x + 2y - 24 = 0. \qquad (3)$$

By deleting θ of (1) and (2):

$$20x - 2y - 2\theta = 0$$
$$-x + 12y - 2\theta = 0$$
$$21x - 14y = 0$$
$$3x - 2y = 0$$

and by replacing $2y$ with $3x$ in (3):

$$0 = x + 2y - 24 = x + 3x - 24$$
$$= 4x - 24$$

we get $x = 6$. By replacing this x value in (3), we get $y = 9$. The critical point has coordinates $(6, 9)$. We compute the partial derivatives of second order to verify whether it is a maximum or a minimum. We get:

$$\frac{\partial^2 F}{\partial x^2} = 10, \quad \frac{\partial^2 F}{\partial y^2} = 12, \quad \frac{\partial^2 F}{\partial x \partial y} = -1,$$

$$\left[\frac{\partial^2 F}{\partial x^2}\right] \cdot \left[\frac{\partial^2 F}{\partial y^2}\right] - \left[\frac{\partial^2 F}{\partial x \partial y}\right]^2$$

$$= 10 \cdot 12 - (-1)^2 = 119 > 0,$$

$$\frac{\partial^2 F}{\partial x^2} > 0, \quad \frac{\partial^2 F}{\partial y^2} > 0.$$

Note that the point $(6, 9)$ corresponds to a minimum for the objective function f subject to the constraints above. This minimum yields $f(6, 9) = 180 + 486 - 54 = 612$.

FURTHER READING
▶ **Linear programming**
▶ **Optimization**

REFERENCES
Dodge, Y. (2004) Optimization Appliquée. Springer, Paris

Laplace Distribution

The **random variable** X follows a Laplace distribution if its **density function** is of the following form:

$$f(x) = \frac{1}{2\Phi} \cdot \exp\left[\left(-\frac{|x - \theta|}{\Phi}\right)\right] \quad \Phi > 0,$$

where Φ is the dispersion **parameter** and θ is the **expected value**.

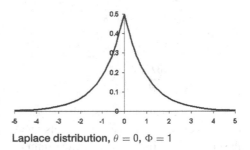

Laplace distribution, $\theta = 0$, $\Phi = 1$

The Laplace distribution is symmetric around its expected value θ, which is also the **mode** and the **median** of the distribution. The Laplace distribution is a **continuous probability distribution**.

HISTORY
This **continuous probability distribution** carries the name of its author. In 1774, **Laplace, P.S.** wrote a fundamental article about symmetric distributions with a view to describing the **errors** of measurement.

MATHEMATICAL ASPECTS
The **expected value** of a **random variable** X following a Laplace distribution is given by:

$$E[X] = \theta$$

and the **variance** is equal to:

$$\text{Var}(X) = 2 \cdot \Phi^2.$$

The standard form of the **density function** of the Laplace distribution is written:

$$f(x) = \frac{1}{\sigma\sqrt{2}} \cdot \exp\left[\left(-\sqrt{2} \cdot \frac{|x - \mu|}{\sigma}\right)\right]$$

$$\sigma > 0,$$

with

$$\mu = \theta \quad \text{and} \quad \sigma = \sqrt{2} \cdot \Phi.$$

This permits one to write the **distribution function** of the Laplace distribution given by:

$$F(x) = \begin{cases} \dfrac{1}{2}\exp\left[\left(\dfrac{x - \theta}{\Phi}\right)\right] \\ \qquad\qquad \text{if } x \leq \theta, \\ 1 - \dfrac{1}{2}\exp\left[\left(-\dfrac{x - \theta}{\Phi}\right)\right] \\ \qquad\qquad \text{if } x \geq \theta. \end{cases}$$

FURTHER READING
▶ **Continuous probability distribution**

REFERENCES
Laplace, P.S. de: Mémoire sur la probabilité des causes par les événements. Mem. Acad. Roy. Sci. (presented by various scientists) **6**, 621–656 (1774) (or Laplace, P.S. de (1891) Œuvres complètes, vol 8. Gauthier-Villars, Paris, pp. 27–65)

Laplace Pierre Simon De

Born in France in 1749 to a middle-class family, Marquis Pierre Simon de Laplace was one of the pioneers of **statistics**. Interested in mathematics, theoretical astronomy, **probability**, and statistics, his first publications appeared in the early 1770s. He became a member of the Academy of Science in 1785, and he also directed the Longitudes Bureau. In 1799, he was named Minister of Interior Affairs by Bonaparte, who in 1806 conferred upon him the title of count. Elected to the French Academy in 1816, he pursued his scientific work until his death, in Paris, in 1827.

Some main works and articles of de Laplace, P.S.:

1774 Mémoire sur la probabilité des causes par les événements. Oeuvres complètes, Vol. 8, pp. 27–65 (1891). Gauthier-Villars, Paris.

1781 Mémoire sur les probabilités. Oeuvres complètes, Vol. 9 pp. 383–485 (1891). Gauthier-Villars, Paris.

1805 Traité de Mécanique Céleste. Vol. 1–4. Duprot, Paris. Vol. 5. Bachelier, Paris.

1812 Théorie analytique des probabilités. Courcier, Paris.

1878–1912 Oeuvres complètes. 14 volumes. Gauthier-Villars, Paris.

Laspeyres Index

The Laspeyres **index** is a **composite index number** of price constructed by the weighted sum method. This index number represents the ratio of the sum of prices in the actual period n to the price sum in the reference period 0, these sums being weighted by the respective quantities of the reference period. Therefore the index number measures the relative price change of the goods, the respective quantities being considered unchanged. The Laspeyres index differs from the **Paasche index** on the choice of the weights: in the Paasche index, the weighting is done by the sold quantities of the current period Q_n, whereas in the Laspeyres index, it is done by the quantities sold in the reference period Q_0.

HISTORY

Laspeyres, Etienne, German economist and statistician of French origin, reformulated the index number developed by Lowe that would become known as the Laspeyres index in the mid-19th century. During the years 1864–1871 Laspeyres, Etienne worked on goods and their prices in Hamburg (he excluded services).

MATHEMATICAL ASPECTS

The Laspeyres **index** is calculated as follows:

$$I_{n/0} = \frac{\sum P_n \cdot Q_0}{\sum P_0 \cdot Q_0},$$

where P_n and Q_n are the prices and quantities sold in the current period and P_0 and Q_0 are the prices and quantities sold in the reference period, expressed in base 100:

$$I_{n/0} = \frac{\sum P_n \cdot Q_0}{\sum P_0 \cdot Q_0} \cdot 100.$$

The sum concerns the considered goods.
The Laspeyres model can also be used to calculate a quantity index (also called volume index). In this case, it is the prices that are constants and the quantities that are variables:

$$I_{n/0} = \frac{\sum Q_n \cdot P_0}{\sum Q_0 \cdot P_0} \cdot 100.$$

EXAMPLES

Consider the following table indicating the respective prices of three consumer goods in reference year 0 and in the current year n, as well as the quantities sold in the reference year:

Product	Quantity sold in 1970 (Q_0) (thousands)	Price (euros) 1970 (P_0)	Price (euros) 1988 (P_n)
Milk	50.5	0.20	1.20
Bread	42.8	0.15	1.10
Butter	15.5	0.50	2.00

From the following table we have

$\sum P_n Q_0 = 138.68$ and $\sum P_0 Q_0 = 24.27$.

Product	$\sum P_n Q_0$	$\sum P_0 Q_0$
Milk	60.60	10.10
Bread	47.08	6.42
Butter	31.00	7.75
Total	138.68	24.27

We can then find the Laspeyres index:

$$\begin{aligned} I_{n/0} &= \frac{\sum P_n \cdot Q_0}{\sum P_0 \cdot Q_0} \cdot 100 \\ &= \frac{138.68}{24.27} \cdot 100 = 571.4. \end{aligned}$$

In other words, according to the Laspeyres index, the price index of these goods has increased 471.4% ($571.4 - 100$) during the considered period.

FURTHER READING
▶ **Composite index number**
▶ **Fisher index**
▶ **Index number**
▶ **Paasche index**

REFERENCES

Laspeyres, E.: Hamburger Warenpreise 1850–1863 und die kalifornisch-australischen Geldentdeckung seit 1848. Jahrb. Natl. Stat. **3**, 81–118, 209–236 (1864)

Laspeyres, E.: Die Berechnung einer mittleren Waarenpreissteigerung. Jahrb. Natl. Stat. **16**, 296–314 (1871)

Latin Square

See **latin square designs**.

Latin Square Designs

A Latin square design is the arrangement of t treatments, each one repeated t times, in such a way that each treatment appears exactly one time in each row and each column in the design. We denote by Roman characters the treatments. Therefore the design is called a Latin square design. This kind of design is used to reduce systematic error due to rows (treatments) and columns.

HISTORY

According to Preece (1983), the history of Latin square dates back to 1624. In **statistics**, **Fisher, Ronald Aylmer** (1925) introduced the Latin square designs.

EXAMPLES

A farmer has in his property two fields where he wants to cultivate corn. He wishes to conduct an experiment involving his four different types of corn (**treatments**). Furthermore, he knows that his fields do not receive the same sunshine and humidity. There are two systematic error sources given that sunshine and humidity affect corn cultivation.

The farmer associates to each small piece of land a certain type of corn in such a way that each one appears once in each line and column:

A: grain 1
B: grain 2
C: grain 3
D: grain 4 .

The following figure shows the 4×4 Latin square where the Roman letters A, B, C, and D represent the four treatments above:

$$
\begin{array}{cccc}
A & B & C & D \\
B & C & D & A \\
C & D & A & B \\
D & A & B & C
\end{array}
$$

An **analysis of variance** will indicate whether, after deleting the lines (sun) and the column (humidity) effects, there exists a significant difference between the treatments.

FURTHER READING
▶ **Analysis of variance**
▶ **Design of experiments**

REFERENCES
Fisher, R.A.: Statistical Methods for Research Workers. Oliver & Boyd, Edinburgh (1925)

Preece, D.A.: Latin squares, Latin cubes, Latin rectangles, etc. In: Kotz, S., Johnson, N.L. (eds.) Encyclopedia of Statistical Sciences, vol. 4. Wiley, New York (1983)

Law of Large Numbers

The law of large numbers stipulates that the **mean** of a sum of independent and identically distributed **random variables** converges toward the **expected value** of their distribution when the **sample size** tends toward infinity.

HISTORY

The law of large numbers was first established by **Bernoulli, J.** in his work entitled *Ars Conjectandi*, published in 1713, in the framework of an empirical definition of **probability**. He stated that the relative **frequency** of an **event** converges, during the repetition of identical tasks (**Bernoulli distribution**), toward a number that consists in its probability.

The inequality of Bienaymé–Tchebychev was initially discovered by Bienaymé, J. (1853), before Tchebychev, P.L. (1867) discovered completely independently a few years later.

The general version of the law of large numbers is attributed to the Russian mathematician Khintchin, A. (1894–1959).

MATHEMATICAL ASPECTS

Let us first present the inequality of Bienaymé–Tchebychev:

Consider X a **random variable** with **expected value** μ and **variance** σ^2, both finite. For every real $\varepsilon > 0$, there is:

$$P(|X - \mu| \geq \varepsilon) \leq \frac{\sigma^2}{\varepsilon^2},$$

which means that the **probability** that a **random variable** X deviates from its expected value by an amount ε its **expected value** cannot be larger than σ^2/ε^2.

The importance of this inequality lies in the fact that it allows one to limit the **value** of certain **probabilities** in cases where only the **expected value** of the distribution is known, ultimately its **variance**. It is obvious that if the distribution itself is known, there is no need to use these limits since the exact value of these probabilities can be computed. With this inequality it is possible to compute the

convergence of the observed **mean** of a **random variable** toward its **expected value**. Consider n independent and identically distributed **random variables** $X_1, X_2, \ldots, X_n$ with **expected value** μ and **variance** σ^2; then for every $\varepsilon > 0$:

$$P\left(\left|\frac{X_1 + X_2 + \ldots + X_n}{n} - \mu\right| > \varepsilon\right) \xrightarrow[n \to \infty]{} 0.$$

DOMAINS AND LIMITATIONS

The law of large numbers underlines the fact that there persists a **probability**, even if very small, of substantial deviation between the empirical **mean** of a series of **events** and its **expected value**. It is for this reason that the term *weak law of large numbers* is used.

In fact, it is possible to demonstrate that this is not the case and that the convergence is quasicertain. This result, stronger than the previous one, is called the *strong law of large numbers* and is written:

$$P\left(\lim_{n \to \infty} \bar{X} = \mu\right) = 1.$$

EXAMPLES

The law of large numbers states that it is always possible to find a **value** for n so that the **probability** that $\bar{X}$ is included in an **interval** $\mu \pm \varepsilon$ is as large as desired.

Let us give an example of a **probability distribution** with **variance** $\sigma^2 = 1$. By choosing an **interval** $\varepsilon = 0.5$ and a **probability** of 0.05, we can write:

$$P(|\bar{X} - \mu| \geq 0.5) = 0.05.$$

By the inequality of Tchebychev, we know that

$$P(|\bar{X} - \mu| \geq \varepsilon) \leq \frac{\sigma^2}{n\varepsilon^2}.$$

We can therefore establish the following inequality:

$$0.05 \leq \frac{1}{n \cdot 0.25},$$

from which:

$$0.05 \cdot 0.25 \le 1/n,$$
$$n \ge 80.$$

FURTHER READING
► **Central limit theorem**
► **Convergence theorem**
(*Théorème de convergence*)

REFERENCES
Bernoulli, J.: Ars Conjectandi, Opus Posthumum. Accedit Tractatus de Seriebus infinitis, et Epistola Gallice scripta de ludo Pilae recticularis. Impensis Thurnisiorum, Fratrum, Basel (1713)

Bienaymé, I.J.: Considérations à l'appui de la découverte de Laplace sur la loi de probabilité dans la méthode des moindres carrés. Comptes Rendus de l'Académie des Sciences, Paris **37**, 5–13 (1853); réédité en 1867 dans le journal de Liouville précédant la preuve des Inégalités de Bienaymé-Chebyshev, Journal de Mathématiques Pures et Appliquées **12**, 158–176

Tchebychev, P.L.: Des valeurs moyennes. J. Math. Pures Appl. **12**(2), 177–184 (1867). Publié simultanément en Russe, dans Mat. Sbornik **2**(2), 1–9

Least Absolute Deviation Regression

The least absolute deviations method (the LAD method) is one of the principal alternatives to the least-squares methods when one seeks to estimate regression parameters. The goal of the LAD regression is to provide a **robust estimator**.

HISTORY
The least absolute deviation regression was introduced around 50 years before the **least-squares method**, in 1757, by **Roger Joseph Boscovich**. He used this procedure while trying to reconcile incoherent measures that were used to estimate the shape of the earth. **Pierre Simon de Laplace** adopted this method 30 years later, yet it was obscured under the shadow of the least-squares method developed by **Adrien Marie Legendre** and **Carl Friedrich Gauss**. The easy calculus of the least-squares method made least squares much more popular than the LAD method. Yet in recent years and with advances in statistical computing, the LAD method can be easily used.

MATHEMATICAL ASPECTS
We treat here only simple LAD regression cases.

Consider the simple **regression** model: $y_i = \beta_0 + \beta_1 x_i + \varepsilon_i$. The LAD method implements the following criterion: choose $\widehat{\beta}_0$ and $\widehat{\beta}_1$ such that they minimize the sum of the residual absolute values, that is,

$$\sum |e_i| = \sum \left| y_i - \widehat{\beta}_0 - \widehat{\beta}_1 x_i \right|,$$

where e_i denotes the ith residual. The particularity of this method is the fact that there is no explicit formula to compute $\widehat{\beta}_0$ and $\widehat{\beta}_1$. LAD estimates can be computed by applying an iterative **algorithm**.

Simple LAD Algorithm
The essential point that is used by the LAD algorithm is the fact that the regression line always crosses at least two data points.

Let (x_1, y_1) be our chosen data point. We search the best line that passes through this point. This line crosses at least one other point, (x_2, y_2).

The next step is to search the best line, with respect to the sum $\sum |e_i|$, that crosses this time (x_2, y_2). This line crosses now at least one other point, (x_3, y_3). We iteratively continue and seek the best line that crosses (x_3, y_3) and so forth. During the iterations, the quantity $\sum |e_i|$ decreases. If, for an iteration, the obtained line is identical to the previous one, we can conclude that this is the best line.

Construction of Best Line Crossing (x_1, y_1) in Terms of $\sum |e_i|$

Since this line crosses another data point, we should find the point (x_k, y_k) for which the line

$$\widehat{y}(x) = y_1 + \frac{y_k - y_1}{x_k - x_1}(x - x_1) \, ,$$

with slope

$$\widehat{\beta}_1 = \frac{y_k - y_1}{x_k - x_1}$$

and intercept

$$\widehat{\beta}_0 = y_1 - \widehat{\beta}_1 x_1 \, ,$$

is the best one in terms of $\sum |e_i|$. To find that point, we rename the $(n - 1)$ candidate points $(x_2, y_2), \ldots, (x_n, y_n)$ in such a way that

$$\frac{y_2 - y_1}{x_2 - x_1} \le \frac{y_3 - y_1}{x_3 - x_1} \le \ldots \le \frac{y_n - y_1}{x_n - x_1} \, .$$

We define $T = \sum_{i=1}^{n} |x_i - x_1|$. The searched point (x_k, y_k) is now determined by the index k for which

$$\begin{cases} |x_2 - x_1| + \ldots + |x_{k-1} - x_1| & < \frac{T}{2} \\ |x_2 - x_1| + \ldots \\ + |x_{k-1} - x_1| + |x_k - x_1| & > \frac{T}{2} \, . \end{cases}$$

This condition guarantees that $\widehat{\beta}_1$ minimizes the quantity

$$\sum_{i=1}^{n} |(y_i - y_1) - \widehat{\beta}_1 (x_i - x_1)|$$

analogously to $\sum |e_i|$ for the regression lines passing through (x_1, y_1). The $\widehat{\beta}_0$ is computed in such a way that the regression line crosses (x_1, y_1). We can equally verify that it passes through the data point (x_k, y_k). We just have to rename the point (x_k, y_k) by (x_2, y_2) and restart.

DOMAINS AND LIMITATIONS

Two particular cases can appear when one runs the LAD regression: the **nonunique solution** and the **degeneracy**. Nonunique solution means that there are at least two lines minimizing the sum of the residual absolute values. Degeneracy means that the LAD regression line passes through more than two data points.

EXAMPLES

Let us take an example involving 14 North and Central American countries. The following table contains the data for each country, the birthrate (number of births per 1000 people) together with the urbanization rate (percentage of population living in cities with more than 100000 inhabitants) in 1980.

Table: Urbanization rate and birthrate

Obs. i	Country	Urbanization rate x_i	Birthrate y_i
1	Canada	55.0	16.2
2	Costa Rica	27.3	30.5
3	Cuba	33.3	16.9
4	USA	56.5	16.0
5	El Salvador	11.5	40.2

Obs. i	Country	Urbaniza-tion rate x_i	Birth-rate y_i
6	Guatemala	14.2	38.4
7	Haiti	13.9	41.3
8	Honduras	19.0	43.9
9	Jamaica	33.1	28.3
10	Mexico	43.2	33.9
11	Nicaragua	28.5	44.2
12	Trinidad/Tobago	6.8	24.6
13	Panama	37.7	28.0
14	Dominican Republic	37.1	33.1

Source: Birkes and Dodge (1993)

We try to estimate the birthrate y_i as a function of the urbanization rate x_i using LAD regression, which is to construct the LAD regression model

$$y_i = \beta_0 + \beta_1 x_i + \varepsilon_i .$$

To use the LAD algorithm to estimate β_0 and β_1, in the fist step we must search for the best line crossing at least one point, for example Canada (randomly chosen). We have $(x_1, y_1) = (55.0, 16.2)$. We compute the slopes $\frac{y_i - 16.2}{x_i - 55.0}$ for all 13 remaining countries. For example, for Costa Rica, the obtained result is $\frac{30.5 - 16.2}{27.3 - 55.0} = -0.5162$. The 13 countries are now reordered in ascending order according to the computed slopes, as can be seen in the table below. The point (x_2, y_2) is Mexico, the point (x_3, y_3) is Nicaragua, and so forth. In what follows we compute the quantity $T = \sum |x_i - 55.0| = 355.9$ and $\frac{T}{2} = 177.95$. This has to do with defining the country k that satisfies the condition

$$\begin{cases} |x_2 - x_1| + \ldots + |x_{k-1} - x_1| & < \frac{T}{2} , \\ |x_2 - x_1| + \ldots \\ + |x_{k-1} - x_1| + |x_k - x_1| & > \frac{T}{2} . \end{cases}$$

Since

$$\sum_{j=2}^{8} |x_j - x_1| = 172.5 < \frac{T}{2}$$

and

$$\sum_{j=2}^{9} |x_j - x_1| = 216.0 > \frac{T}{2} ,$$

we can conclude that $k = 9$ and that the best regression line passes through Canada (x_1, y_1) and El Salvador (x_9, y_9). Hence:

$$\widehat{\beta_1} = \frac{y_9 - y_1}{x_9 - x_1} = \frac{40.2 - 16.2}{11.5 - 55.0} = -0.552 ,$$

$$\widehat{\beta_0} = y_1 - \widehat{\beta_1} x_1 = 16.2 - (-0.552) \cdot 55.0$$
$$= 46.54 ,$$

and the LAD regression line is:

$$\widehat{y}(x) = 46.54 - 0.552x .$$

Table: Computation to find LAD regression line

Obs. i	Country	$\frac{y_i - y_1}{x_i - x_1}$	$|x_i - x_1|$	$\sum_{j=2}^{i} |x_j - x_1|$
2	Mexico	−1.5000	11.8	11.8
3	Nicaragua	−1.0566	26.5	38.3
4	Domin. Rep.	−0.9441	17.9	56.2
5	Honduras	−0.7694	36.0	92.2
6	Panama	−0.6821	17.3	109.5
7	Haiti	−0.6107	41.1	150.6
8	Jamaica	−0.5525	21.9	172.5
9	El Salvador	−0.5517	43.5	216.0
10	Guatemala	−0.5447	40.8	256.8
11	Costa Rica	−0.5162	27.7	284.5
12	Trinidad/Tobago	−0.1743	48.2	332.7
13	USA	−0.1333	1.5	334.2
14	Cuba	−0.0323	21.7	355.9

To improve our first estimate, the next step consists in finding the best line crossing El

L

Salvador. We rename this point $(x_1, y_1) = (11.5, 40.2)$ and reorder the 13 countries in ascending order according to their slopes $\frac{y_i - 11.5}{x_i - 40.2}$, and, similarly, we obtain the best regression line that passes through El Salvador as well as through the United States. Hence,

$$\widehat{\beta_1} = \frac{40.2 - 16.0}{11.5 - 56.5} = -0.538,$$

$$\widehat{\beta_0} = 40.2 - (-0.538) \cdot 11.5 = 46.38,$$

and the regression line is

$$\widehat{y}(x) = 46.38 - 0.538x.$$

The third step consists in finding the best line passing through the United States. Since we find (in a similar way as above) that this regression line passes through El Salvador, our algorithm stops and the regression line $\widehat{y}(x) = 46.38 - 0.538x$ is the LAD regression line of our problem.

We note here that the line computed using the least-squares method for these data is given by

$$\widehat{y}(x) = 42.99 - 0.399x.$$

Note that with this regression line Cuba, Nicaragua, and Trinidad/Tobago have abnormally large residual values. By deleting these three points and recomputing the least-squares regression line, we obtain

$$\widehat{y}(x) = 48.52 - 0.528x,$$

which is very close to the LAD regression line.

FURTHER READING
▶ **Analysis of variance**
▶ **Least squares**
▶ **Multiple linear regression**
▶ **Regression analysis**

▶ **Residual**
▶ **Robust estimation**
▶ **Simple linear regression**

REFERENCES
Birkes, D., Dodge, Y.: Alternative Methods of Regression. Wiley, New York (1993)

Least Significant Difference Test

The least significant difference (LSD) test is used in the context of the **analysis of variance**, when the F-ratio suggests rejection of the **null hypothesis** H_0, that is, when the difference between the population means is significant.

This test helps to identify the populations whose means are statistically different. The basic idea of the test is to compare the populations taken in pairs. It is then used to proceed in a **one-way or two-way analysis of variance**, given that the null hypothesis has already been rejected.

HISTORY
The LSD test was developed by **Fisher, Ronald Aylmer** (1935), who wanted to know which treatments had a significant effect in an **analysis of variance**.

MATHEMATICAL ASPECTS
If in an **analysis of variance** the F-ratio leads to rejection of the null hypothesis H_0, we can perform the LSD test in order to detect which means have led H_0 to be rejected.

The test consists in a pairwise comparison of the means. In general terms, the **standard deviation** of the difference between the mean of group i and the mean of group j, for $i \neq j$, is equal to:

$$\sqrt{s_I^2 \left(\frac{1}{n_i} + \frac{1}{n_j} \right)},$$

where s_I^2 is the **estimation** of the **variance** inside the groups

$$s_I^2 = \frac{SCI}{N-k} = \frac{\sum_{i=1}^{k} \sum_{j=1}^{n_i} \left(Y_{ij} - \bar{Y}_{i.} \right)^2}{N-k},$$

where SCI is the sum of squares inside the groups, N is the total number of observations, k is the number of groups, Y_{ij} is the jth observation of group i, $\bar{Y}_{i.}$ is the mean of group i, n_i is the number of observations in group i, and n_j is the number of observations in group j.

The ratio

$$\frac{\bar{Y}_{i.} - \bar{Y}_{j.}}{\sqrt{s_I^2 \left(\frac{1}{n_i} + \frac{1}{n_j} \right)}}$$

follows the **Student** distribution with $N - k$ degrees of freedom. The difference between a pair of means is significant when

$$\left| \bar{Y}_{i.} - \bar{Y}_{j.} \right| \geq \sqrt{s_I^2 \left(\frac{1}{n_i} + \frac{1}{n_j} \right)} \cdot t_{N-k, 1-\frac{\alpha}{2}},$$

with $t_{N-k, 1-\frac{\alpha}{2}}$ denoting the **value** of a **Student** variate with $N - k$ degrees of freedom for a **significance level** set to α.

The quantity $\sqrt{s_I^2 (\frac{1}{n_i} + \frac{1}{n_j})} \cdot t_{N-k, 1-\frac{\alpha}{2}}$ is called the LSD.

DOMAINS AND LIMITATIONS

Like the **analysis of variance**, the LSD test demands independent and identically distributed normal variables with constant variance.

EXAMPLES

During boiling, all-butter croissants absorb fat in variable quantities. We want to verify whether the absorbed quantity depends on the type of fat content (animal or vegetable fat). We prepare for our experiment four different types of fat, and we boil six all-butter croissants for each croissant type.

Here are the results of our experiment:

Type of fat			
1	2	3	4
64	78	75	55
72	91	93	66
68	97	78	49
77	82	71	64
56	85	63	70
95	77	76	68

We run a **one-way analysis of variance** that permits us to test the **null hypothesis**

$$H_0: \quad \mu_1 = \mu_2 = \mu_3 = \mu_4,$$

which is evidently rejected.

Since H_0 is rejected, we run the LSD test seeking to identify which means caused the rejection of H_0.

In this example, the number of observations is the same for each group ($n_1 = n_2 = n_3 = n_4 = n = 6$). Consequently, $\sqrt{s_I^2 \left(\frac{1}{n_i} + \frac{1}{n_j} \right)}$ is simplified to $\sqrt{\frac{2s_I^2}{n}}$.

The value for the corresponding LSD test is equal to:

$$\sqrt{\frac{2s_I^2}{n}} \cdot t_{N-k, 1-\frac{\alpha}{2}}.$$

The results of the **analysis of variance** provide the computational elements needed to find the value of s_I^2:

$$s_I^2 = \frac{SCI}{N-k} = \frac{\sum\limits_{i=1}^{k}\sum\limits_{j=1}^{n_i}\left(Y_{ij} - \bar{Y}_{i.}\right)^2}{N-k}$$

$$= \frac{2018}{20}$$

$$= 100.9.$$

If we choose a **significance level** $\alpha = 0.05$, the **Student** value for $N - k = 20$ degrees of freedom is equal to:

$$t_{N-k,1-\frac{\alpha}{2}} = t_{20,0.975} = 2.086.$$

Hence the value for the LSD statistic is:

$$LSD = \sqrt{\frac{2s_I^2}{n}} \cdot t_{20,0.975}$$

$$= \sqrt{\frac{2 \cdot 100.9}{6}} \cdot 2.086$$

$$= 12.0976.$$

The difference between a pair of means $\bar{Y}_i$ and $\bar{Y}_j$ is significant if

$$\left|\bar{Y}_{i.} - \bar{Y}_{j.}\right| \geq 12.0976.$$

We can summarize the results in a table containing the mean differences, the values for the LSD statistic, and the conclusion.

Difference		LSD	Significant
$\left|\bar{Y}_{1.}-\bar{Y}_{2.}\right|$	$= |72 - 85| = 13$	12.0976	Yes
$\left|\bar{Y}_{1.}-\bar{Y}_{3.}\right|$	$= |72 - 76| = 4$	12.0976	No
$\left|\bar{Y}_{1.}-\bar{Y}_{4.}\right|$	$= |72 - 62| = 10$	12.0976	No
$\left|\bar{Y}_{2.}-\bar{Y}_{3.}\right|$	$= |85 - 76| = 9$	12.0976	No
$\left|\bar{Y}_{2.}-\bar{Y}_{4.}\right|$	$= |85 - 62| = 23$	12.0976	Yes
$\left|\bar{Y}_{3.}-\bar{Y}_{4.}\right|$	$= |76 - 62| = 14$	12.0976	Yes

We verify that only the differences $\bar{Y}_{1.} - \bar{Y}_{2.}$, $\bar{Y}_{2.} - \bar{Y}_{4.}$, and $\bar{Y}_{3.} - \bar{Y}_{4.}$ are significant.

FURTHER READING

▶ **Analysis of variance**
▶ **Student distribution**
▶ **Student table**

REFERENCES

Dodge, Y., Thomas, D.R.: On the performance of non-parametric and normal theory multiple comparison procedures. Sankhya **B42**, 11–27 (1980)

Fisher, R.A.: The Design of Experiments. Oliver & Boyd, Edinburgh (1935)

Miller, R.G., Jr.: Simultaneous Statistical Inference, 2nd edn. Springer, Berlin Heidelberg New York (1981)

Least Squares

See **least-squares method**.

Least-Squares Method

The least-squares method consists in minimizing the sum of the squared residuals. The latter correspond to the squared deviations between estimated and observed values.

HISTORY

The least-squares method was introduced by **Legendre, Adrien-Marie** (1805).

MATHEMATICAL ASPECTS

Consider the **multiple linear regression model**:

$$Y_i = \beta_0 + \sum_{j=1}^{p-1} \beta_j \cdot X_{ji} + \varepsilon_i, \quad i = 1,\dots,n,$$

where Y_i is the **dependent variable**, $X_{ji}, j = 1,\dots,p - 1$ are the explanatory or independent variables, $\beta_j, j = 0,\dots,p - 1$ are the parameters to be estimated, and ε_i is the **error** term.

Parameter **estimation** using the least-squares method consists in minimizing the sum of the squared errors (residuals):

$$\text{minimum} f\left(\beta_0, \ldots, \beta_{p-1}\right),$$

where:

$$f\left(\beta_0, \ldots, \beta_{p-1}\right) = \sum_{i=1}^{n} \varepsilon_i^2.$$

The residual sum of squares is equal to:

$$f\left(\beta_0, \ldots, \beta_{p-1}\right) = \sum_{i=1}^{n} \varepsilon_i^2$$

$$= \sum_{i=1}^{n} \left(Y_i - \beta_0 - \sum_{j=1}^{p-1} \beta_j \cdot X_{ji}\right)^2.$$

We find the estimated parameters of the function by equating to zero the partial derivatives of this function with respect to each parameter:

$$\frac{\partial f\left(\beta_0, \ldots, \beta_{p-1}\right)}{\beta_0} = 0,$$

$$\frac{\partial f\left(\beta_0, \ldots, \beta_{p-1}\right)}{\beta_1} = 0,$$

$$\vdots$$

$$\frac{\partial f\left(\beta_0, \ldots, \beta_{p-1}\right)}{\beta_{p-1}} = 0.$$

Suppose we have n observations and that $p = 2$:

$$\left(X_1, Y_1\right), \left(X_2, Y_2\right), \ldots, \left(X_n, Y_n\right).$$

The equation relating Y_i and X_i can be defined from the linear model:

$$Y_i = \beta_0 + \beta_1 \cdot X_i + \varepsilon_i, \quad i = 1, \ldots, n.$$

The sum of the squared deviations from the estimated regression line is:

$$f\left(\beta_0, \beta_1\right) = \sum_{i=1}^{n} \varepsilon_i^2$$

$$= \sum_{i=1}^{n} \left(y_i - \beta_0 - \beta_1 \cdot X_i\right)^2.$$

Mathematically, we determine β_0 and β_1 by taking the partial derivatives of function f with respect to parameters β_0 and β_1 and setting them equal to zero.

Partial derivative with respect to β_0:

$$\frac{\partial f}{\partial \beta_0} = -2 \cdot \sum_{i=1}^{n} \left(Y_i - \beta_0 - \beta_1 \cdot X_i\right).$$

Partial derivative with respect to β_1:

$$\frac{\partial f}{\partial \beta_1} = -2 \cdot \sum_{i=1}^{n} X_i \cdot \left(y_i - \beta_0 - \beta_1 \cdot X_i\right).$$

Hence, the estimated **values** for β_0 and β_1, denoted as $\hat{\beta}_0$ and $\hat{\beta}_1$, are given as the solution to the following equations:

$$\sum_{i=1}^{n} \left(Y_i - \hat{\beta}_0 - \hat{\beta}_1 \cdot X_i\right) = 0,$$

$$\sum_{i=1}^{n} X_i \cdot \left(Y_i - \hat{\beta}_0 - \hat{\beta}_1 \cdot X_i\right) = 0.$$

Developing these two equations, we obtain:

$$\sum_{i=1}^{n} Y_i = n \cdot \hat{\beta}_0 + \hat{\beta}_1 \cdot \sum_{i=1}^{n} X_i$$

and

$$\sum_{i=1}^{n} \left(X_i \cdot Y_i\right) = \hat{\beta}_0 \cdot \sum_{i=1}^{n} X_i + \hat{\beta}_1 \cdot \sum_{i=1}^{n} X_i^2,$$

which we call the **normal equations** of the least-squares regression line.

Given

$$\bar{X} = \frac{\sum_{i=1}^{n} X_i}{n} \quad \text{and} \quad \bar{Y} = \frac{\sum_{i=1}^{n} Y_i}{n},$$

the solution of the normal equations provides the estimates of the following parameters:

$$\hat{\beta}_0 = \bar{Y} - \hat{\beta}_1 \cdot \bar{X},$$

$$\hat{\beta}_1 =$$

$$\frac{n \cdot \sum_{i=1}^{n} (X_i \cdot Y_i) - \left(\sum_{i=1}^{n} X_i\right) \cdot \left(\sum_{i=1}^{n} Y_i\right)}{n \cdot \sum_{i=1}^{n} X_i^2 - \left(\sum_{i=1}^{n} X_i\right)^2}$$

$$= \frac{\sum_{i=1}^{n} (X_i - \bar{X})(Y_i - \bar{Y})}{\sum_{i=1}^{n} (X_i - \bar{X})^2}.$$

The least-squares estimates for the dependent variable are:

$$\hat{Y}_i = \hat{\beta}_0 + \hat{\beta}_1 \cdot X_i, \quad i = 1, \ldots, n$$

or

$$\hat{Y}_i = \bar{Y} + \hat{\beta}_1 \cdot (X_i - \bar{X}), \quad i = 1, \ldots, n.$$

According to this equation, the least-squares regression line passes through the point $(\bar{X}, \bar{Y})$, which is called the barycenter or center of gravity for the scatter cloud of the data points.

We can, equally, express the multiple linear regression model in terms of vectors and matrices:

$$\mathbf{Y} = \mathbf{X} \cdot \boldsymbol{\beta} + \boldsymbol{\varepsilon},$$

where $\mathbf{Y}$ is the $(n \times 1)$ response **vector** (the dependent variable), $\mathbf{X}$ is the $(n \times p)$ independent variables **matrix**, $\boldsymbol{\varepsilon}$ is the $(n \times 1)$ vector of the error term, and $\boldsymbol{\beta}$ is the $(p \times 1)$ vector of the parameters to be estimated.

The application of the least-squares method consists in minimizing the following equation:

$$\boldsymbol{\varepsilon}' \cdot \boldsymbol{\varepsilon} = (\mathbf{Y} - \mathbf{X} \cdot \boldsymbol{\beta})' \cdot (\mathbf{Y} - \mathbf{X} \cdot \boldsymbol{\beta})$$
$$= \mathbf{Y}' \cdot \mathbf{Y} - \boldsymbol{\beta}' \cdot \mathbf{X}' \cdot \mathbf{Y} - \mathbf{Y}' \cdot \mathbf{X} \cdot \boldsymbol{\beta}$$
$$\quad + \boldsymbol{\beta}' \cdot \mathbf{X}' \cdot \mathbf{X} \cdot \boldsymbol{\beta}$$
$$= \mathbf{Y}' \cdot \mathbf{Y} - 2 \cdot \boldsymbol{\beta}' \cdot \mathbf{X}' \cdot \mathbf{Y}$$
$$\quad + \boldsymbol{\beta}' \cdot \mathbf{X}' \cdot \mathbf{X} \cdot \boldsymbol{\beta}.$$

The normal equations are given, in matrix form, by:

$$\mathbf{X}' \cdot \mathbf{X} \cdot \hat{\boldsymbol{\beta}} = \mathbf{X}' \cdot \mathbf{Y}.$$

If $\mathbf{X}' \cdot \mathbf{X}$ is invertible, the parameters $\boldsymbol{\beta}$ are estimated by:

$$\hat{\boldsymbol{\beta}} = (\mathbf{X}' \cdot \mathbf{X})^{-1} \cdot \mathbf{X}' \cdot \mathbf{Y}.$$

EXAMPLES
See **simple linear regression**.

FURTHER READING
▶ **Estimation**
▶ **Multiple linear regression**
▶ **Regression analysis**
▶ **Simple linear regression**

REFERENCES
Legendre, A.M.: Nouvelles méthodes pour la détermination des orbites des comètes. Courcier, Paris (1805)

Legendre Adrien Marie

Born in 1752 in Paris, French mathematician Legendre, Adrien Marie succeeded **de Laplace, P.S.** as professor in mathematics, first at the Ecole Militaire and then at the Ecole Normale of Paris. He was interested in the theory and practice of astronomy and geodesy. From 1792 he was part of the

French commission in charge of measuring the length of a meridian quadrant from the North Pole to the Equator passing through Paris.

He died in 1833, but in the history of **statistics** his name remained attached to the publication of his work in 1805, which included an appendix titled "On the **least-squares** method." This publication was at the root of the controversy that opposed him to **Gauss, C.F.**, who vied with him over credit for the discovery of this method.

Main work of Legendre, A.M.:

1805 Nouvelles méthodes pour la détermination des orbites des comètes. Courcier, Paris.

REFERENCES

Plackett, R.L.: Studies in the history of probability and statistics. In: Kendall, M., Plackett, R.L. (eds.) The discovery of the method of least squares. vol. II. Griffin, London (1977)

Lehmann Erich

Lehmann, Erich was born in Strasbourg, France, in 1917 and went to the United States in 1940. He has made fundamental contributions to the theory of statistics. His research interests are the statistical theory and history and philosophy of statistics. He is the author of *Basic Concepts of Probability and Statistics*, *Elements of Finite Probability*, *Nonparametrics: Statistical Methods Based on Ranks*, *Testing Statistical Hypotheses*, and *Theory of Point Estimation*. He is a member of the American Academy of Arts and Sciences and of the National Academy of Science (NAS), having been elected to the NAS in 1978. He has been editor of the Annals of Mathematical Statistics and President of the Institute of Mathematical Statistics.

Selected works and publications of Erich Lehmann:

1983 Theory of Point Estimation, 2nd edn. Wiley, NY.

1986 Testing Statistical Hypotheses. 2nd edn. In: Series in Probability and Mathematical Statistics. Wiley, NY.

1993 (with Hodges, J.L.) Testing Statistical Hypotheses. Chapman & Hall, NY.

FURTHER READING
▶ **Estimation**
▶ **Hypothesis testing**

Level of Significance

See **significance level**.

Leverage Point

In **regression analysis**, we call the leverage point an **observation** i for which the **estimated** response value $\hat{Y}_i$ is influenced by the **value** of the corresponding **independent variable** X_i. The notion of leverage point is equivalent to that of leverage in physics. That is, even a small modification to the value of observation i can seriously affect the estimation of $\widehat{y}_i$.

MATHEMATICAL ASPECTS

The diagonal elements of **matrix H** in regression analysis reflect the influence of the **observation** on the estimation of the response. We associate in principle the diagonal elements h_{ii} of matrix $\mathbf{H}$ with the notion of the leverage point. Since the **trace** of matrix $\mathbf{H}$ is equal to the number of parameters to estimate, we expect to have h_{ii} close to p/n. The h_{ii} should be close to p/n if all the observations have the same influence on the estimates. If for an observation i we obtain h_{ii} larger than p/n, then observation i is considered a leverage point. In practice, an observation i is called a high leverage point if:

$$h_{ii} > \frac{2p}{n}.$$

Consider the following linear regression model in matrix notation:

$$\mathbf{Y} = \mathbf{X} \cdot \boldsymbol{\beta} + \boldsymbol{\varepsilon},$$

where $\mathbf{Y}$ is the $(n \times 1)$ response **vector** (the dependent variable), $\mathbf{X}$ is the $(n \times p)$ independent variables **matrix**, $\boldsymbol{\varepsilon}$ is the $(n \times 1)$ vector of the error term, and $\boldsymbol{\beta}$ is the $(p \times 1)$ vector of the parameters to be estimated. The **estimation** $\hat{\boldsymbol{\beta}}$ of **vector** $\boldsymbol{\beta}$ is given by:

$$\hat{\boldsymbol{\beta}} = \left(\mathbf{X}'\mathbf{X}\right)^{-1}\mathbf{X}'\mathbf{Y}.$$

Matrix $\mathbf{H}$ is defined by:

$$\mathbf{H} = \mathbf{X}\left(\mathbf{X}'\mathbf{X}\right)^{-1}\mathbf{X}'.$$

In particular, the diagonal elements h_{ii} are defined by:

$$h_{ii} = x_i \left(\mathbf{X}'\mathbf{X}\right)^{-1} x_i',$$

where x_i is the ith line of $\mathbf{X}$.
We call the leverage point each observation i for which:

$$h_{ii} = x_i \left(\mathbf{X}'\mathbf{X}\right)^{-1} x_i' > 2 \cdot \frac{p}{n}.$$

DOMAINS AND LIMITATIONS

Huber (1981) refers to leverage points when the diagonal element of the **matrix** $\mathbf{X}\left(\mathbf{X}'\mathbf{X}\right)^{-1}\mathbf{X}'$ is larger than 0.2, the critical limit not being dependent either on the number of the parameters p or on the number of observations n.

EXAMPLES

Consider the following table, where Y is a **dependent variable** explained by the **independent variable** X:

X	Y
50	6
52	8
55	9
75	7
57	8
58	10

The linear regression model in matrix notation is:

$$\mathbf{Y} = \mathbf{X} \cdot \boldsymbol{\beta} + \boldsymbol{\varepsilon},$$

where

$$\mathbf{Y} = \begin{bmatrix} 6 \\ 8 \\ 9 \\ 7 \\ 8 \\ 10 \end{bmatrix}, \quad \mathbf{X} = \begin{bmatrix} 1 & 50 \\ 1 & 52 \\ 1 & 55 \\ 1 & 75 \\ 1 & 57 \\ 1 & 58 \end{bmatrix},$$

$\boldsymbol{\varepsilon}$ is the (6×1) residual **vector**, and $\boldsymbol{\beta}$ is the (2×1) parameter vector.
We get **matrix H** by:

$$\mathbf{H} = \mathbf{X}\left(\mathbf{X}'\mathbf{X}\right)^{-1}\mathbf{X}'.$$

We find:

$$\mathbf{H} = \begin{bmatrix} 0.32 & 0.28 & 0.22 & -0.17 & 0.18 & 0.16 \\ 0.28 & 0.25 & 0.21 & -0.08 & 0.18 & 0.16 \\ 0.22 & 0.21 & 0.19 & 0.04 & 0.17 & 0.17 \\ -0.17 & -0.08 & 0.04 & 0.90 & 0.13 & 0.17 \\ 0.18 & 0.18 & 0.17 & 0.13 & 0.17 & 0.17 \\ 0.16 & 0.16 & 0.17 & 0.17 & 0.17 & 0.17 \end{bmatrix}.$$

We write the diagonal of matrix $\mathbf{H}$ on the vector $\mathbf{h}$:

$$\mathbf{h} = \begin{bmatrix} 0.32 \\ 0.25 \\ 0.19 \\ 0.90 \\ 0.17 \\ 0.17 \end{bmatrix}.$$

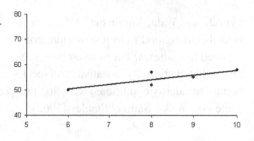

In this case there are two parameters ($p = 2$) and six observations ($n = 6$), that is, the **critical value** equals $\frac{4}{6} = 0.67$.

Comparing the components of $\mathbf{h}$ to this **value**, we notice that, with the exception of observation $i = 4$, all observations have $h_{ii} < 0.67$. The fourth observation is a leverage point for the problem under study. The following figure, where variables X and Y are represented on the two axes, illustrates the fact that the point $(75, 7)$ is a high leverage point.

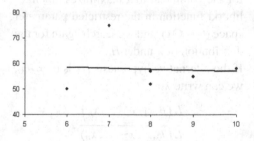

Inspecting this figure we notice that the regression line is highly influenced by this leverage point and does not represent correctly the data set under study.

In the following figure we have deleted this point from the data set. We can see that the new regression line gives a much better representation of the data set.

FURTHER READING
▶ **Hat matrix**
▶ **Matrix**
▶ **Multiple linear regression**
▶ **Regression analysis**
▶ **Simple linear regression**

REFERENCES
Huber, P.J.: Robust Statistics. Wiley, New York (1981)

Mosteller, F., Tukey, J.W.: Data Analysis and Regression: A Second Course in Statistics. Addison-Wesley, Reading, MA (1977)

Likelihood Ratio Test

The likelihood ratio test is a **hypothesis test**. It allows to test general hypotheses concerning the parameters of interest of a parametric family as well as to test two different models built on the same data. The main idea of this test is the following: compute the **probability** of observing the data under the null hypothesis H_0 and under the **alternative hypothesis** using the likelihood function.

HISTORY
It is **Neyman, Jerzy** and **Pearson, Egon Sharpe** (1928) who came up with the idea of using the likelihood ratio **statistic** to test

hypotheses. Wald, Abraham (1941) generalized the likelihood ratio test to more complicated hypotheses. The asymptotic results on the distribution of this statistic (subject to certain regularity conditions) were first presented by Wilks, Samuel Stanley (1962).

MATHEMATICAL ASPECTS

In the likelihood ratio test, the **null hypothesis** is rejected if the likelihood under the **alternative hypothesis** is significantly larger than the likelihood under the null hypothesis. Hence, the problem is not finding the most adequate parameter but knowing if under the alternative hypothesis we get a significantly better estimate.

The Neyman–Pearson theorem states that, for all **significant levels** α, the likelihood ratio test (for testing the simple hypothesis) has more power than any other test.

Let $x_1, x_2, \ldots, x_n$ be n independent observations ($n > 1$) of a **random variable** X following a probability distribution with **parameter** θ (or a parameter vector $\boldsymbol{\theta}$). We denote by $L(\theta; x_1, x_2, \ldots, x_n)$ the likelihood function that, for fixed x_i, depends only upon θ, and we denote by Θ the space of all possible values of θ and Θ_0, with Θ_1 are subspace of Θ.

We distinguish between:

Simple Null Hypothesis vs. a Simple Alternative Hypothesis

$$H_0: \quad \theta = \theta_0$$
$$\text{vs. } H_1: \quad \theta = \theta_1.$$

In this case, the likelihood ratio is:

$$\begin{aligned}
\lambda &= \lambda(x_1, x_2, \ldots, x_n) \\
&= \frac{L(\theta_0; x_1, x_2, \ldots, x_n)}{L(\theta_1; x_1, x_2, \ldots, x_n)}.
\end{aligned}$$

The likelihood ratio for a **significance level** α is defined by the decision rule:

$$\text{Reject } H_0 \quad \text{if} \quad \lambda > \lambda_0,$$

where λ_0 is the critical value defined by the significance level (α) and the distribution of λ under the null hypothesis:

$$\alpha = P(\lambda(x_1, x_2, \ldots, x_n) \geq \lambda_0 | H_0).$$

The problem here is to find the **critical region** since there is no generally accepted procedure to do so. The idea is to search for a statistic whose distribution is known.

Simple Null Hypothesis vs. a General Alternative

$$H_0: \quad \theta = \theta_0$$
$$\text{vs. } H_1: \quad \theta \in \Theta_1 (\text{for example } \theta < \theta_0).$$

We pose:

$$\lambda = \frac{\sup_{\theta \in \Theta_1} L(\theta_0; x_1, x_2, \ldots, x_n)}{L(\theta_1; x_1, x_2, \ldots, x_n)},$$

where $\sup_{\theta \in \Theta_1} L(\theta; x_1, x_2, \ldots, x_n)$ represents the value of the likelihood function for the parameter that maximizes this likelihood function in the restricted parameter space ($\theta \in \Theta_1$), and we search again for the distribution of λ under H_0.

If the alternative hypothesis H_1 is $\theta \neq \theta_0$, we can write λ as:

$$\frac{L(\hat{\theta}; x_1, x_2, \ldots, x_n)}{L(\theta_0; x_1, x_2, \ldots, x_n)},$$

where $\hat{\theta}$ is the maximum likelihood estimate for θ.

General Hypothesis $H_0: \theta \in \Theta_0$ vs. Alternative Hypothesis $H_1: \theta \in \Theta \setminus \Theta_0 = \{\theta \in \Theta; \theta \notin \Theta_0\}$

Consider the following **statistic**:

$$\lambda = \frac{\sup_{\theta \in \Theta_0} L(\theta; x_1, x_2, \ldots, x_n)}{\sup_{\theta \in \Theta} L(\theta; x_1, x_2, \ldots, x_n)}$$

and search the rejection region as in the previous cases.

The $\sup_{\theta \in \Theta_0} L(\theta; x_1, x_2, \ldots, x_n)$ represents the value of the likelihood function for the parameter that maximizes the likelihood function (the maximum likelihood estimate) on the restricted space of the parameters. The $\sup_{\theta \in \Theta} L(\theta; x_1, x_2, \ldots, x_n)$ represents the same thing but for the whole parameter space. The larger the difference between these two values, the more inappropriate will be the restricted model (and H_0 will be rejected).

Under H_0, the distribution of $-2 \log(\lambda)$ follows asymptotically a **chi-square** distribution with u degrees of freedom, with $u = \dim \Theta - \dim \Theta_0$ (that is, the number of parameters specified by H_0). In other words:

$$-2 \lim_{n \to \infty} \log \left(\frac{\sup_{\theta \in \Theta} L(\theta; x_1, x_2, \ldots, x_n)}{\sup_{\theta \in \Theta_0} L(!0; x_1, x_2, \ldots, x_n)} \right)$$
$$= \chi^2(u) .$$

For the particular cases treated earlier, the result is:

$$-2 \log \lambda \sim \chi^2 \text{ on 1 degree of freedom}$$
$$\text{if } n \to \infty .$$

DOMAINS AND LIMITATIONS

If the data distribution is normal, the likelihood ratio test is equivalent to the **Fisher test**. The advantage of the latter (when it can be explicitly constructed) is that it guarantees the best performance and power [according to the Neyman–Pearson theorem] to test the simple hypothesis.

EXAMPLES

Example 1

Let X be a normal random variable with mean μ and variance σ^2. We want to test the following hypothesis:

$$H_0: \quad \mu = 0, \quad \sigma^2 > 0$$
$$H_1: \quad \mu \neq 0, \quad \sigma^2 > 0.$$

By considering the parameter space $\Theta = \{(\mu, \sigma^2); -\infty < \mu < \infty, 0 < \sigma^2 < \infty\}$ and a subspace $\Theta_1 = \{(\mu, \sigma^2) \in \Theta; \mu = 0\}$, the hypothesis can be now written as:

$$H_0: \quad \theta = \left(\mu, \sigma^2 \right) \in \Theta_0$$
$$H_1: \quad \theta = \left(\mu, \sigma^2 \right) \in \Theta \backslash \Theta_0 .$$

Consider now $x_1, \ldots, x_n$ n observations of variable X ($n > 1$). The likelihood function is (see example under **maximum likelihood**):

$$L(\theta; x_1, \ldots, x_n) = L\left(\mu, \sigma^2; x_1, \ldots, x_n \right)$$
$$= \left(\frac{1}{2\pi\sigma^2} \right)^{n/2} \cdot \exp \left(-\frac{\sum_{i=1}^{n} (x_i - \mu)^2}{2\sigma^2} \right)$$

We estimate the sample mean and variance by using the maximum likelihood estimates:

$$\widehat{\mu} = \frac{1}{n} \cdot \sum_{i=1}^{n} x_i = \overline{x},$$
$$\widehat{\sigma}^2 = \frac{1}{n} \cdot \sum_{i=1}^{n} (x_i - \mu)^2 .$$

Hence, the likelihood functions are, for the null hypothesis:

$$L\left(0, \widehat{\sigma}^2; x_1, \ldots, x_n \right)$$
$$= \left(\frac{n}{2\pi \sum_{i=1}^{n} x_i^2} \right)^{n/2} \cdot \exp \left(-\frac{n \sum_{i=1}^{n} x_i^2}{2 \sum_{i=1}^{n} x_i^2} \right)$$
$$= \left(\frac{n}{2\pi \sum_{i=1}^{n} x_i^2} \right)^{n/2} \exp \left(-\frac{n}{2} \right)$$

L

$$= \left(\frac{n \exp(-1)}{2\pi \sum\limits_{i=1}^{n} x_i^2} \right)^{n/2} ,$$

and for the alternative hypothesis:

$$L\left(\widehat{\mu}, \widehat{\sigma}^2; x_1, \ldots, x_n \right)$$

$$= \left(\frac{n}{2\pi \sum\limits_{i=1}^{n} (x_i - \overline{x})^2} \right)^{n/2}$$

$$\cdot \exp\left(-\frac{n \sum\limits_{i=1}^{n} (x_i - \overline{x})^2}{2 \sum\limits_{i=1}^{n} (x_i - \overline{x})^2} \right)$$

$$= \left(\frac{1}{2\pi \sum\limits_{i=1}^{n} (x_i - \overline{x})^2} \right)^{n/2} \cdot \exp\left(-\frac{n}{2} \right)$$

$$= \left(\frac{n \exp(-1)}{2\pi \sum\limits_{i=1}^{n} (x_i - \overline{x})^2} \right)^{n/2} ,$$

so the likelihood ratio is:

$$\lambda = \lambda(x_1, x_2, \ldots, x_n)$$

$$= \frac{L\left(0, \widehat{\sigma}^2; x_1, x_2, \ldots, x_n \right)}{L\left(\widehat{\mu}, \widehat{\sigma}^2; x_1, x_2, \ldots, x_n \right)}$$

$$= \left(\frac{n \exp(-1)}{2\pi \sum\limits_{i=1}^{n} x_i^2} \right)^{n/2} \Bigg/$$

$$\left(\frac{n \exp(-1)}{2\pi \sum\limits_{i=1}^{n} (x_i - \overline{x})^2} \right)^{n/2}$$

$$= \left(\frac{\sum\limits_{i=1}^{n} (x_i - \overline{x})^2}{\sum_{i=1}^{n} x_i^2} \right)^{n/2} .$$

One must find at this point the distribution of λ. To do so we recall that $\sum\limits_{i=1}^{n} x_i^2 = \sum\limits_{i=1}^{n} (x_i - \overline{x})^2 + n\overline{x}^2$, and we obtain:

$$\lambda = \left(\frac{\sum\limits_{i=1}^{n} (x_i - \overline{x})^2}{\sum\limits_{i=1}^{n} (x_i - \overline{x})^2 + n\overline{x}^2} \right)^{n/2}$$

$$= 1 \Bigg/ \left(1 + \frac{n\overline{x}^2}{\sum\limits_{i=1}^{n} (x_i - \overline{x})^2} \right)^{n/2} .$$

We search for λ_0 for which the null hypothesis is not rejected if $\lambda \leq \lambda_0$, which we can write as:

$$\lambda \leq \lambda_0$$

$$\text{or } 1 \Bigg/ \left(1 + \frac{n\overline{x}^2}{\sum\limits_{i=1}^{n} (x_i - \overline{x})^2} \right)^{n/2} \leq \lambda_0$$

$$\text{or } \left(1 + \frac{n\overline{x}^2}{\sum\limits_{i=1}^{n} (x_i - \overline{x})^2} \right)^{n/2} \geq \lambda_0^{-1}$$

$$\text{or } \frac{n\overline{x}^2}{\sum\limits_{i=1}^{n} (x_i - \overline{x})^2} \geq \lambda_0^{-2/n} - 1 .$$

By multiplying each term of the inequality by $(n-1)$, we obtain:

$$\lambda \leq \lambda_0 \quad \text{or}$$

$$\frac{\sqrt{n}\,|\overline{x}|}{\sqrt{\sum\limits_{i=1}^{n} (x_i - \overline{x})^2 / (n-1)}}$$

$$\geq \sqrt{(n-1)\left(\lambda_0^{-2/n} - 1 \right)} .$$

Since the random variable

$$T = \frac{\overline{X} - \mu}{\widehat{\sigma}/(n-1)},$$

where $\widehat{\sigma}^2$ is the sample variance, follows a t distribution with $(n-1)$ degrees of freedom; we can equivalently write:

$$T = \frac{\sqrt{n}\,(\overline{X} - \mu)}{\sqrt{n\widehat{\sigma}^2/(n-1)}},$$

which, for mean set to zero, gives:

$$T = \frac{\sqrt{n}\overline{X}}{\sqrt{n\widehat{\sigma}^2/(n-1)}}$$

$$= \frac{\sqrt{n}\overline{X}}{\sqrt{\sum_{i=1}^{n} x_i^2/(n-1)}}.$$

One must search now for the value λ_0 for which $c := \sqrt{(n-1)\left(\lambda_0^{-2/n} - 1\right)}$ is the value of a Student variable for a chosen significance level α:

$$\lambda_0 = \left(\frac{c^2}{n-1} + 1\right)^{-n/2}.$$

For example, given a sample of size $n = 6$, and for $\alpha = 0.05$, we find $z_{5,0.05/2} = 2.751$, so it is sufficient to see that:

$$\lambda \le \left(\frac{(2.751)^2}{5} + 1\right)^{-3} = 0.063$$

for the null hypothesis to be true.

Example 2

Two independent random samples $X_1, \ldots, X_n$ and $Y_1, \ldots, Y_m$ $(m + n > 2)$ follow normal distributions with parameters

(μ_1, σ^2) and (μ_2, σ^2), respectively. We wish to test the following hypothesis:

$$H_0: \quad \mu_1 = \mu_2$$
$$H_1: \quad \mu_1 \ne \mu_2.$$

We define the sets:

$$\Theta = \{(\mu_1, \mu_2, \sigma^2); -\infty < \mu_1, \mu_2$$
$$< \infty, 0 < \sigma^2 < \infty\},$$
$$\Theta_1 = \{(\mu_1, \mu_2, \sigma^2) \in \Theta; \mu_1 = \mu_2\}.$$

We can now write the hypothesis as follows:

$$H_0: \quad \left(\mu_1, \mu_2, \sigma^2\right) \in \Theta_1$$
$$H_1: \quad \left(\mu_1, \mu_2, \sigma^2\right) \in \Theta \backslash \Theta_1.$$

So the likelihood functions are:

$$L\left(\mu_1, \mu_1, \sigma^2; x_1, \ldots, x_n, y_1, \ldots, y_m\right)$$
$$= \left(\frac{1}{\sqrt{2\pi}\sigma}\right)^{n+m}$$
$$\cdot \exp\left(-\frac{\sum_{i=1}^{n}(x_i - \mu_1)^2 + \sum_{i=1}^{m}(y_i - \mu_1)^2}{2\sigma^2}\right),$$

$$L\left(\mu_1, \mu_2, \sigma^2; x_1, \ldots, x_n, y_1, \ldots, y_m\right)$$
$$= \left(\frac{1}{\sqrt{2\pi}\sigma^2}\right)^{n+m}$$
$$\cdot \exp\left(-\frac{\sum_{i=1}^{n}(x_i - \mu_1)^2 + \sum_{i=1}^{m}(y_i - \mu_2)^2}{2\sigma^2}\right).$$

By using the maximum likelihood estimates, which are for the null hypothesis:

$$\widehat{\mu}_1 = \frac{\sum_{i=1}^{n} x_i + \sum_{i=1}^{m} y_i}{n+m} = \frac{n\overline{x} + m\overline{y}}{n+m},$$

$$\widehat{\sigma}^2 = \frac{\sum_{i=1}^{n}(x_i - \widehat{\mu}_1)^2 + \sum_{i=1}^{m}(y_i - \widehat{\mu}_1)^2}{n+m}$$

and for the alternative hypothesis:

$$\widehat{\mu}_1' = \frac{\sum\limits_{i=1}^{n} x_i}{n} = \bar{x},$$

$$\widehat{\mu}_2 = \frac{\sum\limits_{i=1}^{m} y_i}{m} = \bar{y},$$

$$\widehat{\sigma}'^2 = \frac{\sum\limits_{i=1}^{n} (x_i - \bar{x})^2 + \sum\limits_{i=1}^{m} (y_i - \bar{y})^2}{n + m},$$

we get the likelihood ratio:

$$\lambda = \frac{L\left(\widehat{\mu}_1, \widehat{\mu}_1, \widehat{\sigma}^2; x_1, \ldots, x_n, y_1, \ldots, y_m\right)}{L\left(\widehat{\mu}_1', \widehat{\mu}_2', \widehat{\sigma}'^2; x_1, \ldots, x_n, y_1, \ldots, y_m\right)}$$

$$= \left(\frac{2\pi}{n+m}\right)^{-(n+m)/2} \exp\left(-\frac{n+m}{2}\right)$$

$$\cdot \left[\sum\limits_{i=1}^{n}\left(x_i - \frac{n\bar{x} + m\bar{y}}{n+m}\right)^2\right.$$

$$\left. + \sum\limits_{i=1}^{m}\left(y_i - \frac{n\bar{x} + m\bar{y}}{n+m}\right)^2\right]^{-(n+m)/2}$$

$$\cdot \left(\frac{2\pi}{n+m}\right)^{(n+m)/2} \exp\left(\frac{n+m}{2}\right)$$

$$\left[\sum\limits_{i=1}^{n}(x_i - \bar{x})^2 + \sum\limits_{i=1}^{m}(y_i - \bar{y})^2\right]^{(n+m)/2}$$

$$= \left(\sum\limits_{i=1}^{n}\left(x_i - \frac{n\bar{x} + m\bar{y}}{n+m}\right)^2\right.$$

$$\left. + \sum\limits_{i=1}^{m}\left(y_i - \frac{n\bar{x} + m\bar{y}}{n+m}\right)^2\right)^{-(n+m)/2}$$

$$\cdot \left(\sum\limits_{i=1}^{n}(x_i - \bar{x})^2 + \sum\limits_{i=1}^{m}(y_i - \bar{y})^2\right)^{(n+m)/2}$$

$$= \left(\sum\limits_{i=1}^{n}(x_i - \bar{x})^2 + \sum\limits_{i=1}^{m}(y_i - \bar{y})^2\right)^{(n+m)/2}$$

$$\left/ \left(\sum\limits_{i=1}^{n}\left(x_i - \frac{n\bar{x} + m\bar{y}}{n+m}\right)^2\right.\right.$$

$$\left.\left. + \sum\limits_{i=1}^{m}\left(y_i - \frac{n\bar{x} + m\bar{y}}{n+m}\right)^2\right)^{(n+m)/2}\right..$$

The statistic $\lambda^{2/(n+m)}$ is given by:

$$\frac{\sum\limits_{i=1}^{n}\left(X_i - \bar{X}\right)^2 + \sum\limits_{i=1}^{m}\left(Y_i - \bar{Y}\right)^2}{\sum\limits_{i=1}^{n}\left(X_i - \frac{n\bar{X}+m\bar{Y}}{n+m}\right)^2 + \sum\limits_{i=1}^{m}\left(Y_i - \frac{n\bar{X}+m\bar{Y}}{n+m}\right)^2}.$$

To find the distribution of this statistic, we use the following arguments (note that $\sum\limits_{i=1}^{n}(X_i - \bar{X}) = 0$):

$$\sum\limits_{i=1}^{n}\left(X_i - \frac{n\bar{X} + m\bar{Y}}{n+m}\right)^2$$

$$= \sum\limits_{i=1}^{n}\left((X_i - \bar{X}) + \left(\bar{X} - \frac{n\bar{X} + m\bar{Y}}{n+m}\right)\right)^2$$

$$= \sum\limits_{i=1}^{n}(X_i - \bar{X})^2 + n\left(\bar{X} - \frac{n\bar{X} + m\bar{Y}}{n+m}\right)^2$$

$$= \sum\limits_{i=1}^{n}(X_i - \bar{X})^2 + \frac{nm^2}{(n+m)^2}\left(\bar{X} - \bar{Y}\right)^2.$$

The same holds for Y:

$$\sum\limits_{i=1}^{m}\left(Y_i - \frac{n\bar{X} + m\bar{Y}}{n+m}\right)^2$$

$$= \sum\limits_{i=1}^{m}(Y_i - \bar{Y})^2 + \frac{mn^2}{(n+m)^2}\left(\bar{X} - \bar{Y}\right)^2.$$

We can now write:

$$\lambda^{2/(n+m)}$$

$$= \left(\sum_{i=1}^{n} \left(X_i - \overline{X}\right)^2 + \sum_{i=1}^{m} \left(Y_i - \overline{Y}\right)^2 \right)$$

$$\cdot \left(\sum_{i=1}^{n} \left(X_i - \overline{X}\right)^2 + \sum_{i=1}^{m} \left(Y_i - \overline{Y}\right)^2 \right.$$

$$\left. + \frac{nm}{n+m} \left(\overline{X} - \overline{Y}\right)^2 \right)^{-1}$$

$$= \left(1 + \frac{\frac{nm}{n+m}\left(\overline{X} - \overline{Y}\right)^2}{\sum_{i=1}^{n}\left(X_i - \overline{X}\right)^2 + \sum_{i=1}^{m}\left(Y_i - \overline{Y}\right)^2} \right)^{-1}.$$

If the null hypothesis is true, then the random variable

$$T = \frac{\sqrt{\frac{nm}{n+m}}\left(\overline{X} - \overline{Y}\right)}{\sqrt{\frac{\sum_{i=1}^{n}(X_i - \overline{X})^2 + \sum_{i=1}^{m}(Y_i - \overline{Y})^2}{n+m-2}}}$$

follows the Student distribution with $n+m-2$ degrees of freedom. Therefore, we obtain:

$$\lambda^{2/(n+m)} = \frac{1}{1 + (n+m-2)\,T^2}$$

$$= \frac{n+m-2}{n+m-2+T^2}.$$

We search for λ_0 for which the null hypothesis is rejected given $\lambda \le \lambda_0 \le 1$:

$$\lambda = \left(\frac{n+m-2}{n+m-2+T^2} \right)^{2/(n+m)} \le \lambda_0$$

$$\Leftrightarrow \quad |T| \ge \sqrt{\frac{\left(\lambda_0^{-\frac{n+m}{2}} - 1\right)}{(n+m-2)}} := c.$$

It is now sufficient to take the critical value c from the table of a Student random variable with α fixed (the one we want) and n the sample size. For example, with $n = 10$, $m = 6$,

and $\alpha = 0.05$, we find:

$$c = z_{14,0.025} = 2.145,$$

$$\lambda_0 = \left(\frac{c^2}{n+m-2} - 1 \right)^{-2/(n+m)}$$

$$= \left(\frac{(2.145)^2}{14} - 1 \right)^{-1/8} = -1.671.$$

FURTHER READING
▶ **Alternative hypothesis**
▶ **Fisher test**
▶ **Hypothesis**
▶ **Hypothesis testing**
▶ **Maximum likelihood**
▶ **Null hypothesis**

REFERENCES
Cox, D.R., Hinkley, D.V.: Theoretical Statistics. Chapman & Hall, London (1973)

Edwards, A.W.F.: Likelihood. An account of the statistical concept of likelihood and its application to scientific inference. Cambridge University Press, Cambridge (1972)

Kendall, M.G., Steward, A.: The Advanced Theory of Statistics, vol. 2. Griffin, London (1967)

Neyman, J., Pearson, E.S.: On the use and interpretation of certain test criteria for purposes of statistical inference, Parts I and II. Biometrika **20**A, 175–240, 263–294 (1928)

Wald, A.: Asymptotically Most Powerful Tests of Statistical Hypotheses. Ann. Math. Stat. **12**, 1–19 (1941)

Wald, A.: Some Examples of Asymptotically Most Powerful Tests. Ann. Math. Stat. **12**, 396–408 (1941)

L

Wilks, S.S.: Mathematical Statistics. Wiley, New York (1962)

Line Chart

The line chart is a special type of frequency graph. It is useful in representing a **frequency distribution** of a discrete **random variable**. The **values** of the **variable** are given in abscissa and the frequencies corresponding to these values are displayed in ordinate.

HISTORY
See **graphical representation**.

MATHEMATICAL ASPECTS
The line chart has two **axes**. The **values** x_i of the discrete **random variable** are given in abscissa. A line is drawn over each x_i value with a length that is equal to the **frequency** n_i (or relative frequency f_i) corresponding to this value.

EXAMPLES
The distribution of the shoe sizes of a **sample** of 84 men are given in the following table.

Shoe size	Frequency
5	3
6	1
7	13
8	16
9	21
10	19
11	11
Total	84

Here is the line chart established from this table:

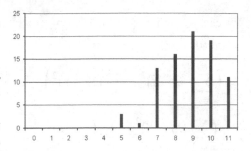

FURTHER READING
▶ **Dot plot**
▶ **Frequency distribution**
▶ **Graphical representation**
▶ **Histogram**

Linear Programming

Mathematical programming concerns the optimal allocation of limited resources in order to reach the desired objectives. In linear programming the mathematical model under study is expressed by using linear relations. A linear programming problem appears in the form of a set of linear equations (or inequalities), called constraints, and a linear function that states the objective, which is called the objective function.

HISTORY
Mathematical programming appeared in the field of economics around 1930. Some remarkable works on linear programming were published by John Von Neumann (1935–1936). In 1947, George B. Dantzig presented the simplex method for solving linear programming problems.

MATHEMATICAL ASPECTS
A linear programming problem can be defined as follows.

Find a set of values for the variables $x_1, x_2, \ldots, x_n$ that maximizes (or minimizes) linearly the objective function:

$$z = c_1 \cdot x_1 + c_2 \cdot x_2 + \ldots + c_n \cdot x_n$$

subject to the linear constraints:

$$a_{11} \cdot x_1 + a_{12} \cdot x_2 + \ldots$$
$$+ a_{1n} \cdot x_n \left\{ \begin{matrix} \leq \\ = \\ \geq \end{matrix} \right\} b_1 ,$$
$$a_{21} \cdot x_1 + a_{22} \cdot x_2 + \ldots$$
$$+ a_{2n} \cdot x_n \left\{ \begin{matrix} \leq \\ = \\ \geq \end{matrix} \right\} b_2 ,$$
$$\vdots$$
$$a_{m1} \cdot x_1 + a_{m2} \cdot x_2 + \ldots$$
$$+ a_{mn} \cdot x_n \left\{ \begin{matrix} \leq \\ = \\ \geq \end{matrix} \right\} b_m ,$$

and to the nonnegative constraints:

$$x_j \geq 0 , \quad j = 1, 2, \ldots, n ,$$

where a_{ij}, b_i, and c_j are known constants. In other words, this means that in a linear programming problem we seek a nonnegative solution that satisfies the posed constraints and optimizes the objective function.

We can write a linear programming problem using the following matrix notation:

$$\text{Maximize (or minimize) } \mathbf{z} = \mathbf{c}' \cdot \mathbf{x}$$
$$\text{subject to } \mathbf{A} \cdot \mathbf{x} \left\{ \begin{matrix} \leq \\ = \\ \geq \end{matrix} \right\} \mathbf{b}$$
$$\text{and } \mathbf{x} \geq \mathbf{0} ,$$

where $\mathbf{c}'$ is a $(1 \times n)$ line vector, $\mathbf{x}$ is an $(n \times 1)$ column vector, $\mathbf{A}$ is an $(m \times n)$ **matrix**, $\mathbf{b}$ is an $(m \times 1)$ column vector, and $\mathbf{0}$ is the null vector with n components.

The principal tool for solving a linear programming problem is the simplex method. This consists of a set of rules that should be followed to obtain the solution to a given problem. It is an iterative method that provides an exact solution in a finite number of iterations.

DOMAINS AND LIMITATIONS

Linear programming seeks an optimal allocation of these resources that can produce one or more goods. This is done by maximizing (or minimizing) functions such as the profit (or cost). Therefore, the objective function's coefficients are often called the "prices" associated to the variables.

EXAMPLES

A firm manufactures tables and chairs using two machines, A and B. Each product is made on both machines. To make a chair, 2 h of machine A and 1 h of machine B are needed. To make a table, 1 h of machine A and 2 h of machine B are needed. The firm makes a profit of 3.00 CHF on each chair and 4.00 CHF on each table. The two machines are available for a maximum of 12 h daily. Given the above data, we search for the daily production schedule that guarantees the maximum profit. In other words, the firm wants to know the optimal production program, that is, the amount of chairs and tables that must be produced per day to reach the maximum profit.

We denote by x_1 the number of chairs and by x_2 the number of tables that must be produced per day. The daily profit is given by the following function:

$$z = 3 \cdot x_1 + 4 \cdot x_2 .$$

This objective function has to be maximized. We know that the product x_1 requires 2 h of machine A, while product x_2 requires 1 h of machine A. Hence, the total daily hours for machine A is:

$$2 \cdot x_1 + 1 \cdot x_2$$

since x_1 chairs and x_2 tables are produced. Yet the total time for machine A cannot exceed 12 h. In mathematical terms this means:

$$2x_1 + x_2 \leq 12.$$

In the same manner for machine B, we find:

$$x_1 + 2x_2 \leq 12.$$

We have two constraints. Furthermore, we cannot produce negative values of chairs and tables, so two additional constraints are added:

$$x_1 \geq 0 \quad \text{and} \quad x_2 \geq 0.$$

We wish to find the values for the variables x_1 and x_2 that satisfy the constraints and maximize the profit. We can finally represent the mathematical problem in the following way:

$$\max z = 3x_1 + 4x_2,$$
$$\text{s.t. } 2x_1 + x_2 \leq 12,$$
$$x_1 + 2x_2 \leq 12,$$
$$\text{and } x_1, x_2 \geq 0$$

(s.t. means subject to). The maximum profit is given by the pair $(x_1, x_2) = (4, 4)$; 28.00 CHF of profit by the production of 4 chairs and 4 tables. For this problem, the result can be found by computing the profit for each possible pair (there are fewer than 36 of them.)

FURTHER READING

▶ **Operations research**
▶ **Optimization**

REFERENCES

Arthanari, T.S., Dodge, Y.: Mathematical Programming in Statistics. Wiley, New York (1981)

Dantzig, G.B.: Applications et prolongements de la programmation linéaire. Dunod, Paris (1966)

Hadley, G.: Linear programming. Addison-Wesley, Reading, MA (1962)

Von Neumann, J.: Über ein ökonomisches Gleichungssystem und eine Verallgemeinerung des Brouwerschen Fixpunktsatzes. Ergebn. math. Kolloqu. Wien **8**, 73–83 (1937)

Logarithmic Transformation

One of the most useful and common **transformations** is the logarithmic transformation. Indeed, before running a linear regression, it might be wise to replace the **dependent variable** Y by its logarithm $\log(Y)$ or by a linear combination of the latter $a + b \cdot \log(Y)$ (a and b being constants adjusted by the **least squares**). Such an operation would stabilize the **variance** of Y and would make the distribution of the transformed variable closer to normal.

A logarithmic transformation would also allow one to shift from a multiplicative to an additive **model** with estimates that are easier to define and simpler to interpret. On the other hand, the logarithmic transformation of a **gamma** random variable Y results in a new random variable that is closer to the **normal** distribution. The latter could be easily adapted in a **regression analysis**.

FURTHER READING
▶ **Analysis of variance**
▶ **Dependent variable**
▶ **Regression analysis**

Logistic Regression

Given a binary dependent variable, logistic regression offers the possibility of fitting a regression model. It permits one to study the **relation** between a proportion and a set of explanatory variables, either quantitative or qualitative.

MATHEMATICAL ASPECTS

The logistic regression model with dependent variable Y and independent variables $X_1, X_2, \ldots, \_p$ is expressed by:

$$\text{logit } \pi \left((X_{ij})_{i=1,\ldots,p} \right) = \log \left(\frac{P(Y_j = 1)}{P(Y_j = 0)} \right)$$

$$= \beta_0 + \sum_{i=1}^{p} \beta_i X_{ij},$$

$$j = 1, \ldots, n,$$

where Y_j $(j = 1, \ldots, n)$ corresponds to the observations of variables Y and X_{ij} to the observations of the independent variables and π is the probability (of success) function of the logistic regression:

$$\pi \left((X_{ij})_{i=1,\ldots,p} \right) =$$

$$\frac{\exp \left(\beta_0 + \sum_{i=1}^{p} \beta_i X_{ij} \right)}{1 + \exp \left(\beta_0 + \sum_{i=1}^{p} \beta_i X_{ij} \right)}.$$

Equivalently, we can write:

$$P(Y_j = 1) = \frac{\exp \left(\beta_0 + \sum_{i=1}^{p} \beta_i X_{ij} \right)}{1 + \exp \left(\beta_0 + \sum_{i=1}^{p} \beta_i X_{ij} \right)},$$

which permits one to compute the probability that variable Y takes a value of 1 depending upon the realization of the independent variables $X_1, X_2, \ldots, X_p$.
An interesting property is that the model parameters $(\beta_0, \beta_1, \ldots, \beta_p)$ can take any value, since all values of the logit transformation $logit (\pi (Y)) = \log \left(\frac{\pi(Y)}{1 - \pi(Y)} \right)$ correspond to a number between 0 and 1 for $P(Y = 1)$.
The model parameters can be estimated using the **maximum likelihood method**, thereby maximizing the following function:

$$L(\beta_0, \beta_1, \ldots, \beta_n) = \prod_{j=1}^{n} \pi \left((X_{ij})_{i=1,\ldots,p} \right)^{Y_i} \cdot$$

$$\cdot \left(1 - \pi \left((X_{ij})_{i=1,\ldots,p} \right) \right)^{1-Y_j}$$

or, in an easier way, by maximizing the log-likelihood function:

$$\log L(\beta_0, \beta_1, \ldots, \beta_n)$$

$$= \sum_{j=1}^{n} Y_j \log \pi \left((X_{ij})_{i=1,\ldots,p} \right)$$

$$+ (1 - Y_j) \log \left(1 - \pi \left((X_{ij})_{i=1,\ldots,p} \right) \right).$$

We can therefore do hypothesis testing on the parameter values (for example $\beta_0 = 0, \beta_3 = \beta_2$, etc.) by using the **likelihood ratio test**. The interpretation of the regression coefficients corresponding to binary independent variables is described in the next paragraph. When an independent variable X is continuous, the regression coefficient reflects the relative risk of a one-unit increase in X. When the independent variable X has more than one class, we must transform it into binary variables. The coefficient's interpretation is then clear and straightforward.

DOMAINS AND LIMITATIONS

Logistic regression is often used in **epidemiology** since:

- Its sigmoidal shape (of the previously given expression $P(Y_j = 1)$) fits very well the relation usually observed between a dose X and the frequency of a disease Y: $\frac{\exp(x)}{1+\exp(x)}$.
- It is easy to interpret: the association measure between a disease and a risk factor M (corresponding to a binary variable X_i) is expressed by the **odds ratio**. This is a good approximation of the **relative risk** when the probabilities $P(Y = 1 | X_i = 1)$ and $P(Y = 1 | X_i = 0)$ are small and it is computed simply as the exponential of the parameter that is associated with the variable X_i; thus:

$$odds - ratio \text{ of the factor } M = \exp(\beta_i).$$

EXAMPLES

We take an example given by Hosmer and Lemeshow. The dependent variable indicates the presence or absence of heart disease for 100 subjects, and the independent variable is age.

The resulting logistic regression model is:

$$\pi(X) = -5.3 + 0.11 \cdot X$$

with a standard deviation for the constant and the β (coefficient of the independent variable X) equal to 1.1337 and 0.241, respectively. The two estimated parameters are statistically significant.

The risk of having heart disease increases by $\exp(0.11) = 1.12$ with every year.

It is also possible to compute the **confidence interval** by taking the exponential of the interval for β:

$$(\exp(1.11 - 1.96 \cdot 0.241),$$
$$\exp(1.11 + 1.96 \cdot 0.241))$$
$$= (1.065, 1.171).$$

The following table gives the values for the log-likelihood as well as the parameter values that are estimated by maximizing the log-likelihood function in an iterative algorithm:

	Log-likelihood	Constant	β
1	−54.24	−4.15	0.872
2	−53.68	−5.18	1.083
3	−53.67	−5.31	1.109
4	−53.67	−5.31	1.109

FURTHER READING

- ▶ **Binary data**
- ▶ **Contingency table**
- ▶ **Likelihood ratio test**
- ▶ **Maximum likelihood**
- ▶ **Model**
- ▶ **Odds and odds ratio**
- ▶ **Regression analysis**

REFERENCES

Bishop, Y.M.M., Fienberg, S.E., Holland, P.W.: Discrete Multivariate Analysis: Theory and Practice. MIT Press, Cambridge, MA (1975)

Cox, D.R., Snell, E.J.: The Analysis of Binary Data. Chapman & Hall (1989)

Hosmer, D.W., Lemeshow S.: Applied logistic regression. Wiley Series in Probability and Statistics. Wiley, New York (1989)

Altman, D.G.: Practical Statistics for Medical Research. Chapman & Hall, London (1991)

Lognormal Distribution

A **random variable** Y follows a lognormal distribution if the **values** of Y are a function of the values of X according to the equation:

$$y = \exp(x) \,,$$

where the **random variable** X follows a **normal distribution**.

The **density function** of the lognormal distribution is given by:

$$f(y) = \frac{1}{\sigma y \sqrt{2\pi}}$$
$$\cdot \exp\left(\left[\frac{-(\log y - \mu)^2}{2\sigma^2}\right]\right).$$

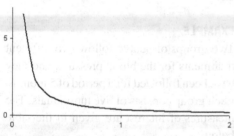

Lognormal distribution, $\mu = 0, \sigma = 1$

The lognormal distribution is a **continuous probability distribution**.

HISTORY

The lognormal distribution is due to the works of **Galton, F.** (1879) and McAlister, D. (1879), who obtained expressions for the **mean**, **median**, **mode**, **variance**, and certain **quantiles** of the resulting distribution. Galton, F. went from n independent positive **random variables** and constructed the product and, with the help of the logarithm, passed from the product to a sum of new random variables.

Kapteyn, J.C. (1903) reconsidered the construction of a random variable following a lognormal distribution with van Uven, M.J. (1916). They developed a graphical method for the estimation of **parameters**. Since that time, many works relative to the lognormal distribution have been published. An exhaustive list is given by Johnson, N.L. and Kotz, S. (1970).

MATHEMATICAL ASPECTS

The **expected value** of the lognormal distribution is given by:

$$E[Y] = \exp\left(\mu + \frac{1}{2\sigma^2}\right).$$

The **variance** is equal to:

$$\text{Var}(X) = \exp(2\mu)$$
$$\cdot \left[\exp\left(2\sigma^2\right) - \exp\left(\sigma^2\right)\right].$$

If the **random variable** Y follows a lognormal distribution, the random variable

$$X = \ln Y$$

follows a **normal distribution**.

DOMAINS AND LIMITATIONS

The lognormal distribution is largely applied in common statistical practice. For example, the critical proportioning in pharmaceutical applications, the length of a visit to the doctor, and the distribution of economic forces all follow a lognormal distribution.

The lognormal distribution can be used as an approximation of the **normal distribution**. As σ gets small, the lognormal distribution tends toward the **normal distribution**.

FURTHER READING

▶ **Continuous probability distribution**
▶ **Normal distribution**

REFERENCES

Galton, F.: The geometric mean in vital and social statistics. Proc. Roy. Soc. Lond. **29**, 365–367 (1879)

Johnson, N.L., Kotz, S.: Distributions in Statistics: Continuous Univariate Distributions, vols. 1 and 2. Wiley, New York (1970)

Kapteyn, J.: Skew frequency curves in biology and statistics. Astronomical Laboratory Noordhoff, Groningen (1903)

Kapteyn, J., van Uven, M.J.: Skew frequency curves in biology and statistics. Hoitsema Brothers, Groningen (1916)

McAlister, D.: The law of the geometric mean. Proc. Roy. Soc. Lond. **29**, 367–376 (1879)

Longitudinal Data

Longitudinal data are data collected on subjects over time. These correspond to multiple observations per subject as the same units are repeatedly measured on more than one occasion. Repeated measures, in biology and medicine, as well as **panel** studies, in social and economic sciences, are characteristic examples of longitudinal data studies.

HISTORY

See history of **panel data**.

MATHEMATICAL ASPECTS

A very general way to write the linear model in longitudinal data analysis is the following:

$$y_i = X_i\beta + W_i\gamma_i + \varepsilon_i ,$$

where y_i is the $n_i \times 1$ column response vector for the subject i, X_i is an $n_i \times b$ design matrix usually consisting of dummy coded variables, β is the $n_i \times b$ regression coefficient vector, W_i is an $b \times 1$ design matrix for random effects γ_i, and ε_i are the subject error terms.

The general model above is a mixed model since it includes both fixed (β) and random effects (γ_i).

Parameter estimation is often done by means of **maximum likelihood estimation** with the likelihood function strongly dependent on the assumptions made on the covariance structure of γ_i and ε_i. It is straightforward to construct statistical tests on the estimated parameters, to obtain confidence intervals, and to measure prediction. Model-selection techniques are also used to test nested models.

EXAMPLE

Two groups of people follow two different treatments for the blood pressure, and they have been followed for a period of 5 months. Each group consists of five individuals. The measures of interest are given in the table below:

	Month (Jan to May)				
	105	68	95	86	93
	100	92	98	74	35
1st group	102	85	95	62	82
	70	81	72	35	39
	98	34	34	37	40
	83	25	31	42	38
	69	44	41	30	32
2nd group	84	64	77	80	82
	63	54	55	47	44
	93	77	44	38	40

Each row corresponds to observations taken over time for the same individual.

Below is the figure that plots the group mean curves ± 1 standard error.

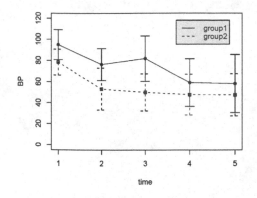

The simplest approach in this case would be to ignore the random effects ($\gamma_i = 0$) and to construct a simple model with fixed effect only, that is, the group where a subject belongs. In such a case the model reduces to

$$y_{ij} = \beta_j + \varepsilon_{ij},$$

with j indicating the group and being a simple dummy variable. If there is a significant difference between individuals, then random effect γ_i is added to the model above.

DOMAINS AND LIMITATIONS

Longitudinal data are mostly used in epidemiology and medical statistics, as well as in socioeconomic studies. In the former cases longitudinal data represent repeated measurements for different treatment groups in **clinical trials**. In the latter we note the **panel** studies that usually measure social and eco-nomic indexes, such as income, consumption, and job satisfaction, in different time periods.

There are many possible complications in longitudinal data analysis. Firstly, data are correlated since the same subjects are measured over time. This must be accounted for in the final analysis in order to provide proper results. Moreover, missing data often appear in such data analysis since not all of the subjects are measured during the whole time period. The latter becomes more problematic, for example, in survival analysis and followup studies where the subjects (usually the patients) die off or quit the trials and abandon the treatments.

FURTHER READING
▶ **Panel**

REFERENCES

Twisk, J.W.R.: Applied Longitudinal Data Analysis for Epidemiology. A Practical Guide. Cambridge University Press, Cambridge (2003)

Fitzmaurice, G., Laird, N., Ware, J.: Applied Longitudinal Analysis. Wiley, New York (2004)

Molenberghs, G., Geert, V.: Models for Discrete Longitudinal Data Series. Springer Series in Statistics. Springer, Berlin Heidelberg New York (2005)

Mahalanobis Distance

The Mahalanobis distance is the distance from X to the quantity μ defined as:

$$d_M^2(\mathbf{X}, \mu) = (\mathbf{X} - \mu)^t \sum{}^{-1} (\mathbf{X} - \mu).$$

This distance is based on the correlation between variables or the variance–covariance matrix.

It differs from the Euclidean distance in that it takes into account the correlation of the data set and does not depend on the scale of measurement. Mahalanobis distance is widely used in **cluster analysis** and other **classification** methods.

HISTORY

The Mahalanobis distance was introduced by Mahalanobis, P.C. in 1936.

MATHEMATICAL ASPECTS

If μ_i denotes $E(X_i)$, then by definition the expected value of $\mathbf{X} = (X_1, \ldots, X_p)$ vector:

$$E[\mathbf{X}] = \mu = \begin{bmatrix} \mu_1 \\ \mu_2 \\ \vdots \\ \vdots \\ \mu_p \end{bmatrix}.$$

The variance–covariance matrix Σ of $\mathbf{X}$ is defined by:

$$\Sigma = \begin{bmatrix} \sigma_1^2 & cov(X_1, X_2) & \ldots & cov(X_1, X_p) \\ & \sigma_2^2 & & \\ \vdots & & \ddots & \vdots \\ & & & \sigma_p^2 \end{bmatrix}$$

$$= E[\mathbf{XX}'] - \mu\mu'.$$

It is a square symmetrical matrix of order p. If the X_i variables are standardized, then Σ is identical to the correlation matrix:

$$\begin{bmatrix} 1 & \rho_{12} & \rho_{13} & \ldots & \rho_{1p} \\ & 1 & \rho_{23} & \ldots & \rho_{2p} \\ & & 1 & \ldots & \rho_{3p} \\ & & & \ddots & \vdots \\ & & & & 1 \end{bmatrix}.$$

Then $\Sigma^{-1/2} \cdot \mathbf{Y} = \Sigma^{-1/2}(\mathbf{X} - \mu)$ is a random standardized vector with noncorrelated components. The random variable $(\mathbf{X} - \mu)' \Sigma^{-1}(\mathbf{X} - \mu) = D^2$ has expected value of p. In fact, $D^2 = \sum_{i=1}^{p} Y_i^2$, where the Y_i are of random variables with mean 0 and 1. D is called a Mahalanobis distance from $\mathbf{X}$ to μ.

FURTHER READING
► **Classification**
► **Cluster analysis**
► **Correspondence analysis**
► **Covariance analysis**
► **Distance**

REFERENCES

Mahalanobis, P.C.: On Tests and Meassures of Groups Divergence. Part I. Theoretical formulae. J. Asiatic Soc. Bengal **26**, 541–588 (1930)

Mahalanobis, P.C.: On the generalized distance in statistics, Proc. natn. Inst. Sci. India **2**, 49–55 (1936)

Mahalanobis Prasanta Chandra

Prasanta Chandra Mahalanobis (1893–1972) studied in school at Calcutta until 1908. In 1912, he received his B.Sc. in physics at the Presidential College of Calcutta. In 1915, he went to England to earn a Tripos (B.Sc. received at Cambridge) in mathematics and physics at King's College of Cambridge. Before beginning his research, he returned to Calcutta on holiday, but he did not return to England because of the war.

As a statistician, Mahalanobis did not use statistics as an end in itself but mostly as a tool for comprehending and interpreting scientific data as well as for making decisions regarding society's well-being. He used statistics first in his study of anthropology (about 1917) and then floods. In 1930, he published an article on the D^2 statistic called "Test and Measures of Group Divergence". His greatest contribution to statistics is surely the large-scale census. He made many contributions in this area such as the opti-

mal choice of sampling using variance and cost functions. He was named president of the *United Nations Subcommission on Statistical Sampling* in 1947, a position he held until 1951.

His friendship with **Ronald Aylmer Fisher**, which lasted from 1926 until Fisher's death, is also notable.

Principal works and articles of Prasanta Chandra Mahalanobis:

1930 On tests and meassures of groups divergence. Part I. Theoretical formulae. Journal of the Asiatic Society of Bengal, 26 pp. 541–588.

1936 On the generalized distance in statistics, Proc. Natn. Inst. Sci. India 2, 49–55.

1940 A sample survey of the acreage under jute in Bengal. Sankhya 4, 511–530.

1944 On large-scale sample surveys. Philos. Trans. Roy. Soc. Lond. Ser. B 231329-451.

1963 The Approach of Operational Research to Planning in India. Asia Publishing House, Bombay.

FURTHER READING
► **Mahalanobis distance**

Main Effect

In a **factorial experiment**, a main effect is the contribution of a **factor level** specific to the response **variable**.

MATHEMATICAL ASPECTS

The linear **model** that only takes into account the main effects of factors and that ignores the **interactions** among factors is called an additive model. Consider the additive **model** of a double classification design:

$$Y_{ij} = \mu + \alpha_i + \beta_j + \varepsilon_{ij},$$
$$i = 1, 2, \ldots, a \text{ and } j = 1, 2, \ldots, b,$$

where Y_{ij} is the response receiving the **treatment** ij, μ is the **mean** common to all the treatments, α_i is the main effect of the ith **level** of **factor** A, β_j is the main effect of the jth level of factor B, and ϵ_{ij} is the experimental **error** of the **observation** Y_{ij}.

If $a = b = 2$, then the principal effects of factors A and B, denoted a and b, respectively, are given by:

$$a = \frac{(\alpha_1 - \alpha_0)(\beta_1 + \beta_0)}{2},$$
$$b = \frac{(\alpha_1 + \alpha_0)(\beta_1 - \beta_0)}{2}.$$

FURTHER READING
▶ **Analysis of variance**
▶ **Estimation**
▶ **Factor**
▶ **Factorial experiment**
▶ **Hypothesis testing**
▶ **Interaction**

REFERENCES
Box, G.E.P., Hunter, W.G., Hunter, J.S.: Statistics for Experimenters. An Introduction to Design, Data Analysis, and Model Building. Wiley, New York (1978)

Mann–Whitney Test

The Mann–Whitney test is a **nonparametric test** that aims to test the equality of two populations. It is used when we have two samples coming from two populations.

HISTORY
The Mann–Whitney test was first introduced for the case $m = n$ by **Wilcoxon, Frank** (1945). Then, the **Wilcoxon test** was extended to the case of samples of different dimensions by White, C. (1952).

Note that a test equivalent to those of the Wilcoxon test was developed independently and introduced by Festinger, L. (1946). Mann, H.B. and Whitney, D.R. (1947) were the first to consider the samples of different dimensions and to provide the tables adapted to small samples.

MATHEMATICAL ASPECTS
Let $(X_1, X_2, \ldots, X_n)$ be a **sample** of dimension n coming from a **population** 1, and let $(Y_1, Y_2, \ldots, Y_m)$ be a sample of dimension m coming from a population 2.

We get $N = n + m$ observations that we class in increasing order noting each time the sample to which a given observation belongs. The Mann and Whitney **statistic**, denoted by U, is defined as the total number of times that an X_i precedes a Y_j in the **classification** in increasing order of the N observations. When the dimension of the sample is big enough, the determination of U becomes long, but we can use the following relation that gives identical results:

$$U = mn + \frac{n(n+1)}{2} - T,$$

where T is the sum of ranks attributed to the X from the **Wilcoxon test**.

Hypotheses
The hypotheses corresponding to the Mann–Whitney test can be formulated as follows, according to the one-tail or two-tail case:

M

A: Two-sided case:

$$H_0: \quad P(X < Y) = \tfrac{1}{2}$$
$$H_1: \quad P(X < Y) \neq \tfrac{1}{2}.$$

B: One-sided case:

$$H_0: \quad P(X < Y) \leq \tfrac{1}{2}$$
$$H_1: \quad P(X < Y) > \tfrac{1}{2}.$$

C: One-sided case:

$$H_0: \quad P(X < Y) \geq \tfrac{1}{2}$$
$$H_1: \quad P(X < Y) < \tfrac{1}{2}.$$

In case A, the **null hypothesis** corresponds to the situation where there is no difference between the populations. In case B, the null hypothesis means that population 1 is greater than population 2. In case C, the null hypothesis indicates that population 1 is smaller than population 2.

Decision Rules

If $m, n < 12$, we compare **statistic** U with the **value** found in the Mann–Whitney table to test H_0. On the other hand, if $m, n \geq 12$, then the **sampling distribution** of U approaches very quickly the normal distribution with **mean** $\mu = \dfrac{m \cdot n}{2}$ and **standard deviation**

$$\sigma = \sqrt{\frac{mn\,(m + n + 1)}{12}}.$$

Thus we can determine the *meaning* of the observed value of U by:

$$Z = \frac{U - \mu}{\sigma},$$

where Z is a random standardized variable. In other words, we can compare the value of Z thus obtained with the **normal table**. Moreover, the decision rules are different and depend on the posed hypotheses. In this way we have rules A, B, and C relative to the previous cases A, B, and C.

Decision rule A

We reject the null hypothesis H_0 at the **significance level** α if U is smaller than the value in the Mann–Whitney table with parameters n, m, and $\frac{\alpha}{2}$ denoted by $t_{n,m,1-\frac{\alpha}{2}}$, or if U is greater than the value in the Mann–Whitney table for n, m, and $1 - \frac{\alpha}{2}$, denoted $t_{n,m,\frac{\alpha}{2}}$, that is, if

$$t_{n,m,1-\frac{\alpha}{2}} < U < t_{n,m,\frac{\alpha}{2}}.$$

Decision rule B

We reject the null hypothesis H_0 at the significance level α if U is smaller than the value in the Mann–Whitney table with parameters n, m, and α, denoted by $t_{n,m,\alpha}$, that is, if

$$U < t_{n,m,\alpha}.$$

Decision rule C

We reject the null hypothesis H_0 at the significance level α if U is greater than the value in the Mann–Whitney table with parameters n, m, and $1 - \alpha$, denoted by $t_{n,m,1-\alpha}$, that is, if

$$U > t_{n,m,1-\alpha}.$$

DOMAINS AND LIMITATIONS

As there exists a linear relation between the **Wilcoxon test** and the Mann–Whitney test ($U = mn + n(n + 1)/2 - T$), these two tests are equivalent.

Thus they have two common properties, and both must be respected:

1. Both samples are random samples taken from their respective **populations**.
2. In addition to the **independence** inside each sample, there is mutual independence between samples.
3. The scale of measurement is at least ordinal.

EXAMPLES

In a class, we count nine pupils: five boys and four girls. The class takes a test on mental calculation to see if the boys are, in general, better than the girls in this domain.

The data are in the following table, the highest score corresponding to the best test result.

Boys (X_i)	Girls (Y_i)
9.2	11.1
11.3	15.6
19.8	20.3
21.1	23.6
24.3	

We class the score in increasing order noting each time the **sample** to which a given **observation** belongs.

Score	Sample	Rank	Number of times X precedes Y
9.2	X	1	4
11.1	Y	2	–
11.3	X	3	3
15.6	Y	4	–
19.8	X	5	2
20.3	Y	6	–
21.1	X	7	1
23.6	Y	8	–
24.3	X	9	0

We find $U = 4 + 3 + 2 + 1 = 10$. In the same way, if we use the relation $U = mn + \frac{n(n+1)}{2} - T$, then we have:

$$T = \sum_{i=1}^{n} R(X_i) = 1 + 3 + 5 + 7 + 9 = 25$$

and $U = 5 \cdot 4 + \frac{5 \cdot 6}{2} - 25 = 10$.

The hypotheses to be tested are as follows:

H_0: Boys are not generally better than girls in mental calculation.

H_1: Boys are generally better than girls in mental calculation.

The null hypothesis may be restated as follows: H_0: $P(X < Y) \geq 1/2$ and corresponds to case C. The decision rule is then as follows: Reject H_0 at **significance level** α if

$$U > t_{n,m,1-\alpha} ,$$

where $t_{n,m,1-\alpha}$ is the **value** in the Mann–Whitney table with parameters n, m, and $1 - \alpha$. Recall that in our case, $m = 4$ and $n = 5$. As we have a **relation** between the **Wilcoxon test** and the Mann–Whitney test, instead of using the U table, we use the **Wilcoxon table** for T. If we choose $\alpha = 0.05$, then the **value** of the Wilcoxon table equals 32. As $T = 25 < 32$, we do not reject H_0 and we conclude that boys are not generally better than girls in mental calculation.

FURTHER READING

▶ **Hypothesis testing**
▶ **Nonparametric test**
▶ **Wilcoxon test**

REFERENCE

Festinger, L.: The significance of difference between means without reference to the frequency distribution function. Psychometrika **11**, 97–105 (1946)

Mann, H.B., Whitney, D.R.: On a test whether one of two random variables is stochastically larger than the other. Ann. Math. Stat. **18**, 50–60 (1947)

White, C.: The use of ranks in a test of significance for comparing two treatments. Biometrics **8**, 33–41 (1952)

Wilcoxon, F.: Individual comparisons by ranking methods. Biometrics **1**, 80–83 (1945)

M

Marginal Density Function

In the case of a **pair of random variables** (X, Y), when **random variable** X (or Y) is considered by itself, its **density function** is called the marginal density function.

HISTORY
See **probability**.

MATHEMATICAL ASPECTS
In the case of a **pair of continuous random variables** (X, Y) with a **joint density function** $f(x, y)$, the marginal **density function** of X is obtained by integration:

$$P(X \le a, -\infty < Y < \infty)$$
$$= \int_{-\infty}^{a} \int_{-\infty}^{\infty} f(x, y) \, dy \, dx$$
$$= \int_{-\infty}^{a} f_X(x) \, dx,$$

where $f_X(x) = \int_{-\infty}^{\infty} f(x, y) \, dy$ is the density of X.

The marginal **density function** of Y is obtained in the same way:

$$f_Y(y) = \int_{-\infty}^{\infty} f(x, y) \, dx.$$

FURTHER READING
▶ **Density function**
▶ **Joint density function**
▶ **Pair of random variables**

REFERENCES
Hogg, R.V., Craig, A.T.: Introduction to Mathematical Statistics, 2nd edn. Macmillan, New York (1965)

Ross, S.M.: Introduction to Probability Models, 8th edn. John Wiley, New York (2006)

Marginal Distribution Function

In the case of a **pair of random variables** (X, Y), when **random variable** X (or Y) is considered by itself, its **distribution function** is called the marginal distribution function.

MATHEMATICAL ASPECTS
Consider a **pair of random variables** (X, Y) and their **joint distribution function**:

$$F(a, b) = P(X \le a, Y \le b).$$

The marginal distribution function of X is equal to:

$$
\begin{aligned}
F_X(a) &= P(X \le a) \\
&= P(X \le a, Y < \infty) \\
&= P[\lim_{b \to \infty} (X \le a, Y \le b)] \\
&= \lim_{b \to \infty} P(X \le a, Y \le b) \\
&= \lim_{b \to \infty} F(a, b) \\
&= F(a, \infty).
\end{aligned}
$$

In a similar way, the marginal distribution function of Y is equal to:

$$F_Y(b) = F(\infty, b).$$

FURTHER READING
▶ **Distribution function**
▶ **Joint distribution function**
▶ **Pair of random variables**

Markov, Andrei Andreevich

Markov, Andrei Andreevich was born in 1856 in Ryazan and died in 1922 in St. Petersburg.

This student of **Tchebychev, Pafnutii Lvovich** made important contributions to the field of probability calculus. He gave a rigorous proof of the **central limit theorem**. Through his works on Markov chains, the concept of Markovian dependence evolved into the modern theory and application of random processes. The globalism of his work has influenced the development of probability and international statistics.

He studied mathematics at St. Petersburg University, where he received his bachelor degree, and his doctorate in 1878. He then taught, continuing the courses of Tchebychev, Pafnutii Lvovich after Tchebychev, P.L. left the university. In 1886, he was elected member of the School of Mathematics founded by Tchebychev, P.L. at the St. Petersburg Academy of Sciences. He became a full member of the Academy in 1886 and in 1905 retired from the university, where he continued teaching.

Markov and Liapunov were famous students of Tchebychev, P.L. and were ever striving to establish probability as an exact and practical mathematical science.

The first appearance of Markov chains in Markov's work happened in Izvestiia (Bulletin) of the Physico-Mathematical Society of Kazan University. Markov completed the proof of the central limit theorem that **Tchebychev** started but did not finish. He approached it using the method of moments, and the proof was published in the third edition of *Ischislenie Veroiatnostei* (Probability Calculus).

Principal work of Markov, Andrei Andreevich:

1899 Application des functions continues au calcul des probabilités, Kasan. Bull. **9**(2), 29–34.

FURTHER READING
▶ **Central limit theorem**
▶ **Gauss–Markov theorem**

REFERENCES
Markov, A.A.: Application des functions continues au calcul des probabilités. Kasan. Bull. **9**(2), 29–34 (1899)

Matrix

A rectangular table with m lines and n columns is called an $(m \times n)$ matrix. The elements figuring in the table are called coefficients or components of the matrix.

MATHEMATICAL ASPECTS
A matrix can be represented as follows:

$$
A = \begin{bmatrix}
a_{11} & a_{12} & \cdots & a_{1n} \\
a_{21} & a_{22} & \cdots & a_{2n} \\
\vdots & \vdots & \ddots & \vdots \\
a_{m1} & a_{m2} & \cdots & a_{mn}
\end{bmatrix}.
$$

An element is characterized by its **value** and position. Therefore a_{ij} is an element of matrix A where i indicates the line number and j the column number. A matrix with m lines and n columns is of order $(m \times n)$. There are different types of matrices:

- Square matrix of order n
  A matrix is called a square matrix of order n if it has as many lines as columns (denoted by n).
- Null matrix
  The matrix on which all the elements are null is called the null matrix. It will be denoted by O.
- Identity matrix
  For any square matrix of order n, the elements that have the same index for the line

as for the column [meaning the elements in position (i, i)] form the main diagonal. An **identity matrix** is a square matrix that has "1" on the main diagonal and "0" everywhere else. It is denoted by I or sometimes by I_n, where n is the order of the identity matrix.

- Symmetric matrix
  A square matrix A is symmetric if $A = A'$, where A' is the **transpose** of matrix A.
- Antisymmetric matrix
  A square matrix A is antisymmetric if $A = -A'$, where A' is the **transpose** of matrix A; this condition leads to the elements on the main diagonal a_{ij} for $i = j$ being null.
- Scalar matrix
  A square matrix A is called a scalar matrix if
  $$a_{ij} = \beta \quad \text{for } i = j,$$
  $$a_{ij} = 0 \quad \text{for } i \neq j,$$
  where β is a scalar.
- Diagonal matrix
  A square matrix A is called a diagonal matrix if $a_{ij} = 0$ for $i \neq j$ and where a_{ij} is any number.
- Triangular matrix
  A square matrix A is called upper triangular if $a_{ij} = 0$ for $i > j$, or lower triangular if $a_{ij} = 0$ for $i < j$.

Operations on Matrices

- Addition of matrices
  If $A = (a_{ij})$ and $B = (b_{ij})$ are two matrices of order $(m \times n)$, then their sum $A + B$ is defined by the matrix $C = (c_{ij})$ of order $(m \times n)$ of which each element is the sum of corresponding elements of A and B:
  $$C = (c_{ij}) = (a_{ij} + b_{ij}).$$
- Multiplication of a matrix by a scalar
  If α is a scalar and A a matrix, the product αA is the matrix of the same order as A

obtained by multiplying each element a_{ij} of A by the scalar α:
$$\alpha \mathbf{A} = \mathbf{A}\alpha = (\alpha \cdot a_{ij}).$$

- Multiplication of matrices
  Let A and B be two matrices; the product AB is defined if and only if the number of columns of A is equal to the number of lines of B. If A is of order $(m \times n)$ and B of order $(n \times q)$, then the product AB is defined by the matrix C of order $(m \times q)$ whose elements are obtained by calculating:
  $$c_{ij} = \sum_{k=1}^{n} a_{ik} \cdot b_{kj}$$
  for $i = 1, \ldots, m$ and $j = 1, \ldots, q$.

  This means that the elements of the ith line of A are multiplied by the corresponding elements of the jth column of B, and the results are added.
- Integer power of square matrices
  Let A be a square matrix; if p is a positive integer, the pth power of A is defined by:
  $$A^p = A \cdot A \cdot \ldots \cdot A \quad p \text{ times}.$$

  Normally, for $p = 0$, we pose $A^0 = I$, where I is the **identity matrix** of the same order as A.
- Square root of a diagonal matrix
  Let A be a diagonal matrix of nonnegative coefficients. We denote by $\sqrt{A}$ the square root of A, defined by the diagonal matrix of the same order as A, in which we replace each element by its square root:
  $$\sqrt{\mathbf{A}} = \begin{bmatrix} \sqrt{a_{11}} & 0 & \cdots & 0 \\ 0 & \sqrt{a_{22}} & \cdots & 0 \\ \vdots & \vdots & \ddots & \vdots \\ 0 & \cdots & 0 & \sqrt{a_{nn}} \end{bmatrix}.$$

DOMAINS AND LIMITATIONS

The use of matrices, or generally matrix calculus, is widely used in regression analysis, in **analysis of variance**, and in many other domains of **statistics**. It allows to describe in the most concise manner a system of equations and facilitates greatly the notations. For the same reasons, the matrices appear in **analyses of data** where we generally encounter large data tables.

EXAMPLES

Let **A** and **B** be two square matrices of order 2:

$$\mathbf{A} = \begin{bmatrix} 1 & 2 \\ 1 & 3 \end{bmatrix} \text{ and } \mathbf{B} = \begin{bmatrix} 6 & 2 \\ -2 & 1 \end{bmatrix}.$$

The sum of two matrices is defined by:

$$\mathbf{A} + \mathbf{B} = \begin{bmatrix} 1 & 2 \\ 1 & 3 \end{bmatrix} + \begin{bmatrix} 6 & 2 \\ -2 & 1 \end{bmatrix}$$

$$= \begin{bmatrix} 1+6 & 2+2 \\ 1-2 & 3+1 \end{bmatrix}$$

$$= \begin{bmatrix} 7 & 4 \\ -1 & 4 \end{bmatrix}.$$

The multiplication of matrix **A** by 3 gives:

$$3 \cdot \mathbf{A} = 3 \cdot \begin{bmatrix} 1 & 2 \\ 1 & 3 \end{bmatrix} = \begin{bmatrix} 3 \cdot 1 & 3 \cdot 2 \\ 3 \cdot 1 & 3 \cdot 3 \end{bmatrix}$$

$$= \begin{bmatrix} 3 & 6 \\ 3 & 9 \end{bmatrix}.$$

The matrix product of **A** and **B** equals:

$$\mathbf{A} \cdot \mathbf{B} = \begin{bmatrix} 1 & 2 \\ 1 & 3 \end{bmatrix} \cdot \begin{bmatrix} 6 & 2 \\ -2 & 1 \end{bmatrix}$$

$$= \begin{bmatrix} 1 \cdot 6 + 2 \cdot (-2) & 1 \cdot 2 + 2 \cdot 1 \\ 1 \cdot 6 + 3 \cdot (-2) & 1 \cdot 2 + 3 \cdot 1 \end{bmatrix}$$

$$\mathbf{A} \cdot \mathbf{B} = \begin{bmatrix} 2 & 4 \\ 0 & 5 \end{bmatrix}.$$

Remark: $\mathbf{A} \cdot \mathbf{B}$ and $\mathbf{B} \cdot \mathbf{A}$ are generally not equal, which illustrates the noncommutativity of the matrix multiplication.

The square of matrix A is obtained by multiplying A by A. Thus:

$$\mathbf{A}^2 = \mathbf{A} \cdot \mathbf{A} = \begin{bmatrix} 1 & 2 \\ 1 & 3 \end{bmatrix} \cdot \begin{bmatrix} 1 & 2 \\ 1 & 3 \end{bmatrix}$$

$$= \begin{bmatrix} 1^2 + 2 \cdot 1 & 1 \cdot 2 + 2 \cdot 3 \\ 1^2 + 3 \cdot 1 & 1 \cdot 2 + 3 \cdot 3 \end{bmatrix}$$

$$\mathbf{A}^2 = \begin{bmatrix} 3 & 8 \\ 4 & 11 \end{bmatrix}.$$

Let **D** be the following diagonal matrix of order 2:

$$\mathbf{D} = \begin{bmatrix} 9 & 0 \\ 0 & 2 \end{bmatrix}.$$

We define the square root of square matrix **D**, denoted $\sqrt{\mathbf{D}}$, by:

$$\sqrt{\mathbf{D}} = \begin{bmatrix} \sqrt{9} & 0 \\ 0 & \sqrt{2} \end{bmatrix} = \begin{bmatrix} 3 & 0 \\ 0 & \sqrt{2} \end{bmatrix}.$$

FURTHER READING

▶ **Correspondence analysis**
▶ **Data analysis**
▶ **Determinant**
▶ **Hat matrix**
▶ **Inertia matrix**
▶ **Inverse matrix**
▶ **Multiple linear regression**
▶ **Regression analysis**
▶ **Trace**
▶ **Transpose**

REFERENCES

Balestra, P.: Calcul matriciel pour économistes. Editions Castella, Albeuve, Switzerland (1972)

M

Graybill, F.A.: Theory and Applications of the Linear Model. Duxbury, North Scituate, MA (Waldsworth and Brooks/Cole, Pacific Grove, CA) (1976)

Searle, S.R.: Matrix Algebra Useful for Statistics. Wiley, New York (1982)

Maximum Likelihood

The term maximum likelihood refers to a method of estimating parameters of a **population** from a random **sample**. It is applied when we know the general form of distribution of the population but when one or more parameters of this distribution are unknown. The method consists in choosing an estimator of unknown parameters whose values maximize the **probability** of obtaining the observed sample.

HISTORY

Generally the maximum likelihood method is attributed to **Fisher, Ronald Aylmer**. However, the method can be traced back to the works of the 18th-century scientists Lambert, J.H. and Bernoulli, D. Fisher, R.A. introduced the method of maximum likelihood in his first statistical publications in 1912.

MATHEMATICAL ASPECTS

Consider a **sample** of n random variables $X_1, X_2, \ldots, X_n$, where each random variable is distributed according to a **probability distribution** giving a **density function** $f(x, \theta)$, with $\theta \in \Omega$, where Ω is the parameter space. The **joint density function** of $X_1, X_2, \ldots, X_n$ is

$$f(x_1, \theta) \cdot f(x_2, \theta) \cdot \ldots \cdot f(x_n, \theta).$$

For fixed x, this function is a function of θ and is called likelihood function L:

$$L(\theta; x_1, x_2, \ldots, x_n) = f(x_1, \theta) \cdot f(x_2, \theta) \\ \cdot \ldots \cdot f(x_n, \theta).$$

An **estimator** is the maximum likelihood estimator of $\widehat{\theta}$ if it maximizes the likelihood function. Note that this function L can be maximized by setting to zero the first partial derivative by θ and resolving the equations thus obtained. Moreover, when each function L and lnL are maximized for each value of θ, it is often possible to resolve:

$$\frac{\partial \ln L(\theta; x_1, x_2, \ldots, x_n)}{\partial \theta} = 0.$$

Finally, the maximum likelihood estimators are not always unique and this method does not always give unbiased estimators.

EXAMPLES

Let $X_1, X_2, \ldots, X_n$ be a random **sample** coming from the normal distribution with the **mean** μ and the **variance** σ^2. The method of **estimation** of maximum likelihood consists in choosing as estimators of unknown parameters μ and σ^2 the values that maximize the **probability** of having the observed sample.

Thus maximizing the likelihood function entails the following:

$$L\left(\mu, \sigma^2; x_1 \ldots, x_n\right)$$

$$= \frac{1}{\left(\sigma\sqrt{2\pi}\right)^n} \exp\left(\frac{-\sum_{i=1}^{n}(x_i - \mu)^2}{2\sigma^2}\right),$$

or, which is the same thing, its natural logarithm:

$$\ln L = \ln \left[\frac{1}{\left(\sigma \sqrt{2\pi}\right)^n} \cdot \exp\left(\frac{-\sum_{i=1}^{n}(x_i - \mu)^2}{2\sigma^2} \right) \right]$$

$$= \ln \left(\sigma \sqrt{2\pi}\right)^{-n} - \frac{\sum_{i=1}^{n}(x_i - \mu)^2}{2\sigma^2}$$

$$= \ln (2\pi)^{-\frac{n}{2}} + \ln (\sigma)^{-n}$$

$$- \frac{\sum_{i=1}^{n}(x_i - \mu)^2}{2\sigma^2}$$

$$= -\frac{n}{2} \cdot \ln (2\pi) - n \cdot \ln(\sigma)$$

$$- \frac{\sum_{i=1}^{n}(x_i - \mu)^2}{2\sigma^2}.$$

We set equal to zero the partial derivatives $\frac{\partial \ln L}{\partial \mu}$ and $\frac{\partial \ln L}{\partial \sigma}$:

$$\frac{\partial lnL}{\partial \mu} = \frac{1}{\sigma^2} \cdot \sum_{i=1}^{n}(x_i - \mu) = 0, \quad (1)$$

$$\frac{\partial lnL}{\partial \sigma} = -\frac{n}{\sigma} + \frac{1}{\sigma \cdot \sigma^2} \cdot \sum_{i=1}^{n}(x_i - \mu)^2 \quad (2)$$

$$= 0.$$

From (1) we deduce:

$$\frac{1}{\sigma^2} \cdot \sum_{i=1}^{n}(x_i - \mu) = 0,$$

$$\sum_{i=1}^{n}(x_i - \mu) = 0,$$

$$\sum_{i=1}^{n} x_i - n \cdot \mu = 0,$$

$$\hat{\mu} = \frac{1}{n} \cdot \sum_{i=1}^{n} x_i = \bar{x}.$$

Thus for the **normal distribution**, the **arithmetic mean** is the maximum likelihood estimator.

From (2) we get:

$$-\frac{n}{\sigma} + \frac{1}{\sigma \cdot \sigma^2} \cdot \sum_{i=1}^{n}(x_i - \mu)^2 = 0,$$

$$\frac{1}{\sigma \cdot \sigma^2} \cdot \sum_{i=1}^{n}(x_i - \mu)^2 = \frac{n}{\sigma},$$

$$\hat{\sigma}^2 = \frac{1}{n} \cdot \sum_{i=1}^{n}(x_i - \mu)^2.$$

We find in this case as maximum likelihood estimator the empirical variance of the sample.

Note that the maximum likelihood estimation for the **mean** is without **bias** and those of the variance are biased (divided by n instead of $n - 1$). We find that for the normal distribution, the mean and variance of the sample are the maximum likelihood estimators of the mean and variance of the **population**. This is not necessarily true for other distributions: consider, for example, the case of the **uniform distribution** in the interval $[a, b]$, where a and b are unknown and where the parameter to be estimated is the mean of the distribution $\mu = (a + b)/2$. We can verify that the estimator of the maximum likelihood of μ equals:

$$\hat{\mu} = \frac{x_1 + x_n}{2},$$

where x_1 and x_n are the smallest and the largest **observations** of the sample $(x_1, x_2, \ldots, x_n)$, respectively. This estimator is the middle point of the interval with the limits x_1 and x_n. We see that in this case, the maximum likelihood estimator is different from the mean of the sample:

$$\hat{\mu} \neq \bar{x} .$$

FURTHER READING
▶ **Estimation**
▶ **Estimator**
▶ **Likelihood ratio test**
▶ **Robust estimation**

REFERENCES
Fisher, R.A.: On an absolute criterium for fitting frequency curves. Mess. Math. **41**, 155–160 (1912)

DOMAINS AND LIMITATIONS
Empirically, the determination of the **arithmetic mean** of a set of observations gives us a measure of the central **value** of this set. This measure gives us an *estimation* of the mean μ of **random variable** X from which the observations are given.

FURTHER READING
▶ **Arithmetic mean**
▶ **Expected value**
▶ **Measure of central tendency**
▶ **Measure of kurtosis**
▶ **Measure of skewness**
▶ **Weighted arithmetic mean**

Mean

The mean of a **random variable** is a **central tendency measure** of this **variable**. It is also called the **expected value**. In practice, the term mean is often used in the meaning of the **arithmetic mean**.

HISTORY
See **arithmetic mean**.

MATHEMATICAL ASPECTS
Let X be a continuous **random variable** whose **density function** is $f(x)$. The mean μ of X, if it exists, is given by:

$$\mu = \int x \cdot f(x) \, dx .$$

The mean of a discrete random variable X of **probability function** $P(X = x)$, if it exists, is given by:

$$\mu = \sum x \cdot P(X = x) .$$

Mean Absolute Deviation

The mean deviation of a set of quantitative **observations** is a **measure of dispersion** indicating the **mean** of the absolute values of the **deviations** of each observation with respect to a **measure of position** of the distribution.

HISTORY
See L_1 **estimation**.

MATHEMATICAL ASPECTS
Consider $x_1, x_2, \dots, x_n$ a set of n quantities or of n **observations** *related* to a **quantitative variable** X. According to the definition, the mean deviation is calculated as follows:

$$EM = \frac{\sum_{i=1}^{n} |x_i - \bar{x}|}{n} ,$$

where $\bar{x}$ is the **arithmetic mean** of the observations. When the **observations** are ordered in a **frequency distribution**, the mean deviation is calculated as follows:

$$EM = \frac{\sum_{i=1}^{k} f_i \cdot |x_i - \bar{x}|}{\sum_{i=1}^{k} f_i},$$

where

x_i represents the different **values** of the **variable**,

f_i is the **frequencies** associated to these values, and

k is the number of different **values**.

DOMAINS AND LIMITATIONS

The mean absolute deviation is a measure of dispersion more robust than the standard deviation.

EXAMPLES

Five students successively passed two exams for which they obtained the following grades:

Exam 1:

$$\overline{3.5 \ 4 \ 4.5 \ 3.5 \ 4.5} \qquad \bar{x} = \frac{20}{5} = 4$$

Exam 2:

$$\overline{2.5 \ 5.5 \ 3.5 \ 4.5 \ 4} \qquad \bar{x} = \frac{20}{5} = 4$$

The **arithmetic mean** $\bar{x}$ of the grades is identical for both exams. Nevertheless, the variability of the grades (or the dispersion) is not the same.

For exam 1, the mean deviation is equal to:

$$\begin{aligned}
\text{Mean} &= \frac{|3.5 - 4|}{5} + \frac{|4 - 4|}{5} \\
\text{deviation} &\quad + \frac{|4.5 - 4|}{5} + \frac{|3.5 - 4|}{5} \\
&\quad + \frac{|4.5 - 4|}{5}
\end{aligned}$$

$$= \frac{2}{5} = 0.4.$$

For exam 2, the mean deviation is much higher:

$$\begin{aligned}
\text{Mean} &= \frac{|2.5 - 4|}{5} + \frac{|5.5 - 4|}{5} \\
\text{deviation} &\quad + \frac{|3.5 - 4|}{5} + \frac{|4.5 - 4|}{5} \\
&\quad + \frac{|4 - 4|}{5}
\end{aligned}$$

$$= \frac{4}{5} = 0.8.$$

The grades of the second exam are therefore more dispersed around the **arithmetic mean** than the grades of the first exam.

FURTHER READING
▶ L_1 **estimation**
▶ **Measure of dispersion**
▶ **Standard deviation**
▶ **Variance**

Mean Squared Error

Let $\hat{\theta}$ be an **estimator** of **parameter** θ. The squared **error** in estimating θ by $\hat{\theta}$ is $\left(\theta - \hat{\theta}\right)^2$.

The mean squared error *(MSE)* is the **mathematical expectation** of this value:

$$MSE(\hat{\theta}) = E\left[\left(\theta - \hat{\theta}\right)^2\right].$$

We can prove that the mean squared **error** equals the sum of the **variance** of the estimator and of the square of the **bias**:

$$MSE(\hat{\theta}) = \text{Var}\left(\hat{\theta}\right) + \left(E\left[\hat{\theta}\right] - \theta\right)^2.$$

Thus when the considered estimator is without bias, the mean squared error corresponds exactly to the variance of the estimator.

HISTORY

In the early 19th century, **de Laplace, Pierre Simon** (1820) and **Gauss, Carl Friedrich** (1821) compared the **estimation** of an unknown quantity, based on observations with random errors, with a chance game. They drew a parallel between the error of the estimated **value** and the loss due to such a game. Gauss, Carl Friedrich proposed a measure of inexactitude, the square of error. He recognized that his choice was random, but he justified it by the mathematical simplicity of the function "make the square".

MATHEMATICAL ASPECTS

Let $X_1, X_2, \ldots, X_n$ a **sample** of dimension n and $\hat{\theta}$ an **estimator** of **parameter** θ; the deviation (squares) of the **estimation** is given by $\left(\theta - \hat{\theta}\right)^2$.

The mean squared **error** is defined by:

$$MSE\left(\hat{\theta}\right) = E\left[\left(\theta - \hat{\theta}\right)^2\right].$$

We can prove that:

$$MSE\left(\hat{\theta}\right) = \text{Var}\left(\hat{\theta}\right) + \left(E\left[\hat{\theta}\right] - \theta\right)^2,$$

that is, the mean squared error equals the sum of the **variance** of the estimator and the square of the **bias**.

In fact:

$$MSE(\hat{\theta}) = E\left[\left(\theta - \hat{\theta}\right)^2\right]$$

$$= E\left[\left(\theta - E\left[\hat{\theta}\right] + E\left[\hat{\theta}\right] - \hat{\theta}\right)^2\right]$$

$$= E[\left(\theta - E\left[\hat{\theta}\right]\right)^2 + \left(E\left[\hat{\theta}\right] - \hat{\theta}\right)^2$$

$$+ 2 \cdot \left(\theta - E\left[\hat{\theta}\right]\right) \cdot \left(E\left[\hat{\theta}\right] - \hat{\theta}\right)]$$

$$= E\left[\left(\theta - E\left[\hat{\theta}\right]\right)^2\right] + E\left[\left(E\left[\hat{\theta}\right] - \hat{\theta}\right)^2\right]$$

$$+ 2 \cdot \left(\theta - E\left[\hat{\theta}\right]\right) \cdot (E\left[\hat{\theta}\right] - E\left[\hat{\theta}\right])$$

$$= E\left[\left(E\left[\hat{\theta}\right] - \hat{\theta}\right)^2\right] + \left(\theta - E\left[\hat{\theta}\right]\right)^2$$

$$= \text{Var}\left(\hat{\theta}\right) + (bias)^2,$$

where the bias is $E[\hat{\theta}] - \theta$.
Thus when the estimator $\hat{\theta}$ is without bias, that is, $E[\hat{\theta}] = \theta$, we get:

$$MSE\left(\hat{\theta}\right) = \text{Var}\left(\hat{\theta}\right).$$

The efficiency of an estimator is inversely proportional to its mean squared error. We say that the estimator $\hat{\theta}$ is consistent if

$$\lim_{n \to \infty} MSE(\hat{\theta}) = 0.$$

In this case the variance and the bias of the estimator tend to 0 when the number of observations n tends to infinity.

DOMAINS AND LIMITATIONS

Certain authors prefer to use the **mean deviation** in absolute value defined by $E[|\theta - \hat{\theta}|]$, where $\hat{\theta}$ is the **estimator** of **parameter** θ. The disadvantage of the last definition is at the level of mathematical properties, because the function of absolute value cannot be differentiated everywhere.

EXAMPLES

We consider a **multiple linear regression** in matrix form:

$$\mathbf{Y} = \mathbf{X} \cdot \boldsymbol{\beta} + \boldsymbol{\varepsilon},$$

where
Y is a **vector** of order n (n observations of the **independent variable**),
X is a given **matrix** $n \times p$,
$\boldsymbol{\beta}$ is the vector of order p of the coefficients to be determined, and

$\boldsymbol{\varepsilon}$ is the vector of errors that we suppose are independent and each following a **normal distribution** with **mean** 0 and **variance** σ^2.

The vector $\hat{\boldsymbol{\beta}} = (\mathbf{X}' \cdot \mathbf{X})^{-1} \cdot \mathbf{X}' \cdot \mathbf{Y}$ gives an **estimator** without **bias** of $\boldsymbol{\beta}$.

We then obtain:

$$MSE\left(\hat{\boldsymbol{\beta}}\right) = \mathrm{Var}\left(\hat{\boldsymbol{\beta}}\right).$$

FURTHER READING

▶ **Bias**
▶ **Estimator**
▶ **Expected value**
▶ **Variance**

REFERENCES

Gauss, C.F.: In: Gauss's Work (1803–1826) on the Theory of Least Squares. Trans. Trotter, H.F. Statistical Techniques Research Group, Technical Report No. 5. Princeton University Press, Princeton, NJ (1821)

Laplace, P.S. de: Théorie analytique des probabilités, 3rd edn. Courcier, Paris (1820)

Measure of Central Tendency

A measure of central tendency is a statistic that summarizes a set of data relative to a **quantitative variable**. More precisely, it allows to determine a fixed **value**, called a central value, around which the set of data has the tendency to group. The principal measures of central tendencies are:

- **Arithmetic mean**
- **Median**
- **Mode**

MATHEMATICAL ASPECTS

The **mean** μ, **median** M_d, and **mode** M_o are measures of the center of distribution.

In a symmetric distribution, the mean, median, and mode coincide in a common **value**. This value gives the center of symmetry of the distribution. For a symmetric unimodal distribution, we have:

$$\mu = M_d = M_o.$$

In this case, an **estimation** of the center of the distribution can be given by the mean, median, or mode.

In an asymmetric distribution, the different measures of central tendency have different values. (Their comparison can be used as a **measure of skewness**.)

For a unimodal distribution stretched to the right, we generally have:

$$\mu \geq M_d \geq M_o.$$

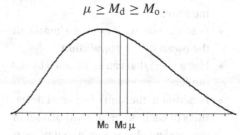

This relation is inverted when the distribution is unimodal and stretched to the left:

$$\mu \leq M_d \leq M_o.$$

For a unimodal distribution, moderately asymmetric, it was empirically proved that

M

the mode M_o, the median M_d, and the mean μ often satisfy the following relation, in an approximate manner:

$$M_o - \mu = 3 \cdot (M_d - \mu) \, .$$

DOMAINS AND LIMITATIONS

In practice, the choice of the measure allowing one to characterize at best the center of a set of observations depends on the specific features of the statistical studies and the information we wish to obtain.

Let us compare the **arithmetic mean**, the **mode**, and the **median** depending on different criteria:

1. The *arithmetic mean*:
   - Depends on the **value** of all observations.
   - Is simple to interpret.
   - Is the most familiar and the most used measure.
   - Is frequently used as an **estimator** of the **mean** of the **population**.
   - Has a value that can be falsified by the outliers.
   - In addition, the sum of squared deviations of each observation x_i of a set of data and a value α is minimal when α equals the arithmetic mean:

$$\min_\alpha \sum_{i=1}^{n} (x_i - \alpha)^2$$
$$\Rightarrow \quad \alpha = \text{arithmet. mean}$$

2. The *median*:
   - Is easy to determine because only one data classification is needed.
   - Is easy to understand but less used than the arithmetic mean.
   - Is not influenced by outliers, which gives it an advantage over the arith-

metic mean, if the series really have outliers.
   - Is used as an **estimator** of central values of a distribution, especially when it is asymmetric or has outliers.
   - The sum of squared deviations in absolute value between each observation x_i of a set of data and a value α is minimal when α equals the median:

$$\min_\alpha \sum_{i=1}^{n} |x_i - \alpha| \quad \Rightarrow \quad \alpha = M_d \, .$$

3. The *mode*:
   - Has practical interest because it is the most represented value of a set.
   - Is in any event a rarely used measure.
   - Has a value that is little influenced by outliers.
   - Has a value that is strongly influenced by the fluctuations of a **sampling**. It can strongly vary from one **sample** to another.

In addition, there can be many (or no) modes in a data set.

FURTHER READING

▶ **Arithmetic mean**
▶ **Expected value**
▶ **Mean**
▶ **Measure of skewness**
▶ **Median**
▶ **Mode**
▶ **Weighted arithmetic mean**

REFERENCES

Stuart, A., Ord, J.K.: Kendall's advanced theory of statistics. Vol. I. Distribution theory. Wiley, New York (1943)

Dodge, Y.: Premiers pas en statistique. Springer, Paris (1999)

Measure of Dispersion

A measure of dispersion allows to describe a set of data concerning a particular **variable**, giving an indication of the variability of the **values** inside the data set. The measure of dispersion completes the description given by the **measure of central tendency** of a distribution.

If we observe different distributions, we can say that for some of them, all the data are grouped in a more or less short **distance** from the central value; for others the distance is much greater.

We can class the measures of dispersion into two groups:

1. Measures defined by the distance between two representative values of the distribution:
   - **Range**, also called the **interval** of variation
   - **Interquartile Range**
2. Measures calculated depending on deviations of each datum from a central value:
   - **Geometric deviation**
   - **Median deviation**
   - **Mean absolute deviation**
   - **Standard deviation**

Among measures of dispersion, the most important and most used is the standard deviation.

DOMAINS AND LIMITATIONS

Except for the descriptive aspect allowing one to characterize a distribution or compare many distributions, the knowledge of the dispersion or of the variability of a distribution can have a considerable practical interest. What would happen if our decisions were based on the **mean** alone.

- Roads would be constructed to get only the mean traffic, and the traffic jams during holidays would be immeasurable.
- Houses would be constructed in such as way as to resist the mean force of wind with the consequences that would result in the case of a strong storm.

These examples show that in certain cases, having a **measure of central tendency** is not enough to be able to make a reliable decision. Moreover, the measure of dispersion is an indication of the character of the central tendency measure. The smaller the variability of a given set, the more the value of the measure of central tendency will be representative of the data set.

FURTHER READING

- ▶ **Coefficient of variation**
- ▶ **Interquartile range**
- ▶ **Mean absolute deviation**
- ▶ **Measure of central tendency**
- ▶ **Range**
- ▶ **Standard deviation**
- ▶ **Variance**

REFERENCES

Dodge, Y.: Premiers pas en statistique. Springer, Paris (1999)

Measure of Dissimilarity

The measure of dissimilarity is a distance measure that assigns to pair of objects a real positive (or zero) number.

If we suppose, for example, that each object is characterized by n variables, a measure of dissimilarity between two objects would consist in giving the number of different points that represent two considered objects.

This number is a whole number between 0 and n and equals n when two objects have no common point.

MATHEMATICAL ASPECTS

Let Ω be a **population** on which we would like to define a measure of dissimilarity s'. We call a measure of dissimilarity s' any mapping of $\Omega \times \Omega$ to the real positive (or zero) numbers, satisfying the following two properties:

- For each w, w' in Ω, $s'(w, w') = s'(w', w)$; that is, s' is symmetric.
- For each w in Ω, $s'(w, w) = 0$, that is, the measure of dissimilarity between an object and itself is zero.

DOMAINS AND LIMITATIONS

As a measure of dissimilarity s' on a finite **population**, Ω is a limited application, that is:

$$\max \left\{ s'\left(w, w'\right) ; w, w' \in \Omega \right\} = m < \infty .$$

We can define a **measure of similarity** on the population Ω by:

$$s\left(w, w'\right) = m - s'\left(w, w'\right) .$$

These two notions, measure of similarity and measure of dissimilarity, can be considered complementary.

We use the measure of dissimilarity to resolve the problems of **classification**, instead of a **distance** that is a particular case of it and that has the so-called property of triangular inequality.

EXAMPLES

We consider the **population** Ω composed of three flowers

$$\Omega = \{w_1, w_2, w_3\}$$

on which the following three variables were measured:

- v_1: color of flower
- v_2: number of petals
- v_3: number of leaves

where

v_1 is summarized here as a **dichotomous variable** because we consider only red (R) and yellow (Y) flowers and

v_2 and v_3 are discrete random variables taking their values in cardinal numbers.

We have observed:

		Variables		
		v_1	v_2	v_3
	w_1	R	5	2
Ω	w_2	Y	4	2
	w_3	R	5	1

We determine for each couple of flowers the number of points that distinguish them. We see that we have a measure of dissimilarity for the set of three flowers:

$$s'(w_1, w_1) = s'(w_2, w_2) = s'(w_3, w_3) = 0$$

because a flower is not different from itself;

$$s'(w_1, w_2) = s'(w_2, w_1) = 2$$

because the color and the number of petals are different at w_1 and w_2;

$$s'(w_1, w_3) = s'(w_3, w_1) = 1$$

as only the number of leaves allows to distinguish w_1 from w_3;

$$s'(w_2, w_3) = s'(w_3, w_2) = 3$$

as w_2 and w_3 are different in each of the three studied characteristics.

With the help of these possible values, we can deduce that s' satisfies the two properties of a measure of dissimilarity.

FURTHER READING
▶ **Classification**
▶ **Distance**
▶ **Measure of similarity**
▶ **Population**

REFERENCES
Celeux, G., Diday, E., Govaert, G., Lecheval-lier, Y., Ralambondrainy, H.: Classification automatique des données—aspects statistiques et informatiques. Dunod, Paris (1989)

Hand, D.J.: Discrimination and classification. Wiley, New York (1981)

Measure of Kurtosis

The measures of kurtosis are a part of the measures of form and characterize an aspect of the form of a given distribution. More precisely, they characterize the degree of kurtosis of the distribution toward a normal distribution. Certain distributions are close to the **normal distribution** without being totally identical. It is very useful to be able to test if the form of the distribution represents a deviation from a kurtosis of the normal distribution.

We talk about a platicurtical distribution if the curve is more flattened than the normal curve and of a leptocurtical distribution if it is sharper than the normal curve.

To test the kurtosis of a curve, we use the **coefficient of kurtosis**.

MATHEMATICAL ASPECTS

Given a random variable X with probability distribution f_x the coefficient of kurtosis is defined as:

$$\gamma_2 = \frac{\mu_4}{\sigma^4} - 3,$$

where μ_4 stands for the fourth moment about the mean, and σ^4 is the squared variance. The sample kurtosis is obtained by using sampling analogues, that is:

$$\hat{\gamma} = \frac{\hat{\mu}_4}{\hat{\sigma}^4} - 3.$$

FURTHER READING
▶ **Coefficient of kurtosis**
▶ **Measure of shape**
▶ **Measure of skewness**

REFERENCES
Joanes, D.N. & Gill, C.A.: Comparing measures of samp skewness and kurtosis, JRSS.D **47**(1), 183–189 (1998)

Measure of Location

A measure of location (or of position) is a measure that proposes to synthesize a set of statistical data by only one fixed **value**, highlighting a particular point of the studied **variable**. The most frequently used measures of position are:

- Measures of central tendency (**mean, mode, median**) that tend to determine the center of a set of data.
- The **quantiles** (**quartile, decile, centile**), which tend to determine not the center, but a particular position of a subset of data.

DOMAINS AND LIMITATIONS

The measures of location, as well as those of dispersion and of form, are used especially to compare many sets of data or many frequency distributions.

FURTHER READING
▶ **Measure of central tendency**
▶ **Quantile**

Measure of Shape

We can try to characterize the form of a **frequency distribution** with the help of appropriate coefficients. Certain frequency distributions are close to the **normal distribution** ("bell curve"). These distributions can also represent an asymmetry or a kurtosis from the normal curve. The two measures of form are:

- **Measure of asymmetry**
- **Measure of kurtosis.**

HISTORY

Pearson, Karl (1894, 1895) was the first to test the differences between certain distributions and the **normal distribution**. We have shown that deviations from the normal curve can be characterized by moments of order 3 and 4 of a distribution.

Before 1890, Gram, Jorgen Pedersen and Thiele, Thorvald Nicolai of Denmark developed a theory about the symmetry of frequency curves. After Pearson, Karl published his sophisticated and extremely interesting system (1894, 1895), many articles were published on this subject.

FURTHER READING

▶ **Measure of dispersion**
▶ **Measure of kurtosis**
▶ **Measure of location**
▶ **Measure of skewness**

REFERENCES

Pearson, K.: Contributions to the mathematical theory of evolution. I. In: Karl Pearson's Early Statistical Papers. Cambridge University Press, Cambridge, pp. 1–40 (1948). First published in 1894

as: On the dissection of asymmetrical frequency curves. Philos. Trans. Roy. Soc. Lond. Ser. A **185**, 71–110

Pearson, K.: Contributions to the mathematical theory of evolution. II: Skew variation in homogeneous material. In: Karl Pearson's Early Statistical Papers. Cambridge University Press, Cambridge, pp. 41–112 (1948). First published in 1895 in Philos. Trans. Roy. Soc. Lond. Ser. A **186**, 343–414

Measure of Similarity

The measure of similarity is a distance measure assigns a real positive (or zero) number correspond to a pair of objects of the studied **population**. If we suppose, for example, that each object is characterized by n variables, a measure of similarity between two objects would consist in giving the number of common points to two considered objects. This number is a whole one between 0 and n and equals n when the two objects are identical.

MATHEMATICAL ASPECTS

Let Ω be a **population** on which we would like to define a measure of similarity s. We call a measure of similarity s any mapping from $\Omega \times \Omega$ to the real positive (or zero) numbers, satisfying:

- For each w, w' in Ω, $s\left(w, w'\right) = s\left(w', w\right)$, that is, s is symmetric;
- For each w, w' in Ω, with $w \neq w'$

$$s(w, w) = s\left(w', w'\right) > s\left(w, w'\right),$$

that is, an object is more "similar" to itself than to another object.

DOMAINS AND LIMITATIONS

As a measure of similarity s on a **population**, Ω is a limited application, that is:

$$\max \left\{ s\left(w, w'\right); w, w' \in \Omega \right\} =$$
$$s\left(w, w\right) < \infty,$$

where w is any object of Ω; if we pose $m = s(w, w)$, then we can define a **measure of dissimilarity** of the population Ω by:

$$s'\left(w, w'\right) = m - s\left(w, w'\right).$$

These two notions, measure of similarity and measure of dissimilarity, can be considered complementary.

We use the measures of similarity, as well as the measures of dissimilarity, for **classification** problems.

EXAMPLES

Let the **population** Ω be the forms of three flowers

$$\Omega = \{w_1, w_2, w_3\}$$

on which we measure the three variables:

- v_1: color of flower
- v_2: number of petals
- v_3: number of leaves

where

v_1 is summarized here as a **dichotomous variable** because we consider only red (R) and yellow (Y) flowers and

v_2 and v_3 are discrete random variables taking their values in cardinal numbers.

We have observed:

Variables

	v_1	v_2	v_3
w_1	R	5	2
w_2	Y	4	2
w_3	R	5	1

Ω

We determine for each couple of flowers the number of points that are common to them. We see that we have a measure of similarity of the set of three flowers:

$$s\left(w_1, w_1\right) = s\left(w_2, w_2\right) = s\left(w_3, w_3\right) = 3,$$

which corresponds to the number of variables;

$$s\left(w_1, w_2\right) = s\left(w_2, w_1\right) = 1$$

because the color and number of petals are different at w_1 and w_2;

$$s\left(w_1, w_3\right) = s\left(w_3, w_1\right) = 2$$

as only the number of leaves allows us to distinguish w_1 from w_3;

$$s\left(w_2, w_3\right) = s\left(w_3, w_2\right) = 0$$

as w_2 and w_3 have no common points.

With the help of these possible values, we can deduce that s satisfies the two properties of a measure of similarity.

FURTHER READING

- ► **Distance**
- ► **Measure of dissimilarity**
- ► **Population**

REFERENCES

See **measure of dissimilarity**.

Measure of Skewness

In a symmetric distribution, the **median**, **arithmetic mean**, and **mode** are in the same central point. This is not true when the distribution is skewed. In this case, the mode is separate from the arithmetic mean, and the median is between two of them. In consequence, it is necessary to develop the measures of skewness to study the degree of deviation of the form of the distribution from symmetric distribution. The principal skewness measures are:

- **Yule and Kendall coefficient**
- **Coefficient of skewness**

M

MATHEMATICAL ASPECT

Given a random variable X with prob. distribution f_x, the coefficient of skewness is defined as:

$$\gamma_1 = \frac{\mu_3}{\sigma^3}$$

where μ_3 stands for the third moment about the mean, and σ^3 is the stand deviation.

HISTORY

See **coefficient of skewness**.

FURTHER READING

▶ **Coefficient of skewness**
▶ **Measure of kurtosis**
▶ **Measure of shape**
▶ **Yule and Kendall coefficient**

Median

The median is a **measure of central tendency** defined as the **value** that is in the center of a set of ordered observations when it is in increasing or decreasing order.

For a random variable X, the median, denoted by M_{d}, has the property that:

$$P(X \leq M_{\mathrm{d}}) \leq \tfrac{1}{2} \text{ and } P(X \geq M_{\mathrm{d}}) \leq \tfrac{1}{2}.$$

We find then 50% of the observations on each side of the median:

If we have an odd number of observations, the median corresponds to the values of the middle observation. If we have an even number of observations, there is no unique observation in the middle. The median will be given by the **arithmetic mean** of the values of

two observations of the middle of ordered observations.

HISTORY

In 1748, Euler, Leonhard and Mayer, Johann Tobias proposed, independently, a method that consists in dividing the observations of a data set into two equal parts.

MATHEMATICAL ASPECTS

When the observations are given individually, the process of calculating the median is simple:

1. Arrange the n observations in increasing or decreasing order.
2. If the number of observations is odd:
   - Find the observation of the middle $\frac{n+1}{2}$.
   - The median equals the **value** of the middle observation.
3. If the number of observations is even:
   - Find the two observations of the middle $\frac{n}{2}$ and $\frac{n}{2} + 1$.
   - The median equals the **arithmetic mean** of the values of these two observations.

When we do not have individual observations but they are grouped into classes, the median is determined as follows:

1. Determine the median class (the one that contains the $\frac{n}{2}$-th observation).
2. Calculate the median using the following formula:

$$M_{\mathrm{d}} = L + \left[\frac{\frac{n}{2} - \sum f_{\text{inf}}}{f_{M_{\mathrm{d}}}} \right] \cdot c,$$

where

L is the lower limit of the median class,

n is the total number of observations,

$\sum f_{inf}$ is the sum of the frequencies smaller than the median class,

f_{M_d} is the frequency of the median class, and

c is the length of the **interval** of the median class.

This formula is based on the **hypothesis** according to which the observations are uniformly distributed inside each class. It also supposes that the lower and upper limits of the median class are defined and known.

Properties

The sum of deviations in absolute values between each observation x_i of a data set and a value α is minimal when α equals the median:

$$\text{minimize}_{\alpha} \sum_{i=1}^{n} |x_i - \alpha| \implies \alpha = M_d.$$

DOMAINS AND LIMITATIONS

In the calculation of the **arithmetic mean**, we take into account the **value** of each **observation**. Thus an **outlier** strongly influences the arithmetic mean. On the other hand, in the calculation of the median, we take into account only the rank of the observations. Thus the outliers do not influence the median. When we have a strongly skewed **frequency distribution**, we are interested in choosing the median as a **measure of central tendency** and not the arithmetic mean in order to neutralize the effect of the extreme values.

EXAMPLES

The median of a set of five numbers 42, 25, 75, 40, and 20 is 40 because 40 is the middle

number when the numbers are ordered:

$$20 \quad 25 \quad \underbrace{40}_{\text{median}} \quad 42 \quad 75 \,.$$

If we add the number 62 to this set, the number of observations is even, and the median equals the **arithmetic mean** of the two middle observations:

$$20 \quad 25 \quad \underbrace{40 \quad 42}_{\frac{40+42}{2}=41=\text{median}} \quad 62 \quad 75 \,.$$

We consider now the median of data grouped into classes. Let us take, for example, the following **frequency table** that represents the daily revenue of 50 grocery stores:

Class (revenue in euros)	Frequency f_i (number of grocery stores)	Cumulated frequency
500–550	3	3
550–600	12	15
600–650	17	32
650–700	8	40
700–750	6	46
750–800	4	50
Total	50	

The median class is the class 600–650 as it contains the middle observation (the 25th observation).

Supposing that the observations are uniformly distributed in each class, the median equals:

$$M_d = 600 + \frac{\frac{50}{2} - 15}{17} \cdot (650 - 600)$$

$$= 629.41 \,.$$

This result means that 50% of the grocery stores have daily revenues in excess of 629.41 euros and the other 50% have revenues totaling less than 629.41 euros.

FURTHER READING
► **Mean**
► **Measure of central tendency**
► **Mode**
► **Quantile**

REFERENCES

Boscovich, R.J.: De Litteraria Expeditione per Pontificiam ditionem, et Synopsis amplioris Operis, ac habentur plura eius ex exemplaria etiam sensorum impressa. Bononiensi Scientiarium et Artium Instituto Atque Academia Commentarii, vol. IV, pp. 353–396 (1757)

Euler, L.: Recherches sur la question des inégalités du mouvement de Saturne et de Jupiter, pièce ayant remporté le prix de l'année 1748, par l'Académie royale des sciences de Paris. Republié en 1960, dans Leonhardi Euleri, Opera Omnia, 2ème série. Turici, Bâle, 25, pp. 47–157 (1749)

Mayer, T.: Abhandlung über die Umwälzung des Monds um seine Achse und die scheinbare Bewegung der Mondflecken. Kosmographische Nachrichten und Sammlungen auf das Jahr 1748 **1**, 52–183 (1750)

Median Absolute Deviation

The median absolute deviation of a set of quantitative observations is a **measure of dispersion**. It corresponds to the **mean** of the absolute values of deviation of each observation relative to the **median**.

HISTORY

See L_1 **estimation**.

MATHEMATICAL ASPECTS

Let $x_1, x_2, \ldots, x_n$ be a set of n observations relative to a **quantitative variable** X.

The median absolute deviation, denoted by E_{MAD}, is calculated as follows:

$$E_{\text{MAD}} = \frac{\sum_{i=1}^{n} |x_i - Md|}{n} \, ,$$

where Md is the median of the observations.

DOMAINS AND LIMITATIONS

The median absolute deviation is a fundamental notion of the L_1 **estimation**. The L_1 **estimation** of a **parameter** of the **measure of central tendency** θ is really a method of **estimation** based on the minimum absolute deviations of observations x_i from parameter θ:

$$\sum_{i=1}^{n} |x_i - \theta| \, .$$

The value that minimizes this expression is the **median** of the **sample**. In the case of data containing outliers, this method is more efficient than the **least-squares** method.

FURTHER READING
► L_1 **estimation**
► **Measure of dispersion**
► **Standard deviation**
► **Variance**

Method of Moments

The method of moments is used for estimating the parameters of a distribution from a **sample**. The idea is as follows. If the number of parameters to be estimated equals k, we pose the k first moments of the distribution (whose expression depends on unknown parameters) equal the corresponding empirical moments, that is, to the estimators of the moments of order k calculated on the sample. We should solve the system of k equations with k unknown variables.

HISTORY

In the late 19th century, **Pearson, Karl** used the method of moments to estimate the parameters of a normal mixed model (with parameters $p_1, p_2, \mu_1, \mu_2, \sigma_1^2, \sigma_2^2$) given by

$$P(x) = \sum_{i=1}^{2} p_i \cdot g_i(x) \, ,$$

where the g_i are normally distributed with the mean μ_i and variance σ_i^2. The method was further developed and studied by Chuprov, A.A. (1874–1926), Thiele, Thorvald Nicolai, **Fisher, Ronald Aylmer**, and **Pearson, Karl**, among others.

MATHEMATICAL ASPECTS

Let $\theta_1, \ldots, \theta_k$ be the k parameters to be estimated for the random variable X with a given distribution.

We denote by $\mu_j((\theta_1, \ldots, \theta_k)) = E[X^j]$ the moment of order j of X expressed as a function of the parameters to be estimated.

The method of moments consists in solving the following system of equations:

$$\mu_j(\theta_1, \ldots, \theta_k) = m_j \ \text{ for } j = 1, \ldots, k \, .$$

To distinguish between real parameters and estimators, we denote the latter by $\hat{\theta}_1, \ldots, \hat{\theta}_k$.

EXAMPLES

1. Let X be a random variable following an **exponential distribution** with unknown parameter θ. We have n observations $x_1, \ldots, x_n$ of X, and we want to estimate θ.

As the moment of order 1 is the mean of the distribution, to estimate θ we should equalize the theoretical and the empirical mean. We know that (see exponential distribution) the (theoretical) mean equals $\frac{1}{\theta}$. The desired estimator $\hat{\theta}$ satisfies

$$\frac{1}{\hat{\theta}} = \frac{1}{n} \sum_{i=1}^{n} x_i.$$

2. According to the method of moments, the **mean** μ and the **variance** σ^2 of a random variable X following a **normal distribution** are estimated by the empirical mean $\bar{x}$ and the variance

$$\frac{1}{n} \sum_{i=1}^{n} (x_i - \bar{x})^2$$

of a sample $\{x_1, x_2, \ldots, x_n\}$ of X.

DOMAINS AND LIMITATIONS

First, we should observe if the system of equations given previously has solutions. It cannot have solutions if, for example, the parameters to be estimated have one or more particular constraints. In any case, the method does not guarantee particularly efficient estimators. For example, the estimator found for the variance of a normal distribution is biased.

FURTHER READING

▶ **Chi-square distance**
▶ **Estimator**
▶ **Least squares**
▶ **Maximum likelihood**
▶ **Moment**
▶ **Parameter**
▶ **Sample**

REFERENCES

Fisher, R.A.: Moments and product moments of sampling distributions. Proc. Lond. Math. Soc. **30**, 199–238 (1929)

Fisher, R.A.: The moments of the distribution for normal samples of measures of departure from normality. Proc. Roy. Soc. Lond. Ser. A **130**, 16–28 (1930)

Kendall, M.G., Stuard, A.: The Advanced Theory of Statistics. Vol. 1. Distribution Theory, 4th edn. Griffin, London (1977)

Pearson, K.: Contributions to the mathematical theory of evolution. I. In: Karl Pearson's Early Statistical Papers. Cambridge University Press, Cambridge, pp. 1–40 (1948). First published in 1894 as: On the dissection of asymmetrical frequency curves. Philos. Trans. Roy. Soc. Lond. Ser. A **185**, 71–110

Thiele, T.N.: Theory of Observations. C. and E. Layton, London; Ann. Math. Stat. **2**, 165–308 (1903)

Missing Data

Missing data occur in statistics when certain outcome are not present. This happens when data are not available (the data were destroyed by a saving error, an animal died during an **experiment**, etc.).

HISTORY

Allan, F.E. and Wishart, J. (1930) were apparently the first to consider the problem of missing data in the analysis of experimental design. Their works were followed by those of Yates, Frank (1933).

Allan, F.E. and Wishart, J., as well as Yates, F., tried to find an **estimate** of missing data by the method of iteration. In a situation where many observations are missing, the proposed method does not give a satisfactory solution. Bartlett, M.S. (1937) used the **analysis of covariance** to resolve the same problem.

Since then, many other methods have been proposed, among them the method of Rubin, D.B. (1972). Birkes, D. et al. (1976) presented an exact method for solving the problem of missing data using the **incidence matrix**. The generalized matrix approach was proposed by Dodge, Y. and Majumdar, D. (1979). More details concerning the historical aspects of missing data from 1933 to 1985 can be found in Dodge, Y. (1985).

MATHEMATICAL ASPECTS

We can solve the problem of missing data by repeating the experiment under the same conditions and so obtain the values for the missing data. Such a solution, despite the fact that it is ideal, is not always economically and materially possible.

There exists also other approaches to solve the problem of missing data in experimental design. We mention the approach by the **generalized inverse** or by the incidence **matrix**, which indicates which observations are missing in the design.

FURTHER READING
▶ **Data**
▶ **Design of experiments**
▶ **Experiment**
▶ **Observation**

REFERENCES
Allan, F.E., Wishart, J.: A method of estimating the yield of a missing plot in field experimental work. J. Agricult. Sci. **20**, 399–406 (1930)

Bartlett, M.S.: Some examples of statistical methods of research in agriculture and applied biology. J. Roy. Stat. Soc. (Suppl.) **4**, 137–183 (1937)

Birkes, D., Dodge, Y., Seely, J.: Spanning sets for estimable contrasts in classification models. Ann. Stat. **4**, 86–107 (1976)

Dodge, Y.: Analysis of Experiments with Missing Data. Wiley, New York (1985)

Dodge, Y., Majumdar, D.: An algorithm for finding least square generalized inverses for classification models with arbitrary patterns. J. Stat. Comput. Simul. **9**, 1–17 (1979)

Rubin, D.B.: A non-iterative algorithm for least squares estimation of missing values in any analysis of variance design. Appl. Stat. **21**, 136–141 (1972)

Yates, F.: The analysis of replicated experiments when the field results are incomplete. Empire J. Exp. Agricult. **1**, 129–142 (1933)

Missing Data Analysis

See **missing data**.

Mode

The mode is a **measure of central tendency**. The mode of a set of observations is the **value** of the observation that have the highest **frequency**. According to this definition, a distribution can have a unique mode (called the unimodal distribution). In some situations a distribution may have many modes (called the bimodal, trimodal, multimodal, etc. distribution).

MATHEMATICAL ASPECTS

Consider a **random variable** X whose **density function** is $f(x)$. M_0 is a mode of the distribution if $f(x)$ has a local maximum at M_0.

Empirically, the mode of a set of observations is determined as follows:

1. Establish the **frequency distribution** of the set of observations.
2. The mode or modes are the values whose **frequency** is greater or equals to the frequency of the other observations.

For the distribution of observations grouped into classes of the same range ($= maximum - minimum$) whose limits are known, the mode is determined as follows:

1. Determine the modal classes—the classes with a frequency greater than or equal to the frequencies of other classes. (If many classes with the same frequency, we group these classes in such a way as to calculate only one mode of this set of classes).
2. Calculate the mode M_0 taking into account the joint classes:

$$M_0 = L + \left(\frac{d_1}{d_1 + d_2} \right) \cdot c,$$

where

L is the smaller limit of the modal class,

d_1 is the difference between the frequency of the modal class and the frequency of the previous class,

d_2 is the difference between the frequency of the modal class and the frequency of the next class, and

c is the length of the modal class, common to all the classes.

M

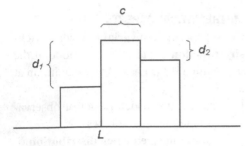

If the length of the classes is not identical, the mode must be calculated by modifying the division of the classes in order to obtain, if possible, the classes of equal length.

DOMAINS AND LIMITATIONS

For a discrete **variable**, the mode has the advantage of being easy to determine and interpret. It is used especially when the distribution is not symmetric.

For a continuous variable, the determination of the mode is not always precise because it depends on the grouping into classes. Generally, the mode is a good **indicator** of the center of the data only if there is one dominant **value** in the set of data. If there are many dominant values, the distribution is called plurimodal. In this case, the modes are not the measure of central tendency. A bimodal distribution generally indicates that the considered **population** is in reality heterogeneous and is composed of two subpopulations having different central values.

EXAMPLES

Let the set be the following numbers:

1 2 2 3 3 3 4 4 4 5 5 5 5 6 6 7 7 8 9.

The **frequency table** derived from this set is:

Values	1	2	3	4	5	6	7	8	9
Frequency	1	2	3	3	4	2	2	1	1

The mode of this set of numbers is 5 because it appears with the greatest **frequency** (4). Consider now the following frequency table, which represents the daily revenues of 50 shops, grouped by class of revenue:

Class (revenue in euros)	Frequency f_i (number of shops)
500–550	3
550–600	12
600–650	17
650–700	8
700–750	6
750–800	4
Total	50

We can immediately see that the modal class is 600–650 because it has the greatest frequency.

The mode is calculated as follows:

$$M_0 = L_1 + \left(\frac{d_1}{d_1 + d_2} \right) \cdot c$$

$$= 600 + \left(\frac{5}{5 + 9} \right) \cdot 50 = 617.86.$$

Let us take once more the same example but grouping the numbers into different classes:

Class (revenue in euros)	Frequency f_i (number of shops)
500–600	15
600–650	17
650–700	8
700–800	10
Total	50

The classes do not have the same length. We will do a new grouping to obtain classes of the same length.

Case 1

Let us take, for example, as reference the class 600–650, whose length is 50, and suppose that the revenues are uniformly distributed in a class.

The table producing this new grouping is the following:

Class (revenue in euros)	Frequency f_i (number of shops)
500–550	7.5
550–600	7.5
600–650	17
650–700	8
700–750	5
750–800	5
Total	50

The modal class is the class 600–650, and the values of the mode are:

$$M_o = 600 + \left(\frac{9.5}{9.5 + 9}\right) \cdot 50 = 625.67 \,.$$

Case 2

We could also base our calculation on the length of the first class; this yields the following table:

Class (revenue in euros)	Frequency f_i (number of shops)
500–600	15
600–700	25
700–800	10
Total	50

The modal class is the class 600–700, and the values of the mode are:

$$M_o = 600 + \left(\frac{10}{10 + 15}\right) \cdot 100 = 640 \,.$$

These results show that we should be careful to the use of the mode because the obtained results are different depending on how the data are grouped into classes.

FURTHER READING

▶ **Mean**
▶ **Measure of central tendency**
▶ **Measure of skewness**
▶ **Median**

Model

A model is a theoretical representation of a real situation. It is composed of mathematical symbols. A mathematical model, based on a certain number of observations and hypotheses, tries to give the best possible description of the phenomenon under study. A mathematical model contains essentially two types of elements: the directly or indirectly observable variables that concern the studied phenomenon and the parameters, which are fixed quantities, generally unknown, that relate the variables among them inside the model.

HISTORY

The notion of "model" first appeared in the 18th century, in three major problems:

1. Understanding the observed inequality between the movements of Jupiter and Saturn.
2. Determining and mathematically describing the movements of the moon.
3. Determining the shape of the earth.

These three problems used observations from astronomy and some theoretical knowledge about gravitation. They occupied many scientists, especially Euler, Leonhard (1749), who first treated the problem, and

Mayer, Tobias (1750), who was interested in the movements of the moon. The third problem was treated by **Boscovich, Roger Joseph** and Maire, Christopher (1755).

MATHEMATICAL ASPECTS

There exist many types of models. Among them, the determinist model is a model where the relations are considered to be exact. On the other hand, the stochastic model presupposes that the relations are not exact but established by a random process.

A stochastic model is principally determined by the nature of independent variables (exogenous or explanatory). We speak about continuous models when the independent variables X_j are the quantitative continuous variables. We try to express the **dependent variable** Y (also called the endogenous or response) by a function of the type:

$$Y = f\left(X_1, \ldots, X_p | \beta_0, \beta_1, \ldots, \beta_{p-1}\right) + \varepsilon,$$

where the β_j are the parameters (or coefficients) to be estimated and ε is a nonobservable random **error**. Of this type of model, we can distinguish linear models and nonlinear models; the linearity is relative to the coefficients. A linear model can be written in the form of a polynomial. For example:

$$Y = \beta_0 + \beta_1 X + \beta_2 X^2 + \varepsilon$$

is a linear model called "of the second order" (the order comes from the greatest power of variable X). The model

$$\log V = \beta_0 + \frac{\beta_1}{W}$$

is also a linear model for parameters β_0 and β_1. This model also becomes linear for variables posing:

$$Y = \log V \quad \text{and} \quad X = \frac{1}{W}.$$

It gives:

$$Y = \beta_0 + \beta_1 \cdot X.$$

On the other hand, the model

$$Y = \frac{\beta_0}{X + \beta_1} + \varepsilon$$

is a nonlinear model for the parameter β_1. There are many methods for estimating the parameters of a model. Among these methods, we may cite the **least-squares** and the L_1 **estimation** methods.

Except for the continuous models, we talk about discrete models when the studied **population** can be divided into subpopulations on the basis of qualitative categorical variables (or factors) where each can have a certain number of **levels**. The type of model depends on the established **experimental design**. The objective of such a design is to determine how the experimental conditions influence the observed values of the response variable. We can distinguish different types of discrete models:

- A model with fixed effects (model I) is a linear model for which the factors are fixed, that is, take a finite number of levels all of which are represented in the established experimental design. In this case, the model can be written:

$$\text{observed value} = \begin{pmatrix} \text{linear function} \\ \text{of unknown} \\ \text{parameters} \end{pmatrix} + \text{error};$$

the errors are independent random variables **mean** equal to zero.

- A model with variable effects (model II) is a linear model for which the studied levels of each factor were randomly chosen from a finite or nonfinite number of possible levels. In this case, we can write:

observed value = constant

$$+ \left(\begin{array}{c} \text{linear function} \\ \text{of random variables} \end{array} \right) + \text{error} \,;$$

the random variables are independent with **mean** equal to zero.

- A model with mixed effects (model III) is a linear model that contains at the same time fixed and random factors. Thus we can write:

$$\text{observed value} = \left(\begin{array}{c} \text{linear function} \\ \text{of unknown} \\ \text{parameters} \end{array} \right)$$

$$+ \left(\begin{array}{c} \text{linear function} \\ \text{of random} \\ \text{variables} \end{array} \right) + \text{error}.$$

When the experimental design contains a single factor, denoted by α, the model can be written as:

$$Y_{ij} = \mu + \alpha_i + \varepsilon_{ij} \,,$$

$$i = 1, \ldots, I \text{ and } j = 1, \ldots, n_i \,,$$

where I is the number of levels of a factor α, n_i is the number of observations at each level, μ is the **mean** of the **population**, α_i are the effects associated to each level of the factor α, and ε_{ij} are the independent random errors, normally distributed with **mean** to zero and homogeneous **variances** σ^2.

Such a model can be of type I or type II, depending on the nature of the factor α: it is of type I if the levels of the factor α are fixed and if they all appear on the entire design. It is of type II if α is a random variable independent of ϵ, distributed according to any distribution of zero mathematical expectancy and vari-

ance σ_α^2 (if the levels of α are randomly chosen from a finite or infinite number of possible levels).

When the experimental design contains many factors, for example:

$$Y_{ijk} = \mu + \alpha_i + \beta_j + (\alpha\beta)_{ij} + \varepsilon_{ijk} \,,$$

$$i = 1, \ldots, I \,,$$

$$j = 1, \ldots, J \,,$$

$$\text{and } k = 1, \ldots, n_{ij} \,,$$

the model can be of type I (the level of all the factors are fixed), II (all the factors are random variables with the same restrictions as in the previous model), or III (certain factors have fixed levels and others are variable). The term $(\alpha\beta)_{ij}$ represents here the effect of the **interaction** between the I levels of the factor α and the J levels of the factor β.

A model of type I can be analyzed using an **analysis of variance**.

M

FURTHER READING

▶ **Analysis of variance**
▶ **Dependent variable**
▶ **Design of experiments**
▶ **Independent variable**
▶ **Parameter**
▶ **Random variable**
▶ **Regression analysis**
▶ **Variable**

REFERENCES

Anderson, V.L., McLean, R.A.: Design of experiments: a realistic approach. Marcel Dekker, New York (1974)

Boscovich, R.J. et Maire, C.: De Litteraria Expeditione per Pontificiam ditionem ad dimetiendas duas Meridiani gradus. Palladis, Rome (1755)

Euler, L.: Recherches sur la question des inégalités du mouvement de Saturne et de Jupiter, pièce ayant remporté le prix de l'année 1748, par l'Académie royale des sciences de Paris. Republié en 1960, dans Leonhardi Euleri, Opera Omnia, 2ème série. Turici, Bâle, 25, pp. 47–157 (1749)

Mayer, T.: Abhandlung über die Umwälzung des Monds um seine Achse und die scheinbare Bewegung der Mondflecken. Kosmographische Nachrichten und Sammlungen auf das Jahr 1748 **1**, 52–183 (1750)

De Moivre Abraham

Abraham de Moivre (1667–1754) first studied in Sedan and Saumur (both in France) before continuing his studies in mathematics and physics at the Sorbonne. Coming from a Protestant family, he was forced to leave France in 1685 due to the revocation of the Edict of Nantes and was exiled to London. After a rough start in his new homeland, he was able to study mathematics. He was elected member of the Royal Society of London in 1697.

His most important works were *The Doctrine of Chances*, which appeared in 1718 and was dedicated to questions that continue to play a fundamental role in modern probability theory, and *Miscellanea Analytica de Seriebus et Quadraturis*, where for the first time appeared the **density function** of the **normal distribution** in the context of probability calculations for games of hazard. **Carl Friedrich Gauss** took the same density function and applied it to problems of measurement in astronomy. He also made contributions to actuarial sciences and mathematical analysis.

Principal works of Abraham De Moivre:

1718 The Doctrine of Chances: or, A Method of Calculating the Probability of Events in Play. Pearson, London.

1725 Annuities upon Lives. Pearson, London.

1730 Miscellanea Analytica de Seriebus et Quadraturis. Tonson and Watts, London.

FURTHER READING
▶ **De Moivre–Laplace theorem**
▶ **Normal distribution**

De Moivre–Laplace Theorem

This theorem established the relation between the **binomial distribution** with parameters n and p and the **normal distribution** for n that tends to infinity. Its utility consists especially in the estimation of the **distribution function** $F(k)$ of the binomial distribution by the distribution function of the normal distribution, which is found to be easier.

HISTORY

This theorem dates back to 1870, when **Pierre Simon de Laplace** studied, with others (among them **de Moivre, Abraham**), problems related to the approximation of the binomial distribution and to the theory of errors.

MATHEMATICAL ASPECTS

Let $X_1, X_2, \ldots, X_n, \ldots$ be a sequence of random variables, independent and identically distributed, following a **binomial distribution** with parameters n and p, where $0 < p < 1$.

We consider the corresponding adjusted variable Z_n, $n = 1, 2, \ldots$ such as:

$$Z_n = \frac{X_n - np}{\sqrt{npq}},$$

where np represents the **mean** of a binomial distribution and $\sqrt{npq}$ its **variance** (the variable Z_n corresponds to the **normalization** of X_n).

The de Moivre–Laplace theorem is expressed as follows:

$$P(Z_n \leq x) \longrightarrow_{n \to \infty} \Phi(x),$$

where Φ is the **distribution function** of the standard **normal distribution**.

The approximation can also be improved using the **factor** of correction, "a half". Hence, the following approximation:

$$P(X_n \geq x) \cong 1 - \Phi\left(\frac{x - \frac{1}{2} - np}{\sqrt{npq}}\right)$$

is better than those obtained without taking into account the corrective factor $\frac{1}{2}$.

DOMAINS AND LIMITATIONS

This theorem is a particular case of the **central limit theorem**. A **binomial distribution** (such as was defined earlier) can be written as a sum of n random variables distributed according to a **Bernoulli distribution** of **parameter** p. The **mean** of a random Bernoulli variable is equal to p and its **variance** to $p(1 - p)$. The de Moivre–Laplace theorem is easily deduced from the central limit theorem.

EXAMPLES

We want to calculate the **probability** of more than 27 successes in a binomial **experiment** of 100 trials, each trial having a probability of success of $p = 0.2$. The desired binomial probability can be expressed by the sum:

$$P(X_n > 27)$$
$$= \sum_{k=27}^{100} \binom{100}{k} \cdot (0.2)^k (0.8)^{100-k}.$$

The direct calculus of this probability demands an evaluation of 74 terms, each of the form

$$\binom{100}{k}(0.2)^k (0.8)^{100-k}.$$

Using the de Moivre–Laplace theorem, this probability can be approximately evaluated taking into account the relation between the binomial and normal distributions. Thus:

$$P(X_n > 27) \cong 1 - \Phi\left(\frac{27 - np}{\sqrt{npq}}\right)$$
$$= 1 - \Phi\left(\frac{27 - 100 \cdot 0.2}{\sqrt{100 \cdot 0.2 \cdot 0.8}}\right)$$
$$= 1 - \Phi(1.75).$$

Referring to the **normal table**, we obtain $\Phi(1.75) = 0.9599$, and so:

$$P(X_n > 27) \cong 0.0401.$$

The exact **value** to 4 decimal places of the probability is: $P(X_n > 27) = 0.0558$. A comparison of the two values 0.0401 and 0.0558 shows that the approximation by the normal distribution gives a result that is very close to the exact value.

We obtain the following approximation using the corrective factor:

$$P(X_n > x) \cong 1 - \Phi(1.625) = 0.0521.$$

The result is closer to the exact value 0.0558 than those obtained without using the corrective factor (0.0401).

M

FURTHER READING
- ▶ **Binomial distribution**
- ▶ **Central limit theorem**
- ▶ **Normal distribution**
- ▶ **Normalization**

REFERENCE

Laplace, P.S. de: Mémoire sur les approximations des formules qui sont fonctions de très grands nombres et sur leur application aux probabilités. Mémoires de l'Académie Royale des Sciences de Paris, 10. Reproduced in: Œuvres de Laplace **12**, 301–347 (1810)

Moivre, A. de: Approximatio ad summam terminorum binomii $(a + b)^n$, in seriem expansi. Supplementum II to Miscellanae Analytica, pp. 1–7 (1733). Photographically reprinted in a rare pamphlet on Moivre and some of his discoveries. Published by Archibald, R.C. Isis **8**, 671–683 (1926)

Sheynin, O.B.: P.S. Laplace's work on probability, Archive for History of Exact Sciences **16**, 137–187 (1976)

Sheynin, O.B.: P.S. Laplace's theory of errors, Archive for History of Exact Sciences **17**, 1–61 (1977)

Moment

We call the moment of order k of the **random variable** X relative to any **value** x_0 the mean difference between the random variable and x_0, to the power k. If $x_0 = 0$, we refer to the moment (or initial moment) of order k. If $x_0 = \mu$, where μ is the **expected value** of random variable X, we refer to the central moment of order k.

HISTORY

The concept of "moment" in statistics comes from the concept of moment in physics which is derived from Archimedes (287–212 BC) discovery of the operating of the lever, where we speak of the moment of forces. The corrected formulas for the calculation of the moments of grouped data were adapted by Sheppard, W.F. (1886).

MATHEMATICAL ASPECTS

The initial moment of order k of a discrete **random variable** X is expressed by:

$$E\left[X^k\right] = \sum_{i=1}^{n} x_i^k P(x_i) \,,$$

if it exists, that is, if $E\left[X^k\right] < \infty$, where $P(x)$ is the **probability function** of the discrete random variable X taking the n values $x_1, \ldots, x_n$.

In the case of a continuous random variable, if the moment exists, it is defined by:

$$E\left[X^k\right] = \int_{-\infty}^{\infty} x^k f(x)\, dx \,,$$

where $f(x)$ is the **density function** of the continuous random variable.

For $k = 1$, the initial moment is the same as the **expected value** of the random variable X (if this $E(X)$ exists):

$$(X \text{ discrete}) \ E[X] = \sum_{i=1}^{n} x_i P(x_i) \,,$$

$$(X \text{ continuous}) \ [X] = \int_{-\infty}^{\infty} x f(x)\, dx \,.$$

The central moment of order k of a discrete random variable X, if it exists, is expressed by:

$$E\left[(X - E[X])^k\right] =$$
$$\sum_{i=1}^{n} (x_i - E[X])^k P(x_i) \,.$$

In the same way, in the case of a continuous random variable, we have:

$$E\left[(X - E[X])^k\right]$$

$$= \int_{-\infty}^{\infty} (x - E[X])^k f(x)\, dx$$

if the integral is well defined.

It is evident that with any random variable X having a finite mathematical expectancy, the centered moment of order 1 is zero:

$$E[X - \mu] = \sum_{i=1}^{n} (x_i - \mu) P(x_i)$$

$$= \sum_{i=1}^{n} x_i P(x_i) - \mu \sum_{i=1}^{n} P(x_i)$$

$$= \mu - \mu = 0.$$

On the other hand, if it exists, the central moment of order 2 is the same as the **variance** of the random variable:

$(X$ discrete$) E\left[(X - \mu)^2\right]$

$$= \sum_{i=1}^{n} (x_i - \mu)^2 P(x_i) = \text{Var}(X) ,$$

$(X$ continuous$) E\left[(X - \mu)^2\right]$

$$= \int_{-\infty}^{\infty} (x - \mu)^2 f(x)\, dx = \text{Var}(X) .$$

DOMAINS AND LIMITATIONS

The **estimator** of the moment of order k calculated for a **sample** $x_1, x_2, \ldots, x_n$ is denoted by m_k. It equals:

$$m_k = \frac{\sum_{i=1}^{n} (x_i - x_0)^k}{n},$$

where n is the total number of observations. If $x_0 = 0$, then we have the moment (or initial moment) of order k.

If $x_0 = \bar{x}$, where $\bar{x}$ is the **arithmetic mean**, we have of the central moment of order k.

For the case where a **random variable** X takes its values x_i with the frequencies f_i, $i = 1, 2, \ldots, h$, the moment of order k is given by:

$$m_k = \frac{\sum_{i=1}^{h} f_i (x_i - x_0)^k}{\sum_{i=1}^{h} f_i} .$$

FURTHER READING

▶ **Arithmetic mean**
▶ **Density function**
▶ **Expected value**
▶ **Probability function**
▶ **Random variable**
▶ **Variance**
▶ **Weighted arithmetic mean**

REFERENCES

Sheppard, W.F.: On the calculation of the most probable values of frequency constants for data arranged according to equidistant divisions of a scale. London Math. Soc. Proc. **29**, 353–380 (1898)

M

Monte Carlo Method

Any numerical technique for solving mathematical problems that uses random or pseudorandom numbers is called a Monte Carlo method. These mathematical or statistical problems have no analytical solutions.

HISTORY

The name Monte Carlo comes from the city of the same name in Monaco, famous for its casinos. The roulette is one of the simplest mechanisms for the **generation of random numbers**.

According to Sobol, I.M. (1975), the Monte Carlo method owes its existence to the American mathematicians von Neumann,

John (1951), Ulam, Stanislaw Marcin, and Metropolis, Nicholas (1949). It was developed in around 1949, but it became practical with the invention of computers.

EXAMPLES

Suppose that we like to compute the surface of a figure S that is situated inside a square; its side equals 1. We will generate N random points inside the square. To do this, we note the square on the system of perpendicular axes whose origin is the lower left angle of the square. This means that all the points that we will generate inside the square will have coordinates between 0 and 1. It is enough to take uniform random variables to obtain a point. The first **random number** will be the abscissa and the second the ordinate.

In the following example, let us choose 40 random points. We will need two samples of 40 random numbers. The surface S that we would like to estimate is the square of 0.75 of the side, in other words the exact **value** equals 0.5625.

After producing of the 40 random points in the unit square, it is sufficient to count the number of points inside figure S (21 in this case). The **estimation** of the searched surface is obtained by the ratio $\frac{21}{40} = 0.525$.

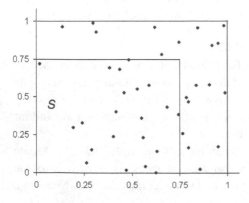

It is possible to improve this estimate by performing the **experiment** many times and taking the **mean** of the obtained surfaces.

FURTHER READING
▶ **Generation of random numbers**
▶ **Jackknife method**
▶ **Random number**
▶ **Simulation**

REFERENCES

Metropolis, N., Ulam, S.: The Monte Carlo method. J. Am. Stat. Assoc. **44**, 335–341 (1949)

Sobol, I.M.: The Monte Carlo Method. Mir Publishers, Moscow (1975)

Von Neumann, J.: Various Techniques Used in Connection with Random Digits. National Bureau of Standards symposium, NBS, Applied Mathematics Series 12, pp. 36–38, National Bureau of Standards, Washington, D.C (1951)

Moving Average

We call the moving average of order k the **arithmetic mean** calculated on k successive values of **time series**. The rest of these arithmetic means give the series of moving averages of order k. The process that consists in replacing the initial series is called smoothing of the time series.

Concerning the time series that are comprised of annual or monthly data, the averages take, respectively, the name of moving averages on k years or k months. When we have monthly data, for example, calculating a moving average over 12 months, the obtained results correspond to the middle of the considered period, at the end of the sixth month (or on the first day of the seventh month), instead of being affected in the middle of a month as in the original data. We correct this by making a centered moving average over 12 months. Generally, a centered moving average of order k is calculated making the moving average of order 2 of the moving average of order k of the initial time series.

HISTORY
See **time series**.

MATHEMATICAL ASPECTS
We have the **time series** $Y_1, Y_2, Y_3, \ldots, Y_N$, and we define the series of moving averages of order k by the series of arithmetic means:

$$\frac{Y_1 + Y_2 + \cdots + Y_k}{k},$$
$$\frac{Y_2 + Y_3 + \cdots + Y_{k+1}}{k}, \ldots,$$
$$\frac{Y_{N-k+1} + Y_{N-k+2} + \cdots + Y_N}{k}.$$

The sum of numerators is called moving sums of order k.

We can use the weighted arithmetic means with previously specified weights; the rest of what is obtained in this way is called a series of weighted moving averages of order k.

DOMAINS AND LIMITATIONS
The moving averages are used in the study of time series, for the **estimation** of the **secular tendency**, of seasonal variations, and of cyclic fluctuations.

The operation "moving average" as applied to a time series allows to:

- Determine the seasonal variations in the limits where they are rigorously periodical,
- Smooth irregular variations, and
- Approximately conserve the extraseasonal movement.

Let us mention the disadvantages of the moving average method:

- The data of the beginning and the end of series are missing.
- Moving averages can cause cycles or other movements not present in the original data.
- Moving averages are strongly influenced by outliers and accidental values.
- When a graph of the studied time series shows an exponential secular tendency or, more generally, a strong curve, the moving average method does not give very precise results for the estimation of extraseasonal movement.

EXAMPLES
With the **time series** 7, 2, 3, 4, 5, 0, 1 data, we obtain the moving averages of order 3:

$$\frac{7+2+3}{3}, \frac{2+3+4}{3}, \frac{3+4+5}{3},$$
$$\frac{4+5+0}{3}, \frac{5+0+1}{3},$$

that is, 4, 3, 4, 3, 2.

In a moving average series, it is common to localize each number on its relative position to the original data. In this example, we write:

Original data	7, 2, 3, 4, 5, 0, 1
Moving average of order 3	4, 3, 4, 3, 2

In the graph below each number of the moving average series is the **arithmetic mean** of the three numbers immediately below it.

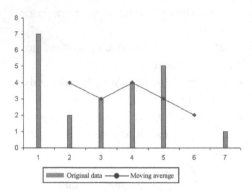

We propose to calculate the centered moving average over 4 years of the mean monthly production of oil in a country. In the first step, we calculate the moving sums over 4 years, adding the data of 4 successive years. Then, dividing each moving sum by 4, we obtain moving averages over 4 years, the last corresponding to 1 January of the third year considered in different sums. Finally, we obtain moving averages over 4 years, calculating the arithmetic mean of two moving averages over 4 successive years.

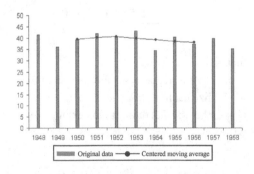

Average monthly oil production in a country in millions of tons

Year	Datum	Moving sum over 4 years	Moving average over 4 years	Centered moving average over 4 years
1948	41.4			
1949	36.0			
		158.8	39.7	
1950	39.3			39.6
		158.0	39.5	
1951	42.1			40.4
		165.2	41.3	
1952	40.6			40.7
		160.4	40.1	
1953	43.2			39.9
		158.8	39.7	
1954	34.5			39.3
		155.6	38.9	
1955	40.5			38.5
		152.4	38.1	
1956	37.4			38.2
		153.2	38.3	
1957	40.0			
1958	35.3			

FURTHER READING
▶ **Arithmetic mean**
▶ **Time series**
▶ **Weighted arithmetic mean**

Multicollinearity

Multicollinearity is a term used to describe a situation where a **variable** is almost equal to a linear combination of other variables. Multicollinearity is sometimes present at regression problems when there is a strong corellation between the explanatory variables.

The existence of linear or almost linear relations among independent variables causes a big imprecision in the estimation of coefficients of **regression**. If the relations are rigorously linear, the coefficients would be biased.

MATHEMATICAL ASPECTS
See **ridge regression**.

FURTHER READING
▶ **Correlation coefficient**
▶ **Multiple linear regression**

REFERENCES
Belsley, D., Kuh, E., Welsch, R.E.: Regression diagnistics Idendifying Influential data and sources of collinearity. Wiley, New York (1976)

Multimodal Distribution

See **frequency distribution**.

Multinomial Distribution

Consider a **random experiment** with n independent trials. Suppose that each trial can only give one of **event** E_i, with a **probability** p_i.

Random variable X_i represents the number of appearances of **event** E_i.

The multinomial distribution with **parameters** n and p_i is defined by the following **joint density function**:

$$P(X_1 = x_1, X_2 = x_2, \ldots, X_s = x_s)$$

$$= \frac{n!}{\prod\limits_{i=1}^{s} x_i!} \prod_{i=1}^{s} p_i^{x_i} \,,$$

with the integers $x_i \geq 0$ such that $\sum\limits_{i=1}^{s} x_i = n$.

Indeed, the **probability** of obtaining the sequence

$$\underbrace{E_1, \ldots, E_1}_{x_1}; \underbrace{E_2, \ldots, E_2}_{x_2}; \ldots; \underbrace{E_s, \ldots, E_s}_{x_s}$$

is equal to $\prod\limits_{i=1}^{s} p_i^{x_i}$.

The number of existing possibilities is the number of **permutations** of n objects of which x_1 is **event** E_1, x_2 is event $E_2, \ldots, x_s$ is event E_s, meaning that:

$$\frac{n!}{\prod\limits_{i=1}^{s} x_i!} \,.$$

The multinomial distribution is a **discrete probability distribution**.

MATHEMATICAL ASPECTS
For the multinomial distribution

$$\sum_{i=1}^{s} p_i = 1 \quad \text{and} \quad \sum_{i=1}^{s} x_i = n \,,$$

with $0 < p_i < 1$.

The **expected value** is calculated for each **random variable** X_i.

We therefore have:

$$E[X_i] = n \cdot p_i \,.$$

Indeed each **random variable** can be considered individually as following a **binomial distribution**.

We can make the same remark for the **variance**. It is then equal to:

$$\text{Var}(X_i) = n \cdot p_i \cdot q_i \,.$$

DOMAINS AND LIMITATIONS
The multinomial distribution has many applications in the analysis of statistical

M

data. It is used especially in situations where the data must be classified into several **categories** of **events** (for example, in the analysis of categorical data).

EXAMPLES

Consider the following **random experiment**:

An urn contains nine balls, two white, three red, and four black. A ball is chosen at random; its color is noted and it is put back in the urn. This is repeated 5 times.

We can describe this **random experiment** as follows:

We have three possible **events**:

E_1: choosing a white ball

E_2: choosing a red ball

E_3: choosing a black ball

and three **random variables**:

X_1: number of white balls chosen

X_2: number of red balls chosen

X_3: number of black balls chosen

The **probabilities** associated to these three **events** are:

$$p_1 = \frac{2}{9},$$

$$p_2 = \frac{3}{9},$$

$$p_3 = \frac{4}{9}.$$

We will calculate the **probability** that on five chosen balls ($n = 5$), one will be white, two will be red, and two will be black.

The number of possibilities is equal to:

$$5!/(1!2!2!).$$

The **probability** of obtaining the sequence (w, r, r, b, b), where w, r, and b are, respectively, white, red, and black balls, is:

$$\prod_{i=1}^{3} p_i^{x_i} = \frac{2}{9} \cdot \left(\frac{3}{9}\right)^2 \cdot \left(\frac{4}{9}\right)^2 = 0.0049 .$$

Hence,

$$P(X_1 = x_1, X_2 = x_2, X_3 = x_3)$$
$$= \frac{5!}{\prod\limits_{i=1}^{3} x_i!} \cdot \left(\frac{2}{9}\right)^{x_1} \cdot \left(\frac{3}{9}\right)^{x_2} \cdot \left(\frac{4}{9}\right)^{x_3} ,$$

where

$$P(X_1 = 1, X_2 = 2, X_3 = 2) = 30 \cdot 0.0049$$
$$= 0.147 .$$

FURTHER READING

▶ **Binomial distribution**

▶ **Discrete probability distribution**

Multiple Correlation Coefficient

See **correlation coefficient**.

Multiple Linear Regression

A **regression analysis** where **independent variable** Y linearly depends on many independent variables $X_1, X_2, \ldots, X_k$ is called multiple linear regression.

A multiple linear regression equation is of the form:

$$Y = f(X_1, X_2, \ldots, X_k) ,$$

where $f(X_1, X_2, \ldots, X_k)$ is a linear function of $X_1, X_2, \ldots, X_k$.

HISTORY

See **regression analysis**.

MATHEMATICAL ASPECTS

A general **model** of multiple linear regression containing $k = (p - 1)$ independent variables (and n observations) is written as:

$$Y_i = \beta_0 + \sum_{j=1}^{p-1} X_{ji}\beta_j + \varepsilon_i , \quad i = 1, \ldots, n ,$$

where Y_i is the **dependent variable**, $X_{ji}, j = 1, \ldots, p - 1$, are the independent variables, $\beta_j, j = 0, \ldots, p - 1$, are the parameters to be estimated, and ε_i is the term of random nonobservable **error**.

In the matrix form, this **model** is written as:

$$\mathbf{Y} = \mathbf{X}\boldsymbol{\beta} + \boldsymbol{\varepsilon},$$

where $\mathbf{Y}$ is the vector $(n \times 1)$ of observations related to the dependent variable (n observations), $\boldsymbol{\beta}$ is the vector $(p \times 1)$ of parameters to be estimated, $\boldsymbol{\varepsilon}$ is the vector $(n \times 1)$ of errors,

and $\mathbf{X} = \begin{pmatrix} 1 & X_{11} & \cdots & X_{1(p-1)} \\ \vdots & \vdots & & \vdots \\ 1 & X_{n1} & \cdots & X_{n(p-1)} \end{pmatrix}$ is the

$(n \times p)$ **matrix** of the independent variables.

Estimation of Vector β

Starting from the model

$$\mathbf{Y} = \mathbf{X}\boldsymbol{\beta} + \boldsymbol{\varepsilon},$$

we obtain the estimate $\hat{\boldsymbol{\beta}}$ of the vector $\boldsymbol{\beta}$ by the **least-squares** method:

$$\hat{\boldsymbol{\beta}} = (\mathbf{X'X})^{-1}\mathbf{X'Y},$$

and we calculate an estimated value $\widehat{\mathbf{Y}}$ for $\mathbf{Y}$:

$$\widehat{\mathbf{Y}} = \mathbf{X} \cdot \hat{\boldsymbol{\beta}}.$$

At this step, we can calculate the residuals, denoted by vector $\mathbf{e}$, that we find in the following manner:

$$\mathbf{e} = \mathbf{Y} - \widehat{\mathbf{Y}}.$$

Measure of Reliability of Estimation of Y

To know which measure to trust for the chosen linear model, it is useful to conduct an **analysis of variance** and to test the hypotheses on vector $\boldsymbol{\beta}$ of the regression model. To conduct these tests, we must make the following assumptions:

- For each **value** of X_{ji} and for all $i = 1, \ldots, n$ and $j = 1 \ldots, p-1$, there is a **random variable Y** distributed according to the **normal distribution**.
- The **variance** of $\mathbf{Y}$ is the same for all X_{ji}; it equals σ^2 (unknown).
- The different observations on $\mathbf{Y}$ are independent of one another but conditioned by the values of X_{ji}.

Analysis of Variance in Matrix Form

In the matrix form, the table of **analysis of variance** for the regression is as follows:

Analysis of variance

Source of variation	Degrees of freedom	Sum of squares	Mean of squares
Regression	$p - 1$	$\hat{\beta}'\mathbf{X'Y} - n\bar{y}^2$	$\dfrac{\hat{\beta}'\mathbf{X'Y} - n\hat{Y}^2}{p - 1}$
Residual	$n - p$	$\mathbf{Y'Y} - \hat{\beta}'\mathbf{X'Y}$	$\dfrac{\mathbf{Y'Y} - \hat{\beta}'\mathbf{X'Y}}{n - p}$
Total	$n - 1$	$\mathbf{Y'Y} - n\bar{y}^2$	

If the **model** is correct, then

$$S^2 = \frac{\mathbf{Y'Y} - \hat{\beta}'\mathbf{X'Y}}{n - p}$$

is an unbiased **estimator** of σ^2.
The analysis of variance allows us to test the **null hypothesis**:

$$H_0: \quad \beta_j = 0 \text{ for } j = 1, \ldots, p - 1$$

against the **alternative hypothesis**:

$H_1:$ at least one of the parameters $\beta_j, j \neq 0$, is different from zero

calculating the **statistic**:

$$F = \frac{EMSE}{RMSE} = \frac{EMSE}{S^2},$$

where $EMSE$ is the mean of squares of the regression, $RMSE$ is the mean of squares of residuals, and $TMSE$ is the total mean of squares.

This F statistic must be compared to the **value** $F_{\alpha,p-1,n-p}$ of the **Fisher table**, where α is the **significance level** of the test:

> $\Rightarrow$ if $F \leq F_{\alpha,p-1,n-p}$, we accept H_0;
>
>   if $F > F_{\alpha,p-1,n-p}$, we reject H_0 for H_1.

The coefficient of determination R^2 is calculated in the following manner:

$$R^2 = \frac{ESS}{TSS} = \frac{\hat{\beta}'\mathbf{X}'\mathbf{Y} - n\bar{y}^2}{\mathbf{Y}'\mathbf{Y} - n\bar{y}^2},$$

where ESS is the sum of squares of the regression and TSS is the total sum of squares.

DOMAINS AND LIMITATIONS
See **analysis of regression**.

EXAMPLES
The following data concern ten companies of the chemical industry. We try to establish a relation between the production, the hours of work, and capital.

Production	Work	Capital
(100 tons)	(h)	(machine hours)
60	1100	300
120	1200	400
190	1430	420
250	1500	400
300	1520	510
360	1620	590
380	1800	600
430	1820	630
440	1800	610
490	1750	630

The matrix **model** is written as:

$$\mathbf{Y} = \mathbf{X}\beta + \varepsilon,$$

where

$$\mathbf{Y} = \begin{bmatrix} 60 \\ 120 \\ 190 \\ 250 \\ 300 \\ 360 \\ 380 \\ 430 \\ 440 \\ 490 \end{bmatrix}, \quad \mathbf{X} = \begin{bmatrix} 1 & 1100 & 300 \\ 1 & 1200 & 400 \\ 1 & 1430 & 420 \\ 1 & 1500 & 400 \\ 1 & 1520 & 510 \\ 1 & 1620 & 590 \\ 1 & 1800 & 600 \\ 1 & 1820 & 630 \\ 1 & 1800 & 610 \\ 1 & 1750 & 630 \end{bmatrix},$$

$$\varepsilon = \begin{bmatrix} \varepsilon_1 \\ \varepsilon_2 \\ \varepsilon_3 \\ \varepsilon_4 \\ \varepsilon_5 \\ \varepsilon_6 \\ \varepsilon_7 \\ \varepsilon_8 \\ \varepsilon_9 \\ \varepsilon_{10} \end{bmatrix}, \quad \beta = \begin{bmatrix} \beta_0 \\ \beta_1 \\ \beta_2 \end{bmatrix}.$$

The equations that form the model are the following:

$$60 = \beta_0 + 1100\beta_1 + 300\beta_2 + \varepsilon_1,$$
$$120 = \beta_0 + 1200\beta_1 + 400\beta_2 + \varepsilon_2,$$
$$\cdots$$
$$490 = \beta_0 + 1750\beta_1 + 630\beta_2 + \varepsilon_{10}.$$

We calculate the estimators $\hat{\beta}$ using the result:

$$\hat{\beta} = \left(\mathbf{X}'\mathbf{X}\right)^{-1}\mathbf{X}'\mathbf{Y},$$

with

$$(\mathbf{X}'\mathbf{X}) = \begin{bmatrix} 10 & 15540 & 5090 \\ 15540 & 24734600 & 8168700 \\ 5090 & 8168700 & 2720500 \end{bmatrix},$$

$$(\mathbf{X}'\mathbf{X})^{-1} =$$
$$\begin{bmatrix} 6.304288 & -0.007817 & 0.011677 \\ -0.007817 & 0.000015 & -0.000029 \\ 0.011677 & -0.000029 & 0.000066 \end{bmatrix},$$

and

$$\mathbf{X'Y} = \begin{bmatrix} 3020 \\ 5012000 \\ 1687200 \end{bmatrix} .$$

That gives:

$$\hat{\beta} = \begin{bmatrix} -439.269 \\ 0.283 \\ 0.591 \end{bmatrix} = \begin{bmatrix} \hat{\beta}_0 \\ \hat{\beta}_1 \\ \hat{\beta}_2 \end{bmatrix} .$$

Thus we obtain the equation for the estimated line:

$$\hat{Y}_i = -439.269 + 0.283X_{i1} + 0.591X_{i2} .$$

Analysis of Variance in Matrix Form

The calculation of different elements of the table gives us

1. Calculation of degrees of freedom:

$$dl_{\text{reg}} = p - 1 = 3 - 1 = 2 ,$$
$$dl_{\text{res}} = n - p = 10 - 3 = 7 ,$$
$$dl_{\text{tot}} = n - 1 = 10 - 1 = 9 ,$$

where dl_{reg} is the degrees of freedom of the regression, dl_{res} is the degrees of freedom of the residuals, and dl_{tot} is the total degrees of freedom.

2. Calculation of the sum of squares:

$$ESS = \hat{\beta}'\mathbf{X'Y} - n\bar{Y}^2$$
$$= [-439.269 \; 0.283 \; 0.591]$$
$$\cdot \begin{bmatrix} 3020 \\ 5012000 \\ 1687200 \end{bmatrix} - 912040$$
$$= 179063.39 .$$

$$TSS = \mathbf{Y'Y} - n\bar{Y}^2$$

$$= [60 \ldots 490] \cdot \begin{bmatrix} 60 \\ \vdots \\ 490 \end{bmatrix} - 912040$$

$$= 187160 .$$

$$RSS = \mathbf{Y'Y} - \hat{\beta}'\mathbf{X'Y}$$
$$= TSS - ESS$$
$$= 187160.00 - 179063.39$$
$$= 8096.61 .$$

We obtain the following table:

Analysis of variance

Source of variation	Degree of free-dom	Sum of squares	Mean of squares
Regression	2	179063.39	89531.70
Residual	7	8096.61	1156.66 $= S^2$
Total	9	187160.00	

This allows us to test the **null hypothesis**:

$$H_0: \quad \beta_1 = \beta_2 = 0 .$$

The calculated **value** of F is:

$$F = \frac{EMSE}{RMSE} = \frac{EMSE}{S^2} = \frac{89531.70}{1156.66}$$
$$= 77.41 .$$

The value of F in the **Fisher table** with a **significance level** $\alpha = 0.05$ is

$$F_{0.05,2,7} = 4.74 .$$

In consequence, the **null hypothesis** H_0 is rejected. The **alternative hypothesis** can be one of the three following possibilities:

$$H_1: \quad \beta_1 \neq 0 \text{ and } \beta_2 \neq 0$$
$$H_2: \quad \beta_1 = 0 \text{ and } \beta_2 \neq 0$$
$$H_3: \quad \beta_1 \neq 0 \text{ and } \beta_2 = 0$$

M

We can calculate the **coefficient of determination** R^2:

$$R^2 = \frac{ESS}{TSS} = \frac{179063.39}{187160.00} \cong 0.96 .$$

FURTHER READING
- ▶ **Analysis of residuals**
- ▶ **Analysis of variance**
- ▶ **Coefficient of determination**
- ▶ **Collinearity**
- ▶ **Correlation coefficient**
- ▶ **Hat matrix**
- ▶ **Least squares**
- ▶ **Leverage point**
- ▶ **Normal equations**
- ▶ **Regression analysis**
- ▶ **Residual**
- ▶ **Simple linear regression**

REFERENCES

Weiseberg, S.: Applied linear regression. Wiley, New York (2006)

Chatterjee, P., Hadi, A.S.: Regression Analysis by Example. Wiley, New York (2006)

Negative Binomial Distribution

A **random variable** X follows a negative binomial distribution with **parameters** k and r if its **probability function** is of the form:

$$P(X = k) = C_{r+k-1}^k \cdot p^r \cdot q^k,$$

where p is the **probability** of success, q is the probability of failure, and C_n^x is the number of combinations of x objects among n.

In the situation where a "success–failure" **random experiment** is repeated in an independent way, the **probability** of success is denoted by p and the probability of failure by $q = 1 - p$. The **experiment** is repeated as many times as required to obtain r successes. The number of obtained failures before attaining this goal is a **random variable** following a negative binomial distribution described above. The negative binomial distribution is a **discrete binomial distribution**.

HISTORY

The first to treat the negative binomial distribution was Pascal, Blaise (1679). Then de Montmort, P.R. (1714) applied the negative binomial distribution to represent the number of times that a coin should be flipped to obtain a certain number of heads. Student (1907) used the negative binomial distribution as an alternative to the **Poisson distribution**.

Greenwood, M. and Yule, G.U. (1920) and Eggenberger, F. and Polya, G. (1923) found applications of the negative binomial distribution. Ever since, there has been an increasing number of applications of this distribution, and the statistical techniques based on this distribution have been developed in a parallel way.

MATHEMATICAL ASPECTS

If $X_1, X_2, \ldots, X_n$ are n independent **random variables** following a **geometric distribution**, then the random variable

$$X = X_1 + X_2 + \ldots + X_r$$

follows a negative binomial distribution. To calculate the **expected value** of X, the following property is used, where Y and Z are **random variables**:

$$E[Y + Z] = E[Y] + E[Z] .$$

We therefore have:

$$
\begin{aligned}
E[X] &= E[X_1 + X_2 + \ldots + X_r] \\
&= E[X_1] + E[X_2] + \ldots \\
&\quad + E[X_r] \\
&= \frac{q}{p} + \frac{q}{p} + \ldots + \frac{q}{p} \\
&= r \cdot \frac{q}{p} .
\end{aligned}
$$

To calculate the **variance** of X, the following property is used, where Y and Z are independent **variables**:

$$\text{Var}(Y + Z) = \text{Var}(Y) + \text{Var}(Z) \, .$$

We therefore have:

$$\begin{aligned}
\text{Var}(X) &= \text{Var}(X_1 + X_2 + \ldots + X_r) \\
&= \text{Var}(X_1) + \text{Var}(X_2) + \ldots \\
&\quad + \text{Var}(X_r) \\
&= \frac{q}{p^2} + \frac{q}{p^2} + \ldots + \frac{q}{p^2} \\
&= r \cdot \frac{q}{p^2} \, .
\end{aligned}$$

DOMAINS AND LIMITATIONS

Among the specific fields for which the negative binomial distribution has been applied are accidents **statistics**, biological sciences, ecology, market studies, medical research, and psychology.

EXAMPLES

A coin is flipped several times. We are interested in the **probability** of obtaining heads a third time on the sixth throw.
We then have:

Number of successes: $r = 3$

Number of failures: $k = 6 - r = 3$

Probability of one success: $p = \dfrac{1}{2}$

(obtaining heads)

Probability of one failure: $q = \dfrac{1}{2}$

(obtaining tails)

The **probability** of obtaining k tails before the third heads is given by:

$$P(X = k) = C_{3+k-1}^k \cdot \left(\frac{1}{2}\right)^3 \cdot \left(\frac{1}{2}\right)^k \, .$$

The **probability** of obtaining a third heads on the sixth throw, meaning the probability of obtaining tails three times before obtaining the third heads, is therefore equal to:

$$\begin{aligned}
P(X = 3) &= C_{3+3-1}^3 \cdot \frac{1}{2^3} \cdot \frac{1}{2^3} \\
&= C_5^3 \cdot \frac{1}{2^3} \cdot \frac{1}{2^3} \\
&= \frac{5!}{3!(5-3)!} \cdot \frac{1}{2^3} \cdot \frac{1}{2^3} \\
&= 0.1563 \, .
\end{aligned}$$

FURTHER READING

▶ **Bernoulli distribution**
▶ **Binomial distribution**
▶ **Discrete probability distribution**
▶ **Poisson distribution**

REFERENCES

Eggenberger, F. und Polya, G.: Über die Statistik verketteter Vorgänge. Zeitschrift für angewandte Mathematische Mechanik **3**, 279–289 (1923)

Greenwood, M., Yule, G.U.: An inquiry into the nature of frequency distributions representative of multiple happenings with particular reference to the occurrence of multiple attacks of disease or of repeated accidents. J. Roy. Stat. Soc. Ser. A **83**, 255–279 (1920)

Montmort, P.R. de: Essai d'analyse sur les jeux de hasard, 2nd edn. Quillau, Paris (1713)

Pascal, B.: Varia Opera Mathematica. D. Petri de Fermat. Tolosae (1679)

Gosset, S.W. ("Student"): On the error of counting with a haemacytometer. Biometrika **5**, 351–360 (1907)

Nelder John A.

John Nelder was born in 1924 at Dulverton (Somerset, England). After receiving his diploma in mathematical statistics in Cambridge, he was named president of the Section of Statistiscs of the National Vegetable Research Station at Wellesbourn, a position he held from 1951 to 1968. He earned the title of Doctor in sciences at Birmingham and was then elected president of the Statistical Department at Rothamsted Experimental Station in Harpenden from 1968 to 1984. He is now invited professor in the Imperial College of Science, Technology and Medicine in London. He was elected member of the Royal Society in 1984 and was president of two prestigious societies: the International Biometric Society and the Royal Statistical Society.

Nelder is the author of statistical computational systems Genstat and GLIM, now used in more than 70 countries. Author of more than 140 articles published in statistical and biological reviews, he is also author of *Computers in Biology* and coauthor with Peter McCullagh of a book treating the generalized linear model. Nelder also developed the idea of generally balanced designs.

Selected works and articles of John A. Nelder:

1965 (with Mead, R.) A simplex method for function minimization. Computational Journal, 7, 303–333.

1971 Computers in Biology. Wykeham, London and Winchester, pp. 149.

1972 (with Wedderburn, R.W.M.) Generalized linear models. J. Roy. Stat. Soc. Ser. A 135, 370–384.

1974 Genstat: a statistical system. In: COMPSTAT: Proceedings in Computational Statistics. Physica, Vienna.

1983 (with McCullagh, P.) Generalized Linear Models. Chapman & Hall, London , pp. 261.

FURTHER READING
▶ **Statistical software**

Newcomb Simon

Newcomb, Simon was born in 1835 in Nova Scotia (Canada) and died in 1909. This astronomer of Canadian origin contributed to the treatment of outliers in statistics, to the application of probability theory for the interpretation of data, and to the development of what we call today robust methods in statistics. Until the age of 16 he studied essentially by consulting the books that his father found for him. After that, he began studying medicine with plans of becoming a doctor. In 1857, he was engaged in the American Ephemeris and Nautical Almanac in Cambridge in the state of Massachussetts. At the same time, he studied in Harvard and he graduated in 1858. In 1861, he became professor of mathematics at the Naval Observatory.

The collection of Notes on the Theory of Probabilities (Mathematics Monthly, 1859–61) constitutes a work that today is still considered modern. Newcomb's most remarkable contribution to statistics is his approach to the treatment of outliers in astronomical data; affirming that the normal distribution does not fit, he invented the normal contaminated distribution.

Selected works and articles of Newcomb, Simon:

1859–61 Notes on the theory of probabilities. Mathematics Monthly, 1, 136–

139, 233–235, 331–355, 349–350; **2**, 134–140, 272–275; **3**, 68, 119–125, 341–349.

Neyman Jerzy

Neyman, Jerzy (1894–1981) was one of the founders of modern **statistics**. One of his outstanding contributions was establishing, with **Pearson, Egon Sharpe**, the basis of the theory of hypothesis testing. Born in Russia, Neyman, Jerzy studied physics and mathematics at Kharkov University. In 1921, he went to Poland, his ancestral country of origin, where he worked for a while as a statistician at the National Institute of Agriculture of Bydgoszcz. In 1924, he spent some time in London, where he could study in University College under the direction of **Pearson, Karl**. At this time he met Pearson, Egon Sharpe **Gosset, William Sealy**, and **Fisher, Ronald Aylmer**. In 1937, at the end of a trip to the United States, where he presented papers at many conferences, he accepted the position of professor at the University of California at Berkeley. He created the Department of Statistics and finished his brilliant career at this university.

Selected articles of Neyman, Jerzy:

1928   (with Pearson, E.S.) On the use and interpretation of certain test criteria for purposes of statistical inference, I, II. Biometrika **20A**, 175–240, pp. 263–295.

1933   (with Pearson, E.S.) Testing of statistical hypotheses in relation to probabilities a priori. Proc. Camb. Philos. Soc. **29**, 492–510.

1934   On the two different aspects of the representative method: the method of stratified sampling and the method of purposive selection. J. Roy. Stat. Soc. **97**, 558–606, discussion pp. 607–625.

1938   Contribution to the theory of sampling human populations. J. Am. Stat. Assoc. **33**, 101–116.

FURTHER READING

▶ Hypothesis testing

Nonlinear Regression

An **analysis of regression** where the **dependent variable** Y depends on one or many independent variables $X_1, \ldots, X_k$ is called a nonlinear regression if the equation of regression $Y = f(X_1, \ldots, X_k; \beta_0, \ldots, \beta_p)$ is not linear in parameters $\beta_0, \ldots, \beta_p$.

HISTORY

Nonlinear regression dates back to the 1920s to **Fisher, Ronald Aylmer** and Mackenzie, W.A. However, the use and more detailed investigation of these models had to wait for advances in automatic calculations in the 1970s.

MATHEMATICAL ASPECTS

Let $Y_1, Y_2, \ldots, Y_r$ and $X_{11}, \ldots, X_{1r}$, $X_{21}, \ldots, X_{2r}, \ldots, X_{k1}, \ldots, X_{kr}$ be the r observations (respectively) of the **dependent variable** Y and the independent variables $X_1, \ldots, X_k$.

The goal is to estimate the parameters $\beta_0, \ldots, \beta_p$ of the model:

$$Y_i = f(X_{1i}, \ldots, X_{ki}; \beta_0, \ldots, \beta_p) + \varepsilon_i,$$
$$i = 1, \ldots, r.$$

By the **least-squares** method and under the corresponding hypothesis concerning errors (see **analysis of residuals**), we estimate the parameters $\beta_0, \ldots, \beta_p$ by the values $\hat{\beta}_0, \ldots, \hat{\beta}_p$ that minimize the sum of squared errors, denoted by $S(\beta_0, \ldots, \beta_p) = \sum_{i=1}^{r} (Y_i - f(X_{1i}, \ldots, X_{ki}; \beta_0, \ldots, \beta_p))^2$:

$$\min_{\beta_0, \ldots, \beta_p} \sum_{i=1}^{r} \varepsilon_i^2 = \min_{\beta_0, \ldots, \beta_p} S(\beta_0, \ldots, \beta_p).$$

If f is nonlinear, the resolution of the **normal equations** (neither of which is linear):

$$\sum_{i=1}^{r} (Y_i - f(X_{1i}, \ldots, X_{ki}; \beta_0, \ldots, \beta_p))$$

$$\left[\frac{\partial f(X_{1i}, \ldots, X_{ki}; \beta_0, \ldots, \beta_p)}{\partial \beta_j} \right] = 0$$

($j = 0, \ldots, p$) can be difficult.

The problem is that the given equations do not necessarily have a solution or have more than one.

In what follows, we discuss the different approaches to the resolution of the problem.

1. *Gauss–Newton*

   The procedure, iterative in nature, consists in the linearization of f with the help of the Taylor expansion in the successive estimation of parameters by linear **regression**.

   We will develop this method in more detail in the following **example**.

2. *Steepest descent*

   The idea of this method is to fix, in an approximate way, the parameters β_j for estimates and then to add an approximation of

   $$-\frac{\partial S(\beta_0, \ldots, \beta_p)}{\partial \beta_j}$$

that corresponds to the maximum of descent from function S.

Iteratively, we determine the parameters that minimize S.

3. *Marquard–Levenburg* This method, not developed here, integrates the advantages of two mentioned procedures forcing and accelerating the convergence of the approximations of parameters to be estimated.

DOMAINS AND LIMITATIONS

In **multiple linear regression**, there are functions that are not linear in parameters but that become linear after a transformation. If this is not the case, we call these functions nonlinear and treat them with nonlinear methods.

An example of a nonlinear function is given by:

$$Y = \frac{\beta_1}{\beta_1 - \beta_2} \left[\exp(-\beta_2 X) - \exp(-\beta_1 X) \right] + \varepsilon.$$

EXAMPLES

Let $P_1 = (x_1, y_1)$, $P_2 = (x_2, y_2)$, and $P_3 = (x_3, y_3)$ be three points on a graph and d_1, d_2, and d_3 the approximative distances (the "approximations" being distributed according to the Gauss distribution of the same variance) between the data points and a point $P = (x, y)$ of unknown and searched coordinates.

A possible regression **model** to estimate the coordinates of P is as follows:

$$d_i = f(x_i, y_i; x, y) + \varepsilon_i, \quad i = 1, 2, 3,$$

where the function f represents the distance between point P_i and point P, that is:

$$f(x_i, y_i; x, y) = \sqrt{(x - x_i)^2 + (y - y_i)^2}.$$

The distance function is clearly nonlinear and cannot be transformed in a linear form, so the appropriate regression is nonlinear. We apply the method of *Gauss–Newton*.

Considering f as a function of parameters x and y to be estimated, the Taylor development of f, until it becomes the linear term in (x_0, y_0), is given by:

$$f(x_i, y_i; x, y)$$
$$\simeq f(x_i, y_i; x_0, y_0)$$
$$+ \left. \frac{\partial f(x_i, y_i; x, y)}{\partial x} \right|_{\substack{x=x_0 \\ y=y_0}} \cdot (x - x_0)$$
$$+ \left. \frac{\partial f(x_i, y_i; x, y)}{\partial x} \right|_{\substack{x=x_0 \\ y=y_0}} \cdot (y - y_0) \, ,$$

where

$$\left. \frac{\partial f(x_i, y_i; x, y)}{\partial x} \right|_{\substack{x=x_0 \\ y=y_0}} = \frac{x_i - x_0}{f(x_i, y_i; x_0, y_0)}$$

and

$$\left. \frac{\partial f(x_i, y_i; x, y)}{\partial x} \right|_{\substack{x=x_0 \\ y=y_0}} = \frac{y_i - y_0}{f(x_i, y_i; x_0, y_0)} \, .$$

So let (x_0, y_0) be a first estimation of P, found, for example, geometrically.

The linearized regression model is thus expressed by:

$$d_i = f(x_i, y_i; x_0, y_0)$$
$$+ \frac{x_i - x_0}{f(x_i, y_i; x_0, y_0)} \cdot (x - x_0)$$
$$+ \frac{y_i - y_0}{f(x_i, y_i; x_0, y_0)} \cdot (y - y_0) + \varepsilon_i \, ,$$
$$i = 1, 2, 3 \, .$$

With the help of a **multiple linear regression** (without constants), we can estimate the parameters $(\Delta x, \Delta y) = (x - x_0, y - y_0)$ of

the following model (the "observations" of the model appear in brackets):

$$\left[d_i - f(x_i, y_i; x_0, y_0) \right]$$
$$= \Delta x \cdot \left[\frac{x_i - x_0}{f(x_i, y_i; x_0, y_0)} \right]$$
$$+ \Delta y \cdot \left[\frac{y_i - y_0}{f(x_i, y_i; x_0, y_0)} \right] + \varepsilon_i \, ,$$
$$i = 1, 2, 3 \, .$$

Let $\left(\hat{\Delta} x, \hat{\Delta} y \right)$ be the estimation of $(\Delta x, \Delta y)$ that was found. Thus we get a better estimation of coordinates of P by $\left(x_0 + \hat{\Delta} x, y_0 + \hat{\Delta} y \right)$.

Using this new point as our point of departure, we can use the same procedure again. The remaining coordinates are found in converging way to the desired point, that is, to the point that approaches the best.

Let us take now a concrete example: $P_1 = (0, 5)$; $P_2 = (5, 0)$; $P_3 = (10, 10)$; $d_1 = 6$; $d_2 = 4$; $d_3 = 6$.

If we choose as the initial point of the *Gauss–Newton* procedure the following point: $P = \left(\frac{0+5+10}{3}, \frac{5+0+10}{3} \right) = (5, 5)$, we obtain the set of the following coordinates:

$$(5, 5)$$
$$(6.2536, 4.7537)$$
$$(6.1704, 4.7711)$$
$$(6.1829, 4.7621)$$
$$(6.1806, 4.7639)$$

that converges to the point $(6.1806, 4.7639)$. The distances between the point found and P_1, P_2, and P_3 are respectively 6.1856, 4.9078, and 6.4811.

FURTHER READING
▶ **Least squares**

▶ **Model**
▶ **Multiple linear regression**
▶ **Normal equations**
▶ **Regression analysis**
▶ **Simple linear regression**

REFERENCES

Bates, D.M., Watts D.G.: Nonlinear regression analysis and its applications. Wiley, New York (1988)

Draper, N.R., Smith, H.: Applied Regression Analysis, 3rd edn. Wiley, New York (1998)

Fisher, R.A.: On the mathematical foundations of theoretical statistics. Philos. Trans. Roy. Soc. Lond. Ser. A **222**, 309–368 (1922)

Fisher, R.A., Mackenzie, W. A.: The manurial response of different potato varieties. J. Agricult. Sci. **13**, 311–320 (1923)

Fletcher, R.: Practical Methods of Optimization. Wiley, New York (1997)

Seber, A.F., Wilt, S.C.: Nonlinear Regression. Wiley, New York (2003)

Nonparametric Statistics

Statistical procedures that allow us to process data from small samples, on variables about which nothing is known concerning their distribution. Specifically, nonparametric statistical methods were developed to be used in cases when the researcher knows nothing about the parameters of the variable of interest in the population (hence the name *nonparametric*). Nonparametric methods do not rely on the estimation of parameters (such as the mean or the standard deviation) describing the distribution of the variable of interest in the population. Nonparametric models differ from parametric models in that the model structure is not specified a priori but is instead determined from data. Therefore, these methods are also sometimes called *parameter-free* methods or *distribution-free methods*.

HISTORY

The term *nonparametric* was first used by Wolfowitz, 1942.

See also **nonparametric test**.

DOMAIN AND LIMITATIONS

The nonparametric method varies from the analysis of a one-way classification model for comparing treatments to regression and curve fitting problems.

The most frequently used nonparametric tests are **Anderson-Darling test**, **Chi-square test**, **Kendall's tau**, **Kolmogorov-Smirnov test**, **Kruskall–Wallis test**, **Wilcoxon rank sum test**, **Spearman's rank correlation coefficient**, and **Wilcoxon sign rank test**.

Nonparametric tests have less power than the appropriate parametric tests, but are more robust when the assumptions underlying the parametric test are not satisfied.

EXAMPLES

A histogram is a simple nonparametric estimate of a probability distribution. A generalization of histogram is kernel smoothing technique by which from a given data set a very smooth probability density function can be constructed.

See also **Nonparametric tests**.

FURTHER READING
▶ **Nonparametric test**

REFERENCES

Parzen, E.: On estimation of a probability density function and mode. Ann. Math. Stat. **33**, 1065–1076 (1962)

Savage, R.: Bibliography of Nonparametric Statistics and Related Topics: Introduction. J. Am. Stat. Assoc. **48**, 844–849 (1953)

Gibbons, J.D., Chakraborti, S.: Nonparametric Statistical Inference, 4th ed. CRC (2003)

Nonparametric Test

A nonparametric test is a type of **hypothesis testing** in which it is not necessary to specify the form of the distribution of the **population** under study. In any event, we should have independent observations, that is, that the selection of individuals that forms the **sample** included must not influence the choice of other individuals.

HISTORY

The first nonparametric test appeared in the works of Arbuthnott, J. (1710), who introduced the **sign test**. But most nonparametric tests were developed between 1940 and 1955.

We make special mention of the articles of Kolmogorov, Andrey Nikolaevich (1933), Smirnov, Vladimir Ivanovich (1939), Wilcoxon, F. (1945, 1947), Mann, H.B. and Whitney, D.R. (1947), Mood, A.M. (1950), and Kruskal, W.H. and Wallis, W.A. (1952). Later, many other articles were added to this list. Savage, I.R. (1962) published a bibliography of about 3000 articles, written before 1962, concerning nonparametric tests.

DOMAINS AND LIMITATIONS

The fast development of nonparametric tests can be explained by the following points:

- Nonparametric methods require few assumption concerning the **population** under study, such as assumption of normality.
- Nonparametric methods are often easier to understand and to apply than the equivalent parametric tests.
- Nonparametric methods are applicable in situations where parametric tests cannot be used, for example, when the variables are measured only on ordinal scales.
- Despite the fact that, at first view, nonparametric methods seem to sacrifice an essential part of information contained in the samples, theoretical researches have shown that nonparametric tests are only slightly inferior to their parametric counterparts when the distribution of the studied population is specified, for example, the **normal distribution**. Nonparametric tests, on the other hand, are superior to parametric tests when the distribution of the population is far away from the specified (normal) distribution.

EXAMPLES
See **Kolmogorov–Smirnov test, Kruskal–Wallis test, Mann–Withney test, Wilcoxon signed test), sign test.**

FURTHER READING
▶ **Goodness of fit test**
▶ **Hypothesis testing**
▶ **Kolmogorov–Smirnov test**
▶ **Kruskal-Wallis test**

► **Mann–Whitney test**
► **Sign test**
► **Spearman rank correlation coefficient**
► **Test of independence**
► **Wilcoxon signed test**
► **Wilcoxon test**

REFERENCES

Arbuthnott, J.: An argument for Divine Providence, taken from the constant regularity observed in the births of both sexes. Philos. Trans. **27**, 186–190 (1710)(3.4).

Kolmogorov, A.N.: Sulla determinazione empirica di una legge di distribuzione. Giornale dell'Instituto Italiano degli Attuari **4**, 83–91 (6.1) (1933)

Kruskal, W.H., Wallis, W.A.: Use of ranks in one-criterion variance analysis. J. Am. Stat. Assoc. **47**, 583–621 and errata, ibid. **48**, 907–911 (1952)

Mann, H.B., Whitney, D.R.: On a test whether one of two random variables is stochastically larger than the other. Ann. Math. Stat. **18**, 50–60 (1947)

Mood, A.M.: Introduction to the Theory of Statistics. McGraw-Hill, New York (1950) (chap. 16).

Savage, I.R.: Bibliography of Nonparametric Statistics. Harvard University Press, Cambridge, MA (1962)

Smirnov, N.V.: Estimate of deviation between empirical distribution functions in two independent samples. (Russian). Bull. Moscow Univ. **2**(2), 3–16 (6.1, 6.2) (1939)

Wilcoxon, F.: Individual comparisons by ranking methods. Biometrics **1**, 80–83 (1945)

Wilcoxon, F.: Some rapid approximate statistical procedures. American Cyanamid, Stamford Research Laboratories, Stamford, CT (1957)

Norm of a Vector

The norm of a **vector** indicates the length of this vector defined from the **scalar product** of the vector by itself. The norm of a vector is obtained by taking the square root of this scalar product.

A vector of norm 1 is called a unit vector.

MATHEMATICAL ASPECTS

For a given **vector** x,

$$
x = \begin{bmatrix} x_1 \\ x_2 \\ \vdots \\ x_n \end{bmatrix},
$$

we define the **scalar product** of x by itself by:

$$
x' \cdot x = \sum_{i=1}^{n} x_i^2.
$$

The norm of vector x thus equals:

$$
\| x \| = \sqrt{x' \cdot x} = \sqrt{\sum_{i=1}^{n} x_i^2}.
$$

DOMAINS AND LIMITATIONS

As the norm of a **vector** is calculated with the help of the **scalar product**, the properties of the norm come from those of the scalar product. Thus:

• The norm of a vector x is strictly positive if the vector is not zero, and zero if it is:

$$
\| x \| > 0 \quad \text{if} \quad x \neq 0 \quad \text{and}
$$
$$
\| x \| = 0 \quad \text{if and only if} \quad x = 0.
$$

- The norm of the result of a multiplication of a vector x by a scalar equals the product of the norm of x and the absolute value of the scalar:

$$\| k \cdot x \| = |k| \cdot \| x \| .$$

- For two vectors x and y, we have:

$$\| x + y \| \leq \| x \| + \| y \| ,$$

that is, the norm of a sum of two vectors is smaller than or equal to the sum of the norms of these vectors.

- For two vectors x and y, the absolute value of the scalar product of x and y is smaller than or equal to the product of norms:

$$\left| x' \cdot y \right| \leq \| x \| \cdot \| y \| .$$

This result is called the Cauchy–Schwarz inequality.

The norm of a vector is also used to obtain a unit vector having the same direction as the given vector. It is enough in this case to divide the initial vector by its norm:

$$y = \frac{x}{\| x \|} .$$

EXAMPLES

Consider the following **vector** defined in the Euclidean space of three dimensions:

$$x = \begin{bmatrix} 12 \\ 15 \\ 16 \end{bmatrix} .$$

The **scalar product** of this vector by itself equals:

$$x' \cdot x = 12^2 + 15^2 + 16^2$$
$$= 144 + 225 + 256 = 625$$

from where we get the norm of x:

$$\| x \| = \sqrt{x' \cdot x} = \sqrt{625} = 25 .$$

The unit vector having the direction of x is obtained by dividing each component of x by 25:

$$y = \begin{bmatrix} 0.48 \\ 0.6 \\ 0.64 \end{bmatrix} ,$$

and we want to verify that $\| y \| = 1$. Hence:

$$\| y \|^2 = 0.48^2 + 0.6^2 + 0.64^2$$
$$= 0.2304 + 0.36 + 0.4096 = 1 .$$

FURTHER READING
▶ **Least squares**
▶ **L_1 estimation**
▶ **Vector**

REFERENCES
Dodge, Y.: Mathématiques de base pour économistes. Springer, Berlin Heidelberg New York (2002)

Dodge, Y., Rousson, V.: Multivariate L_1-mean. Metrika **49**, 127–134 (1999)

Normal Distribution

Random variable X is distributed according to a normal distribution if it has a **density function** of the form:

$$f(x) = \frac{1}{\sigma \sqrt{2\pi}} \exp\left(\frac{-(x - \mu)^2}{2\sigma^2} \right) ,$$

$$(\sigma > 0) .$$

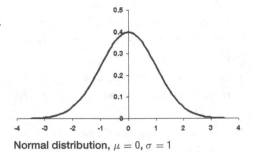

Normal distribution, $\mu = 0, \sigma = 1$

We will say that X follows a normal distribution of mean μ and of variance σ^2. The normal distribution is a **continuous probability distribution**.

HISTORY

The normal distribution is often attributed to **Laplace, P.S.** and **Gauss, C.F.**, whose name it bears. However, its origin dates back to the works of **Bernoulli, J.** who in his work *Ars Conjectandi* (1713) provided the first basic elements of the **law of large numbers**. In 1733, **de Moivre, A.** was the first to obtain the normal distribution as an approximation of the **binomial distribution**. This work was written in Latin and published in an English version in 1738. De Moivre, A. called what he found a "curve"; he discovered this curve while calculating the **probabilities** of gain for different games of hazard.

Laplace, P.S., after **de Moivre, A.**, studied this distribution and obtained a more formal and general result of the de Moivre approximation. In 1774 he obtained the normal distribution as an approximation of the **hypergeometric distribution**.

Gauss, C.F. studied this distribution through problems of measurement in astronomy. His works in 1809 and 1816 established techniques based on the normal distribution that became standard methods during the 19th century. Note that even if the first approximation of this distribution is due to de Moivre, Galileo had already found that the **errors** of **observation** were distributed in a symmetric way and tended to regroup around their true **value**.

Many denominations of the normal distribution can be found in the literature. Quetelet, A. (1796–1874) spoke of the "curve of possibilities" or the "distribution of possibilities".

Note also that **Galton, F.** spoke of the "**frequency** of **error** distribution" or of the "distribution of **deviations** from a **mean**". Stigler, S. (1980) presents a more detailed discussion on the different names of this curve.

MATHEMATICAL ASPECTS

The **expected value** of the normal distribution is given by:

$$E[X] = \mu.$$

The **variance** is equal to:

$$\mathrm{Var}(X) = \sigma^2.$$

If the **mean** μ is equal to 0, and the **variance** σ^2 is equal to 1, then we obtain a centered and reduced normal distribution (or standard normal distribution) whose **density function** is given by:

$$f(x) = \frac{1}{\sqrt{2\pi}} \exp\left(\frac{-x^2}{2}\right).$$

If a **random variable** X follows a normal distribution of mean μ and variance σ^2, then the random variable

$$Z = \frac{X - \mu}{\sigma}$$

follows a centered and reduced normal distribution (of mean 0 and variance 1).

DOMAINS AND LIMITATIONS

The normal distribution is the most famous **continuous probability distribution**. It plays a central role in the theory of **probability** and its statistical applications. Many measurements such as the size or weight of individuals, the diameter of a piece of machinery, the results of an IQ test, etc.

approximately follow a normal distribution. The normal distribution is frequently used as an approximation, either when the normality is attributed to a distribution in the construction of a **model** or when a known distribution is replaced by a normal distribution with the same **expected value** or **variance**. It is used for the approximation of the **chi-square distribution**, the **Student distribution**, and **discrete probability distributions** such as the **binomial distribution** and the **Poisson distribution**.

The normal distribution is also a fundamental element of the theory of **sampling**, where its role is important in the study of correlation, **regression** analysis, **variance analysis**, and **covariance analysis**.

FURTHER READING

▶ **Continuous probability distribution**
▶ **Lognormal distribution**
▶ **Normal table**

REFERENCES

Bernoulli, J.: Ars Conjectandi, Opus Posthumum. Accedit Tractatus de Seriebus infinitis, et Epistola Gallice scripta de ludo Pilae recticularis. Impensis Thurnisiorum, Fratrum, Basel (1713)

Gauss, C.F.: Theoria Motus Corporum Coelestium. Werke, 7 (1809)

Gauss, C.F.: Bestimmung der Genauigkeit der Beobachtungen, vol. 4, pp. 109–117 (1816). In: Gauss, C.F. Werke (published in 1880). Dieterichsche Universitäts-Druckerei, Göttingen

Laplace, P.S. de: Mémoire sur la probabilité des causes par les événements. Mem. Acad. Roy. Sci. (presented by various scientists) **6**, 621–656 (1774) (or Laplace,

P.S. de (1891) Œuvres complètes, vol 8. Gauthier-Villars, Paris, pp. 27–65)

Moivre, A. de: Approximatio ad summam terminorum binomii $(a + b)^n$, in seriem expansi. Supplementum II to Miscellanae Analytica, pp. 1–7 (1733). Photographically reprinted in a rare pamphlet on Moivre and some of his discoveries. Published by Archibald, R.C. Isis **8**, 671–683 (1926)

Moivre, A. de: The Doctrine of Chances: or, A Method of Calculating the Probability of Events in Play. Pearson, London (1718)

Stigler, S.: Stigler's law of eponymy, Transactions of the New York Academy of Sciences, 2nd series **39**, 147–157 (1980)

Normal Equations

Normal equations are equations obtained by setting equal to zero the partial derivatives of the sum of squared errors (**least squares**); normal equations allow one to estimate the parameters of a multiple linear regression.

HISTORY

See **analysis of regression**.

MATHEMATICAL ASPECTS

Consider a general **model** of **multiple linear regression**:

$$Y_i = \beta_0 + \sum_{j=1}^{p-1} \beta_j \cdot X_{ji} + \varepsilon_i, \quad i = 1, \dots, n,$$

where Y_i is the **dependent variable**, $X_{ji}, j = 1, \dots, p-1$, are the independent variables, ε_i is the term of random nonobservable **error**,

$\beta_j, j = 0, \ldots, p-1$, are the parameters to be estimated, and n is the number of observations.

To apply the method of **least squares**, we should minimize the sum of squared errors:

$$f\left(\beta_0, \beta_1, \ldots, \beta_{p-1}\right) = \sum_{i=1}^{n} \varepsilon_i^2$$

$$= \sum_{i=1}^{n} \left(Y_i - \beta_0 - \sum_{j=1}^{p-1} \beta_j \cdot X_{ji}\right)^2.$$

Setting the p partial derivatives equal to zero relative to the parameters to be estimated, we get the p normal equations:

$$\begin{cases} \frac{\partial f(\beta_0, \beta_1, \ldots, \beta_{p-1})}{\partial \beta_0} = 0, \\ \frac{\partial f(\beta_0, \beta_1, \ldots, \beta_{p-1})}{\partial \beta_1} = 0, \\ \vdots \\ \frac{\partial f(\beta_0, \beta_1, \ldots, \beta_{p-1})}{\partial \beta_{p-1}} = 0. \end{cases}$$

We can also express the same model in matrix form:

$$\mathbf{Y} = \mathbf{X}\boldsymbol{\beta} + \boldsymbol{\varepsilon},$$

where $\mathbf{Y}$ is the **vector** $(n \times 1)$ of observations related to the dependent variable (n observations), $\mathbf{X}$ is the **matrix** $(n \times p)$ of the plan related to the independent variables, $\boldsymbol{\varepsilon}$ is the $(n \times 1)$ vector of errors, and $\boldsymbol{\beta}$ is the $(p \times 1)$ vector of the parameters to be estimated.

By the **least-squares** method, we should minimize:

$$\boldsymbol{\varepsilon}'\boldsymbol{\varepsilon} = (\mathbf{Y} - \mathbf{X}\boldsymbol{\beta}')(\mathbf{Y} - \mathbf{X}\boldsymbol{\beta})$$

$$= \mathbf{Y}'\mathbf{Y} - \boldsymbol{\beta}'\mathbf{X}'\mathbf{Y} - \mathbf{Y}'\mathbf{X}\boldsymbol{\beta}$$

$$+ \boldsymbol{\beta}'\mathbf{X}'\mathbf{X}\boldsymbol{\beta}$$

$$= \mathbf{Y}'\mathbf{Y} - 2\boldsymbol{\beta}'\mathbf{X}'\mathbf{Y} + \boldsymbol{\beta}'\mathbf{X}'\mathbf{X}\boldsymbol{\beta}.$$

Setting to zero the derivatives relative to $\boldsymbol{\beta}$ (in matrix form), corresponding to the partial derivatives but written in vector form, we find the normal equations:

$$\mathbf{X}'\mathbf{X}\hat{\boldsymbol{\beta}} = \mathbf{X}'\mathbf{Y}.$$

DOMAINS AND LIMITATIONS

We can generalize the concept of normal equations in the case of a **nonlinear regression** (with r observations and $p + 1$ parameters to estimate).

They are expressed by:

$$\sum_{i=1}^{r} \frac{\left(Y_i - f\left(X_{1i}, \ldots, X_{ki}; \beta_0, \ldots, \beta_p\right)\right)^2}{\partial \beta_j}$$

$$= \sum_{i=1}^{r} \left(Y_i - f\left(X_{1i}, \ldots, X_{ki}; \beta_0, \ldots, \beta_p\right)\right)$$

$$\cdot \left[\frac{\partial f\left(X_{1i}, \ldots, X_{ki}; \beta_0, \ldots, \beta_p\right)}{\partial \beta_j}\right] = 0,$$

where $j = 0, \ldots, p$. Because f is nonlinear in the parameters to be estimated, the normal equations are also nonlinear and so can be very difficult to solve. Moreover, the system of equations can have more than one solution that corresponds to the possibility of many minima of f. Normally, we try to develop an iterative procedure to solve the equations or refer to the techniques explained in nonlinear regression.

EXAMPLES

If the **model** contains, for example, 3 parameters β_0, β_1, and β_2, the normal equations are the following:

$$\begin{cases} \frac{\partial f(\beta_0, \beta_1, \beta_2)}{\partial \beta_0} = 0, \\ \frac{\partial f(\beta_0, \beta_1, \beta_2)}{\partial \beta_1} = 0, \\ \frac{\partial f(\beta_0, \beta_1, \beta_2)}{\partial \beta_1} = 0. \end{cases}$$

That is:

$$
\begin{cases}
\beta_0 \cdot n + \beta_1 \cdot \sum_{i=1}^{n} X_{1i} + \beta_2 \cdot \sum_{i=1}^{n} X_{2i} = \sum_{i=1}^{n} Y_i, \\[2ex]
\beta_0 \cdot \sum_{i=1}^{n} X_{1i} + \beta_1 \cdot \sum_{i=1}^{n} X_{1i}^2 + \beta_2 \cdot \sum_{i=1}^{n} X_{1i} \cdot X_{2i} \\[1ex]
\qquad\qquad\qquad\qquad = \sum_{i=1}^{n} X_{1i} \cdot Y_i, \\[2ex]
\beta_0 \cdot \sum_{i=1}^{n} X_{2i} + \beta_1 \cdot \sum_{i=1}^{n} X_{2i} \cdot X_{1i} + \beta_2 \cdot \sum_{i=1}^{n} X_{2i}^2 \\[1ex]
\qquad\qquad\qquad\qquad = \sum_{i=1}^{n} X_{2i} \cdot Y_i.
\end{cases}
$$

See **simple linear regression**.

FURTHER READING
▶ **Error**
▶ **Least squares**
▶ **Multiple linear regression**
▶ **Nonlinear regression**
▶ **Parameter**
▶ **Simple linear regression**

REFERENCES
Legendre, A.M.: Nouvelles méthodes pour la détermination des orbites des comètes. Courcier, Paris (1805)

Normal Probability Plot

A normal probability plot allows one to verify if a **data** set is distributed according to a **normal distribution**. When this is the case, it is possible to estimate the **mean** and the **standard deviation** from the plot.

The cumulated **frequencies** are transcribed in **ordinate** on a Gaussian **scale**, graduated by putting the **value** of the **distribution function** $F(t)$ of the **normal distribution** (centered and reduced) opposite to the point located at a **distance** t from the origin. In **abscissa**, the **observations** are represented

on an arithmetic scale. In practice, the normal probability papers sold on the market are used.

MATHEMATICAL ASPECTS
On an **axis** system the choice of the origin and size of the unity length is made according to the **observations** that are to be represented. The **abscissa** is then arithmetically graduated like millimeter paper.

In **ordinate**, the values of the **distribution function** $F(t)$ of the **normal distribution** are transcribed at the height t of a fictive origin placed at mid-height of the sheet of paper. Therefore:

$F(t)$	t
0.9772	2
0.8412	1
0.5000	0
0.1588	−1
0.0	

will be placed, and only the **scale** on the left, which will be graduated from 0 to 1 (or from 0 to 100 if written in %), will be kept.

Consider a set of n point **data** that are supposed to be classified in increasing order. Let us call this series $x_1, x_2, \ldots, x_n$, and for each **abscissa** x_i ($i = 1, \ldots, n$), the **ordinate** is calculated as

$$
100 \cdot \frac{i - \frac{3}{8}}{n + \frac{1}{4}}
$$

and represented on a normal probability plot. It is also possible to transcribe

$$
100 \cdot \frac{i - \frac{1}{2}}{n}
$$

as a function of x_i on the plot. This second approximation of the cumulated **frequency** is explained as follows. If the surface located under the normal curve (with area equal to

Papier gausso-arithmetique. Echelle verticale : 0.01 - 99.99 % (made with Mathematica^{TM})

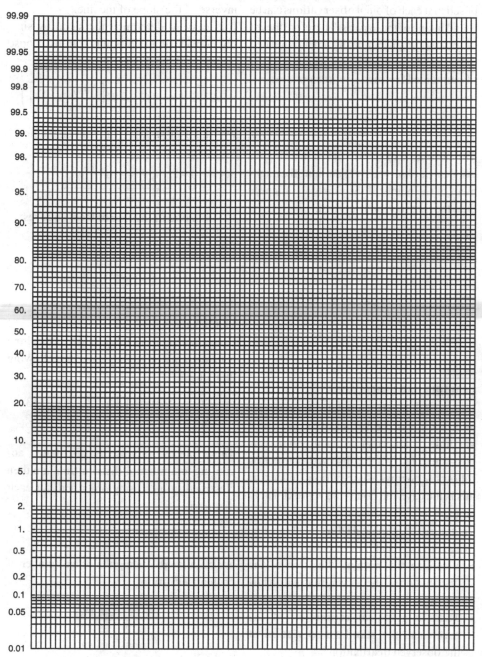

Normal probability paper. Vertical scale: 0.01 - 99.99 %

1) is divided into n equal parts, it can be supposed that each of our n **observations** can be found in one of these parts. Therefore the ith observation x_i (in increasing order) is located in the middle of the ith part, which in terms of cumulated **frequency** corresponds to

$$\frac{i - \frac{1}{2}}{n}.$$

The normality of the observed distribution can then be observed by examining the alignment of the points: if the points are aligned, then the distribution is normal.

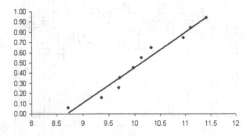

DOMAINS AND LIMITATIONS

By representing the cumulated **frequencies** as a function of the **values** of the statistical **variable** on a normal probability plot, it is possible to verify if the observed distribution follows a **normal distribution**. Indeed, if this is the case, the points are approximately aligned.

Therefore the graphical adjustment of a line to a set of points allows to:

- Visualize the normal character of the distribution in a better way than with a **histogram**.
- Make a graphical **estimation** of the **mean** and the **standard deviation** of the observed distribution.

The line of **least squares** cuts the **ordinate** 0.50 (or 50%) at the **estimation** of the mean (which can be read on the **abscissa**). The estimated standard deviation is given by the inverse of the slope of the line.

It is easy to determine the slope by simply considering two points. Let us choose $(\hat{\mu}, 0.5)$ and the point at the **ordinate**

$$0.9772 \left(= F(2) = F\left[\frac{X_{0.9772} - \hat{\mu}}{\hat{\sigma}}\right]\right)$$

whose **abscissa** $X_{0.9772}$ is read on the line. We have:

$$\frac{1}{\hat{\sigma}} = \frac{F^{-1}(0.9772) - F^{-1}(0.5)}{X - \mu}$$

$$= \text{slope of the line of \textbf{least squares}},$$

from which we finally obtain:

$$\hat{\sigma} = \frac{X_{0.9772} - \hat{\mu}}{2}.$$

EXAMPLES

Consider ten pieces fabricated by a machine X taken randomly: They have the following diameters:

$$9.36, \quad 9.69, \quad 11.10, \quad 8.72, \quad 10.13,$$
$$9.98, \quad 11.42, \quad 10.33, \quad 9.71, \quad 10.96.$$

These pieces, supposedly distributed according to a **normal distribution**, are then classified in increasing order and, for each diameter

$$f_i = \frac{i - \frac{3}{8}}{10 + \frac{1}{4}},$$

are calculated, i being the order number of each **observation**:

x_i	f_i
8.72	5/82
9.36	13/82
9.69	21/82
9.71	29/82
9.98	37/82

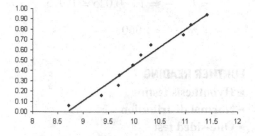

x_i	f_i
10.13	45/82
10.33	53/82
10.96	61/82
11.10	69/82
11.42	77/82

The **graphical representation** of f_i with respect to x_i on a piece of normal probability paper provides for our random **sample**: estimated **mean** of diameters: $\hat{\mu} = 10.2$, estimated **standard deviation**: $\hat{\sigma} = 0.9$.

FURTHER READING
▶ **Distribution function**
▶ **Frequency**
▶ **Frequency distribution**
▶ **Normal distribution**

Normal Table

The normal table, also called a Gauss table or a **normal distribution** table, gives the values of the **distribution function** of a **random variable** following a standard normal distribution, that is, of the **mean** equalling 0 and of **variance** equalling 1.

HISTORY
de Laplace, Pierre Simon (1774) obtained the **normal distribution** from the **hypergeometric distribution** and in a second work (dated 1778 but published in 1781) he gave a normal table. **Pearson, Egon Sharpe** and Hartley, H.O. (1948) edited a normal table based on the values calculated by Sheppard, W.F. (1903, 1907). Exhaustive lists of normal tables on the market was given by the National Bureau of Standards (until 1952) and by Greenwood, J.A. and Hartley, H.O. (until 1958).

DOMAINS AND LIMITATIONS
See **central limit theorem**.

MATHEMATICAL ASPECTS
Let **random variable** Z follow a **normal distribution** of **mean** 0 and of **variance** 1. The **density function** $f(t)$ thus equals:

$$f(t) = \frac{1}{\sqrt{2\pi}} \cdot \exp\left(-\frac{t^2}{2}\right).$$

The **distribution function** of random variable Z is defined by:

$$F(z) = P(Z \leq z) = \int_{-\infty}^{z} f(t)\,dt.$$

The normal table represented here gives the values of $F(z)$ for the different values of z between 0 and 3.5.

With a **normal distribution** symmetric relative to the mean, we have:

$$F(z) = 1 - F(-z),$$

which allows to determine $F(z)$ for a negative value of z.

Sometimes we must use the normal table, in inverse direction, to find the value of z corresponding to a given probability. We generally denote by $z = z_\alpha$ the value of random variable Z for which

$$P(Z \leq z_\alpha) = 1 - \alpha.$$

N

EXAMPLES
See Appendix G.
Examples of normal table use:

$$P(Z \leq 2.5) = 0.9938,$$

$$P(1 \leq Z \leq 2)$$
$$= P(Z \leq 2) - P(Z \leq 1)$$
$$= 0.9772 - 0.8413 = 0.1359,$$

$$P(Z \leq -1)$$
$$= 1 - P(Z \leq 1)$$
$$= 1 - 0.8413 = 0.1587,$$

$$P(-0.5 \leq Z \leq 1.5)$$
$$= P(Z \leq 1.5) - P(Z \leq -0.5)$$
$$= P(Z \leq 1.5) - [1 - P(Z \leq 0.5)]$$
$$= 0.9332 - [1 - 0.6915]$$
$$= 0.9332 - 0.3085 = 0.6247.$$

Conversely, we can use the normal table to determine the limit z_α for which the **probability** that the **random variable** Z takes a value smaller to that limit (z_α), is equal to a fixed value $(1 - \alpha)$, that is:

$$P(Z \leq z_\alpha) = 0.95 \quad \Rightarrow \quad z_\alpha = 1.645,$$
$$P(Z \leq z_\alpha) = 0.975 \quad \Rightarrow \quad z_\alpha = 1.960.$$

This allows us, in **hypothesis testing**, to determine the **critical value** z_α relative to a given **significance level** α.

Example of Application
In the case of a **one-tailed test**, if the significance level α equals 5%, then we can determine the critical value $z_{0.05}$ in the following manner:

$$P(Z \leq z_{0.05}) = 1 - \alpha$$
$$= 1 - 0.05$$
$$= 0.95$$
$$\Rightarrow z_{0.05} = 1.645.$$

In the case of a **two-tail test**, we should find the value $z_{\alpha/2}$. With a significance level of 5%, the critical value $z_{0.025}$ is obtained in the following manner:

$$P(Z \leq z_{0.025})$$
$$= 1 - \frac{\alpha}{2} = 1 - 0.025 = 0.975$$
$$\Rightarrow z_{0.025} = 1.960.$$

FURTHER READING
▶ **Hypothesis testing**
▶ **Normal distribution**
▶ **One-sided test**
▶ **Statistical table**
▶ **Two-sided test**

REFERENCES
Greenwood, J.A., Hartley, H.O.: Guide to Tables in Mathematical Statistics. Princeton University Press, Princeton, NJ (1962)

Laplace, P.S. de: Mémoire sur la probabilité des causes par les événements. Mem. Acad. Roy. Sci. (presented by various scientists) **6**, 621–656 (1774) (or Laplace, P.S. de (1891) Œuvres complètes, vol 8. Gauthier-Villars, Paris, pp. 27–65)

Laplace, P.S. de: Mémoire sur les probabilités. Mem. Acad. Roy. Sci. Paris, 227–332 (1781) (or Laplace, P.S. de (1891) Œuvres complètes, vol. 9. Gauthier-Villars, Paris, pp. 385–485.)

National Bureau of Standards.: A Guide to Tables of the Normal Probability Integral. U.S. Department of Commerce. Applied Mathematics Series, 21 (1952)

Pearson, E.S., Hartley, H.O.: Biometrika Tables for Statisticians, 1 (2nd edn.). Cambridge University Press, Cambridge (1948)

Sheppard, W.F.: New tables of the probability integral. Biometrika **2**, 174–190 (1903)

Sheppard, W.F.: Table of deviates of the normal curve. Biometrika **5**, 404–406 (1907)

Normalization

Normalization is the transformation from a normally distributed **random variable** to a random variable following a standard normal distribution. It allows to calculate and compare the values belonging to the normal curves of the **mean** and of different **variances** on the basis of a reference normal distribution, that is, the standard normal distribution.

MATHEMATICAL ASPECTS

Let the **random variable** X follow the **normal distribution** $N\left(\mu, \sigma^2\right)$. Its normalization gives us a random variable

$$Z = \frac{X - \mu}{\sigma}$$

that follows a standard normal distribution $N(0, 1)$.

Each **value** of the distribution $N\left(\mu, \sigma^2\right)$ can be transformed into standard **variable** Z, each Z representing a deviation from the **mean** expressed in units of **standard deviation**.

EXAMPLES

The students of a professional school have had two exams. Each exam was graded on a scale of 1 to 60, and the grades are considered the realizations of two random variables of a **normal distribution**. Let us compare the grades received by the students on these two exams.

Here are the means and the standard deviations of each exam calculated on all the students:

$$\text{Exam 1: } \mu_1 = 35, \quad \sigma_1 = 4,$$
$$\text{Exam 2: } \mu_2 = 45, \quad \sigma_2 = 1.5,$$

A student named Marc got the following results:

$$\text{Exam 1: } X_1 = 41,$$
$$\text{Exam 2: } X_2 = 48.$$

The question is to know which exam Marc scored better on relative to the other students. We cannot directly compare the results of two exams because the results belong to distributions with different means and standard deviations.

Is we simply examine the difference of each note and compare that to the mean of its distribution, we get:

$$\text{Exam 1: } X_1 - \mu_1 = 41 - 35 = 6,$$
$$\text{Exam 2: } X_2 - \mu_2 = 48 - 45 = 3.$$

Note that Marc's score was 6 points higher than the mean on exam 1 and only 3 points higher than the mean on exam 2. A hasty conclusion would suggest to us that Marc did better on exam 1 than on exam 2, relative to the other students.

This conclusion takes into account only the difference of each result from the mean. It does not consider the dispersion of student grades around the mean. We divide the difference from the mean by the standard deviation to make the results comparable:

$$\text{Exam 1: } Z_1 = \frac{X_1 - \mu_1}{\sigma_1} = \frac{6}{4} = 1.5,$$
$$\text{Exam 1: } Z_2 = \frac{X_2 - \mu_2}{\sigma_2} = \frac{3}{1.5} = 2.$$

By this calculation we have normalized the scores X_1 and X_2. We can conclude that the normalized **value** is greater for exam 2 ($Z_2 = 2$) than for exam 1 ($Z_1 = 1.5$) and that Marc did better on exam 1, relative to other students.

FURTHER READING
▶ **Normal distribution**

Null Hypothesis

In the fulfillment of **hypothesis testing**, the null hypothesis is the hypothesis that is to be tested. It is designated by H_0. The opposite **hypothesis** is called the **alternative hypothesis**. It is the alternative hypothesis that will be accepted if the test leads to rejecting the null hypothesis.

HISTORY
See **hypothesis** and **hypothesis testing**.

MATHEMATICAL ASPECTS
During the fulfillment of **hypothesis testing** on a **parameter** of a **population**, the null hypothesis is usually a supposition on the presumed **value** of this parameter. The null hypothesis will then be presented as:

$$H_0: \quad \theta = \theta_0,$$

θ being the unknown parameter of the population and θ_0 the presumed value of this parameter. This **parameter** can be, for example, the **mean** of the population.

When **hypothesis testing** aims at comparing two **populations**, the null hypothesis generally supposes that the **parameters** are equal:

$$H_0: \quad \theta_1 = \theta_2,$$

where θ_1 is the parameter of the **population** where the first sample came from and θ_2 the population that the second sample came from.

EXAMPLES
We are going to examine the null hypothesis on three examples of **hypothesis testing**:

1. *Hypothesis testing on the percentage of a sample*

   A candidate running for office wants to know if he will receive more than 50% of the vote.

   The **null hypothesis** for this problem can be posed thus:

   $$H_0: \quad \pi = 0.5,$$

   where π is the **percentage** of the **population** to be estimated.

2. *Hypothesis testing on the mean of the population*

   A manufacturer wants to test the precision of a new machine that should produce pieces 8 mm in diameter.

   We can pose the following **null hypothesis**:

   $$H_0: \quad \mu = 8,$$

   where μ is the **mean** of the **population** to be estimated.

3. *Hypothesis testing on the comparison of means of two populations*

   An insurance company has decided to equip its offices with computers. It wants to buy computers from two different suppliers if there is no significant difference between the reliability of the two brands. It draws two samples of PC's comming from each brand. Then it records how much time was consumed for each sample to have a problem on PC.

   According to the null hypothesis, the mean of the time passed since the first problem is the same for each brand:

   $$H_0: \quad \mu_1 - \mu_2 = 0,$$

where μ_1 and μ_2 are the respective means of two populations.

This **hypothesis** can also be written as:

$$H_0: \quad \mu_1 = \mu_2.$$

FURTHER READING
▶ **Alternative hypothesis**
▶ **Analysis of variance**
▶ **Hypothesis**
▶ **Hypothesis testing**

N

O

Observation

An observation is the result of a scientific study assembling information on a statistical unit belonging to a given **population**.

FURTHER READING

► **Data**
► **Outlier**
► **Population**
► **Sample**

Odds and Odds Ratio

Odds are defined as the ratio of the probability of an event and the probability of the complementary event, that is, the ratio of the probability of an event occurring (in the particular case of epidemiology, an illness) and the probability of its not occurring.

The odds ratio is, like **relative risk**, a measure of the relative effect of a cause. More precisely, it is the ratio of the odds calculated within populations exposed to given risk factor to varying degrees.

MATHEMATICAL ASPECTS

Odds attached to an event "having an illness" are, formally, the following ratio, where p denotes the probability of having the illness:

$$\text{Odds} = \frac{p}{1-p}.$$

Let there be a factor of risk to which a group E is exposed and a group NE is not exposed. The corresponding odds ratio, OR, is the ratio of the odds of group E and the odds of group NE:

$$\text{OR} = \frac{\text{odds of group E}}{\text{odds of group NE}}.$$

DOMAINS AND LIMITATIONS

Let us take as an example the calculation of the odds ratio from the data in the following table comparing two groups of individuals attacked by the same potentially dangerous illness; one group ("treated" group) receives medical treatment while the other does not ("reference" group):

Group	Deaths	Survivors	Odds	OR
Treated	5%	95%	0.053	0.47
Reference	10%	90%	0.111	1.0*

*) By definition the OR equals 1 for the group of reference.

If we call here "**risk**" the risk attached to the event "death", then the OR is a good approximation of the relative risk, on the condition that the risks of death are similar or, if they are strongly different, are small. We generally consider a risk smaller than 10%, as in this case, as being small. The odds ratio does not remain an approximation of relative risk because there is a **bias** of the OR which appears for small risks. That is:

$$OR = \frac{\frac{\text{risk of treated group}}{\text{survival rate in treated group}}}{\frac{\text{risk in reference group}}{\text{survival rate in reference group}}}$$

$$= \frac{\text{risk in treated group}}{\text{risk in reference group}}$$

$$\times \frac{\text{survival rate in reference group}}{\text{survival rate in treated group}}$$

$$= \text{relative risk} \times \text{bias of odds ratio}$$

with:

$$\text{bias of OR} =$$
$$\frac{\text{survival rate in reference group}}{\text{survival rate in treated group}} .$$

The estimations of the relative and odds ratios are used to measure the force of causal associations between the exposure to a risk factor and an illness. The two ratios can take values between 0 and infinity that are justified by the same general interpretation. We can distinguish two clearly different cases:

1. An OR value greater than 1 indicates a risk or an odds of illness much greater when the studied group is exposed to a specific risk factor. In such a case there is a positive association.
2. A smaller value implies a reduced exposure to a risk factor. In this case there is a negative association.

Confidence Interval at 95% of Odds Ratio

The technique and notation used for a confidence interval of 95% of the odds ratio are the same as those for **relative risk**:

$$\exp\left\{ (\ln OR) \pm 1.96\sqrt{\frac{1}{a} - \frac{1}{n} + \frac{1}{c} - \frac{1}{m}} \right\} .$$

a, b, c, and d must be large. If the confidence interval includes the value 1.0, then the odds ratio is not statistically significant.

EXAMPLES

Consider a study of the relation between the risk factor "smoking" and breast cancer. The odds of breast cancer is the ratio of the probability of developing this cancer and the probability of not having this illness during a given period. The odds of developing breast cancer for each group are defined by the group's level of exposure to smoking. The following table illustrates this example:

Group	Cancer (100000/ 2 years)	No cancer (100000/ 2 years)	Odds of breast cancer (/2 years)	OR
Non-exposed	114	99886	0.00114	1.0
Passive smokers	252	99748	0.00252	2.2
Active smokers	276	99724	0.00276	2.4

In this example, the odds of breast cancer of the nonexposed group are 0.00114 (114 : 99886), those of the passive smokers 0.00252 (252 : 99748), and those of the active smokers 0.00276 (276 : 99274). Taking as the reference group the nonexposed subjects, the odds ratio of the breast cancer is, for the active smokers:

$$\frac{\frac{276}{99724}}{\frac{114}{99886}} = 2.4 = \frac{276}{114} \cdot \frac{99886}{99724} .$$

The risk of breast cancer over 2 years being very small, the survival of active smokers as well as that of nonexposed subjects is close to 100%, and so the survival ratio (the bias of the odds ratio) is practically equal to 1.0, which is interpreted as almost the absence of any OR bias. In this example, the odds ratio of breast cancer is identical to the relative risk of breast cancer in the passive smokers and the active smokers. Since breast cancer is a rare illness, the odds ratio is a good approximation of the relative risk.

FURTHER READING
► **Attributable risk**
► **Avoidable risk**
► **Cause and effect in epidemiology**
► **Incidence**
► **Incidence rate**
► **Odds and odds ratio**
► **Prevalence**
► **Prevalence rate**
► **Relative risk**
► **Risk**

REFERENCES

Lilienfeld, A.M., Lilienfeld, D.E.: Foundations of Epidemiology, 2nd edn. Clarendon, Oxford (1980)

MacMahon, B., Pugh, T.F.: Epidemiology: Principles and Methods. Little Brown, Boston, MA (1970)

Morabia, A.: Epidemiologie Causale. Editions Médecine et Hygiène, Geneva (1996)

Morabia, A.: L'Épidémiologie Clinique. Editions "Que sais-je?". Presses Universitaires de France, Paris (1996)

Official Statistics

The adjective *official* means *coming from the government, the administration; authorized by a proper authority*. Thus the word *official*, in conjunction with the word *statistics*, indicates a relation with the state. Moreover, the etymology of the word *statistics*, from the Latin *status* (state), meaning to be governmental statistics of data collected by government stablishement or private agencies working for the goverment. The term *official statistics* refers to a very general concept that can be obtained in different ways.

HISTORY

The **census**, the most widely known of official statistics, is a practice almost as old as the social organization of people. Since antiquity, the kings of Babylon and Egypt, as well as Chinese, Greek, Hebrew, Persian, and Roman chiefs, tried to determine the number of soldiers under their command and the goods they possessed. The first statistical trials date back more than 6000 years, but they were only sporadic trials of counting people and goods. Statisticians, in the modern sense of the term, appear only in the late 17th century, with the establishment of the modern state and the needs of colonial administration in France and England. In the 18th century, in Diderot and Alembert's encyclopedia, "arithmetical politics" is defined as "the application of arithmetical calculations to political subjects and usage". The French Revolution, and then the birth of democracy in the United States, justified the new developments in official statistics. The new states, to know the political weight of their different entities, needed to know about their populations. That is why the first regular decennial censuses started in the US in 1790.

Since the Second World War, empires, colonies, and tribal societies have been gradually replaced by national states. To organize elections and assemblies and facilitate the development of modern corporations, these new states must establish statistical systems that were essential conditions of their development. The United Nations, and in particular its Statistical Commission, expended considerable effort in promoting and developing official statistics in different states. Under their auspices, the first world population census was organized in 1950. Another followed in 1960.

Now official statistics are ubiquitous. The Organization for Economic Cooperation and Development, OECD, wants to produce statistics that could help in the formulation of a precise, comprehensive, and balanced assessment regarding the state of the principal sectors of society. These statistics, called indicators, have since the 1960s been the subject of much research in economics, health, and education. The goal is to create, for politicians, a system of information allowing them to determine the effects of their economic, social, and educational policies.

EXAMPLES

Official statistics treat many domains, of which the most important are:

- *Population* (families households, births, adoptions, recognitions, deaths, marriages, divorces, migrations, etc.)
- *Environment* (land, climate, water, fauna, flora, etc.)
- *Employment and active life* (companies and organizations, employment, work conditions, etc.)
- *National accounts* (national accounting, balance of payments, etc.)
- *Price* (prices and price *indices*)
- *Production, commerce, and consumption* (performance of companies, production, turnover, foreign trade, etc.)
- *Agriculture and silviculture* (resource use, labor, machines, forests, fishing, etc.)
- *Energy* (balance, production, consumption, etc.)

- *Construction and housing* (building and housing structures, living conditions, etc.)
- *Tourism* (hotels, trips, etc.)
- *Transportation and communications* (vehicles, installations, traffic, highway accidents, etc.)
- *Financial markets, and banks* (banking system, financial accounting, etc.)
- *Insurance* (illness insurance, private insurance, etc.)
- *Health* (healthcare system, mortality and morbidity causes, healthcare workers, equipment, healthcare costs, etc.)
- *Education and science* (primary and secondary education, higher education, research and development, etc.)
- *Sports, culture, and standard of living* (language, religion, cultural events, sports, etc.)
- *Politics* (elections, participation in political life, etc.)
- *Public finance* (federal and local state revenues and expenditures, taxation, etc.)
- *Legal system* (penal code, courts, recidivism, etc.)

DOMAINS AND LIMITATIONS

History shows the mutual interdependence between statistics and the evolution of the state. This interdependence explains, at least partly, the difficulty that statisticians have in defining clear limits in their discipline. Throughout history and even today, political power determines the limits of the study of official statistics, expanding or limiting the contents depending on its needs.

Political choice comes into play in many areas of official statistics: in definitions of such concepts as unemployment, poverty, wellness, etc.; in the definitions of variables such as professional categories, level of salary, etc. This problem of choice leads to two other problems:

1. The comparability of statistics

   The needs and objectives of the state dictate choice in this domain. Because needs are not always the same, choice will be different in different places and international statistics will not always be comparable. The difference in different states' means for data collection and use also explains the difference in the results among countries.

2. Objectivity

   Official statistics require objectivity. Even if political will cannot be completely eliminated, the evolution of statistical theories and improvements in tools allow for much more important and more reliable data collection. The volume of treated data has increased greatly, and information has improved in terms of objectivity.

FURTHER READING

► **Census**
► **Data**
► **Index number**
► **Indicator**
► **Population**

REFERENCES

Duncan, J.W., Gross, A.C.: Statistics for the 21st Century: Proposals for Improving Statistics for Better Decision Making. Irwin Professional Publishing, Chicago (1995)

Dupâquier, J. et Dupâquier, M.: Histoire de la démographie. Académie Perrin, Paris (1985)

INSEE: Pour une histoire de la statistique. Vol. 1: Contributions. INSEE, Economica, Paris (1977)

INSEE: Pour une histoire de la statistique. Vol. 2: Matériaux. J. Affichard (ed.) INSEE, Economica, Paris (1987)

ISI: The Future of Statistics: An International Perspective. Voorburg (1994)

Stigler, S.: The History of Statistics, the Measurement of Uncertainty Before 1900. Belknap, London (1986)

United Nations Statistical Commission Fundamental Principles of Official Statistics

Ogive

An ogive is a **graphical representation** of a cumulated **frequency distribution**. This is a type of frequency graphic and is also called a cumulated **frequency polygon**. It serves to give the number (or proportion) of observations smaller than or equal to a particular **value**.

MATHEMATICAL ASPECTS

An ogive is constructed on a system of perpendicular axes. We place on the horizontal axis the limits of the class intervals previously determined. Based on each of these limit values, we determine the ordinate of the height equal to the cumulated **frequency** corresponding to this value. Joining by line segments the successive points thus determined, we obtain a line called an ogive. We normally place the cumulated frequency in the left vertical axis and the **percentage** of cumulated frequency on the right vertical axis.

DOMAINS AND LIMITATIONS

Ogives are especially useful for estimating centiles in a distribution. For example, we can know the central point so that 50% of the observations would be below this point and 50% above. To do this, we draw a line from the point of 50% on the axis of percentage until it intersects with the curve. Then we vertically project the intersection onto the horizontal axis. The last intersection gives us the desired **value**. The **frequency polygon** and ogive are used to compare two statistical sets whose number could be different.

EXAMPLES

From the table representing the distribution of the daily turnover of 75 stores we will construct an ogive.

Distribution of daily turnover of 75 stores (in dollars):

Turnover (class intervals)	Fre- quency	Cumu- lated frequen- cy	Percen- tage of cumulated frequency
215–235	4	4	5.3
235–255	6	10	13.3
255–275	13	23	30.7
275–295	22	45	60.0
295–315	15	60	80.0
315–335	6	66	88.0
335–355	5	71	94.7
355–375	4	75	100.0
Total	75		

In this example, we obtain the **estimation** that 50% of stores have a turnover smaller than or equal to about US&289.

FURTHER READING

▶ **Frequency distribution**
▶ **Frequency polygon**
▶ **Graphical representation**

Olkin Ingram

Olkin, Ingram was born in 1924 in Waterbury, CT, USA. He received his diploma in 1947 at the City College of New York and continued his studies at Columbia University, where he received a master's degree in statistics (1949). He received his doctorate in statistical mathematics at the University of North Carolina in 1951. He then taught at Michigan State University and at the University of Minnesota before being named professor of statistics at Stanford University, a position he currently occupies.

Ingram Olkin has served as editor of the *Annals of Statistics*, associate editor of *Psychometrika*, *Journal of Educational Statistics*, *Journal of the American Statistical Association*, and *Linear Algebra and Its Applications*, among other mathematical and statistical journals. His research is focused on multivariate analysis, inequalities, and the theory and application of meta-analysis in the social sciences and biology. He is coauthor of many works including *Inequality Theory of Majorization and Its Applications* (1979), *Selecting and Ordering Populations* (1980), *Probability Models and Applications* (1980), and, more recently, *Statistical Methods in Meta-Analysis* (1985).

Selected works and articles of Ingram Olkin:

1979   (with Marshall, A.) Inequalities: Theory of Majorization and Its Applications. Academic, New York.

1984   (with Marsaglia, G.) Generating Correlation Matrices. SIAM Journal of Scientific and Statistical Computing 5, pp. 470–475.

1985 (with Hedges, L.V.) Statistical Methods for Meta-Analysis. Academic, New York.

1994 (with Gleser, L.J. and Derman, C.) Probability Models and Applications. Macmillan, New York.

1995 (with Moher, D.) Meta-analysis of randomised controlled trials. A concern for standards. J. Am. Med. Assoc. 274, 1962–1963.

1996 (with Begg C., Cho M., Eastwood S., Horton R., Moher D., Pitkin R., Rennie D., Schulz KF., Simel D., Stroup D.F.) Improving the quality of reporting of randomized controlled trials. The CONSORT statement. J. Am. Med. Assoc. 276(8), 637–639.

One-Sided Test

A one-sided or a one-tailed test on a **population** parameter is a type of **hypothesis test** in which the values for which we can reject the null hypothesis, denoted H_0 are located entirely in one tail of the probability distribution.

MATHEMATICAL ASPECTS

A one-sided test on a **sample** is a type of **hypothesis test** where the **hypotheses** are of the form:

$$(1) \quad H_0: \quad \theta \geq \theta_0$$
$$H_1: \quad \theta < \theta_0$$

or

$$(2) \quad H_0: \quad \theta \leq \theta_0$$
$$H_1: \quad \theta > \theta_0,$$

where H_0 is the **null hypothesis**, H_1 the **alternative hypothesis**, θ an unknown **parameter** of the **population**, and θ_0 the presumed **value** of this parameter.

In the case of a one-sided test on two samples, the hypotheses are as follows:

$$(1) \quad H_0: \quad \theta_1 \geq \theta_2$$
$$H_1: \quad \theta_1 < \theta_2$$

or

$$(2) \quad H_0: \quad \theta_1 \leq \theta_2$$
$$H_1: \quad \theta_1 > \theta_2,$$

where θ_1 and θ_2 are the unknown parameters of two populations being compared.

The critical region which is defined to be the set of values of the test statistics for which the null hypothesis, H_0 is rejected, for a one-sided test is the set of values less than the **critical value** of the test, or the set of values greater than **critical value** of the test.

EXAMPLES
One-sided Test on a Population Mean

A perfume manufacturer wants to make sure that their bottles really contain a minimum of 40 ml of perfume. A **sample** of 50 bottles gives a **mean** of 39 ml, with a **standard deviation** of 4 ml.

The hypotheses are as follows:

Null hypothesis H_0: $\mu \geq 40$,

Alternative hypothesis H_1: $\mu < 40$.

If the dimension of the sample is large enough, the **sampling distribution** of the mean can be approximated by a **normal distribution**. For a **significance level** $\alpha = 1\%$, we obtain the **value** of z_α in the **normal table**: $z_\alpha = 2.33$. The **critical value** of the **null hypothesis** is calculated by the expression:

$$\mu_{\bar{x}} - z_\alpha \cdot \sigma_{\bar{x}},$$

where $\mu_{\bar{x}}$ is the mean of the sampling distribution of the means

$$\mu_{\bar{x}} = \mu$$

and $\sigma_{\bar{x}}$ is the standard deviation of the sampling distribution of the means, or **standard error** of the mean

$$\sigma_{\bar{x}} = \frac{\sigma}{\sqrt{n}}.$$

We obtain thus the following critical value:

$$40 - 2.33 \cdot \frac{4}{\sqrt{50}} = 38.68.$$

As the sampling mean $\bar{x} = 39$ is greater than the critical value for the one-tailed left test, the manufacturer cannot reject the null hypothesis and so he can consider that his bottles of perfume do indeed contain at least 40 ml of perfume, the mean of the sample being randomly distributed.

One-sided Test on Percentage of Two Populations

The proposer of a bill on road traffic believes that his bill will be more readily accepted by urban populations than by rural ones. An inquiry was made on two samples of 100 people from urban and rural environments. In the urban environment (population 1), 82 people responded favorably to his bill, while in the rural environment (population 2), only 69 people responded favorably.

To confirm or to invalidate the proposer's suspicion, we formulate a **hypothesis test** posing the following hypotheses:

Null hypothesis H_0: $\pi_1 \leq \pi_2$

or $\pi_1 - \pi_2 \leq 0$

Alt. hypothesis H_1: $\pi_1 > \pi_2$

or $\pi_1 - \pi_2 > 0$

where π_1 and π_2 represent the favorable proportion of the urban and rural populations, respectively.

Depending on the measured percentage of the two samples ($p_1 = 0.82$ and $p_2 = 0.69$),

we can estimate the **value** of the standard deviation of the sampling distribution of the percentage difference or standard error of the percentage difference:

$$\hat{\sigma}_{p_1 - p_2} = \sqrt{\frac{p_1 \cdot (1 - p_1)}{n_1} + \frac{p_2 \cdot (1 - p_2)}{n_2}}$$

$$= \sqrt{\frac{\frac{0.82 \cdot (1 - 0.82)}{100}}{+ \frac{0.69 \cdot (1 - 0.69)}{100}}}$$

$$= 0.06012.$$

If we design a test with a **significance level** α of 5%, then the **value** of z_α in the **normal table** equals 1.645. The critical value of the null hypothesis is calculated by:

$$\pi_1 - \pi_2 + z_\alpha \cdot \sigma_{p_1 - p_2}$$

$$= 0 + 1.645 \cdot 0.06012 = 0.0989.$$

Because the difference between the observed proportions $p_1 - p_2 = 0.82 - 0.69 = 0.13$ is greater than the critical value, we have to reject the null hypothesis in favor of the alternative hypothesis, thus confirming the supposition of the proposer.

FURTHER READING
▶ **Acceptance region**
▶ **Alternative hypothesis**
▶ **Hypothesis testing**
▶ **Null hypothesis**
▶ **Rejection region**
▶ **Sampling distribution**
▶ **Significance level**
▶ **Two-sided test**

REFERENCES
Miller, R.G., Jr.: Simultaneous Statistical Inference, 2nd edn. Springer, Berlin Heidelberg New York (1981)

One-Way Analysis of Variance

One-way analysis of variance for a **factor** with t **levels** (or t **treatments**) is a technique that helps to determine if there exists a significant difference between the t treatments (**populations**). The principle of one-way analysis of variance is to compare the variability within **samples** taken from the **populations** with the variability between these samples. The source of variation called "**error**" corresponds to the variability within the **samples**, and the source of variation called "effect" corresponds to the variability between the samples.

HISTORY

See **analysis of variance**.

MATHEMATICAL ASPECTS

The linear **model** for a **factor** with t **levels** (**treatments**) is as follows:

$$Y_{ij} = \mu + \tau_i + \varepsilon_{ij},$$
$$i = 1, 2, \ldots, t, \quad j = 1, 2, \ldots, n_i,$$

where Y_{ij} represents **observation** j receiving **treatment** i, μ is the general **mean** common to all treatments, τ_i is the actual effect of treatment i on the observations, and ε_{ij} is the experimental **error** of observation Y_{ij}.

This **model** is subjected to the basic **hypotheses** associated to **analysis of variance** if we suppose that the errors ϵ_{ij} are independent **random variables** following a **normal distribution** $N(0, \sigma^2)$.

To see if there exists a difference between the t **treatments**, a **test of hypothesis** is done. The following **null hypothesis**:

$$H_0: \tau_1 = \tau_2 = \ldots = \tau_t$$

means that the t treatments are identical.

The **alternative hypothesis** is formulated as follows:

$$H_1: \quad \text{not all } \textbf{values} \text{ of } \tau_i \ (i = 1, 2, \ldots, t)$$
$$\text{are identical.}$$

The principle of **analysis of variance** is to compare the variability within each **sample** with the variability among the samples. For this, the **Fisher test** is used, in which a ratio is formed whose numerator is an **estimation** of the **variance** among treatments (samples) and whose denominator is an estimation of the variance within treatments.

This ratio, denoted F, follows a **Fisher distribution** with $t - 1$ and $N - t$ **degrees of freedom** (N being the total number of **observations**). The **null hypothesis** H_0 will be rejected at the **significance level** α if the F ratio is superior or equal to the **value** in the Fisher table, meaning if

$$F \geq F_{t-1, N-t, \alpha}.$$

If the **null hypothesis** H_0 cannot be rejected, then we can consider that the t **samples** come from the same **population**.

Calculation of Variance Among Treatments

For the **variance** among treatments, the sum of squares among treatments is calculated as follows:

$$SS_{Tr} = \sum_{i=1}^{t} n_i (\bar{Y}_{i.} - \bar{Y}_{..})^2,$$

where $\bar{Y}_{i.}$ is the mean of the ith treatment.

The number of **degrees of freedom** associated to this sum is equal to $t - 1$, t being the number of **samples** (or groups).

The **variance** among treatments is then equal to:

$$s_{Tr}^2 = \frac{SS_{Tr}}{t - 1}.$$

Calculation of Variance Within Treatments

For the **variance** within treatments the sum of squares within treatments also called error, must be calculated as follows:

$$SS_E = \sum_{i=1}^{t} \sum_{j=1}^{n_i} (Y_{ij} - \bar{Y}_{i.})^2 .$$

The number of **degrees of freedom** associated to this sum is equal to $N - t$, N being the total number of **observations** and t the number of **samples**. The **variance** within treatments is then equal to:

$$s_E^2 = \frac{SS_E}{N - t} .$$

The total sum of squares (SS_T) is the sum of squares of the **deviations** of each **observation** Y_{ij} from the general **mean** $\bar{Y}_{..}$:

$$SS_T = \sum_{i=1}^{t} \sum_{j=1}^{n_i} (Y_{ij} - \bar{Y}_{..})^2 .$$

The number of **degrees of freedom** associated to this sum is equal to $N - 1$, N being the total number of **observations**. We can also find SS_T in the following way:

$$SS_T = SS_{Tr} + SS_E .$$

Fisher Test

We can now calculate the F ratio and determine if the **samples** can be considered as having been taken from the same **population**. The F ratio is equal to the **estimation** of the **variance** among treatments divided by the estimation of the variance within treatments, meaning

$$F = \frac{s_{Tr}^2}{s_E^2} .$$

If F is larger than or equal to the **value** in the **Fisher table** for $t - 1$ and $N - t$ **degrees**

of freedom, then we can conclude that the difference between the **samples** is due to the experimental **treatments** and not only to randomness.

Table of Analysis of Variance

It is customary to summarize the information of an **analysis of variance** in a table called an analysis of variance table, presented as follows:

Source of variation	Degrees of freedom	Sum of squares	Mean of squares	F
Among treatments	t-1	SS_{Tr}	s_{Tr}^2	$\frac{s_{Tr}^2}{s_E^2}$
Within treatments	N-t	SS_E	s_E^2	
Total	N-1	SS_T		

Pairwise Comparison of Means

When the Fisher test rejects the **null hypothesis**, meaning there is a significant difference between the **means** of **samples**, we can wonder where this difference is. Several tests have been developed to answer this question. There is the **least significant difference** (LSD) **test** (in French the *test de la différence minimale*) that makes comparisons of **means** taken in pairs. For this test to be used, the F ratio must indicate a significant difference between the means.

EXAMPLES

During cooking, croissants absorb grease in variable quantities. We want to see if the quantity absorbed depends on the type of grease. Four different types of grease are prepared, and six croissants are cooked per type of grease.

The **data** are in the following table (the numbers are quantities of grease absorbed per croissant):

Grease			
1	2	3	4
64	78	75	55
72	91	93	66
68	97	78	49
77	82	71	64
56	85	63	70
95	77	76	68

The **null hypothesis** is that there is no difference among **treatments**, meaning that the quantity of grease absorbed during cooking does not depend on the type of grease:

$$H_0: \quad \tau_1 = \tau_2 = \tau_3 = \tau_4.$$

To test this **hypothesis**, the **Fisher test** is used, for which the F ratio has to be determined. Therefore the **variance** among treatments and the variance within treatments must be calculated.

Variance Among Treatments

The general **mean** $\bar{Y}_{..}$ and the means of the four sets $\bar{Y}_{i.} = \frac{1}{n_i} \sum_{j=1}^{n_i} Y_{1i}$ ($i = 1, 2, 3, 4$) have to be calculated. Since all n_i are equal to 6 and N is equal to 24, we obtain:

$$\bar{Y}_{..} = \frac{64 + 72 + \ldots + 70 + 68}{24}$$
$$= \frac{1770}{24}$$
$$= 73.75.$$

We also calculate:

$$\bar{Y}_{1.} = \frac{1}{6} \sum_{j=1}^{6} Y_{1j} = \frac{432}{6} = 72,$$

$$\bar{Y}_{2.} = \frac{1}{6} \sum_{j=1}^{6} Y_{2j} = \frac{510}{6} = 85,$$

$$\bar{Y}_{3.} = \frac{1}{6} \sum_{j=1}^{6} Y_{3j} = \frac{456}{6} = 76,$$

$$\bar{Y}_{4.} = \frac{1}{6} \sum_{j=1}^{6} Y_{4j} = \frac{372}{6} = 62.$$

We can then calculate the sum of squares among treatments:

$$SS_{Tr} = \sum_{i=1}^{4} n_i (\bar{Y}_{i.} - \bar{Y}_{..})^2$$
$$= 6(72 - 73.75)^2 + 6(85 - 73.75)^2$$
$$\quad + 6(76 - 73.75)^2 + 6(62 - 73.75)^2$$
$$= 1636.5.$$

The number of **degrees of freedom** associated to this sum is equal to $4 - 1 = 3$. The **variance** among treatments is then equal to:

$$s_{Tr}^2 = \frac{SS_{Tr}}{t - 1} = \frac{1636.5}{3} = 545.5.$$

Variance Within Treatments

The sum of squares within treatments or error is equal to:

$$SS_E = \sum_{i=1}^{4} \sum_{j=1}^{6} (Y_{ij} - \bar{Y}_{i.})^2$$
$$= (64 - 72)^2 + (72 - 72)^2 + \ldots +$$
$$\quad (70 - 62)^2 + (68 - 62)^2$$
$$= 2018.$$

The number of **degrees of freedom** associated to this sum is equal to $24 - 4 = 20$.

The **variance** within treatments is then equal to:

$$s_E^2 = \frac{SS_E}{N-t} = \frac{2018}{20} = 100.9 \, .$$

Fisher Test

All the elements necessary for the calculation of F are now known:

$$F = \frac{s_{Tr}^2}{s_E^2} = \frac{545.5}{100.9} = 5.4063 \, .$$

We have to find the **value** of the **Fisher table** with 3 and 20 **degrees of freedom** with a **significance level** $\alpha = 0.05$:

$$F_{3.20,0.05} = 3.10 \, .$$

We notice that

$$F \geq F_{3,20,0.05} \, ,$$
$$5.4063 \geq 3.10 \, ,$$

which means that H_0 must be rejected. This means that there is a significant difference between the **treatments**. Therefore the quantity of grease absorbed depends on the type of grease used.

Table of Analysis of Variance

This information is summarized in the following table of **analysis of variance**:

Source of variation	Degrees of freedom	Sum of squares	Mean of squares	F
Among treatments	3	1636.5	545.5	5.4063
Within treatments	20	2018	100.9	
Total	23	3654.5		

FURTHER READING

▶ **Analysis of variance**
▶ **Contrast**
▶ **Fisher table**
▶ **Fisher test**
▶ **Least significant difference test**
▶ **Two-way analysis of variance**

REFERENCES

Cox, D.R., Reid, N.: Theory of the design of experiments. Chapman & Hall, London (2000)

Montgomery, D.C.: Design and analysis of experiments, 4th edn. Wiley, Chichester (1997)

Sheffé, H.: The analysis of variance. C'n'H (1959)

Operations Research

Operations research is a domain of applied mathematics that uses scientific methods to provide a necessary basis for decision making. It is generally applied to the complex problems of people and equipment organization in order to find the best solution to reach some goal. Included in operations research methods are **simulation**, **linear programming**, mathematical programming (nonlinear), game theory, etc.

HISTORY

The original objective of operations research was to find military applications in the United Kingdom in the 1930s. These military applications were then largely extended, and since the Second World War the use of operations research has been extended to non-military public applications and even to the private sector. The impact on operations research of the rapid development of computers was very important.

EXAMPLES

Optimization, **linear programming**, and **simulation** are examples of operations research.

FURTHER READING
▶ **Linear programming**
▶ **Optimization**
▶ **Simulation**

REFERENCES

Bealle, E.M.L.: Introduction to Optimization. Wiley, New York (1988)

Collatz, L., Wetterling, W.: Optimization problems. Springer, Berlin Heidelberg New York (1975)

Optimal Design

An optimal design is an **experimental design** that satisfies certain criterion of optimality (for example the minimization of errors associated to the estimations that we must make). As the treatments applied to different experimental units are fixed by the experimenter, we can try to improve the quality of the results. We can also increase the precision of the estimators, reduce the confidence intervals, or increase the power of the hypothesis testing, choosing the optimal design inside the **category** to which it belongs.

HISTORY

The theory of optimal designs was developed by **Kiefer, Jack** (1924–1981). One of the principal points of his work is based on the statistical efficiency of design. In his article of 1959, he presents the principal criteria for obtaining the optimal design. Kiefer's collected articles in the area of experimental design was published by Springer with the collaboration of the Institute of Mathematical Statistics.

DOMAINS AND LIMITATIONS

As we cannot minimize all **errors** at the same time, the optimal design will be determined by the choice of optimality criteria. These different criteria are revealed in Kiefer's 1959 article.

FURTHER READING
▶ **Design of experiments**

REFERENCES

Brown, L.D., Olkin, I., Sacks, J., Wynn, H.P.: Jack Carl Kiefer, Collected Papers III: Design of Experiments. Springer, Berlin Heidelberg New York (1985)

Fedorov, V.V.: Theory of Optimal Experiments. Academic, New York (1972)

Kiefer, J.: Optimum experimental designs. J. Roy. Stat. Soc. Ser. B **21**, 272–319 (1959)

Optimization

A large number of statistical problems can be solved using optimization techniques. We use optimization techniques in statistical procedures such as **least squares**, **maximum likelihood**, L_1 **estimation**, etc. In many other areas of **statistics** such as regression analysis, hypothesis testing, experimental design, etc., optimization plays a hidden, but important, role.

Optimization methods can be classed in the following manner:
1. Classical optimization methods (differential calculation, Lagrange multipliers)
2. Mathematical programming methods (**linear programming**, nonlinear programming, dynamic programming)

HISTORY

The problems of optimization were formulated by Euclid, but only with the development of the differential calculus and the calculus of variations in the 17th and 18th centuries were mathematical tools capable of resolving such problems available. These methods were employed for the first time in the resolution of certain problems of optimization in geometry and physics.

Problems of this type were formulated and studied by well-known mathematicians such as Euler, Leonhard **Bernoulli, Jakob**, Jacobi, Carl Gustav Jacob, and Lagrange, Joseph-Louis, to whom we owe the Lagrange multiplier.

The discovery of important applications of the optimization technique in military problems during the Second World War and advances in technology (computer) allowed for the resolution of increasingly complex problems of optimization in the most varied fields.

FURTHER READING
▶ **Lagrange multiplier**
▶ **Linear programming**
▶ **Operations research**

REFERENCES
Arthanari, T.S., Dodge, Y.: Mathematical Programming in Statistics. Wiley, New York (1981)

Outlier

Outlier is an observation which is well seperated from the rest of the data.

Outliers are defined with respect to a supposed underlying distribution or a theoretical model. If these change, the observation may be no more outlying.

HISTORY

The problem of outliers has been addressed by many scientists since 1750 by **Boscovich, Roger Joseph**, **de Laplace, Pierre Simon**, and **Legendre, Adrien Marie**.

According to Stigler, Stephen M. (1973), Legendre, A.M. proposed, in 1805, the rejection of outliers; and in 1852, Peirce, Benjamin established a first criterion for determining outliers. Criticisms appeared, especially from Airy, George Biddell (1856), against the use of such a criterion.

DOMAINS AND LIMITATIONS

In the framework of statistical study, we can formulate the question of outliers in two different ways:
- The first consists of adapting methods that are resistent to the presence of outliers in the **sample**.
- the Second tries to eliminate the outliers. After identification, delete them from the sample before making a statistical analysis.

Deletion of outliers from a data set is a controversial issue especially in a small data set. The problem of outliers led to the research of new statistical methods called robust methods. Estimators not sensitive to outliers are said to be robust.

FURTHER READING

REFERENCES
Airy, G.B.: Letter from Professor Airy, Astronomer Royal, to the editor. Astronom. J. **4**, 137–138 (1856)

Barnett, V., Lewis, T.: Outliers in statistical data. Wiley, New York (1984)

Peirce, B.: Criterion for the Rejection of Doubtful Observations. Astron. J. **2**, 161–163 (1852)

Stigler, S.: Simon Newcomb, Percy Daniell, and the History of Robust Estimation 1885–1920. J. Am. Stat. Assoc. **68**, 872–879 (1973)

O

Paasche Index

The Paasche **index** is a **composite index number** of price arrived at by the weighted sum method. This index number corresponds to the ratio of the sum of the prices of the actual period n and the sum of prices of the reference period 0, these sums being weighted by the respective quantities of the actual period.

The Paasche index differs from the **Laspeyres index** only by the choice of weights method. Indeed, in the Laspeyres index, the weights are given using quantities of the reference period Q_0 rather than those of the current period Q_n.

HISTORY

In the mid-19th century German statistician Paasche, Hermann developed a formula for the **index number** that carries his name. Paasche, H. (1874) worked on prices recorded in Hamburg.

MATHEMATICAL ASPECTS

The Paasche **index** is calculated as follows:

$$I_{n/0} = \frac{\sum P_n \cdot Q_n}{\sum P_0 \cdot Q_n},$$

where P_n and Q_n are, respectively, the prices and sold quantities in the current period and P_0 and Q_0 are the prices and sold quantities in the reference period. The sums relate to the considered goods and are expressed in base 100.

$$I_{n/0} = \frac{\sum P_n \cdot Q_n}{\sum P_0 \cdot Q_n} \cdot 100.$$

The Paasche model can also be applied to calculate a quantity index (also called volume index). In this case, it is the prices that are constant and the quantities that are variable:

$$I_{n/0} = \frac{\sum Q_n \cdot P_n}{\sum Q_0 \cdot P_n} \cdot 100.$$

EXAMPLES

Consider the following table indicating the respective prices of three food items in reference year 0 and in the current year n, as well as the quantities sold in the current year:

Product	Quantity sold in 1988 (Q_n) (thousands)	Price (euros) 1970 (P_0)	Price (euros) 1988 (P_n)
Milk	85.5	0.20	1.20
Bread	50.5	0.15	1.10
Butter	40.5	0.50	2.00

From the following table we have:

$$\sum P_n Q_n = 239.15 \quad \text{and}$$
$$\sum P_0 Q_n = 44.925.$$

Product	$\sum P_n Q_n$	$\sum P_0 Q_n$
Milk	102.60	17.100
Bread	55.55	7.575
Butter	81.00	20.250
Total	239.15	44.925

We can then find the Paasche index:

$$I_{n/0} = \frac{\sum P_n \cdot Q_n}{\sum P_0 \cdot Q_n} \cdot 100$$

$$= \frac{239.15}{44.925} \cdot 100 = 532.3 .$$

In other words, according to the Paasche index, the price index number of these products has risen by 432.3% (532.3 − 100) during the considered period.

FURTHER READING
▶ **Composite index number**
▶ **Fisher index**
▶ **Index number**
▶ **Laspeyres index**
▶ **Simple index number**

REFERENCES
Paasche, H.: Über die Preisentwicklung der letzten Jahre nach den Hamburger Borsen-notirungen. Jahrb. Natl. Stat. **23**, 168–178 (1874)

Pair of Random Variables

A pair of **variables** whose **values** are determined by a **random experiment** is called a pair of random variables. There are two types of pairs:

- A pair of **random variables** is *discrete* if the set of values taken by each of the random variables is a finite or infinite countable set.

- A pair of random variables is *continuous* if the set of values taken by each of the random variables is an infinite noncountable set.

In the case of a pair of random variables, the set of all the possible results of the experiment comprises the **sample space**, which is located in a two-dimensional space.

MATHEMATICAL ASPECTS
A pair of **random variables** is generally denoted by the capital letters such as X and Y. It is a two-variable function with true **values** in $\mathbb{R}^2$.

If A is a subset of the **sample space**, then:

$$P(A) = \textbf{Probability that } (X, Y) \in A$$
$$= P[(X, Y) \in A] .$$

The case of a pair of **random variables** can be generalized to the case of n random variables: If n numbers are necessary to describe the realization of a **random experiment**, such a realization is represented by n random variables $X_1, X_2, \ldots, X_n$. The **sample space** is then located in an n-dimensional space. If A is a subset of the sample space, then we can say that:

$$P(A) = P[(X_1, X_2, \ldots, X_n) \in A] .$$

DOMAINS AND LIMITATIONS
Certain random situations force one to consider not one but several numerical entities simultaneously. Consider, for example, a system of two elements in a series and of random life spans X and Y; the study of the life span of the whole set must consider events concerning X and Y at the same time. Or consider a cylinder, fabricated by a machine, of diameter X and height Y, with

these values being random. The imposed tolerances usually consider both entities simultaneously; the structure of their probabilistic dependence should be known.

FURTHER READING
▶ **Covariance**
▶ **Joint density function**
▶ **Marginal density function**
▶ **Random variable**

REFERENCES
Hogg, R.V., Craig, A.T.: Introduction to Mathematical Statistics, 2nd edn. Macmillan, New York (1965)

Paired Student's T-Test

The paired Student's test is a hypothesis test that is used to compare the means of two populations when each element of a population is related to an element from the other one.

MATHEMATICAL ASPECTS
Let x_{ij} be **observation** j for the pair i ($j = 1, 2$ and $i = 1, 2, \ldots, n$). For each pair of observations we calculate the difference

$$d_i = x_{i2} - x_{i1}.$$

We are looking for the estimated **standard error** of the **mean** of d_i, denoted by $\bar{d}$:

$$s_{\bar{d}} = \frac{1}{\sqrt{n}} \cdot S_d,$$

where S_d is the **standard deviation** of d_i:

$$S_d = \sqrt{\frac{\sum\limits_{i=1}^{n} (d_i - \bar{d})^2}{n - 1}}.$$

The resulting statistical test is defined by:

$$T = \frac{\bar{d}}{s_{\bar{d}}}.$$

Hypotheses
The Paired Student's Test is a **two-sided test**. The hypotheses are:

$H_0: \quad \delta = 0$ (there is no difference among the treatments)

$H_1: \quad \delta \neq 0$ (there is a difference among the treatments),

where δ is the difference between the means of two populations ($\delta = \mu_1 - \mu_2$).

Decision Rules
We reject the **null hypothesis** at **significance level** α if

$$|T| > t_{n-1,\frac{\alpha}{2}},$$

where $t_{n-1,\frac{\alpha}{2}}$ is the **value** of the **Student table** with $n - 1$ degrees of freedom.

EXAMPLES
Suppose that two treatments are applied to ten pairs of observations. The obtained data and the corresponding differences denoted by d_i are presented in the following table:

Pair i	Treatment 1	Treatment 2	$d_i = x_{i2} - x_{i1}$
1	110	118	8
2	99	104	5
3	91	85	−6
4	107	108	1
5	82	81	−1
6	96	93	−3
7	100	102	2
8	87	101	14
9	75	84	9
10	108	111	3

P

The **mean** equals:

$$\bar{d} = \frac{1}{10} \sum_{i=1}^{10} d_i = \frac{32}{10} = 3.2 \,.$$

The **standard deviation** is calculated thus:

$$S_d = \sqrt{\frac{\sum_{i=1}^{10} (d_i - \bar{d})^2}{10 - 1}} = \sqrt{\frac{323.6}{9}} = 6.00 \,,$$

and the **standard error**:

$$s_{\bar{d}} = \frac{1}{\sqrt{n}} \cdot S_d$$

$$= \frac{1}{\sqrt{10}} \cdot 6.00 = 1.90 \,.$$

The statistical test then equals:

$$T = \frac{\bar{d}}{s_{\bar{d}}} = \frac{3.2}{1.90} = 1.69 \,.$$

If we choose a **significance level** $\alpha = 0.05$, the **value** of $t_{9,0.025}$ is 2.26. Thus the **null hypothesis** $H_0 : \delta = 0$ cannot be rejected because $|T| < t_{9,0.025}$.

FURTHER READING
▶ **Hypothesis testing**
▶ **Student table**
▶ **Student test**

Panel

The panel is a type of **census** that repeats periodically. It is based on a permanent (or semipermanent) **sample** of individuals, households, etc., who are regularly questioned about their behavior or opinion. The panel offers the advantage of following the individual behavior of questioned people or units and measuring any changes in their behavior over time. The information collected in the panel is often richer than in a simple census while retaining limited costs. We adapt the sample renewal rate depending on the goal of the inquiry. When the objective of the panel is to follow the evolution of individuals over time (e. g., following career paths), the sample stays the same; we call this a longitudinal inquiry. When the objective is to serve as a way of estimating characteristics of the **population** in different periods, the sample will be partially renewed from time to time; we call this a changing panel. Thus the data of the panel constitute a source of very important information because they contain an individual and time dimension simultaneously.

HISTORY
The history of panel research dates back to 1759, when the French Count du Montbeillard, Philibert Guéneau began recording his sons stature at six-month intervals from birth to age 18. His records have little in common with contemporary panel studies, aside from the systematic nature of his repeated observations. The current concept of panel research was established in around the 1920s and 1930s, when several studies of human growth and development began. Since the mid-20th century panel studies have proliferated across the social sciences. Several factors have contributed to the growth of panel research in the social sciences. Much growth in the 1960s and 1970s came in response to increased federal funding for panel studies in the United States. For example, policy concerns regarding work and economic behavior led to funding for ongoing studies like the National Longitudinal Surveys (NLS) in 1966 and the Panel Study of Income

Dynamics in 1968. The latter, for example, began with 4802 families with a list of variables exceeding 5000; poor households were oversampled. The NLS survey included 5020 elderly men, 5225 young men, 5083 mature women, 5159 young women, and 12686 youth.

Technological advances have facilitated data collection at a national scale and increased researchers' capacity for managing, analyzing, and sharing large, complex data sets. In addition, advances in statistical methods for analyzing longitudinal data have encouraged researchers not only to collect panel data but also to ask new questions about existing data sets. The growth of panel studies is evident in other industrialized countries, too.

DOMAINS AND LIMITATIONS

In practice, it is rare to have panels in the strict meaning of the term. We should thus take into account the following problems of the panel: difficulty of recruiting certain categories of panelists, fatigue of panelists, which is one of the main reasons of non-responses, and evolution of the **population** causing the deformation of the **sample** from the population it should represent. In practice, we often try to eliminate this problem by preferring a changing panel to a strict one.

There are also specific errors of measure in panels:

(1) *Telescope effect*: error of dating, often made by new panelists when indicating the date of the event about which they are being asked.

(2) *Panel effect* or *bias of conditioning*: change in behavior of panelists that can appear over time as a result of repeated interviews. The panelists are asked about their behavior, and finally they change it.

EXAMPLES

Inquiry about health and medical care: inquiry conducted by INSEE in about 10000 households. Each household is followed for 3 months during which every 3 weeks members of the household are questioned by a researcher about their medical care.

Panel of Sofres consumers: The Metascope panel of Sofres is based on a **sample** of 20000 households. Each month the panelists receive by mail a self-administered questionnaire on their different purchases. These questionnaires are essentially used by companies who want to know what their clients want.

Panel of audience: The Mediamat panel is based on a sample of 2300 households equipped with audimeters with push buttons. It allows researchers to measure the audience of TV stations.

FURTHER READING
► **Bias**
► **Population**
► **Sample**
► **Survey**

REFERENCES

Duncan, G.J.: Household Panel Studies: Prospects and Problems. Social Science Methodology, Trento, Italy (1992)

Kasprzyk, D., Duncan, G. et al. (eds.): Panel Surveys. New York, Chichester, Brisbane, Toronto, Singapore. Wiley, New York (1989)

Rose, D.: Household panel studies: an overview. Innovation **8**(1), 7–24 (1995)

Diggle, P. J., P. Heagerty, K.-Y. Liang, and S.L. Zeger.: Analysis of longitudinal data, 2nd edn. Oxford University Press, Oxford (2002)

Parameter

A parameter characterizes a quantitative aspect of a **population**.

DOMAINS AND LIMITATIONS

The parameters of a **population** are often unknown. However, we can estimate a parameter by a **statistic** calculated from a **sample** using a method of **estimation**.

EXAMPLES

A parameter is generally designated by a Greek letter:

μ: **mean** of **population**
σ: **standard deviation** of population
π: **percentage** relative to population

FURTHER READING

▶ **Estimation**
▶ **Estimator**
▶ **Least squares**
▶ **Maximum likelihood**
▶ **Moment**
▶ **Statistics**

REFERENCES

Lehmann, E.L.: Theory of Point Estimation, 2nd edn. Wiley, New York (1983)

Parametric Test

A parametric test is a form of **hypothesis testing** in which assumptions are made about the underlying distribution of observed data.

HISTORY

One of the first parametric tests was the chi-square test, introduced by Pearson, K. in 1900.

EXAMPLES

The **Student test** is an example of a parametric test. It aims to compare the means of two normally distributed populations.

Among the best-known parametric tests we mention the **Student t-test** and the **Fisher test**.

FURTHER READING

▶ **Binomial test**
▶ **Fisher test**
▶ **Hypothesis testing**
▶ **Student test**

Partial Autocorrelation

The partial autocorrelation at lag k is the autocorrelation between Y_t and Y_{t-k} that is not accounted for by lags 1 through $k - 1$.

See **autocorrelation**.

Partial Correlation

The partial correlation between two variables is defined as correlation of two variables while controlling for a third or more other variables. A *measure of partial correlation* between variables X and Y has a third variable Z as a measure of the direct relation between X and Y that does not take into account the consequences of the linear relations of these two variables with Z. The calculation of the partial correlation between X and Y given Z serves to identify in what measure the linear relation between X and Y can be due to the correlation of two variables with Z.

HISTORY

Interpreting analyses of correlation coefficients between two variables became an important question in statistics during the increasingly widespread use of correlation methods in the early 1900s. Pearson, Karl knew that a large correlation between two variables could be due to their correlation with a third variable. This phenomenon was not recognized until 1926 when Yule, George Udny proved it through an example by getting the coefficients of correlation between time series.

MATHEMATICAL ASPECTS

From a sample of n observations, (x_1, y_1, z_1), $(x_2, y_2, z_2), \ldots, (x_n, y_n, z_n)$ from an unknown distribution of three random variables X, Y, and Z, the *coefficient of partial correlation*, which we denote by $r_{xy \cdot z}$, is defined as the **coefficient of correlation** calculated between $\widehat{x}_i$ and $\widehat{y}_i$ with

$$\widehat{x}_i = \widehat{\beta}_{0x} + \widehat{\beta}_{1x} z_i ,$$

$$\widehat{y}_i = \widehat{\beta}_{0y} + \widehat{\beta}_{1y} z_i ,$$

where $\widehat{\beta}_{0x}$ and $\widehat{\beta}_{1x}$ are the **least-squares** estimators obtained by making a regression of x_i on z_i, and $\widehat{\beta}_{0y}$ and $\widehat{\beta}_{1y}$ are the least-squares estimators obtained by making a regression of y_i on z_i. Thus by definition we have:

$$r_{xy \cdot z} = \frac{\sum (\widehat{x}_i - \overline{x})(\widehat{y}_i - \overline{y})}{\sqrt{\sum (\widehat{x}_i - \overline{x})^2} \sqrt{\sum (\widehat{y}_i - \overline{y})^2}} .$$

We can prove the following result:

$$r_{xy \cdot z} = \frac{r_{xy} - r_{xz} \cdot r_{yz}}{\sqrt{1 - r_{xz}^2} \cdot \sqrt{1 - r_{yz}^2}} ,$$

where r_{xy}, r_{xz}, and r_{yz} are the coefficients of correlation respectively between x_i and y_i, x_i and z_i, and y_i and z_i. Note that when x_i and y_i are not correlated with z_i, that is, when $r_{xz} = r_{yz} = 0$, we have:

$$r_{xy \cdot z} = r_{xy} ,$$

that is, the coefficient of correlation equals the coefficient of partial correlation, or, in other words, the value of the correlation between x_i and y_i is not due at all to the presence of z_i. On the other hand, if the correlations with z_i are important, then we have $r_{xy \cdot z} \cong 0$, which indicates that the observed correlation between x_i and y_i is only due to the correlation between these variables with z_i. Thus we say that there is no direct relation between x_i and y_i (but rather an indirect relation, through z_i).

More generally, we can calculate the partial correlation between x_i and y_i with z_{i1} and z_{i2} relative to two variables Z_1 and Z_2 in the following manner:

$$r_{xy \cdot z_1 z_2} = \frac{r_{xy \cdot z_1} - r_{xz_2 \cdot z_1} \cdot r_{yz_2 \cdot z_1}}{\sqrt{1 - r_{xz_2 \cdot z_1}^2} \cdot \sqrt{1 - r_{yz_2 \cdot z_1}^2}} .$$

DOMAINS AND LIMITATIONS

We must insist on the fact that a correlation measures only the linear relation between two variables, without taking into account functional models or the predictive or forecasting capacity of a given model.

EXAMPLES

We have observed, for example, a strong positive correlation between the number of registered infarcts during a certain period and the amount of ice cream sold in the same period. We should not conclude that high consumption of ice cream provokes infarcts or, conversely, that an infarct provokes a particular desire for ice cream. In this example, the large number of infarcts does not come from the amount of ice cream sold but from the heat. Thus we have a strong correlation between the number of infarcts and the temperature. The large amount of ice cream sold is also explained by the heat, which implies a strong correlation between the amount of ice cream sold and the temperature. The strong correlation observed between the number of infarcts and the amount of ice cream sold is then the consequence of two strong correlations. A third variable is hidden behind the apparent relation of the two previous variables.

FURTHER READING
▶ Correlation coefficient

REFERENCES
Yule, G.U. (1926) Why do we sometimes get nonsense-correlations between time-series? A study in sampling and the nature of time-series. J. Roy. Stat. Soc. (**2**) 89, 1–64

Partial Least Absolute Deviation Regression

The partial least absolute deviation (partial LAD) regression is a regression method that linearly relates a response vector to a set of predictors using derived components. It is an L_1 regression modeling technique well suited to situations where the number of parameters in a linear regression model exceeds the number of observations. It is mostly used for prediction purposes rather than inference on parameters. The partial LAD regression method extends the **partial least-squares regression** to the L_1 norm associated with LAD regression instead of the L_2 norm, which is associated with partial least squares. The partial LAD regression follows the structure of the univariate partial least-squares regression algorithm and extracts components (denoted by **t**) from directions (denoted by **w**) that depend upon the

response variable. The directions are determined by a Gnanadesikan–Ketterning (GK) covariance estimate that replaces the usual variance based on the L_2 norm with MAD, the median absolute deviation, based on L_1. Therefore we use the notation $\mathbf{w}_k^{\mathrm{mad}}$ for partial LAD directions.

HISTORY
The partial LAD regression method was introduced in Dodge, Y. et al. (2004). It was further developed in Dodge, Y. et al. (2004) and tested using the bootstrap in Kondylis, A. and Whittaker, J. (2005).

MATHEMATICAL ASPECTS
For $i = 1, \ldots, n$ and $j = 1, \ldots, p$, denoting observation units and predictors, the partial LAD regression algorithm is given below:
1. Center or standardize both **X** and **y**.
2. For $k = 1, \ldots, k_{\max}$:
   Compute $\mathbf{w}_k^{\mathrm{mad}}$ according to

$$w_{j,k}^{\mathrm{mad}} = \frac{1}{4}\left(\mathrm{mad}^2(\mathbf{x}_{j,k-1} + \mathbf{y})\right.$$
$$\left. - \mathrm{mad}^2(\mathbf{x}_{j,k-1} - \mathbf{y})\right),$$

and scale $\mathbf{w}_k^{\mathrm{mad}}$ to 1.
   Extract component

$$\mathbf{t}_k = \sum_{j=1}^{p} w_{j,k}^{\mathrm{mad}} \mathbf{x}_{j,k-1},$$

where

$$\mathbf{w}_k^{\mathrm{mad}} = \left(w_{1,k}^{\mathrm{mad}}, \ldots, w_{j,k}^{\mathrm{mad}}\right).$$

Orthogonalize each $\mathbf{x}_{j,k-1}$ with respect to $\mathbf{t}_k$: $\mathbf{x}_{j,k} = \mathbf{x}_{j,k-1} - E(\mathbf{x}_{j,k-1}|\mathbf{t}_k)$.
3. Give the resulting sequence of the fitted vectors $\widehat{\mathbf{y}}_k^{\mathrm{plad}} = \mathbf{T}_k\widehat{\mathbf{q}}_k$, where $\mathbf{T}_k = (\mathbf{t}_1, \ldots, \mathbf{t}_k)$ is the score matrix and $\widehat{\mathbf{q}}_k = (\widehat{q}_1, \ldots, \widehat{q}_k)$ the LAD regression coefficient vector.

4. Recover the implied partial LAD regression coefficients according to $\widehat{\boldsymbol{\beta}}_k^{\text{plad}} = \widetilde{\mathbf{W}}_k^{\text{mad}}\widehat{\mathbf{q}}_k$, where the matrix $\widetilde{\mathbf{W}}_k^{\text{mad}}$ pools in its columns the vectors $\mathbf{w}_k^{\text{mad}}$ expressed in terms of the original $\mathbf{x}_j$.

The univariate partial least-squares regression and the partial LAD method share the following properties:

1. $\mathbf{y} = q_1\mathbf{t}_1 + \ldots + q_k\mathbf{t}_k + \epsilon$,
2. $\mathbf{X} = \mathbf{p}_1\mathbf{t}_1 + \ldots + \mathbf{p}_k\mathbf{t}_k + f$,
3. $\text{cor}(\mathbf{t}_i, \mathbf{t}_j) = 0$ for $i \neq j$,

where $\text{cor}(,)$ denotes the Pearson correlation coefficient, ϵ and f correspond to residual terms, and $\mathbf{p}_k$ the X-loading.

The partial LAD method builds a regression model that relates the predictors to the response according to:

$$\widehat{\mathbf{y}}_k^{\text{plad}} = \sum_{j=1}^{p} \widehat{\boldsymbol{\beta}}_j^{\text{plad}} \mathbf{x}_j.$$

EXAMPLE

We give here an example from near infrared experiments in spectroscopy. We use the octane data set that consists of 39 gasoline samples for which the octanes have been measured at 225 wavelengths (in nanometers). The regressors are highly multicollinear spectra at different numbers of wavelengths, and their number exceeds the sample size n.

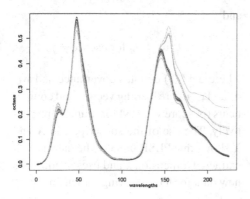

In the figure, we give the lines for the 39 gasoline samples throughout their 225 wavelengths.

We use the partial LAD regression method to build a linear model based on two derived components. The resulting plot of the response $\mathbf{y}$ vs. the fitted values $\widehat{\mathbf{y}}$ is given in the right panel of the following figure under the main title PLAD (partial LAD regression). The right panel of the same figure is the corresponding plot of the response $\mathbf{y}$ vs. the fitted values $\widehat{\mathbf{y}}$ given for a univariate partial least-squares regression model based on two components.

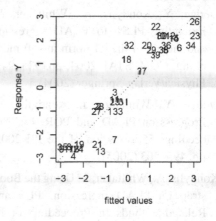

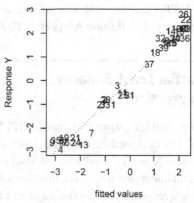

The partial least-squares regression model will require an additional component in the final model in order to provide good predictive results as the partial LAD regression model. This is due to a group of six outliers in the octane data set. These are observations 25, 26, 36, 37, 38, and 39, which contained alcohol. They are visible for wavelengths higher than 140nm in the lineplot above.

FURTHER READING

▶ **Least absolute deviation regression**
▶ **Partial least-squares regression**

REFERENCES

Dodge, Y., Kondylis, A., Whittaker, J.: Extending PLS1 to PLAD regression and the use of the L1 norm in soft modelling. COMPSTAT 2004, pp. 935–942, Physica/Verlag-Springer (2004)

Dodge, Y., Whittaker, J., Kondylis, A.: Progress on PLAD and PQR. In: Proceedings 55th Session of the ISI 2005, pp. 495–503 (2005)

Kondylis, A., Whittaker, J.: Using the Bootstrap on PLAD regression. PLS and Related Methods. In: Proceedings of the PLS'05 International Symposium 2005, pp. 395–402. Edition Aluja et al. (2005)

Partial Least-Squares Regression

The partial least-squares regression (PLSR) is a statistical method that relates two data matrices $\mathbf{X}$ and $\mathbf{Y}$, usually called *blocks*, via a latent linear structure. It is applied either on a single response vector $\mathbf{y}$ (univariate PLSR) or a response matrix $\mathbf{Y}$ (multivariate PLSR). PLSR is commonly used when the recorded variables are highly correlated. It is a very suitable method in cases where the number of the variables exceeds the number of the available observations. This is due to the fact that it uses orthogonal derived components instead of the original variables. The use of a few orthogonal components instead of numerous correlated variables guarantees the reduction of the regression problem on a small subspace that often stabilizes the variability of the estimated coefficients and provides better predictions.

HISTORY

The PLSR was initially developed within the NIPALS algorithm, though it has been implemented in various algorithms including the orthogonal scores and the orthogonal-loading PLSR, the SIMPLS algorithm for PLSR, the Helland PLSR algorithm, and the Kernel PLSR algorithm.

MATHEMATICAL ASPECTS

PLSR methods solve the following maximization problem:

$$\max_{\mathbf{w},\mathbf{q}}\{\mathrm{cov}(\mathbf{X}_{k-1}\mathbf{w}_k, \mathbf{Y}_{k-1}\mathbf{q}_k)\}$$

subject to

$$\mathbf{w}_k^{\mathrm{T}}\mathbf{w}_k = 1, \mathbf{q}_k^{\mathrm{T}}\mathbf{q}_k = 1$$

and

$$\mathbf{t}_k \perp \mathbf{t}_j, \mathbf{u}_k \perp \mathbf{u}_j \text{ for each } k \neq j,$$

where $\mathrm{cov}(\cdot, \cdot)$ denotes covariance and $\mathbf{w}_k$, $\mathbf{q}_k$ and $\mathbf{t}_k$, $\mathbf{u}_k$ are loading vectors and components (or score vectors) for X and Y, respectively. The use of the subscript k in $\mathbf{X}$ and $\mathbf{Y}$ shows that PLSR deflates the data at each extracted dimension k and uses residuals as new data for the following dimension.

Univariate PLSR is much easier to interpret. It is in fact a generalization of **multiple linear regression** that shrinks regression coefficients on directions of low covariance between variables and responses. Given the following relation:

$$\text{cov}(\mathbf{Xw}, \mathbf{y}) \propto \text{cor}(\mathbf{Xw}, \mathbf{y})\text{var}(\mathbf{X})^{1/2},$$

with *cor* and *var* denoting correlation and variance, respectively, it is easy to verify that the maximization criterion in the PLSR is a compromise between **least-squares regression** and regression on **principal components**. The latter methods maximize cor($\mathbf{Xw}, \mathbf{y}$), and var($\mathbf{X}$), respectively.

The univariate PLSR involves the following steps:

1. For $(\mathbf{X}, \mathbf{y})$ *commonly centered*:
2. For $k = 1, \ldots, p$:
   Store the cov($\mathbf{x}_{j,k-1}, \mathbf{y}_{k-1}$) on vector $\mathbf{w}_k$.
   Pool $\mathbf{w}_k$ in matrix $\mathbf{W}_{(p \times k)}$.
   Extract component $\mathbf{t}_k$ as $\mathbf{t}_k = \mathbf{X}_{j,k-1}\mathbf{w}_k$.
   Orthogonalize $\mathbf{X}_{k-1}$ and $\mathbf{y}_{k-1}$ with respect to $\mathbf{t}_k$.
3. Take least-squares residuals as new data for $k \leftarrow k + 1$.

Once the components are extracted, the **least-squares regression** is used to regress them on response vector $\mathbf{y}$. The number of components that should be ultimately retained is defined using **bootstrap** and cross-validation methods.

EXAMPLE

The PLSR method is a generalization of **multiple linear regression**. This is justified using a regression problem where no correlation between the predictors occurs, that is, an **orthogonal design**. The table below contains the results of an orthogonal design with two explanatory variables and one response.

Table: Orthogonal Design

y	x_1	x_2
18	−2	4
12	1	3
10	0	−6
16	−1	−5
11	2	3
9	0	−3
11	−1	6
8	1	−1
7	1	0
12	−1	−1

Multiple linear regression analysis for the data in Table 1 ($\mathbf{y}$ as the response and two explanatory variables $\mathbf{x}_1$ and $\mathbf{x}_2$) lead to the following least-squares estimates:

$$\widehat{\mathbf{y}} = 11.40 - 1.8571\mathbf{x}_1 + 0.1408\mathbf{x}_2.$$

Using the PLSR method (data were initially centered and scaled) and regressing the response on the derived components we are led to exactly the same estimates with only one component used, that is, by $\widehat{\mathbf{y}} = \mathbf{t}_1 \cdot \widehat{\mathbf{q}}_1$. We finally get

$$\widehat{\mathbf{y}} = \mathbf{T}_1\widehat{\mathbf{q}}_1 = \mathbf{Xw}_0\widehat{\mathbf{q}}_1 = \mathbf{X}\widehat{\beta}_{\text{PLS}}.$$

Thus for standardized data the implied regression coefficients are

$$q\widehat{\beta}_{\text{PLS}} = (-0.68, 0.164),$$

which, when transformed back into the original scale, equals $(11, 40, -1.8571, 0.1408)$. The latter are the multiple linear regression estimates.

DOMAINS AND LIMITATIONS

PLSR methods have been extensively used in near infrared experiments in spectroscopy. They have long been used by chemometricians to do multivariate calibration and to measure chemical concentrations. PLSR methods have been equally used in many statistical applications with more variables than observations, for example in environmetrics, in microarray experiments in biostatistics, and in functional data analysis.

PLSR methods focus mainly on prediction accuracy and are mainly used in order to construct good predictive models.

FURTHER READING
▶ **Bootstrap**
▶ **Least-squares method**

REFERENCES

Tenenhaus, M.: La régression PLS. Théorie et pratique. Technip, Paris (1998)

Frank, I., Friedman, J.: A statistical view of some chemometrics regression tools. Technometrics **35**, 109–135 (1993)

Martens, H., and Naes, T.: Multivariate Calibration. Wiley, New York (1989)

Helland: On the structure of PLS regression, Comm. in Stat-Simul. and Comp., 17, 581–607 (1988)

Pearson Egon Shape

Pearson, Egon Shape (1895–1980), son of statistician **Pearson, Karl**, studied mathematics at Trinity College, Cambridge. In 1921, he entered the Department of **Statistics** of the University College in London. He met **Neyman, Jerzy**, with whom he started to collaborate during the visit of the latter to London in 1924–1925, and also collaborated with **Gosset, William Sealy**. When his father left University College in 1933, he transferred to the Applied Statistics Department and in 1936 became director of the journal *Biometrika* following the death of his father, the journal's foundor.

Besides the works that he published in collaboration with Neyman, J. and that led to the development of the theory of hypothesis testing, Pearson, E.S. also touched on problems of quality control and **operations research**.

Selected articles of Pearson, E.S.:

1928 (with Neyman, J.) On the use and interpretation of certain test criteria for purposes of statistical inference. Biometrika 20A, 175–240, pp. 263–295.

1931 The test of significance for the correlation coefficient. J. Am. Stat. Assoc. 26, 128–134.

1932 The percentage limits for the distribution of range in samples from a normal population ($n \leq 100$). Biometrika 24, 404–417.

1933 Neyman, J. and Pearson, E.S. On the testing of statistical hypotheses in relation to probability a priori. Proc. Camb. Philos. Soc. 29, 492–510.

1935 The Application of Statistical Methods to Industrial Standardization and Quality Control. Brit. Stant. 600. British Standards Institution, London.

FURTHER READING
▶ **Hypothesis testing**

Pearson Karl

Born in London, Pearson, Karl (1857–1936) is known for his numerous contributions to statistics. After studying at Kings College, Cambridge, he was named in 1885 chair of the Applied Mathematics Department of University College in London. He spent his entire career at this university, where he was made chair of the Eugenics Department in 1911. In 1901, with the help of **Galton, Francis**, he founded the journal *Biometrika* and was its editor in chief until his death in 1936.

In 1906, he welcomed for 1year in his laboratory **Gosset, William Sealy**, with whom he resolved problems related to samples of small dimension. He retired in 1933, and his department at the university was split into two: the Eugenics Department was entrusted to **Fisher, Ronald Aylmer** and the **Statistics** Department to his own son **Pearson, Egon Shape**.

Selected articles of Pearson, Karl:

1894 1948. Contributions to the mathematical theory of evolution. I. pp. 1–40. In: Karl Pearson's Early Statistical Papers. Cambridge University Press, Cambridge . First published as: On the dissection of asymmetrical frequency curves, Philos. Trans. Roy. Soc. Lond. Ser. A 185, 71–110.

1895 1948. Contributions to the mathematical theory of evolution. II. Skew variation in homogeneous material. In: Karl Pearson's Early Statistical Papers. Cambridge University Press, Cambridge , pp. 41–112. First published in Philos. Trans. Roy. Soc. Lond. Ser. A 186, 343–414.

1896 1948. Mathematical contributions to the theory of evolution. III. Regression, heredity and panmixia. In: Karl Pearson's Early Statistical Papers. Cambridge University Press, Cam-

bridge , pp. 113–178. First published in Philos. Trans. Roy. Soc. Lond. Ser. A 187, 253–318 in the Philosophical Magazine, 5th series, 50, pp. 157–175.

FURTHER READING
▶ **Chi-square distribution**
▶ **Correlation coefficient**

Percentage

Percentage is the notion of a measure allowing one to describe the proportion of individuals or statistical units having a certain characteristic in a collection, evaluated on the basis of 100.

MATHEMATICAL ASPECTS

Percentage is generally denoted by p when it is measured on a **sample** and by π when it concerns a **population**.

Let there be a sample of size n, if k individuals have certain characteristic; the percentage of individuals is given by the **statistic** p:

$$p = \frac{k}{n} \cdot 100\%.$$

The calculation of the same percentage on a population of size N gives us **parameter** π:

$$\pi = \frac{k}{N} \cdot 100\%.$$

Percentile

Percentiles are **measurements of location** computed in a data set. We call percentiles the **values** that divide a distribution into 100 equal parts (each part contains the same number of **observations**). The xth percentile is the value C_x such that $x\%$ of the observations are lower and $(100-x)\%$ of the observations are greater than C_x. For example, 20th percentile is the value that separates the 20% of values that are lower than it from the 80% that are higher than it. We will then have 99 percentiles for a given distribution:

Note for example:
Percentile 10 = first **decile**
Percentile 25 = first **quartile**
Percentile 50 = **median**
This notion is part of the family of **quantiles**.

MATHEMATICAL ASPECTS

The process of calculation is similar to that of the **median**, **quartiles**, and **deciles**. When all the raw **observations** are given, the percentiles are calculated as follows:
1. Organize the n **observations** in the form of a **frequency distribution**.
2. The percentiles correspond to the **observations** for which the relative cumulated **frequency** exceeds respectively 1%, 2%, 3%, ..., 98%, 99%.
   Some authors propose the following formula, which permits one to determine with precision the **value** of the different percentiles:
   Calculation of the jth percentile:
   Consider i the integer part of $\frac{j \cdot (n+1)}{100}$ and k the fractional part of $\frac{j \cdot (n+1)}{100}$.
   Consider x_i and x_{i+1} the data **values** of the **observations** respectively classified in the ith and $(i+1)$th position (when the n observations are sorted in increasing order).

The jth percentile is equal to:

$$C_j = x_i + k \cdot (x_{i+1} - x_i).$$

When we have **observations** that are grouped into classes, the percentiles are determined as follows:
1. Determine the class in which the desired percentile is:
   - 1st percentile: first class for which the relative cumulated **frequency** is over 1%.
   - 2nd percentile: first class for which the relative cumulated frequency is over 2%.
   - 99th percentile: first class for which the relative cumulated frequency is over 99%.
2. Calculate the **value** of the percentiles as a function of **the hypothesis** according to which the **observations** are uniformly distributed in each class:

$$\text{percentile} = L_1 + \left[\frac{(n \cdot q) - \sum f_{\text{inf}}}{f_{\text{centile}}} \right] \cdot c,$$

where L_1 is the lower limit in the class of the percentile, n is the total number of **observations**, q is $\frac{1}{100}$ for the 1st percentile, q is $\frac{2}{100}$ for the 2nd percentile, ..., q is $\frac{99}{100}$ for the 99th percentile, $\sum f_{\text{inf}}$ is the sum of the **frequencies** lower than the percentile class, $f_{\text{percentile}}$ is the frequency of the percentile class, and c is the dimension of the **interval** of the percentile class.

DOMAINS AND LIMITATIONS

The calculation of the percentiles only has true meaning for a **quantitative variable** that can take its **values** in a given interval. In practice, percentiles can be calculated only when there is a large number of **observations** since this calculation consists in dividing the set of observations into 100 parts. This notion is seldom applied because in most descriptive analyses the determination of the **deciles** or even the **quartiles** is sufficient for the interpretation of the results.

EXAMPLES

Consider an example of the calculation of percentiles on the **frequency distribution** of a continuous **variable** where the **observations** are grouped into classes.

The following **frequency table** represents the profits (in thousands of euros) of 2000 bakteries:

Profit (thousands of euros)	Fre-quen-cies	Cumu-lated frequen-cies	Relative cumulat-ed fre-quency
100–150	160	160	0.08
150–200	200	360	0.18
200–250	240	600	0.30
250–300	280	880	0.44
300–350	400	1280	0.64
350–400	320	1600	0.80
400–450	240	1840	0.92
450–500	160	2000	1.00
Total	2000		

The class containing the first percentile is the class 100–150 [the one for which the relative cumulated **frequency** is over 0.01 (1%)]. Considering that the **observations** are uniformly distributed in each class, we obtain for the first percentile the following **value**:

1st percentile

$$= 100 + \left[\frac{\left(2000 \cdot \frac{1}{100}\right) - 0}{160} \right] \cdot 50$$

$$= 106.25 \, .$$

The class containing the 10th percentile is the class 150–200. The value of the 10th percentile is equal to

10th percentile

$$= 150 + \left[\frac{\left(2000 \cdot \frac{10}{100}\right) - 160}{200} \right] \cdot 50$$

$$= 160 \, .$$

We can calculate all the percentiles in the same way. We can then conclude, for example, that 1% of the 2000 bakeries have a profit between 100000 and 106250 euros, or that 10% have a profit between 100000 and 160000 euros, etc.

FURTHER READING

▶ **Decile**
▶ **Measure of location**
▶ **Median**
▶ **Quantile**
▶ **Quartile**

Permutation

The term permutation is a subject of **combinatory analysis**. It refers to an arrangement or an ordered of n objects. Theoretically, we should distinguish between the case where n objects are fully identified and those where they are only partially identified.

HISTORY

See **combinatory analysis**.

MATHEMATICAL ASPECTS

1. *Number of permutations of different objects*

The number of possible permutations of n objects equals n factorial:

Permutations
$$= n! = n \cdot (n-1) \cdot \ldots \cdot 3 \cdot 2 \cdot 1 \,.$$

2. *Number of permutations of partially distinct objects*

The number of possible permutations of n objects among which $n_1, n_2, \ldots, n_r$ are not distinguished among them equals:

$$\text{number of permutations} = \frac{n!}{n_1! \cdot n_2! \cdot \ldots \cdot n_r!},$$

where $n_1 + n_2 + \ldots + n_r = n$.

EXAMPLES

If we have three objects A, B, and C, then the possible permutations are as follows:

A	B	C
A	C	B
B	A	C
B	C	A
C	A	B
C	B	A

Thus we have six possible permutations. Without enumerating them, we can find the number of permutations using the following formula:

$$\text{Number of permutations} = n! \,,$$

where n is the number of objects. This gives us in our example:

Number of permutations $= 3!$
$$= 3 \cdot 2 \cdot 1 = 6 \,.$$

Imagine that we compose the signals aligning some flags. If we have five flags of different colors, the number of signals that we can compose will be:

$$5! = 5 \cdot 4 \cdot 3 \cdot 2 \cdot 1 = 120 \,.$$

On the other hand, if among the flags there are two red, two white, and one black flag, we can compose:

$$\frac{5!}{2! \cdot 2! \cdot 1!} = 30$$

different signals.

FURTHER READING
▶ **Arrangement**
▶ **Combination**
▶ **Combinatory analysis**

Pictogram

A pictogram is a symbol representing a concept, object, place or event by illustration. Like the **bar chart**, the pictogram is used either to compare the **categories** of a **qualitative variable** or to compare **data** sets coming from different years or different places. Pictograms are mostly used in journals as advertisement graphs for comparisons and are of little statistical interest because they only provide rough approximations.

MATHEMATICAL ASPECTS

The figure that is chosen to illustrate the different quantities in a pictogram should if possible be divisible into four parts, so that they can represent halves and quarters. The **scale** is chosen as a function of the available space. Therefore the figure is not necessarily equal to unity.

EXAMPLES

Consider wheat production in two countries A and B. The production of country A is equal to 450.000 quintals and that of country B is of 200.000 quintals (1 quintal is equal to 100kg).

We obtain the following pictogram:

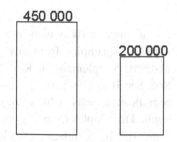

The surface of each figure is proportional to the represented quantity.

FURTHER READING
▶ **Bar chart**
▶ **Graphical representation**
▶ **Quantitative graph**

Pie Chart

The pie chart is a type of **quantitative graph**. It is made of a circle divided into sectors, each sector having an angle that is proportional to the represented magnitude.

HISTORY
See **graphical representation**.

MATHEMATICAL ASPECTS
To establish a pie chart, the total of the represented **frequencies** is calculated and then the relative frequencies representing the different sectors are calculated. To draw these sectors on a circle, the relative **frequencies** are converted into degrees by the following transformation:

$$\alpha = f \cdot 360°,$$

where α represents the angle in the center and f the relative **frequency**.

DOMAINS AND LIMITATIONS
Pie charts are used to give a visual representation of the **data** that form the different parts of a whole population. They are often found in the media (television, journals, magazines, etc.), where they serve to explain in a simple and synthetic way a concept or situation that is difficult to understand with just numbers.

EXAMPLES
We will represent the distribution of the marital status in Australia on 30 June 1981 using a pie chart. The **data** are the following:

Marital status in Australia on 30 June 1981 (in thousands)

Marital status	Frequency	Relative frequency
Bachelor	6587.3	0.452
Married	6836.8	0.469
Divorced	403.5	0.028
Widow	748.7	0.051
Total	14576.3	1.000

Source: Australian Bureau of Statistics, Australian Pocket Year Book 1984, p. 11

First the relative **frequencies** are transformed into degrees:

a) $\alpha = 0.452 \cdot 360° = 162,72°$
b) $\alpha = 0.469 \cdot 360° = 168,84°$
c) $\alpha = 0.028 \cdot 360° = 10,08°$
d) $\alpha = 0.051 \cdot 360° = 18,36°$

The pie chart is shown on the next page.

FURTHER READING
▶ **Bar chart**
▶ **Graphical representation**
▶ **Pictogram**
▶ **Quantitative graph**

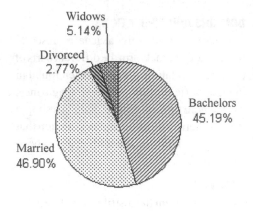

Widows
5.14%

Divorced
2.77%

Bachelors
45.19%

Married
46.90%

Pitman Edwin James George

Pitman, Edwin James George was born in Melbourne (Victoria), Australia in 1897. He went to school at Kensington State School and South Melbourne College. During his last year of study, he was awarded a scholarship to study mathematics in Wyselaskie and in Dixson, as well as at Ormond College. He obtained a degree in letters (similar to a B.A.) (1921), a degree in the sciences (similar to a B.S.) (1922), and a master's degree (1923). He was then named temporary professor of mathematics at Canterbury College, University of New Zealand (1922–1923). He returned to Australia where he was named assistant professor at Trinity and Ormond Colleges as well as part-time lecturer in physics at Melbourne University (1924–1925). In 1926, Pitman, E.J.G. was named professor of mathematics at Tasmania University, where he stayed until his retirement in 1962. He died in 1993, in Kingston (Tasmania).

A major contribution of Pitman, E.J.G. to probability theory concerns the study of the behavior of a characteristic function in a neighborhood of zero.

Selected works of Pitman, Edwin James George:

1937 Significance tests which may be applied to samples from any populations. Supplement, J. Roy. Stat. Soc. Ser. B 4, 119–130.

1937 Significance tests which may be applied to samples from any populations. II. The correlation coefficient test. Supplement, J. Roy. Stat. Soc. Ser. B 4, 225–232.

1938 Significance tests which may be apply to samples from any populations. III. The analysis of variance test. Biometrika 29, 322–335.

1939 The estimation of location and scale parameters. Biometrika 30, 391–421.

1939 Tests of hypotheses concerning location and scale parameters. Biometrika 31, 200–215.

Point Estimation

Point estimation of a **population parameter** allows to obtain a unique **value** calculated from a **sample**. This value will be considered the **estimate** of the unknown parameter.

MATHEMATICAL ASPECTS

Consider a **population** in which trait X is studied; suppose that the shape of the distribution of X is known but that the **parameter** θ on which this distribution depends is unknown.

To estimate θ, a **sample** of size n $(X_1, X_2, \ldots, X_n)$ is taken from the **population** and a function $G(X_1, X_2, \ldots, X_n)$ is created that, for a particular realization $(x_1, x_2, \ldots, x_n)$, will provide a unique **value** as an **estimate** of the value of the **parameter** θ.

DOMAINS AND LIMITATIONS
In general, the point estimation of a **parameter** of a **population** does not exactly correspond to the **value** of this parameter. There exists a **sampling error** that derives from the fact that part of the population has been omitted. It is possible to measure this error by calculating the **variance** or **standard deviation** of the **estimator** that has been used to evaluate the parameter of the population.

EXAMPLES
A company that manufactures light bulbs wants to study the average lifetime of its bulbs. It takes a **sample** of size n $(X_1, X_2, \ldots, X_n)$ from the production of light bulbs.
The function $G(X_1, X_2, \ldots, X_n)$ that it will use to estimate the **parameter** θ corresponding to the **mean** μ of the **population** is as follows:

$$\bar{X} = \sum_{i=1}^{n} \frac{X_i}{n} \,.$$

This **estimator** $\bar{X}$ of μ is a **random variable**, and for a particular realization $(x_1, x_2, \ldots, x_n)$ of the **sample** it takes a precise **value** denoted by $\bar{x}$. Therefore, for example, for $n = 5$, the following sample can be obtained (in hours):

$$(812, \ 1067, \ 604, \ 918, \ 895) \,.$$

The **estimator** $\bar{x}$ becomes:

$$\bar{x} = \frac{812 + 1067 + 604 + 918 + 895}{5}$$
$$= \frac{4296}{5} = 859.2 \,.$$

The **value** 859.2 is a point estimation of the average lifetime of these light bulbs.

FURTHER READING
▶ **Error**
▶ **Estimation**
▶ **Estimator**
▶ **Sample**
▶ **Sampling**

REFERENCES
Lehmann, Erich L., Casella, G.: Theory of point estimation. Springer-Verlag, New York (1998)

Poisson Distribution

The Poisson distribution is a **discrete probability distribution**. It is particularly useful for phenomena of counting in unit of time or space.
The **random variable** X corresponding to the number of elements observed per unit of time or space follows a Poisson distribution of parameter θ, denoted $P(\theta)$, if its **probability function** is:

$$P(X = x) = \frac{\exp(-\theta) \cdot \theta^x}{x!} \,.$$

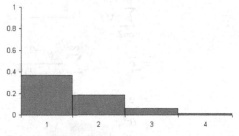

Poisson distribution, $\theta = 1$

HISTORY
In 1837 **Poisson, S.D.** published the distribution that carries his name. He reached this distribution by considering the limits of the **binomial distribution**.

MATHEMATICAL ASPECTS

The **expected value** of the Poisson distribution is by definition:

$$E[X] = \sum_{x=0}^{\infty} x \cdot P(X = x)$$

$$= \sum_{x=0}^{\infty} x \cdot \frac{e^{-\theta}\theta^x}{x!}$$

$$= \sum_{x=1}^{\infty} x \cdot \frac{e^{-\theta}\theta^x}{x!}$$

$$= \theta \cdot e^{-\theta} \sum_{x=1}^{\infty} \frac{\theta^{x-1}}{(x-1)!}$$

$$= \theta \cdot e^{-\theta} \cdot e^{\theta}$$

$$= \theta.$$

The **variance** of X is equal to:

$$\text{Var}(X) = E[X^2] - (E[X])^2$$

$$= E[X(X-1) + X] - (E[X])^2$$

$$= E[X(X-1)] + E[X] - (E[X])^2.$$

Since

$$E[X(X-1)] = \sum_{x=0}^{\infty} x(x-1)\frac{e^{-\theta} \cdot \theta^x}{x!}$$

$$= \sum_{x=2}^{\infty} x(x-1)\frac{e^{-\theta} \cdot \theta^x}{x!}$$

$$= e^{-\theta} \cdot \theta^2 \cdot \sum_{x=2}^{\infty} \frac{\theta^{x-2}}{(x-2)!}$$

$$= e^{-\theta} \cdot \theta^2 \cdot e^{\theta}$$

$$= \theta^2,$$

we obtain:

$$\text{Var}(X) = \theta^2 + \theta - \theta^2 = \theta.$$

DOMAINS AND LIMITATIONS

The Poisson distribution is used to determine the following events, among others:

- Number of particles emitted by a radioactive substance.
- Number of phone calls recorded by a center.
- Number of accidents happening to an insured person.
- Number of arrivals at a counter.
- Number of bacteria in a microscopic preparation.
- Number of plants or animals in a surface in nature determined by an observer.

When θ tends toward infinity, the Poisson distribution can be approximated by the **normal distribution** with **mean** θ and **variance** θ.

EXAMPLES

A secretary makes on average two mistakes per page. What is the probability of having three mistakes on one page?

The number of mistakes per page, X, follows a Poisson distribution of **parameter** $\theta = 2$, and its **probability function** is then:

$$P(X = x) = \frac{\exp(-\theta) \cdot \theta^x}{x!}$$

$$= \frac{\exp(-2) \cdot 2^x}{x!}.$$

The **probability** of having three mistakes on a page is equal to:

$$P(X = 3) = \frac{e^{-2} \cdot 2^3}{3!} = 0.1804.$$

FURTHER READING

▶ **Binomial distribution**
▶ **Discrete probability distribution**
▶ **Normal distribution**

REFERENCES

Poisson, S.D.: Recherches sur la probabilité des jugements en matière criminelle et en matière civile. In: Procédés des Règles Générales du Calcul des Probabilités. Bachelier, Imprimeur-Libraire pour les Mathématiques, Paris (1837)

Poisson Siméon-Denis

Poisson, Siméon-Denis (1781–1840), despite coming from a modest provincial family, was able to pursue his studies at the Ecole Polytechnique in Paris, where he later became a professor.

From 1815, he also taught at the Sorbonne and was elected to the Academy of Sciences that same year.

Poisson was interested in research in different fields: mechanics, physics, and probability theory. In 1837 he published an article called "Recherches sur la probabilité des jugements en matière criminelle et en matière civile, précédées des règles générales du calcul des probabilités". In his numerous writings he presented the basis of a statistical method for the social sciences.

Selected works of Siméon-Denis Poisson:

1824   Sur la probabilité des résultats moyens des observations. Connaissance des temps pour l'an 1827, pp. 273–302.

1829   Suite du mémoire sur la probabilité du résultat moyen des observations, inséré dans la connaissance des temps de l'année 1827. Connaissance des temps pour l'an 1832, pp. 3–22.

1836a   Note sur la loi des grands nombres. Comptes rendus hebdomadaires des séances de l'Académie des sciences 2: pp. 377–382.

1836b   Note sur le calcul des probabilités. Comptes rendus hebdomadaires des

séances de l'Académie des sciences 2: pp. 395–400.

1837   Recherches sur la probabilité des jugements en matière criminelle et en matière civile, précédées des règles générales du calcul des probabilités. Bachelier, Imprimeur-Libraire pour les Mathématiques, Paris .

FURTHER READING
▶ Poisson distribution

Pooled Variance

The pooled **variance** is used to estimate the **value** of the variance of two or more populations when the respective variances of each population are unknown but can be considered as equal.

MATHEMATICAL ASPECTS

On the basis of k samples of dimensions n_1, $n_2, \ldots, n_k$, the pooled **variance** S_p^2 is estimated by:

$$S_p^2 = \frac{\sum_{i=1}^{k} \sum_{j=1}^{n_i} \left(x_{ij} - \bar{x}_{i.} \right)^2}{\left(\sum_{i=1}^{k} n_i \right) - k} \,,$$

where x_{ij} is the jth **observation** of sample i and $\bar{x}_{i.}$ is the **arithmetic mean** of sample i. Knowing the respective variances of each sample, the pooled variance can also be defined by:

$$S_p^2 = \frac{\sum_{i=1}^{k} (n_i - 1) \cdot S_i^2}{\left(\sum_{i=1}^{k} n_i \right) - k} \,,$$

where S_i^2 is the variance of sample i:

$$S_i^2 = \frac{\sum_{j=1}^{n_i} \left(x_{ij} - \bar{x}_{i.}\right)^2}{n_i - 1}.$$

EXAMPLES

Consider two samples of computers of different brands for which we noted the time (in hours) before the first problem:

Brand 1	Brand 2
2800	2800
2700	2600
2850	2400
2650	2700
2700	2600
2800	2500
2900	
3000	

The number of samples k equals 2, and the dimension of the sample brands equals, respectively:

$$n_1 = 8,$$
$$n_2 = 6.$$

The calculation of the **variance** of the first sample gives us

$$S_1^2 = \frac{\sum_{j=1}^{n_1} \left(x_{1j} - \bar{x}_{1.}\right)^2}{n_1 - 1},$$

where

$$\bar{x}_{1.} = \frac{\sum_{j=1}^{n_1} x_{1j}}{n_1} = \frac{22400}{8} = 2800$$

$$S_1^2 = \frac{95000}{8-1} = 13571.428.$$

For the second sample we get:

$$S_2^2 = \frac{\sum_{j=1}^{n_2} \left(x_{2j} - \bar{x}_{2.}\right)^2}{n_2 - 1}$$

with

$$\bar{x}_{2.} = \frac{\sum_{j=1}^{n_2} x_{2j}}{n_2} = \frac{15600}{6} = 2600$$

$$S_2^2 = \frac{70000}{6-1} = 14000.$$

If we assume that the unknown variances of two populations σ_1^2 and σ_2^2 are identical, then we can calculate an **estimation** of $\sigma^2 = \sigma_1^2 = \sigma_2^2$ by the pooled variance:

$$
\begin{aligned}
S_p^2 &= \frac{\sum_{i=1}^{2}(n_i - 1) \cdot S_i^2}{\sum_{i=1}^{2} n_i - 2} \\
&= \frac{(8-1) \cdot 13571.428 + (6-1) \cdot 14000}{8+6-2} \\
&= 13750.
\end{aligned}
$$

FURTHER READING
▶ **Variance**

Population

A population is defined as a collection of statistical units of the same nature whose quantifiable information we are interested in. The population constitutes the reference universe during the study of a given statistical problem.

EXAMPLES

The citizens of a state, a group of trees in a forest, workers in a factory, and prices of consumer goods all form discrete populations.

FURTHER READING
▶ Sample

Prevalence

See prevalence rate.

Prevalence Rate

The **prevalence** of a disease is the number of individuals affected in the statistical population at a given time. This notion is close to the notion of "stock".
The prevalence rate of an illness is the proportion of affected individuals in the population at a given moment.

HISTORY
Farr, William, pioneer in the use of statistics in epidemiology and creator of the concepts of mortality rate and question and answer, showed that the **prevalence** of an illness is equivalent to the product of the **incidence** and duration of the illness.
MacMahon, B. and Pugh, T.F. (1970) illustrated the relation of the incidence rate and the prevalence rate using data collected between 1948 and 1952 on acute leukemia among the white population of Brooklyn, New York. The incidence rate was 32.5 cases per million inhabitants per year, and the prevalence rate was 6.7 cases per million inhabitants. The duration of acute leukemia thus had to be 0.21 years or 2.5 months:

$$\text{Duration of acute leukemia} = \frac{\text{prevalence ate}}{\text{incidence rate}}$$
$$= \frac{6.7}{32.5} \text{ year} = 0.21 \text{ year} .$$

MATHEMATICAL ASPECTS
The prevalence rate is defined as follows:

$$\text{Prevalence rate} = \frac{\text{number of afflicted}}{\text{dimension of population}} .$$

The prevalence rate and the **incidence rate** of an illness are approximately associated to one another by the mean duration of the illness (that is, the mean duration of survival or the mean duration of the cure). This relation is written:

Prevalence rate

$$\simeq (\text{incidence rate}) \cdot \left(\begin{array}{c} \text{mean duration} \\ \text{of illness} \end{array} \right) .$$

DOMAINS AND LIMITATIONS
Prevalence is different from **risk**. A risk is a quantity that has a predictive value, as it concerns a fixed period of observation and contains, because of this, information homogeneous in time. In contrast, prevalence is tangential to the history of an illness in a population: it cannot give any indication about the risk insofar as it aggregates, at a given moment, recently diagnosed cases and others in the distant past. Note that the greater the mean survival, if the illness is incurable, or the longer the convalescence, if the illness is curable, the greater the prevalence of the illness.
Finally, note that the prevalence rate often refers not to a well-defined illness, but to an interval of values of a biological parameter. For example, the prevalence rate of hypercholesterolemy can be defined as the proportion of individuals having a cholesterolemy greater than 6.5mmol/l in given a population at a given moment. Thus the percentiles of a biological **parameter**, considered as measuring the proportion of individuals for which the parameters take a value

P

within a certain interval, can be interpreted as the prevalence rate. For example, the prevalence rate relative to values smaller than or equal to the **median** (or percentile 50) of a biological parameter is 50%.

FURTHER READING
► **Attributable risk**
► **Avoidable risk**
► **Cause and effect in epidemiology**
► **Incidence rate**
► **Odds and odds ratio**
► **Relative risk**
► **Risk**

REFERENCES
Cornfield, J.: A method of estimating comparative rates from clinical data. Applications to cancer of the lung, breast, and cervix. J. Natl. Cancer Inst. **11**, 1269–75 (1951)

Lilienfeld, A.M., Lilienfeld, D.E.: Foundations of Epidemiology, 2nd edn. Clarendon, Oxford (1980)

MacMahon, B., Pugh, T.F.: Epidemiology: Principles and Methods. Little Brown, Boston, MA (1970)

Morabia, A.: Epidemiologie Causale. Editions Médecine et Hygiène, Geneva (1996)

Morabia, A.: L'Épidémiologie Clinique. Editions "Que sais-je?". Presses Universitaires de France, Paris (1996)

Probability

We can define the probability of an **event** either by using the relative frequencies or through an axiomatic approach.

In the first approach, we suppose that a **random experiment** is repeated many times in the same conditions. For each event A defined in the **sample space** Ω, we define n_A as the number of times that event A occurred during the first n repetitions of the **experiment**. In this case, the probability of event A, denoted by $P(A)$, is defined by:

$$P(A) = \lim_{n \to \infty} \frac{n_A}{n},$$

which means that $P(A)$ is defined as the limit relative to the number of times event A occurred relative to the total number of repetitions.

In the second approach, for each event A, we accept that there exists a probability of A, $P(A)$, satisfying the following three axioms:
1. $0 \le P(A) \le 1$,
2. $P(\Omega) = 1$,
3. For each sequence of mutually exclusive events $A_1, A_2, \dots$ (that is of events $A_i \cap A_j = \phi$ if $i \ne j$):

$$P\left[\bigcup_{i=1}^{\infty} A_i\right] = \sum_{i=1}^{\infty} P(A_i).$$

HISTORY
The first hazard games mark the beginning of the history of probability, and we can affirm that they date back to the emergence of *Homo sapiens*.

The origin of the word "hazard" offers less certitude. According to Kendall, Maurice George (1956), it was brought to Europe at the time of the third Crusade and derived from the Arabic word "al zhar", meaning a die.

Benzécri, J.P. (1982) gives a reference for Franc d'Orient, named Guillaume de Tyr, according to whom the word hazard was related to the name of a castle in Syria. This castle was an object of a siege during which the aggressors invented a game of dice, giving its name to these games.

Traces of hazard games can be found in ancient civilizations. Drawings and objects found in Egyptian graves from the First Dynasty (3500 B.C.) show that these games were already being played at that time.

According to the Greek historian Herodotus, hazard games were invented by Palamedes during the siege of Troy.

The game of astragals for dogs was one of the first hazard games. The astragal is a small foot bone symmetric to the vertical axis. Much appreciated by the Greeks, and later by the Romans, it is possibly, according to Kendall, M.G. (1956), the original game of craps. Very old astragals were found in Egypt, but more ancient ones, produced in terra cotta, were discovered in northern Iraq. They date back to the third millennium B.C. Another, also in terra cotta, dating to the same time, was found in India.

There is a wealth of evidence demonstrating how games involving dice were played in ancient times. But nobody at the time thought to establish the equiprobable property to get any side.

For David, F.N. (1955), two hypotheses help to explain this gap. According to the first hypothesis, the dice generally were deformed. According to the second hypothesis, the rolling of dice was of a religious nature. It was a way to ask the gods questions: the result of a given roll of the dice revealed the answer to questions that had been posed to the gods.

Among games of hazard, playing cards are also of very ancient origin. We know that they were used in China, India, Arabia, and Egypt. In the West, we find the first evidence of cards in Venice in 1377, in Nuremberg in 1380,

and in Paris in 1397. Tarot cards are the most ancient playing cards.

Until the 15th century, according to Kendall, M.G. (1956), the Church and kings fought against the practice of dice games. As evidence of this we may mention the laws of Louis IX, Edward III, and Henry VIII. Card games were added to the list of illegal activities, and we find a note on the prohibition of card games issued in Paris in 1397. Despite the interdictions, games of hazard were present without interruption, from the Romans to the Renaissance, in all social classes.

It is not surprising to learn that they are at the origin of mathematical works on probability. In the West, the first writings on this subject come from Italy. They were written by Cardano in his treatise *Liber de Ludo Aleae*, published posthumously in 1663, and by Galileo Galilei in his *Sopra le Scoperte dei Dadi*. In the works of Galileo we find the principle of equiprobability of dice games.

In the rest of Europe, these hazard-related problems also interested mathematicians. Huygens, from the Netherlands, published in 1657 a work entitled *De Ratiociniis in Aleae Ludo* that, according to Holgate, P. (1984), extended the interest in probabilities to other domains; game theory would be developed in parallel to this in relation to body dynamics. The works of Huygens considerably influenced those of two other well-known mathematicians: **Bernoulli, Jakob** and **de Moivre, Abraham**.

But it was in France, with Pascal, Blaise (1623–1662) and de Fermat, Pierre (1601–1665), that probability theory really took shape. According to Todhunter, I. (1949), Pascal, B. was asked by a famous gambler, Gombauld, Antoine, Knight of Méré, to resolve the following problem: In a game there are two adverseries The first one wins n of m plays while the second wins p of m plays, which are the chances for one of them to win the game (given that it wins the first to take m plays).

Pascal, B. contacted de Fermat, P., who found the solution. Pascal himself discovered the recurrence formula showing the same result. According to Benzécri, J.P. (1982), at a time when the West was witnessing the fall of the Roman Empire and the advent of the Christian era, the East was engaged in intense scientific and artistic activity. In these circumstances, Eastern scholars, such as Omar Khayam, probably discovered probability rules.

There is an aspect of probability that is of particular interest to mathematicians: the problems of **combinatory analysis**.

MATHEMATICAL ASPECTS
Axiomatic Basis of Probability

We consider a **random experiment** with the **sample space** Ω containing n elements:

$$\Omega = \{x_1, x_2, \ldots, x_n\} \, .$$

We can associate to each simple **event** x_i a probability $P(x_i)$ having the following properties:

1. The probabilities $P(x_i)$ are nonnegative:

$$P(x_i) \geq 0 \quad \text{for} \quad i = 1, 2, \ldots, n \, .$$

2. The sum of the probabilities of x_i for i going from 1 to n equals 1:

$$\sum_{i=1}^{n} P(x_i) = 1 \, .$$

3. The probability P is a function that assigns a number between 0 and 1 to each simple event of a random experiment:

$$P \colon \Omega \to [0, 1] \, .$$

Properties of Probabilities

1. The probability of a **sample space** is the highest probability that can be associated to an event:

$$P(\Omega) = 1 \, .$$

2. The probability of an impossible event equals 0:

$$\text{If} \quad A = \phi, \quad \text{then} \quad P(A) = 0 \, .$$

3. Let $\bar{A}$ be the **complement** of A in Ω, then the probability of $\bar{A}$ equals 1 minus the probability of A:

$$P(\bar{A}) = 1 - P(A) \, .$$

4. Let A and B be two incompatible events ($A \cap B = \phi$). The probability of $A \cup B$ equals the sum of the probabilities of A and B:

$$P(A \cup B) = P(A) + P(B) \, .$$

5. Let A and B be two events. The probability of $A \cup B$ equals:

$$P(A \cup B) = P(A) + P(B) - P(A \cap B) \, .$$

FURTHER READING

► **Conditional probability**
► **Continuous probability distribution**
► **Discrete probability distribution**
► **Event**
► **Random experiment**
► **Sample space**

REFERENCES

Benzécri, J.P.: Histoire et préhistoire de l'analyse des données. Dunod, Paris (1982)

David, F.N.: Dicing and gaming (a note on the history of probability). Biometrika **42**, 1–15 (1955)

Holgate, P.: The influence of Huygens' work in Dynamics on his Contribution to Probability. Int. Stat. Rev. **52**, 137–140 (1984)

Kendall, M.G.: Studies in the history of probability and statistics: II. The beginnings of a probability calculus. Biometrika **43**, 1–14 (1956)

Stigler, S.: The History of Statistics, the Measurement of Uncertainty Before 1900. Belknap, London (1986)

Todhunter, I.: A history of the mathematical theory of probability. Chelsea Publishing Company, New York (1949)

Probability Distribution

Synonym for **probability function**, but used much more frequently than "**probability function**" to describe the distribution giving the probability occurance of a value of a random variable X at the value of x.

HISTORY

It is in the 17th century that the systematic study of problems related to random phenomena began. The eminent physicist Galileo had already tried to study the **errors** of physical measurements, considering these as random and estimating their **probability**. During this period, the theory of insurances also appeared and was based on the analysis of the laws that rule random phenomena such as morbidity, mortality, accidents, etc. However, it was first necessary to study simpler phenomena such as games of chance. These provide particularly simple and clear **models** of random phenomena, allowing one to observe and study the specific laws that rule them; moreover, the possibility of repeating the same **experiment** many times allows for experimental verification of these laws.

MATHEMATICAL ASPECTS

The function $P(b) = P(X = b)$, where b varies according to the possible **values** of the discrete **random variable** X, is called the probability function of X. Since $P(X = b)$ is always positive or zero, the probability function is also positive or zero.

The probability function is represented on an axis system. The different **values** b of X are plotted as abscissae, the images $P(b)$ as ordinates. The **probability** $P(b)$ is represented by rectangles with a width equal to unity and a height equal to the probability of b. Since X must take at least one of the **values** b, the sum of the $P(b)$ must be equal to 1, meaning that the sum of the surfaces of the rectangles must be equal to 1.

EXAMPLES

Consider a **random experiment** that consists in rolling a fixed die. Consider the **random variable** X corresponding to the number of obtained points. The probability function $P(X = b)$ is given by the **probabilities** associated to each **value** of X:

b	1	2	3	4	5	6
$P(b)$	$\frac{1}{6}$	$\frac{1}{4}$	$\frac{1}{12}$	$\frac{1}{12}$	$\frac{1}{4}$	$\frac{1}{6}$

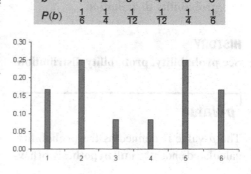

FURTHER READING
▶ **Continuous probability distribution**
▶ **Density function**
▶ **Discrete distribution function**
▶ **Discrete probability distribution**
▶ **Distribution function**
▶ **Probability**
▶ **Probability function**
▶ **Random variable**

REFERENCES
Johnson, N.L., Kotz, S.: Distributions in Statistics: Discrete Distributions. Wiley, New York (1969)

Johnson, N.L., Kotz, S.: Distributions in Statistics: Continuous Univariate Distributions, vols. 1 and 2. Wiley, New York (1970)

Johnson, N.L., Kotz, S., Balakrishnan, N.: Discrete Multivariate Distributions, John Wiley (1997)

Rothschild, V., Logothetis, N.: Probability Distributions. Wiley, New York (1986)

Probability Function

The probability function of a discrete **random variable** is a function that associates each **value** of this random variable to its **probability**.

See **probability distribution**.

HISTORY
See **probability, probability distribution**.

p-Value

The p-value is defined as the **probability**, calculated under the **null hypothesis**, of having outcome as extreme as the observed **value** in the **sample** or is the probability of obtaining a result at least as extreme as a given data point, assuming the data point was a result of chance alone.

HISTORY
The p-value was introduced by Gibbens and Pratt in 1975.

MATHEMATICAL ASPECTS
Let us illustrate the case of a **hypothesis test** made on the **estimator** of **mean**; the same principle is applied for any other estimator; only the notations are different.

Suppose that we want to test the following hypotheses:

$$H_0: \quad \mu = \mu_0$$
$$H_1: \quad \mu > \mu_0,$$

where μ represents the mean of a normally distributed **population** with a known **standard deviation** σ. A **sample** of dimension n gives an observed mean $\bar{x}$.

Consider the case where $\bar{x}$ exceeds μ_0 and calculate the **probability** of obtaining an **estimation** $\hat{\mu}$ greater than or equal to $\bar{x}$ under the **null hypothesis** $\mu = \mu_0$. The value p corresponds to this probability. Thus:

$$p = P(\hat{\mu} \geq \bar{x} | \mu = \mu_0).$$

The standard **random value** Z given by $Z = \dfrac{\hat{\mu} - \mu_0}{\frac{\sigma}{\sqrt{n}}}$ follows normal distribution of mean 0 with **standard deviation** 1. Introducing this **variable** into the expression of p, we find:

$$p = P\left(Z \geq \frac{\bar{x} - \mu_0}{\frac{\sigma}{\sqrt{n}}}\right),$$

which in this form can be read in the **normal table**.

For a **significance level** α, the comparison of p to α allows to make a decision about an eventual rejection of the **null hypothesis**. If:

- $p \leq \alpha$: we reject the null hypothesis H_0 in favour of the **alternative hypothesis** H_1;
- $p > \alpha$: we do not reject the null hypothesis H_0.

We also can calculate the upper limit of the **acceptance region** of H_0, knowing the **value** C_α such as

$$P(\hat{\mu} \geq C_\alpha | \mu = \mu_0) = \alpha.$$

Depending on the standard normal random variable, z_α, the value C_α is expressed by:

$$C_\alpha = \mu_0 + z_\alpha \cdot \frac{\sigma}{\sqrt{n}}.$$

DOMAINS AND LIMITATIONS

We frequently use the p value to understand the result of a **hypothesis test**.

It is used for a **one-tailed test**, when the hypotheses are of the form:

$$H_0: \quad \mu = \mu_0$$
$$H_1: \quad \mu > \mu_0,$$

or

$$H_0: \quad \mu = \mu_0$$
$$H_1: \quad \mu < \mu_0,$$

or

$$H_0: \quad \mu \leq \mu_0$$
$$H_1: \quad \mu > \mu_0,$$

or

$$H_0: \quad \mu \geq \mu_0$$
$$H_1: \quad \mu < \mu_0.$$

The value p can be interpreted as the smallest **significance level** for which the **null hypothesis** cannot be rejected.

EXAMPLES

Suppose that we want to conduct a one-tailed **hypothesis test**

$$H_0: \quad \mu = 30$$
$$\text{against} \quad H_1: \quad \mu > 30$$

for a normally distributed **population** with a **standard deviation** $\sigma = 8$. A **sample** of **size** $n = 25$ gives an observed **mean** of 34. As $\bar{x}$ (= 34) clearly exceeds $\mu = 30$, we should try to accept the **alternative hypothesis**, which is $\mu > 30$.

Calculate the **probability** of obtaining an observed mean of 34 under the **null hypothesis** $\mu = 30$. This is the p value:

$$p = P(\bar{X} \geq 34 | \mu = 30)$$
$$= P\left(\frac{\bar{X} - 30}{\frac{8}{5}} \geq \frac{34 - 30}{\frac{8}{5}} \right)$$
$$= P(Z \geq 2.5),$$

where $Z = \dfrac{\bar{X} - \mu}{\frac{\sigma}{\sqrt{n}}}$ is normally distributed with mean 0 and standard deviation 1; $p = 0.0062$ according to the **normal table**. For a **significance level** $\alpha = 0.01$, we reject the null hypothesis for the alternative hypothesis $\mu > 30$. We can calculate the upper limit of the **acceptance region** of H_0, C_α such as

$$P(\bar{X} \geq C_\alpha | \mu = 30) = 0.01$$

and find $C_\alpha = 30 + 1.64 \cdot \dfrac{8}{5} = 32.624$.

This confirms that we really must reject the null hypothesis $\mu = 30$ when the observed mean equals 34.

FURTHER READING
► Acceptance region
► Hypothesis testing

► Normal table
► Significance level

REFERENCES

Gibbens, R.J., Pratt, J.W.: *P*-values Interpretation and Methodology. Am. Stat. **29**(1), 20–25 (1975)

Q-Q Plot
(Quantile to Quantile Plot)

The Q-Q plot, or quantile to quantile plot, is a graph that tests the conformity between the empirical distribution and the given theoretical distribution.

One of the methods used to verify the normality of errors of a regression model is to construct a Q-Q plot of the residuals. If the points are aligned on the line $x = y$, then the data are normally distributed.

HISTORY

The method of **graphical representation** known as the Q-Q plot appeared in the early 1960s. Since then, it has become a necessary tool in the analysis of data and/or residuals because it is a very provides a wealth of information and is easy to interpret.

MATHEMATICAL ASPECTS

To construct a Q-Q plot we follow two steps.

1. Arrange the data $x_1, x_2, \ldots, x_n$ in increasing order:

$$x_{[1]} \leq x_{[2]} \leq \cdots \leq x_{[n]} \,.$$

2. Associate to each data point $x_{[i]}$ the $i/(n+1)$-**quantile** q_i of the standard normal distribution. Plot on a graph as ordinates the ordered data x_i and as abscissae the quantiles q_i.

If the variables have the same distribution, then the graphical representation between the quantiles of the first variable relative to the quantiles of the second distribution will be a line of slope 1. So if the data x_i are normally distributed, the points on the graph must be almost aligned on the line of equation $x_i = q_i$.

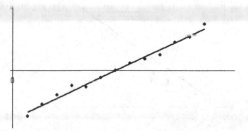

DOMAINS AND LIMITATIONS

The Q-Q plot is used to verify if data follow a particular distribution or if two given data sets have the same distribution. If the distributions are the same, the graph is a line. The further the obtained result is from the 45° diagonal, the further is the empirical distribution from the theoretical one. The extreme points have a greater variability than those in the center of the distribution. Thus a U-shaped graph means that one distribution is skewed relative to another. An S-shaped graph indicates that the distributions represent a greater influence of extreme values on another distribution (long tail).

A frequent use of the Q-Q plots is when one wants to know the behavior of residuals during a simple or **multiple linear regression**, more specifically, to know if their distribution is normal.

i	e_i	q_i
5	−0.17	−0.37
6	−0.12	−0.18
7	0.04	0.00
8	0.18	0.18
9	0.34	0.37
10	0.42	0.57
11	0.64	0.79
12	1.33	1.07
13	2.54	1.47

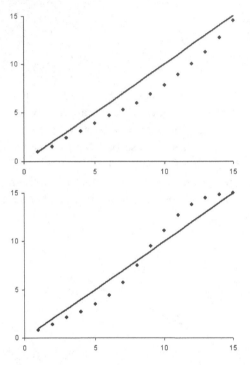

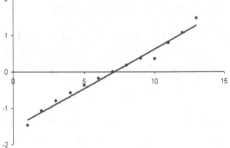

EXAMPLES

The following table contains the residuals (in increasing order) of a linear regression with 13 observations. We associate them to the quantiles q_i as described above. The obtained Q-Q plot is represented by the following figure. We see that except for two points that are probably outliers, the points are well aligned.

Table: Residuals and quantiles

i	e_i	q_i
1	−1.62	−1.47
2	−0.66	−1.07
3	−0.62	−0.79
4	−0.37	−0.57

FURTHER READING
▶ **Analysis of residuals**
▶ **Data analysis**
▶ **Exploratory data analysis**
▶ **Multiple linear regression**
▶ **Probability distribution**
▶ **Quantile**
▶ **Residual**
▶ **Simple linear regression**

REFERENCES

Chambers, J.M., Cleveland, W.S., Kleiner, B., Tukey, P.A.: Graphical Methods for Data Analysis. Wadsworth, Belmont, CA (1983)

Hoaglin, D.C., Mosteller, F., Tukey, J.W. (eds.): Understanding Robust and Exploratory Data Analysis. Wiley, New York (1983)

Wilk, M.B., Ganadesikan, R.: Probability plotting methods for the analysis of data. Biometrika **55**, 1–17 (1968)

Qualitative Categorical Variable

A qualitative categorical **variable** is a variable with modalities in the form of categories, such as, for example, "man" and "woman" of the variable "sex"; or the categories "red", "orange", "green", "blue", "indigo", and "violet" of the variable "color". The modalities of qualitative categorical variable can be represented on a nominal scale or on an ordinal scale.

An example of qualitative categorical variable having an ordinal scale is the qualificative professional variable with categories "qualified", "semiqualified", and "nonqualified".

FURTHER READING
▶ **Category**
▶ **Variable**

Quantile

Quantiles measure position (or the **central tendency**) and do not necessarily try to determine the center of a distribution of observations, but to describe a particular position.

This notion is an extension of the concept of the **median** (which divides a distribution of observation into two parts). The most frequently used quantiles are:

• Quartiles, which separate a collection of observations into four parts,
• Deciles, which separate a collection of observations into ten parts,
• Centiles, which separate a collection of observations into a hundred parts.

DOMAINS AND LIMITATIONS
The calculation of quantiles makes sense only for a **quantitative variable** that can take values on a determined **interval**.

The concept of quantile indicates the separation of a distribution of observations in an arbitrary number of parts. Note that the greater the number of observations, the more sophisticated the separation of the distribution can be.

Quartiles can generally be used for any distribution. The calculation of deciles and, a fortiori, centiles requires a relatively large number of observations to obtain a valid interpretation.

FURTHER READING
▶ **Decile**
▶ **Measure of location**
▶ **Median**
▶ **Percentile**
▶ **Quartile**

Quantitative Graph

Quantitative graphs are used to present and summarize numerical information coming from the study of a **categorical quantitative variable**.

The most frequently used types of quantitative graphs are:

• **Bar chart**
• **Pictogram**
• **Pie chart**

HISTORY
Beniger, J.R. and Robyn, J.L. (1978) retrace the historical development of quantitative charts since the 17th century. Schmid, C.F. and Schmid, S.E. (1979) discuss the numerous forms of quantitative graphs.

FURTHER READING

▶ **Bar chart**
▶ **Graphical representation**
▶ **Pictogram**
▶ **Pie chart**

REFERENCES

Beniger, J.R., Robyn, D.L.: Quantitative graphics in statistics: a brief history. Am. Stat. **32**, 1–11 (1978)

Schmid, C.F., Schmid, S.E.: Handbook of Graphic Presentation, 2nd edn. Wiley, New York (1979)

Quantitative Variable

A quantitative variable is a **variable** with numerical modalities. For example, weight, dimension, age, speed, and time are quantitative variables. We distinguish discrete variables (e. g., number of children per family) from continuous variables (e. g., length of a jump).

FURTHER READING

▶ **Variable**

Quartile

Quartiles are location measures of a distribution of observations. Quartiles separate a distribution into four parts. Thus there are three quartiles for a given distribution. Between each quartile we find 25% of the total observations:

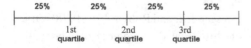

Note that the second quartile equals the **median**.

MATHEMATICAL ASPECTS

Calculation of the quartile is similar to that of the **median**. When we have all the observations, quartiles are calculated as follows:

1. The n observations must be arranged in the form of a **frequency distribution**.
2. Quartiles correspond to observations for which the relative cumulated **frequency** exceeds 25%, 50%, and 75%. Certain authors propose the following formula, which allows to determine with precision the **value** of different quartiles:

Computation of jth quartile:
Let i be the integer part of $\frac{j\cdot(n+1)}{4}$ and k the fraction part of $\frac{j\cdot(n+1)}{4}$.
Let x_i and x_{i+1} be the values of the observations respectively in the ith and $(i+1)$th position (when the observations are arranged in increasing order).
The jth quartile is

$$Q_j = x_i + k \cdot (x_{i+1} - x_i).$$

When we have observations grouped into classes, quartiles are determined as follows:

1. Determine the class where the quartile is found:
   - *First quartile*: class for which the relative cumulative frequency exceeds 25%.
   - *Second quartile*: class for which the relative cumulative frequency exceeds 50%.
   - *Third quartile*: class for which the relative cumulative frequency exceeds 75%.
2. Calculate the value of the quartile depending on the assumption according to which the observations are uniformly distributed in each class; the jth quartile Q_j is:

$$Q_j = L_j + \left[\frac{n \cdot \frac{j}{4} - \sum f_{\inf}}{f_j} \right] \cdot c_j,$$

where L_j is the lower limit of the class of quartile Q_j, n is the total number of observations, $\sum f_{inf}$ is the sum of frequencies lower than the class of the quartile, f_j is the frequency of the class of quartile Q_j, and c_j is the size of the **interval** of the class of quartile Q_j.

DOMAINS AND LIMITATIONS

The calculation of quartiles makes sense only for a **quantitative variable** that can take values on a determined **interval**.

The quartile is similar to the **median**. It is also based on observation rank, not values. An **outlier** will have only a small influence on the quartile values.

EXAMPLES

Let us first take an example where we have ten observations ($n = 10$):

$$1 \; 2 \; 4 \; 4 \; 5 \; 5 \; 5 \; 6 \; 7 \; 9$$

Despite the fact that quartiles cannot be calculated for a small number of observations (we should be very cautious when interpreting them), we will study this case in order to understand the principles behind the calculation.

The first quartile Q_1 is found at the position $\frac{n+1}{4} = 2.75$. Q_1 is three quarters of the **distance** between the second and third observations (which we will call x_2 and x_3). We can calculate Q_1 as follows:

$$Q_1 = x_2 + 0.75 \cdot (x_3 - x_2)$$
$$= 2 + 0.75 \cdot (4 - 2)$$
$$= 3.5 .$$

The second quartile Q_2 (which is **median**) is found at the position $2 \cdot \frac{n+1}{4}$ (or $\frac{n+1}{2}$), which

equals for our example 5.5.

$$Q_2 = x_5 + 0.5 \cdot (x_6 - x_5)$$
$$= 5 + 0.5 \cdot (5 - 5)$$
$$= 5 .$$

The third quartile Q_3 is found at the position $3 \cdot \frac{n+1}{4} = 8.25$. Q_3 equals:

$$Q_3 = x_8 + 0.25 \cdot (x_9 - x_8)$$
$$= 6 + 0.25 \cdot (7 - 6)$$
$$= 6.25 .$$

The values 3.5, 5, and 6.25 separate the observations into four equal parts.

For the second example, let us consider the following **frequency table** representing the number of children per family in 200 families:

Value (number of children)	Relative (number of families)	Relative frequency	Cumulated frequency
0	6	0.03	0.03
1	38	0.19	0.22
2	50	0.25	0.47
3	54	0.27	0.74
4	42	0.21	0.95
5	8	0.04	0.99
6	2	0.01	1.00
Total	200	1.00	

The first quartile equals those observations that have a relative cumulative **frequency** exceeding 25%, which corresponds to two children (because the relative cumulative frequency for two children goes from 22 to 47%, which includes 25%). The second quartile equals three children because the relative cumulated frequency for three children goes from 47 to 74%, which includes

50%. The third quartile equals four children because the relative cumulated frequency for four children goes from 74 to 95%, which includes 75%.

Quartiles 2, 3, and 4 separate the 200 families into quarters. We can group 50 of the 200 families into the first quarter with zero, one, or two children, 50 into the second quarter with two or three children, 50 into the third quarter with three or four children, and 50 into the fourth quarter with four, five, or six children.

Now consider an example involving the calculation of quartiles from the **frequency distribution** of a continuous **variable** where the observations are grouped into classes.

The following frequency table represents the profits (in thousands of francs) of 100 stores:

Profit (thousands of francs)	Cumulated frequency	Cumulated frequency	Relative frequency
100–200	10	10	0.1
200–300	20	30	0.3
300–400	40	70	0.7
400–500	30	100	1.0
Total	100		

The class containing the first quartile is the class 200–300 (the one with a relative cumulative frequency of 25%).

Considering that the observations are uniformly distributed in each class, we obtain for the first quartile the following **values**:

1st quartile

$$= 200 + \left[\frac{\left(100 \cdot \frac{1}{4}\right) - 10}{20} \right] \cdot 100$$

$$= 275 .$$

The class containing the second quartile is the class 300–400. The **value** of the second quartile equals:

2nd quartile

$$= 300 + \left[\frac{\left(100 \cdot \frac{2}{4}\right) - 30}{40} \right] \cdot 100$$

$$= 350 .$$

The class containing the third quartile is the class 400–500. The value of the third quartile equals:

3rd quartile

$$= 400 + \left[\frac{\left(100 \cdot \frac{3}{4}\right) - 70}{30} \right] \cdot 100$$

$$= 416.66 .$$

We can conclude that 25 of the 100 stores have profits between 100000 and 275000 francs, 25 have a profits between 275000 and 350000 francs, 25 have profits between 350000 and 416066 francs, and 25 have profits between 416066 and 500000 francs.

FURTHER READING
▶ **Decile**
▶ **Measure of location**
▶ **Median**
▶ **Percentile**
▶ **Quantile**

Quetelet Adolphe

Quetelet, Adolphe was born in 1796 in Ghent, Belgium. In 1823, the minister of public education asked him to supervise the construction of an observatory for the city of Brussels. To this end, he was sent to Paris to study astronomy and to prepare plans for the construction of the observatory. In France, Bouvard, Alexis was charged with his education. Quetelet, Adolphe met, among others, **Poisson, Siméon-Denis** and **de Laplace, Pierre Simon**, who convinced him of the importance of probabilistic calculations. Over the next 4 years, he traveled widely to visit different observatories and to collect instruments. He wrote popular works on science: *Astronomie élémentaire*, *Astronomie populaire et Positions de physique*, and *Instructions populaires sur le calcul des probabilités* in 1828. Moreover, he founded with Garnier (his thesis advisor) a journal: Correspondance mathématique et physique. In 1828, he was named astronomer of the Brussels observatory and rose quickly to become director of the observatory, a position he would occupy for the next 25 years. In 1836, he became the tutor of Princes Ernest and Albert of Saxe-Cobourg and Gotha, which allowed him to pursue his interest in the calculation of probabilities. His lessons to the boys were published in 1846 under the title *Lettres à S.A.R. le Duc régnant de Saxe-Cobourg et Gotha*; they addressed the theory of probability as applied to ethical and political sciences. Quetelet died in 1874.

Principal works of Quetelet, Adolphe:

1826   Astronomie élémentaire. Malther, Paris.

1828   Positions de physique. Brussels.

1828   Instructions populaires sur le calcul des probabilités.

1846   Lettres à S.A.R. le Duc régnant de Saxe-Cobourg et Gotha sur la théorie

des probabilités, appliquée aux sciences morales et politiques, Brussels.

Quota Sampling

Quota sampling is a nonrandom **sampling** method. A **sample** is chosen in such a way that it reproduces an image that is as close as possible to the **population**.

The quota method is based on the known distribution of a **population** for certain characteristic (gender, age, socioeconomic class, etc.).

HISTORY

See **sampling**.

MATHEMATICAL ASPECTS

Consider a **population** composed by N individuals. The distribution of the population is known to have the characteristics $A, B, \ldots, Z$ with the respective modalities $A_1, A_2, \ldots, A_a$; $B_1, B_2, \ldots, B_b$; $\ldots$; $Z_1, Z_2, \ldots, Z_z$, the desired **sample** is obtained by multiplying the number of the different modalities of all the control traits by the **sounding** rate q, meaning the **percentage** of the population that one wants to sound.

The **sample** will be of size $n = N \cdot q$ and will have the following distribution:

$$q \cdot X_{A_1}, q \cdot X_{A_2}, \ldots, q \cdot X_{A_a},$$
$$q \cdot X_{B_1}, q \cdot X_{B_2}, \ldots, q \cdot X_{B_b}, \ldots,$$
$$q \cdot X_{Z_1}, q \cdot X_{Z_2}, \ldots, q \cdot X_{Z_z},$$

where $X_{\ldots}$ represents the number of the different modalities for all the control traits.

DOMAINS AND LIMITATIONS

A control trait that will be used as a basis for the distribution of the **sample** must obey the following rules: (1) It must be in close correlation with the under study variables. (2) It must have a known distribution for the whole **population**.

Commonly used control traits are gender, age, socioprofessional class, region, etc.

The advantages of quota **sampling** are the following:

- Unlike random sampling, quota **sampling** does not require the existence of a **sounding** basis.
- The cost is much lower than for random sampling.

The main disadvantages are:

- The quota method does not have a sufficient theoretical foundation. It is based on the following **hypothesis**: a correct distribution of the control traits ensures the representativity of the distribution of the observed traits.
- The quota method cannot calculate the precision of the **estimations** obtained from the **sample**. Because the investigators choose the people to be surveyed, it is impossible to know with what **probability** any given individual of the **population** may be part of the sample. It is therefore impossible to apply the calculation of probabilities that, in the case of random sampling, can associate to each estimation a measure of the **error** that could have been committed.

EXAMPLES

Suppose we want to study the market possibilities for a consumer product in town X. The chosen control variables are gender, age, and socioprofessional class. The **population**

distribution (in thousands of persons) is the following:

Gender		Age		Socioprofessional class	
Male	162	0–20	70	Chief	19.25
Female	188	21–60	192.5	Independent Prof.	14.7
		≥ 61	87.5	Managers	79.1
				Workers	60.9
				Unemployed	176.05
Total	350	Total	350	Total	350

We choose a sounding rate equal to $\frac{1}{300}$. The **sample** will therefore have the following size and distribution:

$$n = N \cdot q = 350000 \cdot \frac{1}{300} = 1167 \,.$$

Gender		Age		Socioprofessional class	
Male	540	0–20	233	Chief	64
Female	627	21–60	642	Independent Prof.	49
		≥ 61	292	Managers	264
				Workers	203
				Unemployed	587
Total	1167	Total	1167	Total	1167

The choice of individuals is then made by **simple random sampling** in each category.

FURTHER READING

▶ **Estimation**
▶ **Simple random sampling**
▶ **Sampling**

REFERENCES

Sukhatme, P.V.: Sampling Theory of Surveys with Applications. Iowa State University Press, Ames, IA (1954)

Random Experiment

An **experiment** in which the outcome is not predictable in advance is called a random experiment.

A random experiment can be characterized as follows:

1. It is possible to describe the set of all possible results (called the **sample space** of the random experiment).
2. It is not possible to predict the result with certainty.
3. It is possible to associate each possible outcome to a **probability** of appearing.

DOMAINS AND LIMITATIONS

The random phenomena to which **probabilities** apply can be encountered in many different fields such as economic sciences, social sciences, physics, medicine, psychology, etc.

EXAMPLES

Here are some examples of random experiments, characterized by their action and by the set of the possible results (their **sample space**).

Action	Sample space
Flipping a coin	$\Omega = \{Heads, Tails\}$
Rolling a die	$\Omega = \{1, 2, 3, 4, 5, 6\}$
Drawing a ball from a container with one green ball, one blue ball, and one yellow ball	$\Omega = \{Blue, Green, Yellow\}$
Rainfall in a determined period	$\Omega =$ set of nonnegative real numbers (infinite noncountable set)

FURTHER READING
▶ **Event**
▶ **Probability**
▶ **Sample space**

Random Number

A random number is called the realization of a uniformly distributed random variable in the interval [0, 1].

DOMAINS AND LIMITATIONS

If the conditions imposed above are not rigorously verified, that is, if the distribution from where the sample is taken is not exactly the **uniform distribution** and the successive draws are not necessarily independent, then we say that the set of randomly obtained numbers form a series of pseudorandom numbers.

EXAMPLES

A container holds 10 balls from 0 to 9. If our goal is to obtain a random number between 0 and 1 to m decimal places, we need to draw m balls with replacement seeking to form a number of m digits. Then we consider these draws as if there were only one number, dividing it by 10^m.

As an example, we suppose that we draw successively the following balls: 8, 7, 5, 4, and 7. The corresponding random number is 0.87547.

FURTHER READING

▶ **Generation of random numbers**

▶ **Uniform distribution**

Random Number Generation

See **generation of random numbers**.

Random Variable

A **variable** whose value is determined by the result of a **random experiment** is called a random variable.

MATHEMATICAL ASPECTS

A random variable is generally denoted by one of the last letters of the alphabet (in uppercase). It is a real valued function defined on the **sample space** Ω. In other words, a real random variable X is a mapping of Ω onto $\mathbb{R}$:

$$X : \Omega \to \mathbb{R}.$$

We refer to:

- Discrete random variables, when the set of all possible values for the random variable is either finite or infinite countable;

- Continuous random variables, when the set of all possible values for the random variable is infinite uncountable.

EXAMPLES
Examples of Discrete Random Variable

If we roll two dice simultaneously, the 36 possible results that comprise Ω are the following:

$$\Omega = \{(1, 1), (1, 2), \ldots, (1, 6),$$
$$(2, 1), (2, 2), \ldots, (2, 6), \ldots,$$
$$(6, 1), (6, 2), \ldots, (6, 6)\}.$$

The random variable $X =$ "sum of two dice" is a discrete random variable. It takes its values in the finite set $E = \{2, 3, 4, \ldots, 11, 12\}$, that is, the sum of two dice can be 2, 3, 4, ... or 12. In this example, X is an entirely positive random variable and the operation is from Ω in E.

It is possible to attribute a **probability** to the different values that the random variable X can take.

The value "sum equal to 2" is then obtained only by the result $(1, 1)$. On the other hand, the value "sum equal to 8" is obtained by the results $(2, 6), (3, 5), (4, 4), (5, 3)$, and $(6, 2)$. Since Ω contains 36 results with equal probability, we can claim that the "chance" or the "probability" of obtaining a "sum equal to 2" is $\frac{1}{36}$. The probability of obtaining a "sum equal to 8" is $\frac{5}{36}$, etc.

Example of Continuous Random Variable

Such random variables characterize measures like, for example, the time period of a telephone call defined in the interval $]0, \infty[$, wind direction defined in the interval $[0, 360[$, etc.

FURTHER READING
▶ Density function
▶ Probability
▶ Random experiment
▶ Sample space
▶ Value

Randomization

Randomization is a procedure by which treatments are randomly assingned to experimental units. It is used to avoid subjective treatment affectations. It allows to eliminate a systematic error caused by noncontrollable factors that persist even in repeated experiments. Randomization can be used to avoid any bias that might corrupt the data. Finally, randomization is necessary if we seek to estimate the experimental error.

HISTORY
In September 1919, **Fisher, Ronald Aylmer** began his career at the Rothamsted Experimental Station, where agricultural research was taking place. He started with historical data and then took into account data from agricultural trials, for which he developed the **analysis of variance**. These trials led him to develop experimental designs. He developed highly efficient experimental designs using the principles of randomization, Latin square, randomized block, and factorial arrangements.

All these methods can be found in his work *Statistical Methods for Research Workers* (1925). His article in 1926 on the arrangement of his agricultural experiments became the book *The Design of Experiments* (1935).

FURTHER READING
▶ Completely randomized design
▶ Design of experiments
▶ Experiment
▶ Experimental unit
▶ Treatment

REFERENCES
Fisher, R.A.: Statistical Methods for Research Workers. Oliver & Boyd, Edinburgh (1925)

Fisher, R.A.: The arrangement of field experiments. J. Ministry Agric. **33**, 503–513 (1926)

Fisher, R.A.: The Design of Experiments. Oliver & Boyd, Edinburgh (1935)

Fisher, R.A.: Statistical Methods, Experimental Design and Scientific Inference. A re-issue of Statistical Methods for Research Workers, "The Design of Experiments', and "Statistical Methods and Scientific Inference." Bennett, J.H., Yates, F. (eds.) Oxford University Press, Oxford (1990)

Randomized Block Design

R

A randomized block design is an **experimental design** where the experimental units are in groups called blocks. The treatments are randomly allocated to the experimental units inside each block. When all treatments appear at least once in each block, we have a completely randomized block design. Otherwise, we have an incomplete randomized block design.

This kind of design is used to minimize the effects of systematic error. If the experimenter focuses exclusively on the differences between treatments, the effects due to variations between the different blocks should be eliminated.

HISTORY
See **experimental design**.

EXAMPLES
A farmer possesses five plots of land where he wishes to cultivate corn. He wants to run an experiment since he has two kinds of corn and two types of fertilizer. Moreover, he knows that his plots are quite heterogeneous regarding sunshine, and therefore a systematic error could arise if sunshine does indeed facilitate corn cultivation.

The farmer divides the land into five blocks and randomly attributes to each block the following four treatments:

> A: corn 1 with fertilizer 1
> B: corn 1 with fertilizer 2
> C: corn 2 with fertilizer 1
> D: corn 2 with fertilizer 2

He eliminates therefore the source of the systematic **error** due to the **factor** sun. By following the same procedure randomly inside each block, he does not favor a special corn variety with respect to the others, nor a special type of fertilizer with respect to the others. The following table shows the random arrangement of the four treatments in the five blocks:

Block 1	D	C	A	B
Block 2	C	B	A	D
Block 3	A	B	C	D
Block 4	A	C	B	D
Block 5	D	C	A	B

An **analysis of variance** permits one to draw conclusions, after block effects are removed, as to whether there exists a significative difference between the treatments.

FURTHER READING
▶ **Analysis of variance**
▶ **Design of experiments**

Range

The range is the easiest **measure of dispersion** to calculate.

Consider a set of **observations** relative to a **quantitative variable** X. The range is defined as the difference between the highest observed value and the lowest observed value.

MATHEMATICAL ASPECTS
Consider a set of **observations** relative to a **quantitative variable** X.

If we denote by X_{max} the **value** of the highest observation in a set of **observations** and by X_{min} the lowest value, then the range is given by:

$$\text{range} = X_{max} - X_{min} .$$

When **observations** are grouped into classes, the range is equal to the difference between the center of the two extreme classes. Let δ_1 be the center of the first class and δ_k the center of the last class. The range is equal to:

$$\text{range} = \delta_k - \delta_1 .$$

DOMAINS AND LIMITATIONS
The range is a **measure of dispersion** that is not used very often because it only provides an indication on the span of the **values** but does not provide any information on the distribution within this span.

It nevertheless has the great advantage of being easy to calculate. However, one should be very careful with its interpretation.

The range is a **measure of dispersion** that only takes into account the two extreme **values**. This method can therefore lead to wrong interpretations if the number of children in seven families is 0, 0, 1, 1, 2, 3, and 6, the range is 6, whereas all but one of the families have between 0 and 3 children. In this case, the variability is relatively low whereas the range is high.

Using the range to measure dispersion (or variability) presents another inconvenience. It can only increase when the number of **observations** increases. This is a disadvantage because ideally a measure of variability should be independent of the number of **observations**.

One must exercise extreme caution when interpreting a range because there may be **outliers**. These have a direct influence on the **value** of the range.

EXAMPLES

Here are the ages of two groups of people taking an evening class:

Group 1:

33	34	34	37	40
40	40	43	44	45

Group 2:

22	25	31	40	40
41	44	45	50	52

The **arithmetic mean** $\bar{x}$ of these two sets of **observations** is identical:

$$\bar{x} = \frac{390}{10} = 39 .$$

Even if the **arithmetic mean** of the ages of these two groups is the same, 39 years, the dispersion around the mean is very different. The range for the first group is equal to 12 $(45 - 33)$, whereas the range for the second group is equal to $52 - 22 = 30$ years old. Notice that the range only gives an indicative measure on the variability of the **values**; it does not indicate the shape of the distribution within this span.

FURTHER READING
▶ **Interquartile range**
▶ **Measure of dispersion**

REFERENCES
Anderson, T.W., Sclove, S.L.: An Introduction to the Statistical Analysis of Data. Houghton Mifflin, Boston (1978)

Rank of a Matrix

The rank of a **matrix** is the maximum number of linearly independent rows or columns.

MATHEMATICAL ASPECTS
Let A be an $m \times n$ **matrix**. We define its rank, denoted by r_A, as the maximum number of linearly independent rows or columns. In particular:

$$r_A \leq \min (m, n) .$$

If $r_A = \min (m, n)$, then we say that matrix A is of full rank.

If the rank of a square matrix of order n equals n, then its **determinant** is nonzero; in other words, this matrix is invertible.

EXAMPLES
Let A be a 2×3 matrix:

$$A = \begin{bmatrix} -7 & 3 & 2 \\ 1 & 0 & 4 \end{bmatrix} .$$

Its rank is less than or equal to $\min(2, 3)$. The number of linearly independent lines or columns equals 2. Consequently, $r_A = 2$.

FURTHER READING
▶ **Determinant**
▶ **Matrix**

Rao Calyampudi Radhakrishna

Rao, Radhakrishna was born in 1920 at Hadagali, Karnata, India. He received his Ph.D. in 1948 at Cambridge University. Since 1989 he has received honorary doctorates from many universities around the world, notably the University of Neuchâtel, in Switzerland.

His contribution to the development of statistical theory and its applications is comparable to the works of **Fisher, Ronald Aylmer**, Neyman, Jerzy, and other important modern statisticians. Among his contributions, the most important are the Fisher–Rao theorem, the Rao–Blackwell theorem, the Cramer–Rao inequality, and Rao's U-test. He is the coauthor of 12 books and more than 250 scientific articles. His book *Statistics and Truth: Putting Chance to Work* deals with fundamental logic and statistical applications and has been translated into many languages.

Selected works and publications of Rao, Calyampudi Radhakrishna:

1973 Linear Statistical Inference and Its Applications, 2nd edn. Wiley, New York.

1989 (with Shanbhag, D.N.) Further extensions of the Choquet–Deny and Deny theorems with applications in characterization theory. Q. J. Math., 40, 333–350.

1992 (with Zhao, L.C.) Linear representation of R-estimates in linear models. Can. J. Stat., 220:359–368.

1997 Statistics and Truth: Putting Chance to Work, 2nd edn. World Scientific, Singapore.

1999 (with Toutenburg, H.) Linear Models. Least Squares and Alternatives, 2nd edn. Springer Series in Statistics. Springer, Berlin Heidelberg New York.

FURTHER READING
▶ **Gauss–Markov theorem**
▶ **Generalized inverse**

Regression Analysis

Regression analysis is a technique that permits one to study and measure the **relation** between two or more variables. Starting from data registered in a **sample**, regression analysis seeks to determine an estimate of a mathematical relation between two or more variables. The goal is to estimate the value of one variable as a function of one or more other variables. The estimated variable is called the **dependent variable** and is commonly denoted by Y. In contrast, the variables that explain the variations in Y are called independent variables, and they are denoted by X.

When Y depends on only one X, we have simple regression analysis, but when Y depends on more than one independent variable, we have multiple regression analysis. If the relation between the dependent and the independent variables is linear, then we have linear regression analysis.

HISTORY

The pioneer in linear regression analysis, **Boscovich, Roger Joseph**, an astronomer as well as a physician, was one of the first to find a method for determining the coefficients of a regression line. To obtain a line passing as close as possible to all observations, two parameters should be determined in such a way that two Boscovich, R.J. conditions are respected:

1. The sum of the deviations must equal zero.
2. The sum of the deviations absolute value must be minimal.

Boscovich, R.J. established these conditions around 1755–1757. He used a method of geometrical resolution that he applied when measuring the length of five of the earth's meridians.

In 1789, **de Laplace, Pierre Simon**, in his work *Sur les degrés mesurés des méridiens, et sur les longueurs observées sur pendule*, adopted the two conditions imposed by Boscovich, R.J. and established an algebraic method for the resolution of Boscovich's **algorithm**.

Two other mathematicians, **Gauss, Carl Friedrich** and **Legendre, Adrien Marie**, seem to have discovered, without conferring with each other, the **least-squares** method. It seems that Gauss, C.F. had been using it since 1795. But it is Legendre, A.M. who published the least-squares method in 1805 in his work *Nouvelles méthodes pour la détermination des orbites des comètes*, where we find the appendix *Sur la méthode des moindres carrés*. A controversy about the true author of the discovery erupted between the two men in 1809. Gauss, C.F. published his method citing references to his previous works. Note that in the same period, an American named Adrian, Robert, who had no knowledge of Gauss' and Legendre's work, was also working with least squares.

The 18th century marked the development of the least-squares method. According to Stigler, S. (1986), this development is strongly associated with three problems from that period. The first concerned the mathematical representation and determination of the movements of the moon. The second was related to explaining the secular inequality observed in the movement of Jupiter and Saturn. Finally, the third concerned the determination of the shape of the earth.

According to the same author, one finds in the works of **Galton, Francis** the origin of the term linear regression. His research focused on heredity: the notion of regression arose from an analysis of heredity among family members.

In 1875, Galton, C.F. conducted an **experiment** with small peas distributed in seven groups by weight. For each group, he took ten seeds and asked seven friends to cultivate them. He thus cultivated 490 germs. He discovered that each seed group classified by weight followed a normal distribution and the curves, centered on different weights, were dispersed in an equally normal distribution; the variability of the different groups was identical. Galton, F. talked about "reversion" (in biology "reversion" refers to a return to a primitive type). It was a linear reversion that Galton, F. had witnessed since the seven seed groups were normally distributed, compared not to their parents' weight, but to a value close to the overall population mean.

Human heredity was a research topic for Galton, F. as well, who was trying to define the relation between the length of parent seeds and that of their offspring. In 1889, he published his work under the title *Natural Inheritance*. In this field, from 1885 he used the term regression, and in his last works he ended up with the concept of **correlation**.

MATHEMATICAL ASPECTS

See **LAD regression**, **multiple linear regression**, and **simple linear regression**.

DOMAINS AND LIMITATIONS

The goal of regression analysis is not only to determine the relation between the **dependent variable** and the **independent variable(s)**. Regression analysis seeks as well to establish the reliability of estimates and consequently the reliability of the obtained predictions.

Regression analysis allows furthermore to examine whether the results are statistically significant and if the **relation** between the variables is real or only apparent.

Regression analysis has various applications in all scientific fields. In fact, from the moment we determine a relation among variables, we can predict future values, provided that the conditions remain unchanged and that there always exists an **error**. Commonly we check the validity of a **regression analysis** by residual analysis.

EXAMPLES

See **linear regression** (**multiple** or **simple**), **LAD regression**, and **ridge**.

FURTHER READING

- ► **Analysis of residuals**
- ► **Analysis of variance**
- ► **Coefficient of determination**
- ► **Hat matrix**
- ► **Least squares**
- ► **Leverage point**
- ► **Multiple linear regression**
- ► **Normal equations**
- ► **Residual**
- ► **Simple linear regression**

REFERENCES

Eisenhart, C.: Boscovich and the Combination of Observations. In: Kendall, M., Plackett, R.L. (eds.) Studies in the History of Statistics and Probability, vol. II. Griffin, London (1977)

Galton, F.: Natural Inheritance. Macmillan, London (1889)

Gauss, C.F.: Theoria Motus Corporum Coelestium. Werke, 7 (1809)

Laplace, P.S. de: Sur les degrés mesurés des méridiens, et sur les longueurs observées sur pendule. Histoire de l'Académie royale des inscriptions et belles lettres, avec les Mémoires de littérature tirées des registres de cette académie. Paris (1789)

Legendre, A.M.: Nouvelles méthodes pour la détermination des orbites des comètes. Courcier, Paris (1805)

Plackett, R.L.: Studies in the history of probability and statistics. In: Kendall, M., Plackett, R.L. (eds.) The discovery of the method of least squares. vol. II. Griffin, London (1977)

Stigler, S.: The History of Statistics, the Measurement of Uncertainty Before 1900. Belknap, London (1986)

Rejection Region

The rejection region is the **interval**, measured in the **sampling distribution** of the **statistic** under study, that leads to rejection of the **null hypothesis** H_0 in a **hypothesis test**. The rejection region is **complementary** to the **acceptance region** and is associated to a **probability** α, called the **significance level** of the test or **type I error**.

MATHEMATICAL ASPECTS

Consider a **hypothesis test** on a **parameter** θ. We utilize the **statistic** T in order to estimate this parameter.

Case 1: One-sided Test (Right Tail)

The hypothesis is the following:

$$H_0: \quad \theta = \theta_0$$
$$H_1: \quad \theta > \theta_0 .$$

where θ_0 is the preassigned value for parameter θ.

In a one-sided test (test for the right tail), the rejection region corresponds to an **interval** limited on the right side of the **critical value**:

Rejection region = [critical value; ∞[, and Acceptance region =]$-\infty$; critical value[,

given that the critical value equals:

$$\mu_T + z_\alpha \cdot \sigma_T ,$$

where μ_T is the **mean** of the **sampling distribution** for statistic T, σ_T is the **standard error** for the statistic T, and α is the **significance level** of the test.

The value z_α is obtained from **statistical tables** of distribution T. If the value of T computed in the **sample** lies in the rejection region, that is, if T is greater than or equal to the critical value, the **null hypothesis** H_0 must be rejected in favor of the **alternative hypothesis** H_1. Otherwise, the null hypothesis H_0 cannot be rejected.

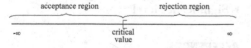

acceptance region   rejection region

$-\infty$ critical value ∞

Cas 2: One-sided Test (Left Tail)

The hypothesis is as follows:

$$H_0: \quad \theta = \theta_0$$
$$H_1: \quad \theta < \theta_0 .$$

In a one-sided test (test for the right tail), the rejection region is limited on the left side by the critical value:

$$\text{rejection region} =]-\infty; \text{ critical value}] ,$$

and the acceptance region is:

$$\text{acceptance region} =]\text{critical value}; \infty[,$$

provided the critical value equals:

$$\mu_T - z_\alpha \cdot \sigma_T .$$

If the value of T computed in the **sample** lies in the rejection region, that is, if T is less than or equal to the critical value, the **null hypothesis** H_0 must be rejected in favor of the **alternative hypothesis** H_1. Otherwise, the null hypothesis H_0 cannot be rejected.

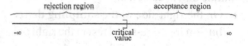

rejection region   acceptance region

$-\infty$ critical value ∞

Case 3: Two-sided Test

The hypothesis is as follows:

$$H_0: \quad \theta = \theta_0$$
$$H_1: \quad \theta \neq \theta_0$$

In a **two-sided test** the rejection region is divided into two intervals:

$$\text{rejection region} = \mathbb{R} \backslash]\text{lower critical value}; \text{ upper critical value}[$$

and the acceptance region:

$$\text{acceptance region} =]\text{lower critical value}; \text{ upper critical value}[,$$

where the critical values are equal to:

$$\mu_T - z_{\frac{\alpha}{2}} \cdot \sigma_T \quad \text{and} \quad \mu_T + z_{\frac{\alpha}{2}} \cdot \sigma_T .$$

R

If the value of T computed in the **sample** lies in the rejection region, that is, if T is greater than or equal to the upper critical value or if T is less than or equal to the lower critical value, the **null hypothesis** H_0 must be rejected in favor of the **alternative hypothesis** H_1. Otherwise, the null hypothesis H_0 cannot be rejected.

The following figure illustrates the acceptance and rejection regions for the cases mentioned above.

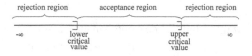

DOMAINS AND LIMITATIONS

In a **one-sided test**, the rejection region is unique and is detected:

- On the right side of the **sampling distribution** for a one-sided test in the right tail.
- On the left side of the sampling distribution for a one-sided test in the left tail.

In contrast, the rejection region is divided into two parts in the **two-sided test**, which are found on the extreme left and right sides of the sampling distribution. To each one of these regions is associated a **probability** $\frac{\alpha}{2}$, with α being the **significance level** of the test.

EXAMPLES

A company produces steel cables. Based on a sample of $n = 100$ units, it wants to verify whether the diameter of the cables is 0.9 cm. The **standard deviation** of the population is known to be equal to 0.05 cm.

The **hypothesis test** in this case is a **two-sided test**. The hypothesis is as follows:

Null hypothesis H_0: $\mu = 0.9$
Alternative hypothesis H_1: $\mu \neq 0.9$

For a **significance level** set to $\alpha = 5\%$, the value $z_{\frac{\alpha}{2}}$ of the **normal table** equals 1.96. The critical values are then:

$$\mu_{\bar{x}} \pm z_{\frac{\alpha}{2}} \cdot \sigma_{\bar{x}} \,;$$

where $\mu_{\bar{x}}$ is the **mean** of the **sampling distribution** for the means:

$$\mu_{\bar{x}} = \mu = 0.9$$

and $\sigma_{\bar{x}}$ is the **standard error** of the mean:

$$\sigma_{\bar{x}} = \frac{\sigma}{\sqrt{n}} = \frac{0.05}{\sqrt{100}} = 0.005$$

$$\Rightarrow \quad \mu_{\bar{x}} \pm z_{\frac{\alpha}{2}} \sigma_{\bar{x}} = 0.9 \pm 1.96 \cdot 0.005$$
$$= 0.9 \pm 0.0098\,.$$

The **interval** $]0.8902; 0.9098[$ corresponds to the **acceptance region** of the null hypothesis.

The rejection region is the complementary interval and is divided into two parts:

$$]-\infty; 0.8902] \text{ and } [0.9098; \infty[\,.$$

If the obtained sampling mean lies in this interval, the **null hypothesis** H_0 must be rejected in favor of the **alternative hypothesis** H_1.

FURTHER READING

▶ **Acceptance region**
▶ **Critical value**
▶ **Hypothesis testing**
▶ **Significance level**

REFERENCES

Bickel, R.D.: Selecting an optimal rejection region for multiple testing. Working paper, Office of Biostatistics and Bioinformation, Medical College of Georgia (2002)

Relation

The notion of relation expresses the rapport that exists between two random variables. It is one of the most important notions in **statistics**. It appears in concepts like **classification, correlation, dependence, regression analysis, time series**, etc.

FURTHER READING
- ▶ **Classification**
- ▶ **Correlation coefficient**
- ▶ **Dependence**
- ▶ **Model**
- ▶ **Regression analysis**
- ▶ **Time series**

Relative Risk

The **relative risk** is defined as the ratio between the **risk** relative to the individuals who are exposed to a factor. It measures the relative effect of a certain cause, that is, the degree of the association between a disease and its cause.

HISTORY
See **risk**.

MATHEMATICAL ASPECTS
The relative risk is a probability ratio that is estimated by:

$$\text{relative risk} = \frac{\text{risk of exposed}}{\text{risk of nonexposed}}$$
$$= \frac{\text{incidence rate for exposed}}{\text{incidence rate for nonexposed}}.$$

Note that the relative risk is a measure without a unit since it is defined as the ratio of two quantities measured in the same unit.

DOMAINS AND LIMITATIONS
The relative risk expresses the risk of the units exposed to a factor as a multiple of the risk of the units not exposed to this factor:

risk of units exposed
$$= \text{risk of units nonexposed}$$
$$\cdot \text{ relative risk}.$$

We want to study the relation between a risk factor and a disease. When there is no association, the relative risk should theoretically be equal to 1. If exposure to the risk factor increases the number of cases of the disease, then the relative risk becomes larger than 1. If, on the other hand, exposure to the factor decreases the number of cases of the disease, then the relative risk lies between 0 and 1. It is easy to compute a $P = (1 - \alpha)\%$ **confidence interval** for the relative risk. We consider here the following general case:
The confidence interval is computed in the logarithmic scale and then transformed back into the arithmetic scale. The lowercase letters refer to the following table:

	Disease	Nondisease	Total
Exposed	a	n − a	n
Nonexposed	c	m − c	m

The $P = (1 - \alpha)\%$ confidence interval for the relative risk, RR, is then given by:

$$\exp\left\{ \ln(RR) \pm z_\alpha \sqrt{\frac{1}{a} - \frac{1}{n} + \frac{1}{c} - \frac{1}{m}} \right\},$$

with "exp" the exponent for the expression in brackets and z_α the **value** in the normal **table**. It is assumed that a, n, c, and m are large enough. If the confidence interval for the relative risk includes the value 1.0, then we conclude that the exposure factor does not play any statistically significant role in the probability of having the disease.

EXAMPLES

We use here example 1 from the section on **attributable risk**. The relative risk of breast cancer associated to exposure to smoke corresponds to the **risk** (or **incidence rate**) for each exposure level divided by the risk (or incidence rate) for the nonexposed:

Group	Incidence rate (100000 year) (A)	Relative risk ($A/57.0$)
Nonexposed	57.0	1.0
Passive smokers	126.2	2.2
Active smokers	138.1	2.4

The relative risk is 2.2 for passive smokers and 2.4 for active smokers. Compared to the risk of the nonexposed, the risk for active smokers and passive smokers is now 2.4 and 2.2 times larger. There is a roughly modest association that underlines the fact that breast cancer depends also on factors other than smoking. The relative risk provides thus a partial description of the relation between smoking and breast cancer. By itself it does not allow one to draw any firm conclusions and indicates that smoking is probably responsible for 50% of breast cancer cases in the population.

FURTHER READING

▶ **Attributable risk**
▶ **Avoidable risk**
▶ **Cause and effect in epidemiology**
▶ **Incidence rate**
▶ **Odds and odds ratio**
▶ **Prevalence rate**
▶ **Risk**

REFERENCES

Cornfield, J.: A method of estimating comparative rates from clinical data. Applications to cancer of the lung, breast, and cervix. J. Natl. Cancer Inst. **11**, 1269–75 (1951)

Lilienfeld, A.M., Lilienfeld, D.E.: Foundations of Epidemiology, 2nd edn. Clarendon, Oxford (1980)

MacMahon, B., Pugh, T.F.: Epidemiology: Principles and Methods. Little Brown, Boston, MA (1970)

Morabia, A.: Epidemiologie Causale. Editions Médecine et Hygiène, Geneva (1996)

Morabia, A.: L'Épidémiologie Clinique. Editions "Que sais-je?". Presses Universitaires de France, Paris (1996)

Resampling

While using a **sample** to estimate a certain **parameter** concerning the **population** from which the sample is drawn, we can use the same sample more than once to improve our statistical analysis and inference. This happens by sampling inside the initial sample. This is the idea behind resampling methods. Using resampling we may compute the bias or the standard deviation of an estimated parameter, or we can also draw conclusions from construction confidence interval and testing hypotheses for the estimated parameter. Among the techniques that use resampling, **bootstrap**, **jackknife**, and cross validation can be mentioned, for example. Their special interest is in nonparametric techniques because:

- The exact distribution of the data is not known and
- The data arise from a nonlinear **design**, which is very difficult to model.

On the other hand, a problem related to such procedures arises when the sample does not represent the population and we risk making absolutely false inferences.

HISTORY

The jackknife method was introduced in the 1950s as a way to reduce the **bias** of an **estimate**. In the 1960s and 1970s, resampling was widely used in **sampling** as a way of estimating the **variance**.

The **bootstrap** method was invented by Efron, B. in the late 1970s and has since been developed by numerous researchers in different fields.

MATHEMATICAL ASPECTS

See **bootstrap**, **jackknife**.

FURTHER READING

► **Bootstrap**
► **Jackknife method**
► **Monte Carlo method**
► **Randomization**

REFERENCES

Edington, E. S.: Randomisation Tests. Marcel Dekker, New York (1980)

Efron, B.: The jackknife, the bootstrap, and other resampling plans, CBMS Monograph. No. 38. SIAM, Philadelphia (1982)

Fisher, R.A.: The Design of Experiments. Oliver & Boyd, Edinburgh (1935)

Pitman, E.J.G.: Significance tests which may be applied to samples from any populations. Supplement, J. Roy. Stat. Soc. **4**, 119–130 (1937)

Residual

Residuals are defined as the difference between the observed values and the esti-

mated (fitted) values of a regression **model**. They represent what is not explained by the regression equation.

HISTORY
See **error**.

MATHEMATICAL ASPECTS

The residuals e_i are given by:

$$e_i = Y_i - \hat{Y}_i, \quad i = 1, \ldots, n,$$

where Y_i denotes an **observation** and $\hat{Y}_i$ its fitted **value**, obtained by fitting a regression **model**.

The following graph illustrates the residuals in a **simple linear regression**:

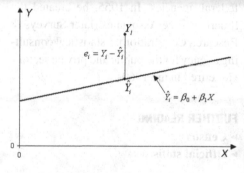

EXAMPLES
See **residual analysis**.

FURTHER READING

► **Analysis of residuals**
► **Error**
► **Hat matrix**
► **Leverage point**
► **Multiple linear regression**
► **Regression analysis**
► **Simple linear regression**

R

Rice Stuart Arthur

Rice, Stuart Arthur (1889–1969) studied at the University of Washington (obtaining his bachelor of science degree in 1912 and his master's degree in 1915). His academic work was devoted to the development of a methodology in the social sciences.

In the early 1930s, Rice, S.A. joined the administration of Franklin D. Roosevelt. He became the assistant director of the census office from 1933 until the early 1950s. His major contribution to statistics consists in the modernization of the use of statistics in government. He promoted the development of COGSIS (Committee on Government Statistics and Information Services) and worked toward adapting modern techniques of sampling as well as mathematical statistics in federal agencies. In 1955, he created the Stuart A. Rice Associates (later Surveys & Research Corporation), a statistical consulting agency for the public and private sectors. He retired in the 1960s.

FURTHER READING
▶ **Census**
▶ **Official statistics**

Ridge Regression

The main goal of the ridge method is to minimize the mean squared error of the estimates, that is, to compromise bias and variance. By considering collinearity as a problem of quasisingularity of $\mathbf{W}'\mathbf{W}$, where $\mathbf{W}$ denotes the matrix of the explanatory variables $\mathbf{X}$ suitably transformed (centered and/or scaled), we use the ridge regression method, which modifies the matrix $\mathbf{W}'\mathbf{W}$ in order to remove singularities.

HISTORY

The ridge regression was introduced in 1962 by Hoerl, A.E., who stated that the existence of correlation between explanatory variables can cause errors in estimation while applying the least-squares method. As an alternative, he developed the ridge regression, which allows to compute estimates that are biased but of lower variance than in the least-squares estimates. While the least-squares method provides nonbiased estimates, the ridge minimizes the **mean squared error** of the estimates.

MATHEMATICAL ASPECTS

Let us consider the **multiple linear regression** model:

$$\mathbf{Y} = \mathbf{X}\boldsymbol{\beta} + \boldsymbol{\varepsilon},$$

where $\mathbf{Y} = (y_1, \ldots, y_n)'$ is an $(n \times 1)$ vector of the observations related to the **dependent variable** (n observations), $\boldsymbol{\beta} = (\beta_0, \ldots, \beta_{p-1})'$ is the $(p \times 1)$ parameter vector to be estimated, $\boldsymbol{\varepsilon} = (\varepsilon_1, \ldots, \varepsilon_n)'$ is the $(n \times 1)$ error vector, and

$$\mathbf{X} = \begin{pmatrix} 1 & X_{11} & \ldots & X_{1(p-1)} \\ \vdots & \vdots & & \vdots \\ 1 & X_{n1} & \ldots & X_{n(p-1)} \end{pmatrix}$$

is an $(n \times p)$ **matrix** with independent variables.

The **estimation** of $\boldsymbol{\beta}$ by means of **least squares** is given by:

$$\widehat{\boldsymbol{\beta}} = (\widehat{\beta}_0, \ldots, \widehat{\beta}_{p-1})' = (\mathbf{X}'\mathbf{X})^{-1}\mathbf{X}'\mathbf{Y}.$$

Consider the standardized data **matrix**:

$$\mathbf{W} = \begin{pmatrix} X_{11}^s & X_{12}^s & \cdots & X_{1p-1}^s \\ X_{21}^s & X_{22}^s & \cdots & X_{2p-1}^s \\ \vdots & \vdots & \ddots & \vdots \\ {}_{n1}^s & X_{n2}^s & \cdots & X_{np-1}^s \end{pmatrix}.$$

Hence the model becomes:

$$Y = \mu + W\gamma + \varepsilon,$$

where

$$\mu = (\gamma_0, \ldots, \gamma_0)' = \left(\beta_0^s, \ldots, \beta_0^s\right)' \text{ and}$$

$$\gamma = (\gamma_1, \ldots, \gamma_{p-1})' = \left(\beta_1^s, \ldots, \beta_{p-1}^s\right)'.$$

The least-squares estimates are given by $\widehat{\gamma}_0 = \widehat{\beta}_0^s = \overline{Y}$ and by:

$$\widehat{\gamma} = (\widehat{\gamma}_1, \ldots, \widehat{\gamma}_{p-1})'$$

$$= \left(\widehat{\beta}_1^s, \ldots, \widehat{\beta}_{p-1}^s\right)'$$

$$= (W'W)^{-1} W'Y.$$

These estimates are unbiased, that is:

$$E\left(\widehat{\beta}_j\right) = \beta_j, \quad j = 0, \ldots, p-1,$$

$$E\left(\widehat{\gamma}_i\right) = \gamma_i, \quad i = 0, \ldots, p-1.$$

Among all possible linear unbiased estimates, they are the estimates with the minimal **variance**. We know that

$$\text{Var}\left(\widehat{\beta}\right) = \sigma^2 \left(X'X\right)^{-1},$$

$$\text{Var}\left(\widehat{\gamma}\right) = \sigma^2 \left(W'W\right)^{-1},$$

where σ^2 is the error variance. Note that $W'W$ is equal to $(n-1)$ times the correlation matrix for the independent variables. Consequently, in the most favorable case no collinearity arises since $W'W$ is a multiple of I. In order to have a matrix $W'W$ as close as possible to the favorable case, we replace $W'W$ by $W'W + kI$, where k is a positive scalar.

Ridge regression replaces the least-squares estimates

$$\widehat{\gamma} = (W'W)^{-1} W'Y$$

by the ridge estimates

$$\widehat{\gamma}_R = (W'W + kI)^{-1} W'Y$$

$$= \left(\widehat{\gamma}_{R_1}, \ldots, \widehat{\gamma}_{R_{(p-1)}}\right).$$

There exists always a **value** for k for which the **total mean squared error** of the ridge estimates is less than the total mean squared error of the least-squares estimates. In 1975, Hoerl et al. suggested that $k = \frac{(p-1)\sigma^2}{\gamma'\gamma}$ gives a lower total mean squared error, where $(p-1)$ is the number of explanatory variables. To compute k, we must estimate σ and γ by s and $\widehat{\gamma}$. Then, γ is reestimated by $\widehat{\gamma}_R$ using ridge regression with $k = \frac{(p-1)\sigma^2}{\gamma'\gamma}$. Note, though, that the k that has been used is a random variable, since it depends on the data.

DOMAINS AND LIMITATIONS
Properties of Ridge Estimates
Ridge estimates are biased since the expected values $E\left(\widehat{\gamma}_{R_j}\right)$ of the estimates $\widehat{\gamma}_{R_j}$ are in the **vector**:

$$E\left(\widehat{\gamma}_{R_j}\right) = \left(W'W + kI\right)^{-1} W'W\gamma \neq \gamma,$$

$$\forall k > 0.$$

Note that the value $k = 0$ corresponds to the least-squares estimates. Ridge estimates trade some bias in order to reduce the variance of the estimates. We have:

$$\text{Var}\left(\widehat{\gamma}_R\right) = \sigma^2 V_R,$$

with:

$$V_R = \left(W'W + kI\right)^{-1} W'W \left(W'W + kI\right)^{-1}.$$

Note that the diagonal elements of matrix V_R are smaller than the diagonal elements corresponding to $\left(W'W\right)^{-1}$, implying that ridge estimates have lower variance than least-squares estimates.

Thus ridge estimates are less variable than least-squares estimates. Is their bias acceptable? To measure a compromise between the bias and variance of an estimate, we generally compute its mean squared error, denoted by MSE. For an estimate $\widehat{\theta}$ of θ this is defined as:

$$MSE\left(\widehat{\theta}\right) = E\left(\left(\widehat{\theta} - \theta\right)^2\right)$$

$$= \text{Var}\left(\widehat{\theta}\right) + \left(E\left(\widehat{\theta}\right) - \theta\right)^2 .$$

Equivalently, we define the total mean squared error of an estimated vector by the sum of the mean squared errors of its components. We have:

$$TMSE\left(\widehat{\gamma}_R\right) = \sum_{j=1}^{p-1} MSE\left(\widehat{\gamma}_{Rj}\right)$$

$$= \sum_{j=1}^{p-1} \left(E\left(\widehat{\gamma}_{Rj} - \gamma_j\right)^2\right)$$

$$= \sum_{j=1}^{p-1} \left[\text{Var}\left(\widehat{\gamma}_{Rj}\right)\right.$$

$$\left. + \left(E\left(\widehat{\gamma}_{Rj}\right) - \gamma_j\right)^2\right]$$

$$= (p-1)\,\sigma^2 \text{Trace}\,(V_R)$$

$$+ \sum_{j=1}^{p-1}\left(E\left(\widehat{\gamma}_{Rj}\right) - \gamma_j\right)^2 .$$

This method is called adaptive ridge regression. The properties of the ridge estimates hold only when k is a constant, which is not the case for adaptive ridge regression. Nevertheless, the properties for the ridge estimates suggest that adaptive ridge estimates are good estimates. In fact, statistical studies have revealed that for many data sets an adaptive ridge estimation is better in terms of mean squared error than least-squares estimation.

EXAMPLES

In the following example, 13 samples of cement were set. For each one, the percentages of the four chemical ingredients was measured and are tabulated below. The amount of heat emitted was also measured. The goal was to define how the quantities x_{i1}, x_{i2}, x_{i3}, and x_{i4} affect y_i, the amount of heat emitted.

Table: Heat emitted

Samples	Ingredient				Heat
	1	2	3	4	
i	X_{i1}	X_{i2}	X_{i3}	X_{i4}	Y_i
1	7	26	6	60	78.5
2	1	29	15	52	74.3
3	11	56	8	20	104.3
4	11	31	8	47	87.6
5	7	52	6	33	95.9
6	11	55	9	22	109.2
7	3	71	17	6	102.7
8	1	31	22	44	72.5
9	2	54	18	22	93.1
10	21	47	4	26	115.9
11	1	40	23	34	83.9
12	11	66	9	12	113.3
13	10	68	8	12	109.4

where Y_i is the heat evolved by the sample i, X_{i1} is the quantity of ingredient 1 in sample i, X_{i2} is the quantity of ingredient 2 in sample i, X_{i3} is the quantity of ingredient 3 in sample i, and X_{i4} is the quantity of ingredient 4 in sample i.

We initially carry out a multiple linear regression. The linear regression model is:

$$y_i = \beta_0 + \beta_1 X_{i1} + \beta_2 X_{i2} + \beta_3 X_{i3}$$

$$+ \beta_4 X_{i4} + \varepsilon_i .$$

We obtain the following results:

Variable	Coefficient	Std. dev.	t_c
Constant	62.4100	70.0700	0.89
X_1	1.5511	0.7448	2.08
X_2	0.5102	0.7238	0.70
X_3	0.1019	0.7547	0.14
X_4	−0.1441	0.7091	−0.20

The estimated vector obtained by **standardized data** is:

$$\widehat{\gamma} = (9.12; 7.94; 0.65; -2.41)' ,$$

with a standard deviation of 4.381, 11.26, 4.834, and 11.87, respectively. The estimate for s of σ is $s = 2.446$. We follow a ridge analysis using:

$$k = \frac{(p-1)\,\sigma^2}{\gamma'\gamma}$$

$$= \frac{4 \cdot (2.446)^2}{(9.12)^2 + (7.94)^2 + (0.65)^2 + (2.41)^2}$$

$$= 0.157 .$$

This is an adaptive ridge regression since k depends on the data. We use, though, the term ridge regression. We obtain:

$$\widehat{Y}_R = (7.64; 4.67; -0.91; -5.84)' ,$$

with a standard deviation of 1.189, 1.377, 1.169, and 1.391, respectively. We have the estimated values:

$$\widehat{Y}_i = 95.4 + 7.64X_{i1}^s + 4.67X_{i2}^s - 0.91X_{i3}^s - 5.84X_{i4}^s .$$

The estimated value for the constant does not differ from the least-squares estimate.

FURTHER READING
▶ **Collinearity**
▶ **Mean squared error**

▶ **Simulation**
▶ **Standardized data**

REFERENCES

Hoerl A.E.: Application of Ridge Analysis to Regression Problems. Chem. Eng. Prog. **58**(3), 54–59 (1962)

Hoerl A.E., Kennard R.W.: Ridge regression: biased estimation for nonorthogonal Problems. Technometrics **12**, 55–67 (1970)

Hoerl A.E., Kennard R.W., Baldwin K.F.: Ridge regression: some simulations. Commun. Stat., vol. 4, pp. 105–123 (1975)

Risk

In epidemiology, risk corresponds to the probability that a disease appears in a population during a given time interval. The risk is equally defined for either the population or a homogeneous subgroup of individuals. The homogeneity is used here in the sense that individuals are exposed to a certain factor suspected of causing the disease being studied. The notion of risk is strongly related to that of **incidence** (the incidence corresponding to the frequency of a disease, the risk to its probability).

HISTORY
In 1951, Cornfield, Jerome clearly stated for the first time the notion of risk. This resulted from a followup study where risk factors were compared for two groups, one consisting of sick individuals and the other of healthy individuals. Since then risk has been widely used in epidemiologic studies as a measure of association between a disease and the related risk factors.

R

MATHEMATICAL ASPECTS

The risk is the probability of the disease appearing in a fixed time period in a group of individuals exposed to a risk factor under study. We estimate the risk by the ratio:

$$\text{risk} = \frac{\substack{\text{incident cases} \\ \text{(in the exposed population)}}}{\text{population at risk}}$$

during the given time period. The nominator corresponds to the number of new cases of the disease during the fixed time period in the studied population, while the denominator corresponds to the number of persons at risk in the same time period.

DOMAINS AND LIMITATIONS

The notion of risk is intuitively a probability measure, being defined for a given population and a determined time period. Insurance companies use it to predict the frequency of events such as death, disease, etc. in a defined population and according to factors such as sex, age, weight, etc. Risk can be interpreted as the probability for an individual to contract a disease, given that he has been exposed to a certain risk factor. We attribute to the individual the mean value of the risk for the individuals of his group. Insurance companies use risk for each population to determine the contribution of each insured individual. Doctors use risk to estimate the probability that one of their patients develops a certain disease, has a complication, or recovers from the disease.

It is important to note that the notion of risk is always defined for a certain time period: the risk in 2, 3, or 10 years.

The minimal risk is 0%, the maximal risk 100%.

EXAMPLES

Suppose that on 1 January 1992 a group of 1000 60-year-old men is selected, and 10 cases of heart attack are observed during the year. The heart attack risk for this population of 60-year-old men equals $10/1000 = 1\%$ in a year.

We consider here a case of breast cancer over the course of 2 years in a population of 100000 women, distributed in three groups depending on the level of exposure to the factor "smoke". The following contingency table illustrates this example:

Group	Number of cases (incidents)	Number of individuals
Nonexposed	40	35100
Passive smokers	140	55500
Active smokers	164	59400
Total	344	150000

The risk for each group is computed as the ratio of the number of cases divided by the total number of individuals in each group. For example, the risk for the passive smokers is $140/55500 = 0.002523 = 252.3/100000$ in 2 years. We have the following table:

Group	Number of cases (incidents)	Number of individuals	Risk in two years (100000)
Nonexposed	40	35100	114.0
Passive smokers	140	55500	252.3
Active smokers	164	59400	276.1
Total	344	150000	229.3

The notion of risk implies the notion of prediction: among the women of a given population, what proportion of them will develop breast cancer in the next 2 years? From the table above we note that 229.3/100000 may develop this disease in the next 2 years.

FURTHER READING
▶ **Attributable risk**
▶ **Avoidable risk**
▶ **Cause and effect in epidemiology**
▶ **Incidence rate**
▶ **Odds and odds ratio**
▶ **Prevalence rate**
▶ **Relative risk**

REFERENCES
Cornfield, J.: A method of estimating comparative rates from clinical data. Applications to cancer of the lung, breast, and cervix. J. Natl. Cancer Inst. **11**, 1269–75 (1951)

Lilienfeld, A.M., Lilienfeld, D.E.: Foundations of Epidemiology, 2nd edn. Clarendon, Oxford (1980)

MacMahon, B., Pugh, T.F.: Epidemiology: Principles and Methods. Little Brown, Boston, MA (1970)

Morabia, A.: Epidemiologie Causale. Editions Médecine et Hygiène, Geneva (1996)

Morabia, A.: L'Épidémiologie Clinique. Editions "Que sais-je?". Presses Universitaires de France, Paris (1996)

Robust Estimation

An **estimation** is called robust if the **estimators** that are used are not sensitive to **outliers** and small departures from idealized (model) assumptions. These estimation methods include Maximum likelihood type estimation known as M-Estimates, linear combination of order statistics, known as L-Estimates, and methods based on statistical ranks, R-Estimates. Outliers are probably **errors** resulting from a bad reading, a bad recording, or any other cause related to the experimental environment. The **arithmetic mean** and other estimators of the **least squares** are very sensitive to such outliers. That is why it is preferable in these cases to call upon estimators that are more robust.

HISTORY
The term robust, as well as the field to which it is related, was introduced in 1953 by Box, G.E.P. The term was recognized as a field of statistics only in the mid-1960s.

Nevertheless, it is important to note that this concept is not new; in the late 19th century, several scientists already had a clear idea of this concept. In fact, the first mathematical work on robust estimation apparently dates back to 1818 with **Laplace, P.S.** in his work *Le deuxième supplément à la théorie analytique des probabilités*. The distribution of the **median** can be found in this work.

It is from 1964, with the article of Huber, P.J., that this field became a separate field of **statistics** known as robust statistics.

MATHEMATICAL ASPECTS
Consider the following **model**:

$$\mathbf{Y} = \mathbf{X} \cdot \boldsymbol{\beta} + \boldsymbol{\varepsilon},$$

where $\mathbf{Y}$ is the $(n \times 1)$ vector of the observations relative to the **dependent variable** (n observations), $\boldsymbol{\beta}$ is the $(p \times 1)$ vector of the parameters to be estimated, and $\boldsymbol{\varepsilon}$ is the $(n \times 1)$ vector of errors.

There exist several methods of robust estimation that allow one to estimate the **vector** of **parameters** $\boldsymbol{\beta}$.

Some of the methods are:

- L_1 **estimation**:

$$\text{Minimize} \sum_{i=1}^{n} |\varepsilon_i| \ .$$

- In a more general way, the concept of M-estimators based on the idea of replacing the square of the **errors** ϵ_i^2 (**least-squares** method) is defined by a function $\delta(\epsilon_i)$, where δ is a symmetric function having a unique minimum in zero:

$$\text{Minimize} \sum_{i=1}^{n} \delta(\varepsilon_i) \ .$$

In 1964, Huber, P.J. proposed the following function:

$$\delta(x) = \begin{cases} \dfrac{x^2}{2}, & \text{if } |x| \le k \\[2mm] k \cdot |x| - \dfrac{k^2}{2}, & \text{if } |x| > k \end{cases},$$

where k is a positive constant.
Notice that for $\delta(x) = |x|$, the L_1 estimators are obtained.

FURTHER READING

- ▶ **Estimation**
- ▶ **Estimator**
- ▶ **L_1 estimation**
- ▶ **Maximum likelihood**
- ▶ **Outlier**

REFERENCES

Box, G.E.P.: Non-normality and tests on variance. Biometrika **40**, 318–335 (1953)

Huber, P.J.: Robust estimation of a location parameter. Ann. Math. Stat. **35**, 73–101 (1964)

Huber, P.J.: Robust Statistics. Wiley, New York (1981)

Huber, P.J.: Robustness: Where are we now? Student, vol. 1, no. 2, pp. 75–86 (1995)

Laplace, P.S. de: Deuxième supplément à la théorie analytique des probabilités. Courcier, Paris (1818). pp. 531–580 dans les Œuvres complètes de Laplace, vol. 7. Gauthier-Villars, Paris, 1886

Sample

A sample is a subset of a **population** on which statistical studies are made in order to draw conclusions relative to the **population**.

EXAMPLES

A housewife is preparing a soup. To determine if the soup has enough salt, she tastes a spoonful. As a function of what she observed on the sample represented here by a spoonful, she can decide if it is necessary to add salt to the soup, symbolizing the studied **population**.

FURTHER READING

▶ **Chebyshev's Inequality**
▶ **Estimation**
▶ **Population**
▶ **Sample size**
▶ **Sampling**

Sample Size

The sample size is the number of individuals or elements belonging to a **sample**.

HISTORY

Nordin, J.A. (1944) treats the problem of the **estimation** of the sample size within the framework of a theory of decisions through a relative example in the potential sales in a market where a seller wants to have a presence. Cornfield, J. (1951) illustrates in a precise manner the estimation of the sample size to form classes with known proportions. Sittig, J. (1951) discusses the choice of the sample size for a typical economic problem, taking into account the inspection and the cost incurred for defective pieces in an accepted lot and for correct pieces in a rejected lot. Numerous studies were also conducted by Cox, D.R. (1952), who was inspired by the works of Stein, C. (1945), to determine sample size in an ideal way.

MATHEMATICAL ASPECTS

Let $X_1, \ldots, X_n$ be a random **sample** taken without replacement of a **population** of N observations having an unknown **mean** μ. We suppose that the sample distribution of the statistic follows a normal distribution.

The **Chebyshev inequality**, depends on the number n of observations and it allows to estimate the mean μ according to the empirical mean of the obtained values $\overline{X}_n$.

When N is large enough, the size n of the sample can be determined using the Tchebychev inequality, so that the probability that $\overline{X}_n$ would be included in the interval $[\mu - \varepsilon, \mu + \varepsilon]$ ($\varepsilon > 0$ fixed at the required precision) can be as great as we want. If we call

α the **significance level**, we get:

$$P\left(|\bar{x} - \mu| \geq \varepsilon\right) = \alpha \quad \text{or}$$
$$P\left(|\bar{x} - \mu| < \varepsilon\right) = 1 - \alpha,$$

which we can also write as:

$$P\left(-z_{\alpha/2} \leq \frac{\bar{x} - \mu}{\sigma_{\bar{x}}} \leq z_{\alpha/2}\right) = 1 - \alpha,$$

where $\sigma_{\bar{x}}$, the **standard error** of the mean, is estimated by $\frac{\sigma}{\sqrt{n}}$, where σ is the **standard deviation** of the population and $z_{\alpha/2}$ the standard critical value associated to the **confidence level** $1 - \alpha$.

We construct the **confidence interval** with the help of the **normal table** for the confidence level $1 - \alpha$:

$$-z_{\alpha/2} \cdot \frac{\sigma}{\sqrt{n}} \leq \bar{x} - \mu \leq z_{\alpha/2} \cdot \frac{\sigma}{\sqrt{n}}.$$

As the **normal distribution** is symmetric, we are only interested in the inequality on the right. It indicates that $z_{\alpha/2} \cdot \frac{\sigma}{\sqrt{n}}$ is the greatest value that $\bar{x} - \mu$ can take. This limit is also given by the desired precision. From where:

$$\varepsilon = z_{\alpha/2} \cdot \frac{\sigma}{\sqrt{n}},$$

that is:

$$n = \left(z_{\alpha/2} \cdot \frac{\sigma}{\delta}\right)^2.$$

If the desired precision can be expressed in **percentage** relative to the mean μ, or $k\%$, we then have:

$$\delta = k \cdot \mu$$

$$\text{and} \quad n = \left(\frac{z_\alpha}{k} \cdot \frac{\sigma}{\mu}\right)^2,$$

where $\frac{\sigma}{\mu}$ is called the **coefficient of variation**. It measures the relative dispersion of the studied **variable**. In cases where the **sampling distribution** is not normally distributed, these formulas are approximately valid when the sample size, n, is much greater than $(n \geq 30)$.

DOMAINS AND LIMITATIONS

The determination of the sample size is an important step for organizing a statistical inquiry. It implies not only the complex calculations but also the taking into account of the different requirements of the inquiry itself. Thus the desired precision of the inquiry results is to be considered. Generally, the greater the sample size, the greater the precision. The cost of the inquiry constitutes another important constraint. The smaller the budget, the more restricted the sample size. Moreover, the availability of other resources, such as the existence of data comming from a **census** or of a system of inquirers or the duration of the inquiry, as well as the territory to be covered, can have a dramatic impact on the structure of the sample.

Census theory places great importance on samples of "optimal" dimension, especially when optimizing (maximizing or minimizing) an objective function in which the principles are formulated mathematically. Moreover, it is important to underline that generally the optimal sample dimension does not depend (or depends only slightly) on the size of the original **population**.

In the area of experimental design, the problem of optimal sample size is also raised. In this case one should find the minimal number of observations to take into account to detect a significant difference (if there is one) between the different treatments such as many drugs or manufacturing processes.

EXAMPLES

A factory employs several thousand workers. Based on a previous experiment, the researcher knows that weekly salaries are normally distributed with a **standard deviation** of 40 CHF. The researcher wants to estimate the mean weekly salary with a precision of 10 CHF and a **confidence level** of 99%. What should the sample size be?

For a confidence level of 99% ($\alpha = 0.01 = 1 - 0.99$), the $z_{0.01}$ **value** that we find in the **normal table** equals 2.575.

The direct application of the formula gives us:

$$n \geq \left[z_\alpha \cdot \frac{\sigma}{\delta} \right]^2 ,$$

$$n \geq \left[2.575 \cdot \frac{40}{10} \right]^2 ,$$

$$n \geq 106.09 .$$

We will have the desired precision if the sample size is greater than or equal to 107 workers.

FURTHER READING
▶ **Population**
▶ **Sample**

REFERENCES

Cochran, W.G.: Sampling Techniques, 2nd edn. Wiley, New York (1963)

Cornfield, J.: Methods in the Sampling of Human Populations. The determination of sample size. Amer. J. Publ. Health **41**, 654–661 (1951)

Cox, D.R.: Estimation by double sampling. Biometrika **39**, 217–227 (1952)

Harnett, D.L., Murphy, J.L.: Introductory Statistical Analysis. Addison-Wesley, Reading, MA (1975)

Nordin, J.A.: Determining sample size. J. Am. Stat. Assoc. **39**, 497–506 (1944)

Sittig, J.: The economic choice of sampling system in acceptance sampling. Bull. Int. Stat. Inst. **33**(V), 51–84 (1952)

Stein, C.: A two-sample test for a linear hypothesis whose power is independent of the variance. Ann. Math. Stat. **16**, 243–258 (1945)

Sample Space

The sample space of a **random experiment** is the set of all possible outcomes of the **experiment,** usually denoted by Ω.

HISTORY
See **probability**.

MATHEMATICAL ASPECTS
A sample space can be:
- Finite (the outcomes are countable);
- Infinite countable (it is possible to associate each possible outcome to a positive integer in such a way that each possible result has a different number; however, the number of possible outcomes is infinite);
- Infinite noncountable (it is impossible to enumerate all the possible outcomes).

The sample space Ω of a **random experiment** is also called a sure **event**. The **probability** of the sure event is equal to 1:

$$P(\Omega) = 1 .$$

EXAMPLES
Finite sample space:
- Flipping a coin
  $\Omega = \{\text{Heads, Tails}\}$
- Rolling a die
  $\Omega = \{1, 2, 3, 4, 5, 6\}$

Infinite countable sample space:
- Flipping a coin as many times as necessary to obtain "Heads" (H = heads and T = tails)

$$\Omega = \{H, TH, TTH, TTTH, TTTTH,$$
$$TTTTTH, TTTTTTH, \dots \} .$$

Infinite noncountable sample space:
- Location of the **coordinates** of the impact of a bullet on a panel of 10×10 cm. In this case, the sample space Ω is composed of an infinite noncountable set of coordinates because the coordinates are measured on a **scale** of real numbers (it is impossible to give an exhaustive list of the possible results).

FURTHER READING
► **Event**
► **Probability**
► **Random experiment**

REFERENCES
Bailey, D.E.: Probability and statistics: models for research. Wiley, New York (1971)

Sampling

From a general point of view, the term sampling means the selection of part of a **population**, called **sample**, the study of certain characteristics x of this sample, and the fact of making **inferences** relative to the population. Sampling also represents the set of operations required to obtain a **sample** from a given **population**.

HISTORY
The concept of representativity of a **sample** is a very recent one.

Even though the first trials of **extrapolation** appeared in the 13th century, notably in France, where the number of homes was used to estimate the **population**, the **census** remained more popular than sampling until the 19th century.

The principle of sampling with or without replacement appeared for the first time in the work "De ratiociniis in Aleae Ludo", pubished in 1657 by the Dutch scientist Huygens, Christiaan (1629–1695).

It was in 1895, in Bern, during the congress of the International Institute of Statistics, that Kiaer, Anders Nicolai presented a memoir called "Observations et expériences concernant des dénombrements représentatifs", Kiaer, as director of the Central Bureau of Statistics of the Kingdom of Norway at the time, compared in his presentation the structure of a **sample** with the structure of a **population** obtained by **census**. He was therefore defining the term control a posteriori and for the first time used the term "**representative**". Kiaer, A.N. encountered the opposition of almost all congress participants, who were convinced that complete enumeration was the only way to proceed.

Nevertheless, in 1897, Kiaer, A.N. confirmed the notion of representativity and received the support of the Conference of Scandinavian Statisticians. He was given the opportunity to defend his point of view during several conferences: St.-Petersburg (1897), Stockholm (1899), Budapest (1901), and Berlin (1903).

A new step was taken with the works of March, Lucien on the role of randomness during sampling; he was the first to develop the idea of probabilistic sampling, also called random sampling.

Other statisticians got involved with the problem. Von Bortkiewicz, L., a professor from Berlin, suggested the calculation of **probabilities** to test the deviation between the distributions of a **sample** and the total **population** on the key variables. Bowley, A.L. (1906) developed a random sampling procedure called stratification. According to Desrosières, A. (1988), he was also interested in the notion of **confidence** interval, of which he presented the first calculations in 1906 to the Royal Statistical Society.

The year 1925 marked a new step in the history of sampling. It is indeed the year of the congress in Rome of the International Institute of Statistics during which were presented random sampling (choice done by chance) and purposive sampling (judicious choice). Henceforth, the problem would never arise again with respect to the choice between **sampling** and complete enumeration, but it did arise with respect to various sampling methods.

Of the different uses of sampling, the most frequently used in the United States was the one done by the various public opinion institutes and notably preelectoral inquiries. The third day of November 1936, the day the results of the presidential elections were made public, marks one of the most important days in this regard. For this poll, the Readers Digest predicted the victory of Landon, whereas, without consulting each other, pollsters Crossley, Roper, and Gallup announced the victory of his opponent, Roosevelt. Roosevelt got elected. The mistake of the Readers Digest was to base its prediction of the outcome of the election on a telephone survey, but at that time, only rich people owned a telephone, so the sample was biased!

The year 1938 marks in France the birth of the IFOP (Institut Français d'Opinion Publique, the French Institute of Public Opinion) created by Stoetzel, Jean.

It is interesting to go back a little in time to emphasize the part that Russian statisticians played in the evolution of sampling techniques. Indeed, as early as the 19th century, they were known and used in their country. According to Tassi, Ph. (1988), Chuprov, A.I. (1842–1908) was one of the precursors of modern sampling, and from 1910 his son Chuprov, A.A. used random sampling, made reference to cluster sampling, **stratified sampling** with and without replacement, and, under the influence of Markov, A., encouraged the use of **probabilities** in **statistics**. Even if the Russian school was only known very late for its sampling works, the statisticians of this school apparently knew about these techniques long before those of western Europe.

Note that in 1924, Kovalevsky, A.G. had already rigorously worked on **surveys** by stratification and the optimal allocation by stratum, while this result was only rediscovered 10 years later by Neyman, J.

Finally, this history would be incomplete without mentioning the contributions of **Neyman, Jerzy** to sampling. Considered one of the founders of the statistical theory of **surveys**, he was in favor of random sampling on purposeful sampling, and he worked on sampling with or without remittance, **stratified sampling**, and **cluster sampling**. He was already using **estimators** by quotient, difference, and regression, techniques that would be developed later on by Horvitz, D.G., Cochran, W.G., Hansen, M.M., Hurwitz, W.N., and Madow, W.G.

MATHEMATICAL ASPECTS

There are several sampling methods. They can be divided into random methods and nonrandom methods.

The most widely used nonrandom procedure is **quota sampling**.

The random methods (or probabilistic sampling, sometimes also called statistical sampling) call upon a probabilistic mechanism to form a **sample** taken from a **population**.

The **variables** observed on the **sample** are **random variables**. From these variables the corresponding values of the **population** and the error due to sampling can be calculated.

The main random methods are:
- **Simple random sampling**
- **Stratified sampling**
- **Systematic sampling**
- **Cluster sampling**

DOMAINS AND LIMITATIONS

The goal of sampling is to provide enough information to make **inferences** on the characteristics of a **population**. It therefore is necessary to select a **sample** that reproduces these characteristics as accurately as possible.

Attempts are made to reduce the potential **error** due to sampling as much as possible. But even if this sampling error is inevitable, it is tolerated because it is largely compensated by the advantages of the sampling method. These are:

- *Cost reduction*
  In view of all the costs that can be incurred during **data** collection, they can obviously be reduced by sampling.
- *Time savings*
  The speed of decision making can only be ensured if the study is constrained to a **sample**.
- *Increased possibilities*
  In certain fields, a complete analysis on an entire **population** is impossible; the choice, then, is between using sampling or shelving the study.
- *Precise results*
  Since one of the goals of sampling is to recreate a **representative** image of the characteristics of a population, larger

samples would not bring results that are significantly more precise.

Only random sampling methods can evaluate the **error** due to sampling because it is based on the calculation of **probabilities**.

FURTHER READING
- ▶ **Cluster sampling**
- ▶ **Data collection**
- ▶ **Estimation**
- ▶ **Estimator**
- ▶ **Quota sampling**
- ▶ **Simple random sampling**
- ▶ **Stratified sampling**
- ▶ **Systematic sampling**

REFERENCES
Bowley, A.L.: Presidential address to the economic section of the British Association. J. Roy. Stat. Soc. **69**, 540–558 (1906)

Delanius, T.: Elements of Survey Sampling. SAREC, Stockholm (1985)

Desrosières, A. La partie pour le tout: comment généraliser? La préhistoire de la contrainte de représentativité. Estimation et sondages. Cinq contributions à l'histoire de la statistique. Economica, Paris (1988)

Foody, W.N., Hedagat, A.S.: On theory and application of BIB designs with repeated blocks. Annals of Statistics, V=S 932–945 (1977)

Hedayat, A.S., Sinha, B.K.: Design and Inference in Finite Population Sampling. Wiley, New York (1991)

Tassi, Ph.: De l'exhaustif au partiel: un peu d'histoire sur le développement des sondages. Estimation et sondages, cinq contributions à l'histoire de la statistique. Economica, Paris (1988)

Sampling Distribution

A sampling distribution is a distribution of a **statistic** $T(X_1, \ldots, X_n)$, where $T(.)$ is a function of the observed **values** of the **random variables** $X_1, \ldots, X_n$.

HISTORY
See **sampling**.

MATHEMATICAL ASPECTS
Consider, a **population** of size N and of **parameters** μ and σ (μ being the unknown **mean** of the population and σ its **standard deviation**). We take a **sample** of size n from this **population**. The number of possible samples of size n is given by the number of **combinations** of n objects among N:

$$k = C_N^n = \frac{N!}{(N-n)! \cdot n!} .$$

Each sample can be characterized, for example, by its sample mean denoted by:

$$\bar{x}_1, \bar{x}_2, \ldots, \bar{x}_k .$$

The set of these k **means** forms the sampling distribution of the means.
We will adopt the following notations to characterize the **mean** and **standard deviation** of the sampling distribution of the means:

Mean: $\mu_{\bar{x}}$ and standard deviation: $\sigma_{\bar{x}}$.

Mean of Sampling Distribution of Means
Consider $X_1, \ldots, X_n$ independent **random variables** distributed according to a distribution with a mean μ and a **variance** σ^2. By definition of the **arithmetic mean**, the mean of the sampling distribution of the means is

equal to:

$$\mu_{\bar{x}} = \frac{\sum_{i=1}^{k} \bar{x}_i}{k} ,$$

where k is the number of **samples** of size n that it is possible to form from the **population**.
The **expected value** of the **mean** of the sampling distribution of the means is equal to the mean of the **population** μ:

$$E[\mu_{\bar{x}}] = \mu .$$

Because:

$$E[\mu_{\bar{x}}] = E\left[\frac{\sum_{i=1}^{k} \bar{x}_i}{k} \right] = \frac{1}{k} \cdot E\left[\sum_{i=1}^{k} \bar{x}_i \right]$$

$$= \frac{1}{k} \cdot \sum_{i=1}^{k} E[\bar{x}_i]$$

$$= \frac{1}{k} \cdot \sum_{i=1}^{k} E\left[\frac{\sum_{j=1}^{n} x_j}{n} \right]$$

$$= \frac{1}{k} \cdot \sum_{i=1}^{k} \left(\frac{1}{n} \cdot E\left[\sum_{j=1}^{n} x_j \right] \right)$$

$$= \frac{1}{k} \cdot \sum_{i=1}^{k} \left(\frac{1}{n} \cdot \sum_{j=1}^{n} E[x_j] \right)$$

$$= \frac{1}{k} \cdot \sum_{i=1}^{k} \left(\frac{1}{n} \cdot n \cdot \mu \right)$$

$$= \frac{1}{k} \cdot k \cdot \mu = \mu .$$

Standard Deviation of Sampling Distribution of Means
By definition, the standard deviation of the sampling distribution of the means, also

called the **standard error**, is equal to:

$$\sigma_{\bar{x}} = \sqrt{\dfrac{\sum\limits_{i=1}^{k}(\bar{x}_i - \mu_{\bar{x}})^2}{k}}.$$

In the same way that there exists a relation between the **mean** of the sampling distribution of the means and the mean of the **population**, there also exists a relation between the **standard error** and the **standard deviation** of the population:

- In the case of an exhaustive **sampling** (without replacement) on a finite **population**, the **standard error** of the mean is given by:

$$\sigma_{\bar{x}} = \dfrac{\sigma}{\sqrt{n}} \cdot \sqrt{\dfrac{N-n}{N-1}},$$

where σ is the **standard deviation** of the **population**.

- If the **population** is infinite, or if the studied **sampling** is not exhaustive (with replacement), the **standard error** of the **mean** is equal to the **standard deviation** of the **population** divided by the square root of the **sample** size:

$$\sigma_{\bar{x}} = \dfrac{\sigma}{\sqrt{n}}.$$

Characteristics of Sampling Distribution of Means

- If n is large enough ($n \geq 30$), the sampling distribution of the **means** approximately follows a **normal distribution**, whatever the **population** distribution.
- If the **population** is distributed according to a **normal distribution**, then the sampling distribution of the **means** always follows a normal distribution, whatever the size of the **sample**.

- The sampling distribution of the **means** is always symmetric around its mean.

Similar considerations can be made for **statistics** other than the **arithmetic mean**. Consider the distribution of the **variances** obtained from all the possible **samples** of size n, taken from a normal **population** with a variance σ^2. This distribution is called the sampling distribution of the variances. One of the characteristics of this distribution is that it can only take nonnegative **values**, the **variance** s^2 being defined by a sum of squares. The sampling distribution of the variances is related to the **chi-square distribution** by the following relation:

$$s^2 = \dfrac{\chi^2 \sigma^2}{v} = \dfrac{\chi^2 \sigma^2}{n-1},$$

where χ^2 is a **random variable** distributed according to a **chi-square distribution** with $v = n - 1$ **degrees of freedom**. Knowing the **expected value** and the **variance** of a **chi-square distribution**:

$$E\left[\chi^2\right] = v,$$
$$\mathrm{Var}\left(\chi^2\right) = 2v,$$

we can determine these same characteristics for the distribution of s^2 in the case where the **samples** come from an infinite **population**.

Mean of Sampling Distribution of Variances

The **mean** of the distribution of s^2 is equal to the **variance** of the **population**:

$$\mu_{s^2} = E\left[s^2\right] = E\left[\dfrac{\chi^2 \sigma^2}{v}\right]$$
$$= \left(\dfrac{\sigma^2}{v}\right) \cdot E\left[\chi^2\right]$$

(because σ^2/v is a constant)

$$= \left(\frac{\sigma^2}{v} \right) \cdot v$$

$$= \sigma^2 .$$

Variance of Sampling Distribution of Variances

The **variance** of the distribution of s^2 depends not only on the variance of the **population** σ^2, but also on the size of the **samples** $n = v + 1$:

$$\sigma_{s^2}^2 = \text{Var} \left(s^2 \right) = \text{Var} \left[\frac{\chi^2 \sigma^2}{v} \right]$$

$$= \left(\frac{\sigma^4}{v^2} \right) \cdot \text{Var} \left(\chi^2 \right)$$

$$= \left(\frac{\sigma^4}{v^2} \right) \cdot 2v$$

$$= 2 \cdot \frac{\sigma^4}{v} .$$

DOMAINS AND LIMITATIONS

The study of the sampling distribution of a **statistic** allows one to judge the proximity of the statistic measured on a single **sample** with an unknown **parameter** of the **population**. It therefore also allows to specify the **error** margins of the **estimators**, calculated from the sample itself.

The notion of the sampling distribution is therefore at the base of the construction of **confidence intervals**, indicating the degree of precision of an **estimator**, as well as the realization of **hypothesis testing** on the unknown **parameters** of a **population**.

EXAMPLES

Using an example, we will show the relation existing between the **mean** of a sampling distribution and the mean of a **population** and the relation between the **standard error** and the **standard deviation** of that population.

Consider a **population** of five stores whose average price on a certain product we want to know. The following table gives the price of the product for each store belonging to the population:

Store no.	Price
1	30.50
2	32.00
3	37.50
4	30.00
5	33.00

The **mean** of the **population** is equal to:

$$\mu = \frac{\sum_{i=1}^{5} x_i}{5}$$

$$= \frac{30.50 + 32 + 37.50 + 30 + 33}{5}$$

$$= 32.60 .$$

By selecting **samples** of size 3, we can determine the different sampling means that can result from a random selection of the sample. We obtain:

$$k = C_5^3 = \frac{5!}{3! \cdot 2!} = 10 \text{ possible samples.}$$

The following table represents the different **samples** and their respective sampling means:

No. of stores in sample		Sampling means $\bar{x}_i$
1.	1, 2, 3	33.333..
2.	1, 2, 4	30.833..
3.	1, 2, 5	31.833..
4.	1, 3, 4	32.666..
5.	1, 3, 5	33.666..
6.	1, 4, 5	31.166..

S

No. of stores in sample		Sampling means $\bar{x}_i$
7.	2, 3, 4	33.166..
8.	2, 3, 5	34.166..
9.	2, 4, 5	31.666..
10.	3, 4, 5	33.500
	Total	326.000

From the **data** in this table we can calculate the **mean** of the sampling distribution of the means:

$$\mu_{\bar{x}} = \frac{\sum_{i=1}^{k} \bar{x}_i}{k} = \frac{326.00}{10} = 32.60 \,.$$

In this example, we notice that the **mean** of the sampling distribution is equal to the mean of the **population**:

$$\mu_{\bar{x}} = \mu \,.$$

We can proceed in the same way to verify the relation between the **standard error** and the **standard deviation** of the **population**. By definition, the **standard deviation** of the **population** is calculated as follows:

$$\sigma = \sqrt{\frac{\sum_{i=1}^{N}(x_i - \mu)^2}{N}} \,.$$

According to the **data** of the first table, we obtain:

$$\sigma = \sqrt{\frac{35.70}{5}} = 2.6721 \,.$$

By the same definition, we can calculate the **standard deviation** of the sampling distribution of the means:

$$\sigma_{\bar{x}} = \sqrt{\frac{\sum_{i=1}^{k}(\bar{x}_i - \mu_{\bar{x}})^2}{k}}$$

$$= \sqrt{\frac{11.899}{10}} = 1.0908 \,.$$

The following table presents the details of the calculations:

No. of stores in sample		Sampling means $\bar{x}_i$	$(\bar{x}_i - \mu_{\bar{x}})^2$
1.	1, 2, 3	33.333..	0.538
2.	1, 2, 4	30.833..	3.121
3.	1, 2, 5	31.833..	0.588
4.	1, 3, 4	32.666..	0.004
5.	1, 3, 5	33.666..	1.138
6.	1, 4, 5	31.166..	2.054
7.	2, 3, 4	33.166..	0.321
8.	2, 3, 5	34.166..	2.454
9.	2, 4, 5	31.666..	0.871
10.	3, 4, 5	33.500	0.810
	Total	326.000	11.899

According to the obtained results:

$$\sigma = 2.6721 \quad \text{and} \quad \sigma_{\bar{x}} = 1.0908 \,,$$

we can verify the following equality:

$$\sigma_{\bar{x}} = \frac{\sigma}{\sqrt{n}} \cdot \sqrt{\frac{N-n}{N-1}}$$

because $\frac{\sigma}{\sqrt{n}} \cdot \sqrt{\frac{N-n}{N-1}} = \frac{2.6721}{\sqrt{3}} \cdot \sqrt{\frac{5-3}{5-1}} = 1.0908$. These results allow us to conclude that by knowing the **parameters** μ and σ of the **population**, we can evaluate the corresponding characteristics of the sampling distribution of the **means**. Conversely, we can also determine the parameters of the population by knowing the characteristics of the sampling distribution.

FURTHER READING

▶ **Confidence interval**
▶ **Estimation**
▶ **Hypothesis testing**
▶ **Standard error**

Scatterplot

A scatterplot is obtained by transcribing the **data** of a **sample** onto a graphic. In two dimensions, a scatterplot can represent n pairs of **observations** $(x_i; y_i)$, $i = 1, 2, \ldots, n$.

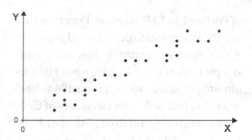

MATHEMATICAL ASPECTS

A scatterplot is obtained by placing the pairs of observed **values** in an **axis** system.

For two **variables** X and Y, the pairs of corresponding points (x_1, y_1), (x_2, y_2), $\ldots$, (x_n, y_n) are placed in a rectangular **axis** system.

The **dependent variable** Y is usually plotted on the vertical **axis** (**ordinate**) and the **independent variable** X on the horizontal axis (**abscissa**).

DOMAINS AND LIMITATIONS

The scatterplot is a very useful and powerful tool often used in **regression analysis**. Each point represents a pair of observed **values** of the **dependent variable** and of the **independent variable**. It allows to determine graphically if there exists a **relation** between two **variables** before choosing an appropriate **model**. Moreover, these scatterplots are very useful in **residual analysis** since they allow one to verify if the **model** is appropriate or not.

EXAMPLES

Consider the following two examples of scatterplots representing pairs of **observations** (x_i, y_i) relative to two **quantitative variables** X and Y:

In this example, the distribution of the observations seems to indicate a linear **relation** between the two **variables**:

The distribution of the **observations** in the following graphic seems to indicate a nonlinear relation between X and Y:

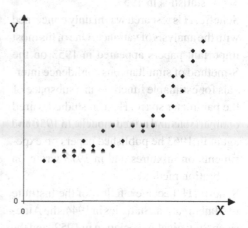

FURTHER READING

► **Analysis of residuals**
► **Regression analysis**

Scheffe Henry

Scheffé Henry was born in 1907 in New York. Scheffé, Henry's doctoral dissertation, "The Asymptotic Solutions of Certain Linear Differential Equations in Which the Coefficient of the Parameter May Have a Zero", was supervised by Langer, Rudolph E. Immediately he began his career as a university professor of mathematics at the University of Wisconsin, then at Oregon State University as well as at Reed College outside of Portland. In 1941 he joined Princeton University, where a statistics team had grown up. A second career started for him as a university professor of statistics. He was on the faculty at Syracuse University in the 1944–1945 academic year and at the University of California at Los Angeles from 1946 until 1948. He left Los Angeles to go to Columbia University, where he became chair of the Statistics Department. After 5 years at Columbia, Scheffe, Henry went to Berkeley as professor of statistics in 1953.

Scheffe, H.'s research was mainly concerned with the analysis of variance. One of his most important papers appeared in 1953 on the S-method of simultaneous confidence intervals for estimable functions in a subspace of the parameter space. He also studied paired comparisons and mixed models. In 1958 and again in 1963 he published papers on experiments on mixtures and in 1973 wrote on calibration methods.

Scheffe, H. became a fellow of the Institute of Mathematical Statistics in 1944, the American Statistical Association in 1952, and the International Statistical Institute in 1964. He was elected president of the International Statistical Institute and vice president of the American Statistical Association. He died in California in 1979.

Selected works of Henry Scheffe:

1959   The analysis of variance. Wiley, New York.

1952   An analysis of variance for paired comparisons. J. Am. Stat. Assoc., 47, 381–400.

1953   A method for judging all contrasts in the analysis of variance. Biometrika, 40, 87–104.

1956   Alternative models for the analysis of variance. Ann. Math. Stat., 27, 251–271.

1973   A statistical theory of calibration. Ann. Stat., 1:1–37.

FURTHER READING
▶ **Analysis of variance**
▶ **Least significant difference test**

Seasonal Index

Consider a **time series** Y_t whose components are:

- **Secular trend** T_t
- **Cyclical fluctuations** C_t
- **Seasonal variations** S_t
- **Irregular variations** I_t

The influence of seasonal variations must be neutral over an entire year, and seasonal variations S_t theoretically repeat themselves in an identical way from period to period (which is not verified in a real case).

To satisfy the requirements of the theoretical **model** and to be able to study the real **time series**, periodical variations that are identical each year (month by month or trimester by trimester), called seasonal variations, must be estimated instead of the observed S_t. They are denoted by S_j, where j varies as follows:

$j = 1$ up to 12 for months (over n years) or

$j = 1$ up to 4 for trimesters (over n years).

Notice that over n years, there only exist 12 seasonal index numbers concerning the months and 4 concerning the trimesters.

MATHEMATICAL ASPECTS
There are many methods for calculating the seasonal indices:

1. *Percentage of mean*

   The data corresponding to each month are expressed as **percentages** of the **mean**. We calculate it using the mean (without taking into account extreme values) or the **median** of the percentages of each month for the different years. Taking the mean for each month, we get 12 percentages for given indices. If the mean does not equal 100%, we should make an adjustment by multiplying the indices by a convenient factor.

   Consider a monthly **time series** over n years, denoted by $Y_{i,j}$, where $j = 1, 2, \ldots, 12$ for the different months and $i = 1, \ldots, n$, indicated by year.

   First, we group the data in a table of the type:

	Month		
	January	...	December
Year	1	...	12
1	$Y_{1,1}$	...	$Y_{1,12}$
...	...	...	...
n	$Y_{n,1}$	...	$Y_{n,12}$

   Then we calculate the annual means for the n years:

   $$M_i = \frac{\bar{Y}_{i,.}}{12}$$

   for $i = 1, \ldots, n$.

   Thus we calculate for each datum:

   $$Q_{i,j} = \frac{Y_{i,j}}{M_i} \quad \text{for } i = 1, \ldots, n$$

   and $j = 1, \ldots, 12$, which we report in the following table:

	Month		
	January	...	December
Year	1	...	12
1	$Q_{1,1}$	...	$Q_{1,12}$
...	...	...	...
n	$Q_{n,1}$	...	$Q_{n,12}$

In each column we include the **arithmetic mean**, denoted by $\bar{Q}_{i,.}$, and we calculate the sum of these 12 means; if it does not equal 1200, then we introduce a correction factor:

$$k = \frac{1200}{\sum_{i=1}^{12} \bar{Q}_{i,.}}.$$

Finally, the seasonal indices are given for each month by the mean multiplied by k.

2. *Tendency ratio*

   The data of each month are expressed as percentages of the monthly value of the **secular tendency**. The index is obtained from an appropriate **mean** of the percentages of the corresponding months. We should adjust the results if the mean does not equal 100% to obtain the desired seasonal indices.

3. *Moving average ratio*

   We calculate a **moving average** on 12 months. As the corresponding results are extended over many months, instead of being in the middle of the month as in the original data, we calculate the moving average over 2 months of the moving average of 12 months. This is what we call a centered moving average over 12 months.

   Then we express the original data of each month as percentages of the corresponding centered moving average over 12 months. We take the mean of the corresponding percentages of each month, which gives the indices. We should then adjust at 100% to obtain the desired seasonal indices.

4. *Relative chains*

   The data of each month are expressed as percentages of the data of the previous month (whence the expression relative chain). We calculate an appropriate mean of the chains for the corresponding months. We can obtain the relative percentages of each month relative to those of January (= 100%). We find then that the following January will ordinarily have a percentage equal to, less than, or greater than 100%. Thus the tendency is either increasing or decreasing. The indices finally obtained, adjusted for their mean equal to 100%, give the desired seasonal indices.

EXAMPLES

We consider the **time series** of any phenomenon; we have trimestrial data over 6 years, summarized in the following table:

Year	1st trim.	2nd trim.	3rd trim.	4th trim.
1	19.65	16.35	21.30	14.90
2	28.15	25.00	29.85	23.40
3	36.75	33.60	38.55	32.10
4	45.30	42.25	47.00	40.65
5	54.15	51.00	55.75	49.50
6	62.80	59.55	64.40	58.05

We determine the seasonal indices using the **percentage** of the **mean**:

First we calculate the arithmetic means for 6 years:

Year	Sum	Mean
1	72.2	18.05
2	106.4	26.6
3	141.0	35.25
4	175.2	43.8
5	210.4	52.6
6	244.8	61.2

We divide the lines of the initial table by the trimestrial mean of the corresponding year and multiply by 100.

Then, if the results show strong enough deviations for the obtained values for the same trimester, it is better to take the **median** instead of the arithmetic mean to obtain the significative numbers.

Year	1st trim.	2nd trim.	3rd trim.	4th trim.
1	108.86	90.58	118.01	82.55
2	105.83	93.98	112.22	87.97
3	104.26	95.32	109.36	91.06
4	103.42	96.46	107.31	92.81
5	102.95	96.96	105.99	94.11
6	102.61	97.30	105.23	94.85
Median	103.84	95.89	108.33	91.94

The median percentage for each trimester is given in the last line of the previous table. The sum of these percentages is exactly 400%, an adjustment is not necessary, and the numbers that we have in the last line represent the desired seasonal indices.

We have seen that there are three other methods to determine the seasonal indices, and they give the following values for the same problem:

Method	Trimester			
	1	2	3	4
Percentage of mean	103.84	95.89	108.33	91.94
Tendency ratio	112.62	98.08	104.85	84.45
Ratio of moving average	111.56	98.87	106.14	83.44
Chains	112.49	98.24	104.99	84.28

As we see, the results agree one with another, despite the diversity of the methods used to determine them.

FURTHER READING

Seasonal Variation

Seasonal variations form one of the basic components of the **time series**. They are variations of a periodic nature that recur regularly. In an economic context, seasonal variations can be caused by the following events:

- Climatic conditions
- Customs appropriate for a **population**
- Religious holidays

These are periodical fluctuations more or less regular that are superimposed on **extraseasonal** movement.

HISTORY

See **time series**.

MATHEMATICAL ASPECTS

The evaluation of seasonal variations is made by determining an **index** of seasonal variation, called the **seasonal index**.

EXAMPLES

We consider the **time series** of any phenomenon; we have trimestrial data taken over a 6 years, summarized in the following table:

Year	1st trim.	2nd trim.	3rd trim.	4th trim.
1	19.65	16.35	21.30	14.90
2	28.15	25.00	29.85	23.40
3	36.75	33.60	38.55	32.10

Year	1st trim.	2nd trim.	3rd trim.	4th trim.
4	45.30	42.25	47.00	40.65
5	54.15	51.00	55.75	49.50
6	62.80	59.55	64.40	58.05

We determine the seasonal indices using the **percentage** of the **mean**:

First, we calculate the trimestrial arithmetic means for the 6 years:

Year	Sum	Mean
1	72.2	18.05
2	106.4	26.6
3	141.0	35.25
4	175.2	43.8
5	210.4	52.6
6	244.8	61.2

We divide the lines of the initial table by the trimestrial mean of the corresponding year and multiply by 100.

Then, if results show large enough deviations for the obtained values for the same trimester, it is better to take the **median** instead to obtain more significant numbers.

Year	1st trim.	2nd trim.	3rd trim.	4th trim.
1	108.86	90.58	118.01	82.55
2	105.83	93.98	112.22	87.97
3	104.26	95.32	109.36	91.06
4	103.42	96.46	107.31	92.81
5	102.95	96.96	105.99	94.11
6	102.61	97.30	105.23	94.85
Median	103.84	95.89	108.33	91.94

The median percentage for each trimester is given by the last line of the previous table. Since these percentages add up to exactly 400%, an adjustment is not necessary, and

S

the numbers in the last line represent the desired seasonal indices.

We have seen that there exist other methods to determine the seasonal indices; they give the following values for the same problem:

Method	Trimester			
	1	2	3	4
Percentage of mean	103.84	95.89	108.33	91.94
Tendency ratio	112.62	98.08	104.85	84.45
Ratio of moving avg.	111.56	98.87	106.14	83.44
Chains	112.49	98.24	104.99	84.28

We see that the results agree with one another, despite the diversity of the methods used to determine them.

FURTHER READING
▶ **Arithmetic mean**
▶ **Cyclical fluctuation**
▶ **Forecasting**
▶ **Index number**
▶ **Moving average**
▶ **Seasonal index**
▶ **Secular trend**
▶ **Time series**

REFERENCES
Fuller, A.W.: Introduction to Statistical Time Series. John Wiley'n'Sas (1976)

Secular Trend

The secular trend forms one of the four basic components of the **time series**. It describes the movement over the long term of a time series that globally can be increasing, decreasing, or stable.

The secular trend can be linear or not. In the last case, we should choose the type of function (exponential, squared, logistic, etc.) that is best adapted to the observed distribution on the **graphical representation** of the time series before estimating the secular trend.

HISTORY
See **time series**.

MATHEMATICAL ASPECTS
Let Y_t be a **time series**. It can be decomposed with the help of four components:
- Secular trend T_t
- **Cyclic fluctuation** C_t
- **Seasonal variations** S_t
- **Irregular variations** I_t

During the study of the time series, we start by analyzing the secular trend, first deciding if this trend can be supposed linear or not. If we can suppose it is linear, the evaluation of the secular trend can be made in many ways:

1. *Graphical method or the raised hand*
   After making a **graphical representation** of the time series, it consists in adjusting a line or a curve of trend at raised hand on the graph. This is the approximation of the secular trend.

2. *Least-squares method*
   This allows to mathematically identify the line representing the best data, that is:

$$Y_t = a + b \cdot t,$$

where Y_t is the **value** of the time series at a given time, a is the value of the line when t is at the origin, b is the slope of the line, and t corresponds to another value in the entire chosen period.

The line of the **least squares** $\hat{Y}_t$ corresponds to the estimation of the secular trend T_t.

Calculation of secular trend

Supposing that the time of the time series is the year, we start from the idea of moving the origin to the corresponding year in the middle of the years in which the measurements were made.

a. *Even number of years n*

Let $j = n - 1/2$, and let us set $x_i = t_i - t_j$ for $i = 1, 2, \ldots, n$; the middle of the year t_j is the new origin and the unit of the new **variable** x is the year.

Thus

$$\hat{a} = \frac{\sum\limits_{i=1}^{n} Y_i}{n} \quad \text{and} \quad \hat{b} = \frac{\sum\limits_{i=1}^{n} x_i Y_i}{\sum\limits_{i=1}^{n} x_i^2}$$

gives us $\hat{Y}_t = \hat{a} + \hat{b}x$ or $\hat{Y}_t = (\hat{a} - \hat{b}t_j) + \hat{b}t$, where t is the year without changing the origin.

b. *Odd number of years n*

Let $j = n/2 + 1$, where t_j is the new origin, set $x_i = 2 \cdot (t_i - t_j) + 1$ for $i = 1, 2, \ldots, n$.

Thus

$$\hat{a} = \frac{\sum\limits_{i=1}^{n} Y_i}{n} \quad \text{and} \quad \hat{b} = \frac{\sum\limits_{i=1}^{n} x_i Y_i}{\sum x_i^2}$$

give us $\hat{Y}_t = \hat{a} + \hat{b}x$. The unity x is the half year counted from 1 January of year t_j.

3. *Moving average method*

Let $Y_1, Y_2, Y_3, \ldots$ be a given time series. Construct new series:

$$\frac{Y_1 + \ldots + Y_n}{n}; \frac{Y_2 + \ldots + Y_{n+1}}{n}, \ldots;$$
$$\frac{Y_j + \ldots + Y_{n+j-1}}{n}, \ldots$$

whose estimation of the secular trend is made by one of the two previous methods.

4. *Semimeans methods*

Let $Y_1, Y_2, \ldots, Y_n$ be a given time series. Denote by k the integer part of $\frac{n}{2}$.

We calculate:

$$y_1 = \frac{Y_1 + \ldots + Y_k}{k} \quad \text{and}$$
$$y_2 = \frac{Y_{k+1} + \ldots + Y_n}{n - k}$$
$$x_1 = \frac{k + 1}{2} \quad \text{and}$$
$$x_2 = \frac{n + k + 1}{2}$$

and determine the line of secular trend passing through two points:

$$(x_1; y_1) \quad \text{and} \quad (x_2; y_2),$$

When the secular trend cannot be supposed linear, we refine the least-squares method. We should apply it replacing $Y_t = a + bt$ by another trend curve as, for example:

- An exponential: $Y_t = a \cdot b^t$
- A modified exponential: $Y_t = c + a \cdot b^t$
- A logistic curve: $Y_t = (c + a \cdot b^t)^{-1}$
- A Gompertz curve: $Y_t = c \cdot a^{b^x}$

DOMAINS AND LIMITATIONS

The analysis of secular trend has the following goals:

- Create a descriptive model of a past situation.
- Make projections of constant structure.
- Eliminate secular trends to study other components of the **time series**.

The four methods of **estimation** of the secular trend have certain disadvantages. Let us mention the principal ones for each:

1. *Graphical method*

   Too subjective.

2. Least-squares method

This method can be used only for a period where the movement is completely in the same direction. When the first part corresponds to an ascending movement and another part to a descending movement, two long lines will have to be created to adjust to these data, each referring to a period with a unique direction.

These lines have some interesting properties:

- We find on the same graph the set of values Y_i of the time series and the trend line. The sum of deviations between the observed values Y_i and the estimate of the secular trend $\hat{Y}_i$ will always equal zero:

$$\sum_{i=1}^{n} \left(Y_i - \hat{Y}_i \right) = 0.$$

- The trend line will minimize the sum of deviations:

$$\sum_{i=1}^{n} \left(Y_i - \hat{Y}_i \right)^2 = \text{minimal value}.$$

3. Moving average method

The data of the beginning and end of the initial time series are "lost". Moving averages can create cycles and other movements not present in the original data; moreover, they are strongly affected by accidental outliers. This method allows to reveal changes in direction.

4. Semimeans method

Even if this method is simple to apply, it can give results without value. Thought valid in the case when data can be classed into two groups where the tendencies are linear, the method is applicable only when the tendency is globally linear or approximately linear. If the procedure is an easy application, it is not very rigorous.

EXAMPLES

Let us estimate the secular trend of the annual sales of the Surfin cookie factory using the following data:

Annual sales of Surfin cookie factory (in millions of euros)

Year	Sales (mill. $)
1975	7.6
1976	6.8
1977	8.4
1978	9.3
1979	12.1
1980	11.9
1981	12.3

Let us use the **least-squares** method:
Let us move the origin to the middle of the 7 years, which is 1978. We complete the following table:

Year	Sales (mill. euros)	Code year		
X_t	Y_t	t	tY_t	t^2
1975	7.6	−3	−22.2	9
1976	6.8	−2	−13.6	4
1977	8.4	−1	−8.4	1
1978	9.3	0	0.0	0
1979	12.1	1	12.1	1
1980	11.9	2	23.8	4
1981	12.3	3	36.9	9
Total	68.4	0	28.0	28

Set $t_i = X_i - 1978$ for $i = 1, 2, \ldots, 7$. The middle of 1978 is the new origin, and the unit of t is the year.
Thus:

$$\hat{a} = \frac{\sum_{t=-3}^{3} Y_t}{7} = \frac{68.4}{7}$$
$$= 9.7714 \text{ (million euros)} \quad \text{and}$$

$$\hat{b} = \frac{\sum\limits_{t=-3}^{3} tY_t}{\sum\limits_{t=-3}^{3} t^2} = \frac{28}{28} = 1.0 \text{ (million euros)},$$

which gives us $T_t = \hat{Y}_t = 9.7714 + t$ for the equation of the line of secular trend.

This gives graphically:

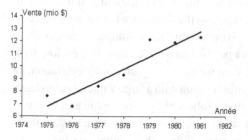

FURTHER READING
▶ **Cyclical fluctuation**
▶ **Graphical representation**
▶ **Least squares**
▶ **Moving average**
▶ **Time series**

REFERENCES
Chatfield, C.: The analysis of Time Series. An introduction. Chapman'n'Hall (2003)

Semilogarithmic Plot

A semilogarithmic plot is a **graphical representation** of a **time series**. It is defined by an arithmetic scale of time t plotted on the abscissa axis and a logarithmic scale plotted on the ordinate axis, meaning that the logarithm of the observed **value** Y_t will be transcribed in ordinate on an arithmetic scale. In general, printed semilogarithmic sheets of paper are used.

MATHEMATICAL ASPECTS
On an axis system, an origin and a unit length are arbitrarily chosen; the abscissa is arithmetically graduated as on millimeter paper.

On the ordinate axis, the unit length chosen by the maker of the semilogarithmic paper, which corresponds to the distance between two successive powers of 10 on the logarithmic scale, can be modified. This distance, which is the unit length of the corresponding arithmetic scale, is called the module.

The **values** printed in the margin of the logarithmic scale can be multiplied by the same number; this multiplicative constant is chosen to be as simple as possible, in such a way that the obtained values are easy to use for the transcription of the points on the plot.

Commercially available semilogarithmic sheets of paper usually have one, two, three, or four modules.

DOMAINS AND LIMITATIONS
The semilogarithmic plot has the following properties.
• The exponential curves Y are represented by a line. If $y(x)$ is an exponential:

$$y(x) = c \cdot a^x,$$

by taking the logarithm:

$$\log y(x) = \log c + x \cdot \log a$$

where $\log y(x) = m \cdot x + h$

where $m = \log a$ and $h = \log c$,

we find a linear expression in x.

• If $a = 1 + i$, where i is the annual interest rate of Y and with an initial sum of money

S

of Y_0, then:

$$Y = Y_0 \cdot (1 + i)^x$$

gives the sum of money after x years.

It is also possible to work in Napierian base (e) since $a^x = e^{x \cdot \ln a}$ and therefore $y(x) = c \cdot e^{x \cdot \ln a}$ or $y(x) = c \cdot e^{r \cdot x}$, where $r = \ln a$ is called the instant growth rate of y. In this case, it is the ratio of the instant variation $\frac{dy}{dx}$ to the **value** of y.

The use of the semilogarithmic plot is judicious in the following situations:

- When there exist great differences in **values** in the **variable** to avoid going outside the plot.
- When one wants to make relative variations appear.
- When cumulations of growth rates, which would make exponentials appear, must be represented in a linear fashion.

FURTHER READING

▶ **Graphical representation**
▶ **Time series**

Serial Correlation

We call serial correlation or **autocorrelation** the dependence between the observations of **time series**.

HISTORY

See **time series**.

MATHEMATICAL ASPECTS

Let $x_1, \ldots, x_n$ be n observations of a random variable X that depends on time (we are speaking here of time series). To measure the dependence between observations and time,

we use the following coefficient:

$$d = \frac{\sum\limits_{t=2}^{n} (x_t - x_{t-1})^2}{\sum\limits_{t=1}^{n} x_t^2},$$

which indicates the relation between the two consecutive observations. This coefficient takes values between 0 and 4. If it is close to 0, so that the difference between two successive observations is minimal, we speak about a positive autocorrelation. If it is close to 4, then we have a negative autocorrelation: an observation with a large value has a tendency to be followed by a low value and vice versa. In the case where observations are independent of time, the coefficient d has a value close to 2.

We can also measure the serial correlation of the observations $x_1, \ldots, x_n$ calculating the **coefficient of correlation** between the series of x_t and those of x_t "moved on k":

$$r_k = \frac{\sum\limits_{t=k+1}^{n} (x_t - \overline{x}) (x_{t-k} - \overline{x})}{\sum\limits_{t=1}^{n} (x_t - \overline{x})^2}$$

and, analyzing in the same way, measure the serial correlation for the coefficient of correlation of two different series.

DOMAINS AND LIMITATIONS
Autoregressive Models

When a serial correlation is observed between the residuals of a model of simple linear regression, convenient models of the domain of time series can be used with success, for example, models of the type

$$Y_t = \beta_0 + \beta_1 X_t + \varepsilon_t,$$

where the terms of error ε_t are not independent but are described by a relation of the type

$$\varepsilon_t = \rho_s \varepsilon_{t-1} + \delta_t,$$

where ρ_s can be estimated by the serial correlation of the ε_t and where the δ_t are independent errors.

EXAMPLES
Graph of Residuals as Functions of Time

This type of graph is especially used in the analysis of time series, that is, when time is an explanatory variable of the model. Residuals relative to time.

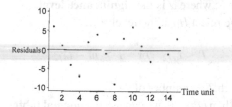

A graphical representation of residuals relative to time always entails the appearance of a serial correlation, positive or negative, as in the case of the figure above. A positive or negative serial correlation implies that the errors are not independent.

Residuals representing a serial correlation.

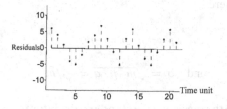

FURTHER READING

▶ **Autocorrelation**
▶ **Hypothesis testing**
▶ **Time series**

REFERENCES

Bartlett, M.S.: On the theoretical specification of sampling properties of autocorrelated time series. J. Roy. Stat. Soc. Ser. B **8**, 27–41 (1946)

Box, G.E.P., Jenkins, G.M.: Time Series Analysis: Forecasting and Control (Series in Time Series Analysis). Holden Day, San Francisco (1970)

Durbin, J., Watson, G.S.: Testing for serial correlation in least squares regression, II. Biometrika **38**, 159–177 (1951)

Sign Test

The sign test is a **nonparametric test** and can be used to test the hypothesis that there is no difference between the distribution of two random variables. We owe its name to the fact that it uses "+" and "−" signs instead of quantitative values. Thus it is applicable even when a series is of an ordinal measure.

HISTORY

In what surely was the first publication about nonparametric tests, Arbuthnott, J. (1710) studied the list of births registered in London in a period of 82 years. For each year he compared the number of children born of each sex. He denoted by "+" the **event** "more boys than girls are born" and by "−" the opposite event. (There was no equality).

To his greatest surprise, 82 times he came up with a "+" sign and never a "−" sign; this made him reject the **null hypothesis** of an equality of births relative to the sex of the child.

MATHEMATICAL ASPECTS

Consider n pairs of observations (x_1, y_1), $(x_2, y_2), \ldots, (x_n, y_n)$. For each pair (x_i, y_i),

S

we make the following comparison according to the dimension of x_i relative to y_i :

$$\text{"+"} \quad \text{if} \ \ x_i < y_i$$
$$\text{"–"} \quad \text{if} \ \ x_i > y_i$$
$$\text{"="} \quad \text{if} \ \ x_i = y_i$$

We subtract from the dimension of the **sample** the number of "=" signs that appear, that is, we take into account only the pairs that represent a positive or negative difference. We denote by m this number of pairs.

We count then the number of "+" signs, which we designate by T.

Hypotheses

According to the test, one-tailed or two-tailed, the null and alternative hypotheses corresponding to the test are:

A: Two-sided case:

$$H_0: \ \ P(X < Y) = P(X > Y)$$
$$H_1: \ \ P(X < Y) \neq P(X > Y)$$

B: One-sided case:

$$H_0: \ \ P(X < Y) \leq P(X > Y)$$
$$H_1: \ \ P(X < Y) > P(X > Y)$$

C: One-sided case:

$$H_0: \ \ P(X < Y) \geq P(X > Y)$$
$$H_1: \ \ P(X < Y) < P(X > Y)$$

In case A, we make the null hypothesis (H_0) that the **probability** that X is smaller than Y is the same as the probability that X is greater than Y.

In case B, we suppose a priori that the probability that X is smaller than Y is smaller than or equal to the probability that X is greater than Y.

Finally, in case C, we suppose a priori that the probability that X is smaller than Y is

greater than or equal to the probability that X is greater than Y.

With the two-tail case, the probability that X is smaller than Y (as that of $X > Y$) equals $\frac{1}{2}$. The sign test is a particular case of the **binomial test** with $p = \frac{1}{2}$, that is, that under the **null hypothesis** H_0 we want to get as many "+" as "–".

Decision Rules

Case A

We use the **binomial table** and we search for the number X with corresponding binomial probability equal to $\frac{1}{2}$.

We should find in the table the **value** closest to $\frac{\alpha}{2}$, where α is the **significance level**. We reject H_0 at the level α if:

$$T \leq t_{\alpha/2} \quad \text{or} \quad T \geq m - t_{\alpha/2},$$

with $T = $ number of "+".

When $m > 20$, we can use the **normal table** as an approximation of the **binomial distribution**.

We transform **statistic** T into a **random variable** Z following a standard normal distribution:

$$Z = \frac{T - \mu}{\sigma},$$

where

$$\mu = m \cdot p = \frac{1}{2}m$$

and $\quad \sigma = \sqrt{m \cdot p \cdot q} = \frac{1}{2}\sqrt{m}$

are the **mean** and the **standard deviation** of the binomial distribution.

Thus we obtain:

$$Z = \frac{T - \frac{1}{2}m}{\frac{1}{2}\sqrt{m}}.$$

The approximation of the value of the binomial table is then:

$$t_{\alpha/2} = \frac{1}{2}\left(m + z_{\alpha/2}\sqrt{m}\right),$$

where $z_{\alpha/2}$ is to to be found in the **normal table** at the level $\frac{\alpha}{2}$. Then, the decision rule is the same.

Case B

We reject H_0 at the level α if:

$$T \geq m - t_\alpha,$$

where t_α is the value of the corresponding binomial table and the **significance level** α (or the closest value).

For $m > 20$, we make the approximation

$$t_\alpha = \frac{1}{2}\left(m + z_\alpha\sqrt{m}\right),$$

where z_α is to be found in the normal table at the level α.

Case C

We reject H_0 at the level α if:

$$T \leq t_\alpha,$$

where the value of t_α is the same as for case B.

DOMAINS AND LIMITATIONS

The requirements of the use of the sign test are the following:

1. The pairs of random variables $(X_i, Y_i), i = 1, 2, \ldots, n$ must be mutually independent.
2. The scale of measure must be at least ordinal, that is, a "+", "−", or "=" sign can be associated to each pair.
3. The pairs (X_i, Y_i), $i = 1, 2, \ldots, n$ must represent a logic between them, that is, if $P\left(\text{"}+\text{"}\right) > P\left(\text{"}-\text{"}\right)$ for one pair (X_i, Y_i), then $P\left(\text{"}+\text{"}\right) > P\left(\text{"}-\text{"}\right)$ for all the other pairs.

The same principle applies if $P\left(\text{"}+\text{"}\right) < P\left(\text{"}-\text{"}\right)$ or $P\left(\text{"}+\text{"}\right) = P\left(\text{"}-\text{"}\right)$.

EXAMPLES

A manufacturer of chocolate is studying a new type of packaging for one of his products. To this end, he proposes to 10 consumers to rate the old and new packaging on a scale of 1 to 6 (with 1 meaning "strongly dislike" and 6 meaning "strongly like").

We perform a **one-sided test** (corresponding to case C) where the **null hypothesis** that we want to test is written:

$$H_0: \ P\left(\text{"}+\text{"}\right) \geq P\left(\text{"}-\text{"}\right)$$

relative to the **alternative hypothesis**:

$$H_1: \ P\left(\text{"}+\text{"}\right) < P\left(\text{"}-\text{"}\right).$$

The "+" sign means that the new packaging is preferred to the old one.

The results obtained are:

| Con-sumer | Rank | | Sign of differ-ence |
	old packaging	new packaging	
1	4	5	+
2	3	5	+
3	1	1	=
4	5	4	−
5	4	4	=
6	2	6	+
7	4	6	+
8	4	5	+
9	2	3	+
10	1	4	+

The table shows that two consumers said there was no difference between the new and old packaging. So there are eight differences ($m = 8$) among which the new packaging was, in seven cases, preferred over the old one. Thus we have $T = 7$ (number of "+"). So we reject H_0 at the level α if

$$T \leq t_\alpha,$$

S

where t_α is the **value** of the **binomial table**. We choose $\alpha = 0.05$; the value of t_α is 1 (for $m = 8$ and $\alpha = 0.0352$). Thus we have $T > t_\alpha$ because $T = 7$ and $t_\alpha = 1$. We do not reject the H_0**hypothesis**: $P\left("+"\right) \geq P\left("-"\right)$, and the chocolate manufacturer may conclude that consumers generally prefer the new packaging.

We take once more the same example, but this time the study is conducted on 100 consumers. The assumed results are the following:

83 people prefer the new packaging.

7 people prefer the old packaging.

10 people are indifferent.

Thus we have:

$$m = 100 - 10 = 90,$$
$$T = 83.$$

The value of t_α can be approximated by the **normal distribution**:

$$t_\alpha = \frac{1}{2}\left(m + z_\alpha \sqrt{m}\right),$$

where z_α is the value of the **normal table** for $\alpha = 0.05$. This gives:

$$t_\alpha = \frac{1}{2}\left(90 - 1.64\sqrt{90}\right)$$
$$= 37.22.$$

If T is greater than t_α ($83 > 37.22$), then the null hypothesis H_0: $P\left("+"\right) \geq P\left("-"\right)$ is not rejected and the chocolate manufacturer draws the same conclusion as before.

FURTHER READING
- ▶ **Binomial table**
- ▶ **Binomial test**
- ▶ **Hypothesis testing**
- ▶ **Nonparametric test**
- ▶ **Normal distribution**

REFERENCES
Arbuthnott, J.: An argument for Divine Providence, taken from the constant regularity observed in the births of both sexes. Philos. Trans. **27**, 186–190 (1710)(3.4).

Significance Level

The significance level is a **parameter** of a **hypothesis test**, and its **value** is fixed by the user in advance.

The significance level of a hypothesis test, denoted by α, is the **probability** of rejecting the **null hypothesis** H_0 when it is true:

$$\alpha = P\{\text{reject } H_0 | H_0 \text{ true}\}.$$

The significance level is also called the probability of **Type I error**.

HISTORY
In 1928, Jerzy Neyman and Egon Pearson discussed the problems related to whether or not a sample may be judged as likely to have been drawn from a certain population. They identified two types of error associated with possible decisions. The probability of the type-I error (rejecting the hypothesis given it is true) corresponds to the significance level of the test.

See **hypothesis testing**.

DOMAINS AND LIMITATIONS
Concerned with the possibility of **error** by rejecting the **null hypothesis** when it is true, statisticians construct a hypothesis test such that the **probability** of error of this type does not exceed a fixed value, given in advance. This **value** corresponds to the significance level of the test.

Note that statisticians conducting a **hypothesis test** must confront the second type of error, called the **Type II error**. It takes place when the null hypothesis is not rejected when in reality it is false.

To decrease the global risk of error in decision making, it is not enough to decrease the significance level, but a compromise should be found between the significance level and the probability of error of the second type. One method is based on the study of the power of the test.

EXAMPLES

In the following example, we illustrate the influence of the choice of the significance level when carrying out a **hypothesis test**. Consider a right **one-sided test** on the **mean** μ of a **population** whose **variance** σ^2 equals 36.

Let us state the following hypothesis:

Null hypothesis H_0: $\qquad \mu = 30$

Alternative hypothesis H_1: $\quad \mu > 30$.

Let us suppose that the mean $\bar{x}$ of a **sample** of dimension $n = 100$ taken from this population equals 31.

Case 1

Choose a significance level $\alpha = 5\%$. We find the **critical value** of $z_\alpha = 1.65$ in the normal table. The upper limit of the **acceptance region** is determined by:

$$\mu_{\bar{x}} + z_\alpha \cdot \sigma_{\bar{x}},$$

where $\mu_{\bar{x}}$ is the mean of the **sampling distribution** of the means and $\sigma_{\bar{x}}$ is the **standard error** (or the **standard deviation** of the sampling distribution of the means):

$$\sigma_{\bar{x}} = \frac{\sigma}{\sqrt{n}}.$$

This upper limit equals:

$$30 + 1.65 \cdot \frac{6}{\sqrt{100}} = 30.99.$$

The **rejection region** of the null hypothesis corresponds to the **interval** $[30.99; \infty[$ and the acceptance region to the interval $] - \infty; 30.99[$.

Case 2

If we fix a smaller significance level, for example $\alpha = 1\%$, the value of z_α in the **normal table** equals 2.33 and the upper limit of the acceptance region now equals:

$$\mu_{\bar{x}} + z_\alpha \cdot \sigma_{\bar{x}} = 30 + 2.33 \cdot \frac{6}{\sqrt{100}}$$

$$= 31.4.$$

The rejection region of the null hypothesis corresponds to the interval $[31.4; \infty[$ and the acceptance region to the interval $] - \infty; 31.4[$.

From the observed means $\bar{x} = 31$ we can see that in the first case ($\alpha = 5\%$) we must reject the null hypothesis because the sampling mean (31) is in the rejection region. In the second case ($\alpha = 1\%$) we must make the converse decision: the null hypothesis cannot be rejected because the sampling mean is in the acceptance region.

We see that the choice of the significance level has a direct influence on the decision about the null hypothesis and, in consequence, on the decision-making process. A smaller significance level (and thus a smaller **probability** of **error of the first type**) restricts the rejection region and increases the acceptance region of the null hypothesis.

A smaller risk of **error** is associated to a smaller precision.

FURTHER READING
▶ **Hypothesis testing**
▶ **Type I error**

REFERENCES
Neyman, J., Pearson, E.S.: On the use and interpretation of certain Test Criteria for purposes of Statistical Inference Part I (1928). Reprinted by Cambridge University Press in 1967

Simple Index Number

A simple **index number** is the ratio of two **values** representing the same **variable**, measured in two different situations or in two different periods. For example, a simple index number of price will give the relative variation of the price between the current period and a reference period. The most commonly used simple index numbers are those of price, quantity, and value.

HISTORY
See **index**.

MATHEMATICAL ASPECTS
The **index number** $I_{n/0}$, which is representative of a **variable** G in situation n with respect to the same variable in situation 0 (reference situation), is defined by:

$$I_{n/0} = \frac{G_n}{G_0} \,,$$

where G_n is the **value** of **variable** G in situation n and G_0 is the value of variable G in situation 0.

Generally, a simple index number is expressed in base 100 in reference situation 0:

$$I_{n/0} = \frac{G_n}{G_0} \cdot 100 \,.$$

Properties of Simple Index Numbers
- *Identity*: If two compared situations (or two periods) are identical, the value of the index number is equal to 1 (or 100):

$$I_{n/n} = I_{0/0} = 1 \,.$$

- A simple index number is *reversible*:

$$I_{n/0} = \frac{1}{I_{0/n}} \,.$$

- A simple index number is *transferable*. Consider two index numbers $I_{1/0}$ and $I_{2/0}$ representing two situations, 1 and 2, with respect to the same base situation 0. If we consider the index number $I_{2/1}$, we can say that:

$$I_{2/0} = I_{2/1} \cdot I_{1/0} \,.$$

Notice that the transferability (also called circularity) implies reversibility:

If $I_{0/1} \cdot I_{1/0} = I_{0/0} = 1$, then $I_{0/1} = \frac{1}{I_{1/0}}$.

DOMAINS AND LIMITATIONS
In economics it is very rare that simple **index numbers** are exploitable in themselves; the information they contain is relatively limited. Nevertheless, they have a fundamental importance in the construction of **composite index numbers**.

EXAMPLES
Consider the following table representing the prices of a certain item and the quantities sold at different periods:

Price		Quantity	
Period 1	Period 2	Period 1	Period 2
$P_0 = 50$	$P_n = 70$	$Q_0 = 25$	$Q_n = 20$

If P_n is the price of the period of interest and P_0 is the price of the reference period, we will have:

$$I_{n/0} = \frac{P_n}{P_0} \cdot 100 = \frac{70}{50} \cdot 100 = 140,$$

which means that in base 100 in period 1, the price index of the item is 140 in period 2. The price of the item increased by 40% (140−100) between the reference period and the actual period.

We can also calculate a simple index number of quantity:

$$I_{n/0} = \frac{Q_n}{Q_0} \cdot 100 = \frac{20}{25} \cdot 100 = 80.$$

The sold quantity therefore has decreased by 20% (100−80) between the reference period and the actual period.

Using these numbers, we can also calculate a value index number, value being defined as price multiplied by quantity:

$$I_{n/0} = \frac{P_n \cdot Q_n}{P_0 \cdot Q_0} = \frac{70 \cdot 20}{50 \cdot 25} \cdot 100 = 112.$$

The value of the considered item has therefore increased by 12% (112 − 100) between the two considered periods.

FURTHER READING
▶ **Composite index number**
▶ **Fisher index**
▶ **Index number**
▶ **Laspeyres index**
▶ **Paasche index**

Simple Linear Regression

The simple linear regression is an **analysis of regression** where the **dependent variable** Y linearly depends on a single **independent variable** X.

The simple linear regression aims to not only estimate the regression function relative to the chosen **model** but also test the reliability of the obtained estimations.

HISTORY
See **analysis of regression**.

MATHEMATICAL ASPECTS
The **model** of simple linear regression is of the form:

$$Y = \beta_0 + \beta_1 X + \varepsilon,$$

where Y is the **dependent variable** (or explicated variable), X is the **independent variable** (or explanatory variable), ε is the term of random non-observable **error**, and β_0 and β_1 are the parameters to estimate.

If we have an set of n observations (X_1, Y_1), ..., (X_n, Y_n), where Y_i is linearly dependent on the corresponding X_i, we can write

$$Y_i = \beta_0 + \beta_1 X_i + \varepsilon_i, \quad i = 1, \ldots, n.$$

The problem with **regression analysis** consists in estimating parameters β_0 and β_1, choosing the values $\hat{\beta}_0$ and $\hat{\beta}_1$ such that the **distance** between Y_i and $(\beta_0 + \beta_1 \cdot X_i)$ is minimal. We should have:

$$\varepsilon_i = Y_i - \beta_0 - \beta_1 \cdot X_i$$

small for all $i = 1, \ldots, n$. To achieve this, we can choose among many criteria:

1. $\min\limits_{\beta_0, \beta_1} \max\limits_{i} |\varepsilon_i|$

2. $\min\limits_{\beta_0, \beta_1} \sum\limits_{i=1}^{n} |\varepsilon_i|$

3. $\min\limits_{\beta_0, \beta_1} \sum\limits_{i=1}^{n} \varepsilon_i^2$

The most used method of **estimation** of parameters β_0 and β_1 is the third, called the **least squares**. It aims to minimize the sum of squared errors.

The estimation of the parameters by the least-squares method gives the following estimators $\hat{\beta}_0$ and $\hat{\beta}_1$:

$$\hat{\beta}_1 = \frac{\sum_{i=1}^{n}(X_i - \bar{X})(\hat{Y}_i - \bar{Y})}{\sum_{i=1}^{n}(X_i - \bar{X})^2}$$

$$\hat{\beta}_0 = \bar{Y} - \hat{\beta}_1\bar{X}.$$

We can then write:

$$\hat{Y}_i = \hat{\beta}_0 + \hat{\beta}_1 X_i,$$

where $\hat{Y}_i$ is the estimated value of Y_i for a given X_i when $\hat{\beta}_0$ and $\hat{\beta}_1$ are known.

Measure of Reliability of Estimation of Y

We have calculated an estimation of the value of Y, or $\hat{Y}$, with the help of the least-squares method, basing our calculation on a linear model to translate the relation that relates Y to X. But how far can we trust this model? To answer this question, it is useful to perform an **analysis of variance** and to test the hypothesis on parameters β_0 and β_1 of the regression line. To carry out these tests, we must make the following suppositions:

- For each value of X, Y is a **random variable** distributed according to the **normal distribution**.
- The **variance** of Y is the same for all X; it equals σ^2 (unknown).
- The different observations on Y are independent of each other but conditioned by the values of X.

Analysis of Variance

The table of the analysis of variance that we must construct is the following:

Analysis of variance

Source of variation	Degree of freedom	Sum of squares	Mean of squares
Regression	1	$\sum_{i=1}^{n}(\hat{Y}_i - \bar{Y})^2$	$\sum_{i=1}^{n}(\hat{Y}_i - \bar{Y})^2$
Residual	$n-2$	$\sum_{i=1}^{n}(\hat{Y}_i - \bar{Y})^2$	$\frac{\sum_{i=1}^{n}(\hat{Y}_i - \bar{Y})^2}{n-2}$
Total	$n-1$	$\sum_{i=1}^{n}(Y_i - \bar{Y})^2$	

If the model is correct, then

$$S^2 = \frac{\sum_{i=1}^{n}(\hat{Y}_i - \bar{Y})^2}{n-2}$$

is an unbiased estimator of σ^2.

The analysis of variance allows us to test the **null hypothesis**:

$$H_0: \quad \beta_1 = 0$$

against the **alternative hypothesis**:

$$H_1: \quad \beta_1 \neq 0$$

calculating the **statistic**:

$$F = \frac{EMSE}{RMSE} = \frac{EMSE}{S^2}.$$

This statistic F must be compared with the value $F_{\alpha,1,n-2}$ of the **Fisher table**, where α is the **significance** of the test.

$\Rightarrow$

If $F \leq F_{\alpha,1,n-2}$, then we accept H_0.

If $F > F_{\alpha,1,n-2}$, then we reject H_0 for H_1.

The **coefficient of determination** R^2 is calculated in the following manner:

$$R^2 = \frac{ESS}{TSS} = \frac{\hat{\beta}'X'Y - n\hat{Y}^2}{Y'Y - n\hat{Y}^2},$$

where ESS is the sum of squares of the regression and TSS is the total sum of squares.

Hypothesis on the Slope β_1

In the case of a simple linear regression, the statistic F allows to test a **hypothesis** using the parameters of the regression equation or β_1, the slope of the line. Another way to carry out the same test is as follows:

If H_0 is true $(\beta_1 = 0)$, then statistic t:

$$t = \frac{\hat{\beta}_1}{S_{\hat{\beta}_1}}$$

follows a **Student distribution** with $(n-2)$ degrees of freedom.

$S_{\hat{\beta}_1}$ is the **standard deviation** of $\hat{\beta}_1$, estimated from the **sample**:

$$S_{\hat{\beta}_1} = \sqrt{\frac{S^2}{\sum_{i=1}^{n}(X_i - \bar{X})^2}}.$$

Statistic t must be compared with the value $t_{\frac{\alpha}{2},n-2}$ of the **Student table**, where α is the significance level and $n-2$ the number of degrees of freedom. The decision rule is the following:

$|t| \leq t_{\frac{\alpha}{2},n-2}$

$\Rightarrow$ we accept H_0 $(\beta_1 = 0)$.

$|t| > t_{\frac{\alpha}{2},n-2}$

$\Rightarrow$ we reject H_0 for H_1 $(\beta_1 \neq 0)$.

It is possible to show that in the case of a simple linear regression, t^2 equals F.

Statistic t allows to calculate a **confidence interval** for β_1.

Hypothesis on Ordinate in Origin β_0

In a similar way, we can construct a confidence interval for β_0 and test the hypothesis:

$$H_0: \quad \beta_0 = 0$$
$$H_1: \quad \beta_0 \neq 0$$

Statistic t:

$$t = \frac{\hat{\beta}_0}{S_{\hat{\beta}_0}}$$

also follows a Student distribution with $(n-2)$ degrees of freedom.

The estimated standard deviation of $\hat{\beta}_0$, denoted by $S_{\hat{\beta}_0}$, is defined by:

$$S_{\hat{\beta}_0} = \sqrt{\frac{S^2 \sum_{i=1}^{n} X_i^2}{n \sum_{i=1}^{n}(X_i - \bar{X})^2}}.$$

The decision rule is the following:

$|t| \leq t_{\frac{\alpha}{2},n-2}$

$\Rightarrow$ we accept H_0 $(\beta_0 = 0)$.

$|t| > t_{\frac{\alpha}{2},n-2}$

$\Rightarrow$ we reject H_0 for H_1 $(\beta_0 \neq 0)$.

$t_{\frac{\alpha}{2},n-2}$ is obtained from the Student table for $(n-2)$ degrees of freedom and a significance level α.

The **coefficient of determination** R^2 is calculated in the following manner:

$$R^2 = \frac{ESS}{TSS} = \frac{\hat{\beta}' X' Y - n\bar{Y}^2}{Y'Y - n\bar{Y}^2},$$

where ESS is the sum of squares of the regression and TSS is the total sum of squares.

DOMAINS AND LIMITATIONS

Simple linear regression is a particular case of **multiple linear regression**. The matrix approach revealed in the multiple linear regression (**model** containing many independent variables) is also valid for the particular case where we have only one independent variable.

See **analysis of regression**.

EXAMPLES

The following table represents the gross national product (GNP) and the demand for staples for the period 1969 to 1980 in certain countries.

Year	GNP	Demand for staple
	X	Y
1969	50	6
1970	52	8
1971	55	9
1972	59	10
1973	57	8
1974	58	10
1975	62	12
1976	65	9
1977	68	11
1978	69	10
1979	70	11
1980	72	14

We want to estimate the demand for staples depending on the GNP according to the **model**:

$$Y_i = \beta_0 + \beta_1 X_i + \varepsilon_i, \quad i = 1, \ldots, n.$$

The **estimation** of parameters β_0 and β_1 by the **least-squares** method gives us the following estimators:

$$\hat{\beta}_1 = \frac{\sum\limits_{i=1}^{12} (X_i - \bar{X})(Y_i - \bar{Y})}{\sum\limits_{i=1}^{12} (X_i - \bar{X})^2} = 0.226$$

$$\hat{\beta}_0 = \bar{Y} - \hat{\beta}_1 \bar{X} = -4.04.$$

The estimated line can be written as:

$$\hat{Y} = -4.04 + 0.226X.$$

Analysis of Variance

We will calculate the degrees of freedom as well as the sum of squares and the mean of squares in order to establish the table of the **analysis of variance**:

$$dl_{\text{reg}} = 1,$$
$$dl_{\text{res}} = n - 2 = 10,$$
$$dl_{\text{tot}} = n - 1 = 11;$$

$$ESS = \sum_{i=1}^{12} \left(\hat{Y}_i - \bar{Y} \right)^2$$
$$= (7.20 - 9.833)^2 + \cdots$$
$$+ (12.23 - 9.833)^2$$
$$= 30.457;$$

$$RSS = \sum_{i=1}^{12} \left(Y_i - \hat{Y}_i \right)^2$$
$$= (6 - 7.20)^2 + \cdots$$
$$+ (14 - 12.23)^2$$
$$= 17.21;$$

$$TSS = \sum_{i=1}^{12} \left(Y_i - \bar{Y} \right)^2$$
$$= (6 - 9.833)^2 + \cdots$$
$$+ (14 - 9.833)^2$$
$$= 47.667;$$

$$EMSE = \frac{ESS}{dl_{\text{reg}}} = 30.457$$

$$S^2 = RMSE = \frac{RSS}{dl_{\text{res}}} = 1.721.$$

Thus we have the following table:

Analysis of variance

Source of variation	Degrees of freedom	Sum of squares	Mean of squares	F
Regression	1	30.457	30.457	17.687
Residual	10	17.211	1.721	
Total	11	47.667		

With a **significance level** of $\alpha = 5\%$, we find in the **Fisher table** the value

$$F_{\alpha,1,n-2} = F_{0.05,1,10} = 4.96.$$

As a result of $F > F_{0.05,1,10}$, we reject the **null hypothesis** H_0: $\beta_1 = 0$, which means that β_1 is significatively different from zero. To calculate the **coefficient of determination** of this example, it is enough to look at the table of analysis of variance because it contains all the required elements:

$$R^2 = \frac{ESS}{TSS} = \frac{30.457}{47.667}$$
$$= 0.6390 = 63.90 \, . \%$$

We can conclude that, according to the chosen model, 63.90% of the variation of the demand for staples is explained by the variation in GNP.

It is evident that the R^2 value of cannot exceed 100; the value 63.90 is large enough, but it is not close enough to 100 to discourage one from trying to improve the model.

This can mean:

- That except for the GNP, other variables should be taken into account for a finer determination of the function of the demand of staples, so the GNP explains only part of the variation.
- We should test another model (one without constant term, nonlinear model, etc).

Hypothesis testing on the parameters will allow us to determine if they are significatively different from zero.

Hypothesis on Slope β_1

In the table of analysis of variance, we have:

$$S^2 = RMSE = 1.721,$$

which allows to calculate the **standard deviation** $S_{\hat{\beta}_1}$ of $\hat{\beta}_1$:

$$S_{\hat{\beta}_1} = \sqrt{\frac{S^2}{\sum_{i=1}^{12}(X_i - \bar{X})^2}}$$

$$= \sqrt{\frac{1.721}{(50 - 61.416)^2 + \cdots + (72 - 61.416)^2}}$$

$$= \sqrt{\frac{1.721}{596.92}} = 0.054.$$

We can calculate the statistic:

$$t = \frac{\hat{\beta}_1}{S_{\hat{\beta}_1}} = \frac{0.226}{0.054} = 4.19.$$

Choosing a significance level of $\alpha = 5\%$, we get:

$$t_{\frac{\alpha}{2},n-2} = t_{0.025,10} = 2.228.$$

Comparing this **value** of t with the value in the table, we get:

$$|t| > t_{\frac{\alpha}{2},n-2},$$

which indicates that the null hypothesis:

$$H_0: \quad \beta_1 = 0$$

must be rejected for the **alternative hypothesis**:

$$H_1: \quad \beta_1 \neq 0.$$

We conclude that there exists a linear relation between the GNP and the demand for staples. If H_0 were accepted, that would mean that there was no linear relation between these two variables.

Remark: We can see that

$$F = t^2,$$
$$17.696 = 4.19^2$$

(without taking into account errors of roundness). In consequence, we can use indifferently test t or test F to test the hypothesis H_0: $\beta_1 = 0$.

Hypothesis on Ordinate in Origin β_0

The **standard deviation** $S_{\hat{\beta}_0}$ of $\hat{\beta}_0$ is:

$$S_{\hat{\beta}_0} = \sqrt{\frac{S^2 \sum_{i=1}^{12} \bar{X}_i^2}{n \sum_{i=1}^{12} \left(X_i - \bar{X}\right)^2}}.$$

$$= \sqrt{\frac{1.721 \left(50^2 + \cdots + 72^2\right)}{(12 \cdot 596.92)}}$$

$$= \sqrt{\frac{1.721 \cdot 45861}{7163.04}} = 3.31.$$

We can calculate the **statistic**:

$$t = \frac{\hat{\beta}_0}{S_{\hat{\beta}_0}} = \frac{-4.04}{3.31} = -1.22.$$

Comparing this **value** of t with the value of the **Student table** $t_{0.025,10} = 2.228$, we get:

$$|t| < t_{\frac{\alpha}{2}, n-2},$$

which indicates the null hypothesis:

$$H_0: \quad \beta_0 = 0$$

must be accepted at the significance level $\alpha = 5\%$. We conclude that the value of β_0 is not significatively different from zero, and the line passes through the origin.

New Model: Regression Passing Through Origin

The new model is the following:

$$Y_i = \beta_1 X_i + \varepsilon_i, \quad i = 1, \ldots, 12.$$

Making the estimation of the parameter β_1 according to the least-squares method, we have:

$$\sum_{i=1}^{12} \varepsilon_i^2 = \sum_{i=1}^{12} (Y_i - \beta_1 X_i)^2.$$

Setting to zero the derivative by β_1, we get:

$$\frac{\partial \sum_{i=1}^{12} \varepsilon_i^2}{\partial \beta_1} = -2 \sum_{i=1}^{12} X_i (Y_i - \beta_1 X_i) = 0$$

or

$$\sum_{i=1}^{12} (X_i Y_i) - \beta_1 \sum_{i=1}^{12} X_i^2 = 0.$$

The value of β_1 that satisfies the equation of the estimator $\hat{\beta}_1$ of β_1 is:

$$\hat{\beta}_1 = \frac{\sum_{i=1}^{12} (X_i Y_i)}{\sum_{i=1}^{12} X_i^2} = \frac{7382}{45861} = 0.161.$$

In consequence, the new regression line is described by:

$$\hat{Y}_i = \hat{\beta}_1 X_i$$
$$\hat{Y}_i = 0.161 X_i.$$

The table of analysis of variance relative to this new model is the following:

Analysis of variance

Source of variation	Degrees of freedom	Sum of squares	Mean of squares	F
Regression	1	1188.2	1188.2	660.11
Residual	11	19.8	1.8	
Total	12	1208.0		

Note that in a model without a constant, the number of degrees of freedom for the total variation equals the number of observations n. The number of degrees of freedom for the residual variation equals $n - 1$.

Thus we can determine R^2:

$$R^2 = \frac{ESS}{TSS} = \frac{1188.2}{1208.0} = 0.9836$$
$$= 98.36\% .$$

We have found with the first model the value $R^2 = 63.90\%$. We can say that this new model without a constant term is much better than the previous one.

FURTHER READING
► **Analysis of residuals**
► **Coefficient of determination**
► **Correlation coefficient**
► **Least squares**
► **Multiple linear regression**
► **Normal equations**
► **Regression analysis**
► **Residual**

REFERENCES
Seber, G.A.F. (1977) Linear Regression Analysis. Wiley, New York

Simple Random Sampling

Simple random sampling is a **sampling** method whereby one chooses n units amongst the N units of a population in such a way that each of the C_N^n possible **samples** has the same **probability** of being selected.

HISTORY
See **sampling**.

MATHEMATICAL ASPECTS
To obtain a simple random **sample** of size n, the individuals of a **population** are first numbered from 1 to N; then n numbers are drawn between 1 and N. The drawings are done in principle without replacement.

Tables of **random numbers** can also be used to obtain a simple random sample. The goal of a simple random **sample** is to provide an **estimation** without a **bias** of the **mean** and of the **variance** of the **population**. Indeed, if we denote by $y_1, y_2, \ldots, y_N$ the characteristics of a population and by $y_1, \ldots, y_n$ the corresponding values in the simple random sample, we obtain the following results:

	Population:	Sample:
Total	$T = \sum_{i=1}^{N} y_i$	$t = \sum_{i=1}^{n} y_i$
Mean	$\bar{Y} = \dfrac{T}{N}$	$\bar{y} = \dfrac{t}{n}$
Variance	$\sigma^2 = \dfrac{\sum_{i=1}^{N}(y_i - \bar{Y})^2}{N-1}$	$s^2 = \dfrac{\sum_{i=1}^{n}(y_i - \bar{y})^2}{n-1}$

DOMAINS AND LIMITATIONS
Simple random sampling can be done with or without replacement. If a drawing is performed with replacement, then the **population** always remains the same. The **sample** is therefore characterized by a series of independent **random variables** that are identically distributed. If the **population** is sufficiently large, and if the size of the **sample** is relatively small with respect to the population, it can be considered that the random variables are **independent** even if the drawings are done without replacement. It is then possible to estimate certain characteristics of the **population** by determining them from this **sample**. Simple random sampling is the basis of the theory of **sampling**.

EXAMPLES
Consider a **population** of 658 individuals. To form a simple random **sample**, we

attribute a three-digit number to every individual: 001, 002, ..., 658. Then with a **random number** table, we obtain a **sample** of ten individuals. We randomly choose a first digit and a reading direction. We then read the first three digits until we obtain ten distinct numbers between 001 and 658 to define our sample.

FURTHER READING
▶ **Cluster sampling**
▶ **Estimation**
▶ **Estimator**
▶ **Sampling**
▶ **Sampling distribution**
▶ **Stratified sampling**
▶ **Systematic sampling**

REFERENCES
Cochran, W.G.: Sampling Techniques, 2nd edn. Wiley, New York (1963)

Simulation

Simulation is a method for analyzing, designing and operating complex systems. Simulation involve designing a model of a system and carrying out experiments on it as it progresses.

The fundamental problem of simulation is in the construction of the artificial samples relative to a statistically known distribution. These distributions are generally known empirically: they result in a statistical study from which we can determine the probability distributions of the random variables characterizing the phenomenon.

Once the distributions are known, the sample is constructed by generating random numbers.

HISTORY
The method of simulation has a long history. It was first introduced by Student (1908), who discovered the sampling distributions of the t **statistic** and the **coefficient of correlation**.

More recently, thanks to computers, simulation methods and the Monte Carlo methods have progressed rapidly.

EXAMPLES
Consider the phenomenon of waiting in the checkout line of a large store. Suppose that we want to know the best "system of cashiers", which is that where the sum of the costs of inactivity of cashiers and the costs of waiting customers are the smallest.

The evolution of the phenomenon essentially depends on the distribution of the arrival of customers at the cashiers and the distribution of the time of service. The simulation consists in constructing a **sampling** of customer arrivals and a sample of time of service. Thus we can calculate the cost of a system. The **experiment** can be repeated for different systems and the best system chosen.

FURTHER READING
▶ **Bootstrap**
▶ **Generation of random numbers**
▶ **Jackknife method**
▶ **Monte Carlo method**
▶ **Random number**

REFERENCES
Gosset, S.W. "Student": The Probable Error of a Mean. Biometrika **6**, 1–25 (1908)pp. 302–310.

Kleijnen, J.P.C.: Statistical Techniques in Simulation (in two parts). Vol. 9 of Statistics: Textbooks and Monographs. Marcel Dekker, New York (1974)

Kleijnen, J.P.C.: Statistical Tools for Simulation Practitioners. Vol. 76 of Statistics: Textbooks and Monographs. Marcel Dekker, New York (1987)

Naylor, T.H., Balintfy, J.L., Burdick, D.S., Chu, K.: Computer Simulation Techniques. Wiley, New York (1967)

Snedecor George Waddel

Snedecor, George Waddel was born in 1881 in Memphis and died in 1974 in Amherst. In 1899, he entered the Alabama Polytechnic Institute in Auburn and stayed there for 2 years. He then spent 2 years earning a teacher's certificate. When his family moved to Tuscaloosa in 1903, Snedecor, George Waddel was transferred to Alabama University, where he received, in 1905, his B.S. in mathematics and physics. He accepted his first academic position at the Selma Military Academy, where he taught from 1905 to 1907. From 1907 and to 1910 he taught mathematics and Greek at Austin College in Sherman.

In 1910, Snedecor, George Waddel moved to Ann Arbor, MI, where he completed his education at Michigan State University in 1913, having earned a master's degree. In the same year, he was hired at the University of Iowa in Ames, where he stayed until 1958.

In 1927, the Mathematics Statistical Service was inaugurated in Ames with Snedecor, G.W. and Brandt, A.E. at the helm. This Service was the precursor of the Statistical Laboratory inaugurated in 1933, where Snedecor, G.W. was director and where **Cox, Gertrude** worked.

In 1931, Snedecor, G.W. was named professor in the Mathematics Department at Iowa State College. During this period, Snedecor, G.W. developed a friendship with **Fisher, Ronald Aylmer**. In 1947, Snedecor, G.W. left the Statistical Laboratory. He became professor of statistics at Iowa State College and remained at this position until his retirement in 1958. He died in February 1974.

Principal work of Snedecor, George Waddel:

1937 Statistical Methods Applied to Experiments in Agriculture and Biology. Collegiate, Ames, USA.

Spatial Data

Spatial data are special kind of data (often called map data, geographic data) that refer to a spatial localization, such as, for example, the concentration of pollutants in a given area, clouds, etc., and all the information that can be represented in the form of a map. Included under spatial data are data with a denumerable collection of spatial sites, for example, the distribution of child mortality in different cities and countries.

We can also consider that spatial data are the realization of a regional variable. The *theory of the regional variable* presupposes that any measure can be modeled (visualized) as being the realization of a random function (or random process). For example, samples taken on the ground can be seen as realizations of a regional variable (see **geostatistics**).

HISTORY

With the development of territory information systems (TIS) or geographic information systems (GIS), the notion of spatial data became widely used in statistics.

DOMAINS AND LIMITATIONS

A GIS is a collection of information programs that facilitate, based on georeferences, the integration of spatial, nonspatial, qualitative, and quantitative data in data bases that can be organized as a unique system. Spatial data come from widely differing fields and often demand a selection and an analysis according to specific criteria for each type of research. Satellites collect enormous quantities of data, and only a tiny fraction is analyzed. This wealth of data must be verified by the selection process. Data selection can be done according to the type of problems to resolve and the models used to make the analysis.

EXAMPLES

Generally, spatial data can be treated as realizations of random variables such as in an ordinary statistical analysis. Below we present, as an example of spatial data, data collected in Swiss Jura. It is a set of 259 data on heavy-metal pollution in a region of 14.5 km^2, 10 of which are represented here:

X	Y	Ground	Rock	Cd	Co
2.386	3.077	3	3	1.740	9.32
2.544	1.972	2	2	1.335	10.00
2.807	3.347	2	3	1.610	10.60
4.308	1.933	3	2	2.150	11.92
4.383	1.081	3	5	1.565	16.32
3.244	4.519	3	5	1.145	3.50
3.925	3.785	3	5	0.894	15.08
2.116	3.498	3	1	0.525	4.20
1.842	0.989	3	1	0.240	4.52

Cr	Cu	Ni	Pb	Zn
38.32	25.72	21.32	77.36	92.56
40.20	24.76	29.72	77.88	73.56

Cr	Cu	Ni	Pb	Zn
47.00	8.88	21.40	30.80	64.80
43.52	22.70	29.72	56.40	90.00
38.52	34.32	26.20	66.40	88.40
40.40	31.28	22.04	72.40	75.20
30.52	27.44	21.76	60.00	72.40
25.40	66.12	9.72	141.00	72.08
27.96	22.32	11.32	52.40	56.40

"Ground", which refers to the use of the ground, and "rock", which describes the type of underlying rock in the sampled location, are the categorical data (**qualitative categorical variables**). The different types of rock are:

1. Argovian,
2. Kimmeridgian,
3. Sequanian,
4. Portlandian,
5. Quaternary.

The use of the ground corresponds to:

1. forest,
2. pasturage,
3. prairie,
4. plowed.

FURTHER READING
▶ **Classification**
▶ **Geostatistics**
▶ **Histogram**
▶ **Sampling**
▶ **Spatial statistics**

REFERENCES

Atteia, O., Dubois, J.-P., Webster, R.: Geostatistical analysis of soil contamination in the Swiss Jura. Environ. Pollut. **86**, 315–327 (1994)

Goovaerts, P.: Geostatistics for Natural Resources Evaluation. Oxford University Press, Oxford (1997)

Matheron, G.: La théorie des variables regionalisées et ses applications. Masson, Paris (1965)

Spatial Statistics

Spatial statistics concern the statistical analysis of **spatial data**. Spatial statistics cover all the techniques used to explore and prove the presence of a spatial dependence between observations distributed in space. Spatial data can come from geology, earth sciences, image treatment, **epidemiology**, agriculture, ecology, astronomy, or forest sciences. Data are assumed to be random, and sometimes their place is also assumed to be random.

HISTORY

The statistical treatment of spatial data date back to Halley, Edmond (1686). Halley, E. superposed on one map the relief of wind directions and monsoon between and around the tropics and tried to explain the causes. The spatial statistics model appeared much later. **Fisher, Ronald Aylmer** (1920–1930), during his researches on the experimental station of Rothamsted in England, formulated the basis of the principles of random choice, of analysis by blocks, and of replication.

MATHEMATICAL ASPECTS

The statistical analysis of spatial data forms the principal subject of spatial statistics. The principal tools of the statistical analysis are the histogram, the curve of frequency cumulation, the **Q-Q plot**, the **median**, the **mean** of the distribution, the coefficients of variation, and the **measure of skewness**. In the case of many variables, the most used statistical approaches are to represent them in the form of a scatterplot and to study their correlation. Spatial analysis has the distinguishing feature that it takes into account spatial information and the relation that exists between two spatial data. Two data close to one another will more likely resemble one another than if they were farther apart. Thus when we analyze a map showing concentrations of pollutants on the ground, we would see not random values but, on the contrary, small values of pollutants grouped as well as large values, and the transition between the classes would be continuous.

DOMAINS AND LIMITATIONS

Spatial representation in the form of maps and drawings and spatial statistics can be united to give a better visual understanding of the influence that the neighboring values have on one another. Superposing statistical information on a map allows to better analyze the information and to draw the appropriate conclusions.

EXAMPLES

See **spatial data**.

FURTHER READING
▶ **Autocorrelation**
▶ **Geostatistics**
▶ **Spatial data**

REFERENCES

Cressie, N.A.C.: Statistics for Spatial Data. Wiley, New York (1991)

Goovaerts, P.: Geostatistics for Natural Resources Evaluation. Oxford University Press, Oxford (1997)

Isaaks, E.H., Srivastava, R.M.: An Introduction to Applied Geostatistics. Oxford University Press, Oxford (1989)

S

Journel, A., Huijbregts, C.J.: Mining Geo-statistics. Academic, New York (1981)

Matheron, G.: La théorie des variables regionalisées et ses applications. Masson, Paris (1965)

Oliver, M.A., Webster, R.: Kriging: a method of interpolation for geographical information system. Int. J. Geogr. Inf. Syst. **4**(3), 313–332 (1990)

Spearman Rank Correlation Coefficient

The Spearman rank correlation coefficient (Spearman ρ) is a nonparametric measurement correlation. It is used to determine the **relation** existing between two sets of **data**.

HISTORY
Spearman, Charles was a psychologist. In 1904 he introduced for the first time the rank correlation coefficient. Often called the ρ of Spearman, it is one of the oldest rank **statistic**.

MATHEMATICAL ASPECTS
Let $(X_1, X_2, \ldots, X_n)$ and $(Y_1, Y_2, \ldots, Y_n)$ be two **samples** of size n. R_{X_i} denotes the rank of X_i compare to the other **values** of the X sample, for $i = 1, 2, \ldots, n$. $R_{X_i} = 1$ if X_i is the smallest value of X, $R_{X_i} = 2$ if X_i is the second smallest value, etc., until $R_{X_i} = n$ if X_i is the largest value of X. In the same way, R_{Y_i} denotes the rank of Y_i, for $i = 1, 2, \ldots, n$. The Spearman rank correlation coefficient, generally denoted by ρ, is defined by:

$$\rho = 1 - \frac{6 \sum_{i=1}^{n} d_i^2}{n(n^2 - 1)} ,$$

where $d_i = R_{X_i} - R_{Y_i}$.

If several **observations** have exactly the same **value**, an average rank will be given to these observations. If there are many average ranks, it is best to make a correction and to calculate:

$$\rho = \frac{S_x + S_y - \sum_{i=1}^{n} d_i^2}{2\sqrt{S_x \cdot S_y}} ,$$

where

$$S_x = \frac{n\left(n^2 - 1\right) - \sum_{i=1}^{g} \left(t_i^3 - t_i\right)}{12} ,$$

with g the number of groups with average ranks and t_i the size of group i for the X **sample**, and

$$S_y = \frac{n\left(n^2 - 1\right) - \sum_{j=1}^{h} \left(t_j^3 - t_j\right)}{12} ,$$

with h the number of groups with average ranks and t_j the size of group j for the Y sample.

(If there are no average ranks, the **observations** are seen as groups of size 1, meaning that $g = h = n$ and $t_i = t_j = 1$ for $i, j = 1, 2, \ldots, n$ and $S_x = S_y = n(n^2 - 1)/12$.)

Hypothesis Testing
The Spearman rank correlation coefficient is often used as a statistical test to determine if there exists a relation between two **random variables**. The test can be a **bilateral test** or a **unilateral test**. The hypotheses are:

A: *Bilateral case*

H_0: X and Y are mutually independent.

H_1: There is either a positive or a negative correlation between X and Y.

There is a positive correlation when the large **values** of X have a tendency to be associated with large values of Y and small values of X with small values of Y. There is a negative correlation when large values of X have a tendency to be associated with small values of Y and vice versa.

B: *Unilateral case*

H_0: X and Y are mutually independent.

H_1: There is a positive correlation between X and Y.

C: *Unilateral case*

H_0: X and Y are mutually independent.

H_1: There is a negative correlation between X and Y.

Decision Rules

The decision rules are different depending on the **hypotheses**. That is why there are decision rules A, B, and C relative to the previous cases.

Decision rule A

Reject H_0 at the **significance level** α if

$$\rho > t_{n,1-\frac{\alpha}{2}} \quad \text{or} \quad \rho < t_{n,\frac{\alpha}{2}},$$

where t is the **critical value** of the test given by the Spearman table.

Decision rule B

Reject H_0 at the **significance level** α if

$$\rho > t_{n,1-\alpha}.$$

Decision rule C

Reject H_0 at the **significance level** α if

$$\rho < t_{n,\alpha}.$$

Remark: The notation t does not mean that the Spearman coefficients are related to those of Student.

DOMAINS AND LIMITATIONS

The Spearman rank correlation coefficient is used as a **hypothesis test** to study the dependence between two **random variables**. It can be considered as a **test of independence**.

As a nonparametric correlation measurement, it can also be used with nominal or ordinal **data**.

A correlation measurement between two **variables** must satisfy the following points:

1. It takes **values** between -1 and $+1$.
2. There is a positive correlation between X and Y if the value of the correlation coefficient is positive; a perfect positive correlation corresponds to a value of $+1$.
3. There is a negative correlation between X and Y if the value of the correlation coefficient is negative; a perfect negative correlation corresponds to a value of -1.
4. There is null correlation between X and Y when the correlation coefficient is close to zero; one can also say that X and Y are not correlated.

The Spearman rank correlation coefficient presents the following advantages:

- The **data** can be nonnumerical **observations** as long as they can be classified according to certain criteria.
- It is easy to calculate.
- The associated statistical test does not formulate a basic **hypothesis** based on the shape of the distribution of the **population** from which the **samples** are taken.

The Spearman table gives the theoretical **values** of the Spearman rank correlation coefficient under the **hypothesis** of the **independence** of two **random variables**.

Here is a sample of the Spearman table for $n = 6, 7$, and 8 and $\alpha = 0.05$ and 0.025:

n	$\alpha = 0.05$	$\alpha = 0.025$
6	0.7714	0.8286
7	0.6786	0.7450
8	0.6190	0.7143

A complete Spearman table can be found in Glasser, G.J. and Winter, R.F. (1961).

EXAMPLES

In this example eight pairs of real twins take intelligence tests. The goal is to see if there is **independence** between the test of the first-born twin and that of the one born second. The data are given in the table below, the highest scores corresponding to the best results.

Pair of twins	Born 1st X_i	Born 2nd Y_i
1	90	88
2	75	79
3	99	98
4	60	66
5	72	64
6	83	83
7	86	86
8	92	95

The X are classified amongst themselves and the Y amongst themselves, and d_i is calculated. This gives:

Pair of twins	Born 1st X_i	R_{X_i}	Born 2nd Y_i	R_{Y_i}	d_i
1	90	6	88	6	0
2	75	3	79	3	0
3	99	8	98	8	0
4	60	1	66	2	−1
5	72	2	64	1	1
6	83	4	83	4	0
7	86	5	86	5	0
8	92	7	95	7	0

The Spearman rank correlation coefficient is then calculated:

$$\rho = 1 - \frac{6 \cdot \sum_{i=1}^{8} d_i^2}{n(n^2 - 1)} = 1 - \frac{6 \cdot 2}{8(8^2 - 1)}$$
$$= 0.9762 .$$

This shows that there is an (almost perfect) positive correlation between the intelligence tests.

Suppose that the results for 7 and 8 are changed. We then obtain the following table:

Pair of twins	Born 1st X_i	R_{X_i}	Born 2nd Y_i	R_{Y_i}	d_i
1	90	6.5	88	5.5	1
2	75	3	79	3	0
3	99	8	98	7.5	−0.5
4	60	1	66	2	−1
5	72	2	64	1	1
6	83	4.5	83	4	0.5
7	83	4.5	88	5.5	−1
8	90	6.5	98	7.5	−1

Since there are average ranks, we use the formula

$$\rho = \frac{S_x + S_y - \sum_{i=1}^{8} d_i^2}{2\sqrt{S_x \cdot S_y}}$$

to calculate the Spearman rank correlation coefficient. We first calculate S_x:

$$S_x = \frac{n\left(n^2 - 1\right) - \sum_{i=1}^{g}\left(t_i^3 - t_i\right)}{12}$$
$$= \frac{8\left(8^2 - 1\right) - \left[2^3 - 2 + 2^3 - 2\right]}{12}$$
$$= 41 .$$

Then S_y:

$$S_y = \frac{n\left(n^2 - 1\right) - \sum_{j=1}^{h}\left(t_j^3 - t_j\right)}{12}$$

$$= \frac{8\left(8^2 - 1\right) - \left[2^3 - 2 + 2^3 - 2\right]}{12}$$

$$= 41 \,.$$

The value of ρ becomes the following:

$$\rho = \frac{41 + 41 - (1 + 0 + \ldots + 1)}{2\sqrt{41 \cdot 41}} = \frac{76.5}{82}$$

$$= 0.9329 \,.$$

We see that in this case there is also an almost perfect positive correlation.

We now carry out the **hypothesis test**:

H_0: There is independence between the intelligence tests of a pair of twins.

H_1: There is a positive correlation between the intelligence tests.

We choose a **significant level** of $\alpha = 0.05$. Since we are in case B, H_0 is rejected if

$$\rho > t_{8,0.95} \,,$$

where $t_{8,0.95}$ is the **value** of the Spearman table, meaning if

$$\rho > 0.6190 \,.$$

In both cases ($\rho = 0.9762$ and $\rho = 0.9329$), H_0 is rejected. We can conclude that there is a positive correlation between the results of the intelligence tests of a pair of twins.

FURTHER READING
▸ **Hypothesis testing**
▸ **Nonparametric test**
▸ **Test of independence**

REFERENCES

Glasser, G.J., Winter, R.F.: Critical values of the coefficient of rank correlation for testing the hypothesis of independance. Biometrika **48**, 444–448 (1961)

Spearman, C.: The Proof and Measurement of Association between Two Things. Am. J. Psychol. **15**, 72–101 (1904)

Spearman Table

The Spearman table gives the theoretical values of the Spearman rank correlation coefficient under the **hypothesis** of **independence** of two random variables.

HISTORY

See **Spearman rank correlation coefficient**.

FURTHER READING
▸ **Spearman rank correlation coefficient**

Standard Deviation

The standard deviation is a **measure of dispersion**. It corresponds to the positive square root of the **variance**, where the variance is the mean of the squared **deviations** of each **observation** with respect to the **mean** of the set of observations.

It is usually denoted by σ when it is relative to a **population** and by S when it is relative to a **sample**.

In practice, the standard deviation σ of a **population** will be estimated by the standard deviation S of a **sample** of this population.

S

HISTORY

The term standard deviation is closely related to the works of two English mathematicians, **Pearson, Karl** and **Gosset, W.S.** It was indeed during a conference that he gave before the London Royal Society in 1893 that Pearson, K. used the term for the first time. He used it again in his article entitled "On the Dissection of Asymmetrical Frequency Curves" in 1894, and it was also Pearson, K. who introduced the symbol σ to denote the standard deviation. Gosset, W.S., (Student), also worked on these problems. He explained why it is important to distinguish S (standard deviation relative to a **sample**) from σ (standard deviation relative to a **population**).

In his article of March 1908, **Gosset, W.S.** defined the standard deviation of a **sample** by:

$$S = \sqrt{\frac{\sum_{i=1}^{n} (x_i - \bar{x})^2}{n}} \ .$$

A question that came up was to know if this expression should be divided by n or by $n - 1$. Pearson, K. asked the opinion of Gosset, W.S., who answered in a letter on 13 March 1927 that both formulas could be found in the literature. The one with an $n - 1$ denominator gives a mean **value** that is independent of the size of the **sample** and therefore of the **population**. On the other hand, with large samples, the difference between both formulas is negligible and the calculations are simpler if the formula with an n denominator is used.

On the whole, **Gosset, W.S.** indicates that the use of $n - 1$ is probably more appropriate for small **samples**, whereas the use of n is preferable for large samples.

Note that the discovery of the standard deviation is to be placed in the context of the theory of the **estimation** and of **hypothesis testing**. Also, at first the study of variability attracted the attention of astronomers because they were interested in discoveries related to the distribution of **errors**.

MATHEMATICAL ASPECTS

Consider a **random variable** X with an **expected value** $E[X]$. The **variance** σ^2 of the **population** is defined by:

$$\sigma^2 = E\left[(X - E[X])^2\right]$$

or

$$\sigma^2 = E\left[X^2\right] - E[X]^2 \ .$$

The standard deviation σ is the positive square root of σ^2:

$$\sigma = \sqrt{E\left[(X - E[X])^2\right]} \ .$$

To estimate the standard deviation σ of a **population**, the distribution of the population should be known, but in practice, it is convenient to estimate σ from a random **sample** $x_1, \ldots, x_n$ with the formula:

$$S = \sqrt{\frac{\sum_{i=1}^{n} (x_i - \bar{x})^2}{n - 1}}$$

where S is the **estimator** of σ and $\bar{x}$ is the **arithmetic mean** of the **sample**:

$$\bar{x} = \frac{\sum_{i=1}^{n} x_i}{n} \ .$$

The standard deviation can also be calculated using the following alternative formula:

$$S = \sqrt{\frac{\sum_{i=1}^{n} x_i^2 - n\bar{x}^2}{n-1}}$$

$$= \sqrt{\frac{n \sum_{i=1}^{n} x_i^2 - \left(\sum_{i=1}^{n} x_i\right)^2}{n(n-1)}}.$$

DOMAINS AND LIMITATIONS

The standard deviation is used as a scale of measurement in **tests of hypotheses** and **confidence intervals**. Other **measures of dispersion** such as the **range** or the **mean deviation** are also used in **descriptive statistics**, but they play a less important role in tests of hypotheses. It can be verified that.

$$\text{mean deviation} \leq S \leq \frac{\text{range}}{2} \cdot \sqrt{\frac{n}{n-1}}.$$

If we consider a **random variable** that follows a **normal distribution** of **mean** μ and standard deviation σ, we can verify that 68.26% of the **observations** are located in the **interval** $\mu \pm \sigma$. Also, 95.44% of the observations are located in the interval $\mu \pm 2\sigma$, and 99.74% of the observations are located in the interval $\mu \pm 3\sigma$.

EXAMPLES

Five students have successively passed two exams on which they obtained the following grades:

Exam 1:

$$\overline{3.5 \quad 4 \quad 4.5 \quad 3.5 \quad 4.5} \quad \bar{x} = \frac{20}{5} = 4$$

Exam 2:

$$\overline{2.5 \quad 5.5 \quad 3.5 \quad 4.5 \quad 4} \quad \bar{x} = \frac{20}{5} = 4$$

The **arithmetic mean** $\bar{x}$ of these two sets of **observations** is identical. Nevertheless, the dispersion of the observations around the mean is not the same.

To calculate the standard deviation, we first have to calculate the **deviations** of each **observation** with respect to the **arithmetic mean** and then square these deviations:

Exam 1:

Grade	$(x_i - \bar{x})$	$(x_i - \bar{x})^2$
3.5	−0.5	0.25
4	0.0	0.00
4.5	0.5	0.25
3.5	−0.5	0.25
4.5	0.5	0.25
$\sum_{i=1}^{5}(x_i - \bar{x})^2$		1.00

Exam 2:

Grade	$(x_i - \bar{x})$	$(x_i - \bar{x})^2$
2.5	−1.5	2.25
5.5	1.5	2.25
3.5	−0.5	0.25
4.5	0.5	0.25
4	0.0	0.00
$\sum_{i=1}^{5}(x_i - \bar{x})^2$		5.00

The standard deviation for each exam is equal to:

Exam 1:

$$S = \sqrt{\frac{\sum_{i=1}^{5}(x_i - \bar{x})^2}{n-1}} = \sqrt{\frac{1}{4}} = 0.5$$

Exam 2:

$$S = \sqrt{\frac{\sum_{i=1}^{5}(x_i - \bar{x})^2}{n-1}} = \sqrt{\frac{5}{4}} = 1.118$$

Since the standard deviation of the grades in the second exam is larger, the grades are more dispersed around the **arithmetic mean** than for the first exam. The variability of the second exam is therefore larger than that of the first.

FURTHER READING
- ▶ **Coefficient of variation**
- ▶ **Mean absolute deviation**
- ▶ **Measure of dispersion**
- ▶ **Variance**

REFERENCES

Barnard, G.A., Plackett, R.L., Pearson, S.: "Student": a statistical biography of William Sealy Gosset: based on writings by E.S. Pearson. Oxford University Press, Oxford (1990)

Pearson, K.: Contributions to the mathematical theory of evolution. I. In: Karl Pearson's Early Statistical Papers. Cambridge University Press, Cambridge, pp. 1–40 (1948). First published in 1894 as: On the dissection of asymmetrical frequency curves. Philos. Trans. Roy. Soc. Lond. Ser. A **185**, 71–110

Gosset, S.W. "Student": The Probable Error of a Mean. Biometrika **6**, 1–25 (1908)

Standard Error

The standard **error** is the square root of the estimated **variance** of a **statistic**, meaning the **standard deviation** of the **sampling distribution** of this statistic.

HISTORY

The standard error notion is attributed to **Gauss, C.F.** (1816), who apparently did not know the concept of **sampling distribution** of a **statistic** used to estimate the **value** of a **parameter** of a **population**.

MATHEMATICAL ASPECTS

The standard **error** of the **mean**, or **standard deviation** of the **sampling distribution** of the means, denoted by $\sigma_{\bar{x}}$, is calculated as a function of the standard deviation of the **population** and of the respective size of the finite population of the **sample**:

$$\sigma_{\bar{x}} = \frac{\sigma}{\sqrt{n}} \cdot \sqrt{\frac{N-n}{N}} \, ,$$

where σ is the standard deviation of the population, N is the size of the population, and n is the sample size.

If the size N of the population is infinite, the correction factor:

$$\sqrt{\frac{N-n}{N}}$$

can be omitted because it tends toward 1 when N tends toward infinity.

If the **population** is infinite, or if the **sampling** is nonexhaustive (with replacement), then the standard **error** of the **mean** is equal to the **standard deviation** of the **population** divided by the square root of the size of the **sample**:

$$\sigma_{\bar{x}} = \frac{\sigma}{\sqrt{n}} \, .$$

If the **standard deviation** σ of the **population** is unknown, it can be estimated by the standard deviation S of the **sample** given by the square root of the **variance** S^2:

$$S^2 = \frac{\sum_{j=1}^{n} \left(x_j - \bar{x}\right)^2}{n-1} \, .$$

The same formulas are valid for the standard **error** of a **proportion**, or **standard deviation** of the **sampling distribution** of the proportions, denoted by σ_p:

$$\sigma_p = \frac{\sigma}{\sqrt{n}} \cdot \sqrt{\frac{N-n}{N}}$$

in the case of a finite **population** and

$$\sigma_p = \frac{\sigma}{\sqrt{n}}$$

in the case of an infinite population, where σ is the **standard deviation** of the population:

$$\sigma = \sqrt{p \cdot (1 - p)} = \sqrt{p \cdot q},$$

where p is the **probability** that an element possesses the studied trait and $q = 1 - p$ is the probability that it does not.

The standard **error** of the difference between two independent quantities is the square root of the sum of the squared standard errors of both quantities. For example, the standard error of the difference between two **means** is equal to:

$$\sigma_{\bar{x}_1 - \bar{x}_2} = \sqrt{\sigma_{\bar{x}_1}^2 + \sigma_{\bar{x}_2}^2} = \sqrt{\frac{\sigma_1^2}{n_1} + \frac{\sigma_2^2}{n_2}},$$

where σ_1^2 and σ_2^2 are the respective **variances** of the two infinite **populations** to be compared and n_1 and n_2 the respective sizes of the two **samples**.

If the **variances** of the two **populations** are unknown, we can estimate them with the variances calculated on the two **samples**, which gives:

$$\sigma_{\bar{x}_1 - \bar{x}_2} = \sqrt{\frac{S_1^2}{n_1} + \frac{S_2^2}{n_2}},$$

where S_1^2 and S_2^2 are the respective **variances** of the two **samples**.

If we consider that the **variances** of the two **populations** are equal, we can estimate the value of these variances with a **pooled variance** S_p^2 calculated as a function of S_1^2 and S_2^2:

$$S_p^2 = \frac{(n_1 - 1) \cdot S_1^2 + (n_2 - 1) \cdot S_2^2}{n_1 + n_2 - 2},$$

which gives:

$$\sigma_{\bar{x}_1 - \bar{x}_2} = S_p \cdot \sqrt{\frac{1}{n_1} + \frac{1}{n_2}}.$$

The standard **error** of the difference between two **proportions** is calculated in the same way as the standard error of the difference between two **means**. Therefore:

$$\sigma_{p_1 - p_2} = \sqrt{\sigma_{p_1}^2 + \sigma_{p_2}^2}$$
$$= \sqrt{\frac{p_1 \cdot (1 - p_1)}{n_1} + \frac{p_2 \cdot (1 - p_2)}{n_2}},$$

where p_1 and p_2 are the **proportions** for infinite **populations** calculated on the two **samples** of size n_1 and n_2.

EXAMPLES

(1) *Mean standard error*

A manufacturer wants to test the precision of a new machine to make bolts 8 mm in diameter. On a lot of $N = 10000$ pieces, a **sample** of $n = 100$ pieces is examined; the **standarbd deviation** S is 1.2 mm.

The standard error of the **mean** is equal to:

$$\sigma_{\bar{x}} = \frac{\sigma}{\sqrt{n}} \cdot \sqrt{\frac{N - n}{N}}.$$

Since the **standard deviation** σ of the **population** is unknown, we can estimate it using the standard deviation of the **sample** S, which gives:

$$\sigma_{\bar{x}} = \frac{S}{\sqrt{n}} \cdot \sqrt{\frac{N - n}{N}}$$
$$= \frac{1.2}{10} \cdot \sqrt{\frac{10000 - 100}{10000}}$$
$$= 0.12 \cdot 0.995 = 0.119.$$

Note that the factor

$$\sqrt{\frac{N - n}{N}} = 0.995$$

S

has little influence on the result as the size of the **population** is large enough.

(2) *Standard error of a proportion*

A candidate conducts a survey on a **sample** of 200 people to know if he will have more than 50% (= π_0) of the vote, π_0 being the presumed value of **parameter** π (**proportion** of the **population**).

The standard error of the proportion σ_p is equal to:

$$\sigma_p = \sqrt{\frac{\pi_0 \cdot (1 - \pi_0)}{n}},$$

where n is the **sample** size (200 in our example). We can consider that the **population** is infinite, which is why we do not take into account the corrective factor:

$$\sigma_p = \sqrt{\frac{0.5 \cdot (1 - 0.5)}{200}} = 0.0354.$$

(3) *Standard error of difference between two quantities*

An insurance company decides to equip its offices with microcomputers. It wants to buy these microcomputers from two different suppliers as long as there is no significant difference in durability between the two brands. It tests a sample of 35 microcomputers of brand 1 and 32 of brand 2, noting the time that passed before the first breakdown. The observed data are the following:

Brand 1: Standard deviation $S_1 = 57.916$

Brand 2: Standard deviation $S_2 = 57.247$

Case 1:

We suppose that the **variances** of the two **populations** are equal. The pooled standard deviation is equal to:

$$S_p = \sqrt{\frac{(n_1 - 1) \cdot S_1^2 + (n_2 - 1) \cdot S_2^2}{n_1 + n_2 - 2}}$$

$$= \sqrt{\frac{34 \cdot 57.916^2 + 31 \cdot 57.247^2}{35 + 32 - 2}}$$

$$= 57.5979.$$

Knowing the weighted **standard deviation**, we can calculate the **value** of the standard deviation of the **sampling distribution** using the following formula:

$$\sigma_{\bar{x}_1 - \bar{x}_2} = S_p \cdot \sqrt{\frac{1}{n_1} + \frac{1}{n_2}}$$

$$= 57.5979 \cdot \sqrt{\frac{1}{35} + \frac{1}{32}} = 14.0875.$$

Case 2:

If we suppose the **variances** of the two **populations** to be unequal, we must start by calculating the standard error of the **mean** for each **sample**:

$$\sigma_{\bar{x}_1} = \frac{S_1}{\sqrt{n_1}} = \frac{57.916}{\sqrt{35}} = 9.79,$$

$$\sigma_{\bar{x}_2} = \frac{S_2}{\sqrt{n_2}} = \frac{57.247}{\sqrt{32}} = 10.12,$$

$$\sigma_{\bar{x}_1 - \bar{x}_2} = \sqrt{\sigma_{\bar{x}_1}^2 + \sigma_{\bar{x}_2}^2}$$

$$= \sqrt{9.79^2 + 10.12^2}$$

$$= 14.08.$$

We notice in this example that the two cases give values of the standard error that are almost identical. This is explained by the very slight difference between the **standard deviations** of the two **samples** S_1 and S_2.

FURTHER READING

▶ **Hypothesis testing**

▶ **Sampling distribution**

▶ **Standard deviation**

REFERENCES

Box, G.E.P., Hunter, W.G., Hunter, J.S.: Statistics for Experimenters. An Introduction to Design, Data Analysis, and Model Building. Wiley, New York (1978)

Cochran, W.G.: Sampling Techniques, 2nd edn. Wiley, New York (1963)

Gauss, C.F.: Bestimmung der Genauigkeit der Beobachtungen, vol. 4, pp. 109–117 (1816). In: Gauss, C.F. Werke (published in 1880). Dieterichsche Universitäts-Druckerei, Göttingen

Gauss, C.F.: Gauss' Work (1803–1826) on the Theory of Least Squares. Trotter, H.F., trans. Statistical Techniques Research Group, Technical Report, No. 5, Princeton University Press, Princeton, NJ (1957)

Standardized Data

Standardized data are data from which we subtract the mean of the observations related to the **random variable** (associated to the data) and then divide by the standard deviation of the observations. Thus standardized data have a mean of 0 and a variance of 1.

MATHEMATICAL ASPECTS

The standardization of a data series X with the terms x_i, $i = 1, \ldots n$ is the same as replacing the term x_i of the series by x_i^s:

$$x_i^s = \frac{x_i - \bar{x}_i}{s},$$

where $\bar{x}_i$ and s are, respectively, the mean and the standard deviation of x_i.

DOMAINS AND LIMITATIONS

Certain statisticians have a habit of standardizing explanatory variables in a linear regression model to simplify the numerical difficulties arising from matrix calculations and to facilitate the interpretation and comparison of the regression coefficients. The application of the least-squares method using standardized data gives results equivalent to those obtained using the original data. A regression is generally modeled by

$$Y_i = \sum_{j=1}^{p} \beta_j X_{ij}, \quad i = 1, \ldots, n,$$

where X_j is the nonstandardized data. If X_{ij} is standardized (mean 0 and variance 1), β_j is called the standardized regression coefficient and is equivalent to the correlation if Y is also standardized. The same terminology can be used if X_j is reduced to having a variance of 1, but not necessarily a zero mean. The coefficient β_0 absorbs the difference.

EXAMPLES

In the following example, the observations relate to 13 mixtures of cement. Each mixture is composed of four ingredients given in the table. The goal of the experiment is to determine how the quantities X_1, X_2, X_3, and X_4 of these four ingredients influence the quantity of heat Y emitted by the hardening of the cement.

Table: Heat emitted by cement

Mix-ture i	Ingredient				Heat
	X_1	X_2	X_3	X_4	Y
1	7	26	6	60	78.5
2	1	29	15	52	74.3
3	11	56	8	20	104.3
4	11	31	8	47	87.6
5	7	52	6	33	95.9
6	11	55	9	22	109.2
7	3	71	17	6	102.7
8	1	31	22	44	72.5

S

Mix-ture i	Ingredient				Heat
	X_1	X_2	X_3	X_4	Y
9	2	54	18	22	93.1
10	21	47	4	26	115.9
11	1	40	23	34	83.9
12	11	66	9	12	113.3
13	10	68	8	12	109.4

Source: Birkes and Dodge (1993)

y_i quantities of heat given by hardening of ith mixture (in joules);

x_{i1} quantity of ingredient 1 (aluminate of tricalcium) in ith mixture;

x_{i2} quantity of ingredient 2 (silicate of tricalcium) in ith mixture;

x_{i3} quantity of ingredient 3 (aluminoferrite of tetracalcium) in ith mixture;

x_{i4} quantity of ingredient 4 (silicate of dicalcium) in ith mixture.

The obtained model of a simple linear regression is:

$$\widehat{Y} = 62.4 + 1.55X_1 + 0.51X_2$$
$$+ 0.102X_3 - 0.144X_4 + \varepsilon.$$

The following table presents data on the cement with data relative to the standardized explanatory variables, which we denote by $X_1^s, X_2^s, X_3^s,$ and X_4^s.

Data on cement with standardized explanatory variables

Mix-ture i	Standardized ingredient				Heat
	X_1^s	X_2^s	X_3^s	X_4^s	Y
1	-0.079	-1.424	-0.901	1.792	78.5
2	-1.099	-1.231	0.504	1.314	74.3
3	0.602	0.504	-0.589	-0.597	104.3
4	0.602	-1.102	-0.589	1.016	87.6
5	-0.079	0.247	-0.901	0.179	95.9
6	0.602	0.440	-0.432	-0.478	109.2
7	-0.759	1.468	0.817	-1.434	102.7
8	-1.099	-1.102	1.597	0.836	72.5

Mix-ture i	Standardized ingredient				Heat
	X_1^s	X_2^s	X_3^s	X_4^s	Y
9	-0.929	0.376	0.973	-0.478	93.1
10	2.302	-0.074	-1.213	-0.239	115.9
11	-1.099	-0.524	1.753	0.239	83.9
12	0.602	1.147	-0.432	-1.075	113.3
13	0.432	1.275	-0.589	-1.075	109.4

The estimation by the least-squares method of the complete model using the standardized explanatory variables gives:

$$\widehat{Y} = 95.4 + 9.12X_1^s + 7.94X_2^s$$
$$+ 0.65X_3^s - 2.41X_4^s.$$

Note that the coefficient $\widehat{\beta}_1^s = 9.12$ of X_1^s can be calculated by multiplying the coefficient $\widehat{\beta}_1 = 1.55$ of X_1 in the initial model by the standard deviation $s_1 = 5.882$ of X_1. Also note that the mean $\overline{X}_1 = 7.46$ of X_1 is considered not in the relation between $\widehat{\beta}_1^s$ and $\widehat{\beta}_1$ but in the relation between $\widehat{\beta}_0^s$ and $\widehat{\beta}_0$. Taking only the first two standardized explanatory variables, we obtain the solution

$$\widehat{Y} = 95.4 + 8.64X_1^s + 10.3X_2^s.$$

FURTHER READING
▶ **Data**
▶ **Ridge regression**

REFERENCES
Birkes, D., Dodge, Y.: Alternative Methods of Regression. Wiley, New York (1993)

Statistical Software

Statistical software is a set of computer programs and procedures for the treatment of statistical **data**.

Libraries

A library is a series of programs or subroutines that are installed under the same operating system (DOS, Windows, or Unix) and can be used by entering individual commands.

IMSL (International Mathematics and Statistics Library): a collection of about 540 subroutines written in FORTRAN specifically concerning mathematics and statistics.

NAG (Numerical Algorithms Group): a library of algorithms written with the help of three languages (ALGOL 60, ALGOL 68, and FORTRAN ANSI). We find here more than 600 advanced subroutines concerning **simulation**, regression analysis, techniques of **optimization**, GLIMs (general linear models), GenStat, time series, and graphical representations. Moreover, we can treat in a simple manner answers obtained by census.

Software

A software is a set of complex algorithms representing a common structure of data and requiring a minimum of previous programming on the part of the user.

BMDP: a set of subroutines written in FORTRAN that allow one to treat enquiries made by **survey**, to generate **graphical representations**, to apply multivariate techniques or **nonparametric tests**, or to carry out a linear or nonlinear **regression** analysis without forgetting the study of **time series**.

IMSL (International Mathematics and Statistics Library): set of subroutines written in FORTRAN specifically concerning the fields of mathematics and **statistics**.

MINITAB: by far the easiest system to use. Since it can be used in an interactive mode, it allows to carry out analyses of various types: **descriptive statistics**, **analysis of variance**, **regression** analysis, **nonparametric tests**, **random-number generation**, or the study of **time series**, among other things.

NAG: a library of **algorithms** written with the help of three languages (ALGOL 60, ALGOL 68, FORTRAN ANSI). Advanced subroutines can be found concerning **simulation**, **regression** analysis, **optimization** techniques, **time series**, and **graphical representations**. Moreover, the answers obtained by **survey** can be easily treated, and multivariate analysis as well as **nonparametric tests** can be applied to them.

P-STAT: an interactive system that treats **designs of experiment**, **graphical representations**, **nonparametric tests**, and **data analysis**.

SAS (Statistical Analysis System): a set of software including a basic SAS language as well as specific programs. Surely the most complete package, it allows to analyze **data** from very diverse fields.

SPSS (Statistical Package for the Social Sciences): one of the most commonly used systems. It offers interactive ways of treating large **data** bases. Also, some **algorithms** of **graphical representations** are included in the program.

BMDP: basically for biomedical applications, BMDP is a complete software kit composed of subroutines written in FORTRAN that allows to treat inquiries made by **census**, to generate graphical representations, to apply multivariate techniques and nonparametric tests, or to perform a linear or nonlinear **regression analysis**, without neglecting studies of time series. Each subprogram is based on highly competitive numerical algorithms.

DataDesk: an interactive software for analyzing statistical treatments of data. It depicts tools graphically, allowing one to see relations among data, tendencies, subgroups, and outliers.

EViews: software essentially for economists, EViews is the tool of choice for analysis, forecasting, and econometric modeling. EViews can be used for general statistical analysis, estimating time series, large-scale simulations, graphics, or simple data management.

Excel: allows the user to represent and analyze data with the help of basic statistical methods: descriptive statistics, analysis of variance, regression and correlation, and hypothesis testing.

Gauss: a statistical program with oriented matrices, Gauss is a powerful econometric and graphic tool.

JMP: statistical software made for experimental designs in particular, JMP also contains an exhaustive suite of statistical tools.

Lisrel: used in the social sciences and, more specifically, for factorial analysis and modeling, Lisrel is particularly useful for modeling the representation of latent variables.

Maple: a complete mathematical environment designed for the manipulation of algebraical expressions and the resolution of equations and integrals; it also has very powerful graphic tools in two and three dimensions. The user can also program in Maple.

Mathematica: statistical suite that allows to perform descriptive univariate and multivariate statistical analysis, data smoothing, classical hypothesis testing, confidence interval estimation, and linear and nonlinear regression.

Matlab: an interactive environment that can be used to analyze scientific and statistical data. The basic objects of Matlab are matrices. The user can perform numerical analyses, treatment of signs and image treatment. The user can program the functions using C and FORTRAN.

MIM: a program for graphic modeling of discrete and continuous data. Among the families of available graphics programs in this software suite, we mention log-linear models, Gaussian graphic models, Manova standard models, and many other models useful for multivariate analysis.

P-STAT: created in 1960 for the treatment of pharmacological data, P-STAT is now used by researchers in the social sciences. It allows to carry out an analysis of censuses. The power of P-STAT lies in its macros, and it is very useful for the statistical applications related to business. It allows to treat large data sets and treats experimental designs, graphical representations, nonparametric tests, and **data analysis**.

R programming language similar to S, R is also freeware widely used in academia. It represents an implementation different from S, but many S commands work in R. R contains a wide variety of statistical commands (linear and nonlinear modeling, execution of classical tests, analysis of time series, classification, segmentation, ...) as well as graphical techniques.

SAS (Statistical Analysis System): a suite of software programs including a basic SAS language as well as specific programs. No doubt the most complete package, it allows to analyze data of various origins: research, engineering, medicine, and business applications. It includes a significant number of statistical tools: analysis of variance, regression analysis, categorical data analysis, multivariate analysis, survival analysis, decision trees, nonparametric analysis, and "data mining".

S-plus: software written in a powerful and flexible object-oriented programming language called S, it allows for matrix, mathematical, and statistical calculations. It includes a tool for exploratory visual analysis that also allows for advanced data analysis as well as their modeling. S-plus allows geostatisticians to put together a block "S+SpatialStats" containing different tools of **spatial statistics**.

Spad-T: based purely on statistical techniques of multivariate analysis. Contingency tables of treated variables are created with the help of multifactorial

methods giving way to graphical representations.

SPSS (Statistical Package for the Social Sciences): first released in 1968, SPSS is actually one of the most widely used programs by statisticians in various domains. It offers interactive means for the treatment of large data bases. It contains algorithms for graphical representations.

Stata: a statistical software allowing one to carry out statistical analysis from linear modeling on a generalized linear model, treat binary data, use nonparametric methods, conduct multivariate analysis and analysis of time series, and utilize methods of graphical data exploration. Moreover, the software allows to program the user-specific commands. Stata contains, among other things, statistical tools for epidemiologists.

Statistica: first released in 1993, Statistica is an extremely complex software suite that contains powerful exploration tools for large data bases. The basic software includes basic statistics, analysis of variance, nonparametric statistics, linear and nonlinear models, techniques of exploratory multivariate analysis, neural nets, calculation of sample dimensions, experimental design analysis tools, quality control tools, etc.

StatView: statistical software for medical scientists and researchers. StatView allows to produce descriptive statistics, perform parametric and nonparametric hypothesis testing, calculate correlations and covariances, carry out linear, nonlinear, and logistic regression, compile contingency tables, and perform factorial analysis, survival analysis,

S

quality control, and Pareto analysis. A good graphical interface is available.

Systat: powerful graphics software. Allows for the visualization of data as well as the use of basic statistics to find an adequate model.

Solas: software treating missing data. Solas integrates algorithms of multiple imputation.

Vista: designed for those starting out in statistics as well as for professors. Vista can be used as a tutorial software in the domain of univariate, multivariate, graphic, and calculatory statistics.

WinBugs: a program of Bayesian data modeling based on Monte Carlo simulation techniques. It includes the representation of Bayesian models with commands that allow to carry out simulations. Moreover, an interface is available allowing one to control the analysis at any moment. The graphic tools allow one to control the convergence of the simulation.

XploRe: a wide-ranging statistical environment made for exploratory data analysis. It is programmed in the matrix-oriented language, and as such it contains a large group of statistical operations and has a powerful interactive interface. The program supports user-defined macros.

HISTORY

The first effort made by statisticians to simplify their statistical calculations was the construction of statistical tables. With the arrival of computers in the mid-1940s, the encoding of statistical subroutines took place. The programming languages at this time were not widely known to researchers, and they had to wait till the beginning of 1960, when the programming language FORTRAN was released, for the development of statistical software programs. Among the first routines introduced by IBM, the best known is the pseudorandom number generator RANDU. The systematic development of statistical software began in 1960 with the creation of the programs BDM and SPSS, followed by SAS.

REFERENCES

Asimov, D.: The grand tour: a tool for viewing multidimensional data. SIAM J. Sci. Stat. Comput. **6**(1), 128–143 (1985)

Becker, R.A., Chambers, J.M., Wilks, A.R.: The new S Language. Waldsworth and Brooks/Cole, Pacific Grove, CA (1988)

Dixon, W.J. chief ed.: BMDP Statistical Software Communications. The University of California Press. Berkeley, CA (1981)

Chambers, J.M and Hastie, T.K (eds.): Statistical models in S. Waldsworth and Brooks/Cole, Pacific Grove, CA (1992)

Dodge, Y., Hand, D.: What Should Future Statistical Software Look like? Comput. Stat. Data Anal. (1991)

Dodge, Y., Whittaker, J. (eds.): Computational statistics, vols. 1 and 2. COMPSTAT, Neuchâtel, Switzerland, 24–28 August 1992 (English). Physica-Verlag, Heidelberg (1992)

Edwards, D.: Introduction to Graphical Modelling. Springer, Berlin Heidelberg New York (1995)

Francis, I.: Statistical software: a comparative rewiew. North Holland, New York (1981)

Hayes, A.R.: Statistical Software: A survey and Critique of its Development. Office of Naval Research, Arlington, VA (1982)

International Mathematical and Statistical Libraries.: IMSL Library Information. IMSLK, Houston (1981)

Ripley, B.D., Venables, W.N. (2000) S programming. Springer, Berlin Heidelberg New York

Statistical Sciences: Statistical Analysis in S-Plus, Version 3.1. Seattle: StatSci, a division of MathSoft (1993)

Venables, W.N., Ripley, B.D.: Modern Applied Statistics with S-PLUS. Springer, Berlin Heidelberg New York (1994)

Wegman, E.J., Hayes, A.R.: Statistical Software. Encycl. Stat. Sci. **8**, 667–674 (1988)

EXAMPLES

The **normal table**, the **Student table**, Fisher's tables, and the chi-square are examples of statistical tables.

FURTHER READING
▶ **Binomial table**
▶ **Chi-square table**
▶ **Distribution function**
▶ **Fisher table**
▶ **Kruskal-Wallis table**
▶ **Normal table**
▶ **Student table**
▶ **Wilcoxon signed table**
▶ **Wilcoxon table**

REFERENCES
Kokoska, S., Zwillinger, D.: CRC Standard Probability and Statistics Tables and formulae. CRC Press (2000)

Statistical Table

A statistical table gives values of the **distribution function** or of the individual **probability** of a **random variable** following a specific **probability distribution**.

DOMAINS AND LIMITATIONS

The necessity of a statistical table comes from the fact that certain distribution functions cannot be expressed in an easy-to-use mathematical form. In other cases, it is impossible (or too difficult) to find a primitive of the **density function** for the direct calculation of the corresponding distribution function.

The values of the distribution function are calculated in advance and represented in a statistical table to simplify use.

Statistic

Statistic is the result of applying a function to a sample of data where the function itself is independent of the sampling distribution.

MATHEMATICAL ASPECTS

A statistic is an observable random variable, this differentiates it from a parameter. Completeness, sufficiency and unbiasedness are important desirable properties of statistics.

EXAMPLES

arithmetic mean, variance, standard deviation.

FURTHER READING
▶ **Chi-square test of independence**
▶ **Percentile**

▶ *p*-value
▶ Quantile
▶ Student test

Statistics

The word statistics, derived from Latin, refers to the notion of state (status): "What is relative to the state". Governments have a great need to count and measure numerous events and activities, such as changing demographics, births, immigration and emigration trends, changes in employment rates, businesses, etc.

In this perspective, the term "statistics" is used to indicate a set of available data about a given phenomenon (e. g., unemployment statistics).

In the more modern and more accurate sense of the word, "statistics" is considered a discipline that concerns itself with numeric data. It is made up of a set of techniques for obtaining knowledge from incomplete data, from a rigorous scientific system for managing data collection, their organization, analysis, and interpretation, when it is possible to present them in numeric form.

In a population of individuals, we would like to know, in terms of statistical theory, if a given individual has a car or if he smokes. On the other hand, we would like to know how many individuals have a car and are smokers, and if there is a **relation** between possessing a car and smoking habits in the studied population.

We would like to know the traits of the population globally, without concerning ourselves with each person or each object in the population.

We distinguish two subsets of techniques: (1) those involving **descriptive statistics** and (2) those involving **inferential statistics**. The essential goal of descriptive statistics is to represent information in a comprehensible and usable format. Inferential statistics, on the other hand, aims to facilitate the generalization of this information or, more specifically, to make inferences (concerning populations) based on samples of these populations.

HISTORY

The term "statistics", derived from the Latin "status" (state), was used for the first time, according to Kendall, M.G. (1960), by a historian named Ghilini, Girolamo in 1589.

According to Kendall, M.G. (1960) the true origin of modern statistics dates back to 1660.

Statistical methods and statistical distributions developed principally in the 19th century thanks to the important role of statistics in experimental and human sciences.

In the 20th century, statistics became a separate discipline because of the wealth and diversity of methods that it uses. This also explains the birth of specialized statistical encyclopedias such as the *International Encyclopedia of Statistics* (1978) published in two volumes and the *Encyclopedia of Statistical Sciences* (1982–1988) in nine volumes.

DOMAINS AND LIMITATIONS

Statistics is used in many domains, all of them very different from one another. Thus we find it being used in industrial production in the environmental and medical Science, as well as in official agencies.

Let us look at some of the areas that might interest a statistician:

- *Fundamental research in statistics*: research in probabilities, statistical methods, and theory.

- *Biology*: fundamental research and experiments related to the principal phenomena of living organisms and biometry.

- *Commerce*: data management, sales volume, management, inventory methods, industrial planning, communication and theoretical control, and accounting procedures.

- **Demography**: study of the increase in human population (birth and death rate, migratory movement), study of the structure of populations (personal, social, and economic characteristics).

- *Economy*: measure of the volume of production, commerce, resources, employment, and standard of living; analysis of the consumer and manufacturer behavior, responses of the market to price changes, impact of advertising, and government policies.

- *Education*: measures and tests related to the process of studying.

- *Engineering*: research and experiments including design and efficiency testing, improvements in test methods, questions related to **inference** and precision tests, improvements in quality control methods.

- *Health*: cost of accidents, medical care, hospitalization.

- *Insurance*: determining mortality and accident rates of insured people and of the general population; determining the prime rate for each **category** of risk; application of statistical techniques to insurance policy management.

- *Marketing and research on consumption*: problems related to markets, the distribution system, detection of trends; study of consumer preferences and behavior.

- *Medicine:* epidemiology, fundamental research and experiments on the causes, diagnosis, treatment, and prevention of illnesses.

- *Management and administration*: problems related to personnel management, equipment, methods of production, and work conditions.

- *Psychology and psychometry*: problems related to the evaluation of the study capacity, intelligence, personal characteristics, and normal and abnormal behavior of an individual as well as the establishment of scales of measurement and instruments of measure.

- *Social sciences*: establishment of techniques of **sampling** and theoretical tests concerning social systems and societal well-being. Analysis of the cost of the social sciences; analysis of cultural differences on the level of values and social behavior.

- *Space research*: interpretation of experiments and analysis of data collected during space missions.

- *Sciences (in general)*: fundamental research in the natural and social sciences.

FURTHER READING
- ▶ **Bayesian statistics**
- ▶ **Descriptive statistics**
- ▶ **Genetic statistics**
- ▶ **Inferential statistics**
- ▶ **Official statistics**
- ▶ **Spatial statistics**

REFERENCES

Kendall, M.G.: Where shall the history of Statistics begin? Biometrika **47**, 447–449 (1960)

Kotz, S., Johnson, N.L.: Encyclopedia of Statistical Sciences, vols. 1–9. Wiley, New York (1982–1988)

Kruskal, W.H., Tanur, J.M.: International Encyclopedia of Statistics, vols. 1–2. The Free Press, New York (1978)

Stigler, S.: The History of Statistics, the Measurement of Uncertainty Before 1900. Belknap, London (1986)

Stem and Leaf

See **stem-and-leaf diagram**.

Stem-And-Leaf Diagram

The stem-and-leaf diagram is a graphical representation of **frequencies**. The basic idea is to provide information on the **frequency distribution** and retain the **values** of the **data** at the same time. Indeed, the stem corresponds to the class **intervals** and the leaf to the number of **observations** in the class represented by the different **data**. It is then possible to directly read the **values** of the data.

HISTORY

The origin of the stem-and-leaf diagram is often associated with Tukey, J.W. (1977). Its concept is based on the **histogram**, which dates back to the 18th century.

MATHEMATICAL ASPECTS

To construct a stem-and-leaf diagram, each number must be cut into a main part (stem) and a secondary part (leaf). For example, 4.2, 42, or 420 can be separated into 4 for the stem and 2 for the leaf. The stems are then listed vertically (one line per stem) in order of magnitude and the leaves are written next to them, also in order of magnitude. If a set of **data** includes the **values** 4.2, 4.4, 4.8, 4.4, and 4.1, then the stem and leaf are as follows:

$$4|12448.$$

DOMAINS AND LIMITATIONS

The stem-and-leaf diagram provides information about distribution, symmetry, concentration, empty sets, and **outliers**.

Sometimes there are too many numbers with the same stem. In this case it is not possible to put them all on the same line. The most common solution consists in separating the ten leaves into two groups of five leaves, e.g., 4|0122444 and 4|567889.

It is possible to compare two series of **data** by putting the two leaves to the left and right of the same stem:

$$012244|4|567889.$$

EXAMPLES

Per-capita GNP – 1970 (in dollars)

AFRICA			
Algeria	324	Mauritius	233
Angola	294	Morocco	224
Benin	81	Mozambique	228
Botswana	143	Niger	90
Burundi	67	Nigeria	145
Central African Rep.	127	Reunion	769
Chad	74	Rwanda	60
Comores	102	Senegal	236

AFRICA			
Congo	238	Sierra Leone	167
Egypt	217	Somalia	89
Equat. Guinea	267	South Africa	758
Ethiopia	72	Rhodesia (South)	283
Gabon	670	Sudan	117
Gambia	101	Swaziland	270
Ghana	257	Togo	135
Guinea	82	Tunisia	281
Guinea-Bissau	259	Uganda	135
Ivory Coast	347	United Rep. of Cameroon	185
Kenya	143	United Rep. of Tanzania	100
Lesotho	74		
Liberia	268		
Madagascar	133	High-Volta	59
Malawi	74	Zaire	88
Mali	54	Zambia	421
Mauritania	165		

Source: Annual Statistics of the UN, 1976

The above table gives the per-capita GNP, in dollars, for most African countries in 1970. To establish a stem-and-leaf diagram from these **data**, they are first classified in order of magnitude. The stem is then chosen: in this example we will use hundreds. The tens number is then placed to the right of the corresponding hundred, which gives the leaf part of the diagram. The units are not taken into account.

The stem and leaf are as follows:

Unit: 1	1 = 110
0	5566777788889
1	00012333444668
2	12233355667889

Unit: 1	1 = 110
3	24
4	2
5	
6	7
7	56

FURTHER READING
▶ **Frequency distribution**
▶ **Graphical representation**
▶ **Histogram**

REFERENCES
Tukey, J.W.: Explanatory Data Analysis, limited preliminary edition. Addison-Wesley, Reading, MA (1970–1971)

Stochastic Process

A stochastic process is a family of random variables. In practice, it serves to model a large number of temporal phenomena where chance comes into play.

We distinguish many types of stochastic processes using certain mathematical properties. The best known are:

• Markov process (or Markov chain)
• Martingale
• Stationary process
• Process of independent increments

HISTORY
Stochastic processes originally resulted from advances in the early 20th century in certain applied branches of statistics such as statistical mechanics (by Gibbs, J.W., Boltzmann, L., Poincaré, H., Smoluchowski, M., and Langevin, P.). The theoretical foundations were formulated later by Doob, J.L., Kolmogorov, A.N., and others (1930–1940). During this period the word "stochastic",

S

from the Greek word for "guess, believe, imagine", started to be used. Other advances were made starting from a Brownian movement in physics (by Einstein, A., Levy, P., and Wiener, N.).

The Markov process was invented by **Markov, Andrei Andreevich** (1856–1922), who laid the foundation for the theory of Markov processes in finite time, generalized by the Russian mathematician Kolmogorov, A. (1936).

MATHEMATICAL ASPECTS

A stochastic process is a family $(X_t)_{t \in I}$ (where I is a discrete or continuous set) of random variables defined on the same probability space. The processes *in discrete time* are represented by the following families $(X_t)_{t \in \mathbb{N}}$ (that is with $I = \mathbb{N}$) and the processes *in continuous time* by the families $(X_t)_{t \in \mathbb{R}_+}$ (that is, with $I = \mathbb{R}_+ = [0, \infty)$).

The term "stochastic process" is used if set I is infinite. In physics, the elements of I represent time.

For an **event** w belonging to the **fundamental set** Ω, we define the *trajectory* of the studied process as being the following family:

$$(X_t(w))_{t \in I} ,$$

so that the "series" (for a process in discrete time) of the values of the random variables is taken for a chosen and fixed element.

The principal objects of the study of random process are the properties of random variables and trajectories.

We characterize the different types of processes relative to the probabilistic conditions to which they are related.

1. *Markov process*

    A Markov process is a random process $(X_t)_{t \in \mathbb{N}}$ in discrete time for which the

probability of an event (in time $t + 1$) depends only on the immediate past (time t).

This is mathematically expressed by the following condition:

For each $t \in \mathbb{N}$ and for each $i_0, i_1, \ldots, i_{t+1} \in \mathbb{R}_+$ we have:

$$P(X_{t+1} = i_{t+1} | X_t = i_t,$$
$$X_{t-1} = i_{t-1}, \ldots, X_0 = i_0)$$
$$= P(X_{t+1} = i_{t+1} | X_t = i_t) .$$

2. *Martingale*

    A martingale is a random process $(X_t)_{t \in \mathbb{N}}$ in discrete time that verifies the following condition:

    For each $t \in \mathbb{N}$:

$$E(X_{t+1} | X_t, X_{t-1}, \ldots, X_0) = X_t .$$

3. *Stationary process*

    A stationary process (in the strict sense of the term) is a random process $(X_t)_{t \in \mathbb{N}}$ for which the joint distributions of the process depend only on the time difference (between the considered variables in the joint distribution) and not on the "absolute" time. Mathematically:

    For each $t_1, t_2, \ldots, t_k \in I (k \geq 2)$ and for each $a \in I$, $X_{t_1}, X_{t_2}, \ldots, X_{t_k}$ have the same joint distribution as $X_{t_1+a}, X_{t_2+a}, \ldots, X_{t_k+a}$.

4. *Process with independent increments*

    A process with independent increments is a random process $(X_t)_{t \in I}$ for which:

    For each $t_1 < t_2 < \ldots < t_k \in I$: $X_{t_2} - X_{t_1}, X_{t_3} - X_{t_2}, \ldots, X_{t_k} - X_{t_{k-1}}$ the random variables are reciprocally independent.

DOMAINS AND LIMITATIONS

Theoretically, there is no difference between the study of stochastic processes and statistics itself. In practice, the study of stochastic processes focuses more on the structure and properties of models than on inferences from real data, which is properly the domain of statistics.

EXAMPLES

1. Here we give an example that shows the close mathematical relations between statistics and stochastic processes. We consider the random variable X_i = age of individual i of a population (potentially infinite). A typical problem is to estimate the mean $E(X)$ with the following hypothesis:
   - Random variable X_i is independent.
   - All X_i are identically distributed.
2. *Random walk*. Let Z_n, $n \geq 2$, be independent and identically distributed random variables taking the value $+1$ with probability p and the value -1 with probability $q = 1 - p$

We define the stochastic process of the random walk as follows:

$$X_0 = 0, X_t = \sum_{n=1}^{t} Z_n \quad \text{if } t > 1.$$

A sample of a random walk is typically represented as follows:

$$(0, 1, 2, 1, 0, -1, 0, -1, -2, -3, -2, \ldots).$$

It is easy to see that the defined stochastic process is a Markov process (this fact is a direct consequence of the independence of X_t) with the following probabilities:

$$P(X_{t+1} = i + 1 \,|\, X_t = i) = p,$$
$$P(X_{t+1} = i - 1 \,|\, X_t = i) = q.$$

We have that:

$$E(X_{t+1} \,|\, X_t) = X_t + (+1) \cdot p + (-1) \cdot q$$
$$= X_t + (p - q).$$

The random walk is a martingale if and only if $p = q = 0.5$.

FURTHER READING
▶ Convergence
▶ Random variable
▶ Statistics
▶ Time series

REFERENCES
Barlett, M.S.: An Introduction to Stochastic Processes. Cambridge University Press, Cambridge (1960)

Billingsley, P.: Statistical Inference for Markov Processes. University of Chicago Press, Chicago (1961)

Cox, D.R., Miller, H.D.: Theory of Stochastic Processes. Wiley, New York; Methuen, London (1965)

Doob, J.L.: What is a stochastic process? Am. Math. Monthly **49**, 648–653 (1942)

Doob, J.L.: Stochastic Processes. Wiley, New York (1953)

Kolmogorov, A.N.: Math. Sbornik. N.S. **1**, 607–610 (1936)

Stratified Sampling

In stratified **sampling**, the **population** is first divided into subpopulations called strata. These must not interpenetrate each other, and the set of these strata must constitute the whole population.

Once the strata have been determined, a random **sample** is taken (not necessarily of the same size) from each stratum, this sampling being done independently in the different strata.

HISTORY
See **sampling**.

MATHEMATICAL ASPECTS

With respect to determining the number individuals to be taken from each stratum, there are mainly two approaches: the **representative** stratified **sample** and the optimal stratified sample.

Representative Stratified Sample

In each stratum, a number of individuals proportional to the magnitude of the stratum is taken.

Let N_i be the magnitude of stratum i, N the number of individuals of the **population**, and n the size of the global **sample**. The size of a subsample n_i is defined by:

$$n_i = \frac{N_i}{N} \cdot n, \quad i = 1, 2, \ldots, k,$$

where k is the number of strata in the **population**.

Optimal Stratified Sample

The goal is to determine the sizes of the subsamples to be taken to obtain the best possible **estimation**. This is done by minimizing the **variance** of the **estimator** m under the constraint

$$\sum_{i=1}^{k} n_i = n,$$

where m is the **mean** of the **sample** defined by:

$$m = \frac{1}{N} \sum_{i=1}^{k} N_i \bar{x}_i,$$

where $\bar{x}_i$ is the **mean** of subsample i. This is a classical **optimization** problem and gives the following result:

$$n_i = \frac{n}{\sum\limits_{i=1}^{k} N_i \cdot \sigma_i} N_i \cdot \sigma_i,$$

where σ_i is the **standard deviation** of stratum i.

Given a constant factor $n / \sum\limits_{i=1}^{k} N_i \cdot \sigma_i$, the size of the subset is proportional to the magnitude of the stratum and to the **standard deviation**.

The larger the dispersion of the studied variable in the stratum and the larger the size of N_i, the larger the size of subsample n_i.

DOMAINS AND LIMITATIONS

Stratified **sampling** is based on the idea of obtaining, as a result of controlling certain variables, a **sample** that recreates an image that is as close as possible to the **population**. The dispersion in the **population** may be strong; in this case, to maintain the **sampling error** at an acceptable level, a very large **sample** must be created, which can increase the cost considerably. To avoid this problem, stratified sampling can be a good alternative. The population can be subdivided into strata in such a way that the variation within the strata is relatively low, which will reduce the required **sample size** for a given sampling error.

Stratified **sampling** may be justified for the following reasons:

1. If the **population** is too heterogeneous, it is preferable to have more homogeneous groups to obtain a **sample** that is more representative of the population.
2. If one wants to obtain information on particular aspects of a **population** (e. g., with respect to different states), stratified sampling is better suited for this.

A control character used to define the strata must obey the following rules:

- It must be in close correlation with the studied variables.

- It must have a known **value** for every unit of the **population**.

Some of the most commonly used control variables are gender, age, region, socioprofessional class, etc.

EXAMPLES

Consider a **population** of 600 students divided into 4 strata. We are interested in the results of a test with scores ranging from 0 to 100.

Suppose we have the following information:

	Size N_i	Standard deviation σ_i
Stratum 1	285	8.6
Stratum 2	150	5.2
Stratum 3	90	2.2
Stratum 4	75	1.4

If we want to take a stratified **sample** of 30 individuals, we obtain the following subsamples:

1. *Representative stratified sample*

$$n_i = \frac{N_i}{N} \cdot n$$

We then obtain the following values, rounded to the nearest integer:

$$n_1 = \frac{N_1}{N} \cdot n$$
$$= \frac{285}{600} \cdot 30$$
$$= 14.25$$
$$\approx 14$$
$$n_2 = 7.5 \approx 7$$
$$n_3 = 4.5 \approx 5$$
$$n_4 = 3.75 \approx 4$$

2. *Optimal stratified sample*

$$n_i = \frac{n}{\sum\limits_{i=1}^{k} N_i \cdot \sigma_i} \cdot N_i \cdot \sigma_i$$

We have:

$$N_1 \cdot \sigma_1 = 2451$$
$$N_2 \cdot \sigma_2 = 780$$
$$N_3 \cdot \sigma_3 = 198$$
$$N_4 \cdot \sigma_4 = 105$$

The constant factor $n / \sum\limits_{i=1}^{4} N_i \cdot \sigma_i$ is equal to:

$$\frac{n}{\sum\limits_{i=1}^{k} N_i \cdot \sigma_i} = \frac{30}{3534} = 0.0085 .$$

We obtain the following values, rounded to the nearest integer:

$$n_1 = 0.0085 \cdot 2451 = 20.81 \approx 21$$
$$n_2 = 0.0085 \cdot 780 = 6.62 \approx 6$$
$$n_3 = 0.0085 \cdot 198 = 1.68 \approx 2$$
$$n_4 = 0.0085 \cdot 105 = 0.89 \approx 1 .$$

If we compare both methods, we can see that, given a more homogeneous stratum, the second method takes a larger number of individuals in the first stratum (the most dispersed) and a smaller number of individuals in the second stratum.

FURTHER READING
▶ **Estimation**
▶ **Estimator**
▶ **Sampling**
▶ **Simple random sampling**

REFERENCES
Yates, F.: Sampling Methods for Censuses and Surveys. Griffin, London (1949)

Student

See **Sealy, Gosset William**.

Student Distribution

Random variable T follows a Student distribution if its **density function** is of the form:

$$f(t) = \frac{\Gamma\left(\frac{v+1}{2}\right)}{\sqrt{v\pi} \cdot \Gamma\left(\frac{v}{2}\right)} \left(1 + \frac{t^2}{v}\right)^{-\frac{v+1}{2}},$$

where Γ is the gamma function (see **gamma distribution**) and v the number of **degrees of freedom**.

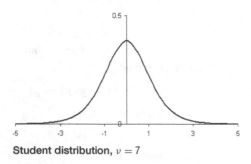

Student distribution, $v = 7$

The Student distribution is a **continuous probability distribution**.

HISTORY

The Student distribution was created by **Gosset, W.S.**, known as "Student", who in 1908 published an article in which he described the **density function** of the difference between the **mean** of a **sample** and the mean of the **population** from which the sample was taken, divided by the **standard deviation** of the sample. He also provided in the same article the first table of the corresponding **distribution function**.

Student continued his research and in 1925 published an article in which he proposed new tables that were more extensive and more precise.

Fisher, R.A. was interested in Gosset's works. He wrote him in 1912 to propose

a geometric demonstration of the Student distribution and to introduce the notion of **degrees of freedom**. In 1925 he published an article in which he defines the ratio t:

$$t = \frac{Z}{\sqrt{\frac{X}{v}}},$$

where X and Z are two independent **random variables**, X is distributed according to a **chi-square distribution** with v **degrees of freedom**, and Z is distributed according to the centered and reduced **normal distribution**. This ratio is called the t of Student.

MATHEMATICAL ASPECTS

If X is a **random variable** that follows a **chi-square distribution** with v **degrees of freedom**, and Z is a random variable distributed according to the standard **normal distribution**, then the random variable

$$T = \frac{Z}{\sqrt{\frac{X}{v}}}$$

follows a Student distribution with v degrees of freedom if X and Z are independent. Consider **random variable** T:

$$T = \frac{\sqrt{n}\,(\bar{x} - \mu)}{S},$$

where $\bar{x}$ is the **mean** of a **sample**, μ is the mean of the **population** from which the sample was taken, and S is the **standard deviation** of the sample given by:

$$S = \sqrt{\frac{1}{n-1} \cdot \sum_{i=1}^{n} (x_i - \bar{x})^2}.$$

Then T follows a Student distribution with $n - 1$ **degrees of freedom**.

The **expected value** of **random variable** T following a Student distribution is given by:

$$E[T] = 0, \quad v > 1,$$

and the **variance** is equal to:

$$\text{Var}(T) = \frac{v}{v-2}, \quad v > 2,$$

where v is the number of **degrees of freedom**.

The Student distribution is related to other **continuous probability distributions**:

- When the number of **degrees of freedom** is sufficiently large, the Student distribution approaches a **normal distribution**.
- When the number of **degrees of freedom** v is equal to 1, the Student distribution is identical to a **Cauchy distribution** with $\alpha = 0$ and $\theta = 1$.
- If **random variable** X follows a Student distribution with v **degrees of freedom**, then random variable X^2 follows a **Fisher distribution** with 1 and v degrees of freedom.

DOMAINS AND LIMITATIONS

The Student distribution is a symmetric distribution with a **mean** equal to 0. The larger its number of **degrees of freedom**, the smaller its **standard deviation** becomes.

The Student distribution is used in **inferential statistics** in relation to the t ratio of Student in **hypothesis testing** and in the construction of **confidence intervals** for the **mean** of a **population**.

In tests of **analysis of variance**, the Student distribution can be used when the sum of squares between the groups (or the sum of squares of the factors) to be compared only has one **degree of freedom**.

The **normal distribution** can be used as an approximation of the Student distribution when the number of **observations** n is large ($n > 30$ according to most authors).

FURTHER READING

- ▶ **Cauchy distribution**
- ▶ **Continuous probability distribution**
- ▶ **Fisher distribution**
- ▶ **Normal distribution**
- ▶ **Student table**
- ▶ **Student test**

REFERENCES

Fisher, R.A.: Applications of "Student's" distribution. Metron **5**, 90–104 (1925)

Gosset, S.W. "Student": The Probable Error of a Mean. Biometrika **6**, 1–25 (1908)

Gosset, S.W. "Student": New tables for testing the significance of observations. Metron **5**, 105–120 (1925)

Student Table

The Student table gives the values of the **distribution function** of a **random variable** following a **Student distribution**.

HISTORY

Gosset, William Sealy, known as "Student", developed in 1908 the **Student test** and the table bearing his name.

MATHEMATICAL ASPECTS

Let **random variable** T follow a **Student distribution** with v degrees of freedom. The **density function** of random variable T is given by:

$$f(s) = \frac{\Gamma\left(\frac{v+1}{2}\right)}{\sqrt{v\pi} \cdot \Gamma\left(\frac{v}{2}\right)} \left(1 + \frac{s^2}{v}\right)^{-(v+1)/2},$$

S

where Γ represents the gamma function (see **gamma distribution**).

The **distribution function** of random variable T is defined by:

$$F(t) = P(T \leq t) = \int_{-\infty}^{t} f(s) \, ds.$$

For each **value** of v, the Student table gives the value of the distribution function $F(t)$. Since the Student distribution is symmetric relative to the origin, we have:

$$F(t) = 1 - F(-t),$$

which allows to determine $F(t)$ for each negative value of t.

The Student table is used in the inverse sense: to find the values of t corresponding to a given **probability**. In this case, the Student table gives the **critical value** for the number of **degrees of freedom** v and the **significance level** α. We generally denote by $t_{v,\alpha}$ the value of random variable T for which:

$$P\left(T \leq t_{v,\alpha}\right) = 1 - \alpha.$$

DOMAINS AND LIMITATIONS
See **Student test**.

EXAMPLES
See Appendix E.

We consider a one-tailed **Student test** on a **sample** of dimension n. For $v = n - 1$ degrees of freedom, the Student table allows to determine the **critical value** $t_{v,\alpha}$ of the test for the given **significance level** α.

For an example using the Student table, see **Student test**.

FURTHER READING
▶ **Statistical table**
▶ **Student distribution**
▶ **Student test**

REFERENCES
Fisher, R.A.: Applications of "Student's" distribution. Metron **5**, 90–104 (1925)

Gosset, S.W. "Student": The Probable Error of a Mean. Biometrika **6**, 1–25 (1908)

Gosset, S.W. "Student": New tables for testing the significance of observations. Metron **5**, 105–120 (1925)

Student Test

The Student test is a parametric **hypothesis test** on the **sample mean** or on the comparison of the means of two samples.

HISTORY
See under history of **Student table**.

MATHEMATICAL ASPECTS
Let X be a **random variable** distributed according to a **normal distribution**. Random variable T defined above follows a **Student distribution** with $n-1$ degrees of freedom:

$$T = \frac{\bar{X} - \mu}{\frac{S}{\sqrt{n}}},$$

where $\bar{X}$ is the **arithmetic mean** of the **sample**, μ is the mean of the **population**, S is the **standard deviation** of the sample, and n is the **sample size**.

The Student test consists in calculating this ratio for the sample data and comparing it with the theoretical **value** of the **Student table**.

A. *Student test for one or single sample*
1. The hypotheses we want to test is:
   – **Null hypothesis**:

$$H_0: \quad \mu = \mu_0.$$

We should test if the sample comes from a population with a specified **mean**, μ_0, or if there is a statistically significant difference between the observed mean in the sample and the hypothetical mean on the population (μ_0) under H_0.

- **Alternative hypothesis**:

  The alternative hypothesis can take three forms:

  H_1: $\mu > \mu_0$ (**one-sided test on the right**)

  H_1: $\mu < \mu_0$ (**one-sided test on the left**)

  H_1: $\mu \neq \mu_0$ (**two-sided test**).

2. Calculation of the t-statistic as:

$$t = \frac{\bar{x} - \mu_0}{\frac{S}{\sqrt{n}}},$$

where μ_0 is the mean of the population specified by H_0, $\bar{x}$ is the sample mean, S is the sample standard deviation, and n is the sample size.

3. Choose of **significance level** α of test

4. Compare of the calculated value t with the appropriate **critical value** t_{n-1} (the value from the Student table with $n - 1$ degrees of freedom). Reject H_0 if the absolute value of t is greater than the critical value t_{n-1}.

The critical values for different degrees of freedom and different significance levels are given by the Student table.

For a one-sided test we take the value $t_{n-1,1-\alpha}$ in the table, and for a two-sided test we take the value $t_{n-1,1-\frac{\alpha}{2}}$.

B. *Student test for two samples*

1. The hypotheses we want to test are:

   - **Null hypothesis**:

$$H_0: \mu_1 = \mu_2.$$

In a test that designed to compare two samples, we want to know if there exists a significative difference between the mean of two populations from which the samples were drawn.

- **Alternative hypothesis**:

  It can take the following three forms:

  H_1: $\mu_1 > \mu_2$ (one-sided test on the right)

  H_1: $\mu_1 < \mu_2$ (one-sided test on the left)

  H_1: $\mu_1 \neq \mu_2$ (two-sided test).

2. Compute the t-statistic by:

$$t = \frac{\bar{x}_1 - \bar{x}_2}{S_p \cdot \sqrt{\frac{1}{n_1} + \frac{1}{n_2}}},$$

where n_1 and n_2 are the respective samples sizes and S_p is the pooled standard deviation:

$$S_p = \sqrt{\frac{(n_1 - 1)\, S_1^2 + (n_2 - 1)\, S_2^2}{n_1 + n_2 - 2}}$$

$$= \sqrt{\frac{\displaystyle\sum_{i=1}^{2}\sum_{j=1}^{n_i}\left(x_{ij} - \bar{x}_i\right)^2}{n_1 + n_2 - 2}},$$

where S_1^2 represents the sample standard deviation 1, S_2 that of sample 2, and x_{ij} **observation** j of sample i.

3. Choose the significance level α of test

4. Compare of the calculated value of t with the critical value $t_{n_1+n_2-2,1-\frac{\alpha}{2}}$ (the value of the Student table with $(n_1 + n_2 - 2)$ degrees of freedom). Reject H_0 if the absolute value of t is greater than that of the critical value.

If the test is one-sided, then take the value $t_{n_1+n_2-2,1-\alpha}$ from the Student table, and if it is two-sided, take the value $t_{n_1+n_2-2,1-\frac{\alpha}{2}}$.

S

Test of Signification of Coefficient of Regression

The Student test can also be applied to regression problems. Consider a simple linear regression **model** for which the data consist of n pairs of observations (x_i, y_i), $i = 1, \ldots, n$, modeled by

$$y_i = \beta_0 + \beta_1 \cdot x_i + \varepsilon_i,$$

where the ε_i are the random independent and normally distributed errors with mean 0 and **variance** σ^2.

The estimators of β_0 and β_1 are given by:

$$\hat{\beta}_0 = \bar{y} - \hat{\beta}_1 \cdot \bar{x},$$

$$\hat{\beta}_1 = \frac{\sum\limits_{i=1}^{n} (x_i - \bar{x})(y_i - \bar{y})}{\sum\limits_{i=1}^{n} (x_i - \bar{x})^2}.$$

The estimator $\hat{\beta}_1$ is normally distributed with mean β_1 and variance

$$\text{Var}(\hat{\beta}_1) = \frac{\sigma^2}{\sum\limits_{i=1}^{n} (x_i - \bar{x})^2}.$$

If σ^2 is estimated by the mean of squares of residuals S^2 and replaced in the expression by the variance of $\hat{\beta}_1$, then the ratio

$$T = \frac{\hat{\beta}_1 - \beta_1}{\sqrt{\widehat{\text{Var}}\left(\hat{\beta}_1\right)}}$$

is distributed according to the Student distribution with $n - 2$ degrees of freedom.

At a certain significance level α we can make a Student test of the coefficient β_1. The hypotheses are the following:

Null hypothesis H_0: $\beta_1 = 0$

Alternative hypothesis H_1: $\beta_1 \neq 0$.

If the absolute value of the calculated ratio T is greater than the value of t in the Student table $t_{n-2, 1-\frac{\alpha}{2}}$, then we can conclude that the coefficient β_1 is significatively different from zero. In the contrary case, we cannot reject the null hypothesis ($\beta_1 = 0$). We conclude in this case that the slope of the regression line is not significatively different from zero.

A similar test can be made for the constant β_0:

The estimator $\hat{\beta}_0$ is normally distributed with mean β_0 and variance

$$\text{Var}(\hat{\beta}_0) = \sigma^2 \cdot \left[\frac{\sum\limits_{i=1}^{n} x_i^2}{\sum\limits_{i=1}^{n} (x_i - \bar{x})^2} \right].$$

If σ^2 is estimated by the mean of squares of residuals S^2 and is replaced by the expression $\text{Var}\left(\hat{\beta}_0\right)$, then the ratio

$$T = \frac{\hat{\beta}_0 - \beta_0}{\sqrt{\widehat{\text{Var}}\left(\hat{\beta}_0\right)}}$$

is distributed according to the Student distribution with $n - 2$ degrees of freedom.

The Student test concerns the following hypotheses:

Null hypothesis H_0: $\beta_0 = 0$

Alternative hypothesis H_1: $\beta_0 \neq 0$.

If the absolute value of the calculated ratio T is greater than the value of t in the Student table $t_{n-2, 1-\frac{\alpha}{2}}$, then we can conclude that the coefficient β_0 is significatively different from zero. In the contrary case, we cannot reject the null hypothesis ($\beta_0 = 0$). We conclude that the regression is not significatively different from zero.

DOMAINS AND LIMITATIONS

Generally, we should be very attentive to the choice of the test to perform. The Student test is often falsely applied. In fact, it means something only if the observations come from a normally distributed **population** or are close to a normally distributed one. If this requirement is not satisfied, the **critical value** of the Student test does not provide an absolute guarantee.

If the normal distribution is applicable, the Student test on a **sample** is used when the **standard deviation** of the population is unknown and when the dimension of the sample is not large ($n < 30$).

Likewise, when there are two samples, the Student test is used when the standard deviations of two populations are unknown and supposed equal and when the sum of dimensions of samples is relatively small ($n_1 + n_2 < 30$).

EXAMPLES

Student Test on a Sample

Consider a **sample** of dimension $n = 10$ where the observations are normally distributed and are the following:

47 51 48 49 48 52 47 49 46 47 .

We want to test the following hypotheses:

Null hypothesis H_0: $\mu = 50$

Alternative hypothesis H_1: $\mu \neq 50$.

The **arithmetic mean** of the observations equals:

$$\bar{x} = \frac{\sum_{i=1}^{10} x_i}{n} = \frac{484}{10} = 48.4 .$$

We summarize the calculation in the following table:

x_i	$(x_i - \bar{x})^2$
47	1.96
51	6.76
48	0.16
49	0.36
48	0.16
52	12.96
47	1.96
49	0.36
46	5.76
47	1.96
Total	32.4

The **standard deviation** equals:

$$S = \sqrt{\frac{\sum_{i=1}^{10} (x_i - \bar{x})^2}{n - 1}} = \sqrt{\frac{32.4}{9}} = 1.90 .$$

We calculate the ratio:

$$t = \frac{\bar{x} - \mu}{\frac{S}{\sqrt{n}}} = \frac{48.4 - 50}{\frac{1.90}{\sqrt{10}}} = -2.67 .$$

The number of degrees of freedom associated to the test equals $n - 1 = 9$.

If we choose a **significance level** α of 5%, then the **value** $t_{n-1,1-\frac{\alpha}{2}}$ of the **Student table** equals:

$$t_{9,0.975} = 2.26 .$$

As $|t| = 2.67 > 2.26$, we reject the null hypothesis and conclude that at a significance level of 5%, the sample does not come from a population with a mean $\mu = 50$.

Student Test on Two Samples

A study is conducted to compare the mean lifetime of two brands of tires. A sample was taken for each brand. The results are:

- For brand 1:
  Mean lifetime: $\bar{x}_1 = 72000 \, \text{km}$
  Standard deviation: $S_1 = 3200 \, \text{km}$
  Sample dimension: $n_1 = 50$

S

- For brand 2:

  Mean lifetime: $\bar{x}_2 = 74400 \, \text{km}$

  Standard deviation: $S_2 = 2400 \, \text{km}$

  Sample size: $n_2 = 40$

We want to know if there exists a significative difference in the lifetime of the two brands of tires, at a significance level α of 1%. This implies the following hypotheses:

Null hypothesis H_0:

$$\mu_1 = \mu_2 \quad \text{or} \quad \mu_1 - \mu_2 = 0$$

Alternative hypothesis H_1:

$$\mu_1 \neq \mu_2 \quad \text{or} \quad \mu_1 - \mu_1 \neq 0.$$

The weighted standard deviation S_p equals:

$$S_p = \sqrt{\frac{(n_1 - 1) \cdot S_1^2 + (n_2 - 1) \cdot S_2^2}{n_1 + n_2 - 2}}$$

$$= \sqrt{\frac{49 \cdot 3200^2 + 39 \cdot 2400^2}{50 + 40 - 2}}$$

$$= 2873.07.$$

We can calculate the ratio t:

$$t = \frac{\bar{x}_1 - \bar{x}_2}{S_p \cdot \sqrt{\frac{1}{n_1} + \frac{1}{n_2}}}$$

$$= \frac{72000 - 74400}{2873.07 \cdot \sqrt{\frac{1}{50} + \frac{1}{40}}}$$

$$= -3.938.$$

The number of degrees of freedom associated to the test equals:

$$v = n_1 + n_2 - 2 = 88.$$

The value of $t_{v, 1-\frac{\alpha}{2}}$ found in the Student table, for $\alpha = 1\%$, is:

$$t_{88, 0.995} = 2.634.$$

As the absolute value of the calculated ratio is greater than this value, $|t| = |-3.938| >$ 2.634, we reject the null hypothesis for the alternative hypothesis and conclude that at the significance level 1%, the two brands of tires do not have the same mean lifetime.

FURTHER READING

► **Hypothesis testing**

► **Parametric test**

► **Student distribution**

► **Student table**

REFERENCES

Fisher, R.A.: Applications of "Student's" distribution. Metron **5**, 90–104 (1925)

Gosset, S.W. "Student": The Probable Error of a Mean. Biometrika **6**, 1–25 (1908)

Gosset, S.W. "Student": New tables for testing the significance of observations. Metron **5**, 105–120 (1925)

Sure Event

See **sample space**.

Survey

An inquiry by survey, or simply survey, is an inquiry made on a restricted part of a **population**. This fraction of the population constitutes the **sample**, and the methods that allow one to construct this sample are called **sampling methods**.

Proceeding with the survey, we obtain information on the sampled units. Then, with the help of **inferential statistics**, we generalize this information to the entire population. This generalization introduces a certain **error**. The importance of this error will depend on the **sample size**, how they were chosen on the sampling scheme used to draw the sample, and the **estimator** that was used. The more representative the sample, the smaller the error. Sampling methods are very useful in order to draw samples which result in a minimal error.

Another technique consists in observing all individuals of a population. We refer here to a **census**. Surveys have cost and speed advantages over censuses that largely compensate for sampling error. Moreover, it is sometimes impossible to carry out a census, for example, when the fact of observing a unit of a population means its destruction or when the population is infinite.

FURTHER READING
- **Census**
- **Data collection**
- **Inferential statistics**
- **Panel**
- **Population**
- **Sample**
- **Sampling**

REFERENCES

Gourieroux, C.: Théorie des sondages. Economica, Paris (1981)

Grosbras, J.M.: Méthodes statistiques des sondages. Economica, Paris (1987)

Systematic Sampling

Systematic sampling is a random type of **sampling**. Individuals are taken from a **population** at fixed intervals according to time, space, or order of occurrence, the first individual being drawn randomly.

HISTORY
See **sampling**.

MATHEMATICAL ASPECTS

Consider a **population** composed of N individuals. A systematic **sample** is made up of individuals whose numbers consitute an arithmetic progression.

A first number b is chosen randomly between 1 and r, where $r = \frac{N}{n}$, n is the size of the **sample**, and r is common difference between successive form of the arithmetic progression. Individuals sampled in this way will have the following numbers:

$$b, b + r, b + 2r, \ldots, b + (n - 1)\, r\,.$$

DOMAINS AND LIMITATIONS

Systematic **sampling** is possible only if the individuals are classified in a certain order. The main advantage of this method lies in its simplicity and, therefore, its low cost.

On the other hand, it has the disadvantage of not taking into account an eventual periodicity of the studied trait. Serious **errors** could occur from this, especially if the period is a submultiple of the ratio of the arithmetic progression.

EXAMPLES

Consider a **population** of 600 students grouped in alphabetical order from which we want to take a **sample** of 30 individuals according to systematic **sampling**.

The ratio of the arithmetic progression is equal to:

$$r = \frac{N}{n} = \frac{600}{30} = 20\,.$$

A first number is chosen randomly between 1 and 20, for example $b = 17$.

The **sample** will be composed of the students with the numbers:

$$17, 37, 57, \ldots, 597\,.$$

FURTHER READING

► **Cluster sampling**
► **Sampling**

REFERENCES

Deming, W.E.: Sample Design in Business Research. Wiley, New York (1960)

Target Population

The concept of the target population is defined as a **population** to which we would like to apply the results of an inquiry.

EXAMPLES

We want to conduct an inquiry on foreigners living in city X. The **population** is represented by the list of foreigners registered in the city.

The target population is defined by the collection of foreigners living in city X, including those not on the list (unannounced, clandestine, etc.).

FURTHER READING

▶ **Population**

Test of Independence

A test of independence is a **hypothesis test** where the objective is to determine if two random variables are independent or not.

HISTORY

See **Kendall rank correlation coefficient**, **Spearman rank correlation coefficient**, and **chi-square test of independence**.

MATHEMATICAL ASPECTS

Let $(X_1, X_2, \ldots, X_n)$ and $(Y_1, Y_2, \ldots, Y_n)$ be two samples of dimension n. The objective is to test the following two hypotheses:

Null hypothesis H_0: The two **variables** X and Y are independent.

Alternative hypothesis H_1: The two variables X and Y are not independent.

The test of independence consists in comparing the empirical distribution with the theoretical distribution by calculating an indicator.

The test of independence applied to two continuous variables is based on the ranks of the observations. Such is the case if the tests are based on the **Spearman rank correlation coefficient** or on the **Kendall rank correlation coefficient**.

In the case of two categorical variables, the most widely used test is the **chi-square test of independence**.

DOMAINS AND LIMITATIONS

To make a test of independence each couple (X_i, Y_i), $i = 1, 2, \ldots, n$ must come from the same bivariate **population**.

EXAMPLES

See **chi-square test of independence**.

FURTHER READING
- ► **Chi-square test of independence**
- ► **Hypothesis testing**
- ► **Kendall rank correlation coefficient**
- ► **Nonparametric test**
- ► **Spearman rank correlation coefficient**

Time Series

A time series is a sequence of observations measured at succesive times. Time series are monthly, trimestrial, or annual, sometimes weekly, daily, or hourly (study of road traffic, telephone traffic), or biennial or decennial. Time series analysis consists of methods that attempt to understand such time series to make predictions.

Time series can be decomposed into four components, each expressing a particular aspect of the movement of the values of the time series.

These four components are:

- **Secular trend**, which describe the movement along the term;
- **Seasonal variations**, which represent seasonal changes;
- **Cyclical fluctuations**, which correspond to periodical but not seasonal variations;
- **Irregular variations**, which are other nonrandom sources of variations of series.

The analysis of time series consists in making mathematical descriptions of these elements, that is, estimating separately the four components.

HISTORY

In an article in 1936 Funkhauser, H.G. reproduced a diagram of the 10th century that shows the inclination of the orbit of seven planets in function of time. According to Kendall this is the oldest time diagram known in the western world.

It is adequate to look to astronomy for the origins of the time series. The **observation** of the stars was popular in antiquity.

In the 16th century, astronomy was again a source of important discoveries. The works of Brahe, Tycho (1546–1601), whose collection of data on the movement of the planets allowed Kepler, Johannes (1571–1630) to formulate his laws on planetary motion. A statistical analysis of data, as empirical as it could be at that time, was at the base of this great scientist's work.

The development of mathematics in the 18th and 19th centuries made it possible to bypass graphical visualization to arrive at the first techniques of time series analysis. Using the frequency approach, scientists developed the first works in this domain.

Frequency analysis (also called harmonic analysis) was originally designed to highlight one or more cyclical "components" of a time series.

In 1919, Persons, W.M. proposed a decomposition of time series in terms of tendency (**secular trends**), cyclical **cyclical fluctuations**), seasonal (**seasonal variation**), and accidental (**irregular variation**) components. Many works have been devoted to the determination and elimination of one or another of these components.

The determination of tendencies has often been related to problems of regression analysis, also developed researchers relative to nonlinear tendencies, as for example the Gompertz curve (established in 1825 by Gompertz, Benjamin) or the logistic curve (introduced in 1844 by Verhulst, Pierre François). Another approach to determining tendencies concerns the use of moving averages because they allow one to treat the problem of tendencies when they serve to eliminate irregular or periodic variations.

The use of moving averages was developed in two directions: for actuarial problems and for the study of the time series in economy. The use of moving averages is particularly developed in frame of the methods of Seasonal Adjustment Methods. More precisely, the separation of seasonal variables from the other components of the time series was studied originally by Copeland, M.T. in 1915 and Persons, W.M. in 1919 and continued by Macauley, F.R. in 1930.

Macauley, F.R. used the method of ratios to the moving average that is still widely used today (see seasonal variation) see also Dufour (2006).

MATHEMATICAL ASPECTS

Let Y_t be a time series that can be decomposed with the help of these four components:

- **Secular trend** T_t
- Seasonal variations S_t
- Cyclical fluctuations C_t
- Irregular variations I_t

We suppose that the values taken by the **random variable** Y_t are determined by a **relation** between the four previous components. We distinguish three models of composition:

1. Additive model:

$$Y_t = T_t + C_t + S_t + I_t.$$

The four components are assumed to be independent of one another.

2. Multiplicative model:

$$Y_t = T_t \cdot C_t \cdot S_t \cdot I_t.$$

With the help of logarithms we pass from the multiplicative model to the additive model.

3. Mixed model:

$$Y_t = S_t + (T_t \cdot C_t \cdot I_t)$$

or

$$Y_t = C_t + (T_t \cdot S_t \cdot I_t).$$

This type of model is very little used.

We choose the additive model when seasonal variations are almost constant and their influence on the tendency does not depend on its level.

When the seasonal variations are of an amplitude almost proportional to that of the secular trends, we introduce the multiplicative model; this is generally the case.

The analysis of time series consists in determining the values taken by each component. We always start with the seasonal variations and end on the cyclical fluctuations. All fluctuations that cannot be attributed to one of the three components will be grouped with the irregular variations.

Based on the used model the analysis can, after the components are determined, adjust subsequent data by substraction or division. Thus when we have estimated the secular trend T_t, at each time t, the time series is adjusted in the following manner:

- If we use the additive **model** $Y_t = T_t + C_t + S_t + I_t$:

$$Y_t - T_t = C_t + S_t + I_t.$$

- For the multiplicative model $Y_t = T_t \cdot C_t \cdot S_t \cdot I_t$:

$$\frac{Y_t}{T_t} = C_t \cdot S_t \cdot I_t.$$

Successively evaluating each of the three components and adjusting the time series among them, we obtain the values attributed to the irregular variations.

T

DOMAINS AND LIMITATIONS

When the time series are not in a long enough period, we group the **secular trend** and the cyclical fluctuations in one component called **extraseasonal movement**.

Graphical Representation

The **graphical representation** of time series helps to analyze observed values depending on time.

To analyze these values, a certain number of precautions must be taken:

1. Each point must be representative to an observation. If the value represents a stock, on a fix date, then the point is found on the reference date. If there is a flow, for example, monthly, then the point is fixed in the middle of the month.
2. Sometimes one has to report, on the same graph, annual data and monthly data. The representative points of the results should be 12 times more frequent than those relative to the annual results. Thus we have to pay attention to the representative points in the right place.
3. If many series are represented on the same graph, they must be identical in nature. Otherwise, we should use different scales or place graphs below one another.

The principal objectives of the study of the time series are:

- Description: determination or climination of different components
- Stochastic modeling
- **Forecasting**
- Filtering: elimination or conversion of certain characteristics
- Control: **observation** of the past of a process to react to future evolution

For forecasting, the analysis of time series is based on the following hypotheses:

- The great historically observed tendencies are maintained in the more or less near future.
- The measurable fluctuations of a **variable** are reproduced in regular intervals.

The analysis of time series will be justified as long as it allows one to reduce the level of uncertainty in the elaboration of forecasts. We distinguish:

1. *Long-term forecast*:
   Based on the secular trend; generally does not take into account cyclical fluctuations.
2. *Middle-term forecast*:
   To take into account the probable effect of the cyclical component, we should multiply the forecasted value of the secular trend by an **estimation** of the relative discounted variation attributable to the cyclical fluctuations.
3. *Short-term forecast:*
   Generally we do not try to forecast irregular variations or make short-term forecasts or forecasts concerning the longer periods. Thus we multiply the forecasted values of the secular trend by estimating the cyclical fluctuations and by the seasonal variation of the corresponding date.

During an analysis of time series, we face the following problems:

1. *Is the classical model appropriate?*
   We expect the quality of the results of analysis to be proportional to the precision and the accuracy of the **model** itself. The multiplicative model $Y_t = T_t \cdot S_t \cdot C_t \cdot I_t$ presupposes that the four components are independent, which is rarely true in reality.
2. *Are our hypotheses of the constancy and regularity valid?*
   The analysis must be adjusted for subjective and qualitative factors that exist in almost any situation.

3. *Can we trust the available data?*
   The problem may arise of breaks of series resulting from insufficiency of data or a change in the quality of variables.

EXAMPLES

A time series is generally represented as follows:

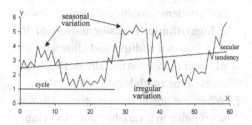

We distinguish four components:

- **Secular trend**, slightly increasing in the present case
- Seasonal variations, readily apparent
- Cyclical fluctuations, in the form of cycles of an approximate amplitude of 27 units of time
- Irregular variations, generally weak enough, except $t = 38$, which represents a very abrupt fall that would not be rational

FURTHER READING
▶ **Cyclical fluctuation**
▶ **Forecasting**
▶ **Graphical representation**
▶ **Irregular variation**
▶ **Seasonal variation**
▶ **Secular trend**

REFERENCES
Copeland, M.T.: Statistical indices of business conditions. Q. J. Econ. **29**, 522–562 (1915)

Dufour, I.M.: Histoire de l'analyse de Séries Chronologues, Université de Montréal, Canada (2006)

Funkhauser, H.G.: A note on a 10th century graph. Osiris **1**, 260–262 (1936)

Kendall, M.G.: Time Series. Griffin, London (1973)

Macauley, F.R.: The smoothing of time series. National Bureau of Economic Research, 121–136 (1930)

Persons, W.M.: An index of general business conditions. Rev. Econom. Stat. **1**, 111–205 (1919)

Trace

For a square matrix A of order n, we define the trace of A as the sum of the terms situated on the diagonal. Thus the trace of a matrix is a scalar quantity.

MATHEMATICAL ASPECTS

Let A be a square **matrix** of order n, $A = (a_{ij})$, where $i, j = 1, 2, \ldots, n$. We define trace A by:

$$\text{tr}(A) = \sum_{i=1}^{n} a_{ii}.$$

DOMAINS AND LIMITATIONS

As we are interested only in diagonal elements, the trace is defined only for square matrices.

1. The trace of the sum of two matrices equals the sum of the traces of the matrices:

   $$\text{tr}(A + B) = \text{tr}(A) + \text{tr}(B).$$

2. The trace of the product of two matrices does not change with the commutations of the matrices. Mathematically:

   $$\text{tr}(A \cdot B) = \text{tr}(B \cdot A).$$

T

EXAMPLES

Let A and B be two square matrices of order 2:

$$A = \begin{bmatrix} 1 & 2 \\ 1 & 3 \end{bmatrix} \text{ and } B = \begin{bmatrix} 5 & 2 \\ -2 & 3 \end{bmatrix}.$$

The calculation of traces gives:

$$\text{tr}(A) = \text{tr}\begin{bmatrix} 1 & 2 \\ 1 & 3 \end{bmatrix} = 1 + 3 = 4,$$

$$\text{tr}(B) = \text{tr}\begin{bmatrix} 5 & 2 \\ -2 & 3 \end{bmatrix} = 5 + 3 = 8,$$

and

$$\text{tr}(A + B) = \text{tr}\begin{bmatrix} 6 & 4 \\ -1 & 6 \end{bmatrix}$$

$$= 6 + 6 = 12.$$

We verify that $\text{tr}(A) + \text{tr}(B) = 12$. On the other hand:

$$A \cdot B = \begin{bmatrix} 1 & 8 \\ -1 & 11 \end{bmatrix} \text{ and}$$

$$B \cdot A = \begin{bmatrix} 7 & 16 \\ 1 & 5 \end{bmatrix},$$

$$\text{tr}(A \cdot B) = \text{tr}\begin{bmatrix} 1 & 8 \\ -1 & 11 \end{bmatrix}$$

$$= 1 + 11 = 12,$$

$$\text{tr}(B \cdot A) = \text{tr}\begin{bmatrix} 7 & 16 \\ 1 & 5 \end{bmatrix}$$

$$= 7 + 5 = 12.$$

Thus $\text{tr}(A \cdot B) = \text{tr}(B \cdot A) = 12$.

FURTHER READING

▶ Matrix

Transformation

A transformation is a change in one or many variables in a statistical study.

We transform variables, for example, by replacing them with their logarithms (**logarithmic transformation**).

DOMAINS AND LIMITATIONS

We transform variables for many reasons:

1. When we are in the presence of a data set relative to many variables and we want to express one of them, Y (**dependent variable**), with the help of others (called independent variables) and we want to linearize the **relation** between the variables. For example, when the relation between independent variables is multiplicative, a **logarithmic transformation** of the data makes it additive and allows the use of linear regression to estimate the parameters of the **model**.

2. When we are in the presence of a data set relative to a variable and its **variance** is not constant, it is generally possible to make it almost constant by transforming this variable.

3. When a **random variable** follows any distribution, an adequate transformation of it allows to obtain a new random variable that approximately follows a **normal distribution**.

The advantages of transforming data set relative to a variable are largely summarized as follows:

- It simplifies existing relations with other variables.
- A remedy for outliers, linearity and homoscedasticity.
- It produces a variable that is approximately normally distributed.

FURTHER READING

▶ **Dependent variable**
▶ **Independent variable**
▶ **Logarithmic transformation**
▶ **Normal distribution**
▶ **Regression analysis**
▶ **Variance**

REFERENCES

Cox, D.R., Box, G.E.P.: An analysis of transformations (with discussion). J. Roy. Stat. Soc. Ser. B **26**, 211–243 (1964)

Transpose

The (matrix) transpose of a **matrix** A of order $(m \times n)$ is a matrix $(n \times m)$, denoted by A', obtained by writing the lines of A in columns of A' and vice versa.

Thus we effect simple "inversion" of the table.

MATHEMATICAL ASPECTS

If $A = (a_{ij})$, $i = 1, 2, \ldots, m$ and $j = 1, 2, \ldots, n$ is a **matrix** of order $(m \times n)$, then the transpose of A is the matrix A' of order $(n \times m)$ given by:

$$A' = (a_{ji}) \quad \text{with } j = 1, 2, \ldots, n$$
$$\text{and } i = 1, 2, \ldots, m.$$

Transposition is a reversible and reciprocal operation; thus taking the transpose of the transpose of a matrix, we find the initial matrix.

DOMAINS AND LIMITATIONS

We use the transpose of a **matrix**, or more precisely the transpose of a **vector**, while calculating the **scalar product**.

EXAMPLES

Let A be the following **matrix** of order (3×2):

$$A = \begin{bmatrix} 1 & 2 \\ 0 & 3 \\ 2 & 5 \end{bmatrix}.$$

The transpose of A is the matrix (2×3):

$$A' = \begin{bmatrix} 1 & 0 & 2 \\ 2 & 3 & 5 \end{bmatrix}.$$

We can also verify that the transpose of A' equals the initial matrix A:

$$\left(A'\right)' = A.$$

FURTHER READING

▶ **Matrix**
▶ **Vector**

Treatment

In an **experimental design**, a treatment is a particular combination of levels of various factors.

EXAMPLES

Experiments are often carried out to compare two or more treatments, for example two different fertilizers on a particular type of plant or many types of drugs to treat a certain illness. Another example consists in measuring the time of coagulation of blood samples of 16 animals having supported different regimes A, B, C, and D.

In this case, different examples of treatments are, respectively, fertilizers, drugs, and regimes.

On the other hand, in experiments that test a particular fertilizer, for example azote, on a harvest of wheat, we can consider different quantities of the same fertilizer as different treatments. Here, we have one **factor** (azote) on different levels (quantities), for example 30 kg, 50 kg, 100 kg, and 200 kg. Each of these levels corresponds to a treatment.

T

The goal of an experiment is to determine if there is a real (significative) difference between these treatments.

FURTHER READING
▶ Design of experiments
▶ Experiment
▶ Experimental unit
▶ Factor

REFERENCES
Senn, St.: Cross-over Trials in Clinical Research. Wiley (1993)

Tukey John Wilder

Tukey, John Wilder was born in Bedford, MA in 1915. He studied chemistry at Brown University and received his doctorate in mathematics at Princeton University in 1939. At the age of 35, he became professor of mathematics at Princeton. He was a key player in the formation in the Dept of Statistics in Princeton University in 1966. He served as its chairman during 1966–1969. He is the author of *Exploratory Data Analysis* (1977) (translated into Russian) and eight volumes of articles. He is coauthor of *Statistical Problems of the Kinsey Report on Sexual Behavior in the Human Male; Data Analysis and Regression* (also translated into Russian), *Index to Statistics and Probability*. He is coeditor of the following books: *Understanding Robust and Exploratory Data Analysis*, *Exploring Data Tables, Trends Shapes, Configural Polysampling*, and *Fundamentals of Exploratory Analysis of Variance*. He died in New Jersey in July 2000.

Selected principal works and articles of Tukey, John Wilder:

1977 Exploratory Data Analysis. 1st edn. Addison-Wesley, Reading, MA .

1980 We need both exploratory and confirmatory. Am. Stat. 34, 23–25.

1986 Data analysis and behavioral science or learning to bear the quantitative man's burden by shunning badmandments. In: Collected Works of John W. Tukey, Vol III: Philosophy and Principles of Data Analysis, 1949–1964. (Jones, L.V., ed.) Wadsworth Advanced Books & Software, Monterey, CA.

1986 Collected Works of John W. Tukey, Vol III: Philosophy and Principles of Data Analysis, 1949–1964. (Jones, L.V., ed.) Wadsworth Advanced Books & Software, Monterey, CA.

1986 Collected Works of John W. Tukey, Vol IV: Philosophy and Principles of Data Analysis, 1965–1986. (Jones, L.V., ed.) Wadsworth Advanced Books & Software, Monterey, CA.

1988 Collected Works of John W. Tukey, Vol V: Graphics, 1965–1985. (Cleveland, W.S., ed.) Waldsworth and Brooks/Cole, Pacific Grove, CA .

FURTHER READING
▶ Box plot
▶ Exploratory data analysis

Two-Sided Test

A two-sided test concerning a **population** is a **hypothesis test** that is applied when we want to compare an estimate of a parameter to a given value against the alternative hypothesis not equal to the statet value.

MATHEMATICAL ASPECTS

A two-tail test is a **hypothesis test** on a **population** of the type:

Null hypothesis H_0: $\theta = \theta_0$

Alternative hypothesis H_1: $\theta \neq \theta_0$,

where θ is a **parameter** of the population whose **value** is unknown and θ_0 is the presumed value of this parameter.

For a test concerning the comparison of two populations, the hypotheses are:

Null hypothesis H_0: $\theta_1 = \theta_2$

Alternative hypothesis H_1: $\theta_1 \neq \theta_2$,

where θ_1 and θ_2 are the unknown parameters of two underlying populations.

DOMAINS AND LIMITATIONS

In a two-sided test, the **rejection region** is divided into two parts, the left and the right sides of the considered **parameter**.

EXAMPLES

A company produces steel cables. It wants to verify if the diameter μ of much of the produced cable conforms to the standard diameter of 0.9 cm.

It takes a **sample** of 100 cables whose mean diameter is 0.93 cm with a **standard deviation** of $S = 0.12$.

The hypotheses are the following:

Null hypothesis H_0: $\mu = 0.9$

Alternative hypothesis H_1: $\mu \neq 0.9$,

As the sample size is large enough, the **sampling distribution** of the **mean** can be approximated by a **normal distribution**.

For a **significance level** $\alpha = 5\%$, we obtain a **critical value** $z_{\alpha/2}$ in the **normal table** of $z_{\alpha/2} = 1.96$.

We find the **acceptance region** of the null hypotheses (see **rejection region**):

$$\mu \pm z_{\alpha/2} \cdot \frac{\sigma}{\sqrt{n}},$$

where σ is the standard deviation of the population. Since the standard deviation σ is unknown, we estimate it using the standard deviation S of the sample ($S = 0.12$), which gives:

$$0.9 \pm 1.96 \cdot \frac{0.12}{\sqrt{100}} = \begin{cases} 0.8765 \\ 0.9235 \end{cases}.$$

The acceptance region of the null hypothesis equals the **interval** $[0.8765, 0.9235]$. Because the mean of the sample (0.93 cm) is outside this interval, we must reject the null hypothesis for the alternative hypothesis. We conclude that at a significance level of 5% and depending on the studied sample, the diameter of the cables does not conform to the norm.

A study is carried out to compare the mean lifetime of two types of tires. A sample of 50 tires of brand 1 gives a mean lifetime of 72000 km, and a sample of 40 tires of brand 2 has a mean lifetime of 74400 km. Supposing that the standard deviation for brand 1 is $\sigma_1 = 3200$ km and that of brand 2 is $\sigma_2 = 2400$ km, we want to know if there is a significative difference in lifetimes between the two brands, at a significance level $\alpha = 1\%$.

The hypotheses are the following:

Null hypothesis H_0: $\mu_1 = \mu_2$

Alternative hypothesis H_1: $\mu_1 \neq \mu_2$.

The dimension of the sample is large enough, and the standard deviations of the populations are known, so that we can approach the

T

sampling distribution using a normal distribution.

The acceptance region of the null hypothesis is given by (see rejection region):

$$\mu_1 - \mu_2 \pm z_{\alpha/2} \cdot \sigma_{\bar{x}_1 - \bar{x}_2},$$

where $\sigma_{\bar{x}_1 - \bar{x}_2}$ is the standard deviation of the sampling distribution of the difference of two means, or the **standard error** of two means. The value $z_{\alpha/2}$ is found in the **normal table**. For a two-tail test with $\alpha = 1\%$, we get $z_{\alpha/2} = 2.575$, which gives:

$$0 \pm 2.575 \cdot \sqrt{\frac{\sigma_1^2}{n_1} + \frac{\sigma_2^2}{n_2}}$$

$$= 0 \pm 2.575 \cdot \sqrt{\frac{3200^2}{50} + \frac{2400^2}{40}}$$

$$= 0 \pm 2.575 \cdot 590.59$$

$$= \pm 1520.77.$$

The acceptance region of the null hypothesis is given by the **interval** $[-1520.77, 1520.77]$. Because the difference in means of the samples ($72000 - 74400 = -2400$) is outside the acceptance region, we must reject the null hypothesis for the alternative hypothesis. Thus we can conclude that there is a significative difference between the lifetimes of the tires of the tested brands.

FURTHER READING
► **Acceptance region**
► **Alternative hypothesis**
► **Hypothesis testing**
► **Null hypothesis**
► **One-sided test**
► **Rejection region**
► **Sampling distribution**
► **Significance level**

Two-Way Analysis of Variance

The two-way analysis of variance is an expansion of one-way analysis of variance in which there are two independent factors (variables). Each factor has two or more levels and treatments are formed by making all possible combinations of levels of two factors.

HISTORY
It is **Fisher, R.A.** (1925) who gave the name "**factorial experiments**" to complex experiments.

Yates, F. (1935, 1937) developed the concept and analysis of these factorial experiments. See **analysis of variance**.

MATHEMATICAL ASPECTS
Given an **experiment** with two **factors**, factor A having a **levels** and factor B having b levels.

If the design associated with this **factorial experiment** is a **completely randomized design**, the **model** is the following:

$$Y_{ijk} = \mu + \alpha_i + \beta_j + (\alpha\beta)_{ij} + \varepsilon_{ijk},$$

$i = 1, 2, \ldots, a$ (**levels** of **factor** A)

$j = 1, 2, \ldots, b$ (**levels** of **factor** B)

$k = 1, 2, \ldots, c$ (number of **observations** receiving the **treatment** ij)

where μ is the general **mean** common to all **treatments**, α_i is the effect of **level** i of **factor** A, β_j is the effect of level j of factor B, $(\alpha\beta)_{ij}$ is the effect of the interaction between

α_i and β_j, and ε_{ijk} is the experimental **error** of **observation** Y_{ijk}.

This **model** is subjected to the basic **assumptions** associated to **analysis of variance** if one supposes that the errors ε_{ijk} are independent **random variables** following a **normal distribution** $N(0, \sigma^2)$.

Three **hypotheses** can then be tested:

1.

$$H_0: \quad \alpha_1 = \alpha_2 = \ldots = \alpha_a$$

$H_1:$ At least one α_i is different from α_j, $i \neq j$.

2.

$$H_0: \quad \beta_1 = \beta_2 = \ldots = \beta_b$$

$H_1:$ At least one β_i is different from β_j, $i \neq j$.

3.

$$H_0: \quad (\alpha\beta)_{11} = (\alpha\beta)_{12} = \ldots = (\alpha\beta)_{1b}$$
$$= (\alpha\beta)_{21} = \ldots = (\alpha\beta)_{ab}$$

$H_1:$ At least one of the interactions is different from the others.

The **Fisher distribution** is used to test the first **hypothesis**. This distribution requires the creation of a ratio whose numerator is an **estimate** of the variance of **factor** A and whose denominator is an estimate of the variance within treatments known also as **error** or **residual**.

This ratio, denoted by F, follows a **Fisher distribution** with $a-1$ and $ab(c-1)$ **degrees of freedom**.

Null hypothesis $H_0:$ $\alpha_1 = \alpha_2 = \ldots = \alpha_a$

will be rejected at the **significant level** α if the ratio F is greater than or equal to the **value** in the **Fisher table**, meaning if

$$F \geq F_{a-1, ab(c-1), \alpha}.$$

To test the second **hypothesis**, we create a ratio whose numerator is an **estimate** of the **variance** of **factor** B and whose denominator is an estimate of the variation within treatments.

This ratio, denoted by F, follows a **Fisher distribution** with $b-1$ and $ab(c-1)$ **degrees of freedom**.

Null hypothesis $H_0:$ $\beta_1 = \beta_2 = \ldots = \beta_b$

will be rejected if the F ratio is greater than or equal to the **value** of the **Fisher table**, meaning if

$$F \geq F_{b-1, ab(c-1), \alpha}.$$

To test the third **hypothesis**, we create a ratio whose numerator is an **estimate** of the **variance** of the interaction between factors A and B and whose denominator is an estimate of the variance within treatments.

This ratio, denoted by F, follows a **Fisher distribution** with $(a-1)(b-1)$ and $ab(c-1)$ **degrees of freedom**.

Null hypothesis $H_0:$

$$(\alpha\beta)_{11} = (\alpha\beta)_{12} = \ldots = (\alpha\beta)_{1b}$$
$$= (\alpha\beta)_{21} = \ldots = (\alpha\beta)_{ab}$$

will be rejected if the F ratio is greater than or equal to the **value** in the **Fisher table**, meaning if

$$F \geq F_{(a-1)(b-1), ab(c-1), \alpha}.$$

T

Variance of Factor A

For the **variance** of **factor** A, the sum of squares must be calculated for factor A (SS_A), which is obtained as follows:

$$SS_A = b \cdot c \sum_{i=1}^{a} (\bar{Y}_{i..} - \bar{Y}_{...})^2,$$

where $\bar{Y}_{i..}$ is the **mean** of all the **observations** of **level** i of factor A and $\bar{Y}_{...}$ is the general mean of all the observations.

The number of **degrees of freedom** associated to this sum is equal to $a - 1$.

The **variance** of **factor** A is then equal to:

$$s_A^2 = \frac{SS_A}{a - 1}.$$

Variance of Factor B

For the **variance** of **factor** B, the sum of squares must be calculated for factor B (SS_B), which is obtained as follows:

$$SS_B = a \cdot c \sum_{j=1}^{b} (\bar{Y}_{.j.} - \bar{Y}_{...})^2,$$

where $\bar{Y}_{i..}$ is the **mean** of all the **observations** of **level** i of factor B and $\bar{Y}_{...}$ is the general mean of all the observations.

The number of **degrees of freedom** associated to this sum is equal to $b - 1$.

The **variance** of **factor** B is then equal to:

$$s_B^2 = \frac{SS_B}{b - 1}.$$

Variance of Interaction AB

For the **variance** of **interaction** AB, the sum of squares must be calculated for interaction AB (SS_{AB}), which is obtained as follows:

$$SS_{AB} = c \sum_{i=1}^{a} \sum_{j=1}^{b} (\bar{Y}_{ij.} - \bar{Y}_{i..} - \bar{Y}_{.j.} + \bar{Y}_{...})^2,$$

where $\bar{Y}_{ij.}$ is the **mean** of all the **observations** of **level** ij of interaction AB, $\bar{Y}_{i..}$ is the mean of all the observations of level i of factor A, $\bar{Y}_{.j.}$ is the mean of all the observations of level j of factor B, and $\bar{Y}_{...}$ is the general mean of all the observations.

The number of **degrees of freedom** associated to this sum is equal to $(a - 1)(b - 1)$.

The **variance** of **interaction** AB is then equal to:

$$s_{AB}^2 = \frac{SS_{AB}}{(a - 1)(b - 1)}.$$

Variance Within Treatments or Error

For the **variance** within treatments, the sum of squares within treatments is obtained as follows:

$$SS_E = \sum_{i=1}^{a} \sum_{j=1}^{b} \sum_{k=1}^{c} (Y_{ijk} - \bar{Y}_{ij.})^2,$$

where a and b are, respectively, the number of **levels** of **factors** A and B, c is the number of observations receiving treatment ij (constant for any i and j), Y_{ijk} is observation k from level i of factor A and from level j of factor B, and $\bar{Y}_{ij.}$ is the mean of all the observations of level ij of interaction AB.

The number of **degrees of freedom** associated to this sum is equal to $ab(c - 1)$.

The **variance** within treatments is thus equal to:

$$s_E^2 = \frac{SS_E}{ab(c - 1)}.$$

The total sum of squares (SS_T) is equal to:

$$SS_T = \sum_{i=1}^{a} \sum_{j=1}^{b} \sum_{k=1}^{c} Y_{ijk}^2.$$

The number of **degrees of freedom** associated to this sum is equal to N, meaning the total number of **observations**.

The sum of squares for the **mean** is equal to:

$$SS_M = N\bar{Y}_{...}^2.$$

The number of **degrees of freedom** associated to this sum is equal to 1.

The total sum of squares (SS_T) can be expressed with all the other sums of squares in the following way:

$$SS_T = SS_M + SS_A + SS_B + SS_{AB} + SS_E.$$

Fisher Tests

We can now calculate the different F ratios.

To test the **null hypothesis**

$$H_0: \quad \alpha_1 = \alpha_2 = \ldots = \alpha_a,$$

the first F ratio is made whose numerator is the **estimation** of the **variance** of **factor** A and denominator is the estimation of the variance within treatments:

$$F = \frac{s_A^2}{s_E^2}.$$

If F is greater than or equal to the **value** in the **Fisher table** for $a - 1$ and $ab(c - 1)$ **degrees of freedom**, the **null hypothesis** is rejected and it is concluded that at least one α_i is different from α_j, $i \neq j$.

To test the **null hypothesis**

$$H_0: \quad \beta_1 = \beta_2 = \ldots = \beta_h,$$

the second F ratio is made whose numerator is the **estimation** of the **variance** of **factor** B and denominator is the estimation of the variance within treatments:

$$F = \frac{s_B^2}{s_E^2}.$$

If F is greater than or equal to the **value** in the **Fisher table** for $b - 1$ and $ab(c - 1)$ **degrees of freedom**, the **null hypothesis** is rejected and it is concluded that H, is true.

To test the **null hypothesis**

$$H_0: \quad (\alpha\beta)_{11} = (\alpha\beta)_{12} = \ldots = (\alpha\beta)_{1b}$$
$$= (\alpha\beta)_{21} = \ldots = (\alpha\beta)_{ab}$$

the third F ratio is made whose numerator is the **estimation** of the **variance** of **interaction** AB and whose denominator is the estimation of the variance within treatments:

$$F = \frac{s_{AB}^2}{s_E^2}.$$

If F is greater than or equal to the **value** in the **Fisher table** for $(a-1)(b-1)$ and $ab(c-1)$ **degrees of freedom**, the **null hypothesis** is rejected and it is concluded that H, is true.

Table of Analysis of Variance

All the information required to calculate the two F ratios can be summarized in a table of **analysis of variance**:

Source of variation	Degrees of freedom	Sum of squares	Mean of squares	F
Mean	1	SS_M		
Factor A	$a - 1$	SS_A	s_A^2	$\dfrac{s_A^2}{s_E^2}$
Factor B	b	1	SS_B	s_B^2	$\dfrac{s_B^2}{s_E^2}$
Interaction AB	$(a - 1) \cdot (b - 1)$	SS_{AB}	s_{AB}^2	$\dfrac{s_{AB}^2}{s_E^2}$
Within treatments	$ab(c-1)$	SS_I	s_E^2	
Total	abc	SS_T		

If $c = 1$, meaning that there is only one **observation** per treatment ij (or per cell ij), it is not possible to attribute part of the **error** "inside the groups" to the **interaction**.

T

EXAMPLES

The director of a school wants to test four brands of typewriters. He asks five professional secretaries to try each brand and repeat the exercise the next day. They must all type the same text, and after 15 min the average number of words typed in 1 min is recorded. Here are the results:

Typewriter	Secretary				
	1	2	3	4	5
1	33	31	34	34	31
	36	31	36	33	31
2	32	37	40	33	35
	35	35	36	36	36
3	37	35	34	31	37
	39	35	37	35	40
4	29	31	33	31	33
	31	33	34	27	33

Let us do a **two-way analysis of variance**, the first factor being the brands of the typewriters and the second the secretaries.
The **null hypotheses** are the following:

1.

H_0: $\alpha_1 = \alpha_2 = \alpha_3 = \alpha_4$

H_1: At least one α_i is different from α_j, $i \neq j$.

2.

H_0: $\beta_1 = \beta_2 = \beta_3 = \beta_4 = \beta_5$

H_1: At least one β_i is different from β_j, $i \neq j$.

3.

H_0: $(\alpha\beta)_{11} = (\alpha\beta)_{12} = \ldots = (\alpha\beta)_{15}$

$\qquad = (\alpha\beta)_{21} = \ldots = (\alpha\beta)_{45}$

H_1: At least one of the interactions is different from the others.

Variance of Factor A "Brands of Typewriters"

Variance of factor A

The sum of squares of **factor A** is equal to:

$$SS_A = 5 \cdot 2 \sum_{i=1}^{4} (\bar{Y}_{i..} - \bar{Y}_{...})^2$$

$$= 5 \cdot 2 \left[(33 - 34)^2 + (35.5 - 34)^2 \right.$$

$$\left. + (36 - 34)^2 + (31.5 - 34)^2 \right]$$

$$= 135 .$$

The number of **degrees of freedom** associated to this sum is equal to $4 - 1 = 3$.
The **variance** of **factor** A (**mean** of squares) is then equal to:

$$s_A^2 = \frac{SS_A}{a - 1}$$

$$= \frac{135}{3}$$

$$= 45 .$$

Variance of Factor B "Secretaries"

The sum of squares of **factor** B is equal to:

$$SS_B = 4 \cdot 2 \sum_{j=1}^{5} (\bar{Y}_{.j.} - \bar{Y}_{...})^2$$

$$= 4 \cdot 2 \left[(34 - 34)^2 + (33.5 - 34)^2 \right.$$

$$+ \ldots + (35.5 - 34)^2$$

$$\left. + (34.5 - 34)^2 \right] = 40 .$$

The number of **degrees of freedom** associated to this sum is equal to $5 - 1 = 4$.
The **variance** of **factor** B (**mean** of squares) is then equal to:

$$s_B^2 = \frac{SS_B}{b - 1}$$

$$= \frac{40}{4}$$

$$= 10 .$$

Variance of Interaction AB "Brands of Typewriters – Secretaries"

The sum of squares of **interaction** AB is equal to:

$$SS_{AB} = 2 \sum_{i=1}^{4} \sum_{j=1}^{5} (\bar{Y}_{ij.} - \bar{Y}_{i..} - \bar{Y}_{.j.} + \bar{Y}_{...})^2$$

$$= 2 \left[(34.5 - 33 - 34 + 34)^2 + \right.$$

$$(31 - 31.5 - 34.5 + 34)^2 + \ldots$$

$$\left. + (33 - 31.5 - 34.5 + 34)^2 \right]$$

$$= 2 \left[1.5^2 + (-1.5)^2 + \ldots + 1^2 \right]$$

$$= 83.$$

The number of **degrees of freedom** associated to this sum is equal to $(4-1)(5-1) = 12$.

The **variance** of **interaction** AB is then equal to:

$$s^2_{AB} = \frac{SS_{AB}}{(a-1)(b-1)}$$

$$= \frac{83}{12}$$

$$= 6.92.$$

Variance Within Treatments (Error)

The sum of squares within treatments is equal to:

$$SS_E = \sum_{i=1}^{4} \sum_{j=1}^{5} \sum_{k=1}^{2} (Y_{ijk} - \bar{Y}_{ij.})^2$$

$$= (33 - 34.5)^2 + (36 - 34.5)^2$$

$$+ (31 - 31)^2 + \ldots + (33 - 33)^2$$

$$= 58.$$

The number of **degrees of freedom** associated to this sum is equal to $4 \cdot 5(2-1) = 20$:

$$s^2_E = \frac{SS_I}{ab(c-1)}$$

$$= \frac{58}{20}$$

$$= 2.9.$$

The total sum of squares is equal to:

$$SS_T = \sum_{i=1}^{4} \sum_{j=1}^{5} \sum_{k=1}^{2} (Y_{ijk})^2$$

$$= 33^2 + 36^2 + 31^2 + \ldots$$

$$+ 27^2 + 33^2 + 33^2$$

$$= 46556.$$

The number of **degrees of freedom** associated to this sum is equal to the number N of observations, meaning 40.

The sum of squares for the **mean** is equal to:

$$SS_M = N \bar{Y}^2_{...}$$

$$= 40 \cdot 34^2$$

$$= 46240.$$

The number of **degrees of freedom** associated to this sum is equal to 1.

Fisher Tests

To test the **null hypothesis**

$$H_0: \quad \alpha_1 = \alpha_2 = \alpha_3 = \alpha_4,$$

we form the first F ratio whose numerator is the **estimation** of the **variance** of **factor** A and denominator is the estimation of the variance within treatments:

$$F = \frac{s^2_A}{s^2_E}$$

$$= \frac{45}{2.9}$$

$$= 15.5172.$$

The **value** of the **Fisher table** with 3 and 20 **degrees of freedom** for a **significant level** $\alpha = 0.05$ is equal to 3.10.

Since $F \geq F_{3,20,0.05}$, we reject the H_0 **hypotheses** and conclude that the brands of typewriters have a significant effect on the number of words typed in 1 min.

To test the **null hypothesis**

$$H_0: \quad \beta_1 = \beta_2 = \beta_3 = \beta_4 = \beta_5,$$

we form the second F ratio whose numerator is the **estimation** of the **variance** of **factor** B and whose denominator is the estimation of

T

the variance within treatments:

$$F = \frac{s_B^2}{s_E^2}$$

$$= \frac{10}{2.9}$$

$$= 3.4483 \, .$$

The **value** of the **Fisher table** with 4 and 20 **degrees of freedom** for $\alpha = 0.05$ is equal to 2.87.

Since $F \geq F_{4,20,0.05}$, we reject the H_0 **hypotheses** and conclude that the secretaries have a significant effect on the number of words typed in 1 min.

To test the **null hypothesis**

$$H_0: \quad (\alpha\beta)_{11} = (\alpha\beta)_{12} = \ldots = (\alpha\beta)_{15}$$
$$= (\alpha\beta)_{21} = \ldots = (\alpha\beta)_{45} \, ,$$

we form the third F ratio whose numerator is the **estimation** of the **variance** of **interaction** AB and whose denominator is the estimation of the variance within treatments:

$$F = \frac{s_{AB}^2}{s_E^2}$$

$$= \frac{6.92}{2.9}$$

$$= 2.385 \, .$$

The **value** of the **Fisher table** with 12 and 20 **degrees of freedom** for $\alpha = 0.05$ is equal to 2.28.

Since $F \geq F_{4,20,0.05}$, we reject the H_0 **hypotheses** and conclude that the effect of the interaction between the secretary and the type of typewriter is significant on the number of words typed in 1 min.

Table of Analysis of Variance

All this information is summarized in the table of **analysis of variance**:

Source of variation	Degrees of freedom	Sum of squares	Mean of squares	F
Mean	1	46240		
Factor A	3	135	45	15.5172
Factor B	4	40	10	3.4483
Interaction AB	12	83	6.92	2.385
Within treatments	20	58	2.9	
Total	40	46556		

FURTHER READING

▶ **Analysis of variance**
▶ **Contrast**
▶ **Fisher table**
▶ **Fisher test**
▶ **Interaction**
▶ **Least significant difference test**
▶ **One-way analysis of variance**

REFERENCES

Cox, D.R., Reid, N.: Theory of the design of experiments. Chapman & Hall, London (2000)

Fisher, R.A.: Statistical Methods for Research Workers. Oliver & Boyd, Edinburgh (1925)

Yates, F.: Complex experiments. Suppl. J. Roy. Stat. Soc. **2**, 181–247 (1935)

Yates, F.: The design and analysis of factorial experiments. Technical Communication of the Commonwealth Bureau of Soils **35**, Commonwealth Agricultural Bureaux, Farnham Royal (1937)

Type I Error

When **hypothesis testing** is carried out, a type I error occurs when one rejects the null hypothesis when it is true, and α is the **probability** of rejecting the **null hypothesis** H_0 when it is true:

$$\alpha = P(\text{reject } H_0 \mid H_0 \text{ is true}) .$$

The probability of type I error α is equal to the **significance level** of the **hypothesis test**.

HISTORY

In 1928, **Neyman, J.** and **Pearson, E.S.** were the first authors to recognize that a rational choice of **hypothesis testing** must take into account not only the **hypothesis** that one wants to verify but also the **alternative hypothesis**. They introduced the type I **error** and the **type II error**.

DOMAINS AND LIMITATIONS

The type I error is one of the two **errors** the statistician is confronted with in **hypothesis testing**: it is the type of error that can occur in decision making if the **null hypothesis** is true.

If the **null hypothesis** is wrong, another type of error arises called the **type II error**, meaning accepting the null hypothesis H_0 when it is false.

The different types of errors can be represented by the following table:

Situation	Decision	
	Accept H_0	Reject H_0
H_0 true	$1 - \alpha$	α
H_0 false	β	$1 - \beta$

with α is the probability of the type I error and β is the probability of the **type II error**.

EXAMPLES

We will illustrate, using an example from everyday life, the different errors that one can commit in making a decision.

When we leave home in the morning, we wonder what the weather will be like. If we think it is going to rain, we take an umbrella. If we think it is going to be sunny, we do not take anything for bad weather.

Therefore we are confronted with the following **hypotheses**:

Null hypothesis H_0: It is going to rain.

Alternative hypothesis H_1: It is going to be sunny.

Suppose we accept the rain hypothesis and take an umbrella.

If it really does rain, we have made the right decision, but if it is sunny, we made an error: the error of accepting a false hypothesis.

In the opposite case, if we reject the rain hypothesis, we have made a good decision if it is sunny, but we will have made an error if it rains: the error of rejecting a true hypothesis.

We can represent these different types of errors in the following table:

Situation	Decision	
	Accept H_0 (I take an umbrella)	Reject H_0 (I do not take an umbrella)
H_0 true (it rains)	Good decision	Type I error
H_0 false (it is sunny)	Type II error β	Good decision

The **probability** α of rejecting a true **hypothesis** is called level of significance.

The **probability** β of accepting a false hypothesis is the probability of **type II error**.

FURTHER READING
▶ **Hypothesis testing**
▶ **Significance level**
▶ **Type II error**

T

REFERENCES

Neyman, J., Pearson, E.S.: On the use and interpretation of certain test criteria for purposes of statistical inference, Parts I and II. Biometrika **20**A, 175–240, 263–294 (1928)

Type II Error

When carrying out a **hypothesis test**, a type II **error** occurs when accepting the null hypothesis given it is false and β is the **probability** of accepting the **null hypothesis** H_0 when it is false:

$$\beta = P(\text{accept } H_0 \mid H_0 \text{ is false}).$$

HISTORY

See **type I error**.

DOMAINS AND LIMITATIONS

The probability of type II **error** is denoted by β, the **probability** of rejecting the **null hypothesis** when it is false is given by $1 - \beta$:

$$\beta = P(\text{reject } H_0 \mid H_0 \text{ is false}).$$

The **probability** $1 - \beta$ is called the power of the test.

Therefore it is always necessary to try and minimize the risk of having a type II **error**, which means increasing the power of the **hypothesis test**.

EXAMPLES

See **type I error**.

FURTHER READING

▶ **Hypothesis testing**
▶ **Type I error**

Uniform Distribution

A uniform distribution is used to describe a **population** that is uniformly distributed in an **interval**.

A **random variable** X is said to be uniformly distributed in the **interval** $[a, b]$ if its **density function** is given by:

$$f(x) = \begin{cases} \dfrac{1}{b-a} & \text{if } a \leq x \leq b \\ 0 & \text{otherwise}. \end{cases}$$

Because of the rectangular shape of its density function, the uniform distribution is also sometimes called the rectangular distribution.

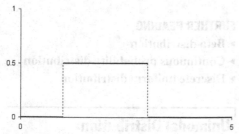

Uniform distribution, $a = 1$, $b = 3$

The uniform distribution is a **continuous probability distribution**.

MATHEMATICAL ASPECTS

The **distribution function** of a uniform distribution is the following:

$$F(x) = \begin{cases} 0 & \text{if } x \leq a \\ \dfrac{x-a}{b-a} & \text{if } a < x < b \\ 1 & \text{otherwise}. \end{cases}$$

The **expected value** of a **random variable** that is uniformly distributed in an **interval** $[a, b]$ is equal to:

$$E[X] = \int_a^b \frac{x}{b-a}\, dx = \frac{b^2 - a^2}{2(b-a)} = \frac{a+b}{2}.$$

This result is intuitively very easy to understand because the **density function** is symmetric around the middle point of the **interval** $[a, b]$.

We can calculate

$$E\left[X^2\right] = \int_a^b x^2 f(x)\, dx = \frac{b^2 + ab + a^2}{3}.$$

The **variance** is therefore equal to:

$$\text{Var}(X) = E\left[X^2\right] - (E[X])^2 = \frac{(b-a)^2}{12}.$$

The uniform distribution is a particular case of the **beta distribution** when $\alpha = \beta = 1$.

DOMAINS AND LIMITATIONS

Applications of the uniform distribution are numerous in the construction of **models** for physical, biological, and social phenomena. The uniform distribution is often used as a **model** for **errors** of approximation during measurements. For example, if the length of some objects is measured with a 1-cm precision, a length of 35 cm can represent all the objects with a length between 34.5 and 35.5 cm. The approximation errors follow a uniform distribution in the interval $[-0.5, 0.5]$.

The uniform distribution is often used to generate **random numbers** of any discrete or continuous **probability distribution**. The method used to generate a continuous distribution is the following.

Consider X a continuous **random variable** and $F(x)$ its **distribution function**. Suppose that F is continuous and strictly increasing. Then $U = F(X)$ is a uniform and continuous random variable. U takes its **values** in the interval $[0,1]$. We then have $X = F^{-1}(U)$, where F^{-1} is the inverse function of the **distribution function**.

Therefore, if $u_1, u_2, \ldots, u_n$ is a series of random numbers, then:

$$x_1 = F^{-1}(u_1),$$
$$x_2 = F^{-1}(u_2),$$
$$\ldots,$$
$$x_n = F^{-1}(u_n)$$

is a series of random numbers generated from the distribution of the **random variable** X.

EXAMPLES

If **random variable** X follows a negative **exponential distribution** of mean equal to 1, then its **distribution function** is

$$F(x) = 1 - \exp(-x).$$

Since $F(x) = u$, we have $u = 1 - e^{-x}$. To find x, we search for F^{-1}:

$$u = 1 - e^{-x},$$
$$u - 1 = -e^{-x},$$
$$1 - u = e^{-x},$$
$$\ln(1 - u) = -x,$$
$$-\ln(1 - u) = x.$$

Notice that if u is a **random variable** that is uniform in $[0,1]$, then $(1 - u)$ is also a **random variable** that is uniform in $[0,1]$.

By applying the aforementioned to a series of **random numbers**, we find a **sample** taken from X:

U	X
0.14	1.97
0.97	0.03
0.53	0.63
0.73	0.31
⋮	⋮

FURTHER READING
▶ **Beta distribution**
▶ **Continuous probability distribution**
▶ **Discrete uniform distribution**

Unimodal Distribution

See **frequency distribution**.

Value

A value is a quantitative or qualitative measure associated to a **variable**.

We speak about a set of values taken by a variable. The values can be cardinal numbers if they are associated to a **quantitative variable** or ordinal numbers if they are associated to a **qualitative categorical variable**

EXAMPLES

The values taken by a certain **variable** can be of the following type:

- Consider the **quantitative variable** "number of children". The values associated to this variable are the natural numbers 0, 1, 2, 3, 4, etc.
- Consider the **qualitative categorical variable** "sex." The associated values are the categories "masculine" and "feminine".

FURTHER READING
► Qualitative categorical variable
► Quantitative variable
► Variable

Variable

A variable is a measurable characteristic to which are attributable many different variables.

We distinguish many types of variables:

- **Quantitative variable**, of which we distinguish:
  - Discrete variable, e.g., the result of a test
  - Continuous variable, e.g., a weight or revenue
- **Qualitative categorical variable**, e.g., marital status

FURTHER READING
► Data
► Dichotomous variable
► Qualitative categorical variable
► Quantitative variable
► Random variable
► Value

Variance

Variance is a **measure of dispersion** of a distribution of a **random variable**.

Empirically, the variance of a **quantitative variable** X is defined as the sum of squared deviations of each **observation** relative to the **arithmetic mean** divided by the number of observations.

HISTORY

The **analysis of variance** such as we understand and practice it today was principally developed by **Fisher, Ronald Aylmer** (1918, 1925, 1935). He also introduced the terms variance and analysis of variance.

MATHEMATICAL ASPECTS

Variance is generally denoted by S^2 when it is relative to a **sample** and by σ^2 when it is relative to a **population**.

We also denote the variance by Var (X) when we speak about the **variance of a random variable**.

Let a population of N observations be relative to a **quantitative variable** X. By definition, the variance of a population is calculated as follows:

$$\sigma^2 = \frac{\sum\limits_{i=1}^{N}(x_i - \mu)^2}{N},$$

where N is the dimension of the population and μ the **mean** of the observation:

$$\mu = \frac{\sum\limits_{i=1}^{N}x_i}{N}.$$

When the observations are ordered in the form of a **frequency distribution**, the calculation of the variance is made in the following manner:

$$\sigma^2 = \frac{\sum\limits_{i=1}^{k}f_i \cdot (x_i - \mu)^2}{\sum\limits_{i=1}^{k}f_i},$$

where x_i are the different values of the **variable**, f_i are the frequencies associated to these values, and k is the number of different values.

To calculate the variance of the frequency distribution of a quantitative X where the values are grouped in classes, we consider that all the observations belonging to a certain class take the values of the center of the class. It is correct only if the **hypothesis** specifying that the observations are uniformly distributed inside each class is verified. If this hypothesis is not verified, the obtained value of the variance will be only approximative.

For the values grouped in classes, we have:

$$\sigma^2 = \frac{\sum\limits_{i=1}^{k}f_i \cdot (\delta_i - \mu)^2}{\sum\limits_{i=1}^{k}f_i},$$

where δ_i are the centers of the classes, f_i are the frequencies associated to each class, and k is the number of classes.

The formula for calculating the variance can be modified to decrease the time of calculation and to increase the precision. Thus it is better to calculate the variance with the following formula:

$$\sigma^2 = \frac{N \cdot \sum\limits_{i=1}^{N}x_i^2 - \left(\sum\limits_{i=1}^{N}x_i\right)^2}{N^2}.$$

The variance corresponds to the centered **moment** of order two.

The variance measured on a sample is an **estimator** of the variance of the population.

Consider a sample of n observations relative to a quantitative variable X and denote by $\bar{x}$ the **arithmetic mean** of this sample:

$$\bar{x} = \frac{\sum\limits_{i=1}^{n}x_i}{n}.$$

To let the variance of the sample be an estimator without **bias** of the variance of the population, it must be calculated by dividing by

$(n - 1)$ and not by n:

$$S^2 = \frac{\sum_{i=1}^{n} (x_i - \bar{x})^2}{n - 1}.$$

EXAMPLES

Five students passed two exams, earning the following grades:

Exam 1:

3.5	4	4.5	3.5	4.5

Exam 2:

2.5	5.5	3.5	4.5	4

Note that the **arithmetic mean** $\bar{x}$ of these two sets of observations is identical:

$$\bar{x}_1 = \bar{x}_2 = 4.$$

The dispersion of the observations around the **mean** is not the same.

To determine the variance, first we calculate the deviations of each observation from the arithmetic mean and then square the deviations:

Exam 1:

Grade	$(x_i - \bar{x}_1)$	$(x_i - \bar{x}_1)^2$
3.5	−0.5	0.25
4	0.0	0.00
4.5	0.5	0.25
3.5	−0.5	0.25
4.5	0.5	0.25
$\sum_{i=1}^{5} (x_i - \bar{x}_1)^2$	1.00	

Exam 2:

Grade	$(x_i - \bar{x}_2)$	$(x_i - \bar{x}_2)^2$
2.5	−1.5	2.25
5.5	1.5	2.25
3.5	−0.5	0.25
4.5	0.5	0.25
4	0.0	0.00
$\sum_{i=1}^{5} (x_i - \bar{x}_2)^2$	5.00	

The variance of the grades of each exam equals:

Exam 1:

$$S_1^2 = \frac{\sum_{i=1}^{5} (x_i - \bar{x}_1)^2}{n - 1} = \frac{1}{4} = 0.25$$

Exam 2:

$$S_2^2 = \frac{\sum_{i=1}^{5} (x_i - \bar{x}_2)^2}{n - 1} = \frac{5}{4} = 1.25$$

Since the variance of the grades of the second exam is greater, on the second exam the grades are more dispersed around the arithmetic mean than on the first exam.

FURTHER READING
▶ **Analysis of variance**
▶ **Measure of dispersion**
▶ **Standard deviation**
▶ **Variance of a random variable**

REFERENCES
Fisher, R.A.: The correlation between relatives on the supposition of Mendelian inheritance. Trans. Roy. Soc. Edinb. **52**, 399–433 (1918)

Fisher, R.A.: Statistical Methods for Research Workers. Oliver & Boyd, Edinburgh (1925)

Fisher, R.A.: The Design of Experiments. Oliver & Boyd, Edinburgh (1935)

V

Variance Analyses

See analysis of variance.

Variance of a Random Variable

The **variance** of a quantitative **random variable** measures the value of the mean deviations of the values of this random variable from the **mathematical expectancy** (that is, from its **mean**).

MATHEMATICAL ASPECTS

Depending on whether the **random variable** is discrete or continuous, we speak about the variance of a discrete or continuous random variable.

In the case where the random variable is discrete X taking n values $x_1, x_2, \ldots, x_n$ with relative frequencies $f_1, f_2, \ldots, f_n$, the **variance** is defined by:

$$\sigma^2 = \text{Var}(X) = E\left[(X - \mu)^2\right]$$
$$= \sum_{i=1}^{n} f_i (x_i - \mu)^2,$$

where μ represents the **mathematical expectancy** of X.

We can establish another formula for which the calculation of the variance is made thus:

$$\text{Var}(X) = E\left[(X - \mu)^2\right]$$
$$= E\left[X^2 - 2\mu X + \mu^2\right]$$
$$= E\left[X^2\right] - E\left[2\mu X\right] + E\left[\mu^2\right]$$
$$= E\left[X^2\right] - 2\mu E\left[X\right] + \mu^2$$
$$= E\left[X^2\right] - 2\mu^2 + \mu^2$$
$$= E\left[X^2\right] - \mu^2$$

$$= E\left[X^2\right] - (E[X])^2$$
$$= \sum_{i=1}^{n} f_i x_i^2 - \left(\sum_{i=1}^{n} f_i x_i\right)^2.$$

In practice, this formula is generally better suited for making the calculations.

If random variable X is continuous in **interval** D, the expression of the variance becomes:

$$\sigma^2 = \text{Var}(X) = E\left[(X - \mu)^2\right]$$
$$= \int_D (x - \mu)^2 f(x)\, dx.$$

We can also transform this formula by the same operations as in the discrete case, and so the variance equals:

$$\text{Var}(X) = \int_D x^2 f(x)\, dx - \left(\int_D x f(x)\, dx\right)^2.$$

Properties of the Variance

1. Let a and b be two constants and X a random variable:

$$\text{Var}(aX + b) = a^2 \cdot \text{Var}(X).$$

2. Let X and Y be two random variables:

$$\text{Var}(X + Y) = \text{Var}(X) + \text{Var}(Y) + 2\text{Cov}(X, Y),$$
$$\text{Var}(X - Y) = \text{Var}(X) + \text{Var}(Y) - 2\text{Cov}(X, Y),$$

where $\text{Cov}(X, Y)$ is the **covariance** between X and Y. In particular, if X and Y are independent:

$$\text{Var}(X + Y) = \text{Var}(X) + \text{Var}(Y)$$
$$\text{Var}(X - Y) = \text{Var}(X) + \text{Var}(Y).$$

EXAMPLES

We consider two examples, one concerning a discrete **random variable**, another a continuous random variable.

We throw a die several times in a row. Suppose we win 1 euro if the result is even and 2 euros if the result is 1 or 3, and that we lose 3 euros if the result is 5.

Random variable X describes the number of euros won or lost. The following table represents different values of X and their respective **probabilities**:

X	-3	1	2
$P(X)$	$\frac{1}{6}$	$\frac{3}{6}$	$\frac{2}{6}$

The variance of discrete random variable X equals:

$$\text{Var}(X) = \sum_{i=1}^{3} f_i x_i^2 - \left[\sum_{i=1}^{3} f_i x_i \right]^2$$

$$= (-3)^2 \frac{1}{6} + (1)^2 \frac{3}{6} + (2)^2 \frac{2}{6}$$

$$- \left(-3\frac{1}{6} + 1\frac{3}{6} + 2\frac{2}{6} \right)^2$$

$$= \frac{26}{9}.$$

Consider a continuous random variable X whose **density function** is uniform on $(0, 1)$:

$$f(x) = \begin{cases} 1 & \text{for } 0 < x < 1 \\ 0 & \text{otherwise}. \end{cases}$$

The **mathematical expectancy** of this random variable equals $\frac{1}{2}$.

We can calculate the variance of random variable X:

$$\text{Var}(X) = \int_0^1 x^2 f(x)\, dx - \left(\int_0^1 x f(x)\, dx \right)^2$$

$$= \int_0^1 x^2 \cdot 1\, dx - \left(\int_0^1 x \cdot 1\, dx \right)^2$$

$$= \frac{x^3}{3} \Big|_0^1 - \left(\frac{x^2}{2} \Big|_0^1 \right)^2$$

$$= \frac{1}{3} - \left(\frac{1}{2} \right)^2$$

$$= \frac{1}{12}.$$

FURTHER READING

▶ **Expected value**
▶ **Random variable**

Variance–Covariance Matrix

The variance–covariance matrix $\mathbf{V}$ between n variables $Y_1, \ldots, X_n$ is a matrix of order $(n \times n)$, where the diagonal contains the **variances** of each variable X_i and outside the diagonals are the **covariances** between the pairs of variables (X_i, X_j) for each $i \neq j$. For n variables, we calculate n variances and $n(n-1)$ covariances. The result is that matrix $\mathbf{V}$ is a matrix of dimension $n \times n$. In the literature, we often encounter this matrix under another name: dispersion matrix or covariance matrix.

If the variables are standardized (see **standardized data**), the variances of the variables equal 1 and the covariances become correlations, and so the matrix becomes a correlation matrix.

HISTORY

See **variance** and **covariance**.

MATHEMATICAL ASPECTS

Let $X_1, \ldots, X_n$ be n random variables. The variance–covariance matrix is a matrix of order $(n \times n)$; it is symmetric and contains

V

on the diagonal the variances of each variable and as the other terms the covariances between the variables. Matrix $\mathbf{V}$ is the following matrix:

$$\begin{bmatrix} \text{Var}(X_1) & \text{Cov}(X_1, X_2) & \cdots & \text{Cov}(X_1, X_n) \\ \text{Cov}(X_2, X_1) & \text{Var}(X_2) & \cdots & \text{Cov}(X_2, X_n) \\ \vdots & \vdots & \ddots & \vdots \\ \text{Cov}(X_n, X_1) & \text{Cov}(X_n, X_2) & \cdots & \text{Var}(X_n) \end{bmatrix},$$

with

$$\text{Var}(X_i) = E\left[X_i^2\right] - (E[X_i])^2 \quad \text{and}$$

$$\text{Cov}\left(X_i, X_j\right)$$
$$= E\left[(X_i - E[X_i])\left(X_j - E[X_j]\right)\right].$$

EXAMPLES

Consider the following three variables X_1, X_2, and X_3 summarized in the following table:

X_1	X_2	X_3
1	2	3
2	3	4
1	2	3
5	4	3
4	4	4

In the case $n = 3$, each random variable consists of five terms, which we denote $N = 5$. We want to calculate the variance–covariance matrix $\mathbf{V}$ of these variables. We complete the following steps.

We estimate the expectancy of each variable by its empirical mean $\bar{x}_i$ and get the following results:

$$\bar{X}_1 = 2.6$$
$$\bar{X}_2 = 3$$
$$\bar{X}_3 = 3.4.$$

The variance for a variable X_j is given by:

$$\frac{\sum_{i=1}^{N} \left(x_{ij} - \bar{x}_j\right)^2}{4}.$$

The covariance between two variables X_r and X_s is calculated as follows:

$$\frac{\sum_{i=1}^{N} (x_{ir} - \bar{x}_r)(x_{is} - \bar{x}_s)}{4}.$$

Then we should subtract the mean $\bar{X}_i$ of each variable X_i. We get a new variable $Y_i = X_i - \bar{X}_i$.

Y_1	Y_2	Y_3
−1.6	−1	−0.4
−0.6	0	0.6
−1.6	−1	−0.4
2.4	1	−0.4
1.4	1	0.6

Note that the mean of each new variable Y_i equals 0. We find these results in matrix form:

$$\mathbf{Y} = \begin{bmatrix} -1.6 & -1 & -0.4 \\ -0.6 & 0 & 0.6 \\ -1.6 & -1 & -0.4 \\ 2.4 & 1 & -0.4 \\ 1.4 & 1 & 0.6 \end{bmatrix}.$$

From this matrix we calculate matrix $\mathbf{Y}' \cdot \mathbf{Y}$, a symmetric matrix, in the following form:

$$\begin{bmatrix} \sum_{i=1}^{N} y_{i1}^2 & \sum_{i=1}^{N} y_{i2} \cdot y_{i1} & \sum_{i=1}^{N} y_{i3} \cdot y_{i1} \\ \sum_{i=1}^{N} y_{i2} \cdot y_{i1} & \sum_{i=1}^{N} y_{i2}^2 & \sum_{i=1}^{N} y_{i3} \cdot y_{i2} \\ \sum_{i=1}^{N} y_{i3} \cdot y_{i1} & \sum_{i=1}^{N} y_{i3} \cdot y_{i2} & \sum_{i=1}^{N} y_{i3}^2 \end{bmatrix}.$$

This matrix contains the values of the sum of products of squares of each variable Y_i (on the diagonal) and the cross product of each pair of variables (Y_i, Y_j) on the rest of the matrix. For the original variables X_i, the following matrix contains on the diagonal the sum of squared deviations from the mean for each variable and on the external terms the cross products of these deviations. Numerically we get:

$$\begin{bmatrix} 13.2 & 7 & 0.8 \\ 7 & 4 & 1 \\ 0.8 & 1 & 1.2 \end{bmatrix}.$$

The mean of the sum of squared deviations from the mean of a random variable X_i of N terms is known as the **variance**. The mean of the crossed products of these deviations is called the **covariance**. Thus to get the variance–covariance matrix V of the variables $X_1, \ldots, X_n$, we should divide the elements of matrix $\mathbf{Y}' \cdot \mathbf{Y}$ by $N - 1$, which is mathematically the same as:

$$\begin{bmatrix} \dfrac{\sum\limits_{i=1}^{N} y_{i1}^2}{N-1} & \dfrac{\sum\limits_{i=1}^{N} y_{i2}\cdot y_{i1}}{N-1} & \dfrac{\sum\limits_{i=1}^{N} y_{i3}\cdot y_{i1}}{N-1} \\ \dfrac{\sum\limits_{i=1}^{N} y_{i2}\, y_{i1}}{N-1} & \dfrac{\sum\limits_{i=1}^{N} y_{i2}^2}{N-1} & \dfrac{\sum\limits_{i=1}^{N} y_{i3}\cdot y_{i2}}{N-1} \\ \dfrac{\sum\limits_{i=1}^{N} y_{i3}\cdot y_{i1}}{N-1} & \dfrac{\sum\limits_{i=1}^{N} y_{i3}\cdot y_{i2}}{N-1} & \dfrac{\sum\limits_{i=1}^{N} y_{i3}^2}{N-1} \end{bmatrix}.$$

The fact of dividing by $N-1$ and not by N for obtaining the means is related to the fact that we generally work with a sample of random variables without knowing the real population and for which we make a nonbiased estimation of the variance and of the covariance. Numerically we get:

$$V = \begin{bmatrix} 3.3 & 1.75 & 1.75 \\ 1.75 & 1 & 0.2 \\ 0.2 & 0.25 & 0.3 \end{bmatrix}.$$

FURTHER READING
▶ **Correlation coefficient**
▶ **Covariance**
▶ **Standardized data**
▶ **Variance**

Vector

We call a vector of dimension n an ($n \times$ 1) **matrix**, that is, a table having only one column and n rows. The elements in this table are called coefficients or vector components.

MATHEMATICAL ASPECTS
Let $\mathbf{x} = (x_i)$ and $\mathbf{y} = (y_i)$ for $i = 1, 2, \ldots, n$ be two vectors of dimension n.
We define:
- The sum of two vectors made component by component:

$$\mathbf{x} + \mathbf{y} = (x_i) + (y_i) = (x_i + y_i).$$

- The product of vector $\mathbf{x}$ by the scalar k:

$$k \cdot \mathbf{x} = (k \; x_i),$$

where each component is multiplied by k.
- The **scalar product** of $\mathbf{x}$ and $\mathbf{y}$:

$$\mathbf{x}' \cdot \mathbf{y} = (x_i)' \cdot (y_i) = \sum_{i=1}^{n} x_i \cdot y_i.$$

- The **norm** of $\mathbf{x}$:

$$\| \mathbf{x} \| = \sqrt{\mathbf{x}' \cdot \mathbf{x}}$$

$$\| \mathbf{x} \| = \sqrt{\sum_{i=1}^{n} x_i^2}.$$

We call a unit vector a vector of norm (or length) 1. We obtain a unit vector by dividing a vector by its norm. Thus for example

$$\frac{1}{\| \mathbf{x} \|} (x_i)$$

is a unit vector having the same direction as $\mathbf{x}$.

V

- The ith vector of the canonic bases in the space of n dimensions is defined by the vector of dimension n having a "1" on line i and "0" for other components.
- A vector is a **matrix** of a certain dimension ($n \times 1$). It is possible to multiply a vector by an ($m \times n$) matrix according to the multiplication rules defined for matrices.

DOMAINS AND LIMITATIONS

A vector can be considered a column of numbers, whence the term vector-column, just as we use the term vector-line to designate the **transpose** of a vector.

Thus the **inner product** of two vectors corresponds to a vector-row multiplied by a vector-column.

EXAMPLES

In a space of three dimensions, we consider the vectors:

$$\mathbf{x} = \begin{bmatrix} 1 \\ 2 \\ 3 \end{bmatrix} \quad \text{and} \quad \mathbf{y} = \begin{bmatrix} 0 \\ 1 \\ 0 \end{bmatrix},$$

where $\mathbf{y}$ is the second vector of the canonic basis.

- The **norm** of $\mathbf{x}$ is given by:

$$\| \mathbf{x} \|^2 = \begin{bmatrix} 1 \\ 2 \\ 3 \end{bmatrix}' \cdot \begin{bmatrix} 1 \\ 2 \\ 3 \end{bmatrix}$$

$$= [1\ 2\ 3] \cdot \begin{bmatrix} 1 \\ 2 \\ 3 \end{bmatrix}$$

$$= 1^2 + 2^2 + 3^2 = 1 + 4 + 9$$

$$= 14,$$

from where $\| \mathbf{x} \| = \sqrt{14} \approx 3.74$.

- In the same way, we get $\| \mathbf{y} \| = 1$.
- The sum of two vectors $\mathbf{x}$ and $\mathbf{y}$ equals:

$$\mathbf{x} + \mathbf{y} = \begin{bmatrix} 1 \\ 2 \\ 3 \end{bmatrix} + \begin{bmatrix} 0 \\ 1 \\ 0 \end{bmatrix}$$

$$= \begin{bmatrix} 1 + 0 \\ 2 + 1 \\ 3 + 0 \end{bmatrix} = \begin{bmatrix} 1 \\ 3 \\ 3 \end{bmatrix}.$$

- The multiplication of vector $\mathbf{x}$ by the scalar product 3 gives:

$$3 \cdot \mathbf{x} = 3 \cdot \begin{bmatrix} 1 \\ 2 \\ 3 \end{bmatrix}$$

$$= \begin{bmatrix} 3 \cdot 1 \\ 3 \cdot 2 \\ 3 \cdot 3 \end{bmatrix} = \begin{bmatrix} 3 \\ 6 \\ 9 \end{bmatrix}.$$

- The **scalar product** of $\mathbf{x}$ and $\mathbf{y}$ equals:

$$\mathbf{x}' \cdot \mathbf{y} = \begin{bmatrix} 1 \\ 2 \\ 3 \end{bmatrix}' \cdot \begin{bmatrix} 0 \\ 1 \\ 0 \end{bmatrix}$$

$$= [1\ 2\ 3] \cdot \begin{bmatrix} 0 \\ 1 \\ 0 \end{bmatrix}$$

$$= 1 \cdot 0 + 2 \cdot 1 + 3 \cdot 0 = 2.$$

FURTHER READING

▶ **Matrix**
▶ **Transpose**

Von Mises Richard

Von Mises, Richard was born in 1883 in Lemberg in the Austro–Hungarian Empire (now L'viv, Ukraine) and died in 1953. Von

Mises studied mechanical engineering at the Technical University in Vienna until 1906. He became assistant to Georg Hamel, professor of mechanics, in Bruenn (now Brno, Czech Republic), where he received the Venia Legendi in 1908. He then worked as associate professor of applied mathematics. After the First World War, in 1920, he became director of the Institute of Applied Mathematics in Berlin.

Von Mises, Richard is principally known for his work on the foundations of probability and statistics, which were rehabilitated in the 1960s. He founded the applied school of mathematics and wrote his first work on philosophical positivism in 1939.

Selected principal works and articles of von Mises, Richard:

1928 Wahrscheinlichkeit, Statistik und Wahrheit. Springer, Berlin Heidelberg New York.

1939 Probability, Statistics and Truth, trans. Neyman, J., Scholl, D., and Rabinowitsch, E. Hodge, Glasgow.

V

Weighted Arithmetic Mean

The **weighted arithmetic mean** is a **measure of central tendency** of a set of quantitative observations when not all the observations have the same importance.

We must assign a weight to each observation depending on its importance relative to other observations.

The weighted arithmetic mean equals the sum of observations multiplied by their weights divided by the sum of their weights.

HISTORY

The weighted arithmetic mean was introduced by Cotes, Roger in 1712. His work was published in 1722, six years after his death.

MATHEMATICAL ASPECTS

Let $x_1, x_2, \ldots, x_n$ be a set of n quantities or n observations relative to a **quantitative variable** X to which we assign the weights $w_1, w_2, \ldots, w_n$.

The weighted arithmetic mean equals:

$$\bar{x} = \frac{\sum_{i=1}^{n} x_i \cdot w_i}{\sum_{i=1}^{n} w_i} \,.$$

DOMAINS AND LIMITATIONS

The weighted arithmetic mean is now used in economics, especially in consumer and producer price indices, etc.

EXAMPLES

Suppose that during a course on the company, management was composed of three relatively different requirements of different importance:

Individual project	30%
Mid-term exam	20%
Final exam	50%

Each student receives a grade on a scale of 1 to 10. One student receives an 8 for his individual project, a 9 on the mid-term exam, and a 4 on the final exam.

His final grade is calculated by the weighted arithmetic mean:

$$\bar{x} = \frac{(8 \cdot 30) + (9 \cdot 20) + (4 \cdot 50)}{30 + 20 + 50} = 6.2 \,.$$

In this example, the weights correspond to the relative importance of the different requirements of the course.

FURTHER READING
▶ **Arithmetic mean**
▶ **Mean**
▶ **Measure of central tendency**

REFERENCES

Cotes, R.: Aestimatio Errorum in Mixta Mathesi, per variationes partium Trianguli plani et sphaerici. In: Smith, R. (ed.) Opera Miscellania, Cambridge (1722)

Weighted Least-Squares Method

The weighted least-squares method is used when the variance of errors is not constant, that is, when the following hypothesis of the least-squares method is violated: the variance of errors is constant (equal to the unknown value σ^2) for any observation i (that is, whatever the value of the concerned x_{ij}). Thus, instead of having, for each $i = 1,\ldots,n$, $\text{Var}(\varepsilon_i) = \sigma^2$, we have:

$$\text{Var}(\varepsilon_i) = \sigma^2 w_i,$$

where the *weights* $w_i > 0$ can be different for each $i = 1,\ldots,n$.

MATHEMATICAL ASPECTS

In the matrix form, we have the model

$$\mathbf{Y} = \mathbf{X}\boldsymbol{\beta} + \boldsymbol{\varepsilon},$$

where $\mathbf{Y}$ is the vector $(n \times 1)$ of the observations relative to the **dependent variable** (n observations), $\boldsymbol{\beta}$ is the vector $(p \times 1)$ of parameters to be estimated, $\boldsymbol{\varepsilon}$ is the vector $(n \times 1)$ of errors, and

$$\mathbf{X} = \begin{pmatrix} 1 & X_{11} & \cdots & X_{1(p-1)} \\ \vdots & \vdots & & \vdots \\ 1 & X_{n1} & \cdots & X_{n(p-1)} \end{pmatrix}$$

is the **matrix** $(n \times p)$ of the plan concerning the independent variables. Moreover, we have:

$$\text{Var}(\boldsymbol{\varepsilon}) = \sigma^2 \mathbf{V},$$

where

$$\mathbf{V} = \begin{pmatrix} 1/w_1 & 0 & \cdots & 0 \\ 0 & 1/w_2 & \cdots & 0 \\ \vdots & \vdots & \ddots & \vdots \\ 0 & 0 & \cdots & 1/w_n \end{pmatrix}.$$

Stating

$$\mathbf{Y}_w = \mathbf{W}\mathbf{Y},$$
$$\mathbf{X}_w = \mathbf{W}\mathbf{X},$$
$$\boldsymbol{\varepsilon}_w = \mathbf{W}\boldsymbol{\varepsilon},$$

with

$$\mathbf{W} = \begin{pmatrix} \sqrt{w_1} & 0 & \cdots & 0 \\ 0 & \sqrt{w_2} & \cdots & 0 \\ \vdots & \vdots & \ddots & \vdots \\ 0 & 0 & \cdots & \sqrt{w_n} \end{pmatrix},$$

such as $\mathbf{W}'\mathbf{W} = \mathbf{V}^{-1}$, we obtain the equivalent model:

$$\mathbf{Y}_w = \mathbf{X}_w \boldsymbol{\beta} + \boldsymbol{\varepsilon}_w,$$

where

$$\text{Var}(\boldsymbol{\varepsilon}_w) = \text{Var}(\mathbf{W}\boldsymbol{\varepsilon}) = \mathbf{W}\text{Var}(\boldsymbol{\varepsilon})\mathbf{W}'$$
$$= \sigma^2 \cdot \mathbf{W}\mathbf{V}\mathbf{W} = \sigma^2 \mathbf{I}_n,$$

where $\mathbf{I}_n$ is the **identity matrix** of dimension n. Since, for this new model, the variance of errors is constant, the least-squares method is used. The vector of estimators is:

$$\hat{\boldsymbol{\beta}}_w = \left(\mathbf{X}_w'\mathbf{X}_w\right)^{-1}\mathbf{X}_w'\mathbf{Y}_w$$
$$= \left(\mathbf{X}'\mathbf{W}'\mathbf{W}\mathbf{X}\right)^{-1}\mathbf{X}'\mathbf{W}'\mathbf{W}\mathbf{Y}$$
$$= \left(\mathbf{X}'\mathbf{V}^{-1}\mathbf{X}\right)^{-1}\mathbf{X}'\mathbf{V}^{-1}\mathbf{Y}.$$

The vectors of estimated values for $\mathbf{Y}_w = \mathbf{W}\mathbf{Y}$ become:

$$\widehat{\mathbf{Y}}_w = \mathbf{X}_w\hat{\boldsymbol{\beta}}_w$$
$$= \mathbf{W}\mathbf{X}\left(\mathbf{X}'\mathbf{V}^{-1}\mathbf{X}\right)^{-1}\mathbf{X}'\mathbf{V}^{-1}\mathbf{Y}.$$

We can obtain a vector of estimated values for $\mathbf{Y} = \mathbf{W}^{-1}\mathbf{Y}_w$ setting:

$$\widehat{\mathbf{Y}} = \mathbf{W}^{-1}\widehat{\mathbf{Y}}_w$$

$$= \mathbf{W}^{-1}\mathbf{W}\mathbf{X}\left(\mathbf{X}'\mathbf{V}^{-1}\mathbf{X}\right)^{-1}\mathbf{X}'\mathbf{V}^{-1}\mathbf{Y}$$

$$= \mathbf{X}\widehat{\boldsymbol{\beta}}_w .$$

The variance σ^2 is estimated by:

$$s_w^2 = \frac{\left(\mathbf{Y}_w - \widehat{\mathbf{Y}}_w\right)'\left(\mathbf{Y}_w - \widehat{\mathbf{Y}}_w\right)}{n - p}$$

$$= \frac{\left(\mathbf{Y} - \widehat{\mathbf{Y}}\right)'\mathbf{W}'\mathbf{W}\left(\mathbf{Y} - \widehat{\mathbf{Y}}\right)}{n - p}$$

$$= \frac{\sum_{i=1}^{n} w_i\left(y_i - \widehat{y}_i\right)^2}{n - p}.$$

In the case of a simple regression, from a sample of n observations (x_i, y_i) and weights w_i we have:

$$\mathbf{X}'\mathbf{V}^{-1}\mathbf{X} = \begin{pmatrix} \sum_{i=1}^{n} w_i & \sum_{i=1}^{n} w_i x_i \\ \sum_{i=1}^{n} w_i x_i & \sum_{i=1}^{n} w_i x_i^2 \end{pmatrix},$$

$$\mathbf{X}\mathbf{V}^{-1}\mathbf{Y} = \begin{pmatrix} \sum_{i=1}^{n} w_i y_i \\ \sum_{i=1}^{n} w_i x_i y_i \end{pmatrix},$$

and thus if we set $\widehat{\boldsymbol{\beta}}_w = \begin{pmatrix} \widehat{\beta}_{0w} \\ \widehat{\beta}_{1w} \end{pmatrix}$, we find:

$$\widehat{\beta}_{0w} = \frac{\sum_{i=1}^{n} w_i y_i \sum_{i=1}^{n} w_i x_i^2 - \sum_{i=1}^{n} w_i x_i \sum_{i=1}^{n} w_i x_i y_i}{\sum_{i=1}^{n} w_i \sum_{i=1}^{n} w_i x_i^2 - \left(\sum_{i=1}^{n} w_i x_i\right)^2},$$

$$\widehat{\beta}_{1w} = \frac{\sum_{i=1}^{n} w_i \sum_{i=1}^{n} w_i x_i y_i - \sum_{i=1}^{n} w_i x_i \sum_{i=1}^{n} w_i y_i}{\sum_{i=1}^{n} w_i \sum_{i=1}^{n} w_i x_i^2 - \left(\sum_{i=1}^{n} w_i x_i\right)^2}.$$

Considering the ponderated weighted means:

$$\overline{x}_w = \frac{\sum_{i=1}^{n} w_i x_i}{\sum_{i=1}^{n} w_i}, \qquad \overline{y}_w = \frac{\sum_{i=1}^{n} w_i y_i}{\sum_{i=1}^{n} w_i},$$

we have the estimators:

$$\widehat{\beta}_{1w} = \frac{\sum_{i=1}^{n} w_i x_i y_i - \sum_{i=1}^{n} w_i \overline{x}_w \overline{y}_w}{\sum_{i=1}^{n} w_i x_i^2 - \sum_{i=1}^{n} w_i \overline{x}_w^2}$$

$$= \frac{\sum_{i=1}^{n} w_i (x_i - \overline{x}_w)(y_i - \overline{y}_w)}{\sum_{i=1}^{n} w_i (x_i - \overline{x}_w)^2}$$

$$\widehat{\beta}_{0w} = \overline{y}_w - \widehat{\beta}_{1w}\overline{x}_w$$

that strongly resemble the least-squares estimators, except that we give more weight to the observations with a large w_i (that is, observation for which the errors have a smaller variance in the initial model). The estimated values are given by:

$$\widehat{y}_i = \widehat{\beta}_{0w} + \widehat{\beta}_{1w} x_i = \overline{y}_w + \widehat{\beta}_{1w}(x_i - \overline{x}_w)$$

and the variance σ^2 is estimated by:

$$s_w^2 = \frac{\sum_{i=1}^{n} w_i (y_i - \widehat{y}_i)^2}{n - 2}$$

$$= \frac{\sum_{i=1}^{n} w_i (y_i - \overline{y}_w)^2 - \widehat{\beta}_{1w}^2 \sum_{i=1}^{n} w_i (x_i - \overline{x}_w)^2}{n - 2}.$$

Examples of weights w_i chosen in simple regression are given by $w_i = 1/x_i$ in cases where the variance of ε_i is proportional to x_i, or where $w_i = 1/x_i^2$ when the variance of ε_i is proportional to x_i^2.

W

DOMAINS AND LIMITATIONS

The greatest disadvantage of the weighted least-squares method, which many people prefer not to know about, is probably the fact that it is based on the assumption that the weight is known exactly. This is almost never the case in practice, and the estimated weights are used instead. Generally, it is difficult to evaluate the effect of the use of estimated weights, but experience indicates that the results of most regression analyses are not very sensitive to the weights used. The advantages of a weighted analysis are often obtained on a large scale, though not always entirely, with the approximate weights. It is important to be aware of and avoid this problem and to use only weights that can be estimated with precision relative to one another. The weighted least-squares regression, as other least-squares methods, is also sensitive to outliers.

EXAMPLES

Consider a data set containing 10 observations with explanatory variable X taking values $X_i = i$ with $i = 1, 2, \ldots, 10$. Variable Y is generated using the model:

$$y_i = 3 + 2x_i + \varepsilon_i,$$

where the ε_i are normally distributed with $E(\varepsilon_i) = 0$ and $\text{Var}(\varepsilon_i) = (0.2x_i)^2$. Thus the variance of error of the first observation corresponds to $[(1)(0.2)]^2 = 0.04$, just as those of the tenth observation correspond to $[(10)(0.2)]^2 = 4$. The data thus generated are presented in the following table:

Values x_i and y_i generated by model $y_i = 3 + 2x_i + \varepsilon_i$

i	x_i	y_i
1	1	4.90
2	2	6.55

i	x_i	y_i
3	3	8.67
4	4	12.59
5	5	17.38
6	6	13.81
7	7	14.60
8	8	32.46
9	9	18.73
10	10	20.27

First we perform a nonweighted regression. We obtain the equation of regression:

$$\widehat{y}_i = 3.49 + 2.09 x_i,$$

$$\widehat{\beta}_0 = 3.49 \quad \text{with} \quad t_c = 0.99,$$

$$\widehat{\beta}_1 = 2.09 \quad \text{with} \quad t_c = 3.69.$$

Analysis of variance

Source of variation	Degrees of freedom	Sum of squares	Mean of squares	F
Regression	1	360.68	360.68	13.59
Residual	8	212.39	26.55	
Total	9	573.07		

The R^2 of this regression is 62.94%. At first glance, the results seem to be good. The estimated coefficient associated to the explanatory variable is clearly significant. This result is not at all surprising because the fact that the variance of errors is not constant rarely influences the estimated coefficients. The mean square of residuals is difficult to explain, considering there is no unique variance to estimate. Moreover, the standard errors and the lengths of the confidence intervals at 95% for the estimated conditional means are relatively constant for all observations. The nonweighted analysis does not take into account the observed inequalities between the variances of the errors.

We now make a weighted regression using as weights the values $1/x^2$. The weights are known because they are proportional to the real variances, which equal $(0.2x)^2$. The results are the following:

$$\widehat{y}_i = 2.53 + 2.28x_i,$$

$$\widehat{\beta}_0 = 2.53 \quad \text{with} \quad t_c = 3.07,$$
$$\widehat{\beta}_1 = 2.28 \quad \text{with} \quad t_c = 7.04.$$

Analysis of variance

Source of variation	Degrees of freedom	Sum of squares	Mean of squares	F
Regression	1	23.29	23.29	49.6
Residual	8	3.76	0.47	
Total	9	27.05		

The R^2 of this regression is 86.11%. On this basis, we can make the following comments:

- All numbers related to the sum of squares of the dependent variable are affected by the weights and are not comparable to those obtained by nonweighted regression.
- The Fisher test and the R^2 can be compared between two models. In our example, two values are greater than the nonweighted regression, but this is not always the case.
- The summarized coefficients are relatively close to those of the nonweighted regression. This is often the case but not a generality.
- The standard errors and confidence intervals for the conditional means reflect the fact that the precision of the estimations

decreases with increasing values of x. This result is the first reason to use the weighted regression.

FURTHER READING
- ▶ **Generalized linear regression**
- ▶ **Least squares**
- ▶ **Regression analysis**

REFERENCES

Seber, G.A.F. (1977) Linear Regression Analysis. Wiley, New York

Wilcoxon Frank

Wilcoxon, Frank (1892–1965) received his master's degree in chemistry at Rutgers University in 1921 and his doctorate in physics in 1924 at Cornell University, where he also did postdoctoral work from 1924 to 1925.

From 1928 to 1929 he was employed by Nichols Copper Company in Maspeth in Queens, New York and was then hired by the Boyce Thompson Institute for Plant Research as chief of a group studying the effects of insecticides and fungicides. He remain at the institute until the start of the Second World War; at this time and for the next 2 years, he moved to the Atlas Powder Company in Ohio where he directed the Control Laboratory.

In 1943, Wilcoxon, Frank was appointed head of the group at the American Cyanamid Company laboratory (in Stamford, CT) that studied insecticides and fungicides. He then developed an interest in statistics: in 1950, he was transferred to Lederle Laboratories division of Cyanamid in Pearl River, NY, where he developed a statistical consultation group. From 1960 until his death in 1965 he taught

W

applied statistics in the Department of Statistics to natural science majors at Florida State University.

Selected principal works and articles of Frank Wilcoxon:

1945 Individual comparisons by ranking methods. Biometrics 1, 80–83.

1957 Some rapid approximate statistical procedures. American Cyanamid, Stamford Research Laboratories, Stamford, CT.

Wilcoxon Signed Table

The Wilcoxon signed table gives theoretical values of **statistic** T of the **Wilcoxon signed test** applied to two paired samples, on the **hypothesis** that two populations follow the same distribution.

HISTORY

Owen, D.B. (1962) published a Wilcoxon signed table for $n \leq 20$ (where n is the number of paired observations) for the **Wilcoxon signed test** developed by Wilcoxon, F. (1945).

The generally used table is that of Harter, H.L. and Owen, D.B. (1970) established for $n \leq 50$.

MATHEMATICAL ASPECTS

Consider a set of n pairs of observations $((X_1, Y_1), \ldots, (X_n, Y_n))$. We calculate the absolute differences $|X_i - Y_i|$ to which we assign a rank (that is, we associate rank 1 to the smallest value, rank 2 to the next highest value, and so on, up to rank n for the greatest value). Then we give to this rank the sign corresponding to the difference $X_i - Y_i$. Thus we calculate the sum of positive ranks:

$$T = \sum_{i=1}^{k} R_i,$$

where k is the number of pairs representing a positive difference $(X_i - Y_i > 0)$.

For different values of m and α (the **significance level**), the Wilcoxon signed table gives the theoretical values of **statistic** T of the **Wilcoxon signed test**, on the **hypothesis** that the two underlying populations are identically distributed.

DOMAINS AND LIMITATIONS

The Wilcoxon signed table is used in nonparametric tests that use rank and especially in the **Wilcoxon signed test**.

EXAMPLES

Below is an extract of a Wilcoxon signed table m going from 10 to 15 and $\alpha = 0.05$ to $\alpha = 0.975$:

m	$\alpha = 0.05$	$\alpha = 0.975$
10	11	46
11	14	55
12	18	64
3	22	73
14	26	83
15	31	94

For an example of how the Wilcoxon signed table is used, see **Wilcoxon signed table**.

FURTHER READING
▶ **Statistical table**
▶ **Wilcoxon signed test**

REFERENCES

Harter, H.L., Owen, D.B.: Selected Tables in Mathematical Statistics, vol. 1. Markham, Chicago (5.7 Appendix) (1970)

Owen, D.B.: Handbook of Statistical Tables. (3.3, 5.7, 5.8, 5.10). Addison-Wesley, Reading, MA (1962)

Wilcoxon, F.: Individual comparisons by ranking methods. Biometrics **1**, 80–83 (1945)(5.1, 5.7).

Wilcoxon Signed Test

The Wilcoxon signed test is a **nonparametric** test. As a **signed test**, it serves to check for any difference between two populations. It tells one if the difference between two pairs of **observations** is positive or negative, taking into account the amplitude of this difference.

HISTORY

The Wilcoxon signed was created by Wilcoxon, F. in 1945. In 1949, he proposed another method to treat the zero differences that can appear between two observations of considered samples.

MATHEMATICAL ASPECTS

Let $(X_1, X_2, \ldots, X_n)$ and $(Y_1, Y_2, \ldots, Y_n)$ be two samples of size n. We consider n pairs of observations $(x_1, y_1), (x_2, y_2), \ldots, (x_n, y_n)$. We denote by $|d_i|$ the difference in absolute value between x_i and y_i:

$$|d_i| = |y_i - x_i|, \quad \text{for } i = 1, 2, \ldots, n.$$

We subtract from the dimension all the pairs of observations that do not represent the difference ($d_i = 0$). We denote by m the number of pairs remaining. We assign then to the pairs of observations a rank from 1 to m depending on the size of $|d_i|$, that is, we give rank 1 to the smallest value of $|d_i|$ and rank m to the largest value of $|d_i|$. We name $R_i, i = 1, \ldots, m$ the rank thus defined.

If many pairs of variables represent the same absolute difference $|d_i|$, we assign to them the mean rank. If there are mean ranks, the **statistic** of the test is calculated by the following expression:

$$T_1 = \frac{\sum_{i=1}^{m} R_i}{\sqrt{\sum_{i=1}^{m} R_i^2}},$$

where R_i is the rank of the pair (x_i, y_i) without the sign of the difference d_i.

If there are no mean ranks, it is more correct to use the sum of ranks associated to the positive difference:

$$T = \sum_{\substack{\text{(for } i \text{ such} \\ \text{as } d_i > 0)}} R_i.$$

Hypotheses

The underlying hypotheses in the Wilcoxon signed test are based on d_k, the **median** of d_i. According to whether it is a **two-tail test** or a **one-tailed test**, we have the following cases:

A: *Two-sided test*:
$$H_0: \quad d_k = 0$$
$$H_1: \quad d_k \neq 0$$

B: *One-sided test*:
$$H_0: \quad d_k \leq 0$$
$$H_1: \quad d_k > 0$$

C: *One-sided test*:
$$H_0: \quad d_k \geq 0$$
$$H_1: \quad d_k < 0$$

W

In case A, the hypothesis represents the situation where there are no differences between the two populations.

In case B, we make the hypothesis that the values of the population of X have a tendency to be greater than those of the population of Y.

In case C, we make the contrary **hypothesis**, i.e., that the values of the **population** of X have a tendency to be smaller than those of the population of Y.

Decision Rules

If T is used and if $m < 50$, then the **Wilcoxon signed table** will be used to test the **null hypothesis** H_0.

The **random variable** T is a linear combination of m independent Bernoulli random variables (but not identically distributed). It has as **mean** and **standard deviation**:

$$\mu = \frac{m(m+1)}{4} \quad \text{and}$$

$$\sigma = \sqrt{\frac{m(m+1)(2m+1)}{24}}.$$

We deduce the **random variable** Z:

$$Z = \frac{T - \mu}{\sigma}.$$

We denote by w_α the value of the Wilcoxon signed table with parameters m and α (where α is the **significance level**).

If $m > 20$, then w_α can be approximated with the help of the corresponding values of the **normal distribution**:

$$w_\alpha = \frac{m(m+1)}{4}$$
$$+ z_\alpha \sqrt{\frac{m(m+1)(2m+1)}{24}},$$

where z_α is the values in the **normal table** at level α.

If T_1 is used, we will base our calculation directly on the normal table to test the null hypothesis H_0. The decision rules are different depending on the hypotheses. We have the decision rules A, B, and C relative to the previous cases A, B, and C. If T_1 is used, we should replace T by T_1 and w_α by z_α in the rules that follow.

Case A

We reject H_0 at the significance level α if T is greater than $w_{1-\alpha/2}$ or smaller than $w_{\alpha/2}$, that is, if

$$T < w_{\alpha/2} \text{ ou } T > w_{1-\alpha/2}.$$

Case B

We reject H_0 at the significance level α if T is greater than $w_{1-\alpha}$, that is, if

$$T > w_{1-\alpha}.$$

Case C

We reject H_0 at the **significance level** α if T is smaller than w_α, that is, if

$$T < w_\alpha.$$

DOMAINS AND LIMITATIONS

To use the Wilcoxon signed test, the following criteria must be met:

1. The distribution of d_i must be symmetric.
2. d_i must be independent.
3. d_i must be measured in true values.

The Wilcoxon signed test is also used as the test of the **median**. The data $X_1, X_2, \ldots, X_n$ consist of a single **sample** of dimension n. The hypotheses corresponding to the previous cases A, B, and C are the following:

A: Two-tail cases:

H_0: The median of X equals a presumed **value** M_d

H_1: The median of X does not equal M_d

B: One-tail case:

H_0: The median of X is $\geq M_d$

H_1: The median of X is $< M_d$

C: One-tail case:

H_0: The median of X is $\leq M_d$

H_1: The median of X is $> M_d$

Note that "median" could be replaced by "**mean**" because we accepted the hypothesis of the symmetry of the distribution of X. In what concerns the calculations, we form n pairs (X_1, M_d), (X_2, M_d), ..., (X_n, M_d) and treat them in the same way as before. The rest of the test is identical.

EXAMPLES

We conduct tests on eight pairs of twins. The goal is to see if the first born is more intelligent than the second born.

The data are presented in the following table, the higher scores corresponding to better test results.

Pair of twins	First born X_i	Second born Y_i	d_i	$|d_i|$	R_i
1	90	88	-2	2	-2
2	75	79	4	4	4
3	99	98	-1	1	-1
4	60	66	6	5	5
5	72	64	-8	6	-6
6	83	83	0	-	-
7	86	86	0	-	-
8	92	95	3	3	3

As pairs 6 and 7 represent no difference ($d_6 = d_7 = 0$), the number of remaining pairs equals $m = 6$. As the d_i are all different, the mean ranks must not be calculated, and we use the sum of positive ranks:

$$T = \sum_{\substack{(\text{for } i \text{ such} \\ \text{as } d_i > 0)}} R_i,$$

where R_i is the rank of the pair (x_i, y_i) representing a positive d_i. Thus we have:

$$T = 4 + 5 + 3 = 12.$$

The hypotheses are the following:

H_0: There is no difference in the level of intelligence between the first and second twin ($d_m = 0$)

H_1: There is a difference ($d_m \neq 0$)

These hypotheses corresponding to case A and the decision rule are the following: Reject H_0 at level α if $T < w_{\alpha/2}$ or $T > w_{1-\alpha/2}$.

If we choose the **significance level** $\alpha = 0.05$, we obtain from the **Wilcoxon signed table** for $m = 6$:

$$w_{1-\alpha/2} = 20 \quad \text{and} \quad w_{\alpha/2} = 1,$$

where T is neither greater than 20 nor smaller than 1, and we cannot reject the **null hypothesis** H_0.

Now we suppose that the results for pairs 7 and 8 have changed with the new hypothesis in the following table:

Pair of twins	First born X_i	Second born Y_i	d_i	$|d_i|$	R_i
1	90	88	-2	4	-4
2	75	79	4	5	5
3	99	98	-1	2	-2
4	60	66	6	6	6
5	72	64	-8	7	-7
6	83	83	0	-	-
7	86	87	1	2	2
8	92	91	-1	2	-2

Now, pair 6 does not represent a difference ($d_6 = 0$); thus m equals 7.

Pairs 3, 7, and 8 represent the same absolute difference: $|d_3| = |d_7| = |d_8| = 1$. We assign a mean rank to the pairs that equal the rank of the ex aequo to which we add half the number of ex aequo minus one. Thus:

$$\text{mean rank} = \text{rank of } |d_3| + \tfrac{1}{2}(3-1)$$
$$= 2 \, .$$

In this case, we use the statistical test T_1 calculated by:

$$T_1 = \frac{\sum\limits_{i=1}^{m} R_i}{\sqrt{\sum\limits_{i=1}^{m} R_i^2}} = \frac{-2}{11.75} = -0.17 \, .$$

The decision rule is the following:
Reject H_0 at level α if $T_1 < z_{\alpha/2}$ or $T_1 > z_{1-\alpha/2}$.
If we choose the level $\alpha = 0.05$, then the values of the **normal table** are:

$$z_{1-\alpha/2} = -1.96 \quad \text{and} \quad z_{\alpha/2} = 1.96 \, .$$

In consequence, T_1 is neither greater than 1.96 nor smaller than -1.96.
We obtain the same result as in the previous case and do not reject the null hypothesis H_0.

FURTHER READING
▶ **Hypothesis testing**
▶ **Nonparametric test**
▶ **Wilcoxon signed table**

REFERENCES
Wilcoxon, F.: Individual comparisons by ranking methods. Biometrics **1**, 80–83 (1945)(5.1, 5.7).

Wilcoxon, F.: Some rapid approximate statistical procedures. American Cyanamid, Stamford Research Laboratories, Stamford, CT (1957)

Wilcoxon Table

The Wilcoxon table gives the theoretical values of **statistic** T of the **Wilcoxon test** under the **hypothesis** that there is no difference between the distribution of two compared populations.

HISTORY
The critical values for $N \le 20$ (where N is the total number of observations) were calculated by Wilcoxon, F. (1947).

MATHEMATICAL ASPECTS
Let $(X_1, X_2, \ldots, X_n)$ be a **sample** of dimension n coming from **population** 1 and $(Y_1, Y_2, \ldots, Y_m)$ a sample of dimension m coming from population 2.
Thus we obtain $N = n+m$ observations that we want to class in increasing order without taking into account their belonging to the samples. We assign a rank to each value: rank 1 to the smallest value, rank 2 to the next value, and so on up to rank N to the largest value.
We define the statistical test T in the following manner:

$$T = \sum_{i=1}^{n} R(X_i) \, ,$$

where $R(X_i)$ is the rank assigned to the observation X_i, $i = 1, 2, \ldots, n$ relative to the set of two samples $(X_1, \ldots, X_n)$ and $(Y_1, \ldots, Y_m)$. For different values of n, m, and α (the **significance level**), the Wilcoxon table gives the theoretical values of **statistic** T of the **Wilcoxon test**, under the **hypothesis** that the two underlying populations are identically distributed.

DOMAINS AND LIMITATIONS

The Wilcoxon table is used in nonparametric tests that use rank, and especially in the **Wilcoxon test**.

EXAMPLES

Here is an extract of a Wilcoxon table for $n = 14$, $m = 11$, and $\alpha = 0.05$ equalling 0.95:

n	m	α	critical
14	11	0.05	152
14	11	0.95	212

For an example of the use of the Wilcoxon table, see **Wilcoxon test**.

FURTHER READING

▶ **Statistical table**
▶ **Wilcoxon test**

REFERENCES

Wilcoxon, F.: Probability tables for individual comparisons by ranking methods. Biometrics **3**, 119–122 (1947)

Wilcoxon Test

The Wilcoxon test is a **nonparametric test**. It is used when we have two samples coming from two populations. The goal is to verify if there is a difference between the populations on the basis of the random samples taken from these populations.

HISTORY

The Wilcoxon test is named for its author and was introduced in 1945.
See also **Mann–Whitney test**.

MATHEMATICAL ASPECTS

Let $(X_1, X_2, \ldots, X_n)$ be a **sample** of dimension n coming from a **population** 1, and let $(Y_1, Y_2, \ldots, Y_m)$ be a sample of dimension m coming from a population 2.

Thus we obtain $N = n + m$ observations that we will class in increasing order without taking into account their belonging to the samples. Then we assign a rank of 1 to the smallest **value**, a rank of 2 to the next highest value, and so on up to rank N, which is assigned to the highest value. If many observations have exactly the same value, then we will assign a mean rank. We denote by $R(X_i)$ the rank assigned to X_i, $i = 1, \ldots, n$.

Statistic W of the test is defined by:

$$W = \sum_{i=1}^{n} R(X_i) .$$

If the dimension of the sample is large ($N \geq 12$), then we use, according to Gibbons, J.D. (1971), an approximation using the statistic W_1, which is supposed to follow a standard **normal distribution** $N(0, 1)$:

$$W_1 = \frac{W - \mu}{\sigma} ,$$

where μ and σ are, respectively, the **mean** and the **standard deviation** of **random variable** W:

$$\mu = \frac{n(N + 1)}{2}$$

and

$$\sigma = \sqrt{\frac{mn(N + 1)}{12}} .$$

If we use the approximation for the large samples W_1, and if we use the fact that there are ranks ex aequo among the N observations, we replace the standard deviation by:

$$\sigma = \sqrt{\frac{mn}{12}\left(N + 1 - \frac{\sum_{j=1}^{g} t_j\left(t_j^2 - 1\right)}{N(N - 1)}\right)} ,$$

W

where g is the number of groups of equal (tie) ranks and t_j the dimension of group j. (If there are no equal (tie) ranks, the observations are seen as many groups of dimension 1. In consequence, $g = N$ and $t_j = 1$ for $j = 1, \ldots, N$, and σ reduces to $\sqrt{\dfrac{mn\,(N+1)}{12}}$.)

Hypotheses

The Wilcoxon test can be made on the basis of a **two-tail test** or **one-tailed test**, according to the type of **hypothesis**:

A: Two-tail test:

$$H_0 : P\,(X < Y) = \tfrac{1}{2}$$
$$H_1 : P\,(X < Y) \neq \tfrac{1}{2}$$

B: One-tail test:

$$H_0 : P\,(X < Y) \leq \tfrac{1}{2}$$
$$H_1 : P\,(X < Y) > \tfrac{1}{2}$$

C: One-tail test:

$$H_0 : P\,(X < Y) \geq \tfrac{1}{2}$$
$$H_1 : P\,(X < Y) < \tfrac{1}{2}$$

Case A expresses the hypothesis that there is no difference between two populations.

Case B represents the hypothesis that population 1 (from where we took the sample of X) generally takes greater values than does population 2 (from where we took the sample of Y).

Case C expresses the contrary hypothesis, i.e., that the values of population 1 have a tendency to be smaller than those of population 2.

Decision Rules

Case A

We reject the **null hypothesis** H_0 at the **significance level** α if W is smaller than the value of the **Wilcoxon table** with parameters n,

m, and $\frac{\alpha}{2}$ denoted by $t_{n,m,\alpha/2}$ or if W is greater than the value in the table for n, m, and $1 - \frac{\alpha}{2}$, by $t_{n,m,1-\alpha/2}$, that is, if

$$W < t_{n,m,\alpha/2} \quad \text{or} \quad W > t_{n,m,1-\alpha/2}\,.$$

If the test uses statistic W_1, a comparison is made with the **normal table**:

$$W_1 < z_{\alpha/2} \quad \text{or} \quad W_1 > z_{1-\alpha/2}\,,$$

with $z_{\alpha/2}$ and $z_{1-\alpha/2}$ the values of the normal table with, respectively, the parameters $\frac{\alpha}{2}$ and $1 - \frac{\alpha}{2}$.

Case B

We reject the null hypothesis H_0 at the significance level α if W is smaller than the value of the Wilcoxon table with parameters n, m, and α, denoted by $t_{n,m,\alpha}$, that is, if

$$W < t_{n,m,\alpha}\,,$$

and in the case where statistic W_1 is used:

$$W_1 < z_\alpha\,,$$

with z_α being the value of the normal table with parameter α.

Case C

We reject the null hypothesis H_0 at the significance level α if W is greater than the value in the Wilcoxon table with parameters n, m, and $1 - \alpha$, denoted by $t_{n,m,1-\alpha}$, that is, if

$$W > t_{n,m,1-\alpha}\,,$$

and with statistic W_1:

$$W_1 > z_{1-\alpha}\,,$$

with $z_{1-\alpha}$ being the value of the normal table with parameter α.

DOMAINS AND LIMITATIONS

To use the Wilcoxon test, the following criteria must be met:

1. The two samples must be from random samples taken from their respective **populations**.
2. In addition to the **independence** inside each sample, there must be mutual independence between the two samples.
3. The scale of measure must be at least ordinal.

EXAMPLES

In a class, we count 25 students: 14 boys and 11 girls. We test them on mental calculation to see if the boys in this class have a tendency to be better than the girls.

The data are in the following table, the highest scores corresponding to the best test results.

Boys (X_i)		Girls (Y_i)	
19.8	17.5	17.7	23.6
12.3	17.9	7.1	11.1
10.6	21.1	21.0	20.3
11.3	16.4	10.6	15.6
14.0	7.7	13.3	
9.2	15.2	8.6	
15.6	16.0	14.1	

We class the scores in increasing order and assign a rank:

Scores	Sample	Rank	$R(X_i)$
7.1	Y	1	–
7.7	X	2	2
8.6	Y	3	–
9.2	X	4	4
10.6	Y	5.5	–
10.6	X	5.5	5.5
11.1	Y	7	–
11.3	X	8	8
12.3	X	9	9

Scores	Sample	Rank	$R(X_i)$
13.3	Y	10	–
14.0	X	11	11
14.1	Y	12	–
15.2	X	13	13
15.6	X	14.5	14.5
15.6	Y	14.5	–
16.0	X	16	16
16.4	X	17	17
17.5	X	18	18
17.7	Y	19	–
17.9	X	20	20
19.8	X	21	21
20.3	Y	22	–
21.0	Y	23	–
21.1	X	24	24
23.6	Y	25	–

We calculate **statistic** W:

$$W = \sum_{i=1}^{14} R(X_i)$$

$$= 2 + 4 + 5.5 + 8 + 9 + 11 + 13$$
$$+ 14.5 + 16 + 17 + 18 + 20$$
$$+ 21 + 24 = 183 .$$

If we choose the approximation for the large samples, taking into account the mean ranks, we make the adjustment:

$$W_1 = \frac{W - \dfrac{n(N+1)}{2}}{\sqrt{\dfrac{mn}{12}\left(m+n+1 - \dfrac{\sum_{j=1}^{g} t_j\left(t_j^2 - 1\right)}{(m+n)(m+n-1)}\right)}} .$$

Here $g = 2$ and $t_j = 2$ for $j = 1, 2$. Thus we have:

$$W_1 = \frac{183 - \dfrac{14(25+1)}{2}}{\sqrt{\dfrac{11\cdot 14}{12}\left(11 + 14 + 1 - \dfrac{2(2^2-1)+2(2^2-1)}{(11+14)(11+14-1)}\right)}}$$

$$= 0.0548 .$$

W

The hypotheses that we want to test are:

H_0: The boys in this class do not have a tendency to be better than the girls in this class in mental calculation.

H_1: The boys in this class have a tendency to be better than the girls in this class in mental calculation.

This is expressed by $H_0: P(X < Y) \geq \frac{1}{2}$ and corresponds to case C. The decision rule is the following:

Reject H_0 at **significance level** α if

$$W > t_{n,m,1-\alpha} \quad \text{or} \quad W_1 > z_{1-\alpha}.$$

If we choose $\alpha = 0.05$, the theoretical **value** in the **Wilcoxon table** is $t_{n,m,1-\alpha} = 212$.
The calculated value of W ($W = 183$) is not greater than $t_{n,m,1-\alpha} = 212$, and we cannot reject the **null hypothesis** H_0.
The value in the **normal table** $z_{1-\alpha}$ equals 1.6449. In consequence, W_1 is not greater than $z_{1-\alpha}$ ($0.0548 < 1.6449$), and we draw the same conclusion on the basis of statistic W: the boys in this class do not have a tendency to be better than the girls in this class in mental calculation.

FURTHER READING
▶ **Hypothesis testing**
▶ **Mann–Whitney test**
▶ **Nonparametric test**
▶ **Wilcoxon signed test**
▶ **Wilcoxon table**

REFERENCES

Gibbons, J.D.: Nonparametric Statistical Inference. McGraw-Hill, New York (1971)

Wilcoxon, F.: Individual comparisons by ranking methods. Biometrics **1**, 80–83 (1945)

Willcox Walter Francis

Willcox, Walter Francis (1861–1964) graduated from Amherst College in Massachusetts (receiving his bachelor's degree in 1884 and his master's degree in 1888). He continued his studies at Columbia University in New York in the late 1880s. He received his doctorate from Columbia University in 1891. His Ph.D. thesis was called *The Divorce Problem: A Study in Statistics*. From 1891, Willcox became professor of economics and of statistics at Cornell University, where he stayed until his retirement in 1931.

His principal contributions in statistics were in the domain of demography as well as in the development of the system of federal statistics.

From 1899 to 1902 he managed the 12th census of the United Nations and presided over the American Statistical Association in 1912, the American Economic Association in 1915, and the International Statistical Institute in 1947.

Selected principal works and articles of Willcox, Walther Francis:

1897 The Divorce Problem: A Study in Statistics. Studies in History, Economics and Public Law. Columbia University Press, New York.

1940 Studies in American Demography, Cornell University Press, Ithaca, NY.

Yates' Algorithm

Yates' **algorithm** is a process used to compute the **estimates** of **main effects** and of the **interactions** in a **factorial experiment**.

HISTORY

Yates' **algorithm** was created by Yates, Franck in 1937.

MATHEMATICAL ASPECTS

Consider a **factorial experiment** with n **factors**, each having two **levels**. We denote this **experiment** by 2^n.

The **factors** are denoted by uppercase letters. We use a lowercase letter for the heigh **level** of the factor and omit the letter for the lower level. When all the factors are at the lower level, we use the symbol (1).

The **algorithm** presented in a table. In the first column we find all the **combinations** of the **levels** of different **factors** in a standard order (the introduction of a letter is followed by combinations with all previous combinations). The second column contains the observations of all the combinations.

For the subsequent n columns, the procedure is the same: the first half of the column is obtained by summing the pairs of successive values of the previous column; the second half of the column is obtained by subtracting the first value from the second value of each pair.

Column n contains the **contrasts** of the **main effects** and **interactions** given in the first column. To find the **estimators** of these effects, we divide the nth column by $m \cdot 2^{n-1}$, where m is the number of repetitions of the **experiment** and n the number of **factors**. Concerning the first line, we find in column n the sum of all the **observations**.

DOMAINS AND LIMITATIONS

Yates' algorithm can be applied to **experiments** with **factors** that have more than two **levels**.

One can also use this **algorithm** to find the sum of squares corresponding to the **main effects** and interactions of the 2^n **factorial experiment**. They are obtained by squaring the nth column and dividing the result by $m \cdot 2^n$ (where m is the number of replications and n the number of **factors**).

EXAMPLES

To compute the **estimates** of the **main effects** and those of the three **interactions**, let us consider a factorial experiment 2^3 (meaning three factors with two levels) relative to the production of the sugar beet. The first factor, denoted A, concerns the type of fertilizer (manure or nonmanure); the second, denoted B, concerns the depth of plowing (20 or 30 cm); and the last factor is C, the variety of sugar beet (variety 1 or 2). Since each factor has two levels, one can speak of a $2 \times 2 \times 2$ experiment.

For example, the **interaction** between **factor** A and factor B is denoted by A × B, and the **interaction** between all three factors is written A × B × C.

The eight **observations** indicate the weight of the harvesting in kilograms for half a hectare, giving the following table:

A		Manure		Nonmanure
B		20 cm	30 cm	20 cm	30 cm
	Var. 1	80	96	84	88
C					
	Var. 2	86	96	89	93

In doing this experiment, we want to quantify not only the effect of fertilizer, plowing, and the variety of beet (simple effects), but also the impact of combined effects (effects of **interaction**) on harvesting.

To establish the table of Yates' algorithm, the eight **observations** must be classified in Yates' order. The classification table is practical in this case; the symbols "−" and "+" represent, respectively, the lower and upper **level** of the **factors**. For example, the first line indicates the observation "80" corresponding to the lower levels for all three factors (A: manure, B: 20 cm, C: variety 1).

Observation classification table in Yates' order

Obs.	Factor			Corresponding
	A	B	C	observation
1	−	−	−	80
2	+	−	−	84
3	−	+	−	96

Obs.	Factor			Corresponding
	A	B	C	observation
4	+	+	−	88
5	−	−	+	86
6	+	−	+	89
7	−	+	+	96
8	+	+	+	93

We can now establish the table of Yates' algorithm:

Table corresponding to Yates' algorithm

Combi-nation	Obs.	Columns			Esti-mation	Average effect
		1	2	3		
(1)	80	164	348	712	89	−
a	84	184	364	−4	−1	A
b	96	175	−4	34	8.5	B
ab	88	189	0	−18	−4.5	A × B
c	86	4	20	16	4	C
ac	89	−8	14	4	1	A × C
bc	96	3	−12	−6	−1.5	B × C
abc	93	−3	−6	6	1.5	A × B × C

The terms in column 2, for example, are calculated in the following way:

- First half of column (sum of elements of each pair of previous column, in this case column 1):

$$348 = 164 + 184$$
$$364 = 175 + 189$$
$$-4 = 4 + (-8)$$
$$0 = 3 + (-3)$$

- Second half of column (subtraction of elements of each pair, in opposite order):

$$20 = 184 - 164$$
$$14 = 189 - 175$$
$$-12 = -8 - 4$$
$$-6 = -3 - 3$$

The estimation for the **main effect** of **factor** A is equal to -1. As one can see, the value -4 (column 3) has been divided by $m \cdot 2^{n-1} = 1 \cdot 2^{3-1} = 4$, where m is equal to 1 because the experiment was done only once (without repetition) and n corresponds to the number of factors, which is three in this case.

FURTHER READING
▶ **Analysis of variance**
▶ **Contrast**
▶ **Design of experiments**
▶ **Factorial experiment**
▶ **Interaction**
▶ **Main effect**

REFERENCES
Yates, F.: The design and analysis of factorial experiments. Technical Communication of the Commonwealth Bureau of Soils **35**, Commonwealth Agricultural Bureaux, Farnham Royal (1937)

Youden William John

Youden, William John (1900–1971) earned his master's degree in chemistry from Columbia University in 1924. From 1924 to 1948 he worked in the Boyce Thompson Institute FOR PLANT RESEARCH in Yonkers, NY. From 1948 until his retirement in 1965, Youden, W.J. was a member of the Applied Mathematics Division in the National Bureau of Standards in Washington, D.C. During this period Youden developed a test and wrote *Statistical Methods for Chemists.*
Principal work of Youden, William John:

1951 Statistical Methods for Chemists. Wiley, New York .

Yule and Kendall Coefficient

The Yule coefficient is used to measure the skewness of a **frequency distribution**. It takes into account the relative positions of the **quartiles** with respect to the **median**, and compares the spreading of the curve to the right and left of the median.

MATHEMATICAL ASPECTS
In a symmetrical distribution, the **quartiles** are at an equal distance on each side of the **median**. This means that

$$(Q_3 - M) - (M - Q_1) = 0,$$

where M is the median, Q_1 the first quartile, and Q_3 the third quartile.
If the distribution is asymmetric, then the previous equality is not true.
The left part of the equation can be rearranged in the following way:

$$Q_3 + Q_1 - 2M.$$

To obtain a coefficient of skewness that is independent of the measure unit, the equation should be divided by the value $(Q_3 - Q_1)$.
The Yule coefficient is then:

$$C_Y = \frac{Q_3 + Q_1 - 2M}{Q_3 - Q_1}.$$

If C_Y is positive, then the distribution spreads toward the right; if C_Y is negative, the distribution spreads toward the left.

DOMAINS AND LIMITATIONS
If the coefficient is positive, then it represents a curve that spreads toward the right. If it is positive, then it represents a curve that spreads toward the left. A coefficient that is

Y

close to zero means that the curve is approximately symmetric.

The Yule and Kendall coefficient can vary between -1 and $+1$. In fact, since the **median** M is always located between the first and third **quartile**, the following two extreme possibilities occur:

1. $M = Q_1$
2. $M = Q_3$

In the first case the coefficient becomes:

$$C_Y = \frac{Q_3 + Q_1 - 2Q_1}{Q_3 - Q_1} = 1,$$

and in the second case it becomes:

$$C_Y = \frac{Q_3 + Q_1 - 2Q_3}{Q_3 - Q_1} = -1.$$

In these conditions, the coefficient should be close to zero for the distribution to be considered symmetric.

This coefficient as well as the other measures of **skewness** are of interest only if they can compare the shapes of two or several distributions. The results vary considerably from one formula to another. It is obvious that comparisons must be made using the same formula.

EXAMPLES

Suppose that we want to compare the shape of the distribution of the daily turnover in 75 bakeries over 2 years. We then calculate in both cases the Yule and Kendall coefficients. The **data** are the following:

Turnover	Frequency year 1	Frequency year 2
215–235	4	25
235–255	6	15
255–275	13	9
275–295	22	8

Turnover	Frequency year 1	Frequency year 2
295–315	15	6
315–335	6	5
335–355	5	4
355–375	4	3

For year 1, the first **quartile**, the **median**, and the third quartile are respectively:

$$Q_1 = 268.46,$$
$$M = 288.18,$$
$$Q_3 = 310.$$

The coefficient is then the following:

$$\begin{aligned} C_Y &= \frac{Q_3 + Q_1 - 2M}{Q_3 - Q_1} \\ &= \frac{310 + 268.46 - 2(288.18)}{310 - 268.46} \\ &= 0.05. \end{aligned}$$

For year 2, these **values** are respectively:

$$Q_1 = 230,$$
$$M = 251.67,$$
$$Q_3 = 293.12.$$

The Yule and Kendall coefficient is then equal to:

$$\begin{aligned} C_Y &= \frac{Q_3 + Q_1 - 2M}{Q_3 - Q_1} \\ &= \frac{293.12 + 230 - 2(251.67)}{293.12 - 230} \\ &= 0.31. \end{aligned}$$

For year 1, since the coefficient gives a result that is close to zero, we can admit that the distribution of the daily turnover in the 75 bakeries is very close to a symmetric distribution. For year 2, the coefficient is higher; this means that the distribution spreads toward the right.

FURTHER READING
▶ **Measure of shape**
▶ **Measure of skewness**

REFERENCES
Yule, G.V., Kendall, M.G.: An Introduction to the Theory of Statistics, 14th edn. Griffin, London (1968)

Yule George Udny

Yule, George Udny was born in 1871 in Beech Hill near Haddington (Ecosse) and died in 1951 in Cambridge.

The Yule coefficient of association, the Yule paradox (also called later the Simpson paradox), and the Yule procedure bear his name in statistics who also contributed to the Mendel theory and the time series.

Selected principal works of Yule, George Udny:

1897 On the Theory of Correlation. J. Roy. Stat. Soc. 60, 812–854.

1900 On the association of attributes in statistics: with illustration from the material of the childhood society. Philos. Trans. Roy. Soc. Lond. Ser. A 194, 257–319.

1926 Why do we sometimes get nonsense-correlations between time-series? A study in sampling and the nature of time-series. J. Roy. Stat. Soc. (2) 89, 1–64.

1968 (with Kendall, M.G.) An Introduction to the Theory of Statistics. 14th edn. Griffin, London.

FURTHER READING
▶ **Measure of skewness**
▶ **Yule and Kendall coefficient**

Y

Appendix

Appendix A: Table of Random Numbers: Decimals of π

Appendix A: Table of Random Numbers, Decimals of π

1415926535	3305727036	5024459455	8583616035	8164706001	9465764078
8979323846	5759591953	3469083026	6370766010	6145249192	9512694683
2643383279	9218611739	4252230825	4710181942	1732172147	9835259570
5028841971	8193261179	3344685035	9555961989	7235014144	9825822620
6939937510	3105118548	2619311881	4676783744	1973568548	5224894077
5820974944	7446237999	7101000313	9448255379	1613611573	2671947826
5923078164	6274956735	7838752886	7747268471	5255213347	8482601476
6286208995	1885752724	5875332083	4047534640	5741849468	9909026401
8628034825	8912279381	8142061717	6208046684	4385233239	3639443745
3421170679	8301194912	7669147303	2590694912	7394143337	5305068203
8214808651	9833673362	5982534904	3313677028	4547762416	4962524517
3282306647	4406566430	2875546873	8989152104	8625189835	4939965143
9384460952	8602139494	1159562863	7521620569	6948556209	1429809190
5058223172	6395224737	8823537875	6602405803	9219222184	6592509372
5359408128	1907021798	9375195778	8150193511	2725502542	2169646151
4811174502	6094370277	1857780532	2533824300	5688767179	5709858387
8410270193	5392171760	1712268066	3558764024	4946016530	4105978859
8521105559	2931767523	1300192787	7496473263	4668049886	5977297549
6446229489	8467481846	6611195909	9141992726	2723279178	8930161753
5493038196	7669405132	2164201989	4269922792	6085784383	9284681382
4428810975	5681271702	3809525720	6782354781	8279679766	6868386894
6659334461	4526356082	1065485863	6360093417	8145410095	2774155991
2847564823	7785771342	2788659361	2164121992	3883786360	8559252459
3786783165	7517896091	5338182796	4586315030	9506800642	5395943104
2712019091	7363717872	8230301952	2861829745	2512520511	9972524680
4564856692	1468440901	3530185292	5570674983	7392984896	8459872736
3460348610	2249534301	6899577362	8505494588	8412848862	4469584865
4543266482	4654958537	2599413891	5869269956	2694560424	3836736222
1339360726	1050792279	2497217752	9092721079	1965285022	6260991246
2491412733	6892589235	8347913151	7509302955	2106611863	8051243888
7245870066	4201995611	5574857242	3211653449	6744278629	4390451244
6315588170	2129021960	4541500959	8720275596	2039194945	1365497627
4881520920	8640344181	5082953311	2364806657	4712371373	8079771569
9628292540	5981362977	6861727855	4991198818	8696095636	1435997700
9171536436	4771309960	8890750983	3479775356	4371917287	1296160894
7892590360	5187072113	8175463746	6369807426	4677646575	4169486855
1133053052	4999999837	4939319255	5425278625	7396241389	5848406353
4882046652	2978049951	6040092774	5181841757	8658326451	4220722258
1384146951	5973173280	1671139001	4672890977	9958133904	2848864815
9415116094	1609631859	9848824012	7727938000	7802759009	8456028506

Appendix B: Binomial Table

Number of trials		Probability of success								
		0.01	0.02	0.03	0.04	0.05	0.1	0.2	0.3	0.5
5	0	0.9510	0.9039	0.8587	0.8154	0.7738	0.5905	0.3277	0.1681	0.0313
	1	0.9990	0.9962	0.9915	0.9852	0.9774	0.9185	0.7373	0.5282	0.1875
	2	1.0000	0.9999	0.9997	0.9994	0.9988	0.9914	0.9421	0.8369	0.5000
	3	1.0000	1.0000	1.0000	1.0000	1.0000	0.9995	0.9933	0.9692	0.8125
	4	1	1.0000	1.0000	1.0000	1.0000	1.0000	0.9997	0.9976	0.9688
	5	1	1	1	1	1	1	1	1	1
10	0	0.9044	0.8171	0.7374	0.6648	0.5987	0.3487	0.1074	0.0282	0.0010
	1	0.9957	0.9838	0.9655	0.9418	0.9139	0.7361	0.3758	0.1493	0.0107
	2	0.9999	0.9991	0.9972	0;9938	0.9885	0.9298	0.6778	0.3828	0.0547
	3	1.0000	1.0000	0.9999	0.9996	0.9990	0.9872	0.8791	0.6496	0.1719
	4	1.0000	1.0000	1.0000	1.0000	0.9999	0.9984	0.0672	0.8497	0.3770
	5	1	1.0000	1.0000	1.0000	1.0000	0.9999	0.9936	0.9527	0.6230
	6	1	1	1.0000	1.0000	1.0000	1.0000	0.9991	0.9894	0.8281
	7	1	1	1	1	1.0000	1.0000	0.9999	0.9984	0.9453
	8	1	1	1	1	1	1.0000	1.0000	0.9999	0.9893
	9	1	1	1	1	1	1	1.0000	1.0000	0.9990
	10	1	1	1	1	1	1	1	1	1
15	0	0.8601	0.7386	0.6333	0.5421	0.4633	0.2059	0.0352	0.0047	0.0000
	1	0.9904	0.9647	0.9270	0.8809	0.8290	0.5490	0.1671	0.0353	0.0005
	2	0.9996	0.9970	0.9906	0.9797	0.9638	0.8159	0.3980	0.1268	0.0037
	3	1.0000	0.9998	0.9992	0.9976	0.9945	0.9444	0.6482	0.2969	0.0176
	4	1.0000	1.0000	0.9999	0.9998	0.9994	0.9873	0.8358	0.5155	0.0592
	5	1.0000	1.0000	1.0000	1.0000	0.9999	0.9978	0.9389	0.7216	0.1509
	6	1	1.0000	1.0000	1.0000	1.0000	0.9997	0.9819	0.8689	0.3036
	7	1	1	1.0000	1.0000	1.0000	1.0000	0.9958	0.9500	0.5000
	8	1	1	1	1.0000	1.0000	1.0000	0.9992	0.9848	0.6964
	9	1	1	1	1	1	1.0000	0.9999	0.9963	0.8491
	10	1	1	1	1	1	1.0000	1.0000	0.9993	0.9408
	11	1	1	1	1	1	1	1.0000	0.9999	0.9824
	12	1	1	1	1	1	1	1.0000	1.0000	0.9963
	13	1	1	1	1	1	1	1.0000	1.0000	0.9995
	14	1	1	1	1	1	1	1	1.0000	1.0000
	15	1	1	1	1	1	1	1	1	1

Number of trials		Probability of success								
		0.01	0.02	0.03	0.04	0.05	0.1	0.2	0.3	0.5
	0	0.8179	0.6676	0.5438	0.4420	0.1585	0.1216	0.0115	0.0008	0.0000
	1	0.9831	0.9401	0.8802	0.8103	0.7358	0.3917	0.0692	0.0076	0.0000
	2	0.9990	0.9929	0.9790	0.9561	0.9245	0.6769	0.2061	0.0355	0.0002
	3	1.0000	0.9994	0.9973	0.9926	0.9841	0.8670	0.4114	0.1071	0.0013
	4	1.0000	1.0000	0.9997	0.9990	0;9974	0.9568	0.6296	0.2375	0.0059
	5	1.0000	1.0000	1.0000	0.9999	0.9001	0.9887	0.8042	0.4164	0.0207
	6	1.0000	1.0000	1.0000	1.0000	1.0000	0.9976	0.9133	0.6080	0.0577
	7	1	1.0000	1.0000	1.0000	1.0000	0.9996	0.9679	0.7723	0.1316
	8	1	1	1.0000	1.0000	1.0000	0.9999	0.9900	0.8867	0.2517
	9	1	1	1	1.0000	1.0000	1.0000	0.9974	0.9520	0.4119
20	10	1	1	1	1	1.0000	1.0000	0.9994	0.9829	0.5881
	11	1	1	1	1	1	1.0000	0.9999	0.9949	0.7483
	12	1	1	1	1	1	1.0000	1.0000	0.9987	0.8684
	13	1	1	1	1	1	1	1.0000	0.9997	0.9423
	14	1	1	1	1	1	1	1.0000	1.0000	0.9793
	15	1	1	1	1	1	1	1.0000	1.0000	0.9941
	16	1	1	1	1	1	1	1.0000	1.0000	0.9987
	17	1	1	1	1	1	1	1	1.0000	0.9998
	18	1	1	1	1	1	1	1	1.0000	1.0000
	19	1	1	1	1	1	1	1	1	1.0000
	20	1	1	1	1	1	1	1	1	1

Number of trials		Probability of success								
		0.01	0.02	0.03	0.04	0.05	0.1	0.2	0.3	0.5
	0	0.7397	0.5455	0.4010	0.2939	0.2146	0.0424	0.0012	0.0000	0.0000
	1	0.9639	0.8795	0.7731	0.6612	0.5535	0.1837	0.0105	0.0003	0.0000
	2	0.9967	0.9783	0.9399	0.8831	0.8122	0.4114	0.0442	0.0021	0.0000
	3	0.9998	0.9971	0.9881	0.9694	0.9392	0.6474	0.1227	0.0093	0.0000
	4	1.0000	0.9997	0.9982	0.9937	0.9844	0.8245	0.2552	0.0302	0.0000
	5	1.0000	1.0000	0.9998	0.9989	0.9967	0.9268	0.4275	0.0766	0.0002
	6	1.0000	1.0000	1.0000	0.9999	0.9994	0.9742	0.6070	0.1595	0.0007
	7	1	1.0000	1.0000	1.0000	0.9999	0.9922	0.7608	0.2814	0.0026
	8	1	1.0000	1.0000	1.0000	1.0000	0.9980	0.8713	0.4315	0.0081
	9	1	1	1.0000	1.0000	1.0000	0.9995	0.9389	0.5888	0.0214
	10	1	1	1.0000	1.0000	1.0000	0.9999	0.9744	0.7304	0.0494
	11	1	1	1	1.0000	1.0000	1.0000	0.9905	0.8407	0.1002
	12	1	1	1	1	1.0000	1.0000	0.9969	0.9155	0.1808
	13	1	1	1	1	1	1.0000	0.9991	0.9599	0.2023
	14	1	1	1	1	1	1.0000	0.9998	0.9831	0.4278
30	15	1	1	1	1	1	1.0000	0.9999	0.9936	0.5722
	16	1	1	1	1	1	1	1.0000	0.9979	0.7077
	17	1	1	1	1	1	1	1.0000	0.9994	0.8192
	18	1	1	1	1	1	1	1.0000	0.9998	0.8998
	19	1	1	1	1	1	1	1.0000	1.0000	0.9506
	20	1	1	1	1	1	1	1.0000	1.0000	0.9786
	21	1	1	1	1	1	1	1	1.0000	0.9919
	22	1	1	1	1	1	1	1	1.0000	0.9974
	23	1	1	1	1	1	1	1	1.0000	0.9993
	24	1	1	1	1	1	1	1	1.0000	0.9998
	25	1	1	1	1	1	1	1	1	1.0000
	26	1	1	1	1	1	1	1	1	1.0000
	27	1	1	1	1	1	1	1	1	1.0000
	28	1	1	1	1	1	1	1	1	1.0000
	29	1	1	1	1	1	1	1	1	1.0000
	30	1	1	1	1	1	1	1	1	1

Appendix C: Fisher Table for $\alpha = 0.05$

ν_2	ν_1									
	1	2	3	4	5	6	7	8	9	10
1	161.4	199.5	215.7	224.6	230.2	234.0	236.8	238.9	240.5	241.9
2	18.51	19.00	19.16	19.25	19.30	19.33	19.35	19.37	19.38	19.40
3	10.13	9.55	9.28	9.12	9.01	8.94	8.89	8.85	8.81	8.79
4	7.71	6.94	6.59	6.39	6.26	6.16	6.09	6.04	6.00	5.96
5	6.61	5.79	5.41	5.19	5.05	4.95	4.88	4.82	4.77	4.74
6	5.99	5.14	4.76	4.53	4.39	4.28	4.21	4.15	4.10	4.06
7	5.59	4.74	4.35	4.12	3.97	3.87	3.79	3.73	3.68	3.64
8	5.32	4.46	4.07	3.84	3.69	3.58	3.50	3.44	3.39	3.35
9	5.12	4.26	3.86	3.63	3.48	3.37	3.29	3.23	3.18	3.14
10	4.96	4.10	3.71	3.48	3.33	3.22	3.14	3.07	3.02	2.98
11	4.84	3.98	3.59	3.36	3.20	3.09	3.01	2.95	2.90	2.85
12	4.75	3.89	3.49	3.26	3.11	3.00	2.91	2.85	2.80	2.75
13	4.67	3.81	3.41	3.18	3.03	2.92	2.83	2.77	2.71	2.67
14	4.60	3.74	3.34	3.11	2.96	2.85	2.76	2.70	2.65	2.60
15	4.54	3.68	3.29	3.06	2.90	2.79	2.71	2.64	2.59	2.54
16	4.49	3.63	3.24	3.01	2.85	2.74	2.66	2.59	2.54	2.49
17	4.45	3.59	3.20	2.96	2.81	2.70	2.61	2.55	2.49	2.45
18	4.41	3.55	3.16	2.93	2.77	2.66	2.58	2.51	2.46	2.41
19	4.38	3.52	3.13	2.90	2.74	2.63	2.54	2.48	2.42	2.38
20	4.35	3.49	3.10	2.87	2.71	2.60	2.51	2.45	2.39	2.35
21	4.32	3.47	3.07	2.84	2.68	2.57	2.49	2.42	2.37	2.32
22	4.30	3.44	3.05	2.82	2.66	2.55	2.46	2.40	2.34	2.30
23	4.28	3.42	3.03	2.80	2.64	2.53	2.44	2.37	2.32	2.27
24	4.26	3.40	3.01	2.78	2.62	2.51	2.42	2.36	2.30	2.25
25	4.24	3.39	2.99	2.76	2.60	2.49	2.40	2.34	2.28	2.24
26	4.23	3.37	2.98	2.74	2.59	2.47	2.39	2.32	2.27	2.22
27	4.21	3.35	2.96	2.73	2.57	2.46	2.37	2.31	2.25	2.20
28	4.20	3.34	2.95	2.71	2.56	2.45	2.36	2.29	2.24	2.19
29	4.18	3.33	2.93	2.70	2.55	2.43	2;35	2.28	2.22	2.18
30	4.17	3.32	2.92	2.69	2.53	2.42	2.33	2.27	2.21	2.16
40	4.08	3.23	2.84	2.61	2.45	2.34	2.25	2.18	2.12	2.08
60	4.00	3.15	2.76	2.53	2.37	2.25	2.17	2.10	2.04	1.99
120	3.92	3.07	2.68	2.45	2.29	2.17	2.09	2.02	1.96	1.91
∞	3.84	3.00	2.60	2.37	2.21	2.10	2.01	1.94	1.88	1.83

ν_2	ν_1								
	12	15	20	24	30	40	60	120	∞
1	243.9	245.9	248.0	249.1	250.1	251.1	252.2	253.3	254.3
2	19.41	19.43	19.45	19.45	19.46	19.47	19.48	19.49	19.50
3	8.74	8.70	8.66	8.64	8.62	8.59	8.57	8.55	8.53
4	5.91	5.86	5.80	5.77	5.75	5.72	5.69	5.66	5.63
5	4.68	4.62	4.56	4.53	4.50	4.46	4.43	4.40	4.36
6	4.00	3.94	3.87	3.84	3.81	3.77	3.74	3.70	3.67
7	3.57	3.51	3.44	3.41	3.38	3.34	3.30	3.27	3.23
8	3.28	3.22	3.15	3.12	3.08	3.04	3.01	2.97	2.93
9	3.07	3.01	2.94	2.90	2.86	2.83	2.79	2.75	2.71
10	2.91	2.85	2.77	2.74	2.70	2.66	2.62	2.58	2.54
11	2.79	2.72	2.65	2.61	2.57	2.53	2.49	2.45	2.40
12	2.69	2.62	2.54	2.51	2.47	2.43	2.38	2.34	2.30
13	2.60	2.53	2.46	2.42	2.38	2.34	2.30	2.25	2.21
14	2.53	2.46	2.39	2.35	2.31	2.27	2.22	2.18	2.13
15	2.48	2.40	2.33	2.29	2.25	2.20	2.16	2.11	2.07
16	2.42	2.35	2.28	2.24	2.19	2.15	2.11	2.06	2.01
17	2.38	2.31	2.23	2.19	2.15	2.10	2.06	2.01	1.96
18	2.34	2.27	2.19	2.15	2.11	2.06	2.02	1.97	1.92
19	2.31	2.23	2.16	2.11	2.07	2.03	1.98	1.93	1.88
20	2.28	2.20	2.12	2.08	2.04	1.99	1.95	1.90	1.84
21	2.25	2.18	2.10	2.05	2.01	1;96	1.92	1.87	1.81
22	2.23	2.15	2.07	2.03	1.98	1.94	1.89	1.84	1.78
23	2.20	2.13	2.05	2.01	1.96	1.91	1.86	1.81	1.76
24	2.18	2.11	2.03	1.98	1.94	1.89	1.84	1.79	1.73
25	2.16	2.09	2.01	1.96	1.92	1.87	1.82	1.77	1.71
26	2.15	2.07	1.99	1.95	1.90	1.85	1.80	1.75	1.69
27	2.13	2.06	1.97	1.93	1.88	1.84	1.79	1.73	1.67
28	2.12	2.04	1.96	1.91	1.87	1.82	1.77	1.71	1.65
29	2.10	2.03	1.94	1.90	1.85	1.81	1.75	1.70	1.64
30	2.09	2.01	1.93	1.89	1.84	1.79	1.74	1.68	1.62
40	2.00	1.92	1.84	1.79	1.74	1.69	1.64	1.58	1.51
60	1.92	1.84	1.75	1.70	1.65	1.59	1.53	1.47	1.39
120	1.83	1.75	1.66	1.61	1.55	1.50	1.43	1.35	1.25
∞	1.75	1.67	1.57	1.52	1.46	1.39	1.32	1.22	1.00

Appendix D: Kruskal-Wallis Table

n_1	n_2	n_3	Critical value	α
2	1	1	2.7000	0.500
2	2	1	3.6000	0.267
2	2	2	4.5711	0.067
			3.7143	0.200
3	1	1	3.2000	0.300
3	2	1	4.2857	0.100
			3.8571	0.133
3	2	2	5.3572	0.029
			4.7143	0.048
			4.5000	0.067
			4.4643	0.105
3	3	1	5.1429	0.043
			4.5714	0.100
			4.0000	0.129
3	3	2	6.2500	0.011
			5.3611	0.032
			5.1389	0.061
			4.5556	0.100
			4.2500	0.121
3	3	3	7.2000	0.004
			6.4889	0.011
			5.6889	0.029
			5.0667	0.086
			4.6222	0.100
4	1	1	3.5714	0.200
4	2	1	4.8214	0.057
			4.5000	0.076
			4.0179	0.114
4	2	2	6.0000	0.014
			5.3333	0.033
			5.1250	0.052
			4.3750	0.100
			4.1667	0.105
4	3	1	5.8333	0.021
			5.2083	0.050
			5.0000	0.057
			4.0556	0.093
			3.8889	0.129

n_1	n_2	n_3	Critical value	α
4	3	2	6.4444	0.009
			6.4222	0.010
			5.4444	0.047
			5.4000	0.052
			4.5111	0.098
			4.4666	0.101
4	3	3	6.7455	0.010
			6.7091	0.013
			5.7909	0.046
			5.7273	0.050
			4.7091	0.094
			4.7000	0.101
4	4	1	6.6667	0.010
			6.1667	0.013
			4.9667	0.046
			4.8667	0.054
			4.1667	0.082
			4.0667	0.102
4	4	2	7.0364	0.006
			6.8727	0.011
			5.4545	0.046
			5.2364	0.052
			4.5545	0.098
			4.4455	0.103
4	4	3	7.1739	0.010
			7.1364	0.011
			5.5985	0.049
			5.5758	0.051
			4.5455	0.099
			4.4773	0.102
4	4	4	7.6538	0.008
			7.5385	0.011
			5.6923	0.049
			5.6538	0.054
			4.6539	0.097
			4.5001	0.104

n_1	n_2	n_3	Critical value	α
5	1	1	3.8571	0.143
5	2	1	5.2500	0.036
			5.0000	0.048
			4.4500	0.071
			4.2000	0.095
			4.0500	0.119
5	2	2	6.5333	0.008
			6.1333	0.013
			5.1600	0.034
			5.0400	0.056
			4.3733	0.090
			4.2933	0.122
5	3	1	6.4000	0.012
			4.9600	0.048
			4.8711	0.052
			4.0178	0.095
			3.8400	0.123
5	3	2	6.9091	0.009
			6.8606	0.011
			5.4424	0.048
			5.3455	0.050
			4.5333	0.097
			4.4121	0.109
5	3	3	6.9818	0.010
			6.8608	0.011
			5.4424	0.048
			5.3455	0.050
			4.5333	0.097
			4.4121	0.109
5	4	1	6.9545	0.008
			6.8400	0.011
			4.9855	0.044
			4.8600	0.056
			3.9873	0.098
			3.9600	0.102
5	4	2	7.2045	0.009
			7.1182	0.010
			5.2727	0.049
			5.2682	0.050
			4.5409	0.098
			4.5182	0.101

n_1	n_2	n_3	Critical value	α
5	4	3	7.4449	0.010
			7.3949	0.011
			5.6564	0.049
			5.6308	0.051
			4.5487	0.099
			4.5231	0.103
5	4	4	7.7604	0.009
			7.7440	0.011
			5.6571	0.049
			5;6116	0.050
			4.6187	0.100
			4.5527	0.102
5	5	1	7.3091	0.009
			6.8364	0.011
			5.1273	0.049
			4.9091	0.053
			4.1091	0.086
			4.0364	0.105
5	5	2	7.3385	0.010
			7.5429	0.010
			5.7055	0.046
			5.6264	0.051
			4.5451	0.100
			4.5363	0.102
5	5	4	7.8229	0.010
			7.1914	0.010
			5.6657	0.049
			5.6429	0.050
			4.5229	0.099
			4.6200	0.101
5	5	5	8.0000	0.09
			7.9800	0.010
			5.7800	0.049
			5.6600	0.051
			4.5600	0.100
			4.5000	0.102

Appendix E: Student Table

v	α					
	0.100	0.050	0.025	0.010	0.005	0.001
1	3.078	6.314	12.706	31.821	63.656	318.289
2	1.886	2.920	4.303	6.965	9.925	22.328
3	1.638	2.353	3.182	4.541	5.841	10.214
4	1.533	2.132	2.776	3.747	4.604	7.173
5	1.476	2.015	2.571	3.365	4.032	5.894
6	1.440	1.943	2.447	3.143	3.707	5.208
7	1.415	1.895	2.315	2.998	3.499	4.785
8	1.397	1.860	2.306	2.896	3.355	4.501
9	1.383	1.833	2.212	2.821	3.250	4.297
10	1.312	1.812	2.228	2.764	3.169	4.144
11	1.363	1.796	2.201	2.718	3.106	4.025
12	1.356	1.182	2.179	2.681	3.055	3.930
13	1.350	1.771	2.160	2.650	3.012	3.852
14	1.345	1.761	2.145	2.624	2.977	3.787
15	1.341	1.753	2.131	2.602	2.947	3.733
16	1.337	1.746	2.120	2.583	2.921	3.686
17	1.133	1.740	2.110	2.567	2.898	3.646
18	1.330	1.734	2.101	2.552	2.878	3.610
19	1.328	1.729	2.093	2.539	2.861	3.579
20	1.325	1.725	2.086	2.528	2.845	3.552
21	1.323	1.721	2.080	2.518	2.831	3.527
22	1.321	1.717	2.074	2.508	2.519	3.505
23	1.319	1.714	2.069	2.500	2.807	3.485
24	1.318	1.711	2.064	2.492	2.797	3.467
25	1.316	1.708	2.060	2.485	2.787	3.450
26	1.315	1.706	2.056	2.479	2.779	3.435
27	1.314	1.103	2.052	2.473	2.771	3.421
28	1.313	1.701	2.048	2.467	2.763	3.408
29	1.311	1.699	2.046	2.462	2.756	3.396
30	1.310	1.697	2.042	2.457	2.750	3.385
40	1.303	1.684	2.021	2.123	2.704	3.301
50	1.299	1.676	2.009	2.403	2.678	3.261
60	1.296	1.571	2.000	2.300	2.660	3.232
70	1.294	1.667	1.994	2.381	2.648	3.211
80	1.292	1.664	1.990	2.374	2.639	3.195
90	1.291	1.662	1.987	2.368	2.632	3.183
100	1.290	1.660	1.984	2.364	2.626	3.174
∞	1.282	1.645	1.960	2.326	2.576	3.090

Appendix F: Chi-Square Table

			Degree of freedeom			
ν	0.990	0.950	0.10	0.05	0.025	0.01
1	0.000	0.004	2.71	3.84	5.020	6.63
2	0.020	0.103	4.61	5.99	7.380	9.21
3	0.115	0.352	6.25	7.81	9.350	11.34
4	0.297	0.711	7.78	9.49	11.140	13.23
5	0.554	1.145	9.24	11.07	12.830	15.09
6	0.872	1.635	10.64	12.53	14.450	16.81
7	1.239	2.107	12.02	14.07	16.010	18.48
8	1.646	2.733	13.36	15.51	17.530	20.09
9	2.088	3.325	14.68	16.92	19.020	21.67
10	2.558	3.940	15.99	18.31	20.480	23.21
11	3.050	4.580	17.29	19.68	21.920	21.72
12	3.570	5.230	18.55	21.03	23.340	20.22
13	4.110	5.890	19.81	21.36	24.740	27.69
14	4.660	6.570	21.06	23.68	26.120	29.14
15	5.230	7.260	22.31	25.00	27.100	30.58
16	5.810	7.960	23.54	26.30	28.850	32.00
17	6.410	8.670	24.77	27.59	30.190	33.41
18	7.020	9.390	25.99	28.87	31.530	34.81
19	7.630	10.120	27.20	30.14	32.850	36.19
20	8.260	10.850	28.41	31.41	34.170	37.57
21	8.900	11.590	29.62	39.67	35.480	38.93
22	9.540	12.340	30.81	33.92	36.780	40.29
23	10.200	13.090	32.01	35.17	38.080	41.64
24	10.860	13.850	33.20	36.42	39.360	42.98
25	11.520	14.610	34.38	37.65	40.650	44.31
26	12.200	15.380	35.56	38.89	41.920	45.64
27	12.880	16.150	36.74	40.11	43.190	46.96
28	13.570	16.930	37.92	41.34	44.160	48.28
29	14.260	17.710	39.09	42.56	45.720	49.59
30	14.950	18.490	40.26	43.77	46.980	50.89

Appendix G: Standard Normal Table

Z	0.00	0.01	0.02	0.03	0.04	0.05	0.06	0.07	0.08	0.09
0.0	0.5000	0.5040	0.5080	0.5120	0.5160	0.5199	0.5239	0.5279	0.5319	0.5359
0.1	0.5398	0.5438	0.5478	0.5517	0.5557	0.5596	0.5636	0.5675	0.5714	0.5753
0.2	0.5793	0.5832	0 5871	0.5910	0.5948	0.5987	0.6026	0.6064	0.6103	0.6141
0.3	0.6179	0.6217	0.6255	0.6293	0.6331	0.6368	0.6406	0.6443	0.6480	0.6517
0.4	0.6501	0.6591	0.6628	0.6664	0.6700	0.6736	0.6772	0.6808	0.6844	0.6879
0.5	0.6915	0.6950	0.6985	0.7019	0.7054	0.7088	0.7123	0.7157	0.7190	0.7224
0.6	0.7257	0.7291	0.7324	0.7357	0.7389	0.7422	0.7454	0.7486	0.7517	0.7549
0.7	0.7580	0.7611	0.7642	0.7673	0.7704	0.7734	0.7764	0.7794	0.7823	0.7852
0.8	0.7881	0.7910	0.7939	0.7967	0.7995	0.8023	0.8051	0.8078	0.8106	0.8133
0.9	0.8119	0.8186	0.8212	0.8238	0.8264	0.8289	0.8315	0.8340	0.8365	0.8389
1.0	0.8413	0.8438	0.8461	0.8485	0.8508	0.8531	0.8554	0.8577	0.8599	0.8621
1.1	0.8643	0.8665	0.8686	0.8708	0.8729	0.8749	0.8770	0.8790	0.8810	0.8830
1.2	0.8849	0.8869	0.8888	0.8907	0.8925	0.8944	0.8962	0.8980	0.8997	0.9015
1.3	0.9032	0.9049	0.9066	0.9082	0.9099	0.9115	0.9131	0.9147	0.9162	0.9177
1.4	0.9192	0.9207	0.9222	0.9236	0.9251	0.9265	0.9279	0.9292	0.9306	0.9319
1.5	0.9332	0.9345	0.9357	0.9370	0.9382	0.9394	0.9406	0.9418	0.9429	0.9441
1.6	0.9452	0.9463	0.9474	0.9484	0.9495	0.9505	0.9515	0.9525	0.9535	0.9545
1.7	0.9554	0.9564	0.9573	0.9582	0.9591	0.9599	0.9608	0.9616	0.9625	0.9633
1.8	0.9641	0.9649	0.9656	0.9664	0.9671	0.9678	0.9686	0.9693	0.9699	0.9706
1.9	0.9713	0.9719	0.9726	0.9732	0.9738	0.9744	0.9750	0.9756	0.9761	0.9767
2.0	0.9772	0.9778	0.9783	0.9788	0.9793	0.9798	0.9803	0.9808	0.9812	0.9817
2.1	0.9821	0.9826	0.9830	0.9834	0.9838	0.9842	0.9846	0.9850	0.9854	0.9857
2.2	0.9861	0.9864	0.9868	0.9871	0.9875	0.9878	0.9881	0.9884	0.9887	0.9890
2.3	0.9893	0.9896	0.9898	0.9901	0.9904	0.9906	0.9909	0.9911	0.9913	0.9916
2.4	0.9918	0.9920	0.9922	0.9925	0.9927	0.9929	0.9931	0.9932	0.9934	0.9936
2.5	0.9938	0.9940	0.9941	0.9943	0.9945	0.9946	0.9948	0.9949	0.9951	0.9952
2.6	0.9953	0.9955	0.9956	0.9957	0.9959	0.9960	0.9961	0.9962	0.9963	0.9964
2.7	0.9965	0.9966	0.9967	0.9968	0.9969	0.9970	0.9971	0.9972	0.9973	0.9974
2.8	0.9974	0.9975	0.9976	0.9977	0.9977	0.9978	0.9979	0.9979	0.9980	0.9981
2.9	0.9981	0.9982	0.9982	0.9983	0.9984	0.9984	0.9985	0.9985	0.9986	0.9986
3.0	0.9987	0.9987	0.9987	0.9988	0.9988	0.9989	0.9989	0.9989	0.9990	0.9990

Bibliography

Abdi, H.: Binomial distribution: Binomical and Sign Tests. In: Salkind, N.J. (ed.) Encyclopedia of Measurement and Statistics. Sage, Thousand Oaks (2007) ► Binomial Test

Abdi, H.: Distance. In: Salkind, N.J. (ed.) Encyclopedia of Measurement and Statistics. Sage, Thousand Oaks (2007) ► Distance

Agresti, A.: Categorical Data Analysis. Wiley, New York (1990) ► Analysis of Categorical Data

Airy, G.B.: Letter from Professor Airy, Astronomer Royal, to the editor. Astronom. J. 4, 137–138 (1856) ► Outlier

Allan, F.E., Wishart, J.: A method of estimating the yield of a missing plot in field experimental work. J. Agricult. Sci. 20, 399–406 (1930) ► Missing Data

Altman, D.G.: Practical Statistics for Medical Research. Chapman & Hall, London (1991) ► Logistic Regression

Anderson, T.W., Darling, D.A.: Asymptotic theory of certain goodness of fit criteria based on stochastic processes. Ann. Math. Stat. 23, 193–212 (1952) ► Anderson–Darling Test

Anderson, T.W., Darling, D.A.: A test of goodness of fit. J. Am. Stat. Assoc. 49, 765–769 (1954) ► Anderson–Darling Test

Anderson, T.W., Sclove, S.L.: An Introduction to the Statistical Analysis of Data. Houghton Mifflin, Boston (1978) ► Range

Anderson, V.L., McLean, R.A.: Design of experiments: a realistic approach. Marcel Dekker, New York (1974) ► Model

Andrews D.F., Hertzberg, A.M.: Data: A Collection of Problems from Many Fields for Students and Research Workers. Springer, Berlin Heidelberg New York (1985) ► C_p Criterion

Anscombe, F.J., Tukey, J.W.: Analysis of residuals. Technometrics 5, 141–160 (1963) ► Analysis of Residuals

Anscombe, F.J.: Examination of residuals. Proc. 4th Berkeley Symp. Math. Statist. Prob. 1, 1–36 (1961) ► Analysis of Residuals

Anscombe, F.J.: Graphs in statistical analysis. Am. Stat. 27, 17–21 (1973) ► Analysis of Residuals

Antille, G., Ujvari, A.: Pratique de la Statistique Inférentielle. PAN, Neuchâtel, Switzerland (1991) ► Data Collection

Arbuthnott, J.: An argument for Divine Providence, taken from the constant regularity observed in the births of both sexes. Philos. Trans. 27, 186–190 (1710) ► Hypothesis Testing, ► Nonparametric Test, ► Sign Test

Armatte, M.: La construction des notions d'estimation et de vraisemblance chez Ronald A. Fisher. In: Mairesse, J. (ed.) Estimation et sondages. Economica, Paris (1988) ► Estimation

Armitage, P., Colton, T.: Encyclopedia of Biostatistics. Wiley, New York (1998) ► Biostatistics

Arthanari, T.S., Dodge, Y.: Mathematical Programming in Statistics. Wiley, New York (1981) ► Linear Programming, ► Optimization

Asimov, D.: The grand tour: a tool for viewing multidimensional data. SIAM J. Sci. Stat. Comput. 6(1), 128–143 (1985) ► Statistical Software

Atteia, O., Dubois, J.-P., Webster, R.: Geostatistical analysis of soil contamination in the Swiss Jura. Environ. Pollut. 86, 315–327 (1994) ► Geostatistics, ► Spatial Data

Bailey, D.E.: Probability and statistics: models for research. Wiley, New York (1971) ► Sample Space

Balestra, P.: Calcul matriciel pour économistes. Editions Castella, Albeuve, Switzerland (1972) ► Matrix

Barlett, M.S.: An Introduction to Stochastic Processes. Cambridge University Press, Cambridge (1960) ► Stochastic Process

Barnard, G.A., Plackett, R.L., Pearson, S.: "Student": a statistical biography of William Sealy Gosset: based on writings by E.S. Pearson. Oxford University Press, Oxford (1990) ► Standard Deviation

Barnett, V., Lewis, T.: Outliers in statistical data. Wiley, New York (1984) ► Outlier

Bartlett, M.S.: On the theoretical specification of sampling properties of autocorrelated time series. J. Roy. Stat. Soc. Ser. B 8, 27–41 (1946) ► Serial Correlation

Bartlett, M.S.: Properties of sufficiency and statistical tests. Proc. Roy. Soc. Lond. Ser. A 160, 268–282 (1937) ► Analysis of Categorical Data

Bartlett, M.S.: Some examples of statistical methods of research in agriculture and applied biology. J. Roy. Stat. Soc. (Suppl.) 4, 137–183 (1937) ► Chi-Square Test, ► Covariance Analysis, ► Missing Data

Basset, G., Koenker, R.: Asymptotic theory of least absolute error regression. J. Am. Stat. Assoc. 73, 618–622 (1978) ► L_1 Estimation

Bates, D.M., Watts D.G.: Nonlinear regression analysis and its applications. Wiley, New York (1988) ► Nonlinear Regression

Bayes, T.: An essay towards solving a problem in the doctrine of chances. Philos. Trans. Roy. Soc. Lond. 53, 370–418 (1763). Published, by the instigation of Price, R., 2 years after his death. Republished with a biography by Barnard, George A. in 1958 and in Pearson, E.S., Kendall, M.G.: Studies in the History of Statistics and Probability. Griffin, London, pp. 131–153 (1970) ► Bayes' Theorem, ► Inferential Statistics

Bealle, E.M.L.: Introduction to Optimization. Wiley, New York (1988) ► Operations Research

Becker, R.A., Chambers, J.M., Wilks, A.R.: The new S Language. Waldsworth and Brooks/Cole, Pacific Grove, CA (1988) ► Statistical Software

Belsley, D., Kuh, E., Welsch, R.E.: Regression diagnostics Idendifying Influential data and sources of collinearity. Wiley, New York (1976) ► Multicollinearity

Belsley, D.A., Kuh, E., Welsch, R.E.: Regression diagnostics. Wiley, New York pp. 16–19 (1980) ► Hat Matrix

Beniger, J.R., Robyn, D.L.: Quantitative graphics in statistics: a brief history. Am. Stat. 32, 1–11 (1978) ► Quantitative Graph

Benzécri, J.P.: Histoire et préhistoire de l'analyse des données, les cahiers de l'analyse des données 1, no. 1–4. Dunod, Paris (1976) ► Correspondence Analysis, ► Data Analysis

Benzécri, J.P.: Histoire et préhistoire de l'analyse des données. Dunod, Paris (1982) ► Probability

Benzécri, J.P.: L'Analyse des données. Vol. 1: La Taxinomie. Vol. 2: L'Analyse factorielle des correspondances. Dunod, Paris (1976) ► Correspondence Analysis, ► Data Analysis

Bernardo, J.M., Smith, A.F.M.: Bayesian Theory. Wiley, Chichester (1994) ► Bayesian Statistics

Bernoulli, D.: Quelle est la cause physique de l'inclinaison des planètes (...). Rec. Pièces Remport. Prix Acad. Roy. Sci. 3, 95–122 (1734) ► Hypothesis Testing

Bernoulli, D.: The most probable choice between several discrepant observations, translation and comments (1778). In: Pearson, E.S. and Kendall, M.G. (eds.) Studies in the History of Statistics and Probability. Griffin, London (1970) ► Estimation

Bernoulli, J.: Ars Conjectandi, Opus Posthumum. Accedit Tractatus de Seriebus infinitis, et Epistola Gallice scripta de ludo Pilae recticularis. Impensis Thurnisiorum, Fratrum, Basel (1713) ► Bernoulli Trial, ► Bernoulli's Theorem, ► Binomial Distribution, ► Estimation, ► Law of Large Numbers, ► Normal Distribution

Berry, D.A.: Statistics, a Bayesian Perspective. Wadsworth, Belmont, CA (1996) ► Bayesian Statistics

Bickel, R.D.: Selecting an optimal rejection region for multiple testing. Working paper, Office of Biostatistics and Bioinformation, Medical College of Georgia (2002) ► Rejection Region

Bienaymé, I.J.: Considérations à l'appui de la découverte de Laplace sur la loi de probabilité dans la méthode des moindres carrés. Comptes Rendus de l'Académie des Sciences, Paris 37, 5–13 (1853); réédité en 1867 dans le journal de Liouville précédant la preuve des Inégalités de Bienaymé-Chebyshev, Journal de Mathématiques

Pures et Appliquées **12**, 158–176 ▶ Law of Large Numbers

Billingsley, P.: Statistical Inference for Markov Processes. University of Chicago Press, Chicago (1961) ▶ Stochastic Process

Birkes, D., Dodge, Y., Seely, J.: Spanning sets for estimable contrasts in classification models. Ann. Stat. **4**, 86–107 (1976) ▶ Missing Data

Birkes, D., Dodge, Y.: Alternative Methods of Regression. Wiley, New York (1993) ▶ Collinearity, ▶ Least Absolute Deviation Regression, ▶ Standardized Data

Bishop, Y.M.M., Fienberg, S.E., Holland, P.W.: Discrete Multivariate Analysis: Theory and Practice. MIT Press, Cambridge, MA (1975) ▶ Binary Data, ▶ Generalized Linear Regression, ▶ Logistic Regression

Bjerhammar, A.: Rectangular reciprocal matrices with special reference to geodetic calculations. Bull. Geod. **20**, 188–220 (1951) ▶ Generalized Inverse

Bock, R.D.: Multivariate Statistical Methods in Behavioral Research. McGraw-Hill, New York (1975) ▶ Collinearity

Boscovich, R.J. et Maire, C.: De Litteraria Expeditione per Pontificiam ditionem ad dimetiendas duas Meridiani gradus. Palladis, Rome (1755) ▶ Model

Boscovich, R.J.: De Litteraria Expeditione per Pontificiam ditionem, et Synopsis amplioris Operis, ac habentur plura eius ex exemplaria etiam sensorum impressa. Bononiensi Scientiarium et Artium Instituto Atque Academia Commentarii, vol. IV, pp. 353–396 (1757) ▶ L_1 Estimation, ▶ Median

Bourbonnais, R.: Econométrie, manuel et exercices corrigés, 2nd edn. Dunod, Paris (1998) ▶ Autocorrelation, ▶ Durbin–Watson Test

Boursin, J.-L.: Les structures du hasard. Seuil, Paris (1986) ▶ Data Collection

Bowerman, B.L., O'Connel, R.T.: Time Series and Forecasting: An Applied Approach. Duxbury, Belmont, CA (1979) ▶ Irregular Variation

Bowley, A.L.: Presidential address to the economic section of the British Association. J. Roy. Stat. Soc. **69**, 540–558 (1906) ▶ Confidence Interval, ▶ Sampling

Box, G.E.P, Draper, N.R.: A basis for the selection of response surface design. J. Am. Stat. Assoc. **54**,

622–654 (1959) ▶ Criterion of Total Mean Squared Error

Box, G.E.P., Hunter, W.G., Hunter, J.S.: Statistics for Experimenters. An Introduction to Design, Data Analysis, and Model Building. Wiley, New York (1978) ▶ Experiment, ▶ Main Effect, ▶ Standard Error

Box, G.E.P., Jenkins, G.M.: Time Series Analysis: Forecasting and Control (Series in Time Series Analysis). Holden Day, San Francisco (1970) ▶ ARMA Models, ▶ Autocorrelation, ▶ Cyclical Fluctuation, ▶ Serial Correlation

Box, G.E.P., Tiao, G.P.: Bayesian Inference in Statistical Analysis. Addison-Wesley, Reading, MA (1973) ▶ Bayesian Statistics

Box, G.E.P.: Non-normality and tests on variance. Biometrika **40**, 318–335 (1953) ▶ Robust Estimation

Brown, L.D., Olkin, I., Sacks, J., Wynn, H.P.: Jack Carl Kiefer, Collected Papers III: Design of Experiments. Springer, Berlin Heidelberg New York (1985) ▶ Optimal Design

Carli, G.R.: Del valore etc. In: Opere Scelte di Carli, vol. I, p. 299. Custodi, Milan (1764) ▶ Index Number

Carlin, B., Louis, T.A.: Bayes & Empirical Bayes Methods. Chapman & Hall /CRC, London (2000) ▶ Bayesian Statistics

Casella, G., Berger, R.L.: Statistical Inference. Duxbury, Pacific Grove (2001) ▶ Joint Probability Distribution Function

Cauchy, A.L.: Sur les résultats moyens d'observations de même nature, et sur les résultats les plus probables. C.R. Acad. Sci. **37**, 198–206 (1853) ▶ Cauchy Distribution

Celeux, G., Diday, E., Govaert, G., Lechevallier, Y., Ralambondrainy, H.: Classification automatique des données—aspects statistiques et informatiques. Dunod, Paris (1989) ▶ Cluster Analysis, ▶ Measure of Dissimilarity

Chambers, J.M and Hastie, T.K (eds.): Statistical models in S. Waldsworth and Brooks/Cole, Pacific Grove, CA (1992) ▶ Statistical Software

Chambers, J.M., Cleveland, W.S., Kleiner, B., Tukey, P.A.: Graphical Methods for Data Analysis. Wadsworth, Belmont, CA (1983) ▶ Q-Q Plot (Quantile to Quantile Plot)

Chambers, J.M.: Computational Methods for Data Analysis. Wiley, New York (1977) ▶ Algorithm

Chatfield, C.: The Analysis of Time Series: An Introduction, 4th edn. Chapman & Hall (1989) ▶ Autocorrelation

Chatfield, C.: The analysis of Time Series. An introduction. Chapman'n'Hall (2003) ▶ Secular Trend

Chatterjee, P., Hadi, A.S.: Regression Analysis by Example. Wiley, New York (2006) ▶ Multiple Linear Regression

Cochran, W.G.: Sampling Techniques, 2nd edn. Wiley, New York (1963) ▶ Sample Size, ▶ Simple Random Sampling, ▶ Standard Error

Collatz, L., Wetterling, W.: Optimization problems. Springer, Berlin Heidelberg New York (1975) ▶ Operations Research

Cook, R.D., Weisberg, S.: An Introduction to Regression Graphics. Wiley, New York (1994) ▶ Analysis of Residuals

Cook, R.D., Weisberg, S.: Applied Regression Including Computing and Graphics. Wiley, New York (1999) ▶ Analysis of Residuals

Cook, R.D., Weisberg, S.: Residuals and Influence in Regression. Chapman & Hall, London (1982) ▶ Analysis of Residuals

Copeland, M.T.: Statistical indices of business conditions. Q. J. Econ. **29**, 522–562 (1915) ▶ Time Series

Cornfield, J.: A method of estimating comparative rates from clinical data. Applications to cancer of the lung, breast, and cervix. J. Natl. Cancer Inst. **11**, 1269–75 (1951) ▶ Attributable Risk, ▶ Avoidable Risk, ▶ Prevalence Rate, ▶ Relative Risk, ▶ Risk

Cornfield, J.: Methods in the Sampling of Human Populations. The determination of sample size. Amer. J. Publ. Health **41**, 654–661 (1951) ▶ Sample Size

Cotes, R.: Aestimatio Errorum in Mixta Mathesi, per variationes partium Trianguli plani et sphaerici. In: Smith, R. (ed.) Opera Miscellania, Cambridge (1722) ▶ Weighted Arithmetic Mean

Cox, D.R., Box, G.E.P.: An analysis of transformations (with discussion). J. Roy. Stat. Soc. Ser. B **26**, 211–243 (1964) ▶ Transformation

Cox, D.R., Hinkley, D.V.: Theoretical Statistics. Chapman & Hall, London (1973) ▶ Likelihood Ratio Test

Cox, D.R., Miller, H.D.: Theory of Stochastic Processes. Wiley, New York; Methuen, London (1965) ▶ Stochastic Process

Cox, D.R., Reid, N.: Theory of the design of experiments. Chapman & Hall, London (2000) ▶ One-Way Analysis of Variance, ▶ Two-way Analysis of Variance

Cox, D.R., Snell, E.J.: Analysis of Binary Data, 2nd edn. Chapman & Hall, London (1990) ▶ Analysis of Categorical Data

Cox, D.R., Snell, E.J.: The Analysis of Binary Data. Chapman & Hall (1989) ▶ Analysis of Binary Data, ▶ Binary Data, ▶ Logistic Regression

Cox, D.R.: Estimation by double sampling. Biometrika **39**, 217–227 (1952) ▶ Sample Size

Cox, D.R.: Interaction. Int. Stat. Rev. **52**, 1–25 (1984) ▶ Interaction

Cox, P.R.: Demography, 5th edn. Cambridge University Press, Cambridge (1976) ▶ Demography

Cressie, N.A.C.: Statistics for Spatial Data. Wiley, New York (1991) ▶ Geostatistics, ▶ Spatial Statistics

Crome, A.F.W.: Geographisch-statistische Darstellung der Staatskräfte. Weygand, Leipzig (1820) ▶ Graphical Representation

Crome, A.F.W.: Ueber die Grösse und Bevölkerung der sämtlichen Europäischen Staaten. Weygand, Leipzig (1785) ▶ Graphical Representation

Dénes, J., Keedwell, A.D.: Latin squares and their applications. Academic, New York (1974) ▶ Graeco-Latin Square Design

Dantzig, G.B.: Applications et prolongements de la programmation linéaire. Dunod, Paris (1966) ▶ Linear Programming

David, F.N.: Dicing and gaming (a note on the history of probability). Biometrika **42**, 1–15 (1955) ▶ Probability

Davison, A.C., Hinkley, D.V.: Bootstrap Methods and Their Application. Cambridge University Press, Cambridge (1997) ▶ Bootstrap

DeLury, D.B.: The analysis of covariance. Biometrics **4**, 153–170 (1948) ▶ Covariance Analysis

Delanius, T.: Elements of Survey Sampling. SAREC, Stockholm (1985) ▶ Sampling

Deming, W.E.: Sample Design in Business Research. Wiley, New York (1960) ▶ Systematic Sampling

Desrosières, A. La partie pour le tout: comment généraliser? La préhistoire de la contrainte de représentativité. Estimation et sondages. Cinq contributions à l'histoire de la statistique. Economica, Paris (1988) ▶ Confidence Interval, ▶ Sampling

Deuchler, G.: Über die Methoden der Korrelationsrechnung in der Pädagogik und Psychologie. Zeitung für Pädagogische Psycholologie und Experimentelle Pädagogik, 15, 114–131, 145–159, 229–242 (1914) ▶ Kendall Rank Correlation Coefficient

Diggle, P. J., P. Heagerty, K.-Y. Liang, and S.L. Zeger.: Analysis of longitudinal data, 2nd edn. Oxford University Press, Oxford (2002) ▶ Panel

Dixon, W.J. chief ed.: BMDP Statistical Software Communications. The University of California Press. Berkeley, CA (1981) ▶ Statistical Software

Dodge, Y. (2004) Optimization Appliquée. Springer, Paris ▶ Lagrange Multiplier

Dodge, Y.: A natural random number generator, Int. Stat. Review, 64, 329–344 (1996) ▶ Generation of Random Numbers

Dodge, Y.: Premiers pas en statistique. Springer, Paris (1999) ▶ Measure of Central Tendency, ▶ Measure of Dispersion

Dodge, Y. (ed.): Statistical Data Analysis Based on the L_1-norm and Related Methods. Elsevier, Amsterdam (1987) ▶ L_1 Estimation

Dodge, Y., Hand, D.: What Should Future Statistical Software Look like? Comput. Stat. Data Anal. (1991) ▶ Statistical Software

Dodge, Y., Kondylis, A., Whittaker, J.: Extending PLS1 to PLAD regression and the use of the L1 norm in soft modelling. COMPSTAT 2004, pp. 935–942, Physica/Verlag-Springer (2004) ▶ Partial Least Absolute Deviation Regression

Dodge, Y., Majumdar, D.: An algorithm for finding least square generalized inverses for classification models with arbitrary patterns. J. Stat. Comput. Simul. 9, 1–17 (1979) ▶ Generalized Inverse, ▶ Missing Data

Dodge, Y., Rousson, V.: Multivariate L_1-mean. Metrika 49, 127–134 (1999) ▶ Norm of a Vector

Dodge, Y., Shah, K.R.: Estimation of parameters in Latin squares and Graeco-latin squares with missing observations. Commun. Stat. Theory A6(15), 1465–1472 (1977) ▶ Graeco-Latin Square Design

Dodge, Y., Thomas, D.R.: On the performance of non-parametric and normal theory multiple comparison procedures. Sankhya B42, 11–27 (1980) ▶ Least Significant Difference Test

Dodge, Y., Whittaker, J. (eds.): Computational statistics, vols. 1 and 2. COMPSTAT, Neuchâtel, Switzerland, 24–28 August 1992 (English). Physica-Verlag, Heidelberg (1992) ▶ Statistical Software

Dodge, Y., Whittaker, J., Kondylis, A.: Progress on PLAD and PQR. In: Proceedings 55th Session of the ISI 2005, pp. 495–503 (2005) ▶ Partial Least Absolute Deviation Regression

Dodge, Y.: Analyse de régression appliquée. Dunod, Paris (1999) ▶ Criterion of Total Mean Squared Error

Dodge, Y.: Analysis of Experiments with Missing Data. Wiley, New York (1985) ▶ Missing Data

Dodge, Y.: Mathématiques de base pour économistes. Springer, Berlin Heidelberg New York (2002) ▶ Eigenvector, ▶ Norm of a Vector

Dodge, Y.: Principaux plans d'expériences. In: Aeschlimann, J., Bonjour, C., Stocker, E. (eds.) Méthodologies et techniques de plans d'expériences: 28ème cours de perfectionnement de l'Association vaudoise des chercheurs en physique, Saas-Fee, 2–8 March 1986 (1986) ▶ Factorial Experiment

Dodge, Y.: Some difficulties involving nonparametric estimation of a density function. J. Offic. Stat. 2(2), 193–202 (1986) ▶ Histogram

Doob, J.L.: Stochastic Processes. Wiley, New York (1953) ▶ Stochastic Process

Doob, J.L.: What is a stochastic process? Am. Math. Monthly 49, 648–653 (1942) ▶ Stochastic Process

Draper, N.R., Smith, H.: Applied Regression Analysis, 3rd edn. Wiley, New York (1998) ▶ Analysis of Residuals, ▶ Nonlinear Regression

Drobisch, M.W.: Über Mittelgrössen und die Anwendbarkeit derselben auf die Berechnung des Steigens und Sinkens des Geldwerts. Berichte über die Verhandlungen der König. Sachs. Ges. Wiss. Leipzig. Math-Phy. Klasse, 23, 25 (1871) ▶ Index Number

Droesbeke, J.J., Tassi, P.H.: Histoire de la Statistique. Editions "Que sais-je?" Presses universitaires de France, Paris (1990) ▶ Geometric Mean

Dufour, I.M.: Histoire de l'analyse de Séries Chronologues, Université de Montréal, Canada (2006) ▶ Time Series

Duncan, G.J.: Household Panel Studies: Prospects and Problems. Social Science Methodology, Trento, Italy (1992)   ► Panel

Duncan, J.W., Gross, A.C.: Statistics for the 21st Century: Proposals for Improving Statistics for Better Decision Making. Irwin Professional Publishing, Chicago (1995)   ► Official Statistics

Dupâquier, J. et Dupâquier, M.: Histoire de la démographie. Académie Perrin, Paris (1985)   ► Official Statistics

Durbin, J., Knott, M., Taylor, C.C.: Components of Cramer-Von Mises statistics, II. J. Roy. Stat. Soc. Ser. B 37, 216–237 (1975)   ► Anderson–Darling Test

Durbin, J., Watson, G.S.: Testing for serial correlation in least squares regression, I. Biometrika 37, 409–428 (1950)   ► Durbin–Watson Test

Durbin, J., Watson, G.S.: Testing for serial correlation in least squares regression, II. Biometrika 38, 159–177 (1951)   ► Durbin–Watson Test, ► Serial Correlation

Durbin, J.: Alternative method to d-test. Biometrika 56, 1–15 (1969)   ► Durbin–Watson Test

Dutot, C.: Réflexions politiques sur les finances, et le commerce. Vaillant and Nicolas Prevost, The Hague (1738)   ► Index Number

Edgeworth F.Y.: A new method of reducing observations relating to several quantities. Philos. Mag. (5th Ser) 24, 222–223 (1887)   ► L_1 Estimation

Edgeworth, F.Y.: Methods of Statistics. Jubilee Volume of the Royal Statistical Society, London (1885)   ► Hypothesis Testing

Edington, E. S.: Randomisation Tests. Marcel Dekker, New York (1980)   ► Resampling

Edwards, A.W.F.: Likelihood. An account of the statistical concept of likelihood and its application to scientific inference. Cambridge University Press, Cambridge (1972)   ► Likelihood Ratio Test

Edwards, D.: Introduction to Graphical Modelling. Springer, Berlin Heidelberg New York (1995)   ► Statistical Software

Efron, B., Tibshirani, R.J.: An Introduction to the Bootstrap. Chapman & Hall, New York (1993)   ► Bootstrap

Efron, B.: Bootstrap methods: another look at the jackknife. Ann. Stat. 7, 1–26 (1979)   ► Bootstrap

Efron, B.: The Jackknife, the Bootstrap and the other Resampling Plans. Society for Industrial and Applied Mathematics, Philadelphia (1982)   ► Jackknife Method

Efron, B.: The jackknife, the bootstrap, and other resampling plans, CBMS Monograph. No. 38. SIAM, Philadelphia (1982)   ► Resampling

Eggenberger, F. und Polya, G.: Über die Statistik verketteter Vorgänge. Zeitschrift für angewandte Mathematische Mechanik 3, 279–289 (1923)   ► Negative Binomial Distribution

Eisenhart, C.: Boscovich and the Combination of Observations. In: Kendall, M., Plackett, R.L. (eds.) Studies in the History of Statistics and Probability, vol. II. Griffin, London (1977)   ► Regression Analysis

Elderton, W.P.: Tables for testing the goodness of fit of theory to observation. Biometrika 1, 155–163 (1902)   ► Chi-Square Table

Euler, L.: Recherches sur la question des inégalités du mouvement de Saturne et de Jupiter, pièce ayant remporté le prix de l'année 1748, par l'Académie royale des sciences de Paris. Republié en 1960, dans Leonhardi Euleri, Opera Omnia, 2ème série. Turici, Bâle, 25, pp. 47–157 (1749)   ► Analysis of Residuals, ► Median, ► Model

Euler, L.: Recherches sur une nouvelle espèce de carrés magiques. Verh. Zeeuw. Gen. Weten. Vlissengen 9, 85–239 (1782)   ► Graeco-Latin Square Design

Evelyn, Sir G.S.: An account of some endeavours to ascertain a standard of weight and measure. Philos. Trans. 113 (1798)   ► Index Number

Everitt, B.S.: Cluster Analysis. Halstead, London (1974)   ► Classification, ► Cluster Analysis, ► Data Analysis

Fechner, G.T.: Kollektivmasslehre. W. Engelmann, Leipzig (1897)   ► Kendall Rank Correlation Coefficient

Federer, W.T., Balaam, L.N.: Bibliography on Experiment and Treatment Design Pre-1968. Hafner, New York (1973)   ► Design of Experiments

Federer, W.T.: Data collection. In: Kotz, S., Johnson, N.L. (eds.) Encyclopedia of Statistical Sciences, vol. 2. Wiley, New York (1982)   ► Data

Fedorov, V.V.: Theory of Optimal Experiments. Academic, New York (1972)   ► Optimal Design

Festinger, L.: The significance of difference between means without reference to the frequency distribution function. Psychometrika **11**, 97–105 (1946) ▶ Mann–Whitney Test

Fienberg, S.E.: Graphical method in statistics. Am. Stat. **33**, 165–178 (1979) ▶ Graphical Representation

Fienberg, S.E.: The Analysis of Cross-Classified Categorical Data, 2nd edn. MIT Press, Cambridge, MA (1980) ▶ Contingency Table

Finetti, B. de: Theory of Probability; A Critical Introductory Treatment. Trans. Machi, A., Smith, A. Wiley, New York (1975) ▶ Bayesian Statistics

Fisher, I.: The Making of Index Numbers. Houghton Mifflin, Boston (1922) ▶ Fisher Index

Fisher, R.A., Mackenzie, W. A.: The manurial response of different potato varieties. J. Agricult. Sci. **13**, 311–320 (1923) ▶ Nonlinear Regression

Fisher, R.A., Yates, F.: Statistical Tables for Biological, Agricultural and Medical Research, 6th edn. Hafner (Macmillan), New York (1963) ▶ Graeco-Latin Square Design

Fisher, R.A., Yates, F.: Statistical Tables for Biological, Agricultural and Medical Research. Oliver and Boyd, Edinburgh and London (1963) ▶ Fisher Table

Fisher, R.A.: Applications of "Student's" distribution. Metron **5**, 90–104 (1925) ▶ Degree of Freedom, ▶ Student Distribution, ▶ Student Table, ▶ Student Test

Fisher, R.A.: Moments and product moments of sampling distributions. Proc. Lond. Math. Soc. **30**, 199–238 (1929) ▶ Method of Moments

Fisher, R.A.: On an absolute criterium for fitting frequency curves. Mess. Math. **41**, 155–160 (1912) ▶ Estimation, ▶ Maximum Likelihood

Fisher, R.A.: On the interpretation of χ^2 from contingency tables, and the calculation of P. J. Roy. Stat. Soc. Ser. A **85**, 87–94 (1922) ▶ Chi-square Test of Independence

Fisher, R.A.: On the mathematical foundations of theoretical statistics. Philos. Trans. Roy. Soc. Lond. Ser. A **222**, 309–368 (1922) ▶ Estimation, ▶ Nonlinear Regression

Fisher, R.A.: Statistical Methods and Scientific Inference. Oliver & Boyd, Edinburgh (1956) ▶ Inferential Statistics

Fisher, R.A.: Statistical Methods for Research Workers. Oliver & Boyd, Edinburgh (1925) ▶ Analysis of Variance, ▶ Block, ▶ Covariance

Analysis, ▶ Fisher Distribution, ▶ Interaction, ▶ Latin Square Designs, ▶ Randomization, ▶ Two-way Analysis of Variance, ▶ Variance

Fisher, R.A.: Statistical Methods, Experimental Design and Scientific Inference. A re-issue of Statistical Methods for Research Workers, "The Design of Experiments', and "Statistical Methods and Scientific Inference." Bennett, J.H., Yates, F. (eds.) Oxford University Press, Oxford (1990) ▶ Randomization

Fisher, R.A.: The Design of Experiments. Oliver & Boyd, Edinburgh (1935) ▶ Hypothesis, ▶ Interaction, ▶ Least Significant Difference Test, ▶ Randomization, ▶ Resampling, ▶ Variance

Fisher, R.A.: The arrangement of field experiments. J. Ministry Agric. **33**, 503–513 (1926) ▶ Design of Experiments, ▶ Randomization

Fisher, R.A.: The correlation between relatives on the supposition of Mendelian inheritance. Trans. Roy. Soc. Edinb. **52**, 399–433 (1918) ▶ Variance

Fisher, R.A.: The moments of the distribution for normal samples of measures of departure from normality. Proc. Roy. Soc. Lond. Ser. A **130**, 16–28 (1930) ▶ Method of Moments

Fisher, R.A.: The precision of discriminant functions. Ann. Eugen. (London) **10**, 422–429 (1940) ▶ Correspondence Analysis, ▶ Data Analysis

Fisher, R.A.: Theory of statistical estimation. Proc. Camb. Philos. Soc. **22**, 700–725 (1925) ▶ Estimation

Fisher-Box, J.: Guinness, Gosset, Fisher, and small samples. Stat. Sci. **2**, 45–52 (1987) ▶ Gosset, William Sealy

Fisher-Box, J.: R.A. Fisher, the Life of a Scientist. Wiley, New York (1978) ▶ Fisher, Ronald Aylmer

Fitzmaurice, G., Laird, N., Ware, J.: Applied Longitudinal Analysis. Wiley, New York (2004) ▶ Longitudinal Data

Flückiger, Y.: Analyse socio-économique des différences cantonales de chômage. Le chômage en Suisse: bilan et perspectives. Forum Helvet. **6**, 91–113 (1995) ▶ Gini Index

Fleetwood, W.: Chronicon Preciosum, 2nd edn. Fleetwood, London (1707) ▶ Index Number

Fletcher, R.: Practical Methods of Optimization. Wiley, New York (1997) ▶ Nonlinear Regression

Foody, W.N., Hedagat, A.S.: On theory and application of BIB designs with repeated blocks.

Annals of Statistics, V=S 932–945 (1977)
► Sampling

Francis, I.: Statistical software: a comparative rewiew.
North Holland, New York (1981)   ► Statistical
Software

Frank, I., Friedman, J.: A statistical view of some
chemometrics regression tools. Technometrics **35**,
109–135 (1993)   ► Partial Least-Squares
Regression

Fredholm, J.: Sur une classe d'équations
fonctionnelles. Acta Math. **27**, 365–390 (1903)
► Generalized Inverse

Freedman, D., Pisani, R., Purves, R.: Statistics.
Norton, New York (1978)   ► Histogram

Fuller, A.W.: Introduction to Statistical Time Series.
John Wiley'n'Sas (1976)   ► Seasonal Variation

Funkhauser, H.G.: A note on a 10th century graph.
Osiris **1**, 260–262 (1936)   ► Time Series

Galton, F.: Co-relations and their measurement,
chiefly from anthropological data. Proc. Roy. Soc.
Lond. **45**, 135–145 (1888)   ► Correlation
Coefficient, ► Genetic Statistics

Galton, F.: Kinship and correlation. North Am. Rev.
150, 419–431 (1890)   ► Correlation Coefficient

Galton, F.: Memories of My Life. Methuen, London
(1908)   ► Correlation Coefficient

Galton, F.: Natural Inheritance. Macmillan, London
(1889)   ► Correlation Coefficient, ► Inferential
Statistics, ► Regression Analysis

Galton, F.: The geometric mean in vital and social
statistics. Proc. Roy. Soc. Lond. **29**, 365–367
(1879)   ► Lognormal Distribution

Galton, F.: Typical laws of heredity. Proc. Roy. Inst.
Great Britain 8, 282–301 (1877) (reprinted in:
Stigler, S.M. (1986) The History of Statistics: The
Measurement of Uncertainty Before 1900. Belknap
Press, Cambridge, MA, p. 280)   ► Genetic
Statistics

Gauss, C.F.: Bestimmung der Genauigkeit der
Beobachtungen, vol. 4, pp. 109–117 (1816). In:
Gauss, C.F. Werke (published in 1880).
Dieterichsche Universitäts-Druckerei, Göttingen
► Normal Distribution, ► Standard Error

Gauss, C.F.: Gauss' Work (1803–1826) on the Theory
of Least Squares. Trotter, H.F., trans. Statistical
Techniques Research Group, Technical Report,
No. 5, Princeton University Press, Princeton, NJ
(1957)   ► Standard Error

Gauss, C.F.: In: Gauss's Work (1803–1826) on the
Theory of Least Squares. Trans. Trotter, H.F.
Statistical Techniques Research Group, Technical
Report No. 5. Princeton University Press,
Princeton, NJ (1821)   ► Mean Squared Error

Gauss, C.F.: Méthode des Moindres Carrés.
Mémoires sur la Combinaison des Observations.
Traduction Française par J. Bertrand.
Mallet-Bachelier, Paris (1855)   ► Bias

Gauss, C.F.: Theoria Combinationis Observationum
Erroribus Minimis Obnoxiae, Parts 1, 2 and suppl.
Werke **4**, 1–108 (1821, 1823, 1826)   ► Bias,
► Gauss–Markov Theorem

Gauss, C.F.: Theoria Motus Corporum Coelestium.
Werke, 7 (1809)   ► Error, ► Estimation,
► Inferential Statistics, ► Normal Distribution,
► Regression Analysis

Gavarret, J.: Principes généraux de statistique
médicale. Beché Jeune & Labé, Paris (1840)
► Hypothesis Testing

Gelman, A., Carlin, J.B., Stern H.S., Rubin, D.B.:
Bayesian Data Analysis. Chapman & Hall /CRC,
London (2003)   ► Bayesian Statistics

Gibbens, R.J., Pratt, J.W.: P-values Interpretation and
Methodology. Am. Stat. **29**(1), 20–25 (1975)
► p-Value

Gibbons, J.D., Chakraborti, S.: Nonparametric
Statistical Inference, 4th ed. CRC (2003)
► Nonparametric Statistics

M Inference. McGraw-Hill, New York (1971)
► Wilcoxon Test

Glasser, G.J., Winter, R.F.: Critical values of the
coefficient of rank correlation for testing the
hypothesis of independance. Biometrika **48**,
444–448 (1961)   ► Spearman Rank Correlation
Coefficient

Goovaerts, P.: Geostatistics for Natural Resources
Evaluation. Oxford University Press, Oxford
(1997)   ► Geostatistics, ► Spatial Data, ► Spatial
Statistics

Gordon, A.D.: Classification. Methods for the
Exploratory Analysis of Multivariate Data.
Chapman & Hall, London (1981)
► Classification, ► Cluster Analysis

Gosset, S.W. ("Student"): On the error of counting
with a haemacytometer. Biometrika **5**, 351–360
(1907)   ► Negative Binomial Distribution

Gosset, S.W. "Student": New tables for testing the
significance of observations. Metron **5**, 105–120

(1925) ▶ Student Distribution, ▶ Student Table, ▶ Student Test

Gosset, S.W. "Student": The Probable Error of a Mean. Biometrika **6**, 1–25 (1908) ▶ Inferential Statistics, ▶ Simulation, ▶ Standard Deviation, ▶ Student Distribution, ▶ Student Table, ▶ Student Test

Gourieroux, C.: Théorie des sondages. Economica, Paris (1981) ▶ Survey

Gower, J.C.: Classification, geometry and data analysis. In: Bock, H.H. (ed.) Classification and Related Methods of Data Analysis. Elsevier, Amsterdam (1988) ▶ Data Analysis

Graunt, J.: Natural and political observations mentioned in a following index, and made upon the bills of mortality: with reference to the government, religion, trade, growth, ayre, diseases, and the several changes of the said city. Tho. Roycroft, for John Martin, James Allestry, and Tho. Dicas (1662) ▶ Biostatistics

Graybill, F.A.: Theory and Applications of the Linear Model. Duxbury, North Scituate, MA (Waldsworth and Brooks/Cole, Pacific Grove, CA) (1976) ▶ Gauss–Markov Theorem, ▶ Generalized Linear Regression, ▶ Matrix

Greenacre, M.: Theory and Applications of Correspondence Analysis. Academic, London (1984) ▶ Correspondence Analysis

Greenwald, D. (ed.): Encyclopédie économique. Economica, Paris, pp. 164–165 (1984) ▶ Gini Index

Greenwood, J.A., Hartley, H.O.: Guide to Tables in Mathematical Statistics. Princeton University Press, Princeton, NJ (1962) ▶ Normal Table

Greenwood, M., Yule, G.U.: An inquiry into the nature of frequency distributions representative of multiple happenings with particular reference to the occurrence of multiple attacks of disease or of repeated accidents. J. Roy. Stat. Soc. Ser. A **83**, 255–279 (1920) ▶ Negative Binomial Distribution

Greenwood, M.: Medical Statistics from Graunt to Farr. Cambridge University Press, Cambridge (1948) ▶ Biostatistics

Grosbras, J.M.: Méthodes statistiques des sondages. Economica, Paris (1987) ▶ Survey

Guerry, A.M.: Essai sur la statistique morale de la France. Crochard, Paris (1833) ▶ Graphical Representation

Haberman, S.J.: Analysis of Qualitative Data. Vol. I: Introductory Topics. Academic, New York (1978) ▶ Analysis of Categorical Data

Hadley, G.: Linear programming. Addison-Wesley, Reading, MA (1962) ▶ Linear Programming

Hall, P.: The Bootstrap and Edgeworth Expansion. Springer, Berlin Heidelberg New York (1992) ▶ Bootstrap

Hand, D.J.: Discrimination and classification. Wiley, New York (1981) ▶ Measure of Dissimilarity

Hansen, M.H., Hurwitz, W.N., Madow, M.G.: Sample Survey Methods and Theory. Vol. I. Methods and Applications. Vol. II. Theory. Chapman & Hall, London (1953) ▶ Cluster Sampling

Hardy, G.H.: Mendelian proportions in a mixed population. Science **28**, 49–50 (1908) ▶ Genetic Statistics

Harnett, D.L., Murphy, J.L.: Introductory Statistical Analysis. Addison-Wesley, Reading, MA (1975) ▶ Irregular Variation, ▶ Sample Size

Harter, H.L., Owen, D.B.: Selected Tables in Mathematical Statistics, vol. 1. Markham, Chicago (5.7 Appendix) (1970) ▶ Wilcoxon Signed Table

Harvard University: Tables of the Cumulative Binomial Probability Distribution, vol. 35. Annals of the Computation Laboratory, Harvard University Press, Cambridge, MA (1955) ▶ Binomial Distribution, ▶ Binomial Table

Harvey, A.C.: The Econometric Analysis of Time Series. Philip Allan, Oxford (Wiley, New York) (1981) ▶ Durbin–Watson Test

Hauser, P.M., Duncan, O.D.: The Study of Population: An Inventory and Appraisal. University of Chicago Press, Chicago, IL (1959) ▶ Demography

Hayes, A.R.: Statistical Software: A survey and Critique of its Development. Office of Naval Research, Arlington, VA (1982) ▶ Statistical Software

Hecht, J.: L'idée du dénombrement jusqu'à la Révolution. Pour une histoire de la statistique, tome 1, pp. 21–82 . Economica/INSEE (1978) ▶ Census

Hedayat, A.S., Sinha, B.K.: Design and Inference in Finite Population Sampling. Wiley, New York (1991) ▶ Sampling

Helland: On the structure of PLS regression, Comm. in Stat-Simul. and Comp., 17, 581–607 (1988) ▶ Partial Least-Squares Regression

Heyde, C.E., Seneta, E.: I.J. Bienaymé. Statistical Theory Anticipated. Springer, Berlin Heidelberg New York (1977) ▶ Chebyshev, Pafnutii Lvovich

Hirschfeld, H.O.: A connection between correlation and contingency. Proc. Camb. Philos. Soc. **31**, 520–524 (1935) ▶ Correspondence Analysis, ▶ Data Analysis

Hoaglin, D.C., Mosteller, F., Tukey, J.W. (eds.): Understanding Robust and Exploratory Data Analysis. Wiley, New York (1983) ▶ Q-Q Plot (Quantile to Quantile Plot)

Hoaglin, D.C., Welsch, R.E.: The hat matrix in regression and ANOVA. Am. Stat. **32**, 17–22 (and correction at **32**, 146) (1978) ▶ Hat Matrix

Hoerl A.E., Kennard R.W., Baldwin K.F.: Ridge regression: some simulations. Commun. Stat., vol. 4, pp. 105–123 (1975) ▶ Ridge Regression

Hoerl A.E., Kennard R.W.: Ridge regression: biased estimation for nonorthogonal Problems. Technometrics **12**, 55–67 (1970) ▶ Ridge Regression

Hoerl A.E.: Application of Ridge Analysis to Regression Problems. Chem. Eng. Prog. **58**(3), 54–59 (1962) ▶ Ridge Regression

Hogg, R.V., Craig, A.T.: Introduction to Mathematical Statistics, 2nd edn. Macmillan, New York (1965) ▶ Estimator, ▶ Joint Distribution Function, ▶ Marginal Density Function, ▶ Pair of Random Variables

Holgate, P.: The influence of Huygens'work in Dynamics on his Contribution to Probability. Int. Stat. Rev. **52**, 137–140 (1984) ▶ Probability

Hosmer, D.W., Lemeshow S.: Applied logistic regression. Wiley Series in Probability and Statistics. Wiley, New York (1989) ▶ Logistic Regression

Huber, P.J.: Robust Statistics. Wiley, New York (1981) ▶ Leverage Point, ▶ Robust Estimation

Huber, P.J.: Robust estimation of a location parameter. Ann. Math. Stat. **35**, 73–101 (1964) ▶ Robust Estimation

Huber, P.J.: Robustness: Where are we now? Student, vol. 1, no. 2, pp. 75–86 (1995) ▶ Robust Estimation

Huitema, B.E.: The Analysis of Covariance and Alternatives. Wiley, New York (1980) ▶ Covariance Analysis

INSEE: Pour une histoire de la statistique. Vol. 1: Contributions. INSEE, Economica, Paris (1977) ▶ Official Statistics

INSEE: Pour une histoire de la statistique. Vol. 2: Matériaux. J. Affichard (ed.) INSEE, Economica, Paris (1987) ▶ Official Statistics

ISI: The Future of Statistics: An International Perspective. Voorburg (1994) ▶ Official Statistics

International Mathematical and Statistical Libraries.: IMSL Library Information. IMSLK, Houston (1981) ▶ Statistical Software

Isaaks, E.H., Srivastava, R.M.: An Introduction to Applied Geostatistics. Oxford University Press, Oxford (1989) ▶ Geostatistics, ▶ Spatial Statistics

Jaffé, W.: Walras, Léon. In: Sills, D.L. (ed.) International Encyclopedia of the Social Sciences, vol. 16. Macmillan and Free Press, New York, pp. 447–453 (1968) ▶ Econometrics

Jambu, M., Lebeaux, M.O.: Classification automatique pour l'analyse de données. Dunod, Paris (1978) ▶ Cluster Analysis

Jevons, W.S.: A Serious Fall in the Value of Gold Ascertained and Its Social Effects Set Forth. Stanford, London (1863) ▶ Index Number

Jevons, W.S.: The Principles of Science: A Treatise on Logic and Scientific Methods (two volumes). Macmillan, London (1874) ▶ Geometric Mean

Joanes, D.N. & Gill, C.A.: Comparing measures of samp skewness and kurtosis, JRSS.D **47**(1), 183–189 (1998) ▶ Measure of Kurtosis

Johnson, N.L., Kotz, S., Balakrishnan, N.: Discrete Multivariate Distributions, John Wiley (1997) ▶ Probability Distribution

Johnson, N.L., Kotz, S.: Distributions in Statistics: Continuous Univariate Distributions, vols. 1 and 2. Wiley, New York (1970) ▶ Continuous Probability Distribution, ▶ Lognormal Distribution, ▶ Probability Distribution

Johnson, N.L., Kotz, S.: Distributions in Statistics: Discrete Distributions. Wiley, New York (1969) ▶ Discrete Probability Distribution, ▶ Probability Distribution

Johnson, N.L., Leone, F.C.: Statistics and experimental design in engineering and the physical sciences. Wiley, New York (1964) ▶ Coefficient of Variation

Journel, A., Huijbregts, C.J.: Mining Geostatistics. Academic, New York (1981) ▶ Geostatistics, ▶ Spatial Statistics

Kaarsemaker, L., van Wijngaarden, A.: Tables for use in rank correlation. Stat. Neerland. **7**, 53 (1953) ▶ Kendall Rank Correlation Coefficient

Kapteyn, J., van Uven, M.J.: Skew frequency curves in biology and statistics. Hoitsema Brothers, Groningen (1916) ▶ Lognormal Distribution

Kapteyn, J.: Skew frequency curves in biology and statistics. Astronomical Laboratory Noordhoff, Groningen (1903) ▶ Lognormal Distribution

Kasprzyk, D., Duncan, G. et al. (eds.): Panel Surveys. New York, Chichester, Brisbane, Toronto, Singapore. Wiley, New York (1989) ▶ Panel

Kaufman, L., Rousseeuw, P.J.: Finding Groups in Data: An Introduction to Cluster Analysis. Wiley, New York (1990) ▶ Classification, ▶ Cluster Analysis

Kendall, M.G., Steward, A.: The Advanced Theory of Statistics, vol. 2. Griffin, London (1967) ▶ Likelihood Ratio Test

Kendall, M.G., Stuard, A.: The Advanced Theory of Statistics. Vol. 1. Distribution Theory, 4th edn. Griffin, London (1977) ▶ Method of Moments

Kendall, M.G.: A new measure of rank correlation. Biometrika **30**, 81–93 (1938) ▶ Kendall Rank Correlation Coefficient

Kendall, M.G.: Rank Correlation Methods. Griffin, London (1948) ▶ Kendall Rank Correlation Coefficient

Kendall, M.G.: Studies in the history of probability and statistics: II. The beginnings of a probability calculus. Biometrika **43**, 1–14 (1956) ▶ Probability

Kendall, M.G.: The early history of index numbers. In: Kendall, M., Plackett, R.L. (eds.) Studies in the History of Statistics and Probability, vol. II. Griffin, London (1977) ▶ Index Number

Kendall, M.G.: Time Series. Griffin, London (1973) ▶ Covariation, ▶ Time Series

Kendall, M.G.: Where shall the history of Statistics begin? Biometrika **47**, 447–449 (1960) ▶ Statistics

Khwarizmi, Musa ibn Meusba (9th cent.). Jabr wa-al-muqeabalah. The algebra of Mohammed ben Musa, Rosen, F. (ed. and transl.). Georg Olms Verlag, Hildesheim (1986) ▶ Algorithm

Kiefer, J.: Optimum experimental designs. J. Roy. Stat. Soc. Ser. B **21**, 272–319 (1959) ▶ Kiefer Jack Carl, ▶ Optimal Design

Kleijnen, J.P.C.: Statistical Techniques in Simulation (in two parts). Vol. 9 of Statistics: Textbooks and Monographs. Marcel Dekker, New York (1974) ▶ Simulation

Kleijnen, J.P.C.: Statistical Tools for Simulation Practitioners. Vol. 76 of Statistics: Textbooks and Monographs. Marcel Dekker, New York (1987) ▶ Simulation

Knight, K.: Mathematical statistics. Chapman & Hall/CRC, London (2000) ▶ Joint Density Function

Kokoska, S., Zwillinger, D.: CRC Standard Probability and Statistics Tables and formulae. CRC Press (2000) ▶ Statistical Table

Kolmogorov, A.N.: Foundations of the Theory of Probability. Chelsea Publishing Company, New York (1956) ▶ Conditional Probability

Kolmogorov, A.N.: Grundbegriffe der Wahrscheinlichkeitsrechnung. Springer, Berlin Heidelberg New York (1933) ▶ Conditional Probability

Kolmogorov, A.N.: Math. Sbornik. N.S. **1**, 607–610 (1936) ▶ Stochastic Process

Kolmogorov, A.N.: Sulla determinazione empirica di una legge di distribuzione. Giornale dell'Instituto Italiano degli Attuari **4**, 83–91 (6.1) (1933) ▶ Kolmogorov–Smirnov Test, ▶ Nonparametric Test

Kondylis, A., Whittaker, J.: Using the Bootstrap on PLAD regression. PLS and Related Methods. In: Proceedings of the PLS'05 International Symposium 2005, pp. 395–402. Edition Aluja et al. (2005) ▶ Partial Least Absolute Deviation Regression

Kotz, S., Johnson, N.L.: Encyclopedia of Statistical Sciences, vols. 1–9. Wiley, New York (1982–1988) ▶ Statistics

Kruskal, W.H., Tanur, J.M.: International Encyclopedia of Statistics, vols. 1–2. The Free Press, New York (1978) ▶ Statistics

Kruskal, W.H., Wallis, W.A.: Use of ranks in one-criterion variance analysis. J. Am. Stat. Assoc. **47**, 583–621 and errata, ibid. **48**, 907–911 (1952) ▶ Kruskal-Wallis Table, ▶ Kruskal-Wallis Test, ▶ Nonparametric Test

L.D. Brown, I. Olkin, J. Sacks and H.P. Wynn (eds.): Jack Karl Kiefer, Collected Papers. I: Statistical inference and probability (1951–1963). New York (1985) ▶ Kiefer Jack Carl

L.D. Brown, I. Olkin, J. Sacks and H.P. Wynn (eds.): Jack Karl Kiefer, Collected Papers. II: Statistical inference and probability (1964–1984). New York (1985) ▶ Kiefer Jack Carl

L.D. Brown, I. Olkin, J. Sacks and H.P. Wynn (eds.): Jack Karl Kiefer, Collected Papers. III: Design of experiments. New York (1985) ▶ Kiefer Jack Carl

Lagarde, J. de: Initiation à l'analyse des données. Dunod, Paris (1983) ▶ Inertia Matrix

Lagrange, J.L.: Mémoire sur l'utilité de la méthode de prendre le milieu entre les résultats de plusieurs observations; dans lequel on examine les avantages de cette méthode par le calcul des probabilités; et où l'on résoud différents problèmes relatifs à cette matière. Misc. Taurinensia 5, 167–232 (1776) ▶ Error

Lancaster, H.O.: Forerunners of the Pearson chi-square. Aust. J. Stat. 8, 117–126 (1966) ▶ Chi-square Distribution, ▶ Gamma Distribution

Laplace, P.S. de: Deuxième supplément à la théorie analytique des probabilités. Courcier, Paris (1818). pp. 531–580 dans les Œuvres complètes de Laplace, vol. 7. Gauthier-Villars, Paris, 1886 ▶ Robust Estimation

Laplace, P.S. de: Mémoire sur l'inclinaison moyenne des orbites des comètes. Mém. Acad. Roy. Sci. Paris 7, 503–524 (1773) ▶ Hypothesis Testing

Laplace, P.S. de: Mémoire sur la probabilité des causes par les événements. Mem. Acad. Roy. Sci. (presented by various scientists) 6, 621–656 (1774) (or Laplace, P.S. de (1891) Œuvres complètes, vol 8. Gauthier-Villars, Paris, pp. 27–65) ▶ Error, ▶ Estimation, ▶ Laplace Distribution, ▶ Normal Distribution, ▶ Normal Table

Laplace, P.S. de: Mémoire sur les approximations des formules qui sont fonctions de très grands nombres et sur leur application aux probabilités. Mémoires de l'Académie Royale des Sciences de Paris, 10. Reproduced in: Œuvres de Laplace 12, 301–347 (1810) ▶ Central Limit Theorem, ▶ De Moivre–Laplace Theorem

Laplace, P.S. de: Mémoire sur les probabilités. Mem. Acad. Roy. Sci. Paris, 227–332 (1781) (or Laplace, P.S. de (1891) Œuvres complètes, vol. 9.

Gauthier-Villars, Paris, pp. 385–485.) ▶ Error, ▶ Normal Table

Laplace, P.S. de: Sur les degrés mesurés des méridiens, et sur les longueurs observées sur pendule. Histoire de l'Académie royale des inscriptions et belles lettres, avec les Mémoires de littérature tirées des registres de cette académie. Paris (1789) ▶ Regression Analysis

Laplace, P.S. de: Théorie analytique des probabilités, 3rd edn. Courcier, Paris (1820) ▶ Mean Squared Error

Laplace, P.S. de: Théorie analytique des probabilités, suppl. to 3rd edn. Courcier, Paris (1836) ▶ Gamma Distribution

Laplace, P.S. de: Théorie analytique des probabilités. Courcier, Paris (1812) ▶ Estimation, ▶ Inferential Statistics

Laspeyres, E.: Die Berechnung einer mittleren Waarenpreissteigerung. Jahrb. Natl. Stat. 16, 296–314 (1871) ▶ Index Number, ▶ Laspeyres Index

1863 und die kalifornisch-australischen Geldentdeckung seit 1848. Jahrb. Natl. Stat. 3, 81–118, 209–236 (1864) ▶ Index Number, ▶ Laspeyres Index

Le Cam, L.M., Yang, C.L.: Asymptotics in Statistics: Some Basic Concepts. Springer, Berlin Heidelberg New York (1990) ▶ Convergence

Le Cam, L.: The Central Limit Theorem around 1935. Stat. Sci. 1, 78–96 (1986) ▶ Central Limit Theorem

Lebart, L., Salem, A.: Analyse Statistique des Données Textuelles. Dunod, Paris (1988) ▶ Correspondence Analysis

Legendre, A.M.: Nouvelles méthodes pour la détermination des orbites des comètes. Courcier, Paris (1805) ▶ Estimation, ▶ Least-Squares Method, ▶ Normal Equations, ▶ Regression Analysis

Lehmann, Erich L., Casella, G.: Theory of point estimation. Springer-Verlag, New York (1998) ▶ Point Estimation

Lehmann, E.I., Romann, S.P.: Testing Statistical Hypothesis, 3rd edn. Springer, New York (2005) ▶ Alternative Hypothesis

Lehmann, E.L.: Testing Statistical Hypotheses, 2nd edn. Wiley, New York (1986) ▶ Hypothesis Testing

Lehmann, E.L.: Theory of Point Estimation, 2nd edn. Wiley, New York (1983) ► Estimator, ► Parameter

Lerman, L.C.: Classification et analyse ordinale des données. Dunod, Paris (1981) ► Cluster Analysis

Liapounov, A.M.: Sur une proposition de la théorie des probabilités. Bulletin de l'Academie Imperiale des Sciences de St.-Petersbourg 8, 1–24 (1900) ► Central Limit Theorem

Lilienfeld, A.M., Lilienfeld, D.E.: Foundations of Epidemiology, 2nd edn. Clarendon, Oxford (1980) ► Attributable Risk, ► Avoidable Risk, ► Cause and Effect in Epidemiology, ► Epidemiology, ► Incidence Rate, ► Odds and Odds Ratio, ► Prevalence Rate, ► Relative Risk, ► Risk

Lindley, D.V.: The philosophy of statistics. Statistician 49, 293–337 (2000) ► Bayesian Statistics

Lipps, G.F.: Die Psychischen Massmethoden. F. Vieweg und Sohn, Braunschweig, Germany (1906) ► Kendall Rank Correlation Coefficient

Lowe, J.: The Present State of England in Regard to Agriculture, Trade, and Finance. Kelley, London (1822) ► Index Number

MacMahon, B., Pugh, T.F.: Epidemiology: Principles and Methods. Little Brown, Boston, MA (1970) ► Attributable Risk, ► Avoidable Risk, ► Cause and Effect in Epidemiology, ► Epidemiology, ► Incidence Rate, ► Odds and Odds Ratio, ► Prevalence Rate, ► Relative Risk, ► Risk

MacMahon, P.A.: Combinatory Analysis, vols. I and II. Cambridge University Press, Cambridge (1915–1916) ► Combinatory Analysis

Macauley, F.R.: The smoothing of time series. National Bureau of Economic Research, 121–136 (1930) ► Time Series

Mahalanobis, P.C.: On the generalized distance in statistics, Proc. natn. Inst. Sci. India 2, 49–55 (1936) ► Mahalanobis Distance

Mahalanobis, P.C.: On Tests and Measures of Groups Divergence. Part I. Theoretical formulae. J. Asiatic Soc. Bengal 26, 541–588 (1930) ► Mahalanobis Distance

Maistrov, L.E.: Probability Theory—A History Sketch (transl. and edit. by Kotz, S.). Academic, New York (1974) ► Independence

Mallows, C.L.: Choosing variables in a linear regression: A graphical aid. Presented at the Central Regional Meeting of the Institute of Mathematical Statistics, Manhattan, KS, 7–9 May 1964 (1964) ► C_p Criterion

Mallows, C.L.: Some comments on C_p. Technometrics 15, 661–675 (1973) ► C_p Criterion

Mann, H.B., Whitney, D.R.: On a test whether one of two random variables is stochastically larger than the other. Ann. Math. Stat. 18, 50–60 (1947) ► Mann–Whitney Test, ► Nonparametric Test

Markov, A.A.: Application des functions continues au calcul des probabilités. Kasan. Bull. 9(2), 29–34 (1899) ► Markov, Andrei Andreevich

Markov, A.A.: Extension des théorèmes limites du calcul des probabilités aux sommes des quantités liées en chaîne. Mem. Acad. Sci. St. Petersburg 8, 365–397 (1908) ► Central Limit Theorem

Martens, H., and Naes, T.: Multivariate Calibration. Wiley, New York (1989) ► Partial Least-Squares Regression

Massey, F.J.: Distribution table for the deviation between two sample cumulatives. Ann. Math. Stat. 23, 435–441 (1952) ► Kolmogorov–Smirnov Test

Matheron, G.: La théorie des variables regionalisées et ses applications. Masson, Paris (1965) ► Geostatistics, ► Spatial Data, ► Spatial Statistics

Mayer, T.: Abhandlung über die Umwälzung des Monds um seine Achse und die scheinbare Bewegung der Mondflecken. Kosmographische Nachrichten und Sammlungen auf das Jahr 1748 1, 52–183 (1750) ► Analysis of Residuals, ► Median, ► Model

Mayr, E., Linsley, E.G., Usinger, R.L.: Methods and Principles of Systematic Zoology. McGraw-Hill, New York (1953) ► Dendrogram

McAlister, D.: The law of the geometric mean. Proc. Roy. Soc. Lond. 29, 367–376 (1879) ► Lognormal Distribution

McMullen, L., Pearson, E.S.: William Sealy Gosset, 1876–1937. In: Pearson, E.S., Kendall, M. (eds.) Studies in the History of Statistics and Probability, vol. I. Griffin, London (1970) ► Gosset, William Sealy

Merrington, M., Thompson, C.M.: Tables of percentage points of the inverted beta (F) distribution. Biometrika 33, 73–88 (1943) ► Fisher Table

Metropolis, N., Ulam, S.: The Monte Carlo method. J. Am. Stat. Assoc. **44**, 335–341 (1949) ► Monte Carlo Method

Meyer, C.D.: Matrix analysis and Applied Linear Algebra, SIAM (2000) ► Eigenvector

Miller, L.H.: Table of percentage points of Kolmogorov statistics. J. Am. Stat. Assoc. **31**, 111–121 (1956) ► Kolmogorov–Smirnov Test

Miller, R.G., Jr.: Simultaneous Statistical Inference, 2nd edn. Springer, Berlin Heidelberg New York (1981) ► Least Significant Difference Test, ► One-Sided Test

Mitrinovic, D.S. (with Vasic, P.M.): Analytic Inequalities. Springer, Berlin Heidelberg New York (1970) ► Harmonic Mean

Moivre, A. de: Approximatio ad summam terminorum binomii $(a + b)^n$, in seriem expansi. Supplementum II to Miscellanae Analytica, pp. 1–7 (1733). Photographically reprinted in a rare pamphlet on Moivre and some of his discoveries. Published by Archibald, R.C. Isis **8**, 671–683 (1926) ► Central Limit Theorem, ► De Moivre–Laplace Theorem, ► Normal Distribution

Moivre, A. de: The Doctrine of Chances: or, A Method of Calculating the Probability of Events in Play. Pearson, London (1718) ► Inferential Statistics, ► Normal Distribution

Molenberghs, G., Geert, V.: Models for Discrete Longitudinal Data Series. Springer Series in Statistics. Springer, Berlin Heidelberg New York (2005) ► Longitudinal Data

Montgomery, D.C.: Design and analysis of experiments, 4th edn. Wiley, Chichester (1997) ► One-Way Analysis of Variance

Montmort, P.R. de: Essai d'analyse sur les jeux de hasard, 2nd edn. Quillau, Paris (1713) ► Negative Binomial Distribution

Mood, A.M.: Introduction to the Theory of Statistics. McGraw-Hill, New York (1950) ► Nonparametric Test

Moore, E.H.: General Analysis, Part I. Mem. Am. Philos. Soc., Philadelphia, PA, pp. 147–209 (1935) ► Generalized Inverse

Morabia, A.: Epidemiologie Causale. Editions Médecine et Hygiène, Geneva (1996) ► Attributable Risk, ► Avoidable Risk, ► Cause and Effect in Epidemiology, ► Epidemiology, ► Incidence Rate, ► Odds and Odds Ratio, ► Prevalence Rate, ► Relative Risk, ► Risk

Morabia, A.: L'Épidémiologie Clinique. Editions "Que sais-je?". Presses Universitaires de France, Paris (1996) ► Attributable Risk, ► Avoidable Risk, ► Incidence Rate, ► Odds and Odds Ratio, ► Prevalence Rate, ► Relative Risk, ► Risk

Mosteller, F., Tukey, J.W.: Data Analysis and Regression: A Second Course in Statistics. Addison-Wesley, Reading, MA (1977) ► Correlation Coefficient, ► Leverage Point

Mudholkar, G.S.: Multiple correlation coefficient. In: Kotz, S., Johnson, N.L. (eds.) Encyclopedia of Statistical Sciences, vol. 5. Wiley, New York (1982) ► Correlation Coefficient

National Bureau of Standards.: A Guide to Tables of the Normal Probability Integral. U.S. Department of Commerce. Applied Mathematics Series, 21 (1952) ► Normal Table

National Bureau of Standards.: Tables of the Binomial Probability Distribution. U.S. Department of Commerce. Applied Mathematics Series 6 (1950) ► Binomial Distribution, ► Binomial Table

Naylor, T.H., Balintfy, J.L., Burdick, D.S., Chu, K.: Computer Simulation Techniques. Wiley, New York (1967) ► Simulation

Neyman, J., Pearson, E.S.: On the problem of the most efficient tests of statistical hypotheses. Philos. Trans. Roy. Soc. Lond. Ser. A **231**, 289–337 (1933) ► Hypothesis Testing

Neyman, J., Pearson, E.S.: On the use and interpretation of certain Test Criteria for purposes of Statistical Inference Part I (1928). Reprinted by Cambridge University Press in 1967 ► Significance Level

Neyman, J., Pearson, E.S.: On the use and interpretation of certain test criteria for purposes of statistical inference, Parts I and II. Biometrika **20**A, 175–240, 263–294 (1928) ► Critical Value, ► Hypothesis Testing, ► Inferential Statistics, ► Likelihood Ratio Test, ► Type I Error

Nordin, J.A.: Determining sample size. J. Am. Stat. Assoc. **39**, 497–506 (1944) ► Sample Size

OCDE: OCDE et les indicateurs internationaux de l'enseignement. Centre de la Recherche et l'innovation dans l'enseignement, OCDE, Paris (1992) ► Indicator

OCDE: Principaux indicateurs de la science et de la technologie. OCDE, Paris (1988) ► Indicator

Oliver, M.A., Webster, R.: Kriging: a method of interpolation for geographical information system.

Int. J. Geogr. Inf. Syst. **4**(3), 313–332 (1990) ► Geostatistics, ► Spatial Statistics

Ostle, B.: Statistics in Research: Basic Concepts and Techniques for Research Workers. Iowa State College Press, Ames, IA (1954) ► Chi-Square Test, ► Contrast

Owen, D.B.: Handbook of Statistical Tables. (3.3, 5.7, 5.8, 5.10). Addison-Wesley, Reading, MA (1962) ► Wilcoxon Signed Table

Paasche, H.: Über die Preisentwicklung der letzten Jahre nach den Hamburger Borsen-notirungen. Jahrb. Natl. Stat. **23**, 168–178 (1874) ► Index Number, ► Paasche Index

Palgrave, R.H.I.: Currency and standard of value in England, France and India etc. (1886) In: IUP (1969) British Parliamentary Papers: Session 1886: Third Report and Final Report of the Royal Commission on the Depression in Trade and Industry, with Minutes of Evidence and Appendices, Appendix B. Irish University Press, Shannon ► Index Number

Pannatier, Y.: Variowin 2.2. Software for Spatial Data Analysis in 2D. Springer, Berlin Heidelberg New York (1996) ► Geostatistics

Parzen, E.: On estimation of a probability density function and mode. Ann. Math. Stat. **33**, 1065–1076 (1962) ► Nonparametric Statistics

Pascal, B.: Mesnard, J. (ed.) Œuvres complètes. Vol. 2. Desclée de Brouwer, Paris (1970) ► Arithmetic Triangle

Pascal, B.: Traité du triangle arithmétique (publ. posthum. in 1665), Paris (1654) ► Arithmetic Triangle

Pascal, B.: Varia Opera Mathematica. D. Petri de Fermat. Tolosae (1679) ► Negative Binomial Distribution

Pascal, B.: Œuvres, vols. 1–14. Brunschvicg, L., Boutroux, P., Gazier, F. (eds.) Les Grands Ecrivains de France. Hachette, Paris (1904–1925) ► Arithmetic Triangle

Pearson, E.S., Hartley, H.O.: Biometrika Tables for Statisticians, 1 (2nd edn.). Cambridge University Press, Cambridge (1948) ► Normal Table

Pearson, E.S., Hartley, H.O.: Biometrika Tables for Statisticians, vols. I and II. Cambridge University Press, Cambridge (1966,1972) ► Coefficient of Skewness

Pearson, E.S.: The Neyman-Pearson story: 1926–34. In: David, F.N. (ed.) Research Papers in Statistics:

Festschrift for J. Neyman. Wiley, New York (1966) ► Hypothesis Testing

Pearson, K., Filon, L.N.G.: Mathematical contributions to the theory of evolution. IV: On the probable errors of frequency constants and on the influence of random selection on variation and correlation. Philos. Trans. Roy. Soc. Lond. Ser. A **191**, 229–311 (1898) ► Estimation

Pearson, K.: Contributions to the mathematical theory of evolution. I. In: Karl Pearson's Early Statistical Papers. Cambridge University Press, Cambridge, pp. 1–40 (1948). First published in 1894 as: On the dissection of asymmetrical frequency curves. Philos. Trans. Roy. Soc. Lond. Ser. A **185**, 71–110 ► Coefficient of Skewness, ► Estimation, ► Measure of Shape, ► Method of Moments, ► Standard Deviation

Pearson, K.: Contributions to the mathematical theory of evolution. II: Skew variation in homogeneous material. In: Karl Pearson's Early Statistical Papers. Cambridge University Press, Cambridge, pp. 41–112 (1948). First published in 1895 in Philos. Trans. Roy. Soc. Lond. Ser. A **186**, 343–414 ► Coefficient of Skewness, ► Frequency Curve, ► Histogram, ► Measure of Shape

Pearson, K.: On the criterion, that a given system of deviations from the probable in the case of a correlated system of variables is such that it can be reasonably supposed to have arisen from random sampling. In: Karl Pearson's Early Statistical Papers. Cambridge University Press, pp. 339–357. First published in 1900 in Philos. Mag. (5th Ser) **50**, 157–175 (1948) ► Chi-square Distribution, ► Chi-square Goodness of Fit Test, ► Chi-square Test of Independence, ► Inferential Statistics

Pearson, K.: On the theory of contingency and its relation to association and normal correlation. Drapers' Company Research Memoirs, Biometric Ser. I., pp. 1–35 (1904) ► Analysis of Categorical Data, ► Contingency Table

Pearson, K.: Studies in the history of statistics and probability. Biometrika **13**, 25–45 (1920). Reprinted in: Pearson, E.S., Kendall, M.G. (eds.) Studies in the History of Statistics and Probability, vol. I. Griffin, London ► Correlation Coefficient

Pearson, K.: Tables of the Incomplete Γ-function. H.M. Stationery Office (Cambridge University

Press, Cambridge since 1934), London (1922)
► Chi-Square Table

Peirce, B.: Criterion for the Rejection of Doubtful
Observations. Astron. J. **2**, 161–163 (1852)
► Outlier

Penrose, R.: A generalized inverse for matrices. Proc.
Camb. Philos. Soc. **51**, 406–413 (1955)
► Generalized Inverse

Persons, W.M.: An index of general business
conditions. Rev. Econom. Stat. **1**, 111–205 (1919)
► Time Series

Petty, W.: Another essay in political arithmetic
concerning the growth of the city of London.
Kelley, New York (1682) (2nd edn. 1963)
► Biostatistics

Petty, W.: Observations Upon the Dublin-Bills of
Mortality, 1681, and the State of That City. Kelley,
New York (1683) (2nd edn. 1963) ► Biostatistics

Petty, W.: Political arithmetic. In: William Petty, The
Economic Writings ..., vol 1. Kelley, New York,
pp. 233–313 (1676) ► Econometrics

Pitman, E.J.G.: Significance tests which may be
applied to samples from any populations.
Supplement, J. Roy. Stat. Soc. **4**, 119–130 (1937)
► Resampling

Plackett, R.L., Barnard, G.A. (eds.): Student:
A Statistical Biography of William Sealy Gosset
(Based on writings of E.S. Pearson). Clarendon,
Oxford (1990) ► Gosset, William Sealy

Plackett, R.L.: Studies in the history of probability
and statistics. In: Kendall, M., Plackett, R.L. (eds.)
The discovery of the method of least squares.
vol. II. Griffin, London (1977) ► Legendre
Adrien Marie, ► Regression Analysis

Plackett, R.L.: Studies in the history of probability
and statistics. VII. The principle of the arithmetic
mean. Biometrika **45**, 130–135 (1958)
► Arithmetic Mean

Playfair, W.: The Commercial and Political Atlas.
Playfair, London (1786) ► Graphical
Representation

Poisson, S.D.: Recherches sur la probabilité des
jugements en matière criminelle et en matière
civile. In: Procédés des Règles Générales du Calcul
des Probabilités. Bachelier, Imprimeur-Libraire
pour les Mathématiques, Paris (1837) ► Poisson
Distribution

Poisson, S.D.: Sur la probabilité des résultats moyens
des observations. Connaissance des temps pour

l'an 1827, pp. 273–302 (1824) ► Central Limit
Theorem

Polyà, G.: Ueber den zentralen Grenzwertsatz der
Wahrscheinlichkeitsrechnung und das
Momentproblem. Mathematische Zeitschrift **8**,
171–181 (1920) ► Central Limit Theorem

Porter, T.M.: The Rise of Statistical Thinking,
1820–1900. Princeton University Press, Princeton,
NJ (1986) ► Correlation Coefficient

Preece, D.A.: Latin squares, Latin cubes, Latin
rectangles, etc. In: Kotz, S., Johnson, N.L. (eds.)
Encyclopedia of Statistical Sciences, vol. 4. Wiley,
New York (1983) ► Latin Square Designs

Py, B.: Statistique Déscriptive. Economica, Paris
(1987) ► Covariation

Quenouille, M.H.: Approximate tests of correlation in
time series. J. Roy. Stat. Soc. Ser. B **11**, 68–84
(1949) ► Jackknife Method

Quenouille, M.H.: Notes on bias in estimation.
Biometrika **43**, 353–360 (1956) ► Jackknife
Method

Quetelet, A.: Letters addressed to H.R.H. the Grand
Duke of Saxe Coburg and Gotha, on the Theory of
Probabilities as Applied to the Moral and Political
Sciences. (French translation by Downs, Olinthus
Gregory). Charles & Edwin Layton, London
(1849) ► Chi-square Test of Independence

Rüegg, A.: Probabilités et statistique. Presses
Polytechniques Romandes, Lausanne, Switzerland
(1985) ► Experiment

Rényi, A.: Probability Theory. North Holland,
Amsterdam (1970) ► Bernoulli's Theorem

Raktoe, B.L. Hedayat, A., Federer, W.T.: Factorial
Designs. Wiley, New York (1981) ► Factor

Rao, C.R., Mitra, S.K., Matthai, A.,
Ramamurthy, K.G.: Formulae and Tables for
Statistical Work. Statistical Publishing Company,
Calcutta, pp. 34–37 (1966) ► Binomial
Distribution

Rao, C.R., Mitra, S.K.: Generalized Inverses of
Matrices and its Applications. Wiley, New York
(1971) ► Generalized Inverse

Rao, C.R.: Advanced Statistical Methods in
Biometric Research. Wiley, New York (1952)
► Analysis of Variance

Rao, C.R.: A note on generalized inverse of a matrix
with applications to problems in mathematical
statistics. J. Roy. Stat. Soc. Ser. B **24**, 152–158
(1962) ► Generalized Inverse

Rao, C.R.: Calculus of generalized inverse of matrices. I. General theory. Sankhya **A29**, 317–342 (1967) ▶ Generalized Inverse

Rao, C.R.: Linear Statistical Inference and Its Applications, 2nd edn. Wiley, New York (1973) ▶ Gauss–Markov Theorem, ▶ Generalized Linear Regression

Rashed, R.: La naissance de l'algèbre. In: Noël, E. (ed.) Le Matin des Mathématiciens. Belin-Radio France, Paris (1985) ▶ Algorithm, ▶ Arithmetic Triangle

Reid, C.: Neyman—From Life. Springer, Berlin Heidelberg New York (1982) ▶ Hypothesis Testing

Ripley, B.D., Venables, W.N. (2000) S programming. Springer, Berlin Heidelberg New York ▶ Statistical Software

Ripley, B.D.: Thoughts on pseudorandom number generators. J. Comput. Appl. Math. **31**, 153–163 (1990) ▶ Generation of Random Numbers

Robert, C.P.: Le choix Bayesien. Springer, Paris (2006) ▶ Bayesian Statistics

Rose, D.: Household panel studies: an overview. Innovation **8**(1), 7–24 (1995) ▶ Panel

Ross, S.M.: Introduction to Probability Models, 8th edn. John Wiley, New York (2006) ▶ Bernoulli Trial, ▶ Joint Distribution Function, ▶ Marginal Density Function

Rothmann, J.K.: Epidemiology. An Introduction. Oxford University Press (2002) ▶ Cause and Effect in Epidemiology, ▶ Epidemiology

Rothschild, V., Logothetis, N.: Probability Distributions. Wiley, New York (1986) ▶ Probability Distribution

Royston, E.: A note on the history of the graphical presentation of data. In: Pearson, E.S., Kendall, M. (eds.) Studies in the History of Statistics and Probability, vol. I. Griffin, London (1970) ▶ Graphical Representation

Rubin, D.B.: A non-iterative algorithm for least squares estimation of missing values in any analysis of variance design. Appl. Stat. **21**, 136–141 (1972) ▶ Missing Data

Saporta, G.: Probabilité, analyse de données et statistiques. Technip, Paris (1990) ▶ Correspondence Analysis

Savage, I.R.: Bibliography of Nonparametric Statistics. Harvard University Press, Cambridge, MA (1962) ▶ Nonparametric Test

Savage, R.: Bibliography of Nonparametric Statistics and Related Topics: Introduction. J. Am. Stat. Assoc. **48**, 844–849 (1953) ▶ Nonparametric Statistics

Scheffé, H.: A method for judging all contrasts in the analysis of variance. Biometrika **40**, 87–104 (1953) ▶ Contrast

Scheffé, H.: The Analysis of Variance. Wiley, New York (1959) ▶ Analysis of Variance

Schmid, C.F., Schmid, S.E.: Handbook of Graphic Presentation, 2nd edn. Wiley, New York (1979) ▶ Quantitative Graph

Schmid, C.F.: Handbook of Graphic Presentation. Ronald Press, New York (1954) ▶ Graphical Representation

Searle, S.R.: Matrix Algebra Useful for Statistics. Wiley, New York (1982) ▶ Matrix

Seber, A.F., Wilt, S.C.: Nonlinear Regression. Wiley, New York (2003) ▶ Nonlinear Regression

Seber, G.A.F. (1977) Linear Regression Analysis. Wiley, New York ▶ Simple Linear Regression, ▶ Weighted Least-Squares Method

Senn, St.: Cross-over Trials in Clinical Research. Wiley (1993) ▶ Treatment

Shao, J., Tu, D.: The Jackknife and Bootstrap. Springer, Berlin Heidelberg New York (1995) ▶ Bootstrap

Sheffé, H.: The analysis of variance. C'n'H (1959) ▶ One-Way Analysis of Variance

Sheppard, W.F.: New tables of the probability integral. Biometrika **2**, 174–190 (1903) ▶ Normal Table

Sheppard, W.F.: On the calculation of the most probable values of frequency constants for data arranged according to equidistant divisions of a scale. London Math. Soc. Proc. **29**, 353–380 (1898) ▶ Moment

Sheppard, W.F.: Table of deviates of the normal curve. Biometrika **5**, 404–406 (1907) ▶ Normal Table

Sheynin, O.B.: On the history of some statistical laws of distribution. In: Kendall, M., Plackett, R.L. (eds.) Studies in the History of Statistics and Probability, vol. II. Griffin, London (1977) ▶ Chi-square Distribution

Sheynin, O.B.: P.S. Laplace's theory of errors, Archive for History of Exact Sciences **17**, 1–61 (1977) ▶ De Moivre–Laplace Theorem

Sheynin, O.B.: P.S. Laplace's work on probability, Archive for History of Exact Sciences **16**, 137–187 (1976) ▶ De Moivre–Laplace Theorem

Simpson, T.: An attempt to show the advantage arising by taking the mean of a number of observations in practical astronomy. In: Miscellaneous Tracts on Some Curious and Very Interesting Subjects in Mechanics, Physical-Astronomy, and Speculative Mathematics. Nourse, London (1757). pp. 64–75 ▶ Arithmetic Mean

Simpson, T.: A letter to the Right Honorable George Earl of Macclesfield, President of the Royal Society, on the advantage of taking the mean of a number of observations in practical astronomy. Philos. Trans. Roy. Soc. Lond. **49**, 82–93 (1755) ▶ Arithmetic Mean, ▶ Error

Simpson, T.: Miscellaneous Tracts on Some Curious and Very Interesting Subjects in Mechanics, Physical-Astronomy and Speculative Mathematics. Nourse, London (1757) ▶ Error

Sittig, J.: The economic choice of sampling system in acceptance sampling. Bull. Int. Stat. Inst. **33**(V), 51–84 (1952) ▶ Sample Size

Smirnov, N.V.: Estimate of deviation between empirical distribution functions in two independent samples. (Russian). Bull. Moscow Univ. **2**(2), 3–16 (6.1, 6.2) (1939) ▶ Kolmogorov–Smirnov Test, ▶ Nonparametric Test

Smirnov, N.V.: Table for estimating the goodness of fit of empirical distributions. Ann. Math. Stat. **19**, 279–281 (6.1) (1948) ▶ Kolmogorov–Smirnov Test

Smith, C.A.B.: Statistics in human genetics. In: Kotz, S., Johnson, N.L. (eds.) Encyclopedia of Statistical Sciences, vol. 3. Wiley, New York (1983) ▶ Genetic Statistics

Sneath, P.H.A., Sokal, R.R.: Numerical Taxonomy: The Principles and Practice of Numerical Classification (A Series of Books in Biology). W.H. Freeman, San Francisco, CA (1973) ▶ Dendrogram

Snedecor, G.W.: Calculation and Interpretation of Analysis of Variance and Covariance. Collegiate, Ames, IA (1934) ▶ Fisher Distribution

Sobol, I.M.: The Monte Carlo Method. Mir Publishers, Moscow (1975) ▶ Monte Carlo Method

Sokal, R.R., Sneath, P.H.: Principles of Numerical Taxonomy. Freeman, San Francisco (1963) ▶ Dendrogram

Spearman, C.: The Proof and Measurement of Association between Two Things. Am. J. Psychol. **15**, 72–101 (1904) ▶ Spearman Rank Correlation Coefficient

Statistical Sciences: Statistical Analysis in S-Plus, Version 3.1. Seattle: StatSci, a division of MathSoft (1993) ▶ Statistical Software

Staudte, R.G., Sheater, S.J.: Robust Estimation and Testing. Wiley, New York (1990) ▶ Convergence

Stein, C.: A two-sample test for a linear hypothesis whose power is independent of the variance. Ann. Math. Stat. **16**, 243–258 (1945) ▶ Sample Size

Stephens, M.A.: EDF statistics for goodness of fit and some comparisons. J. Am. Stat. Assoc. **69**, 730–737 (1974) ▶ Anderson–Darling Test

Stigler, S.: Francis Galton's account of the invention of correlation. Stat. Sci. **4**, 73–79 (1989) ▶ Correlation Coefficient

Stigler, S.: Simon Newcomb, Percy Daniell, and the History of Robust Estimation 1885–1920. J. Am. Stat. Assoc. **68**, 872–879 (1973) ▶ Outlier

Stigler, S.: Stigler's law of eponymy, Transactions of the New York Academy of Sciences, 2nd series **39**, 147–157 (1980) ▶ Normal Distribution

Stigler, S.: The History of Statistics, the Measurement of Uncertainty Before 1900. Belknap, London (1986) ▶ Bernoulli Family, ▶ Combinatory Analysis, ▶ Official Statistics, ▶ Probability, ▶ Regression Analysis, ▶ Statistics

Stuart, A., Ord, J.K.: Kendall's advanced theory of statistics. Vol. I. Distribution theory. Wiley, New York (1943) ▶ Measure of Central Tendency

Sukhatme, P.V.: Sampling Theory of Surveys with Applications. Iowa State University Press, Ames, IA (1954) ▶ Quota Sampling

Takács, L.: Combinatorics. In: Kotz, S., Johnson, N.L. (eds.) Encyclopedia of Statistical Sciences, vol. 2. Wiley, New York (1982) ▶ Combinatory Analysis

Tassi, Ph.: De l'exhaustif au partiel: un peu d'histoire sur le développement des sondages. Estimation et sondages, cinq contributions à l'histoire de la statistique. Economica, Paris (1988) ▶ Sampling

Tchebychev, P.L. (1890–1891). Sur deux théorèmes relatifs aux probabilités. Acta Math. **14**, 305–315 ▶ Central Limit Theorem

Tchebychev, P.L.: Des valeurs moyennes. J. Math. Pures Appl. **12**(2), 177–184 (1867). Publié simultanément en Russe, dans Mat. Sbornik **2**(2), 1–9 ▶ Law of Large Numbers

Tenenhaus, M.: La régression PLS. Théorie et pratique. Technip, Paris (1998) ▶ Partial Least-Squares Regression

Thiele, T.N.: Theory of Observations. C. and E. Layton, London; Ann. Math. Stat. **2**, 165–308 (1903) ▶ Method of Moments

Thorndike, R.L.: Who belongs in a family? Psychometrika **18**, 267–276 (1953) ▶ Classification, ▶ Data Analysis

Todhunter, I.: A history of the mathematical theory of probability. Chelsea Publishing Company, New York (1949) ▶ Probability

Tomassone, R., Daudin, J.J., Danzart, M., Masson, J.P.: Discrimination et classement. Masson, Paris (1988) ▶ Cluster Analysis

Tryon, R.C.: Cluster Analysis. McGraw-Hill, New York (1939) ▶ Classification

Tukey, J.W.: Bias and confidence in not quite large samples. Ann. Math. Stat. **29**, 614 (1958) ▶ Jackknife Method

Tukey, J.W.: Comparing individual means in the analysis of variance. Biometrics **5**, 99–114 (1949) ▶ Contrast

Tukey, J.W.: Explanatory Data Analysis, limited preliminary edition. Addison-Wesley, Reading, MA (1970–1971) ▶ Data Analysis, ▶ Stem-And-Leaf Diagram

Tukey, J.W.: Exploratory Data Analysis. Addison-Wesley, Reading, MA (1977) ▶ Data Analysis, ▶ Exploratory Data Analysis

Tukey, J.W.: Quick and Dirty Methods in Statistics. Part II: Simple Analyses for Standard Designs. Quality Control Conference Papers 1951. American Society for Quality Control, New York, pp. 189–197 (1951) ▶ Contrast

Tukey, J.W.: Some graphical and semigraphical displays. In: Bancroft, T.A. (ed.) Statistical Papers in Honor of George W. Snedecor. Iowa State University Press, Ames, IA , pp. 293–316 (1972) ▶ Box Plot, ▶ Exploratory Data Analysis, ▶ Hat Matrix

Twisk, J.W.R.: Applied Longitudinal Data Analysis for Epidemiology. A Practical Guide. Cambridge University Press, Cambridge (2003) ▶ Longitudinal Data

United Nations Statistical Commission Fundamental Principles of Official Statistics ▶ Official Statistics

Venables, W.N., Ripley, B.D.: Modern Applied Statistics with S-PLUS. Springer, Berlin Heidelberg New York (1994) ▶ Statistical Software

Von Neumann, J.: Various Techniques Used in Connection with Random Digits. National Bureau of Standards symposium, NBS, Applied Mathematics Series 12, pp. 36–38, National Bureau of Standards, Washington, D.C (1951) ▶ Monte Carlo Method

Von Neumann, J.: Über ein ökonomisches Gleichungssystem und eine Verallgemeinerung des Brouwerschen Fixpunktsatzes. Ergebn. math. Kolloqu. Wien **8**, 73–83 (1937) ▶ Linear Programming

Wald, A.: Asymptotically Most Powerful Tests of Statistical Hypotheses. Ann. Math. Stat. **12**, 1–19 (1941) ▶ Likelihood Ratio Test

Wald, A.: Some Examples of Asymptotically Most Powerful Tests. Ann. Math. Stat. **12**, 396–408 (1941) ▶ Likelihood Ratio Test

Wegman, E.J., Hayes, A.R.: Statistical Software. Encycl. Stat. Sci. **8**, 667–674 (1988) ▶ Statistical Software

Weiseberg, S.: Applied linear regression. Wiley, New York (2006) ▶ Multiple Linear Regression

White, C.: The use of ranks in a test of significance for comparing two treatments. Biometrics **8**, 33–41 (1952) ▶ Mann–Whitney Test

Whyte, Lancelot Law: Roger Joseph Boscovich, Studies of His Life and Work on the 250th Anniversary of his Birth. Allen and Unwin, London (1961) ▶ Boscovich, Roger J.

Wilcoxon, F.: Individual comparisons by ranking methods. Biometrics **1**, 80–83 (1945) ▶ Mann–Whitney Test, ▶ Nonparametric Test, ▶ Wilcoxon Signed Table, ▶ Wilcoxon Signed Test, ▶ Wilcoxon Test

Wilcoxon, F.: Probability tables for individual comparisons by ranking methods. Biometrics **3**, 119–122 (1947) ▶ Wilcoxon Table

Wilcoxon, F.: Some rapid approximate statistical procedures. American Cyanamid, Stamford Research Laboratories, Stamford, CT (1957) ▶ Nonparametric Test, ▶ Wilcoxon Signed Test

Wildt, A.R., Ahtola, O.: Analysis of Covariance (Sage University Papers Series on Quantitative Applications in the Social Sciences, Paper 12). Sage, Thousand Oaks, CA (1978)   ▶ Covariance Analysis

Wilk, M.B., Ganadesikan, R.: Probability plotting methods for the analysis of data. Biometrika 55, 1–17 (1968)   ▶ Q-Q Plot (Quantile to Quantile Plot)

Wilks, S.S.: Mathematical Statistics. Wiley, New York (1962)   ▶ Likelihood Ratio Test

Wold, H.O. (ed.): Bibliography on Time Series and Stochastic Processes. Oliver & Boyd, Edinburgh (1965)   ▶ Cyclical Fluctuation

Wonnacott, R.J., Wonnacott, T.H.: Econometrics. Wiley, New York (1970)   ▶ Econometrics

Yates, F.: The design and analysis of factorial experiments, Techn. comm. 35, Imperial Bureau of Soil Science, Harpenden (1937)   ▶ Fractional Factorial Design

Yates, F.: Complex experiments. Suppl. J. Roy. Stat. Soc. 2, 181–247 (1935)   ▶ Two-way Analysis of Variance

Yates, F.: Sampling Methods for Censuses and Surveys. Griffin, London (1949)   ▶ Stratified Sampling

Yates, F.: The analysis of replicated experiments when the field results are incomplete. Empire J. Exp. Agricult. 1, 129–142 (1933)   ▶ Missing Data

Yates, F.: The design and analysis of factorial experiments. Technical Communication of the Commonwealth Bureau of Soils 35, Commonwealth Agricultural Bureaux, Farnham Royal (1937)   ▶ Two-way Analysis of Variance, ▶ Yates' Algorithm

Youschkevitch, A.P.: Les mathématiques arabes (VIIIème-XVème siècles). Partial translation by Cazenave, M., Jaouiche, K. Vrin, Paris (1976)   ▶ Arithmetic Triangle

Yule, G.U. (1926) Why do we sometimes get nonsense-correlations between time-series? A study in sampling and the nature of time-series. J. Roy. Stat. Soc. (2) 89, 1–64   ▶ Correlation Coefficient, ▶ Partial Correlation

Yule, G.U.: On the association of attributes in statistics: with illustration from the material of the childhood society. Philos. Trans. Roy. Soc. Lond. Ser. A 194, 257–319 (1900)   ▶ Chi-square Test of Independence

Yule, G.U.: On the theory of correlation. J. Roy. Stat. Soc. 60, 812–854 (1897)   ▶ Correlation Coefficient

Yule, G.V., Kendall, M.G.: An Introduction to the Theory of Statistics, 14th edn. Griffin, London (1968)   ▶ Yule and Kendall Coefficient

Zubin, J.: A technique for measuring likemindedness. J. Abnorm. Social Psychol. 33, 508–516 (1938)   ▶ Classification, ▶ Data Analysis

List of Entries